Fundamentals of Electric Theory and Circuits

Fundamentals of Electric Theory and Circuits

Sridhar Chitta
Associate Professor (Formerly)
Electronics and Communication Engineering
Matrusri Engineering College
Hyderabad, India

TECHSAR™
NEW DELHI

Note on front cover images: The top set-of-four images depict different circuit elements in circuits with surface charges – resistor, capacitor, inductor and diode from left to right. The lower one is a computer-simulation image of the electric and magnetic field strengths (using vectors E and B) in a waveguide ($TE_{1,0}$ mode); image courtesy, Professor Jacob Dilles, George Mason University, Virginia, USA.

Published and Printed by
TechSar Pvt. Ltd.
4435-36/7, Ansari Road, Daryaganj
New Delhi–110 002 (India)
E-mail: tech_info@techsarworld.com
Website: www.techsarworld.com

ISBN 978-93-85909-46-7

Reprint 2025

Dedicated to

My Father

Late Sri C.S. Suryanarayana

&

My Mother

Smt. C.N. Sita

Foreword

The book by Sridhar Chitta, where electrostatics and electric circuits are treated in a unified way on the basis of surface charges, is one of the rare exceptions in textbooks today. The primary objective of this book is obviously to encourage students to think deeply by themselves and not just to learn and to apply mathematical equations. If students, for instance, just learn about the term potential as "energy per charge" they have not understood much. On the contrary, the majority of students learn such mathematical expressions and unconsciously they feel that they have not understood. As a consequence they might lose interest in further learning. The content offered in Chitta's book can only be "digested" with persistence, activation of spatial imagery and concentrated thinking. For students, properly guided and motivated by faculty or mentors, to easily transcend the limits of merely knowing the circuit and field expressions Ohm's law, Kirchhoff's rules, and Coulomb's law etc., Chitta's book offers the perfect content to deeply understand what they want to and should learn. It explains the nature of electricity in a much deeper manner than almost all the other textbooks. It shows the electrostatic aspect of electric circuits, the behavior of capacitors, the effect of pulses on such elements and many other aspects. Students who have worked through these chapters will leave with an increased self-confidence and the impression that complexity has been reduced, which means something important has been understood.

—Dr Hermann Härtel

Guest Scientist, Institute for Theoretical Physics and Astrophysics (ITAP), University of Kiel, and Author of the seminal work "*THE ELECTRIC VOLTAGE: What do students understand? What can be done for better understanding?*"

Former members of ITAP are Max Planck (founder of Quantum Theory and Nobel Laureate), Heinrich Hertz (proved the existence of electromagnetic waves) and Albrecht Unsöld (renowned Astrophysicist who also developed the theorem of the total electron wavefunction of filled and half-filled sub-shells).

Foreword

(To the Previous Edition Titled "Circuit Process Mechanism")

I feel it a great privilege to introduce the book "Circuit Process Mechanism", a special book written in a style of its own covering the fundamentals of electric and magnetic fields, circuits and electronic processes in a single unit.

There are several books in Physics which deal with the principles and physics of basic electricity and magnetism. Further, at a university level, many books deal with the subject in a detailed manner with a mathematical approach. However, not many books present the real physical processes and implied aspects with suitable examples and engineering applications.

At this juncture, Professor Sridhar Chitta has come out with a lucid presentation of the underlying physical phenomenon and application of electricity and electromagnetic field theory.

Professor Sridhar has vast experience as a practicing instrumentation and controls engineer and has also wide experience in electrical engineering in general. His rich experience, wide knowledge-base and practical touch have been thoroughly used in the writing and development of this book.

He has combined textual descriptions with several illustrative figures full of details and included animated power point slides covering every pertinent principle and phenomenon that will aid readers in visualizing circuit processes. Emphasis has been laid on dealing with a few fundamental topics like "Conduction process" (Chapter 1), electromagnetic fields and its application to electrical power (Chapters 5 and 7), electronic processes of capacitors and semiconductors (Chapters 2, 3 and 9).

The author has given an in-depth and lucid presentation for an applied physicist or engineer to get an accurate and correct viewpoint of these very fundamental electric and magnetic processes. Every case has been dealt so that a present-day electrical and electronic engineer understands these physical and applied aspects at the first level before studying topics of electromagnetic field theory and electrical circuits or networks.

A unified treatment of electrostatics and electric circuits is comprehensively covered. The conventional and basic electrical or electronic engineering textbooks do not touch upon these details and this book fulfils this need.

While "Electric and Magnetic Interactions" by Ruth W. Chabay and Bruce A. Sherwood is a primary book that deals with these principles and processes (which is of an advanced level) this book by Professor Sridhar Chitta forms an excellent fundamental level book on the unified approach especially with respect to ac circuits.

The author has spent more than four years conducting a survey and in-depth search for related literature on the contents. With critical inputs, suggestions and encouragement from renowned Professors Hermann Härtel, Bruce A. Sherwood, David J. Griffiths, Norris W. Preyer, Satyabrata Jit and Dipankar and several other experts in this important field, this book becomes unique by itself.

I feel that anyone who wishes to have a clear and basic idea of the conceptual and physical basis of electricity, magnetism and electric and magnetic fields should read the book. The book can be used for basic electrical and electronic courses, both as a text and a reference.

I congratulate the author in taking a bold step for presenting a tough and difficult part of electrical science in a natural and simple manner to the reader.

I wish all success for the endeavor.

Professor (Dr.) V. Kamaraju
Principal and Professor (Retd.),
Dept. of Electrical Engineering,
JNTU College of Engineering,
Kakinada, A.P., India

Preface

The content of "Fundamentals of Electric Theory and Circuits" is aimed to describe the mechanism of the processes in electrical and electronic circuits and electromagnetic wave propagation. At appropriate places in the text, discussions on the aspects of particle interactions and the underlying Action-at-a-distance, Emission, Classical Electrodynamics, Quantum Electrodynamics, modern Quantum Field and Electromagnetic theories including the postulates of the Theory of Special Relativity and how they predict the relativistic aspect of magnetism, are provided. The presentation of topics on circuit processes and electric theory is organized in a manner such that it will aid the reader in developing an intuitive cognition which will be useful for contextual application of electrodynamics theories in circuit analysis.

The descriptions of circuit processes and electromagnetic propagation, however, are based on Classical Electrodynamics for reasons that are elucidated in the text.

The field concept, which nowadays, is playing little or no role in subjects of circuit analysis in University syllabi is making students lose an important opportunity for additional practice with the electric field and therefore, I have retained most of the topics of "Circuit Process Mechanism", the textbook I authored earlier, which had introduced the role of surface charges and fields in circuit behavior.

Keeping in view the curricular requirements in most University syllabi, apart from those topics that deal with the concept of surface charges which is one aspect of the unified treatment of electrostatics and circuits, I have included topics with discussions on currents in circuits, Gauss's Law, the Superposition Principle and Ohmic contacts in diodes. Student understanding of the field concept may have faded even as they are being taught advanced topics on transmission line theory and electromagnetic waves at a critical juncture in their University education. Keeping this in view, I have expanded the section on Transmission Lines in Chapter 5 and a new Chapter on Wave propagation, Antennas and Waveguides is also included under this new book title.

Based on the lucid descriptions of electric theory, with strong emphasis laid on the field concept, in "Electric and Magnetic Interactions" by Ruth Chabay and Bruce Sherwood (John Wiley & Sons, 1995), I prepared power point presentations with animations. Their textbook includes the role of surface charges in circuit behavior, the unified aspect of electrostatics and circuits, and I learned several aspects from their treatment of this important theory which could be used to explain the physical and applied aspects of not only dc circuits, but also ac circuits. I noted that students were able to easily understand circuit behavior and their learning seemed to be enhanced by the visualizations presented in the slides. Then, sensing that there was a need for a textbook that would be reasonably priced, synchronizing the study of the fundamentals of electric theory with circuit processes I was showing them, involving surface charges, I made a framework for the script for this textbook that would be most suitable to students of electrical and electronic engineering and of applied sciences with majors allied to subjects in electricity and magnetism.

The topics in this new edition are directed towards dispelling common misconceptions about circuit processes and expanded to include topics students are presently taught in circuit theory, field theory, basic electronics, and wave propagation. A few visual aids are available, but these are dispersed in so many locations on the World Wide Web that students find it difficult to piece together a meaningful resource base. This I have done by including a file "**Links_to_animations_articles.pdf**" with selected links to animations, videos and articles available on the internet. At several places in the text, URLs (links) to videos and articles are provided and these URLs (links) are available in the "Links_to_animations_articles.pdf"file in the CD. The links in the pdf file are convenient to copy and paste in Internet browsers for viewing.

The CD accompanying this book also contains **animated power point presentations for all the ten chapters**. Care has been taken to ensure that the slides present the concepts accurately so that they complement the theoretical ideas in the text and afford visualizations of various circuit processes described in the textbook.

Preparing the script was tough and tedious but with the encouragement and the critical inputs from renowned Professors Hermann Härtel, Bruce A. Sherwood, David J. Griffiths, Norris W. Preyer, Satyabrata Jit, Dipankar and several other experts in this field, I grew in confidence. My recent interactions with Prof. A. K. T. Assis, who has over the years brought into sharp focus the theories of William Eduard Weber, gave me added impetus to highlight these in this textbook.

By combining the rigor and formalism of authors Sherwood, Griffiths and Satyabrata Jit in the textual descriptions with several illustrations, I have attempted throughout to strike a balance making use of the power of visual images to convey the fascinating twists and turns of the electric and magnetic fields in circuit operations.

In all chapters, I have made **micro/macro connections** between the underlying microscopic phenomenon and the observed macroscopic behavior of circuits by integrating electrostatics and circuits. As an example, the current (macroscopic) in a circuit is driven by an electric field due to a surface charge gradient (a microscopic phenomenon) which is explained after discussing the atomic structure of conductors and the effects of electric fields on matter. As another example, a discussion of the rise in voltage across a capacitor (macroscopic) charged by a dc voltage using a resistor follows a discussion of transients involved in polarization of conductors and insulators by an electric field (a microscopic phenomenon). Making such micro/macro connections supports the approach to steady-state or quasi-steady state when discussing RC circuits from the field viewpoint.

I urge the reader to view the power point presentations of a chapter provided in the CD before reading the text and to view the web pages whose links appear at appropriate places in the text.

The following gives an overview of each chapter.

The initial sections of **Chapter 1** introduce surface charges and the nature and properties of the electric field, and its effects on matter. Important electrostatic processes of charge separation beginning from neutral atoms, production of resultant electric fields at the atomic level, and emf production in a battery, are presented. This is then followed by rough calculations of the electric field strength due to charged objects such as charged invisible tapes.

The growth of the electric field surrounding a dipole as its constituent charges separate is available as screen shots in "**Growth of Electric Field.pdf**" in the CD.

The large variations in potential far away and between a pair of opposite point charges can be clearly seen in the images provided in "**Potential around a pair of opposite charges.pdf**" in the CD.

The CD also contains a MATLAB program in the Word file "**MATLAB_programs_efield_thin_rod_and_invisible_tape**" that calculates the electric field due to charged thin rods, the expressions for which are derived in Appendix A and running the programs will demonstrate how the result converges when the number of slices is increased. A MATLAB program for the field due to a charged invisible tape is also provided.

The chapter progresses with a description of the structure of insulators and conductors and is followed by a discussion of the nature and speed of surface charge development and the conduction process while addressing the misconception that students have that the electric field inside a metal is always zero (even when the system is not in static equilibrium) and that drifting electrons push each other through a wire just as water molecules push each other through a pipe (despite charge neutrality inside the metal).

One other common misconception that is dispelled is that electrons travel at the speed of light in wires. In fact, according to Härtel, "The kinetic energy of the free electrons is practically zero; the driving force - the EMF - is absolutely dominant".

The resistance property of conductors is discussed with an example of a simple dc circuit and the calculation of the strength of the electric field in conductors. Feedback in circuits affects the surface charge configurations around the circuit to maintain current conservation. A discussion of this aspect is followed by dispelling the student misconception that there cannot be any potential difference across an open switch because V = IR, and there is no I !

Chapter 2 describes the property of capacitance and the discharging and charging process of capacitors with the role of the fringe field and surface charges being given special attention. A detailed description of the process of charging a capacitor by a battery is provided. The CD also contains a file "**What_is_e_in_diffl_equation_solutions.pdf**" which describes the exponent 'e', a constant which appears in the solutions of differential equations, such as the solution to the equation which describes the charging of a capacitor. This is followed by descriptions of the behavior of capacitors to ac inputs and the notion of capacitive reactance is introduced. CR circuits subject to ac inputs are useful as filters and a detailed qualitative analysis of their behavior to inputs at different frequencies is considered. Grounding, a widely misunderstood concept, is discussed towards the end of the chapter. Since, some textbooks of field theory use the concept of electric displacement flux ψ, and electric displacement D, one section is devoted to explain their use in the analysis of distributed charge configurations.

The application of the superposition method in complex circuit analysis is one of the major themes of **Chapter 3**. The chapter begins with a detailed illustrated description of the way a capacitor is used to make ac signals superimpose on a dc voltage. Later sections describe graphical techniques to analyze networks, the application of the superposition principle to analyze circuits with more than one loop, and explain the action of bypass capacitors in transistor amplifiers.

The generation of an ac equivalent circuit to analyze transistor amplifiers is one of the classic applications of the principle of superposition of potential and one section of Chapter 3 is devoted to explain this. Stray capacitance effects limit amplifier high frequency performance, and a description that uses the principle of superposition to throw light on this important aspect is provided. This was done with the aim of preparing students to be able to analyze amplifier circuits in more advanced courses in electronics. The voltage-shunt feedback method finds an important application in a wide class of linear integrated circuits, the operational amplifier, and in a continuing vein of the principle of superposition, the final section describes this method.

The results of simulation runs of circuits that superimpose ac signals on dc voltages "**Simulation_super_ac_sig_on_dc_volt.pdf**" and of CR phase-shifting networks typically used in RC Phase shift oscillators "**RC phase shifters.pdf**" are included in pdf files in the CD. Procedures to obtain the voltages, currents and waveforms for this using TINA simulation software are also provided in the CD.

Chapter 4 was developed to explain the transmission of energy in circuits with an introduction to the important notion that electric and magnetic fields do not travel. The Poynting Theorem is useful to make an energy audit of the various energy exchange processes in circuits and this analysis has been done. The analysis leads to the Poynting Vector that quantifies the energy being delivered to a load by the electric and magnetic fields.

Physicists have developed several theories that explain the forces of attraction (between unlike charges) and repulsion (between like charges). These forces span the region separating the particles whether in a vacuum or not. The classic theory simply calls the region where these forces act, a "field". In attempts to describe the interactions between charged particles and understand the nature of the region between them, several theories were propounded after Michelson-Morley's experiment conclusively ruled out the existence of a luminiferous aether.

Prof. Leigh Page of Yale University had proposed in 1924 an Emission Theory of Electromagnetism which assumes a continuous emission of "moving elements" that retains the formalism of Maxwell's equations accurately. Leigh Page formulated his theory beginning from the postulates of Theory of Special Relativity of massless moving elements emitted by the charges, whose loci are the classical Faraday's lines of electric and magnetic force. Quantum Electrodynamics uses concepts of virtual and real photons that are emitted and absorbed by the charges to form the "Quantum Field". In contrast to these theories,

Weber's Force Law does not subscribe to the concept of a "field" and instead uses simply the term "force" for the interaction between charges. An introduction to these ideas is made in this chapter.

This is then followed by dispelling another common student misconception "a circuit requires two separate wires for operation". Since metallic wires have a colored insulation covering, various colored insulated wires are used in the assembly of circuits. These merely indicate the circuit positive and negative potentials; the circuit does not require separate colored wires for proper operation.

Notes on reactive power in a capacitor circuit are provided to help the student understand how there is a rhythmical flow of power in a capacitive circuit.

Faraday's Law relates the time-varying magnetic field to an electric field and **Chapter 5** discusses this Law beginning with the nature and properties of the magnetic field, it's existence and detection. Progressing with the derivation of expressions for the magnetic field due to different current configurations, the chapter provides demonstrations of this important law and is followed by the use of Lenz's rule to determine the direction of the electric field associated with a time-varying magnetic field. Later sections introduce the concept of inductance. The process of induced emf generation is slightly complex and this has been given due importance while introducing notions of nonconservative curly patterned electric fields. A description of the behavior of inductors to ac inputs is next treated and the concept of inductive reactance is introduced.

LR circuits subject to ac inputs are useful as filters and a detailed qualitative analysis of their behavior to inputs at different frequencies is considered. This is followed by notes on reactive power in an inductor circuit. Detailed descriptions of the behavior of voltages and currents in transmission lines, the wave motion of the electric and magnetic fields including a few common applications, the two-wire or twin-lead cable and the coaxial cable have been provided. Prof. Hermann Härtel prepared an interactive animation of the simulation of a transmission line excited by sine waves, pulse trains and triangular waves. The CD contains in the folder "**Transmission Line Simulation Program**" descriptions of transmission processes in linear systems and an executable jar file "TL" (Java based) to run the simulation.

Descriptions of the process that makes voltage waves slow down in transmission lines are also included in this chapter.

Differential-mode and common-mode signals are important to understand in the design of complex signal-conditioning circuits in medical diagnostic, industrial control and military applications. Topics related to these aspects are discussed in the same section and a power point presentation with animations "**Basic Action of a Differential Amplifer-Heart of the Opamp**" is included in the CD along with this book.

Chapter 6 deals with Magnetic Forces and introduces the Lorentz Force law. A simple example that demonstrates the magnetic force between a pair of current carrying wires is described. The chapter progresses to prove that magnetic forces do no work. A subtlety in the process of the development of a force in a current carrying wire involves Hall Effect surface charges and this effect has also been described.

In **Chapter 7**, a review of Faraday's Law is made and a treatment of its more general application that helps determine the motional emf, the basis of generator operation is provided. Descriptions of the commutation process that enable dc machines generate a unidirectional emf are provided. A study of the energy conservation principle in generator operation is instructive and this has been made in the final section.

Magnetic moments are present in current carrying loops and in atoms as well, and in the presence of magnetic fields, they develop forces. This aspect of magnetic moments is the underlying theme of **Chapter 8**. The theory of magnetic forces in magnetic moments led to its application in the design of motors. A description of the torque developed by a current carrying loop (a magnetic moment) in a magnetic field is followed by details of motor construction and operation. The commutation process in motors enable dc machines to generate a unidirectional torque and this is described with illustrations. Motors are used to drive mechanical loads and the torque imposed by the loads are also described along with the techniques designers use to match load torque to motor torque. This aspect is described in detail. Later sections provide a derivation of an expression for the potential energy of magnetic moments and describe the effects of magnetic fields in matter. Textbook references to the derivation of the Lorentz force law from the Lagrangian and the notion of conservation of energy based on the Hamiltonian are also mentioned and the idea of magnetism as a relativistic effect is presented. Finally, a new consistent approach to visualize the phenomenon of induction based on the circular waves of accelerated charges made by Härtel is introduced.

At the end of the chapter, I have included a reference to an article which contains a set of guidelines that can assist academicians in restructuring the traditional Electricity and Magnetism course in engineering curricula.

In **Chapter 9**, the theory of formation of *pn* junctions which is the basic structure of a diode is provided. A description of the formation of a potential barrier V_o which enables a diode to conduct in one direction and block conduction in the reverse direction is followed by its application to rectify an ac voltage and produce a dc voltage. One important aspect regarding the potential barrier V_o is that it cannot become zero during a forward bias in normal operation, and this has been dealt with in some detail and illustrated with a numerical example.

It is necessary that unwanted rectification is not made by the contacts made to semiconductor diodes and one section describes the technique of making nonrectifying or ohmic contacts beginning with descriptions of the Fermi energy of metals and semiconductors.

The study of Electromagnetic waves has gained importance with the explosive growth of a host of devices like smartphones and wireless-fidelity (wi-fi) networks apart from the sophisticated radar and wireless communication equipment used in the aviation industry and in military equipment. While most of such equipment have been in use for several decades, few textbooks dealing with electromagnetic waves, their properties, production (radiation as by say, antennas) and propagation, treat the new concepts in physics, which in recent years have lent clarity in understanding these processes accurately. I developed **Chapter 10** using a few of these new ideas. The chapter begins with a description of the fundamental notion of accelerated charge and how this kick starts the production of electromagnetic waves and is the basic radiation mechanism of antennas. The process of radiation by bulbs is slightly different and this has been treated.

The processes of light emission by diodes, production of sparks when switches are interrupted which are different in certain aspects from the processes in antennas, are separately described.

Light propagation through media involves reradiation or scattering and this is described in some detail from the viewpoint that these processes are applicable to the interaction of radio frequency waves with matter.

The idea of quantization of energies (emitted by black bodies) was propounded by Planck to explain their distribution with respect to wavelength, and marked an important contribution in Modern Physics. This is also dealt in Chapter 10.

Electromagnetic wave propagation in waveguides is used to couple high frequency radio waves to antennas and to other equipment and this is treated in one section. There are not many animated power point slides in this chapter in the accompanying CD. Yet, I have not kept the student wanting by providing links to some stunning and exciting animations available on the World Wide Web.

Appendix A provides detailed derivations of expressions for the electric field due to dipoles and charge distributions such as, a line charge, a ring of charge, a charged disk, and a pair of charged disks, the capacitor. Appendix A also includes a treatment of Gauss's Law and its applications. One of the important applications described is of its use to determine the neutrality of current-carrying wires in circuits. Also discussed is its use in the determination of the electric field of simple (symmetric) charge configurations, and of the energy stored in spherical shells.

The derivation of expressions for the magnetic field of a short length of current-carrying wire, a long straight wire and of circular loops is also provided. The derivation of an expression for the inductance of inductors is included in Appendix A.

Appendix B includes several important topics such as, a comparison of the electric force with the gravitational force, and the energy levels of outer and inner electrons so that the structure of conductors can be understood more comprehensively. A slightly detailed discussion of various theories of electromagnetism in the topic Fields and Momentum Conservation (the coverage provided in my previous textbook Circuit Process Mechanism is now expanded here), will provide the reader, with the necessary framework to put into perspective ongoing research on the finer aspects of particle and field interactions.

Appendix C provides a description of the field strength of insulators, an often neglected topic in most textbooks on circuit theory, and the processes that make insulators prevent shocks.

Appendix D provides information of the atomic level processes that occur when an object is touched.

The derivation of the *pn* diode equation is described in **Appendix E** after identifying the three distinct processes that makeup the conduction mechanism and provides detailed descriptions of them.

Since the notions of an electric field as an "alteration of space" and it's driving influence on charges was talked about more with reference to charges in and around circuit elements, I felt it necessary to provide details on the aspect of the magnetic field which was described as a "further alteration of space" from the Special Relativity (relativistic) perspective. This has been done in a new **Appendix F** which traces the research work of Coulomb, Ampere, Faraday, Biot-Savart, Maxwell, Grassmann and Weber in electrodynamics. Ampere's force expressions, propounded towards the end of his outstanding pioneering research, have withstood the test of Newton's third law and a comparison with Grassmann's force expression (based on the Biot-Savart law) and the situations where these expressions are valid are discussed. After briefly tracing the history of the "field" concept, a derivation of the Lorentz Force law from Ampere's force law is provided. Then, using the Lorentz transformation set of equations, it is shown how forces are exerted by current carrying wires on moving charges which led to the concept of magnetism as a relativistic phenomenon. A brief introduction to ideas of field theories using Page's massless moving elements, Quantum Field Theory which uses virtual and real photons and an alternative to field theory which deals with the direct interaction between charges without mediatory electric and magnetic fields based on Relational Mechanics (Weber's Electrodynamics) is also provided in this Appendix.

The CD accompanying the book also contains an animated slide presentation "**Regulated_Power_Suppl**y" on the operation of a regulated power supply. Beginning with transformer operation, covering bridge rectifier and filter operation, the slides progress to cover transistor operation with the help of load lines, and then covers design aspects of a series-pass transistor based regulated power supply. The surface charge and field viewpoint detailing the processes are not found in any textbook and the presentation details the dynamics of the regulation function.

Since I observed that students find it difficult to understand the nuances in the design and operation of a transformer coupled audio power amplifier, I have included an animated power point presentation "**Transformer coupled Audio power amplifiers**" in the CD, in which focus had been made on critical design aspects like impedance matching and maximum power transfer.

More practice questions for all the chapters is included in a pdf file in the folder "**Additional practice questions for all chapters**" in the CD along with the book.

The CD contains a list of books in the file "**Recommended_books_for_libraries.pdf**" that can be stocked in college libraries for the ready reference of faculty and students. Numbered references in the text are listed after the last Appendix.

Prof. Hermann Härtel provided constant encouragement over the past several years by his gentle probing questions, and thank him for writing the wonderful foreword for this edition. I wonder oftentimes, "Is Physics dependent on a language like English or German to be understood ?". Words fail to express the immense debt of gratitude I owe Professor Härtel, for making me understand the finer aspects of Electricity, Magnetism and Circuits.

Prof. Hermann Härtel has reminded me on several occasions that the content of my book is based entirely on the classical field interpretation and that readers should be made aware about several possible interpretations of the interactions between objects and charges, apart from the alteration of space mentioned in my content. His friend, Prof. Andre Assis has done extensive research on one of these concepts, namely, the idea of a "force" that mediates interactions between objects and charges (that was first described by Wilhelm Weber) without a field concept. At appropriate places in the text, I have mentioned references to highly resourceful articles, papers and books in which, the interactions between objects (or charges) are explored in-depth.

Prof. Hermann Härtel has prepared a "**A Qualitative Approach to Electricity: A Guide to Visualising Eletrodynamics using symbols and images**" that can help students to make the transition from electrostatics to the study of moving charges and associated magnetic phenomenon consistent without any loss of precision in cognitive assimilation of ideas. The guide is included in the CD along with this book.

The CD also contains in the folder "**Videos by Prof. Härtel**" a set of videos prepared by Prof. Hermann Härtel which shows animated views of the field lines of a charge carrier moving with uniform velocity, accelerated and interacting with another carrier. A few notes about its production are available in the file "Härtel Videos of charge particle in motion" in the CD.

I wish to thank Professor Sherwood who answered several of my queries on various topics with patience and understanding. Professors Ruth W. Chabay and Bruce A. Sherwood were awarded the "Halliday and Resnick Award for Excellence in Undergraduate Teaching" by the American Association of Physics Teachers in 2014. A link to their talk "Thinking Iteratively" on the occasion is here https://www.youtube.com/watch?v=e-shsRZQsi4. Professor David J. Griffiths clarified several points I had raised on the energy stored in charged particles and permanent magnets, for which I express my gratitude. I gratefully acknowledge Professor Norris W. Preyer who permitted the use of images of computer simulation results from his own files and also edited the draft descriptions of the simulation images I made thus making them more accurate and comprehensive.

I thank Prof. A. K. T. Assis, who patiently explained certain aspects in the derivation of Ampere's Law from Weber's Law. Prof. Assis's web page is a treasure trove, which contains numerous articles, papers and links to books. The translations of the original notes and published papers of 19th century scientists such as Ampere, Biot-Savart, Weber, Kirchhoff, Grassmann, Lorentz and several others contain annotations by Assis, which provide deep insights into the ideas and theories of classical electrodynamics. In particular, papers on Weber's research and on Ampere's research provided me with an understanding of interactions between charged particles and fields, and revealed several facets in the theory of electrodynamics that I had not known earlier.

My sincere thanks to Professor Dipankar of Treelabs, Mumbai, India for giving me valuable insights on the variation of the potentials of objects when brought near charged objects.

I express my thanks to John M. Clement and John Denker of Physics Forums for their detailed inputs that provide a continuum to the descriptions of the average field inside matter and the measurement of voltages using probes. I thank Professor Satyabrata Jit for his valuable suggestions on the description of electric field variations within diodes.

Professor (Dr.) V. Kamaraju helped me edit content related to the potentials of objects and the energy of electrons.

I express my sincere gratitude to Prof. Debatosh Guha, for the valuable suggestions he gave me during the development of the new Chapter 10.

I express sincere thanks to Professor (Dr.) R.L. Narasimham, for his continued support and encouragement in the preparation of the manuscript.

I thank Abhay Ravindra Joshi, student in the Dept of Electronics and Communication Engineering, Matrusri Engg College, Hyderabad, India for assisting me during the assembly and trial phase of a voltage shunt-feedback amplifier and for his suggestions when preparing the text of Section 3.9 in Chapter 3.

I am thankful to my wife Mayura and son Sudeendra who were patient and understanding while I was reviewing the literature, preparing the manuscript and editing content over the last so many years.

Sridhar Chitta
Hyderabad, Telangana, India

Contents

1 Surface Charges and Conduction Processes

Chapter 1 introduces *surface charges* broadly and their role in enabling the conduction process. Surface charges maintain the potential around conductors, and provide the electric fields that drive circuit currents, according to Hermann Härtel, a scientist at the Institute for Science Education, University of Kiel.

The chapter begins, by introducing the need for electric fields to be present in conductors to enable the conduction process, and the manner in which this is achieved by the natural phenomenon of surface charge development and their distribution. Simple electrostatic phenomena with examples illustrate how small numbers of surface charges are sufficient to create observable effects. The idea of charge separation that produces voltages and emfs is then presented, followed by a description of the conduction mechanism in conductors and resistors. Electron and conventional currents are then described and expressions to compute them are derived.

The next section is devoted to developing a visualization of current flow in resistors, and continues with sample calculations of the potential profiles inside wires and resistors of simple electric circuits. The chapter concludes with a description of the energy processes during conduction inside wires and resistors. At appropriate places in the text, the student is urged to read the highly instructive notes prepared by Hermann Härtel in "A Qualitative Approach to Electricity: A Guide to Visualization of Electrodynamics using symbols and images" and also learn to depict the Coulomb interactions in the manner prescribed in Härtel's Guide.

1.1 SURFACE CHARGES

Excerpt from a "A Unified treatment of electrostatics and circuits" by Ruth W. Chabay and Bruce A. Sherwood.[1]

Ohm's (German physicist, 1789-1854) steady-state stationary charge distribution consisted of a gradient of volume charge *inside the conductor*, whereas on the basis of what was known about electrostatics there could *not* be a volume charge *inside a* conductor.

"Kirchhoff resolved the remaining fundamental issues, not just with his loop and node rules but more importantly by *unifying* electrostatics and circuits."

Kirchhoff (German physicist, 1824-1887) and Wilhelm Eduard Weber (1804-1891) working independently (Ref. [37]), put the driving charge gradient on the *"surface"* of the conductors and showed that electrostatic and circuit phenomena belonged to one science, not two.

Unfortunately, this unification was later lost sight of as the role of potential came to dominate the analysis of circuits, and the *surface charges.........*

.....disappeared from view.

The above short introduction to the history of developments in the study of surface charge phenomena was given, to put into perspective, some of the reasons why the topic is not treated in modern-day textbooks notwithstanding, its importance in enabling the fundamental process of electrical conduction.

Charges will move, when there is a force acting on them, as in the presence of an electric field, and they will continue to move, unless acted upon by opposing forces such as:

(i) those that impede their motion like collisions with the vibrating lattice of atomic cores in metals, or

(ii) electric fields, produced by charges, which accumulate on the surfaces of capacitor plates, or

(iii) Non-Coulomb electric fields (Faraday's law) or fields due to the motion of conductors in steady magnetic fields.

We will learn more about these opposing fields (ii and iii) and how these are produced, in subsequent chapters.

1.2 CURRENT

Moving charges constitute *a current.*

Current strength in conductors is determined by the charge (electron) collisions with vibrating atomic cores (or lattice ions, which we shall describe later) that physically resist them, and by other electric fields that oppose the driving electric fields of the moving charges (electrons).

Such opposing electric fields are produced by "accumulated" charges on the surfaces of capacitor plates (*fringe fields*), by Faraday's law (*induced emf*), and by the motion of conductors in steady magnetic fields.

The current strength in a conductor of low resistance (large number of free electrons per volume or with a high electron mobility) with a field applied to it, will be more than .....

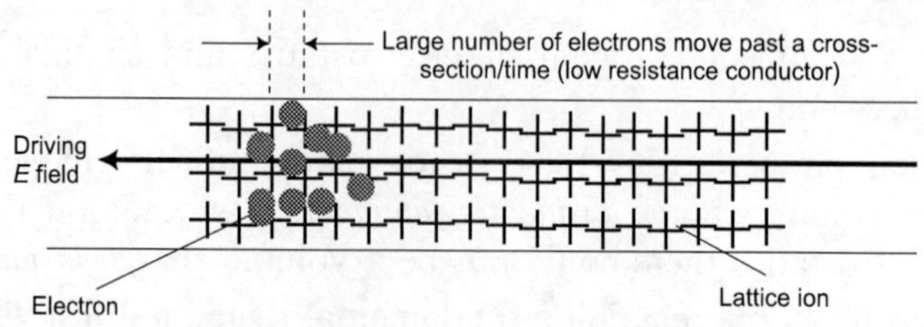

...the current strength in a conductor of high resistance (small number of free electrons per volume or with a low electron mobility), with the same strength of the applied field.

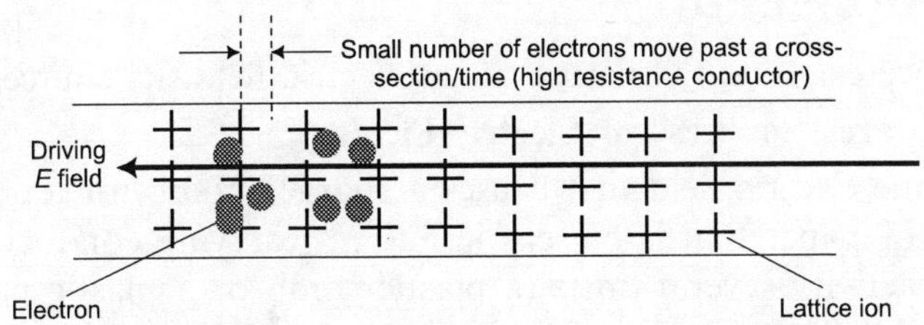

In a conductor with high resistance, only a small number of electrons would have moved past a cross-section/time given the same strength of the driving field. Such conductors are called resistors. Resistance will be discussed in Section 1.19 in more detail.

Current Strength in a Conductor with Opposing Fields

The figure shows an opposing field in the same conductor to which a driving electric field is applied.

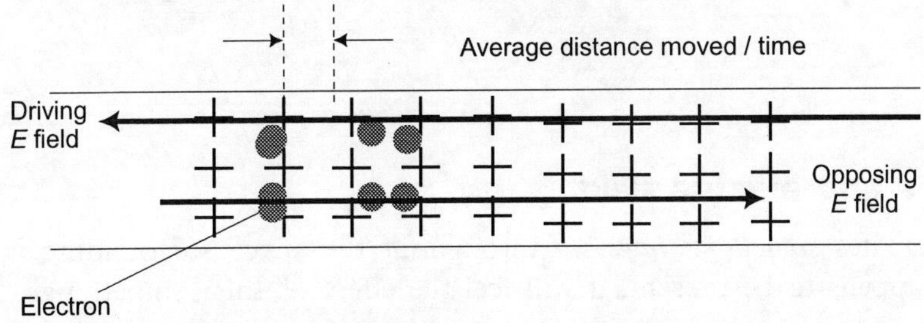

The driving field from a voltage source and the opposing field, say, from a capacitor (fringe) or from an inductor (induced) or in a conductor moving in a steady magnetic field, will combine, to produce a resultant or net field inside the conductor. The driving and the opposing fields are components of the resultant field.

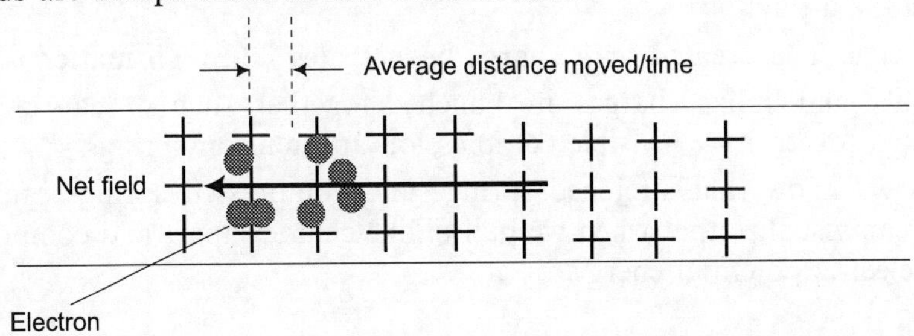

The current strength will be reduced by such opposing fields in the conductor. However, it should be kept in mind that while we consider the 'net' electric field.....in the conductor.......the 'net' field is a result of two opposing component fields and that both these fields are alive and existing like two teams straining in a tug of war competition.

Sometimes, other fields could superimpose on an applied driving field, to *increase* the resultant field strength in conductors, as with the use of precharged capacitors. We will deal with this aspect of the electric field in Chapter 3.

1.3 ELECTRIC FIELD

In this section, we examine the nature of an electric field, its sources and action.

Electric Charges are "Sources" of Electric Fields.

It is the very same electric field that causes a charged comb (or a charged plastic pen) to attract tiny bits of paper that is present in current carrying conductors. I urge the reader to perform this activity several times (a plastic comb or a plastic pen can be charged by rubbing it in one's hair) and make oneself fully *aware* of the *presence* and force that this field exerted on the paper; and then to visualize the force such a field will exert on conduction band electrons inside a conductor. Also observe how the charged plastic comb can make the hair on your hand, stand.

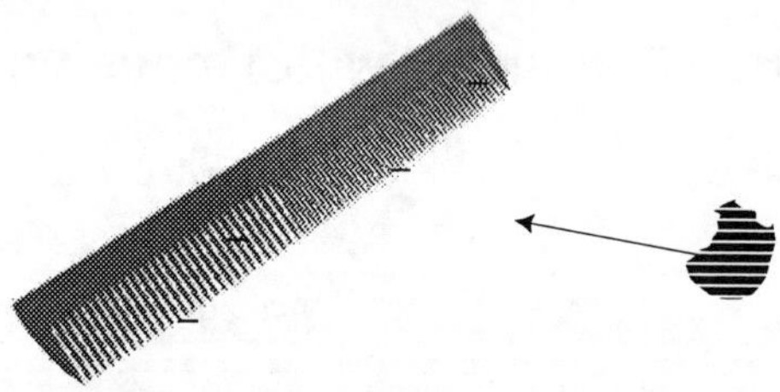

Charges and electric field

Charge creates *something invisible* throughout the space surrounding it, and if another charge happens to be present, it will feel the effect of this "something".

We call this "something" ***electric field***. The electric field created by a charge, is present throughout space at all times, *whether or not there is another charge around to feel its effects.*[2]

A charge *"alters invisibly"* the space surrounding it and we call this altered space, *electric field.* The space between the charged comb and the tiny bit of paper was *altered,* resulting in the attraction.

The electric field created by a charge "penetrates" through matter.

When like and unlike charges are bunched together, then we can imagine that these are connected together by invisible compression and tension springs.

Just as we know that stretched springs and compressed springs can store potential energy, we can visualize that when we pull unlike charges apart and compress like charges, they store electric potential energy.

1.4 CHARGES AND ELECTROSTATIC POTENTIAL ENERGY

A single positive charge is surrounded by, an *electric field.*

If you *bring another* positive charge near this, then you have to do work to bring it near (because like charges repel each other), and this *work done* energy is stored in the *electric field* that extends to the entire universe in this *system* of charges....comprising the 2 charges....

...like compressing a spring.

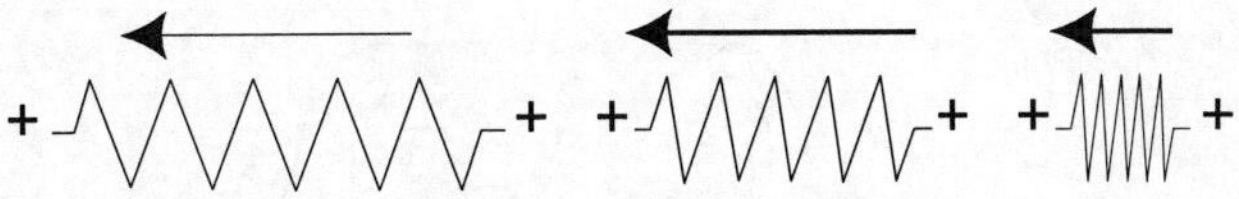

When you *release* the *positive charge*....the electrostatic energy stored in the system of 2 charges, *is released* and is converted to *kinetic energy* in the *positive charge*

...like releasing a coiled spring.

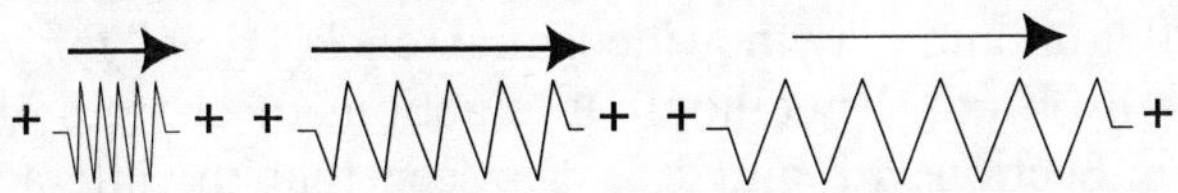

The potential energy is stored in the electric field of the system of charges.

For separating the charges of a neutral atom (electrons and positive atomic cores), you have to do work to separate them (because unlike charges attract each other), and this *work done* energy is stored in the *electric field* of this *system* of charges....comprising the 2 charges...

...like stretching a spring.

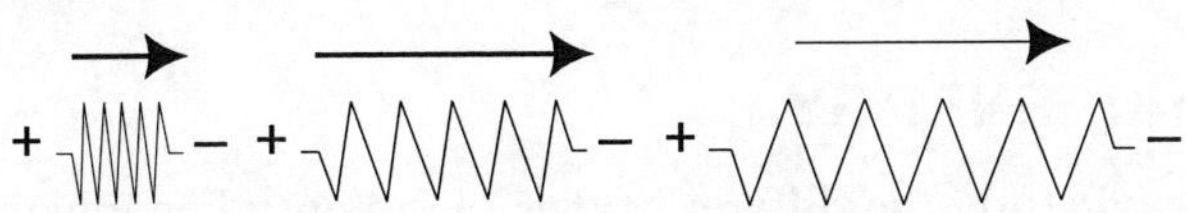

When you *release* the *negative charge*....the *electrostatic energy stored in the system of 2 charges is released* and is converted to *kinetic energy* in the *negative and positive charges...*

...like releasing a stretched spring.

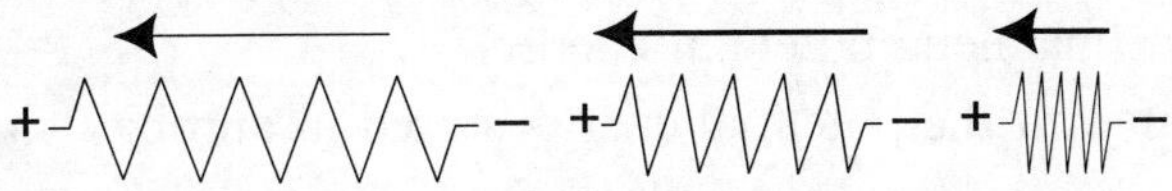

Compressing charges together to make potential energy.

You may imagine that in the palms of your hands, you are holding positive charges and by bringing your palms near one another (compress the charges), you can store energy.... you may release them to make kinetic energy....ever and ever....!

You may repeat this as many times as you please.

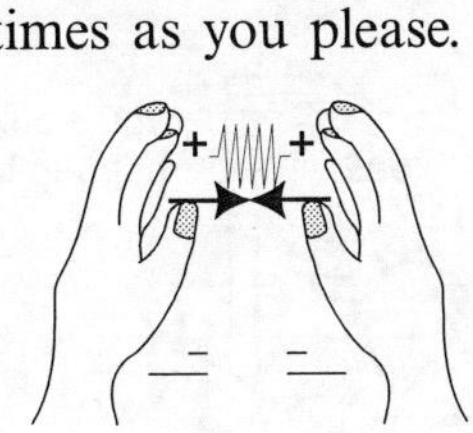

The same may be accomplished with *"opposite"* charges that are attracted and near each other. Separate them to make "potential energy" as in batteries, release them to make "kinetic energy". Do this as many times as you please....over and over....separate opposite charges and make potential energy....*release* and *convert* to kinetic energy.

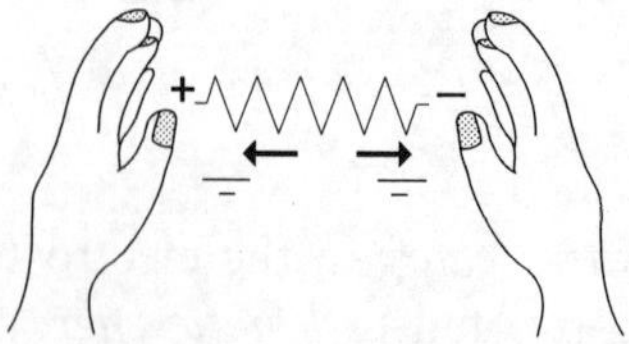

Visit http://ocw.mit.edu/ans7870/8/8.02T/f04/visualizations/electrostatics/38-videogame/38-EMVideoGame320.html and play the videogame (click "Help" for instructions) to get a feel of charge interactions. From the coulomb interactions of charges of repulsion, depicted by the visualizations in Sections 1.3 and 1.4, it is seen that the interaction spans distances in space, unlike objects that collide which we see in everyday life. The theories that describe coulomb interactions were called "action-at-a-distance" theories. The concept of "fields" that mediate the interactions was introduced towards the end of the 19th century, which formed the basis of "Classical Electrodynamics".

In subsequent chapters we will learn about specific aspects of the electric field and it's interactions with charged particles and objects and in general it's conformance to the principles of conservation of energy and momentum.

1.5 POTENTIAL ENERGY

In this section we will study the different types of potential energy and then understand the nature of electric potential energy. Potential energy is *energy*, which results from position or configuration.

An object may have the capacity for doing *work*, as a result of its position, in a gravitational field (gravitational potential energy), an electric field (electric potential energy), or a magnetic field (magnetic potential energy).

We are familiar with the potential energy stored in an object raised above the ground.

You would not like this to happen to you! Do you?

Stored energy, which can be used to do work, is called potential energy. Stored energy, can be made portable as in batteries, and capable of doing work *on demand.*

We consider next the types of potential energy.

Gravitational Potential Energy

When a rock is raised above the earth and released, the gravitational attraction between the rock and the earth, causes the rock to fall to the ground.

A raised object has *gravitational potential energy*, stored in the system of the object and the earth. Potential energy should be associated with PAIRS of interacting objects, not with single objects. When raised, the object does not change in mass, volume, color or shape. It gained potential energy due to its position in the earth's gravitational field. This is the type of potential energy we are most familiar with! Take away the earth, from a raised stone, and gravitational potential energy loses all meaning.

Other Types of Potential Energy

Among the less familiar ones though important types are:

Strain (or Elastic) Potential Energy

If we *stretch* or *compress* a spring and then release it, the spring moves back toward its original length. The stretched or compressed spring has *strain potential energy*, because it has the *potential to move.* It stores potential energy, by straining the bonds between the atoms, of an elastic material.

Examples of strain potential energy

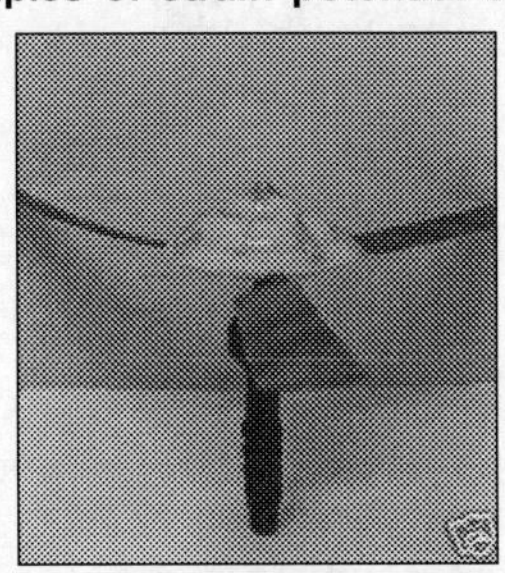

As a child, most of us must have played with this toy!

When the bowstring is released the arrow gains kinetic energy

One more type which is of importance to students of electricity and magnetism is:

Potential Energy of Chemical Reactions

Chemical Potential Energy: Chemical potential energy exists, because atoms and molecules can *take in* or *give off* energy when their chemical bonds are formed or broken.

Batteries store chemical potential energy, which is converted into electrical energy when the battery *operates*.

Potential Energy of Electric Fields

"Electrical potential energy" is stored in electric fields (Refer MIT, USA 8.022 E&M Lecture 6) when positive and negative electric charges *are separated* by an external agent, against their force of attraction. The *amount* of stored energy depends on the number of separated charges *and* on the distance they are separated.

We will learn in Chapter 2 and Appendix A, how energy is "*stored*" in the electric fields.

Potential Energy of Magnetic Moments

Potential energy is stored in magnetic moments, when a current carrying loop is placed in a magnetic field, which we shall discuss in Chapter 8.

Four Fundamental Forces of Nature

The potential energy forms are related to the four fundamental forces of nature from which all other forces are derived. In Appendix D, a description of the four fundamental forces is provided.

1.6 ENERGY IN ELECTRIC FIELDS

An *electric field* surrounds a charge (created by the charge), and its strength falls off as the square of the distance ($1/d^2$) from that charge. The field vector (lengths represent field strength), at representative locations in a two-dimensional slice of space containing the charge, are shown in the diagram below. The vectors are drawn *from* (location $P1$) for positive charges, and *to* (location *P*2, to avoid confusion showing the vector through the negative charge) for negative charges.

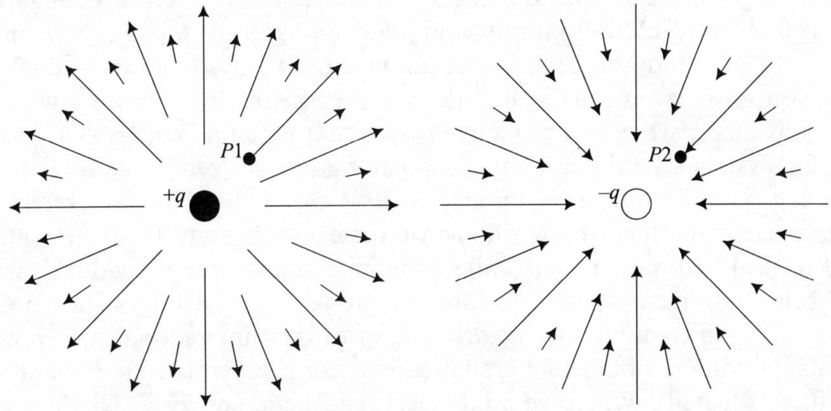

A correct visualization of the electric field surrounding a positive point charge, is afforded in the figure that takes into account the fact that the electric field exists in *all* directions surrounding the charge and fills the entire universe. Visit www.falstad.com for a 3-dimensional view of the field.

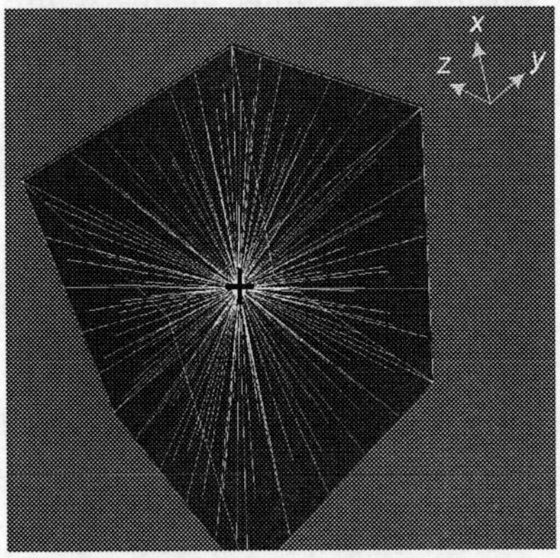

From Einstein's famous E = mc^2 equation, we can show that the energy contained in one gram of matter is 90×10^{12} joules ! This energy is equivalent to that produced by a 100 watt bulb operating for 30,000 years ! But, we do not use tiny grains of matter at home and convert these, to obtain such huge amounts of energy for use in our daily lives!

The picture associating energy with charges is also complex, and is the subject of ongoing study by physicists. Every proton and electron built by nature has an electric field, and the field surrounds the charge. The electric field, whether powerful or weak due to charge in large or small amounts, exists over the entire universe. The field energy of an electron, computed by integrating its' field squared (E^2) over infinite space, is infinite due to the $1/r^2$ form of the electric field.[8] Imagine drawing several thin wires from a block of metal in all directions, making them thinner and thinner as you move away. The electric field around a point charge (such as an electron, or a proton) is like the wires pulled from the block; but, is special, in that it's spread is uniform (*alteration* of space) and not discrete. The rest energy (see Appendix B "Energy stored in a two electron configuration") of an electron is 0.511×10^6 eV, not infinity, and *relating* this to the integral of E^2 is fraught with difficulties. In our study of circuits, the amount of energy of an individual point charge, be it an electron or a proton, is of little consequence, as the following will suggest. Note that for individual particles we associate *energy*; not potential energy. The electric *potential energy* of a system containing *only one* point charge, is "zero", as no energy is required to move the point charge from infinity to its location. But, we are rarely, (I don't feel like "superman", though I have zillions of these in my body !) in the study of circuits, required to deal with how much (finite or infinite) energy the individual particle has, and have to bring our focus to moving around pre-fabricated charges; not one, but several (electrons, say) and compute the *potential energy* of their interactions, which will be in more manageable amounts, obtained by summation using potentials or integration using the electric field (squared)! When we study electromagnetic energy in Chapter 4, we will see that the manageable amounts of energy required in most circuit operations, are *transported* by the electric and magnetic fields.

The electric potential energy of an electron and proton (or nucleus) system, is due to the proximity of the electron to the proton (or nucleus), and the attractive force on the electron by the proton (or nucleus) as shown in the figure.

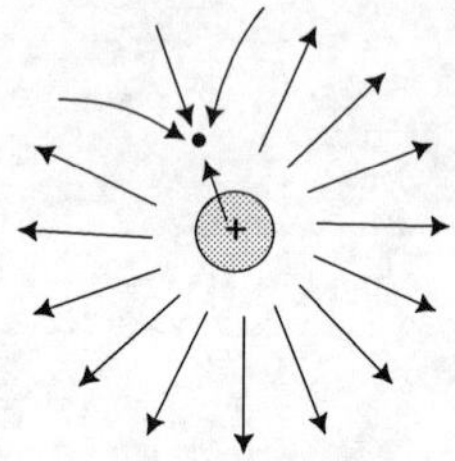

Take away the proton or nucleus, and electric potential energy loses meaning. It is the electric field (negligibly small or large) of the two charges interacting, which connected the two charges, and potential energy is stored in the system of the two charges separated a distance.

Note: Physicists are prone to say the *potential energy of an electron*, when what they *really* mean is the potential energy of the *electron and nucleus (or proton) system of charges*. See Appendix B "Energy stored in a two electron configuration" when two electrons interact at a distance of 10^{-10} meters.

We restrict our discussion of the electric field sourced by charges, to its strength, direction and effect on other point charges, at distances a little away from the charges (idealized as point charges, massless and dimensionless), which is sufficient for the study of most circuits.

The reason for this will become clear in Sections 1.13, 1.18 and 2.26 when we discuss the structure of atoms, how to visualize 'current' and neutrality in conductors carrying currents. Also, see 'Point Charges' in Appendix B.

The electric field which surrounds a stationary point charge *does not* exert a force on that same charge.

In a system of charges that gained or lost potential energy, the charges do not change color, volume or mass !

The potential energy of the system, is due to the relative positions the charges are made to occupy, stretching or compressing the invisible springs between individual charges.

When we bring an uncharged conductor near an isolated positively charged sphere, then charges will be induced on the surface of the conductor, and there will be a net attraction between the sphere and the conductor as seen in the figure.

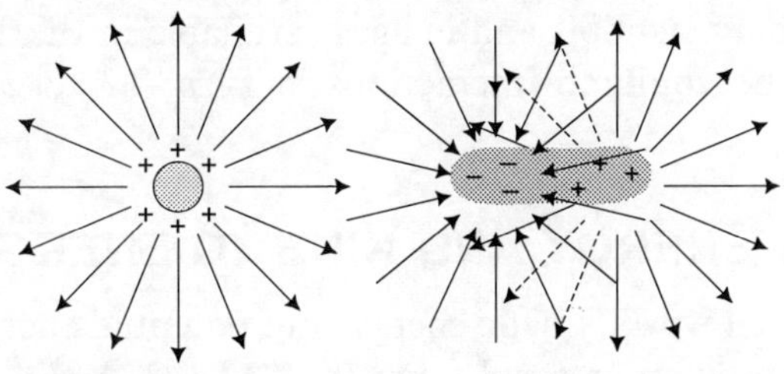

Positively charged sphere   Conductor

When the conductor was uncharged and far away from the positively charged sphere, there was no electric field perceptible surrounding the conductor and we could say that it was *neutral*.

When the positively charged sphere was brought near however, the charges of the conductor separate, and one end charges positive progressively while the other end charges negative progressively. In the figure above, the electric field lines are drawn crossing each other (which are never to be shown this way) merely to illustrate the intermediate stage of development of the resultant field around the system, the final stage of which is shown in the figure below.

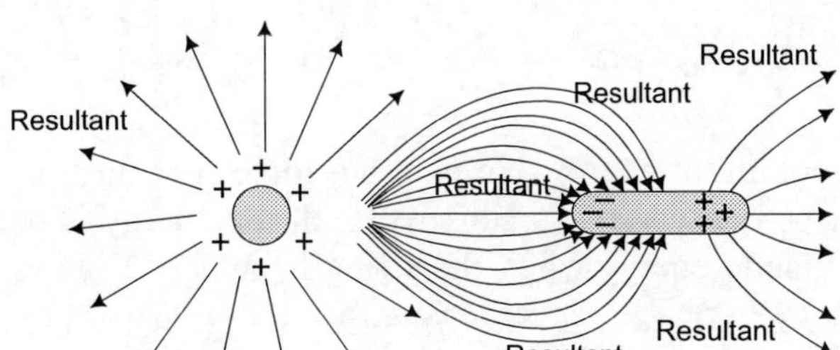

Positively charged sphere   Conductor

A resultant electric field that is non-zero, appears surrounding the conductor.

The system did some work to *separate* the charges when the conductor was brought near. The energy of the system has *gone down*, in comparison to the energy in the field, *before* the conductor was brought near the positively charged sphere.

Note very carefully that the electric field surrounding the system, is due to the electric fields of individual charges, and is a resultant computed from these fields.

The fields of individual charges *ever exist*. They existed, since the time the charges came into existence and will ever remain with the charges.

What is computed (or measured) and shown of the system, is a *resultant* of the fields due to all charges.

It will be instructive for the reader to read Chapter 1 and Chapter 2 (upto and including Sections 2.4.3) in the visualizations provided by Hermann Haertel (written Härtel in German) in the article "A Qualitative Approach to Electricity: A Guide to Visualization of Electrodynamics using symbols and images" available in the Teacher's Guide in the CD. Making sketches of the coulomb interactions in Figs. 2.3 to 2.6 and of Fig. 2.8 will be useful.

1.7 POTENTIAL ENERGY AND KINETIC ENERGY OF CHARGES

The relationship between work, kinetic energy, and potential energy for a system comprising charges is : Work done by an external force $W = \Delta KE + \Delta PE$ where KE = Kinetic energy and PE = Potential energy.[26]

An Example

Two positively charged balls are tied together by a string. One ball has a mass of 30 g and a charge of 1 μC; the other has a mass of 40 g and a charge of 2 μC. The distance between them is 5 cm. The ball with the smaller charge has a mass of 30 g; the other ball has a mass of 40 g. Initially they are at rest, but when the string is cut, they move apart due to mutual repulsion.

When they are a long way away from each other, how fast are they going?

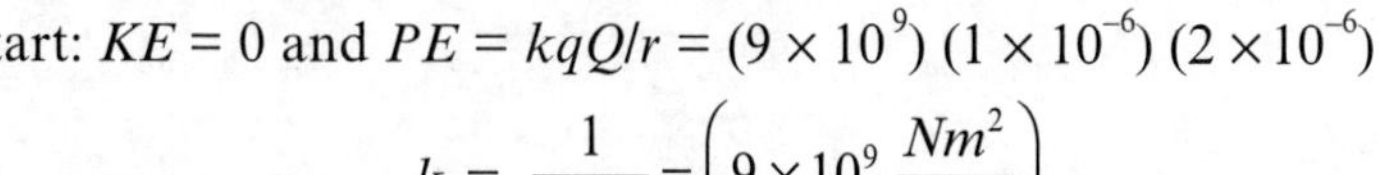

Let's start by looking at energy. No external forces act on this system of two charges, so the energy must be conserved. To start with, all the energy is potential energy; this will be converted into kinetic energy when the string is cut.

Energy at the start: $KE = 0$ and $PE = kqQ/r = (9 \times 10^9)\,(1 \times 10^{-6})\,(2 \times 10^{-6}) / 0.05 = 0.3596\ J$

where $$k = \frac{1}{4\pi\varepsilon_0} = \left(9 \times 10^9\ \frac{Nm^2}{C^2}\right).$$

When the balls are *very far apart* the r in the equation for potential energy will be large, making the potential energy *negligibly small.*

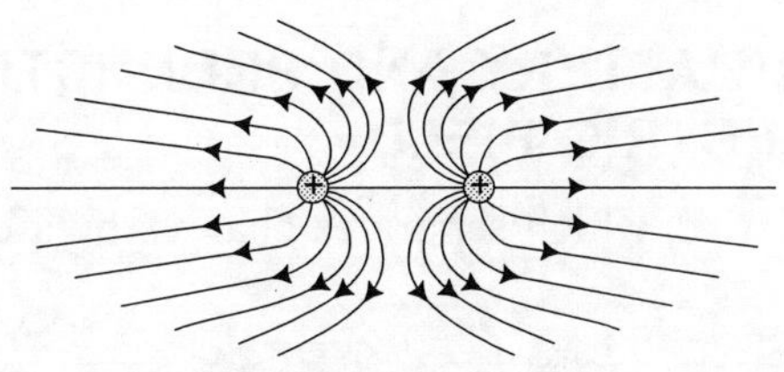

Before the string is cut

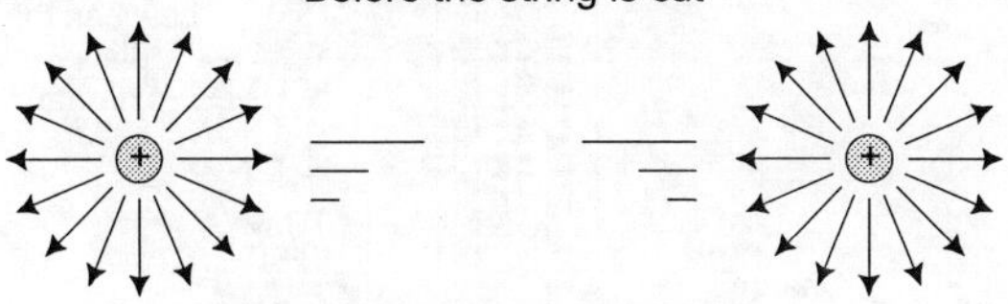

The charged balls are far apart

Energy at the end: $KE = \frac{1}{2}(m_1v_1^2) + \frac{1}{2}(m_2v_2^2)$ and $PE = 0$.

Energy is conserved, so the kinetic energy at the end is equal to the potential energy at the start:

$$\frac{1}{2}(m_1v_1^2) + \frac{1}{2}(m_2v_2^2) = 0.3596\ J.$$

The masses are known, but the two velocities are not. To solve for the velocities, we need another relationship between them.

Because no external forces act on the system, momentum will also be conserved.

Before the string is cut, the momentum is zero, so the momentum has to be zero all the way along.

The momentum of one ball must be equal and opposite to the momentum of the other, so,

$$m_1v_1 = m_2v_2 \text{ and therefore,}$$
$$v_1 = (m_2/m_1)\ v_2 = 4/3\ v_2$$

Plugging this into the energy equation gives:

$$m_1(4/3\ v_2)^2 + (m_2\ v_2)^2 = 2 \times 0.3596$$

This gives $v_2^2 = 0.7192/[1.778\ m_1 + m_2] = 7.7057$ where $m_1 = 30$ g, and $m_2 = 40$ g.

So, $v_2 = 2.78$ m/s and $v_1 = 3.70$ m/s. The speeds are tiny, compared with the speed of light, about 3×10^8 meters per second. We shall see in Chapter 4 that the energy transferred *in electric circuits* does not lie in the kinetic energy of the tiny almost weightless moving electrons though numbering in the millions!

One Newton, in everday life is approximately the force exerted by a medium sized apple on a resting surface, and **one Joule** is approximately the energy required to lift the apple one meter straight up.

In the next section, we shall see how a net electric field is produced between the terminals of a battery.

1.8 ACTION OF A BATTERY AND PRODUCTION OF A NET ELECTRIC FIELD

A battery is used as a device, which separates charges by a chemical reaction as shown in Fig. 1.1.

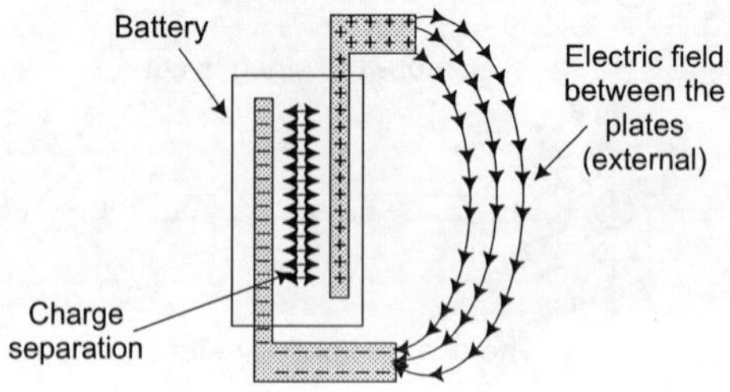

Fig. 1.1 A charged battery.

Charges are *separated* by chemical reactions. The exact process is quite complex and need not concern us. Positive ions form at the positive plate, and negative ions form at the negative plate.

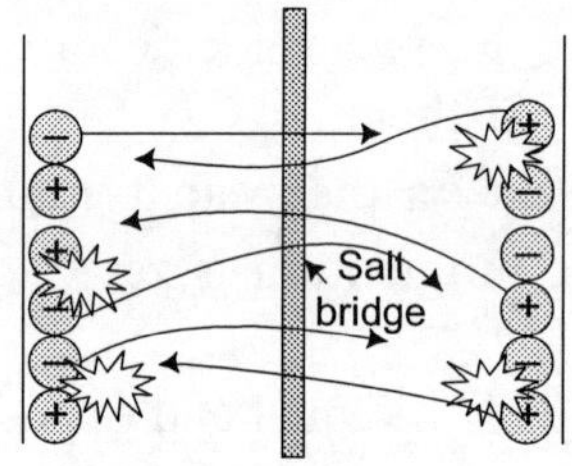

This separation is performed by a chemical process, resulting in a state of equilibrium with a certain number of electrons on the surface of the negative terminal, and the same number of electrons missing on the surface of the positive terminal.

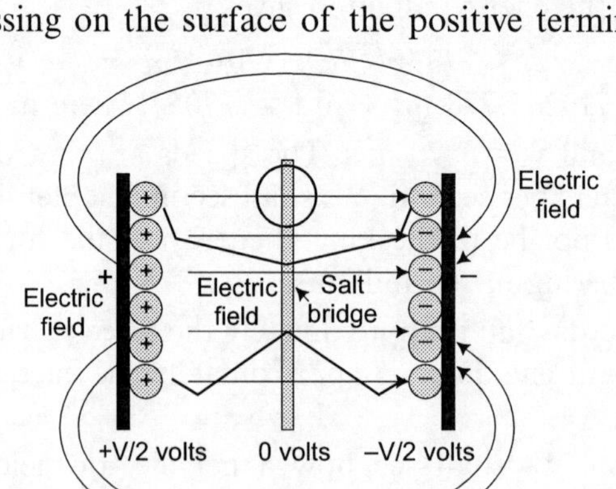

A dynamic equilibrium is reached, when the attracting Coulomb forces between these separated charges are equal to the separating chemical forces. There exists an electric field outside, as well as, inside the battery between the separated charges on the plates.

Potential energy is stored in the electric field of the *configuration* of separated charges!

Conservation of Charge[2]

The net charge of an object is the sum of all the positive charges minus the sum of all the negative charges. Usually, matter has the same number of protons as electrons, so its net charge, is zero. *The net charge of a closed system does not change.* In an extremely wide variety of experiments, no one has ever observed electric charge to be created or destroyed. These observations are summarized by the important principle called *conservation of charge*: the net charge of the universe or of any closed system, does not change. By a closed system, we mean, any region through whose boundaries charged particles neither enter nor leave. For example, the charges in a battery and on its terminals are conserved, and constitute a closed system. For a V-volt battery not having any *external* excess charge, i.e., charge not local to the closed system, at the physical center in the space between its terminals, there will be an equipotential line of zero volts, with the negative terminal at –V/2 volts and the positive terminal at +V/2 volts, produced entirely by internal charge separation. There are high-energy particle reactions that create or destroy charged particles, but always in a way that the *net* charge does not change.

We have introduced a new term "volts". For the purpose of discussing currents in the subsequent sections this quantity is however not important. We will learn about Electro Motive Force, Potential, Potential difference and volts alongwith a working definition for voltage in Sections 2.17, 2.19 and 2.20.

Important: Whenever it is said that a material is electrically neutral, it is merely a statement of account of an equal number of positive and negative charges and that the conservation of charge is preserved in a closed system, and *not* a statement about how these are *distributed* in the system; as dipoles or as electron hole pairs in semiconductors *and* whether there is a resultant electric field around them, unless this is explicitly stated.

1.9 SURFACE CHARGES AND FIELDS IN AND AROUND CIRCUITS

Indicated below are typical questions students raise in the study of electric circuits and circuit behavior, reported by Hermann Härtel, a scientist at the Institute for Science Education, University of Kiel who studied how students attempt to understand the principles and processes of electrical circuits. Later sections develop answers to these questions. (Reproduced by permission, Hermann Härtel).

Student Questions

The analysis of circuits in terms of surface charge provides answers to some profound questions students sometimes raise, *questions which may have no satisfactory answer* within the context of traditional analysis in terms solely of potential. Härtel reports students asking, *"What is the difference in the conditions at the two ends of a resistor in which there is a current"?*

It is quite unsatisfying to say merely that the potential is different.

Consider a battery connected to a bulb with a pair of wires as shown in the figure below from "Unified treatment of electrostatics and circuits" by Chabay and Sherwood.[1] (reproduced from Ref. [1] with permission).

Is electric current due to charges in and around the battery?

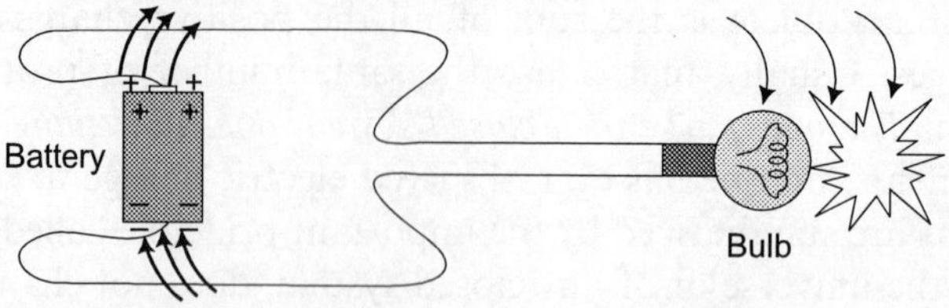

If we bring the bulb closer to the battery, does the bulb glow brighter?

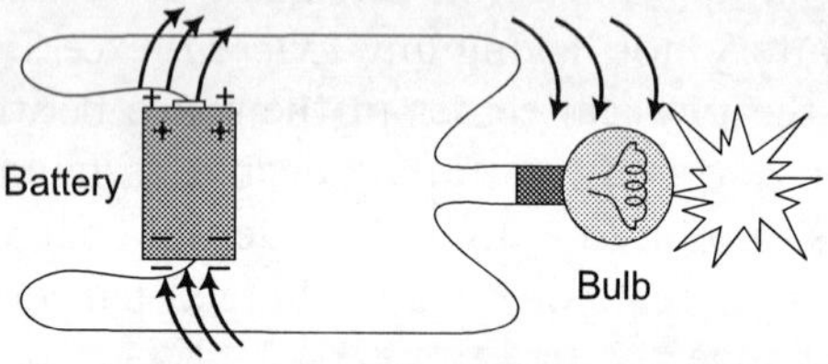

The bulb glows with the *same* brightness even when brought *closer* to the battery.

Coulomb's law predicts that the electric field varies inversely proportional to the square of the distance from the source charge. If the bulb is brought closer to the battery, then there should be a corresponding *increase* in the electric field due to the charges on the battery that would exert more force on the conduction band electrons in the wires and in the filament. But, we observe that there is no increase in the current in the wires and in the brightness of the bulb.

Now rotate the bulb by 90 degrees.

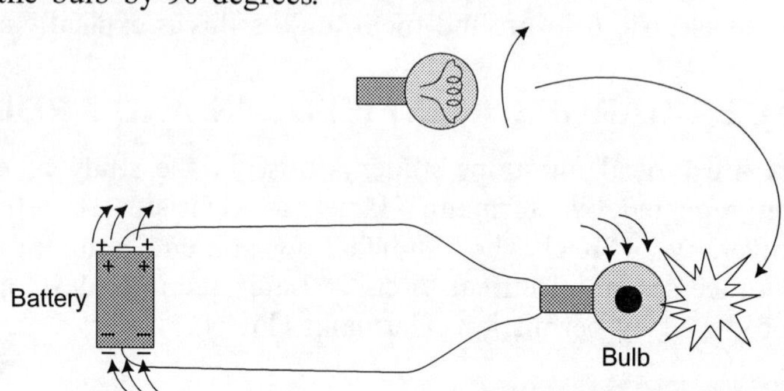

Bulb rotated by 90 degrees

Since the electric field is now perpendicular to the filament, the bulb *should* stop glowing.

The bulb does *not* stop glowing.

Therefore, the electric field from the charges on the battery, has *no significant effect* on the bulb filament.

Evidently, charges in and on the battery *cannot* be the only source of the electric field inside the bulb filament.

It is therefore reasonable to conclude that some *other* charges, in addition to those on the battery, are responsible for producing an electric field that is *pushing* the electrons.

Or,

is our conclusion wrong, and the flowing electrons are simply pushing each other through the wire?

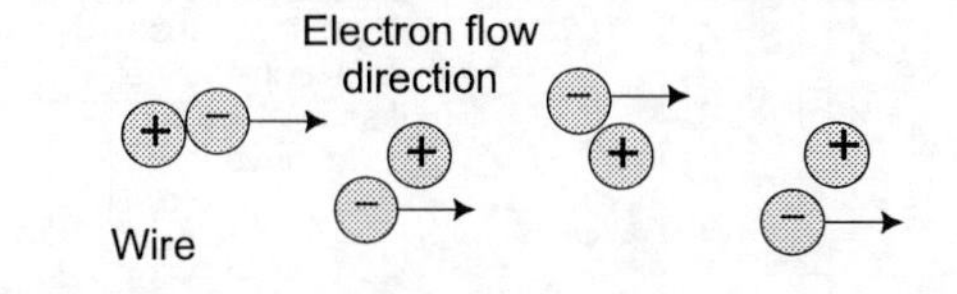

The flowing electrons inside a wire *cannot push each other* (like water molecules that are neutral inside a pipe) through the wire, because on an average, the *repulsion* by any electron, is *canceled* by the *attraction* of a nearby positive atomic core.

One may think that the flowing electrons push each other through the wire as in the figure above, like water molecules pushing neighboring water molecules through a hose.

However, this cannot be true, because averaged over a few atomic diameters, the interior of the metal is everywhere neutral, and on average, the repulsion between flowing electrons is canceled by the attraction to positive atomic cores.

Some *other* charges in addition to those on the battery, are responsible for producing an electric field that *pushes* the electrons.

This is one of the reasons why, an analogy between electric current and the flow of water, can be misleading.

It is easy to understand the constant current, in terms of potential difference in a circuit comprising a battery, wires and a resistor. The potential 'drop' across the resistor, is always nearly equal to the emf of the battery, and the electric field inside the resistor is simply the gradient of the constant potential drop. Is this satisfactory?

We know that charges are the only source of electric field (in the absence of varying magnetic fields). Then, where are the charges, which are responsible for the field inside the resistor?

Hermann Härtel published a paper "Constant Electric Current and the distribution of Surface Charges1"[6] in which he detailed the process of conduction and showed how "surface charges" are responsible for the phenomenon of conduction.

We now introduce surface charges and the nature of the electric field due to surface charges, to gain an intuitive feel for its presence in wires in circuits, and in the next section, go into some detail about the configuration of the surface charges that produce it.

Surface Charges

(Reproduced with permission from Hermann Härtel)

According to Härtel, surface charges are present, whenever a wire carries an electric current, and are necessary for a causal explanation of how an electric current *actually* works.

Consider the case of a battery connected by wires to a resistor as shown in Fig. 1.2.

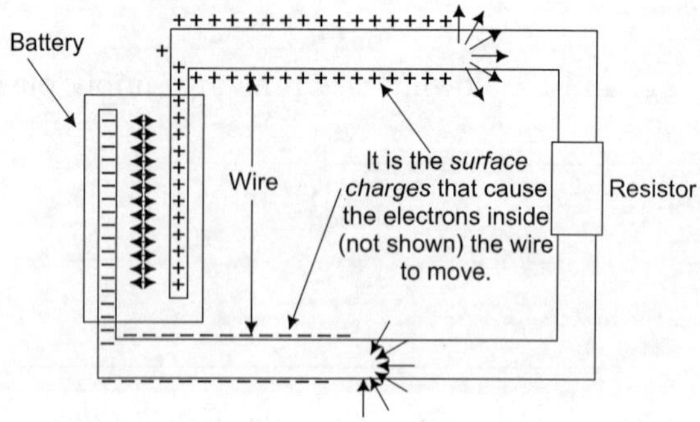

Fig. 1.2 Development of surface charges—intermediate stage.

When the circuit is closed, a potential difference is applied across it, and the charge imbalance that exists between the two terminals of the voltage source, causes a field to appear across the terminals of the circuit.

To deliver energy to the circuit, surface charges appear, and the electric field due to these charges apply forces that cause the movable charges in the wires to begin a slow flowing motion: $F = qE$, where F is the force in Newtons, q is the charge in Coulombs, and E the electric field in Newtons per Coulomb. The electric field is also exactly equal to volts/meter. See the end of Appendix A for a derivation of the equivalence of the two units of the electric field.

The surface charges develop on the wire connecting the battery with the resistor. Shown in Fig. 1.2 is the intermediate development stage.

The way in which these charges are distributed is shown in Fig. 1.3.

Figure 1.3 shows the final stage of development of surface charges. The surface charges develop along the wire; positive surface charges build up at one end of the resistor, while negative surface charges develop at the other end. The surface charge density gradient, produces a powerful resultant electric field inside the resistor, which pushes the conduction electrons to establish a current in the circuit. The surface charges around a circular cross-section wire is in the form of a ring, and in Section 1.10 we shall see how it produces an axially directed electric field inside the wire.

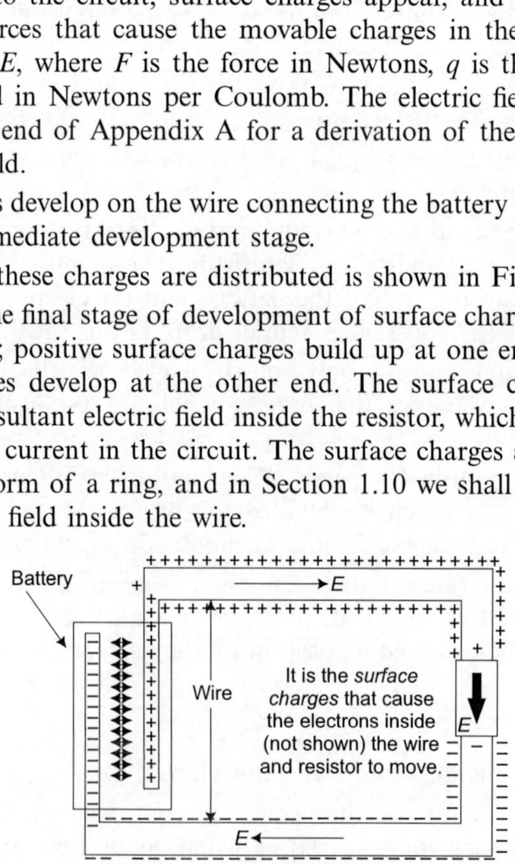

Fig. 1.3 Development of surface charges–final stage.

Another student question reported by Härtel is, *"How does the current know how to split when there are parallel branches"*? This question will be answered in Chapter 2.

Are electric fields really present in and around circuits?

The presence of electric fields in and around circuits can be demonstrated by circuits made out of conductive ink and placed on glass plates, photographs of which are shown in Fig. 1.4. The electric field in the space around the circuit due to surface charges, is demonstrated with the aid of grass seeds strewn upon the glass plate. The grass seeds orient themselves in the neighborhood of charges, because they act like electric dipoles (analog to magnetic compass needles).

These demonstrations were made by Professor Oleg Jefimenko, West Virginia University, USA, and published in a paper, "Demonstration of the electric fields of current-carrying conductors", American Journal of Physics, Vol. 30, No. 1, 1962.

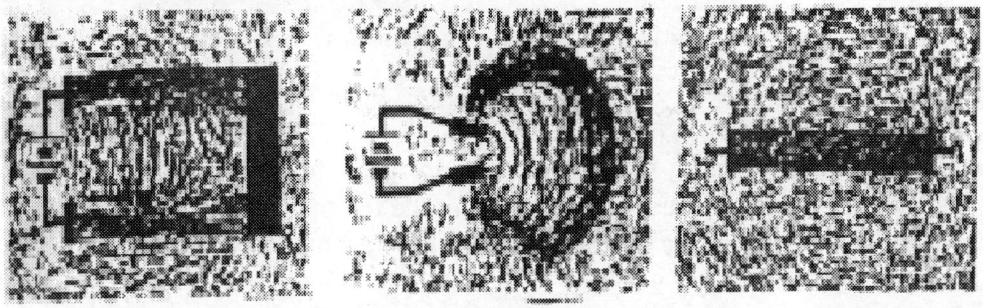

Fig. 1.4 Electric field lines around conductors.

At the end of Section 1.19, we will answer the question as to how the bulb, connected using long wires to a battery, does not increase in brightness, when brought close to the battery, or if the bulb were rotated which would make the electric field from the charges on the battery, perpendicular to the filament.

1.10 ELECTRIC FIELD DUE TO A RING OF CHARGE

It is instructive to get a feel for the electric field, its production and its strength, due to a uniform charge distribution around a ring. Such rings of charges on surfaces of circular cross-section wires used in circuits, produce an electric field *inside* the wires that establish the currents.

In the real world of electricity and electronics, point charges are only electrons (that can never be stationary) and protons. For the current viewpoint of physicists on *point charges* see Appendix B "Point charges". But note how the study of electrostatics using stationary point charges, dipoles and other charge distributions enables us to understand circuit processes and mechanisms that depend on these electrons (though moving that form clouds of charge), protons and *vibrating* lattice ions. We present a derivation of the electric field due to a ring of charge which enables electrical conduction in round uniform wires and resistors.

To derive the expression for the electric field, we make use of an important property of the electric field, the *principle of superposition.*

The Superposition Principle for Electric Fields

The superposition principle is stated thus:

The net electric field at a location in space, is the vector sum of the individual electric fields contributed by all charged particles located elsewhere. The electric field contributed by a charged particle, is unaffected by the presence of other charged particles.

Superposition as a vector sum holds true for forces in general, so it also holds true for fields, because field is a force per unit charge. For a definition and description of this important principle see Appendix A "Superposition of electric fields".

The electric field strength E at a point, is defined as $E = \dfrac{F}{q_{test}}$, where F is the force acting on the test charge q_{test} at the point, where the field E is being measured. The electric field, for points *along the axis* of a ring of charge, is of interest in conduction processes and can be calculated using the principle of superposition, as shown in the figure.

Consider a differential element dl located at the top of the ring. It contains an element of charge given by $dq = q \cdot dl/2\pi r$ where $2\pi r$ is the circumference of the ring and q is the total charge on the ring. The element sets up a differential electric field dE at point P.[13]

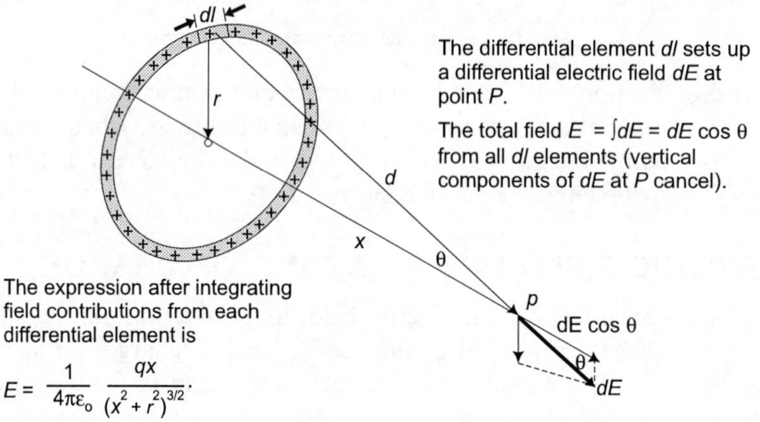

The resultant field E at P, is found by integrating the effects of all the elements that make up the ring. From symmetry, this resultant field must lie along the ring axis. The component established, perpendicular to the axis is canceled out, by an equal but opposite component established by the charge element, on the opposite side of the ring.

Thus, only the component of dE parallel to this axis, contributes to the final result $\dfrac{1}{4\pi\varepsilon_o}\dfrac{qx}{(x^2+r^2)^{3/2}}$.

See Appendix A for details of computing the field due to a ring of charge.

Nature of a Field

The nature of a "field" can be understood by analogy with a gravitational field which acts on all objects.

A field like a **gravitational field,** is a ***special*** kind of field.

It applies a force ***continuously*** or rather relentlessly...on you whether sitting, standing, walking or sleeping for 24 hours and for years...and on all objects...all the time.

It does not stop ever applying a force...***it never gets tired.***

That is the reason for the ***acceleration*** a field produces.

On the other hand, take a simple ***application of a force*** to an object ***once,*** just **once,** like ***hitting an object at rest, once.*** The object will accelerate, but will settle down to move with uniform velocity...and will ***not accelerate*** any more.

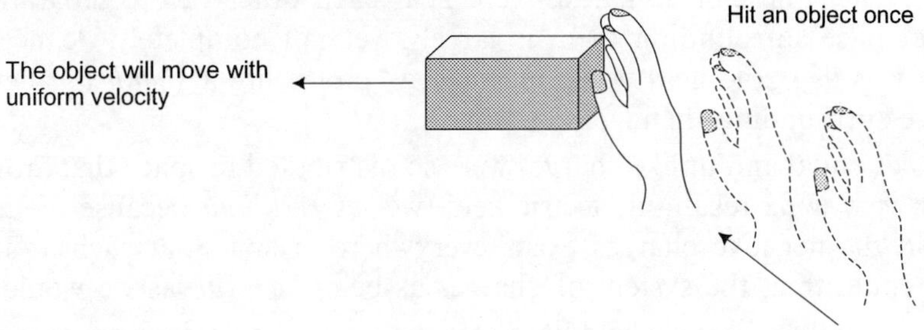

A gravitational field applies a force *continuously*, on all objects and therefore, it accelerates the objects.

The gravitational force acting on the ball released from the top, is continuous. It acts on it when held in the hand *before* release, and also *during* every instant of its movement downward after release, and also *after* it hits the ground. Therefore, the ball *accelerates*, i.e., it's velocity increases continuously. It *does not* remain uniform as it does when an object is hit, just once.

An ***electric field*** like a gravitational field, is another ***special*** kind of field.

It applies a force *continuously*, or rather relentlessly, on electric charges, in its sphere of influence.

That is the reason, for the ***acceleration*** it produces of *electric charges* within its sphere of influence.

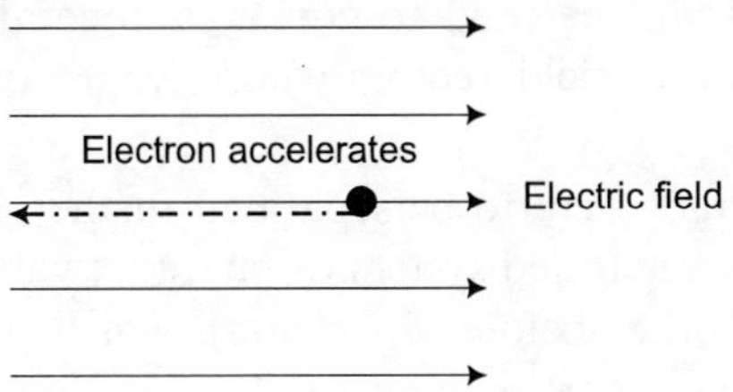

The electric field does not stop ever applying a force. It never "tires", unless if it were reduced to an insignificant force, due to its source charge unifying with an equal and opposite point charge.

Separation of Charges

When unlike charges are separated (as in the formation of dipoles), *three* things happen:

1. A net *non-zero* electric field, is produced in the space surrounding the charges.
2. A potential difference or *voltage* is created.
3. *Energy* (that caused separation) is stored in the electric field.

Every proton and electron built by nature, has an electric field, and the field surrounds the charge. When the *unlike* charges were near each other before separation, the electric field in the space surrounding them is largely, yet not completely, "canceled", and after separation, it is as if the field was "uncanceled" producing a *significant* net non-zero field in the space surrounding them.

If pairs of equal and unlike charges were so distributed in space that around the charges there was a very *weak* resultant electric field (we say *resultant* because the fields due to the positive and the negative charges exists everywhere in space...at each and every point in space), we could treat the system of charges as being *neutral,* as we would a stable atom.

We can draw thick electric field lines originating on the positive charges and terminating on the negative charges, if the charges are near each other. Then, we can draw less thicker and then thinner *appropriately* oriented field lines, representing appropriate magnitudes and field direction, to depict less strong and weak resultant electric fields at points a little away, and at distant points respectively from the charges. Remember, an electric field due to a charge exists at all points in space, and resultant fields are produced at points where there are fields due to other charges.

Depicting an electric field using field lines can be deceptive, because physicists and engineers use these to depict the most *dominant* field strengths. The key to understanding several electrostatic and circuit processes, lies in the ability to visualize the *weaker* fields at points, as either, due to being very far away from source charges, or as a resultant of powerful (sometimes), yet nearly oppositely directed fields which *really should* be depicted by appropriately sized and directed field vectors, but is not done by engineers and physicists, for the sake of clarity.

Note: In this book, field flux lines (continuous) are used to portray the electric field in an expanse of space, and will be used to convey a general feel for the field therein. For more accurate representations, field vectors whose length or thickness represents strength are used.

If now the unlike charges were to be separated further, a *significant* resultant field will begin to appear *around* the separated system of charges (where it was weak when the source charges were nearer each other before separation), which seemingly grows and spreads in the space surrounding the charges. The energy that was given to cause separation, will be stored in the electric field, and the potential difference will increase.

To view the electric field that develops as a dipole is formed by separating charges, visit the Massachusetts Institute of Technology webpage http://ocw.mit.edu/ans7870/18/18.013a/textbook/HTML/chapter28/section04.html and in the first applet, set *all* charges *except* $q1$ and $q2$ equal to zero and click "plot charges and field lines". Then, drag the charges $q1$ and $q2$ apart and observe the development of the resultant electric field around the dipole. If you drag them close to each other, watch the resultant electric field around the dipole collapsing to zero. See screen shots from the MIT webpage in "Growth of electric field.pdf" in the accompanying CD. See the "Links_to_animations_articles" pdf in the accompanying CD for another interactive webpage that shows the dipole and its electric field in 3D.

We shall in subsequent sections, learn that with real electrons and protons of atoms, it is not possible to make the resultant electric field (around them), zero.

1.11 DEVELOPMENT OF SURFACE CHARGE

If a current of 1 amp is flowing through a wire, there are about 10^{19} electrons passing through each cross-section, within one second.

Because of the enormous strength of the Coulomb interaction, and the very high mobility of electrons in metals, it takes only a few electrons at the surface of the wire to produce an electric field, which can overcome the resistance of a metallic wire and push 10^{19} electrons around in a circuit.

We now study qualitatively, various configurations and distribution of excess (local) surface charges of wires, and the resulting electric field inside the wires recollecting from Section 1.10 how a ring of charge produces an axially directed electric field.

Let us first assume that the surface charges have a constant distribution[6] around a wire as shown in Fig. 1.5.

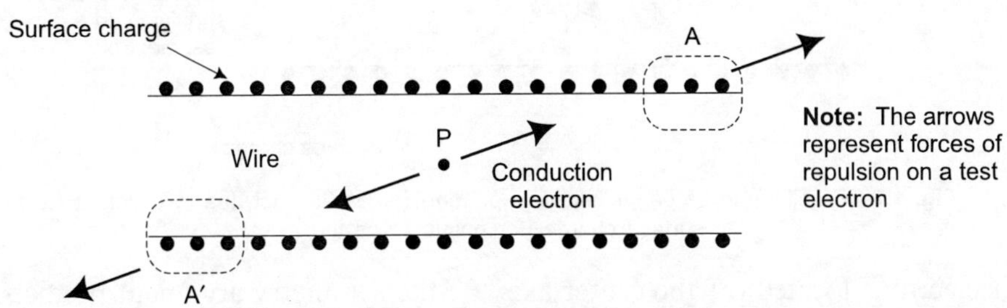

Fig. 1.5 Influence of a constant excess surface charge (for points P on the center axis of the wire).

Along the wire, for every small surface area A containing a certain number of electrons, there exists another area A' on the opposite side with the same number of electrons, which will balance the interaction with a test conduction band electron at point P. Therefore, no resultant force is produced on the conduction electron located at the point P, and a constant excess (local) surface charge cannot produce a current in the wire.

A ring of surface charges has the strongest interaction with the free electrons, within the same cross-section.

But, because of the circular symmetry of the wire however, the interaction of the surface charges cancel each other, and there is no net force on electrons situated at the same cross-section, as shown in Fig. 1.6.

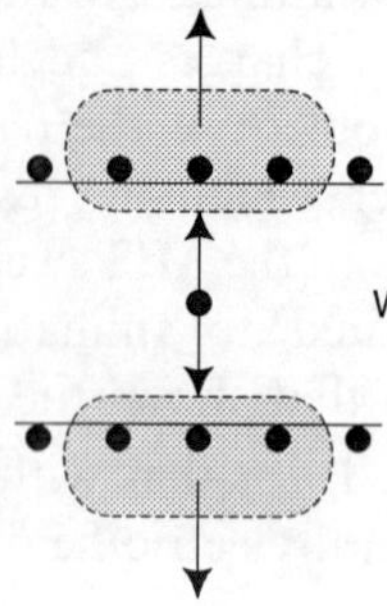

Fig. 1.6 Influence of opposite surface charges.

For reasons of symmetry, it can be concluded that the interaction of equally distributed electrons on the surface of the round wire with centrally located electrons inside the wire, cancels to zero. Therefore, no conduction mechanism is possible for centrally located electrons in the wire, with such a constant distribution of surface charge.

We next consider the effects of a constant excess (local) surface charge distribution on electrons, which are located a little away from the center of the wire, as shown in Fig. 1.7. The forces are resolved into their parallel and perpendicular components.

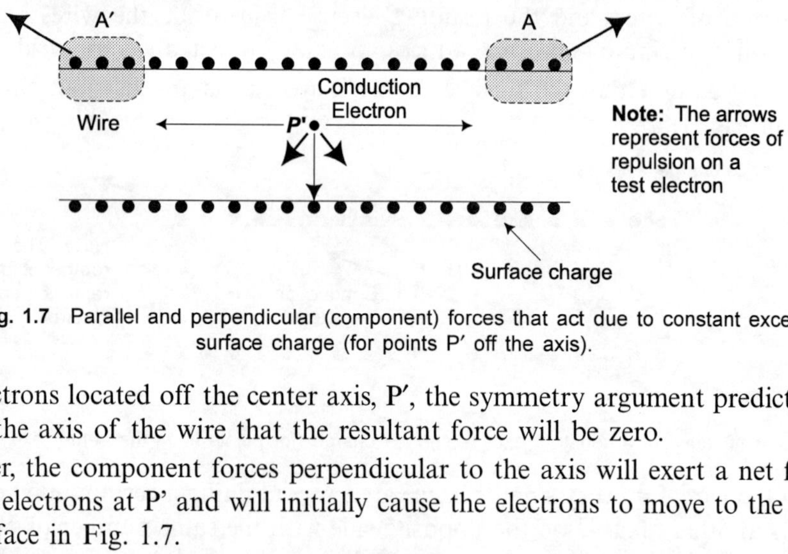

Fig. 1.7 Parallel and perpendicular (component) forces that act due to constant excess surface charge (for points P′ off the axis).

For electrons located off the center axis, P′, the symmetry argument predicts, for forces parallel to the axis of the wire that the resultant force will be zero.

However, the component forces perpendicular to the axis will exert a net force on the conduction electrons at P' and will initially cause the electrons to move to the surface, the bottom surface in Fig. 1.7.

This will cause a redistribution of the excess (local) electrons on the bottom surface of the wire as shown in Fig. 1.8, and the forces due to these excess electrons on the bottom surface will cause the resultant force on the electrons at P′ to become zero. Again, no conduction mechanism is possible, even for electrons located away from the central axis of the wire with a constant excess surface charge distribution.

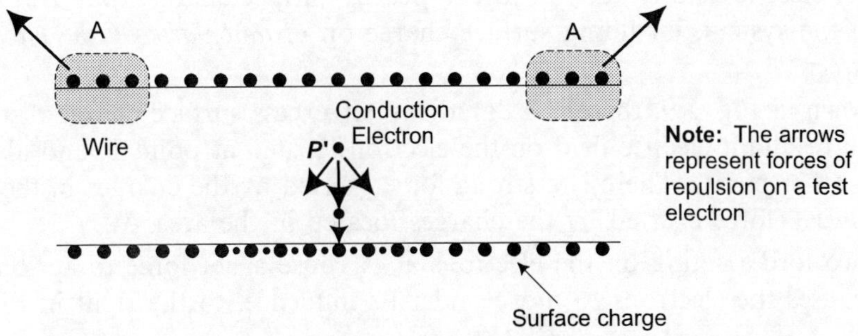

Fig. 1.8 Perpendicular (component) forces that act due to constant surface charge (for points P′ off the axis).

We have seen that *no* "net" electric field in any one direction can be produced, *if* the distribution of surface charge is *spread uniformly* over the surface of conductors.

There has to be some kind of distribution, which *can* produce a *non-zero net* electric field that will *force* the conduction band electrons in conductors into *motion*.

1.12 MECHANISM OF CONDUCTION

We now consider the case when there is a non-uniform distribution of excess (local) surface charges in a circuit due to the applied voltage, which could produce a resultant electric field inside the wires.

In short, to break this equilibrium between the forces generated by the equally distributed surface charges, as was seen in the Figs. 1.5-1.8 and to exert a driving force on the electrons within the wire, it is necessary to "change" this symmetric distribution of surface charges.[6]

On the surfaces of the wires, there have to be more excess (local) surface charges on one side than on the other side, so that the two interactions of opposite areas A and A′ do not balance.

DEVELOPMENT OF SURFACE CHARGE AND THE REAL MECHANISM OF CONDUCTION

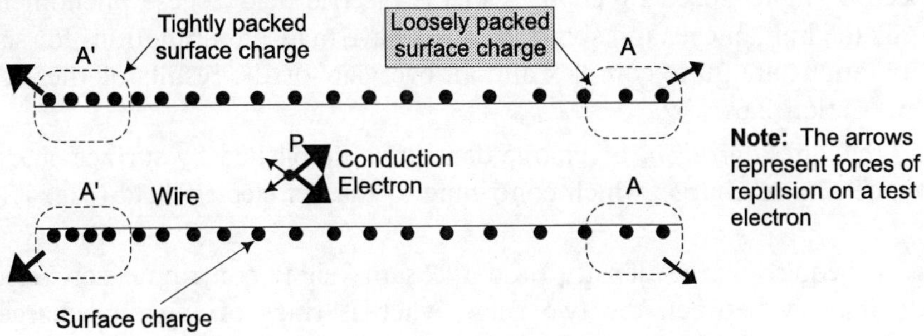

Fig. 1.9 Gradient in excess surface charge distribution.

The *net* electric field is due not just to nearby surface charges, but also to ***all*** the other charges in the system, including surface charge on *distant parts of the circuit and charges on the battery.*

As shown in Fig. 1.9, there is a "gradient" of excess surface charge distribution which produces a resultant electric field on the electron located at point P, and also on electrons located at other points. There is a strong force exerted by the charges in the area A′, while there is a weak force exerted by the charges located in the area A.

The two forces acting on the electrons at **P,** cause a net force to act on the electrons, and this causes the electrons to move under its influence, to the right in Fig. 1.9.

This, is the real mechanism of conduction.

One could imagine positive surface charges in the area A, and negative charges in area A′, in Fig. 1.9, and such a surface charge distribution with charges of opposite signs can also produce the electric fields inside wires.

In Fig. 1.10 is shown a full view of the rings of the same sign distributed with a gradient on surfaces of wires, the electric fields that they produce and the resulting electric field E_{net} that enables conduction.

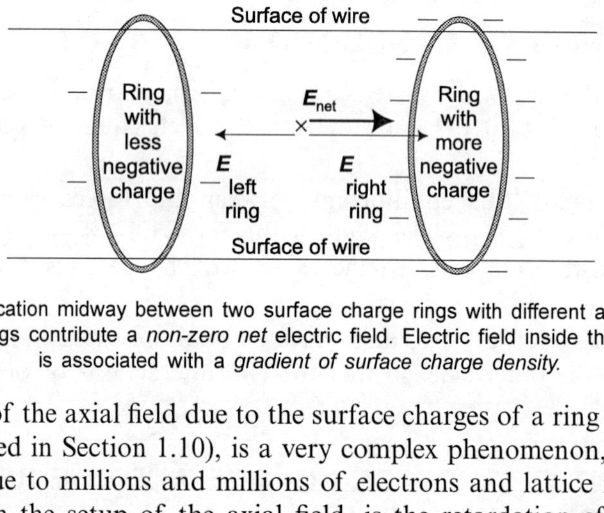

Fig. 1.10 At a location midway between two surface charge rings with different amounts of charge, the rings contribute a *non-zero net* electric field. Electric field inside the wire is associated with a *gradient of surface charge density.*

The setting up of the axial field due to the surface charges of a ring (at points along the axis, as was described in Section 1.10), is a very complex phenomenon, because it involves the electric fields due to millions and millions of electrons and lattice ions. Adding to the complexity to graph the setup of the axial field, is the retardation effect because of the finite (speed of light) speed of changes in the electric field. These phenomena cannot be observed by the human eye, and some scientists have made computations for simple circuits. The computations are quite complex and an overview of the results of these computations is given in Section 2.6.

Along the wire a *gradient* of charge density is established by surface charge rings with different amounts of charge, which contribute to the net electric field (Fig. 1.10) inside the wire.

Rings of equal charge density (and the same sign) contribute zero electric field at a location midway between the two rings, whereas rings of unequal charge density (or different sign), contribute a non-zero field at that location (Fig. 1.9).

In one particularly simple *geometry* of a long straight coaxial cable driven by a battery at one end and short-circuited at the other end, the uniform electric field inside the center conductor, is actually produced by a constant gradient of surface charge density.

In other circuit geometries, the distribution of surface charge may be arbitrarily complicated, even though the pattern of electric field they produce, is simple.

The analysis of surface charge geometries and their resultant electric fields inside wires of varying sizes and arranged in different layouts, was studied by a scientist A. Walz.

According to studies in experiments conducted by him, in some simple cases, the dominant contribution to the electric field inside a wire may be made primarily by nearby surface charges, because the $1/d^2$ character of the Coulomb interaction may make the contributions of distant charges, quite small. To the extent that we can neglect the effects of distant surface charges, we can make qualitative analyses in terms of local causes and local effects.

Sketched in Fig. 1.11 are the results of an analysis by Walz of the relative importance of nearby and far away charges, when there is a constant gradient of surface charge along a straight wire[1]. One finds that the peak contribution dE to the field is made by surface charges at a distance of $1.4r$, from which Walz concluded that E is due to mainly nearby charges.

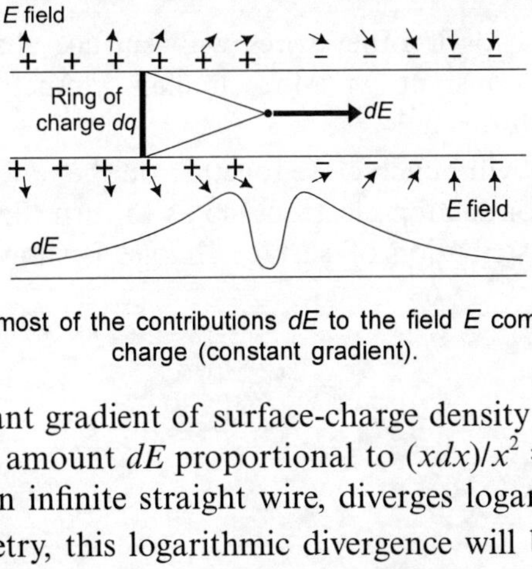

Fig. 1.11 According to Walz, most of the contributions dE to the field E come from nearby rings of surface charge (constant gradient).

However, with a constant gradient of surface-charge density (dq proportional to xdx) a distant ring contributes an amount dE proportional to $(xdx)/x^2 = dx/x$ so, the contribution of distant charges along an infinite straight wire, diverges logarithmically.

In a real circuit geometry, this logarithmic divergence will be cut off due to bends in the wire. While Walz's analysis suggests that nearby charges make a major contribution to the electric field in simple geometries, it is easy to find examples where this approximation breaks down.

For example, if a wire crosses itself or there are multiple loops as in a coil, longitudinally remote but geometrically nearby sections make large contributions to the field.

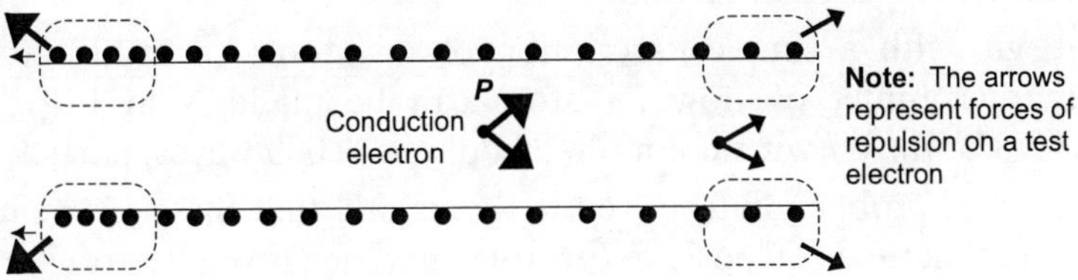

Fig. 1.12 Influence of surface charge at different distances.

Going back to the situation (see Fig. 1.12) where we have a surface charge *gradient*, when moving away from this cross-section, the components in the direction of the wire do not cancel any more; and the resulting forces increase. They reach a maximum at a distance two to five times larger than the diameter of the wire. At larger distances, the interaction vanishes rapidly, because the distance increases.

It is, therefore, the influence of a particular surface charge, at distances comparable to the diameter of the wire for a long straight wire, which exerts maximum force on the free electrons. A detailed discussion of the effect is available in Ref. [37].

For a long and thin wire, it can be shown by calculation that a uniform decrease of surface charges in the direction of the axis (a surface charge with a constant gradient in the X-direction), results in a *uniform and constant force for all the millions and millions of electrons* in the direction parallel to the axis of the wire, *everywhere inside the wire*.

The **electric field** due to the surface charge**s fills the wire** along its entire length.

This is easily proved by Gauss's Law. We will see in Section 1.18 and Appendix A how we can use Gauss's Law to show *neutrality* inside a current carrying wire, and also that the applied electric field fills the wire.

It is interesting to see how the excess (local) surface charges arrange themselves, when there is a bend in the wire.

If the wire is curved, or if other wires with surface charges are closer together, the "gradient" is no longer constant, and the circular symmetry of the charge distribution around the wire may be broken.

A few excess charges will bunch close together on the outer side of the bend, and cause to exert a force on the conduction electrons, so as to turn them and make them follow the bend. This non-linear distribution of surface charges is shown in the figure.

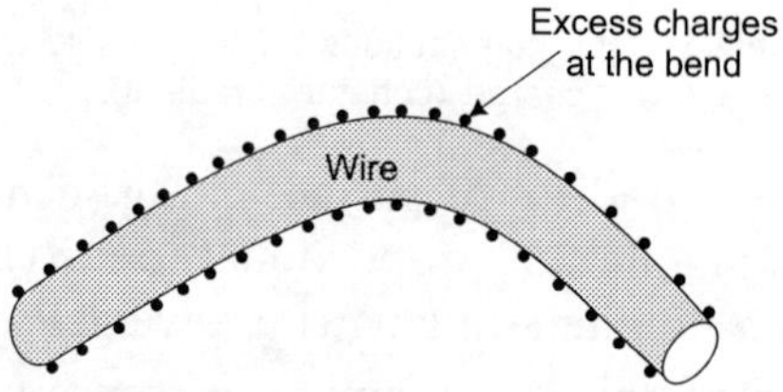

Non-linear distribution of surface charges in a wire with a bend.

1.13 STRUCTURE OF ELECTRICAL CONDUCTORS

In this section, we attempt to visualize the interior of a conductor in order to gain more insight into the process of conduction.

We first begin with a simple description of an atom, "neutral" atom to be precise. For an interactive applet of how an atom can be made, visit http://web.archive.org/web/20060425113223/http://www.colorado.edu/physics/2000/waves_particles/wavpart2.html.

Most of us were probably introduced to atoms and that these are neutral, because they consist of an equal number of positive (protons) and negative (electrons) particles. But that is not the end of the description. In addition, the following are inherent in neutral atoms: (see Section 5.23 also).

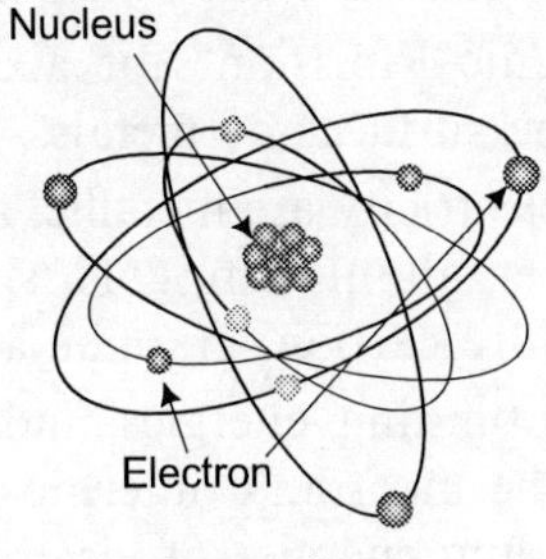

A neutral atom

(i) If the potential energy is set to zero at infinity, the usual convention, then bound electron states have negative potential energy, which means, some amount of energy has to be supplied to the *single* isolated atom to shift out the (outermost) electrons (say, by application of an electric field) from their orbital space and produce a dipole. Also, refer Section 9.7 on potential energy reference convention.

(ii) A nucleus consisting of neutrons and protons with electrons occupying orbital spaces very precisely; the fields due to the protons and electrons always exist, penetrating everything, but they cancel each other so well that there is hardly a net electric field around the atom!

Visualize *thick and hard* springs bonding the inner electrons and protons together in a single isolated atom, making it difficult to stretch the atom. See Appendix B for a comparison of gravitational and electric forces, notes on *stable atom*, and "Is an atom a dipole?". Some of the external forces which can lift or shift an outer electron (visualize *thin* springs holding it with the nucleus) to higher energy levels are heat, light and voltage.

If we visualize, as we had done in Section 1.4, that the unlike charges, positive nucleus and electron of an atom, are connected by a spring, then the analogy with stretching a spring to store potential energy to stretching the atomic bond, is quite appropriate !

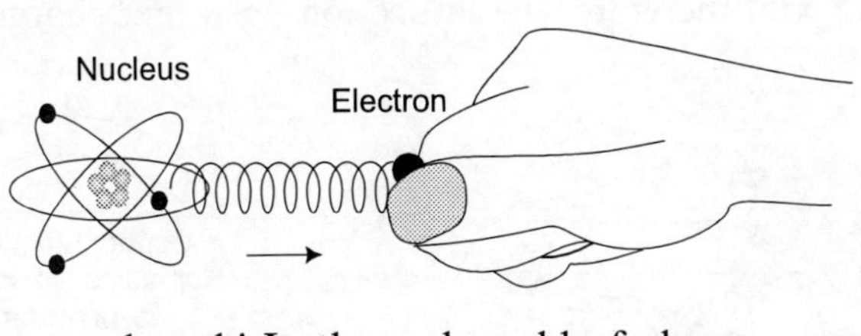

There is one difference though! In the real world of charges, comprising electrons and protons and neutrons, the shifted electrons will continue to occupy slightly different orbital spaces when pulled away, by supplying energy to them, either in the form of heat, light or voltage, and when they are released (by removing the heat, light or voltage), they will return to the stable orbital spaces. We shall see later in this section that it is very difficult to stretch the bonds of inner electrons.

See the end of Appendix A "Energy stored in stretched springs and unlike charges that are separated" for similarities in computation of energy stored in springs that are stretched, and in bonds of attraction between electrons and positive charges that are stretched.

See Appendix B "Thermionic Emission" for a description of how electrons can be extracted, by emission, from the surfaces of metals.

Most good conductors of electricity are metallic, and commonly used ones in electrical and electronic circuits are copper, aluminium, gold, silver and silver-cadmium. Among the good non-metallic conductors, are carbon, graphite and ionic solutions and several more.

For a comparison of the binding energies, indicative of the strength of force of attraction to the nucleus, of innermost and outermost electrons of Copper, see Appendix B "Inner electrons and the binding energies of electrons".

In electrical conductors (metallic), comprising *millions* of very closely spaced-atoms, visualize that the springs holding the *outermost* (valence orbit) electrons and the protons of individual atoms in the conductor, having become ***thin and soft*** (the electron energy level occupancies expand towards *less* negative) due to the strong electric interactions (repulsion between electrons) with adjacent atoms, making it easier for the electrons to shift away from the nucleus. See "Metallic bonding and Metallic structures" in Appendix B.

In electricity and electronics, all that matters is the outermost orbit called the *valence* orbit. The electron, occupying this orbit, controls the electrical properties of the atom.

The outermost valence electrons can easily gain kinetic energy, even at room temperature, or by applied electric fields of moderate magnitudes such as found in electronic and electric circuits, to become free and enable conduction. This is why we call the valence electron, a *free electron*; the inner electrons are *not affected easily*, by moderate electric fields.

The assembly of neutrons, protons *and* the *inner* orbit electrons, is called "the positive atomic core" or "the lattice ion".

Consider the atoms of copper, a metallic conductor. Each atom of copper consists of 29 protons, 29 electrons and 35 neutrons as shown on the left in the figure.

The lattice ion is the assembly of 29 protons, 28 (inner) electrons and 35 neutrons. The sum of the charges on the protons and electrons of the lattice ion is +29 (protons) –28 (electrons) = +1 and therefore, the lattice ion has a net charge of +1 as shown on the right in the figure.

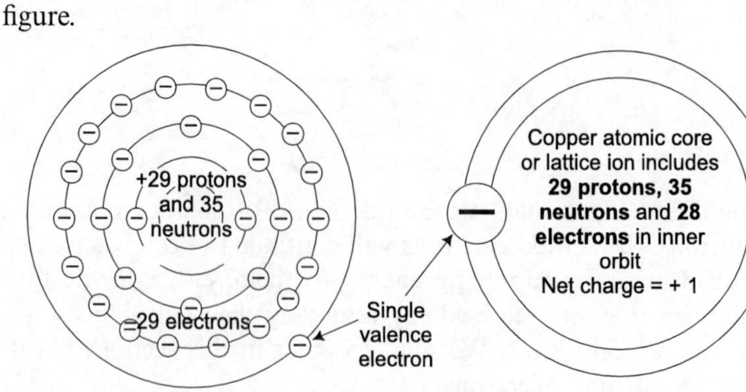

Copper atom

The outermost electron, the valence electron has a charge of –1, as seen on the right in the figure above. If we take the lattice ion and the valence electron together, the net charge is "zero".

The *outermost* (valence electrons) one (in copper), or two or three electrons (in certain other metals), which are quite loosely bound to the nucleus, enabling them to easily gain energy and go into larger orbits in which they become free electrons, roam freely around the lattice ions and form the "electron sea" as seen in the figure.

Atomic Structure of a Conductor

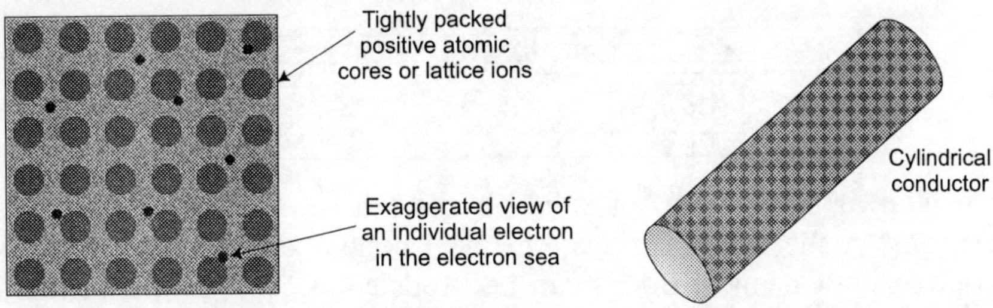

A view of the lattice in a metal and a cylindrical conductor with positive atomic cores and the electron sea.

A simple and useful visualization of a conductor is provided below:

Visualize a long rectangular box or cylinder

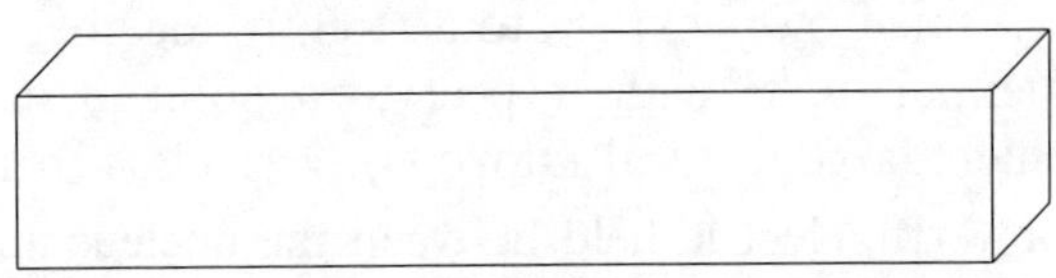

Visualize lattice (window grille) rectangular pieces

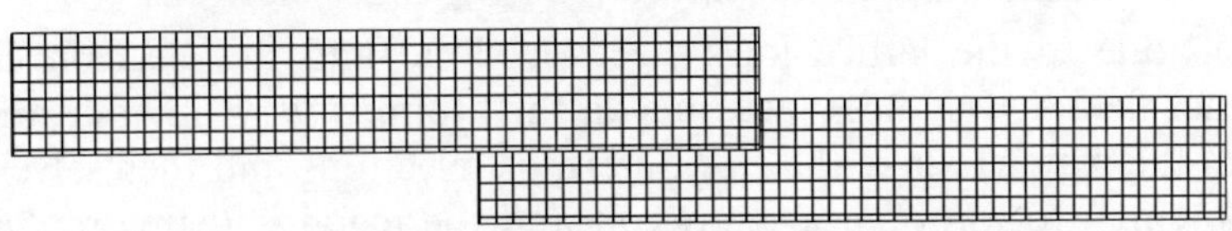

Visualize fixing these lattice strips inside the box arranging them in layers....

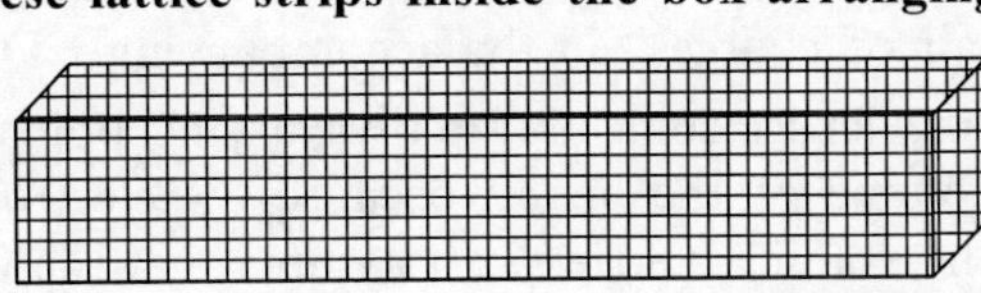

...adding several more until the box is tightly packed.
...What we are visualizing is a conductor with the positive atomic cores (lattice ions) located at the intersection points of the lattice....

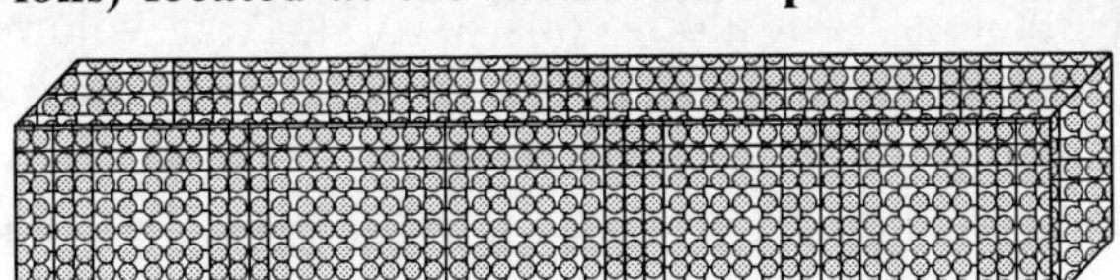

....these are composed of neutrons and protons and include the inner electrons.

Now pour some sand in, and that is the conductor, composed of the lattice ions, and the sand is the *electron sea of the outermost electrons!*

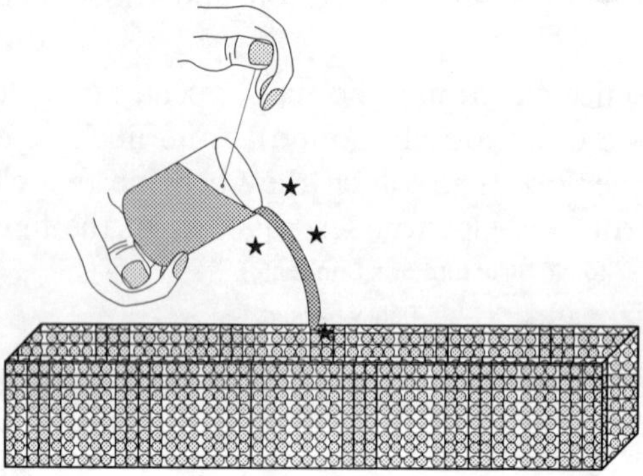

The regular array of atoms of solid objects can be modeled as tiny balls (nuclei with inner electrons) connected to springs (chemical bonds), and a brief description of such a model is given in "Chemical Bonds can be Modeled as Springs", Appendix B.

How is a metal "neutral" despite it being composed of ions (positive atomic cores or lattice ions) and electrons, each ion and electron possessing a huge amount of energy and whose individual potential close to it (volts with respect to a point at infinity) will range in the millions of volts?

When a metal is formed by packing atoms tightly together, then at distances close to a lattice ion, the potential (volts with respect to a point at infinity) varies rapidly with distance and is extremely large (several kilovolts), *very close* to the lattice ion.

There is a very powerful electric field between the nucleus and the electrons occupying orbital spaces, which may be depicted by closely spaced thick field lines originating from the nucleus and terminating almost entirely on the inner electrons.

However, due to the lattice ion's electron cloud and the electron clouds of nearby atoms, the net field at distances just outside the atom, is very tiny and the voltage *rapidly diminishes* in the region, where the free outermost electrons find themselves, and this region is an equipotential region of near zero volt and the metal is therefore, *"neutral"*.

For a pictorial representation of the large potentials near point charges, see the article "Potential around a pair of charges" in the accompanying CD.

The regions near the lattice ions of rapidly varying potential represent a very *small* portion of the total volume of the metal, because the electron cloud occupies the *more significant portion* of the volume, and we can assume that the field distribution within the metal is equipotential and that the free electrons are subject to no forces whatsoever.

While the mass of the proton is 2000 times the mass of an electron, the diameter of the proton is only 1/100000 as big as the diameter of the electron cloud, which is itself only about 2×10^{-10} meters ! An atom consists of a large volume of empty space ! The force of attraction between the *inner* electrons and the nucleus is so huge that even metallic tools *deform* when attempting to separate these!

See Appendix B on metallic structures and how these are held together by the valence or delocalized electrons.

See Appendix B "Density of States" on allowable orbital spaces for the millions and millions of electrons in tiny bands of energy levels.

However, between atoms, in most metals, the forces are more reasonable, permitting metallic ingots and sheets to be sheared and bent by cutting tools and presses.

The Electric Field Inside a Metal

(Chabay & Sherwood, 1995, used by Permission).[2]

When a metal or a conductor is subjected to an electric field then, the electron sea shifts under the influence of this field as shown in Fig. 1.13, and surface charges appear. Electric fields can be present *inside the matter* as well as in empty space.

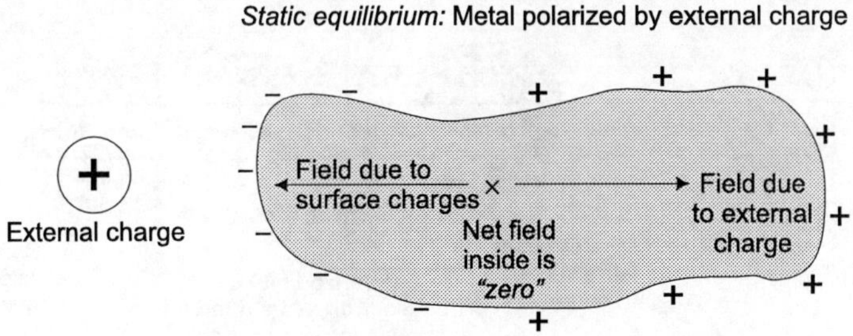

Fig. 1.13 Polarization of a metal.

An external charge polarizes an isolated block of metal *completely*. The free-electron sea **shifts until**, enough surface charge builds up to cancel the influence inside the metal of the external charge.

If this cancellation *weren't perfect*, there would be a force on free electrons inside the metal which would shift the electron sea *until* cancellation would be achieved.

The electric field of the external charge penetrated through the metal block. This is a property of the electric field.

In terms of electric field, we see that at any location inside the metal (marked **x** in the diagram), the electric field of the external charge is cancelled by the electric field produced by the surface charges.

A condition of *static equilibrium* then exists, and the net electric field is zero inside a metal as a result of the field due to external charges and the field due to the surface charges.

Note that there was a movement of the free electrons under the influence of the external electric field, prior to the condition of static equilibrium.

1.14 ELECTRIC FIELDS IN MATTER

Polarization means the displacement of positive and negative electric charge to opposite ends of a nuclear, atomic, molecular, or chemical system, especially by subjection to an electric field or in simple terms, "separation of charges". Atoms and molecules have some inherent polarization. The polarization principle in action in conducting metals is reiterated in the next section. Here, we will study how metals and insulators behave when they are subjected to an external electric field. We will also see the effect of electric fields on atoms.

These are important phenomena to know before piecing together the processes involved in conduction, charging and discharging of capacitors and also in the co-occurrence of non-Coulomb electric fields with time-varying magnetic fields (Faraday's law).

In a solid metal, the atomic cores are packed close together, so that their electron clouds *nearly* touch their neighbors (Chabay & Sherwood, 1995, used by permission).[2] In such metals that are conductors, the conduction electrons can *move freely* throughout the material.

The *sea* of free electrons is nearly *uniform* throughout the solid. When a positive charge is brought near the conductor, its electrons sea *shifts* under the influence of the electric field of the external positive charge as shown in Fig. 1.14.

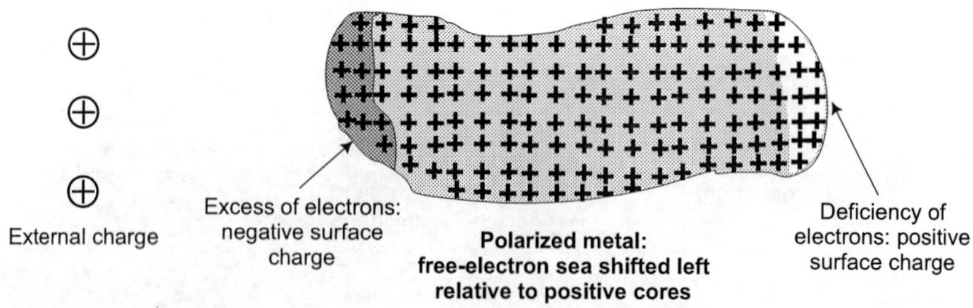

Fig. 1.14 Shifting of the electron sea in a metal under the influence of an electric field.

The shifting stops, when the electric field due to the (separated) shifted sea and the exposed positive cores, cancels the field of the external charge. In the *interior* of the metal, there is essentially *no change and no excess (local) charge*–we still have a uniform sea of electrons filling the space around the positive atomic cores.

At the surface there is some *excess charge*, so we can represent a polarized metal as having thin layers of charge *on its surfaces* but being unpolarized in the interior, unlike an insulator that is polarized *throughout its interior*.

It is important to note that in static equilibrium, the achievable shift in the electron sea is extremely small, typically a tiny fraction of one atomic diameter!

We are unable to apply electric forces so strong as to shift the electron sea very much, because even a tiny shift leads to the appearance of enough surface charge (and accompanying electric fields) to cancel out the forces we're applying.

The configurations of the induced charges is complex, which is discussed in "Induced Charge Configurations" in Appendix C.

Polarization of Atoms under the Influence of an Electric Field

(Chabay & Sherwood, 1995, used by Permission).[2]

In an atom, the electron cloud is *not rigidly connected* to the nucleus. The electron cloud and the nucleus can move relative to each other. If an external charge is nearby, it can exert forces on the electron cloud and on the nucleus which will cause them to *shift position....* creating ....a *dipole.* This is shown in Figs. 1.15 and 1.16.

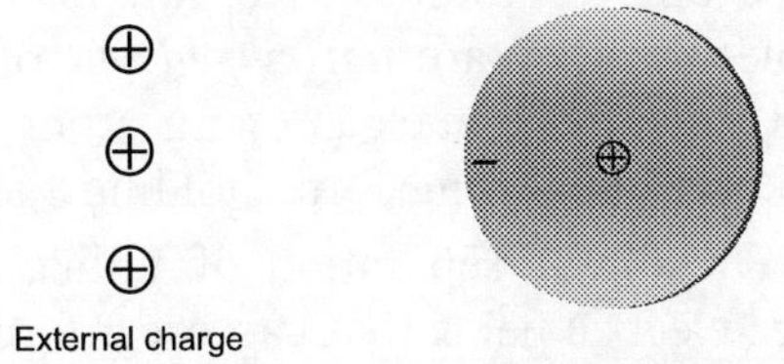

Fig. 1.15 An external charge near a neutral atom.

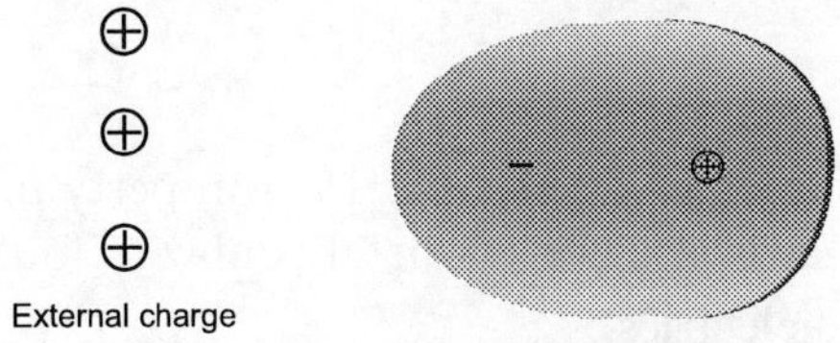

Fig. 1.16 Creating an atomic dipole.

The outer regions of the cloud are affected the most by the external charge, because in the outer regions, the electron is attracted less to the nucleus and can be influenced more by the external charge.

Note: Certain molecules occur in nature as dipoles with inherent polarization and are called "permanent" dipoles, for example, water molecules and HCl molecules.

An *individual* atom or molecule can be *polarized* by an external charge, but this polarization only produces an *induced dipole.*

When the electron cloud of an atom that was neutral, that is, an *insignificant* electric field surrounds it, is shifted away by an external charge, then a *significant* resultant electric field will be produced around it, due to the separation of the positive atomic core and the outermost electron. A computation of the electric field of induced dipoles is given in Appendix A.

In Fig. 1.17 is shown a piece of dielectric or insulator, which has no conduction band electrons, subject to the electric field of an external charge.

In insulators, charged particles are *not free* to move through the material. All of the electrons are firmly bound to the atoms or molecules making up the material. Commonly used insulators in electrical and electronic circuits are, vinyl (polyvinyl chloride or PVC) and rubberised polymers coverings on wires and paper, mica, polycarbonate, polystyrene and Teflon as dielectrics in capacitors.

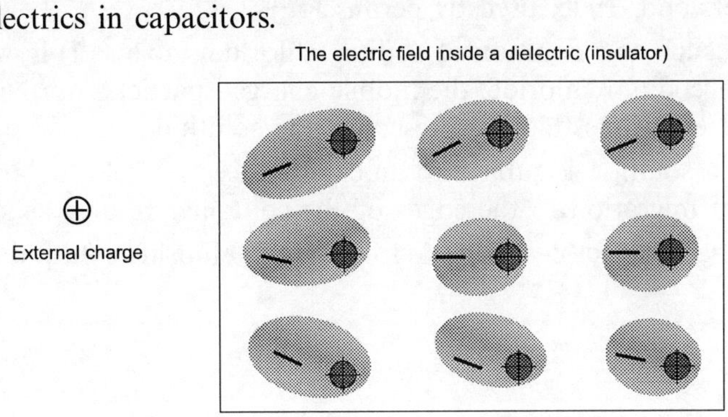

Fig. 1.17 Polarization of atoms in a dielectric.

The electrons in the atoms or molecules of an insulator *shift position slightly*, but remain bound to the molecule; they are *not ripped out* of the molecule. And, of course, protons cannot be ripped out of the nucleus of an atom by an external charge; it would take a *huge force* to overcome the strong force holding the nucleus together.

Induced dipoles are a result of separation of unlike charges by external "inducing" charges, and these produce electric fields that can exert forces on the inducing charges, and the inducing charges also exert forces on the dipole charges.

We can thus summarize the effects of fields due to external charges on conductors and insulators:

1. For metals (conductors):
   One property of metals (and of the $1/d^2$ property of the electric interaction) is that any excess charge on a piece of metal is always found *on an outer or inner surface.*
2. For insulators (dielectrics):
   This is different from an *insulator*, where excess charges can be located *in the interior*, or *bound* to a particular spot on a surface *without spreading out along that surface.*

The effects of electric fields on conductors and insulators are shown in Fig. 1.18.

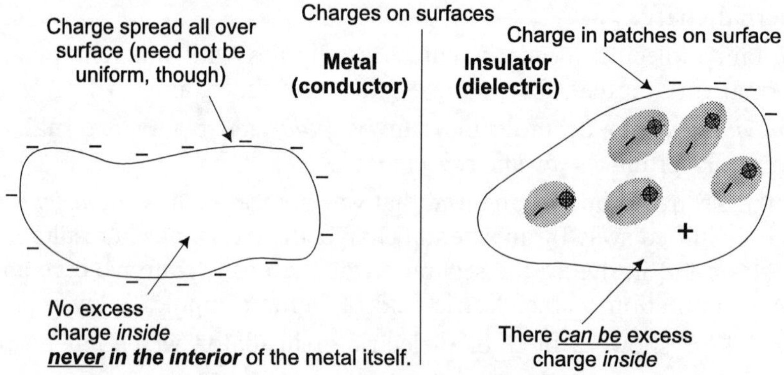

Fig. 1.18 Charges on the surface of conductors and insulators.

Polarization of Ionic Solutions

(Chabay & Sherwood, 1995, used by permission)[2]

Ionic solutions, such as a solution of sodium chloride (table salt) in water, are conductors. In a solution of sodium chloride, the mobile charged particles are Na^+ ions, Cl^- ions, and the few H^+ and OH^- ions that are present in the solution.

When such a solution is subject to an external electric field, polarization into ions and their subsequent migration to the edges of the container results, as shown in Fig. 1.19.

Bring a charged object near a beaker of sodium chloride solution (salt water) as shown in Fig. 1.19a.

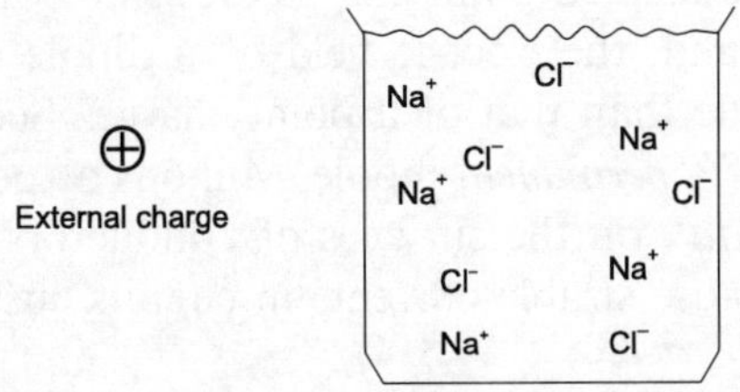

Fig 1.19a External charge brought near an ionic solution.

Polarization of the ions occurs, and the chloride ions move close to the side of the beaker which is near the positive external charge and the sodium ions move towards the opposite side as shown in Fig. 1.19b.

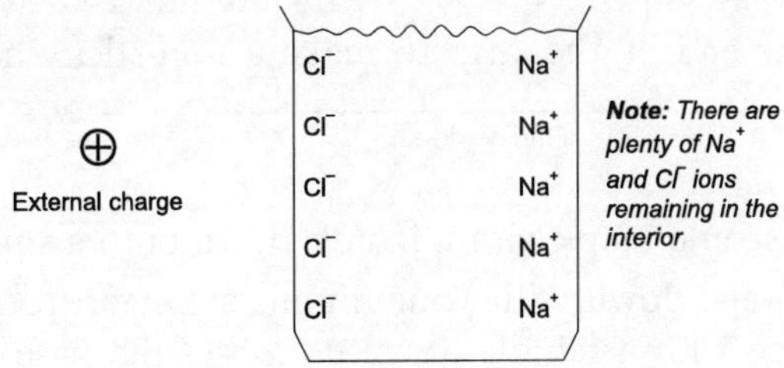

Fig 1.19b Polarization of ionic solutions.

The Electric Field of a Dipole

The figure shows the electric field of a dipole. Picture courtesy: www.falstad.com reproduced with permission.

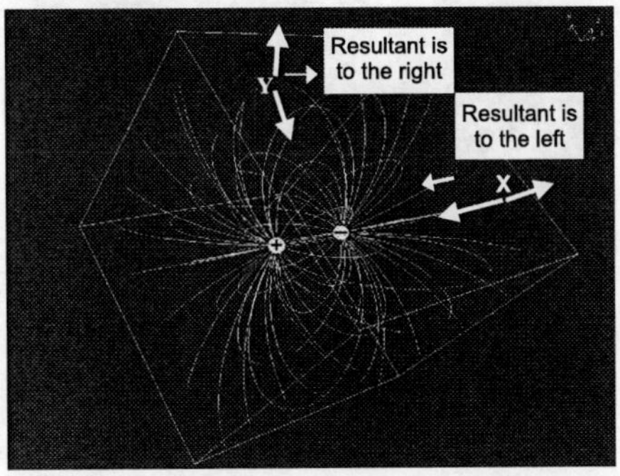

At Y: the vertical components cancel and the resultant E field is to the right

At X: the resultant E field is to the left

The electric field at points **far away** in space where distance **x (or y) >> s** (where s is the separation between the individual charges of a dipole) is important in the study of circuits.

Electric field at point $x = \left[\dfrac{1}{4\pi\varepsilon_o}\right]^2 \dfrac{2qs}{x^3}$, and at point $y = \left[\dfrac{1}{4\pi\varepsilon_o}\right]^2 \dfrac{2qs}{y^3}$

The electric field of an isolated point charge drops away at the rate of $1/d^2$ with distance. Note how on the other hand, the electric field of a dipole drops away at the rate of $1/d^3$, which is a much *steeper* rate than that of a point charge.. See Appendix A for details of the computation of the field of a *permanent* dipole. Millions of point charges whose electric fields drop away at the rate of $1/d^2$, on the surfaces of conductors, add their tiny electric fields to produce large fields that cause sizable currents in circuits, just as millions of tiny droplets of water form the oceans!

The Interactions of Electric Charges

A simple experiment that explains the interactions of charges can be performed using scotch tape (cellotape). Use a strip of tape about 20 cm or 8 inches long (about as long as the width of an A4 size paper) (Chabay & Sherwood, 1995, used by permission).[2]

Fold over one end of the strip to make a non-sticky handle as shown in the figure.

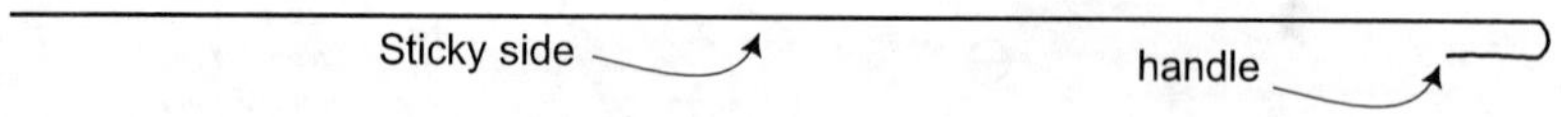

Now, stick a strip of tape with a handle down onto a smooth flat surface such as a desk.

Smooth this tape down with your thumb or fingertips. We will call this the base tape. Stick another tape with a handle down *on top of* the base tape. See the left figure.

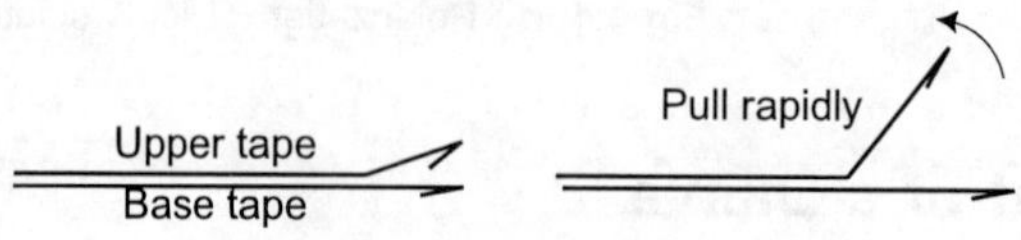

Smooth the upper tape down well with your thumb or fingertips. With a quick motion, pull the upper tape up and off the base tape, leaving the base tape stuck to the desk. See the second figure above. Mark the tape you pulled off "U-tape".

Your Hand and the U Tape

Hang the U tape vertically from the edge of the desk, and bring your hand near the hanging tape.

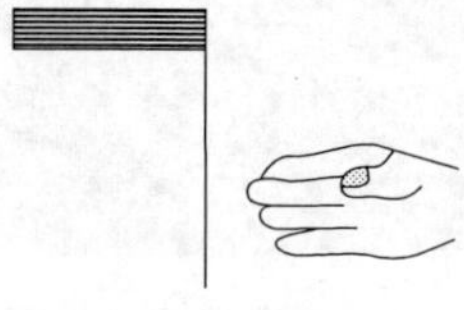

If the tape is in good condition and the room is not too humid, you should find that there is an attraction between the hanging strip and your hand when you get close to either side of the tape.

If you don't see an attraction, remake the tape. You can reuse the same piece of tape by again sticking it down on the base tape, and then running your finger along the base tape that is on the desk, sticking the U-tape down on the base tape, and smoothing the tape down well.

Now, jerk the U-tape very quickly off the base tape.

The tape should attract to your hand when hung vertically from the desk.

Why does the U-tape get attracted to your hand?

Your own body consists mainly of salt water, including the blood and the insides of cells. One effect of polarization of the salt water inside your finger is that there will be a shift of Na^+ and Cl^- ions in the blood and tissues. This shift may be a larger effect than the molecular polarization.

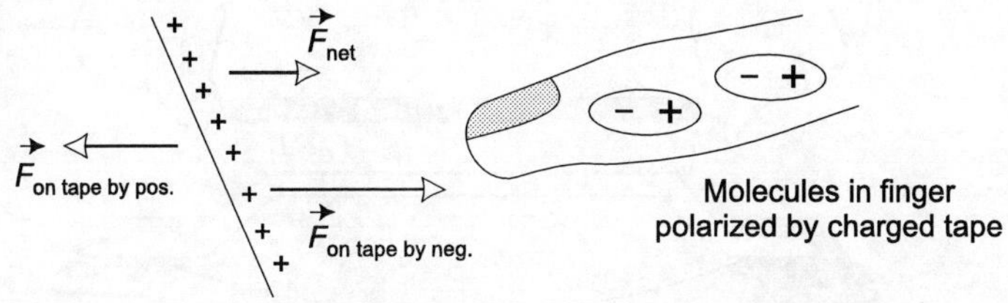

The tape holds a positive charge as seen in the figure and due to these charges, the tape polarizes molecules in the finger. Negative ends of molecules end up closer to the tape. The tape is attracted to the negative ends, and repelled less strongly by the positive ends because they are farther away, and force is proportional to $1/d^2$. The net force on the tape by the finger is the superposition of these forces, and is toward the finger, resulting in the attraction of the tape towards the finger.

Unless the external charge is just a few atomic diameters away, the shift in the electron cloud is typically so tiny that it can't be seen in a picture. But, if all the molecules in your hand are polarized even very slightly by a charged tape, the tiny interactions with the huge number of molecules can add up to an observable effect.

A positively charged object, like a positive piece of invisible tape, is attracted not only to a negative tape and your hand, but also to your desk, your book, and every other nearby neutral object such as objects made of iron, steel, aluminium, copper, zinc and carbon due to the effects of polarization. The same is true for a negative charged object, such as a negative piece of invisible tape.

The tape is shown to hold an excess of positive charge but that could have been negative either.

The polarity of the charge would depend on the type of tape you have used. See Section 1.25 for an explanation of how the tape is charged.

Approximate Amount of Charge on a Piece of Invisible Tape

(Chabay & Sherwood, 2011 reproduced by permission)

We can make a rough estimate of the amount of excess charge on a piece of tape by a simple experiment.

Prepare a pair of oppositely charged tapes using two invisible (Scotch) tapes of 1 cm width and 20 cm length each by pulling the pair of initially neutral tapes apart. Since the tapes started out neutral, one of the tapes must now be negatively charged and the other positively charged, and the absolute value of the charge of each tape must be the same.

If we suspend a (say, positively) charged tape B between two books and lower the negatively charged tape A toward it, when the negative tape gets close enough we will see that the positive tape is lifted upward.

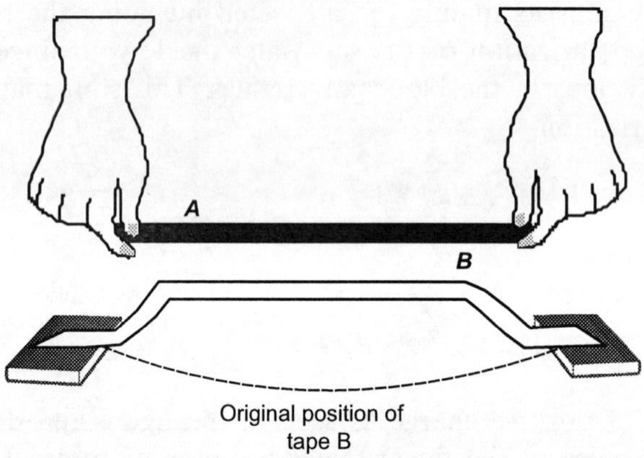

Tapes *A* and *B* are oppositely charged. When *A* gets sufficiently close, the electrical force on *B* is sufficient to lift it up.

The electric field due to A by approximating A as a point charge (a very rough approximation)

$$E_A \approx \frac{1}{4\pi\varepsilon_0}\frac{Q_A}{d^2}$$

So,

$$Q_B \frac{1}{4\pi\varepsilon_o}\frac{Q_A}{d^2} \approx mg$$

If we had a 20 cm long invisible tape with a mass of 0.15 g and the bottom tape starts to be lifted when the top tape is about 2.5 cm away, then since the magnitude of the charge on the tapes is the same,

$$Q \approx \sqrt{\frac{(1.5\times10^{-4}\ \text{kg})\,(9.8\ \text{N/kg})\,(0.025\ \text{m})^2}{\left(9\times10^{9}\ \frac{\text{N.m}^2}{\text{C}^2}\right)}} \approx 1\times10^{-8}\ \text{C} = 10\ \text{nC}$$

where,

$$\frac{1}{4\pi\varepsilon_0} = 9\times10^{9}\ \frac{\text{N.m}^2}{\text{C}^2}$$

This is a rough estimate, but it turns out to be a reasonable one—a small object charged by rubbing usually has a charge on the order of 10 nCs.

The electric field of one tape modeled as a point like object at a location 2.5 cm from the center is

$$\frac{1}{4\pi\varepsilon_o}\frac{Q_A}{d^2} = 9\times10^9 \times \frac{10\times10^{-9}}{0.025^2} = 1.44\times10^5 \text{ N/C}\cdot$$

In Appendix A "Electric field near the centre of a uniformly charged disk", we will show that the field *near* the charged tape can be estimated using an approximate formula to be around 2.8×10^5 N/C.

Since the electric field required to ionize the air itself is about 3×10^6 N/C, the field near the surface of the tape is pretty large. We will learn more about how air ionizes when the field strength is greater than 3×10^6 N/C in Chapter 10.

Fraction of Surface Atoms with Excess Charge

Assuming that the net charge on the negative tape is distributed uniformly over its surface, we can estimate the fraction of those surface atoms that have gained an excess electron or negative ion.

Surface area of a 1-cm-wide, 20-cm-long tape:

$A = (0.2 \text{ m})\ (0.01 \text{ m}) = 2 \times 10^{-3} \text{ m}^2$

Approximate cross-sectional area occupied by an atom whose radius is approximately 1×10^{-10} m:

(assuming its diameter forms one side of a rectangle)

$$A_{\text{atom}} \approx (2 \times 10^{-10}\text{m})^2 = 4 \times 10^{-20}\text{m}^2$$

Number of atoms on the surface:

$$\frac{A}{A_{\text{atom}}} = \frac{2\times10^{-3}\ m^2}{4\times10^{-20}\ m^2} = 5\times10^{16} \text{ atoms}$$

Number of excess electrons (or ions):

$$\frac{10\times10^{-9}\text{C}}{1.6\times10^{-19}\text{C}} = 6.25\times10^{10}$$

Fraction of surface atoms with excess charge:

$$\frac{6.25\times10^{10}}{5\times10^{16}} \approx 1\times10^{-6}$$

Thus only about only one in a million atoms on the surface of the tape has acquired an excess electron or lost an electron—a small fraction.

Electric and Gravitational Fields Penetrate Intervening Matter

Observe that when bringing your hand towards either side of the charged hanging tape, the tape is attracted to your hand, despite the charges being on just one side of the tape. Intervening matter does not "screen" or "shield" the electric field, just as your desk does not "screen" or "shield" your book from the gravitational field of the Earth! Else, you would find the book floating in space! It is evident that the gravitational field penetrates intervening matter.

The gravitational field that penetrates through matter is invisible and may be termed to be an *alteration of space*. It is this property that enables you to stand and sit even on the upper floors of a building. There is a pull exerted by the mass of the concrete floor and the walls of an upper floor. However, it is the combined effect of the gravitational field or alteration of space by the earth, the floor and the walls on you that keep you on the floor. However, the pull of the wall is so weak that it does not affect you significantly. Else, you would be stuck to the wall!

The Electric Field Inside Metallic Wires

During the *polarization* process that we studied previously, in a metal, leading to static equilibrium, charges were *moving* and the electric field is *not zero.*

Non-equilibrium: In an electric circuit powered by a battery, the free electron sea in the body of the wires and resistances is *continually shifted* through the wires by a non-zero applied electric field (*inside the wires*), which applies a force on the electrons.

It is to be noted that this is not a state of static equilibrium.

This is also known as a ***steady-state***.

It will be appropriate for students to become familiar with Chapter 1 and Chapter 2 (upto and including Sections 2.4.3) in the visualizations provided by Hermann Härtel in the article "A Qualtiative Approach to Electricity" which includes a guide to aid in visualizations using symbols and images in the Teacher's Guide in the CD accompanying this book.

1.15 THE NATURE OF SURFACE CHARGES IN CIRCUITS

We shall now see how when we apply a voltage to a simple circuit consisting of a battery and wires, an electric field is established by the development of surface charges due to the applied voltage. Refer to Fig. 1.20 which shows a simple circuit consisting of a battery and wires.

For the purpose of the analysis we shall ignore the fact that if we connect a circuit in this fashion the battery is effectively short-circuited and that the wire will heat up.

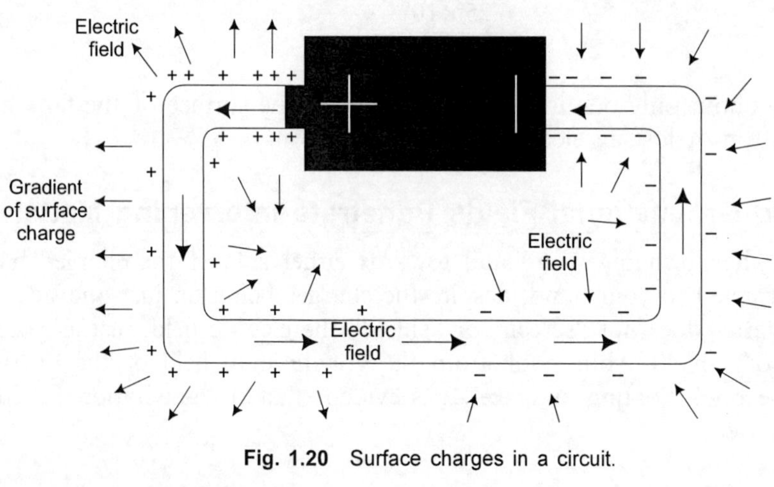

Fig. 1.20 Surface charges in a circuit.

There must be *positive and negative charges* on the surface of the wires near the positive (+) and negative (–) plates of the battery, and in the steady state the surface charge must *vary smoothly* from positive to negative around the circuit, possibly with extra charge on the bends in the circuit, to produce non-zero resultant electric fields inside the wire and in the space surrounding the wire and the circuit (not forgetting that an electric field is an invisible *alteration of* space). If the wire is uniform, the field inside will be uniform too. The field is *not* uniform in the space surrounding the circuit, as there are no mobile carriers there which can produce a uniform field.

Electrons in the free-electron sea inside the wires will move due to the fields from the battery. Imagine that you could take a "snapshot" of the surface charge after the electrons inside the wire have moved just a little bit (that is, after a very short time). Electrons moving toward a positive region of the surface will make the surface charge in that region less positive (negative point charges), while electrons moving away from a negative region of the surface will make the surface charge in that region less negative. A + sign on the surface does not represent some exotic positive charge. Rather, a + sign indicates a region where there is a deficiency of electrons, leaving partially exposed the metal's positive atomic cores, or positive point charges.

Since, surface charges are formed by the expansion or contraction of electrons from the sea of electrons in the conduction band, these are not individual charges of atomic dipoles comprising conduction band electrons and lattice ions. Conduction band electrons are too far away from the lattice for the system to be treated as dipoles, and so surface charges in the conducting wires should be viewed as point charges.

If the ends of a wire are connected to a battery, an electric field E will be set up at every point *within* the wire. If the potential difference maintained by the battery is 10 volts, and if the wire (assumed uniform) is 5 meters long, the strength of this field at *every point* will be 2 volts/meter. This field will act on the electrons and will give them a resultant motion in the direction of $-E$. In Section 1.23, we will verify the field strength value by applying the law of conservation of energy.

Some electrons, will leave the negative terminal because of Coulomb repulsion and *enter the wire*; other electrons will leave the wire and *enter the positive terminal*, charging the wire negative near the negative terminal and positive near the positive terminal.[3]

Thus, a *gradient* of charge density (in the shape of *rings* for circular cross-section wires) will form along the wire, and the field inside the wires from these charges, as well as those on the terminals, will cause a current of density J, given by,

$$J = \sigma E$$

where, J is in amperes per square meter, or current 'I' flowing in a unit cross-sectional area in a conductor whose total area is 'A' ($J = I/A$) of conductivity σ mhos per meter, and E is the applied field due to the battery. The current I is shown in Fig. 1.21.

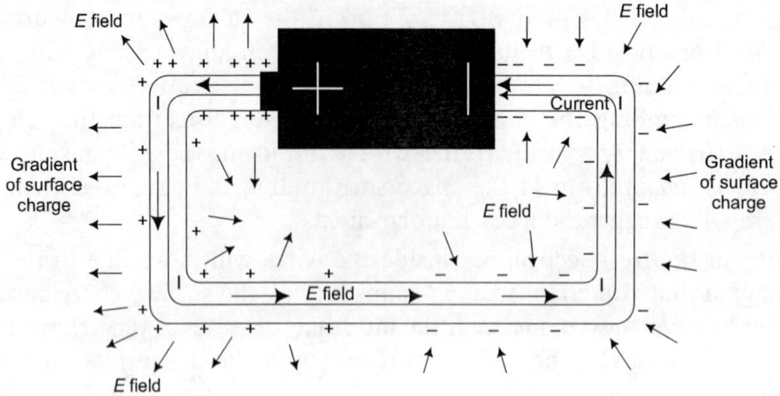

Fig. 1.21 Uniform electric field inside the wire and current I.

We can state a rule thus: When a dc circuit is closed, positive charges will enter the wire near the positive terminal of the source and negative charges will enter the wire near the negative terminal.

"Steady-state" means that some charges are moving, but their velocities at any location aren't changing, and there isn't any change in *the deposits* of excess (local) charge anywhere. The circuit achieves and maintains this condition by a continuous feedback mechanism between the electric field, the surface charge and the moving conduction band electrons in the wire. For an experimental demonstration of the presence of surface charges in circuits, see the video here https://matterandinteractions.org/surface-charge-demo/.

Feedback Leads to Current Conservation

(Chabay & Sherwood, 1995, used by permission)[2]

Consider a straight section of circular cross-section wire that has *more incoming* electron current than outgoing electron current; (shown by the unequal white block arrows at the entrance and exit of the wire in Fig. 1.22) leading *to* a build-up of excess (local) *negative* charge on the surface of this section of wire. This could happen if there was an increase in the applied voltage.

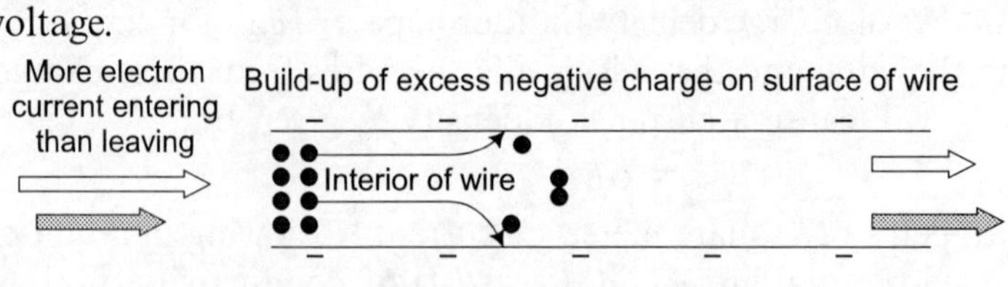

Fig. 1.22 Surface charge and feedback to maintain steady-state.

The surface charge *build-up* (in the shape of rings) will tend to *equalize* the incoming and outgoing electron currents.

The outer surface of this section of wire becomes more negative, and this negative surface charge tends to *impede* the incoming electrons and tends to *speed up* the outgoing electrons, thus making the incoming and outgoing electron currents more nearly equal to each other.

This *feedback process* will continue until the two currents are exactly equal to each other. At that point there will be no further change in surface charge. This is shown by the shaded block arrows.

Now, consider that there is *more outgoing* electron current than incoming electron current say, due to a decrease in the applied voltage (shown by the unequal white block arrows at the entrance and exit of the wire in Fig. 1.23) *leading* to a build-up of excess (local) *positive* charge on the surface of this section of wire.

Positive charge (deficiency of electrons) builds upon surface of wire.

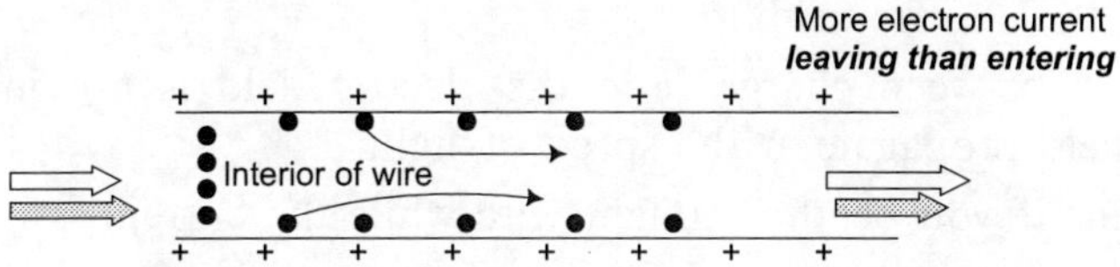

Fig. 1.23 Surface charge and feedback to maintain steady-state.

The resulting *surface charge build-up* (in the shape of rings) will tend to *equalize* the incoming and outgoing electron *current* by speeding up the entry and slowing down the exit of electrons. Eventually, the entry and exit electron currents are equal as shown by the shaded block arrows. So, whichever way there is *an imbalance* between incoming and outgoing currents, the two currents will quickly become equal to each other due to the *automatic feedback mechanism* of changes in the surface charge.

Important note: Observe that surface charge dispersal and accumulation *and* current conservation in wires *and* charge neutrality *in* the wires (Section 1.18), are maintained by *drifting* charges (current up and down variations) due to the source of emf, say a battery (Ref. [37]).

In their paper "A unified treatment of electrostatics and circuits", Chabay and Sherwood indicate that for a straight coaxial cable 30 cm long whose central conductor has a diameter of 1 millimeter, driven by a 3-volt battery at one end and shorted at the other end, the maximum linear surface charge density on the central conductor is *merely a few million* electrons per centimeter.

Note: A 1 centimeter long 1 millimeter diameter copper wire contains 6.6×10^{14} million free electrons and therefore, the surface charges are indeed very few in number compared to the free electrons in the wire.

Connecting a Circuit: The Initial Transient

When you connect a circuit by making the final connection, *feedback* forces a *rapid rearrangement* of surface charges leading to the steady-state.

This *period* of adjustment before establishing the steady-state is called the "initial transient".

Surface Charge and *Feedback* to Maintain Steady-State

The wires in a circuit with steady currents are electrically neutral (neutrality will be explained in Section 1.18) in their interiors, but have *small amounts* of charge on their surfaces.

These surface charges are critical to the behavior we expect from circuits, for they make the *internal* electric field parallel to the wires and ensure that the *same* amount of current flows in different parts of the circuit. *If*, the electric field and surface charges do *not* create the conditions for steady-state currents, rapid changes occur in the surface charge distribution until these conditions *are* satisfied.

There is feedback because the electric fields *change* the surface charges and the surface charges *produce* the electric field.

This *feedback* between charges and the electric field is not instantaneous, because changes in the field propagate at the speed of light *c*.

What happens if you bend a straight wire while the current is running?

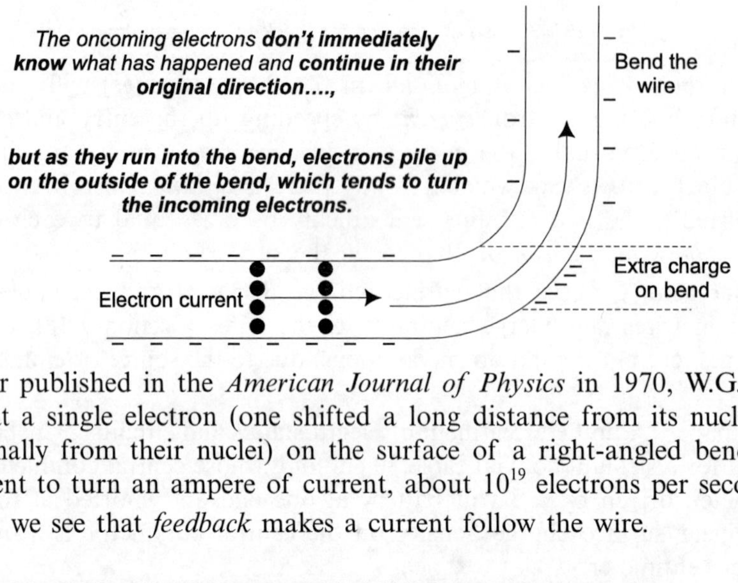

In a paper published in the *American Journal of Physics* in 1970, W.G.V. Rosser has calculated that a single electron (one shifted a long distance from its nucleus or several shifted marginally from their nuclei) on the surface of a right-angled bend in a copper wire is sufficient to turn an ampere of current, about 10^{19} electrons per second !

Therefore, we see that *feedback* makes a current follow the wire.

1.16 THE SPEED OF SURFACE CHARGE BUILD UP

The build up of point charges on the surfaces of conductors in circuits occurs very rapidly of the order of 2 to 5 times the *light crossing time* of the circuit.

The light crossing time of a circuit is the time to cross a physical circuit at the speed of light. In typical electronic circuits the light crossing time will be of the order of a few *nanoseconds*.

The time required to *move* the electrons towards or away from the surface is completely negligible, because the electrons only have to move an *infinitesimal distance* in order to establish a significant amount of surface charge! Surface charges are point charges that are set up to produce a resultant field.

We will see in Section 1.25 and Appendix A "Amount of Cloud Shift in the Carbon Atom......" that the electrons have to move less than 10^{-15} meters away from the atom to produce quite a significant amount of surface charge.

What about the speed of their appearance?

The speed of their appearance (and disappearance) is related to the relaxation time which is of the order of 10^{-14} seconds, and this we will learn in Section 5.22 in connection with the transmission of signal energy at radio frequencies and higher (a few hundreds of Meghahertz; 1 Megahertz = 10^6 cycles per second) by two-wire cables as in broadband communication, is short enough to respond and produce voltages between conductors.

In the simple circuit of Fig 1.24, the electron-sea is *continuously shifting* under the influence of a battery.

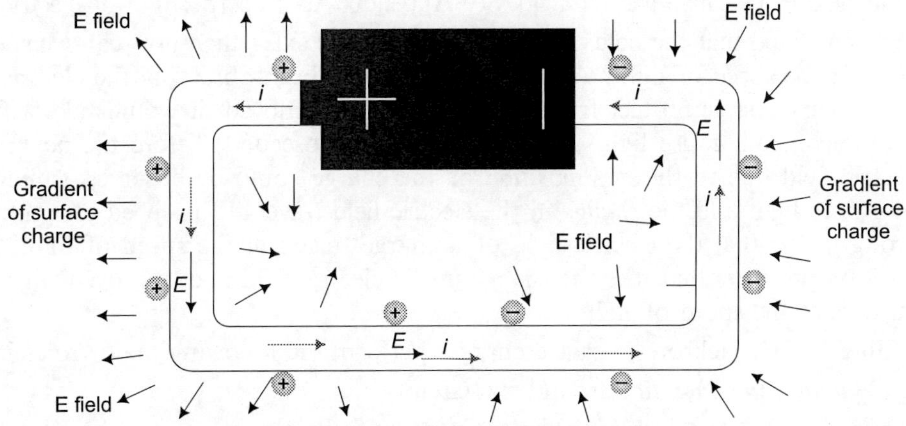

Fig. 1.24 Speed of surface charge "development".

There is *no excess (local) charge* anywhere inside the wire – to every free electron there corresponds a singly charged positive atomic core – an atom minus one electron which has been released to roam freely in the free electron-sea. The conservation of charge principle is upheld when we consider that the battery and wires constitute a larger closed system; never mind that the charges were separating or depleting (inside the battery) and moving as inside the wires.

Averaged over a few atomic diameters, the interior of the metal is neutral, and the repulsion between the free electrons is balanced out on the average by the attractions of the positive cores; at the surface, the electrons *only* have to *move* an *infinitesimal distance* in order to establish a significant amount of surface charge rings, because in reality, surface charges are point charges, which create an electric field in the interior that will extend around the wire.

In Fig. 1.24 surface point charges of both signs (see Section 1.15) are depicted. In a few nanoseconds, the rearrangement of the surface charges will extend all the way around the circuit, through slight expansion or contraction of the mobile-electron sea (compare with the molecular stretch associated with polarizing an insulator, which is less than the radius of a single proton!). When the circuit is completed with the battery, excess negative charge expands the mobile-electron sea so that some of the sea peeks out at the surface, giving a negative surface charge. Excess positive charge contracts the electron sea, allowing positive cores to peek out at the surface. Read the **important note** below Fig. 1.23.

Summarizing the Effects of Feedback

Feedback in *static* electricity situations leads to static equilibrium ($E_{NET} = 0$ inside a metal).

Feedback *in a circuit* leads to surface charges and steady-state current ($E_{NET} \neq 0$ inside a metal).

Never forget that a NET field implies the superposition of all the fields that are alive and existing even as you are computing or visualizing the NET field and that is the reason for the resultant field to be called a NET field. See Appendix A, "Superposition of electric fields".

We had stated that the fields due to charges always exist, then how can we speak about its travel at the speed of light? Assume that you have been observing the electric field due to a stationary charge 30 feet from you. If the charge moved, it would take a finite time (30 nanoseconds, because light travels 1 foot per nanosecond) before the new magnitude of electric field due to the new *position* of the charge could be observed simultaneously, because both light and the *change* in the electric field travel at the speed of light. It would be wrong to say that the electric field of a charge travels at the speed of light. What the observation proves is that it is the *change* in the electric field due to movement of charges that travels at the speed of light.

According to J.D. Jackson[10], surface charges perform the following major roles in circuits:

1. ***Maintain the potential*** around the circuit
2. ***Provide the electric field*** in the space around the circuit
3. Assure the ***confined*** flow of current.

Keeping in view, the conduction mechanism in circuits and the processes which enabled the maintenance of static equilibrium in conductors, it should be possible to answer the question:

Does a voltage develop across an open switch?

We know $V = IR$ therefore, if there is no I there cannot be a V.

Wrong!

The presence of surface charge should dispel that notion.

1.17 CONDUCTION – A MICROSCOPIC AND MACROSCOPIC VIEW

In Section 1.13 it was shown how in metals, the valence electrons are not attached to individual atoms, but are free to move about within the lattice and are called conduction electrons.

In copper, there is one such electron per atom, the other 28 remaining bound to the copper nuclei to form lattice ions or atomic cores.

The figure attempts to depict the random motion of the electron in the metal.[13]

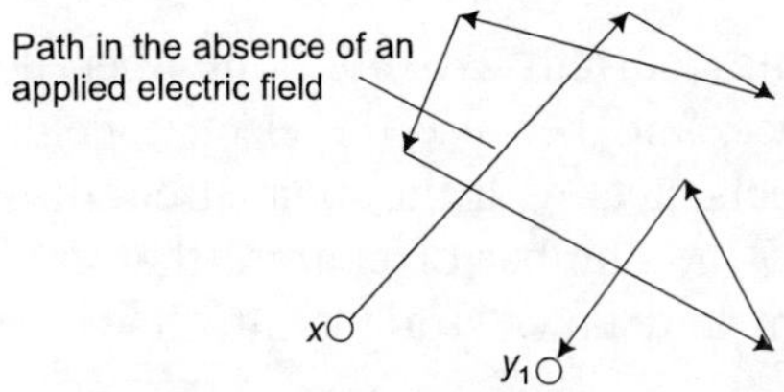

The solid lines depict a possible random path of an electron in the *absence* of an applied electric field, proceeding from x to y_1, making six collisions on the way, bouncing about in an atomic lattice at a speed of about 1/200 the speed of light. The electron collides constantly with the atomic cores (lattice ions), often suffering sudden changes in speed and direction.

In an isolated metallic conductor with no electric field applied, there is no net motion because at any cross-section and for a given interval of time for as many electrons that would have moved to the right, an equal number of electrons would have moved to the left and the current is construed to be zero. For a more rigorous viewpoint of "no net motion" watch the lecture http://freevideolectures.com/Course/2325/Solid-State-Devices-IIT-Madras/12.

The dashed lines in the figure show how this path is altered when an electric field is applied.

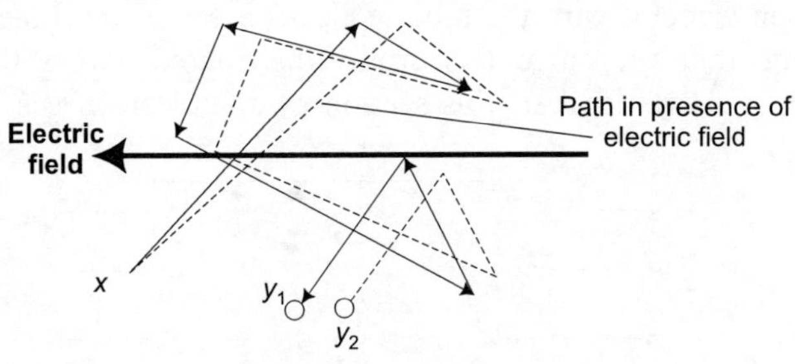

The electron drifts towards the right with a speed much less than a snail's pace. Note that the electron drifts steadily to the right, ending at y_2 rather than at y_1. When a field is applied to an electron in the metal, it will experience a force qE which will impart to it an acceleration. It soon collides with a lattice ion and loses its tendency to drift and will have a truly random direction after the collision. If the applied electric field is steady, the electron accelerates after the collision and again collides and moves to a point a little beyond the point y_2, and so on. A net motion to the right would have been established. In the next section, we shall see how the continuous collisions the electron makes with the lattice results in a slow drifting motion of the electrons. The number of collisions per second is a stupendous *10^{14} per second*!, and the mean free path between two collisions is about 10 diameters of a copper atom.

When the electric field is impressed on a conductor, however, the conduction band electrons are accelerated and their velocity (drift) increases in proportion to mobility. See "Electron Mobility", Appendix B.

The number of collisions remains close to 10^{14} per second but the *intensity* of the collisions increase. At a macroscopic level we say that a current density is established.

The quantity J denotes current *density,* a number flow rate of charges across tiny areas in elemental volumes, and because the electric field penetrates through matter and due to surface charge effects, acts with the *same* strength on *all* the charges in the volume uniformly, it also refers to the number of charges that cross a *unit area* per second. If the current density is uniform and flows straight into a wire of cross-sectional area A, then the current $I = JA$.

A Formula for Electron Current (Chabay & Sherwood, 2011, reproduced by permission)

Consider a section of metal wire through which the mobile electron sea is continuously shifting under the influence of a nonzero electric field, as indicated in the figure.

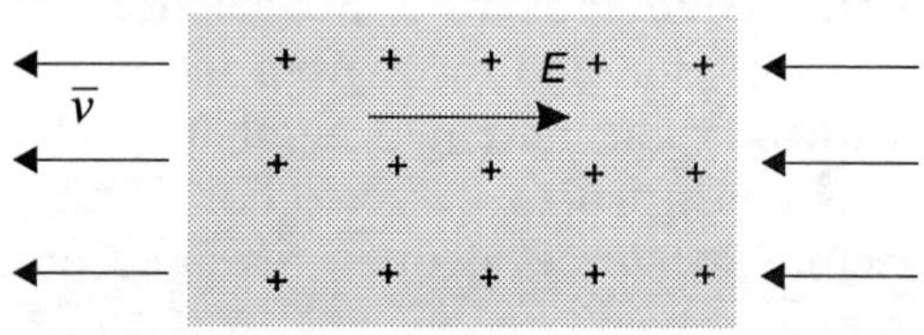

The mobile electron sea drifts to the left with a speed $\bar{v}$.

Suppose that in a metal wire the mobile electron sea as a whole has an average "drift speed" $\bar{v}$. Assume that the current is evenly distributed across the entire cross-section of the wire. After a short time Δt, this section of the electron sea will drift to the left a distance $\bar{v}\,\Delta t$.

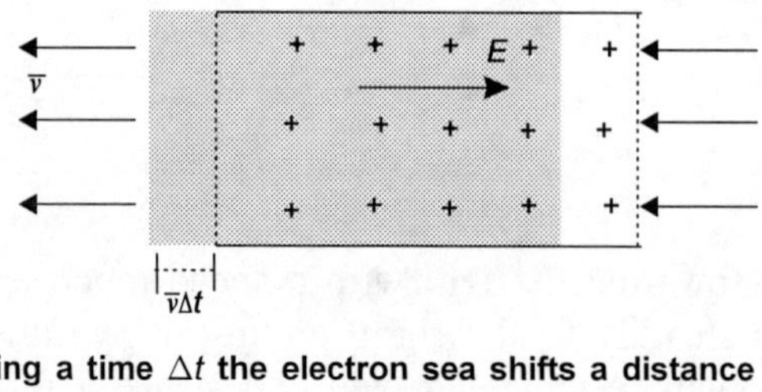

During a time Δt the electron sea shifts a distance $\bar{v}\Delta t$.

If the cross-sectional area of this wire is A, the volume of the disk-shaped portion of the sea that has flowed out of the left of this section of the wire (and into the adjoining section of wire) in the time Δt is $A\bar{v}\,\Delta t$ as seen in the figure.

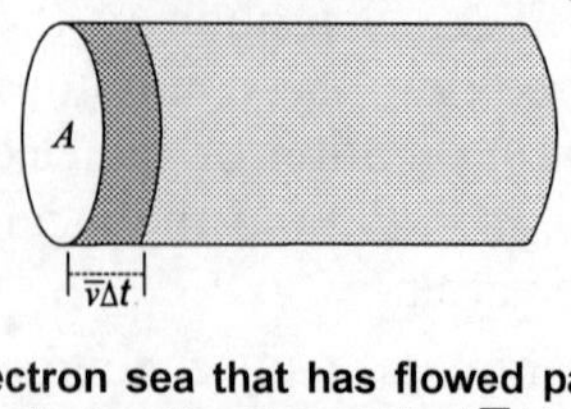

The volume of the electron sea that has flowed past a point in the wire during the time Δt is $A\bar{v}\Delta t$.

If the density of mobile electrons (that is, the number of mobile electrons per unit volume) everywhere in the metal is n, how many electrons are in the disk?

The formula is (with units)

$$\left(n\frac{\text{electrons}}{\text{m}^3}\right)(A\text{m}^2)\left(\overline{v}\frac{m}{s}\right)(\Delta t\ s) = nA\overline{v}\Delta t \ \text{ electrons.}$$

The number of electrons in this disk is the number of electrons that have flowed out of this section of the wire in the short time Δt.

What is the number of mobile electrons per second that flow past the left end of this piece of the wire?

We divide by the time Δt to calculate the number of mobile electrons per second that flow past the left end of this piece of the wire:

$$\frac{\text{Number of electrons}}{s} = \frac{nA\overline{v}\Delta t}{\Delta t} = nA\overline{v}$$

This result, that the number of electrons passing some section of the wire per second is $nA\overline{v}$, is sufficiently important to be worth remembering (though it is even better to remember the simple derivation that gives this result).

ELECTRON CURRENT

The rate i at which electrons pass a section of a wire:

$i = nA\overline{v}$ (number of electrons per second)

n is the mobile electron density (the number of mobile electrons per unit volume), A is the cross-sectional area of wire, and $\overline{v}$ is the average drift speed of electrons.

CONVENTIONAL CURRENT

In most metals, current consists of drifting electrons. In copper, as in most metals, the current is due to moving electrons. However, there are a few materials in which the moving charges act as though they were positive "holes" in the sea of mobile electrons. The positive holes act in every way like real positive particles. For example, holes drift in the direction of the electric field, and they have a charge $+e$.

Electron current moves from the negative end of a battery through a circuit to the positive end of the battery; hole current would go in the opposite direction.

Observations of the magnetic field due to a current-carrying wire are not sufficient to tell the difference, because if the sign of the moving charge is changed, the direction of the drift velocity is also changed: $(+e)\ (+\overline{v}) \rightarrow (-e)\ (-\overline{v})$ as seen in the figures.

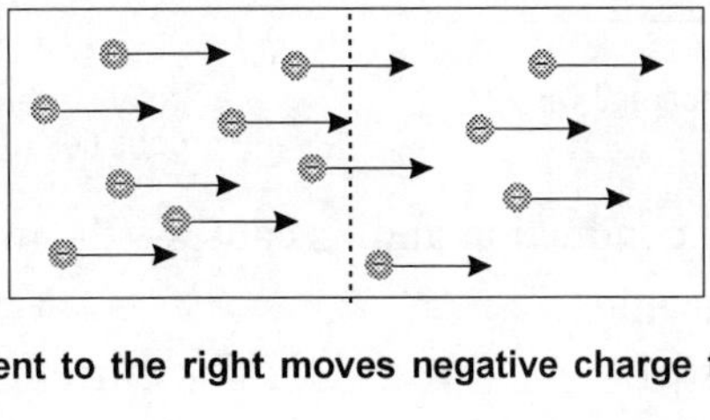

An electron current to the right moves negative charge from left to right.

An electron current to the right moves negative charge from left to right.

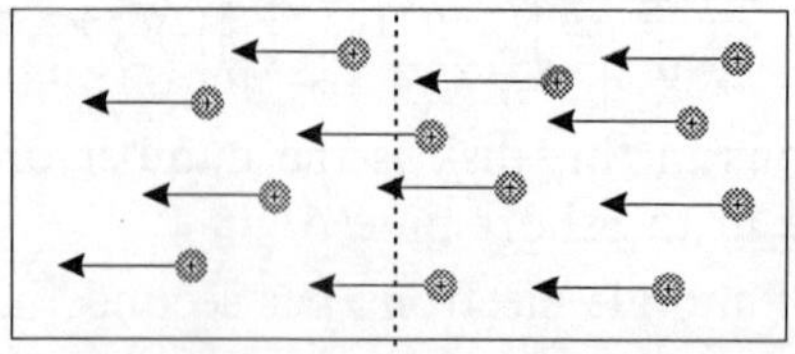

A conventional current to the left would move positive charge from right to left, which would have the same effect as moving negative charge from left to right, as in the previous figure.

The few metals that have hole current include aluminium and zinc. In the doped semiconductors important in electronics, "*p*-type" material (positive type) involves hole current, and "*n*-type" (negative current) involves electron current.

Because most effects (other than the Hall effect) are the same for electron current and hole current, it is traditional to define "conventional current" to run in the direction of the hole current, even if the actual current consists of moving electrons, in which case the "conventional current" runs in the opposite direction to the electrons. (This sign ultimately goes back to a choice made by Benjamin Franklin when he arbitrarily assigned "positive" charge to be the charge we now know to be carried by protons).

In addition, conventional current I is defined not as the number of holes passing some point per second but rather as the amount of charge (in Coulombs) passing that point per second. This is the number of holes per second multiplied by the (positive) charge $|q|$ associated with one hole.

$I = |q|\ nA\bar{v}$ (Coulombs per second, or amperes)

The direction of conventional current is opposite to the direction of electron current. The moving charged particles are assumed to be positive. A battery sends conventional current out of the + end.

In a metal, $|q| = e$, but in an ionic solution the moving ions might have charges that are a multiple of e.

In the figure a pictorial view of the total current I and current density J is given. For ease in visualization, a rectangular cross-sectional wire is assumed.

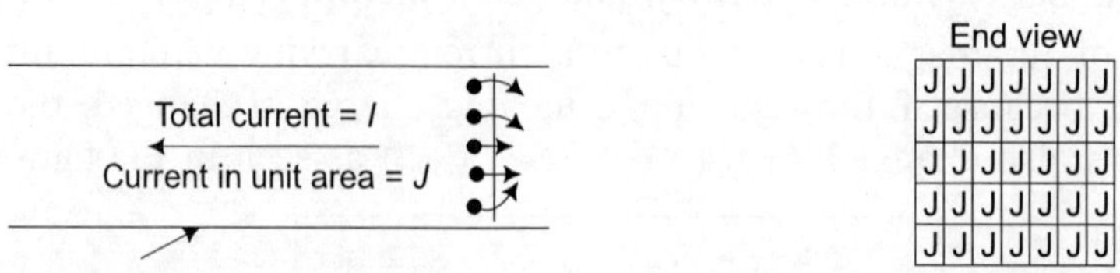

Rectangular cross-sectional wire

Total current I is the sum of the current densities (J) in each elemental area

There are two views of conduction and its causes—the microscopic and the macroscopic.

Microscopic	**Macroscopic**
The drift velocity $v = \mu E$	The current density $J = \sigma E$
and $\quad I = qnAv = qnA\ \mu E$	and $\quad I = qnAv = \dfrac{1}{R}\lvert\Delta V\rvert$

where, v = drift velocity
μ = mobility (cm^2/V-s)
E = applied electric field
A = cross-sectional area
n = number of free electrons per unit volume

where, σ = conductivity = $qn\mu$,
q = magnitude of charge of an electron in coulombs.

In the *microscopic* view, we focus on individual electrons whose drift speed is caused by the electric field produced by external charges (nearby surface charges, surface charge on distant parts of the circuit and charges on the battery).

Typical values of mobility for electrons are: silicon 1350, germanium 3900 and copper 100.

The smaller the mobility, μ, the *tighter* the collisions and resulting movement of the electrons within the lattice (more lattice defects can produce this effect, see "Resistivity of metals" in Appendix B) and larger their mobility μ, the more loose or *freer* their movement, as is evident from the equations for conductivity $\sigma = nq\mu$ and the drift velocity $v_d = \mu E$.

In the *macroscopic* view, we measure conventional current in the every-day units of amperes, and this current is numerically proportional (with proportionality constant $1/R$) to the potential difference, which we measure in the everyday units of volts.

It is useful to gain a feel for the mechanism of current flow. Some scientists have captured the flow and these are given in the next section. These should be only treated as aids to visualization and though not accurate by all means because of the myriads of particles involved at the atomic level, they serve to capture the essence of the movement.

1.18 VISUALISING CURRENT AND NEUTRALITY IN CONDUCTORS

Consider a simple circuit comprising a battery, connecting wires and a resistor that is closed. After the initial transient, lasting a few picoseconds, all the electrons, the millions everywhere in the wires gain a net movement in one preferred direction, and we call this "current".

Since fast running water has kinetic energy which has to leave the system when the waterflow is switched off and the kinetic energy of electrons, however, is negligible compared to all other interaction processes, the analogy of fast flowing water does not hold for conduction processes.

A better and more suitable analogy is a closed water circuit under high pressure but with a very low flowing rate. A current can be switched off with almost no energy dissipation.

Electric current is actually a flowing motion of charged particles.

When charge flows in a battery and resistor circuit, the current is *sustained* by an "applied" e-field in the steady state like a steady and continuous application of force to maintain motion which otherwise will cause the current to stop because of the dissipation or energy exchange in the resistive portion.

Consider a section of metal wire through which the free-electron sea is *continuously shifting to the left*, with a "drift speed" v as shown in the figure. Averaged over a few atomic diameters, the interior of the metal is *neutral.* A detailed discussion of neutrality inside current carrying conductors is provided in a subsequent paragraph in this section.

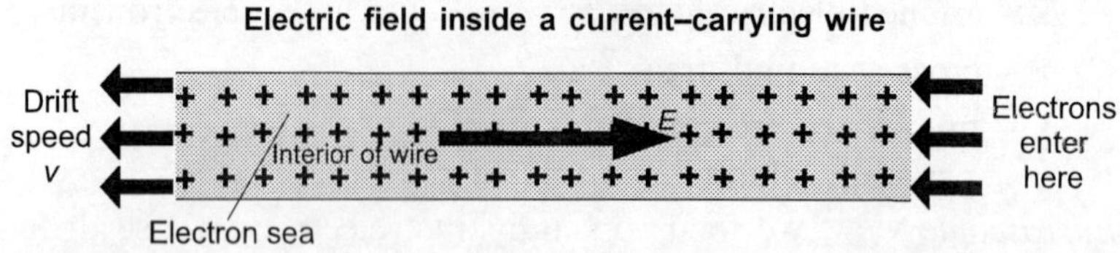

If there is an electric field inside the wire, wouldn't an individual electron accelerate rather than drift at a constant speed? After all, Newton's second law says that an electron should experience,

an acceleration, $$a = \frac{\sum F}{m} = \frac{-eE}{m}$$

The Drude Model

Paul Karl Ludwig Drude (1863 – 1906) formulated a simple classical model of electron motion (called the "Drude" model after Paul Drude). In this model, a mobile electron in the metal, under the influence of the electric field inside the metal, does accelerate and gain energy, but then it loses that energy by colliding with the lattice of atomic cores, which is vibrating because of its own thermal energy. In general, the cooler the metal, the less lattice vibration there is and the fewer collisions between electrons and the lattice. After collision, an electron *again* gets *accelerated*, and *again collides*. This process also called Joule Heating, is what makes the metal filament in a light bulb or the nichrome wire get hot (We will learn in Chapter 10 more about the mechanism of radiation by bulbs). The figure shows a graph of this start-stop sequence.

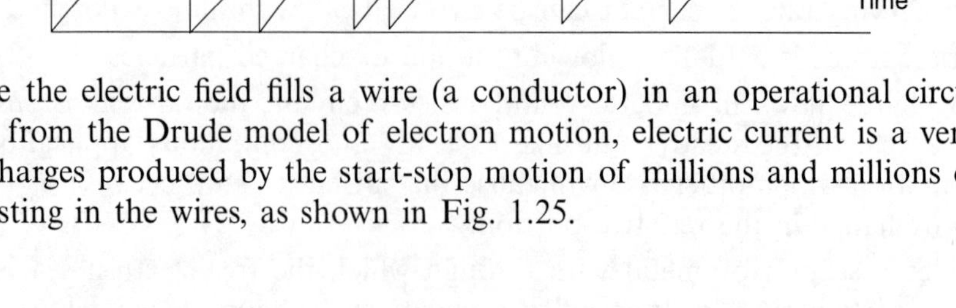

Because the electric field fills a wire (a conductor) in an operational circuit (Section 1.12), then from the Drude model of electron motion, electric current is a very *slow* flow (drift) of charges produced by the start-stop motion of millions and millions of electrons already existing in the wires, as shown in Fig. 1.25.

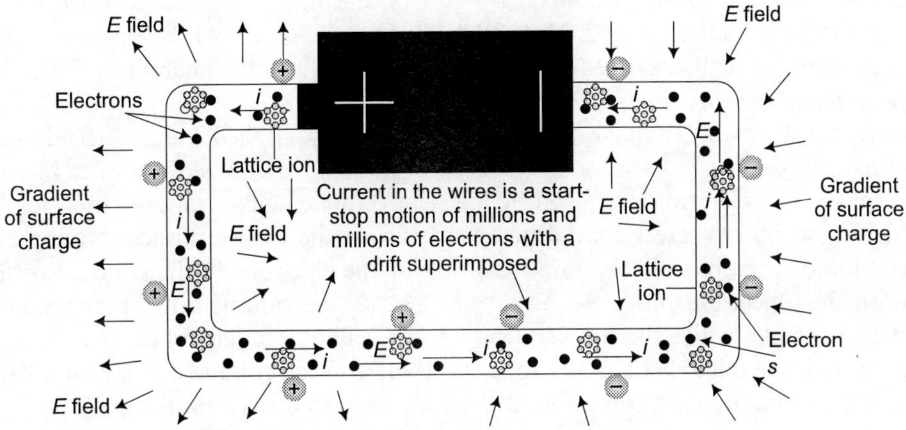

Fig. 1.25 Electric current in a simple circuit.

The average speed in this start-stop motion is called the "drift" speed *v*, and we say that the electrons *"drift"* through the metal. Suppose you hook a battery to a light bulb. Electric current appears in the circuit. There are millions of flowing electrons everywhere in the wires, the battery, the bulb filament and through screw terminals, solder joints, crocodile clips and plug and sockets or whatever that were used to make the circuit. The current in the connections is via overlapping orbitals if these connections are between similar metals or by tunneling if these are between dissimilar metals (see Appendix D and Section 9.15). The current in the connections is due to overlapping orbitals of conduction band electrons. *Inside* the battery there is an electron current which is equal in magnitude to the electron current in the external part of the circuit. The electron current within the battery flows from its positive terminal to its negative terminal. It is that electron current which completes the *circuit*. We will learn more about the action of the battery in Section 1.20.

Now break the circuit. The electrons do not zoom back into the battery; instead they halt where they are, and begin executing the random movement with zero drift in the absence of the electric field as described in the previous section.

If the applied voltage is increased, the start-stop motion of the millions and millions of electrons becomes more vigorous, but remains *tight* within the confines of the lattice making intense collisions, and there is a consequent increase in the drift velocity and the current accordingly increases. For a constant applied voltage (therefore, constant field) and in the steady-state, the drift velocity is constant.

Working Definition of Current

A working definition of current in conductors is: "Current is a start-stop motion of millions and millions of conduction band electrons everywhere within with a drift superimposed".

More details of how electrons in a conductor (metal) gain potential and kinetic energy when subject to an electric field are provided in Section 9.15 under the topic "Fermi Level in metals" with reference to Fig. 9.43.

There are several analogies described in textbooks; for example of the motion of objects dropped in a tank of liquid which attain a constant "terminal velocity", when the gravitational force is balanced out by the viscous forces.[13] The viscous forces of the liquid cancel out the acceleration produced by gravity leaving the object to move with a steady velocity, the "terminal velocity". Note that the viscous forces will be present as long as the object moves, and will disappear if the object is held steady. Therefore, the object moves with a *terminal velocity*.

Analogously, in a conductor subjected to a steady applied voltage (field) the electron collisions with the heavy lattice ions, "cancels" out in a "discretized" manner in a manner of speaking, their acceleration produced by the steady applied field, resulting in a steady velocity, the "drift velocity". There is one difference between the force and acceleration component cancellation of the object in the tank and that of the electrons in the conductor; the viscous forces on the object are produced on its *surface* and yet cancels out the acceleration of the object as a whole, while the acceleration of the millions of electrons is impeded by their collisions with the millions of lattice ions everywhere *in* the conductor, as would ball bearings (electrons) moving down a ramp studded with a forest of nails (lattice).

Such movements are described by equations where velocity and not acceleration (!) is proportional to applied force. However, I strongly recommend visualizing the actual movement of the electrons as described above and this should be sufficient and accurate enough for the visualization of the process of conduction.

If you *take away* the lattice ions and leave a vacuum and apply the field, the electrons will accelerate continuously and their motion is termed "ballistic". This is what happens to liberated electrons in vacuum tubes and oscilloscopes.

If the voltage applied to the conductor is changed, it is the *change* in the electric field that travels at the speed of light and *not the electron*. The *conduction band* electrons (in conductors, which are giant crystals made from atoms) accelerate to very large velocities that are yet small compared to the speed of light, and gain a huge amount of kinetic energy. But they soon decelerate due to the collisions. If they did not gain such large kinetic energies due to their high velocities, their collisions would be analogous to a tiny marble (electron) colliding against a cricket ball (lattice ion) when little energy may be transferred to the cricket ball (lattice ion). At higher velocities, with their large kinetic energies however, they transfer *most* of the energy they gained *between* collisions to the lattice ions and heat them up, like shooting swarms of high velocity marbles at cricket balls to make them move.

Since a coulomb is approximately equal to $6.24150948 \times 10^{18}$ elementary charges, one ampere is approximately equivalent to $6.24150948 \times 10^{18}$ (6.2 million-million-million) elementary charges, such as electrons, moving past a boundary in one second.

Note that in the description of the ampere it is the number of charges that move *past a boundary* in one second that matters and there is *no requirement* for the charges to travel all the way around the circuit to qualify for labeling as a current.

If the electrons in a conventional circuit have a start-stop motion moving a few micrometers, there will be a net transport of electrons past one cross-section (boundary), and there are some other electrons (not the same) that will be crossing a boundary at another cross section (boundary) a few micrometers away and different electrons crossing at any boundary, all around the circuit, at the same instant, with the same number flow rate.

That is sufficient to be registered, as a current….by an ammeter.

Figure 1.26 gives a more detailed view of the electron movement in a small section of a conductor. This view was made based on the above descriptions of the start-stop motions and does not claim to be accurate.

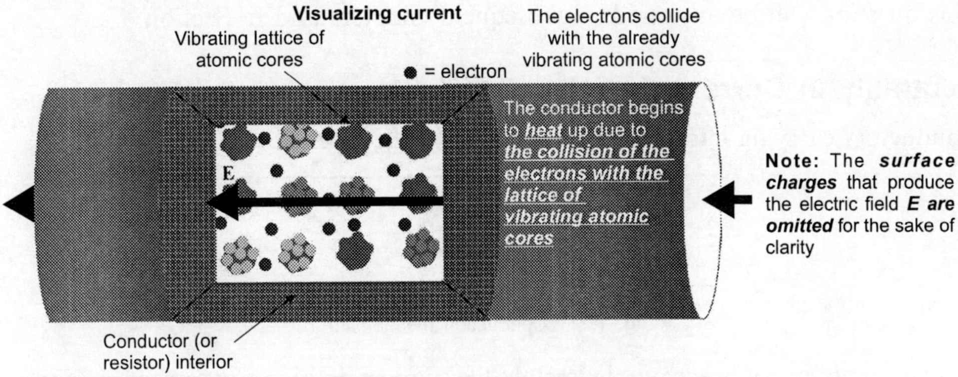

Fig. 1.26 Action in one small slice (a few atomic diameters wide) of a conductor impressed by an *electric field E* in a circuit.

With no electric field applied, the electrons gain energy from their collisions with the rocking lattice ions. The lattice ions, even in the absence of a current, vibrate on their own accord due to temperature. To gain an intuitive understanding of temperature, read "What is temperature" in Appendix B. To view the rocking motion of the lattice ions in a metal, visit http://www.chem.iastate.edu/group/Greenbowe/sections/projectfolder/flashfiles/thermochem/eqilibrium-v1.html and raise its temperature to increase the rocking intensity!

It is instructive to calculate how long it takes an electron to get *from a car battery to the starter motor* after the ignition switch is turned. Assume that the current flowing is 115 A, and that the electrons travel through copper wire with cross-sectional area 31.2 mm^2 and length 85.5 cm.

What is the *current density* in the wire?

$$3.69 \times 10^6 \text{ A/m}^2$$

The number density of the conduction electrons in copper is 8.49×10^{28} $/m^3$.

Given this number density and the current density, what is the *drift speed* of the electrons?

0.00027 meters/second or 0.97 *m/hour* = nearly *forty thousand times slower* than the speed of the fastest runner in the Olympic games.

Is it anywhere near the speed of light? c = 1080 × 1000000 km/hour.

How long does it take for an electron starting at the battery to reach the starter motor? A length of 85.5 cm or nearly a meter.

52.5 minutes (or nearly an hour!)

In that time, the slowest snail has been recorded to move a distance of 12 meters!

Comparing this drift speed (0.00097 km/hr) with the instantaneous speed of electrons in copper at room temperature which is about 5760×10^3 km/hr (only 0.005 times the speed of light), we see that the drift speed is indeed, very very small.

We see that the electron moves only a small distance inside a conductor in which there is a current. Therefore, current in the wires is a start-stop motion of millions and millions of conduction band electrons everywhere within with a drift superimposed (see Fig. 1.25). This raises a question as to how energy is transported in circuits and its speed. This question will be answered in subsequent chapters and in Section 4.2.

Neutrality in Current Carrying Conductors

Conductors carrying currents *remain neutral* in the interior.

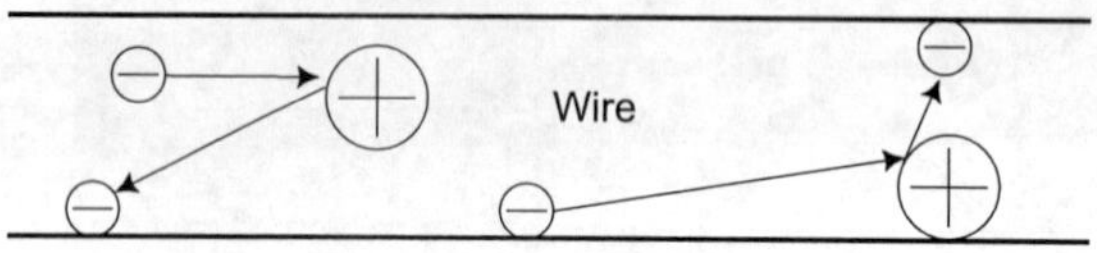

Fig. 1.27 Conduction mechanism in metal.

The mobility of electrons in metals like copper is many orders of magnitude larger than in a resistor. Within the copper wire, the electrons experience a very small force; and it is therefore possible to postulate this concept of a mechanism of conduction, in which the single electrons are accelerated under the influence of the interaction with the surface charges and collide with some atoms within the lattice as shown in Fig. 1.27. The surface charges are not shown in the figure.

As we stated for conductors, the number of collisions per second is 10^{14} per second, and the mean free path between two collisions is about 10 diameters of a copper atom. (**Note:** the world's fastest drummer has been recorded to make 20 strokes per second and the fastest rotating microfabricated gas turbine reached *only* 33,333 revolutions per second). I urge the reader to pause for a moment and visualize the hundred million-million collisions per second each electron makes with the lattice ions!

The electrons can therefore be regarded as independent particles, which are accelerated for a short moment and then stopped by collision. With this approximation, the interaction *between* the free electrons is neglected. In equilibrium, when the distribution of the surface charges is stable and the electric current is constant, this approach is correct because the interaction between the electrons and the enforced increase of the velocity of the electrons is so very small.

Therefore, the density of the electrons is practically not changed, and the demand for **neutrality** is always fulfilled in the interior, despite the presence of the applied electric field. Whenever it is said that a material is *electrically neutral*, it is merely a statement of account of an equal number of positive and negative charges of a closed system. If we treat the interior of the wire (excluding the surface) as a closed system, the neutrality condition is fulfilled *and also* the atomic stretch during the current is very small (no significant dipoles are created) and the interior may be considered to be "truly" neutral. Read the important note below Fig. 1.23.

While the descriptions given above should be sufficient to visualize currents in most circuits, a very thorough and comprehensive description of the flow of the electrons that takes into account, the complex interactions of the driving field on the randomly moving mobile electrons and ions, in the sea of neutral molecules of gases and metals is given in the topic "The physics of electrical conduction" in the Chapter "Electric Currents" in Purcell's "*Electricity and Magnetism*", Ref. [22].

The reader is referred to Gauss's Law and a geometry-based proof of the law in Appendix A. Following that an application of the law is provided which proves that the applied electric field fills the wire and that there is indeed no excess charge in a current carrying wire; it remains "neutral" inside.

Electrical conduction ordinarily involves only a slight systematic drift superimposed on the random motion of the charge carriers. We should not expect the actual average of the velocities of a group of electrons to be exactly zero, for they are statistically independent quantities. One electron doesn't affect the behavior of another. There will in fact be a randomly fluctuating electric current in the *absence* of any driving field, simply as a result of statistical fluctuations in the vector sum of the electron velocities. This spontaneously fluctuating current can be measured. It is a source of *noise* (current) in all electric circuits, and often determines the ultimate limit of sensitivity of devices for detecting weak electric signals[22]. This form of noise in circuits is called thermal, Johnson or Nyquist noise.

It is a fundamental fact that an accelerating or decelerating charge radiates energy in the form of electromagnetic waves. Since, electrons have a start-stop motion in conductors there will be some radiation, due to their accelerations (gained by the applied electric fields in circuits) and their decelerations on collisions, with the ions. However, the amount of this radiation is very small (Section 10.8 "Uniformly moving and sinusoidally oscillating charges").

No scientist has *counted* the number of electrons in a piece of conductor ! Read the interesting article "Determining the Atomic Number and Avogadro's Number" in the folder "Additional Practice Questions for all chapters" in the CD which describes how scientists arrived at the number of electrons of various elements.

1.19 MECHANISM OF CONDUCTION IN RESISTORS

What is a *resistor*? What is *resistance*?

We understand *"resistance"* in several ways. Here are a few samples....

Electrical resistance is a measure of the degree to which *a body opposes the passage of an electric current*. The SI unit of electrical resistance is the ohm. Its reciprocal quantity is electrical conductance measured in Siemens.

Here is another: The *restriction to the flow of electrons*.

And another: The *difficulty* in moving electrical current *through a conductor to which voltage is applied;* expressed in ohms.

Resistor: This is a *circuit element* which *dissipates energy* in the form of heat when current flows through it. Example: an electric duct heater, in which ribbons or coils of *metal having a fairly low conductivity (high resistivity)* are heated to a high temperature by passing current through them.

An appropriate definition of resistance: *Property of a conductor which determines the current produced by a given difference of potential.* The ohm is the practical unit of resistance.

The definition enables us not to forget that a resistor is made from conductive material.

The following description of the function of a resistor was adapted from the article "Constant Electric Current and the Distribution of Surface Charges" by Hermann Härtel and reproduced with permission.

The function of a resistor is, in general, to hinder the flow of charge carriers and to *exchange energy*. Resistors are used in circuits to regulate the strengths or amounts of currents. There are different ways to accomplish this function. One possibility is to reduce the diameter of the conducting wire without changing the material.

In most practical cases, a resistor is introduced by choosing a different material. Such a material possesses a higher density of obstacles (lattice imperfections such as rows of missing atoms and displaced atoms) to electron flow, and/or it has a smaller concentration of free electrons. Both effects normally appear together within manufactured resistors and can cause a dramatic change in mobility for the electrons. The resistivity of a metal can be increased by adding small amounts of impurities, or by straining it severely by drawing it through a die to increase the number of lattice imperfections.

The resistivity of metals, is due to lattice imperfections and the vibrations of the lattice ions, at elevated temperatures. For a description of the phenomenon of resistivity and its variation with temperature, see the Appendix B "Resistivity of metals".

If a uniform conductor and a uniform resistor are soldered together, there exist two thin areas or volumes within which the change of conductivity occurs. These areas are represented in the picture (Fig. 1.28) on an exaggerated scale.

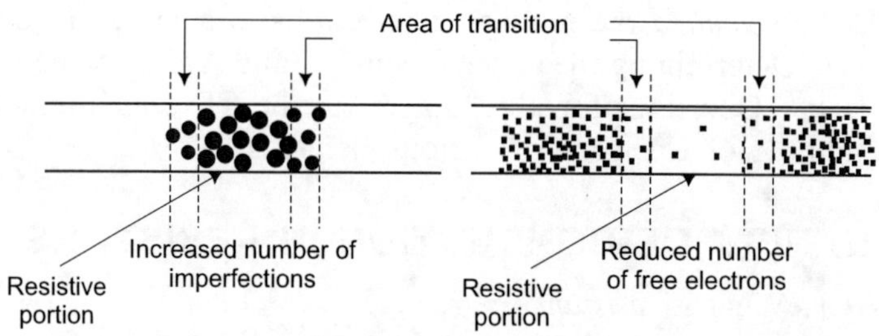

Fig. 1.28 Different mechanisms for electrical resistance.

In the left picture (increased number of imperfections), the filling represents the density of obstacles (lattice imperfections) for the moving electrons in the resistive portion. In the right picture, the reduced number of free electrons is represented in the resistive portion.

In both cases, the function will be the same. Within the area of transition from high to low values for the mobility, there will be one side a slightly higher concentration of electrons than neutral and a slightly lower one on the other side. These extra charges, which are spread out within two thin volumes, will have two effects: on one hand, they oppose the flow in conductors and on the other, they increase the force on the free electrons within the resistor.

In dynamic equilibrium the same number of electrons will flow through any cross-section in any given time. Due to the extra charges and the different gradient of the surface charge distribution, the driving force within the conductor and the resistor are, however, quite different.

According to Härtel, to treat the effect resulting from these transition areas is important for two reasons. First, the principle that nature changes gradually and continuously in space and time is basic for qualitative reasoning. Therefore, the dramatic change in mobility should not be treated as a step function with no further consequences.

Second, this area of transition plays an important role when studying the conduction mechanism in semiconductors, such as diodes and transistors.

A last possibility to resist an electron flow is through magnetic interaction from the outside. In this case, it may be that an additional force (or induced emf which will be discussed in Chapter 5) from outside is holding back parts of the electron distribution within the circuit.

This leads to a redistribution of the surface charges within the other parts just as if an ordinary resistor would have been introduced.

The effect of resistance in respect to electron flow can easily be explained through geometric arrangement. The surface charges A, B, C and D (see Fig. 1.29) which always exist in the presence of a changed cross-section, will oppose the effect of the original charges on the battery terminals and will thus, reduce the intensity of the current flow in the conductor.

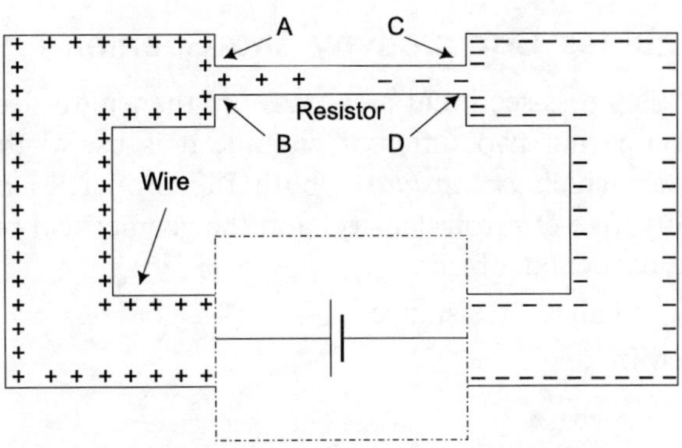

Fig. 1.29 Resistance through reduction of cross-section.

The gradient of the surface charge distribution (point charges set up by the shifting electron sea as described in Sections 1.15 and 1.l6), is larger along the thin wire than along the thick ones. Therefore, a stronger force due to a stronger field is exerted on the free electrons within the resistor, leading to a higher drifting speed. The longer the thin wire, the smaller the gradient and resulting axial electric field will be inside the wire, and this will lead to a smaller current. We will learn in Chapter 4 (Fig. 4.15b) about the presence of interface charges at the two ends of the resistor, which increase the electric field and hence current density through the thin wire, and simultaneously decrease the field and current density in the thicker wires.

When equilibrium is reached, the same number of electrons will pass through any cross-section of the system in any given time. The drifting speed, however, will be different at different cross-sections.

The current in the circuit of Fig. 1.29 is termed "differential" mode, because conventional current leaves the positive terminal and returns to the battery negative terminal. We will learn more about differential mode current in Section 5.22.

It is normal practice to show a circuit zero (commonly called "circuit *zero*" or "signal ground") using the symbol ⏚ or ⏚ in circuit diagrams indicating that voltages or more really the potential *differences* at different points in the circuit, where indicated in a diagram, are measured with respect to the terminal labeled with such a symbol. The battery charge separation process and the polarity of the voltage on its plates described in Section 1.8, are not affected by the inclusion of the symbol.

Every network's voltage source operates by separating charges (Section 1.8), and there will be +V/2 volts on the positive terminal and –V/2 volts on the negative terminal. Then, for example in Fig. 1.29, the negative terminal of the circuit may be shown grounded by adding the circuit *zero* symbol. This is *not* to be interpreted that all currents "terminate" at the circuit zero signal ground. The currents are everywhere in the wires connecting both the terminals of the battery to the resistor and in the battery as well. In Sections 2.26 and 3.4, this will be discussed in more detail. *Real* grounding or making a connection to *earth* will be treated in Section 2.25.

Resistance Combines Conductivity and Geometry

Instead of saying, "the cross-sectional area goes up, the length decreases, and the mobility increases", when comparing two different circuits, it is useful to have a single quantity to *describe a resistor* which *encapsulates* both the properties of the material namely, conductivity (mobility, free-electron density) and the geometrical properties (cross-sectional area, length) of the particular object.

Such a quantity is called "resistance".

And this is given by

$$R = \rho \frac{l}{A}$$

where, ρ is the resistivity of a conductor expressed in ohm-meter; l = the length of the conductor; and A = area of cross-section.

ρ is itself given by: $\frac{E}{J}$ where E = the electric field in the conductor; and J = the current density.

Now consider a circuit consisting of a voltage source, connecting wires and a resistor as shown in Fig. 1.30.

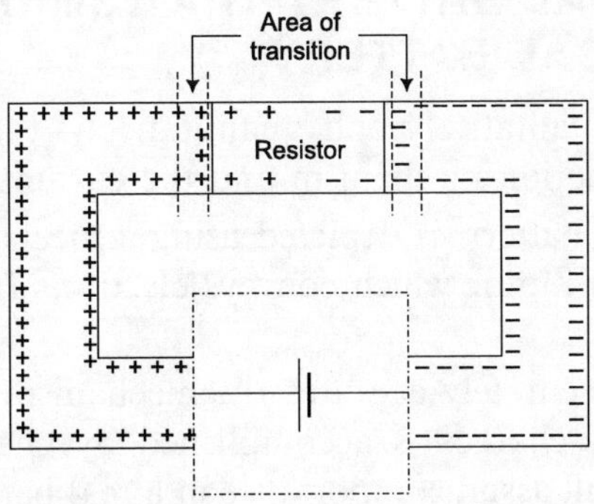

Fig. 1.30 Surface charge distribution (circuit with one resistor).

For any of the different resistors either with more obstacles (more imperfections in the lattice) or fewer movable charges (fewer number of free electrons), a similar surface charge (point charges set up by the shifting electron sea as described in Sections 1.15 and 1.16), distribution will occur. The figure indicates the principal idea, which is the establishment of a *steeper* gradient of surface charge across the resistor, and a more powerful electric field inside it, than inside the wires; it does not represent reality in all respects.

The scale of the wire and resistance in Fig. 1.30 was chosen to optimize visibility. Moreover, the gradient in the surface charge distribution on the copper wires is not shown, because it is too small compared with the gradient along the resistor. To view a live demonstration of surface charge phenomenon and its mechanical effects in circuits, visit https://matterandinteractions.org/videos-of-lecture-demos/

Recollect from Section 1.9 where we sought an answer to the question as to how the electric bulb does not increase in brightness when brought closer to the battery, or when the bulb is twisted so that the filament is perpendicular to the electric field due to charges on the battery. The answer, is the production of a uniform electric field in the filament due to the development of surface charges, in combination with the charges on the battery and at the interface of the wires connecting to the bulb. Continuous feedback between the electric field and surface charges ensures that the brightness of the bulb does not vary as it is moved around or twisted.

Water Circuit and Electric Circuit Analogy Demonstration Provided by Prof. Hermann Härtel

The reader is directed to a very nice demonstration of the analogy between a water circuit and an electric circuit given under the supervision of Prof. Hermann Härtel in this link http://www.astrophysik.uni-kiel.de/~hhaertel/PUB/syringe.htm which brings out clearly the aspect “The kinetic energy of the free electrons is practically zero, the driving force - the EMF - is absolutely dominant”. We will learn more about energy transport and dissipation in circuits in Section 4.4, Chapter 4.

1.20 A MECHANICAL BATTERY THAT SIMULATES THE ACTION OF A CHEMICAL BATTERY

We now describe both qualitatively and quantitatively, the action of sources of emfs in simple circuits, the conduction mechanism and the current determining parameters.

The source of emf, a battery, is depicted using a mechanical battery that consists of a conveyor belt driven by a motor which conveys electrons from one plate of the battery to the other.

This analogy fairly accurately captures the mechanism of a real battery in which the process of separation of charges is accomplished by chemical reactions. The chemical reaction considerations will assume importance in a real battery when it discharges to levels that will affect its emf. The description of the mechanical battery is based on the concept used by Chabay and Sherwood in their book "Electric and Magnetic Interactions", 1995, John Wiley, USA; used by permission.

In the mechanical battery shown in Fig. 1.31, the conveyor belt applies a *non-Coulomb force* F_{NC} to extract an electron from the positive plate of a large parallel-plate capacitor and push it onto the negative plate.

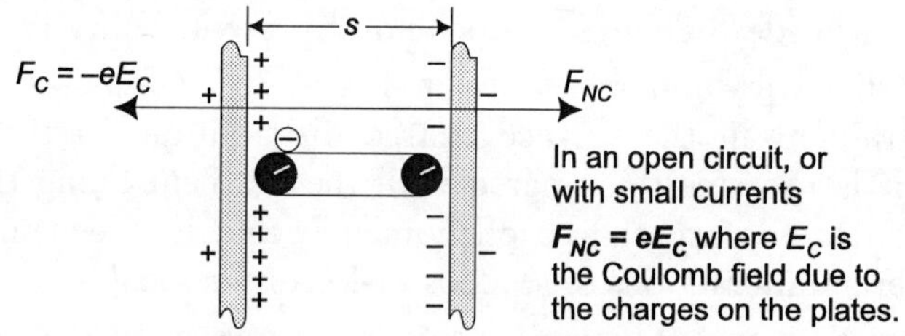

Fig. 1.31 An isolated mechanical battery.

Charge builds up on the plates of the battery until the *Coulomb force* eE_C is equal to the *non-Coulomb force* F_{NC}. The separation s is small compared to the size of the plates.

If we connect a Nichrome wire from one plate of the mechanical battery to the other (Fig. 1.32), a steady-state can be achieved if the motor transports as many electrons per second from the + plate to the – plate as are flowing through the wire. The larger the current, the more electrons per second the motor must transport to keep the plates resupplied. If there is no resistance to the flow of charge inside the battery, we will have $F_C = eE_C = F_{NC}$.

See note on how a real battery operates in Sec. 3.5.

Note on Non-Coulomb Force

We call the force of separation by the conveyor belt a non-coulomb force, because it is due to the *mechanical force* exerted by the belt on the electron as it drags it away from the positive plate.

This is in contrast to the electric force due to electric fields set up by point charges, and which is called a Coulomb force.

Energy Conservation and Electron-current Conservation

In a *circuit* containing a mechanical battery with the plates separated by a distance '*s*' and a series of wires:

Energy conservation leads to an equation relating the amount of work done on an electron by the mechanical battery to the energy dissipated in the wires, where E_i (Newtons/Coulomb) is the electric field in a wire of length L_i.

$F_{NC}s = \Sigma eE_iL_i$ and e = magnitude of the charge of an electron in Coulombs.

A formal derivation of the Loop rule based on energy considerations in a circuit is given in Section 2.20.

Current conservation Because charge is conserved, in the *steady state* (drift speeds and excess charges not changing with time), the amount of current entering any portion of a circuit must equal the amount of current leaving that portion of the circuit.

If there were any *imbalance*, charge would build up inside this portion of the circuit, and steady state would not be achieved.

From the electron current conservation equation:

$$i = nAv = nA\mu E$$

where, n = number of free electrons per unit volume, A = cross-sectional area, v = drift velocity, and μ = mobility in cm^2/V-s.

A battery puts *energy* into a circuit.

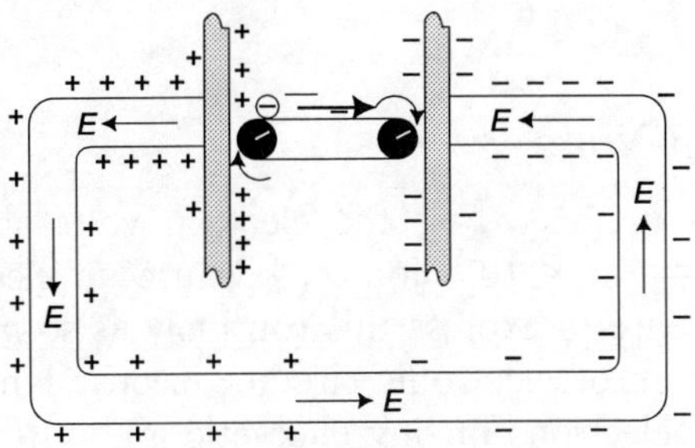

A battery puts energy by separating unlike charges and depositing them on the plates and then holding them apart. If charges are continuously being depleted on the end plates (terminals) after the circuit is closed and a current is established, the battery works to replenish the charges on its plates continuously... thereby maintaining the voltage across its terminals.

In the mechanical battery, the motor puts *energy* into the circuit.

Within the circuit, the moving electrons collide with the atomic cores and heat the metal. The warm metal *heats* the surroundings, and in the *steady-state* the energy input of the battery must be equal to the heat transfer output to the surroundings.

The amount of energy the mechanical battery *must expend* to move one electron from the positive plate to the negative plate is:

$$F_{NC}s$$

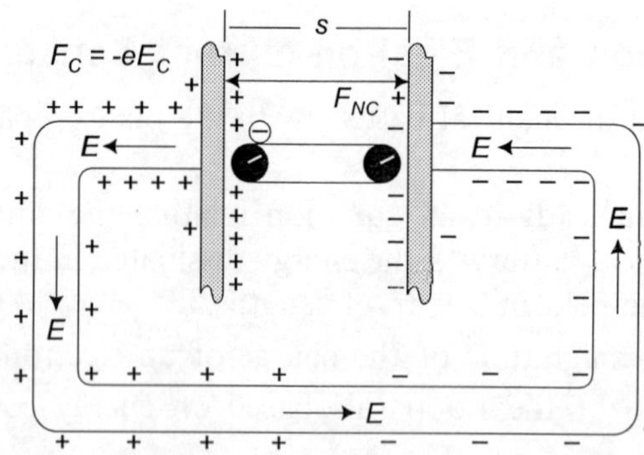

Fig. 1.32 A simple circuit showing energy and current conservation.

This electron is *pushed* through the wire (shown uniform in Fig. 1.32) by the electric force eE inside the wire. The work done on the electron in *forcing it through* the entire length of the wire L is eEL. This work goes into heating the wire.

Due to energy conservation, $F_{NC} \times s = eEL$ in the steady-state.

The above formula is *unaffected* if the wire is *bent* into some complicated shape, because no matter how many twists and bends there are, the surface charge will *arrange itself* to make the electric field be uniform in magnitude, and make the electric field follow the direction of the wire.

Hence, the work done on the electron by the surface charges is eEL, independent of the shape of the wire.

Joules and Electron Volts

Physicists and engineers prefer to use the electron volts abbrev. eV when dealing with electron energies (1 eV = 1.6×10^{-19} joules) because an electron's potential energy is so tiny that it makes little sense to express this routinely as so *many million million millionths* of a joule! In the operation of electronic circuits, electric lamps and motors, millions and millions and millions of electrons in tiny energetic amounts (expressed in electron volts), harmoniously convert energy that is measured in joules! Therefore, physicists and engineers switch between electron volts and joules in the descriptions of microscopic and macroscopic aspects of circuit processes. More details on the mechanism of energy transfer in circuits are provided in Chapter 4.

1.21 DISTINGUISHING INTERNAL ENERGY MECHANISMS AND EXTERNAL ENERGY MECHANISMS

Note that in considering the energy balance, we don't include a term $-F_C s$ corresponding to the Coulomb force F_C acting "on" the electrons which are moved from the "+" ve plate to the "–" plate.

That force, and the associated work or energy, are *internal* to the circuit (but numerically equal to the work done on the electron by the motor $F_{NC}\, s$, if there is no internal resistance of the battery).

Similarly, the work done on the drifting electron by the electric field (due to the charges located on the ends of the battery and on the surfaces of the wires) is *internal* to the circuit but numerically equal to the heat transfer to the surroundings.

The only *external* energy *input* to the circuit is the work done by the motor, and the only *external* energy *output* from the circuit is the heat transfer to the surroundings.

The "emf" of a Battery

What is the significance of the term $F_{NC}s/e$ (e = electron charge in Coulombs) in the energy conservation equation? Chabay & Sherwood, 1995, used by permission.

Since $F_{NC}s$ is the amount of work done by the motor in moving one electron through a mechanical battery, $F_{NC}s$ is the *work done per unit charge* (joules per Coulomb or volts).

This quantity is called the *"emf"*— the *energy input* to the circuit – or the circuit's *"starter"*.

In a chemical battery, the "emf" is a measure of the chemical energy per unit charge expended by the battery in moving charge through the battery and is *nearly constant*, independent of how much current is running through the battery (if the current is not too large) and approximately independent of how long the battery has been used.

It is the approximate constancy of the "emf" that makes this a useful concept, and which makes it useful to anlayze circuits in terms of "emf" and the potential differences in the circuit elements connected to the battery.

1.22 DIFFERENCE BETWEEN EMF AND POTENTIAL DIFFERENCE

It is important to keep in mind that although the units of "emf" are volts, the "emf" is not a potential difference.

"Potential difference" is a path integral of the electric field made by real charges.

We will learn about potential and potential differences and voltage in Sections 2.17, 2.19 and 2.20.

The *"emf"* is the non-Coulomb work per unit charge and might in principle be gravitational or nuclear in nature.

The *"emf"* of a flashlight battery is 1.5 volts, and the associated charge separation is due to chemical reactions in which electrons are reactants or products.

A Simple Circuit and Calculation of the Field Inside the Conductor and Resistor

In Fig. 1.33 is shown a battery connected to a thin wire by using 2 thick wires of equal length. The thin wire constitutes a resistance, because as mentioned in Sec. 1.19 (below Fig. 1.30), when traversing the circuit, there is a portion (thin wire) which has "fewer" mobile charges. Recollect from Section 1.19 that the resistance could be due to fewer charged particles or conduction band electrons like for example the thin diameter wire shown in the circuit of Fig. 1.33.

We have to calculate the electric fields E_1 and E_2 inside the thick and the thin wires respectively.

The joule value $F_{NC}\,s$ given for the battery is for a typical 1.5 volt battery and F_C represents the Coulomb force acting "on" the electrons that are moved from the "+" ve plate to the "–"plate.

Also given for the circuit are n = number of free electrons per unit volume of the wire, A_1 and A_2 cross-sectional areas of the thick and thin wires respectively, and μ = mobility in cm^2/V-s.

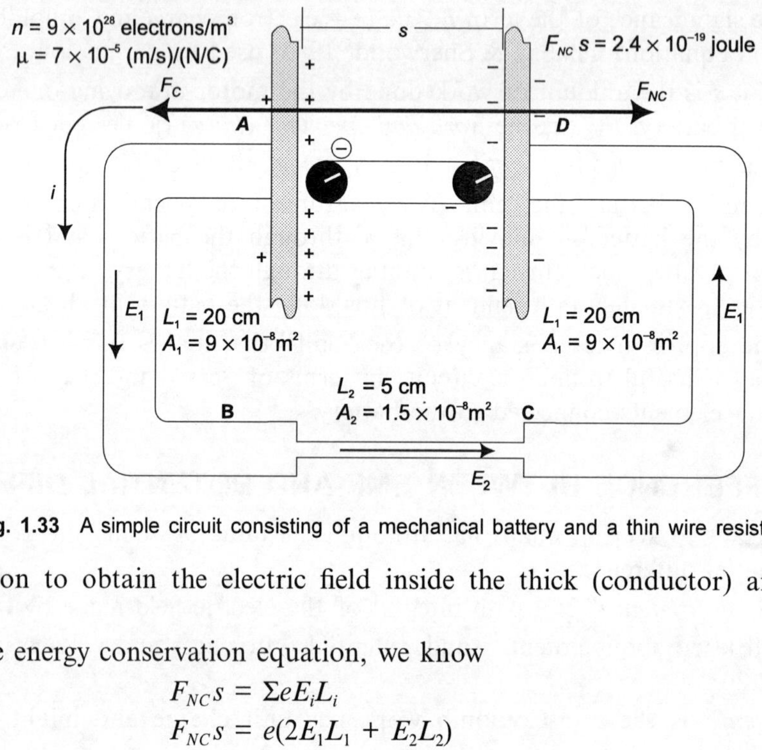

Fig. 1.33 A simple circuit consisting of a mechanical battery and a thin wire resistor.

Calculation to obtain the electric field inside the thick (conductor) and thin wire (resistor)

From the energy conservation equation, we know

$$F_{NC}s = \Sigma eE_iL_i \tag{1}$$

$$\therefore \quad F_{NC}s = e(2E_1L_1 + E_2L_2) \tag{2}$$

From the electron current conservation equation

$$i = nA_1\mu E_1 = nA_2\mu E_2 \tag{3}$$

From (3):
$$\frac{E_1}{E_2} = \frac{A_2}{A_1} = \frac{1.5\times10^{-8}}{9\times10^{-8}} = \frac{1}{6} \tag{4}$$

Or, $\quad E_2 = 6 \times E_1$

From (2): $\quad 2.4 \times 10^{-19} = 1.6 \times 10^{-19}\ (0.4 \times E_1 + 0.05 \times E_2)$

Substituting $\quad E_2 = 6 \times E_1$ from (4):

$$1.5 = 0.4E_1 + 0.05 \times 6 \times E_1 = 0.7E_1$$

Therefore, $\quad E_1 = 2.1$ V/m and $E_2 = 12.6$ V/m

Verification that the result conforms to Ohm's law, i.e., voltage drop across the thick and thin wires = 1.5 V.

The potential difference between points B and C (across the resistor) = $E_2 \times L_2 = 12.6 \times .05 = 0.63$ V, and the drop across the wires = $E_1 \times 2 \times \boldsymbol{L_1} = 2.1 \times 2 \times 0.2 = 0.84$ V

Since, 0.84 V + 0.63 V = **1.47 V = (approx. 1.5 V)** (5)

the relation (5) establishes the conformance of the circuit to Ohm's law.

Note that the electric field in the thin conductor (resistor) is *6 times stronger* than that of the field in the thick conductor.

It is this powerful field in the resistive portion that guarantees a constant current through the entire circuit.

Calculations Based on Resistance and Voltage

Let us check our results by calculating the resistance of the wires given that the battery whose energy is 2.4×10^{-19} joule corresponds to a 1.5 volt battery (1.5 V = 2.4×10^{-19} joules/1.6×10^{-19} coulombs).

Conductivity of the wires $\sigma = qn\mu = 1.6 \times 10^{-19} \times 9 \times 10^{28} \times 7 \times 10^{-5}$ $= 100.8 \times 10^4$ mho per meter.

Resistivity ρ = 1/ Conductivity = $1/\sigma$

Therefore, the resistivity of the material is $1/\ 100.8 \times 10^4$ $= 9.9 \times 10^{-7} = 990 \times 10^{-9}$ = (approx) $1000 \times 10^{-6} \times 100 \times 10^{-3} = 100 \times 10^{-6}$ = **100 microohm-cm.**

Consider the wire of length L = 20 cm and area of cross-section A = 9×10^{-8} m^2.

Resistance of the wire = resistivity × L/A = $990 \times 10^{-9} \times 20 \times 10^{-2}/9 \times 10^{-8} = 19800 \times 10^{-11}/9 \times 10^{-8} = 2200 \times 10^{-3}$ = **2.2 ohms.**

Consider the wire of length L = 5 cm and area of cross-section A = 1.5×10^{-8} m^2

Resistance of the wire = resistivity × L/A = $990 \times 10^{-9} \times 5 \times 10^{-2}/1.5 \times 10^{-8} = 4950 \times 10^{-11}/1.5 \times 10^{-8} = 3300 \times 10^{-3}$ = **3.3 ohms.**

The total current will be 1.5 V/resistance of the wires.

Therefore, the current = 1.5 V/(2 × 2.2 + 3.3) ohms = 0.19 Amps.

The voltage in the thick sections is 0.19 A × 2 × 2.2 = 0.85 V.

The voltage in the thin section is 0.19 A × 3.3 = 0.627 V.

The sum of the two voltage drops is = 0.85 V + 0.627 V = 1.477 V = **approx 1.5 V** (6)

This is the same result which we obtained with calculations (see relation (5)) based on electric fields in the wires.

1.23 FIELD STRENGTH INSIDE THE WIRES AND THE CONSERVATION OF CURRENT

It is useful to keep in mind that the electric field strengths inside wires in most applications where currents do not exceed a few amperes will be in the range 0.1 V/m to 20 V/m.

If we plot a graph as in Fig. 1.34, of the electric field and drift velocity inside the wires and the resistor, we will observe a large drift velocity of the electrons inside the resistor portion in comparison with the low resistance wire portions.

This is rightly so because if it were not, then excess (local) charges (neglecting surface charges) will build up inside the circuit in violation of the charge neutrality condition that should always be maintained.

Because charge is conserved, in the steady state (drift speeds and excess charges not changing with time) the amount of current entering any portion of a circuit must *equal* the amount of current leaving that portion of the circuit and is simply called the *conservation of current*.

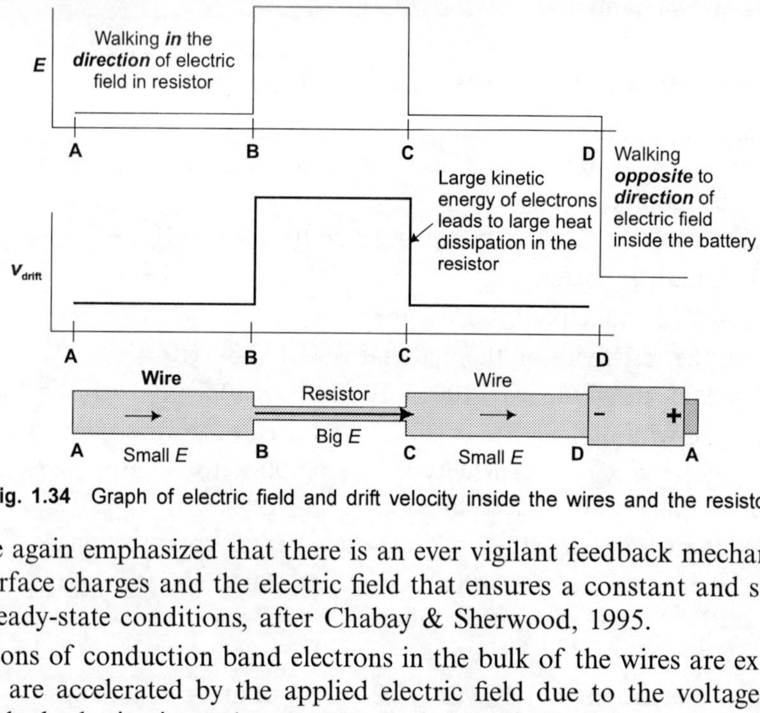

Fig. 1.34 Graph of electric field and drift velocity inside the wires and the resistor.

It is once again emphasized that there is an ever vigilant feedback mechanism between the excess surface charges and the electric field that ensures a constant and steady current leading to steady-state conditions, after Chabay & Sherwood, 1995.

The millions of conduction band electrons in the bulk of the wires are extremely busy: even as they are accelerated by the applied electric field due to the voltage source, only to collide with the lattice ions, they move only infinitesimal distances and this maintains charge *neutrality* in the bulk of the wire.

In response to the applied field, which seems to have disturbed their otherwise pre-programmed movement, they make adjustments to the *surface* charges to preserve voltage conditions on the surface and within the wire, and more importantly, ensure that the conservation of current law is not violated at all times.

A large amount of kinetic energy is imparted to the electrons by the electric field in the resistor leading to large heat dissipation in it.

How is heat produced by an electron collision with a lattice ion? See Appendix B for a description of the mechanism of heat production in conductors and resistors.

The conduction band electrons are busy in the resistor as well. In here, they continuously make adjustments to the surface charges to ensure a uniform voltage drop along its length (provided, of course, the resistance is uniform along its length) (see Fig. 1.35 and the potential graph between points B and C), maintaining charge neutrality in elemental volumes and that the required field strength is available that will ensure that the conservation of current law is not violated at all times. Neutrality is maintained *because* the interaction between the electrons and the enforced increase of the velocity of the electrons due to the applied field is so very *small*.

In the bulk of the resistor (conductor) not counting the surface charges, the conduction band electrons **do not** create *significantly strong* **dipoles** (charge center separation between free electrons and lattice ions) that could produce *opposing* internal electric fields of significance. But the discrete nature of the electrons (charge carriers which are not the ideally approximated finely ground *dimensionless* point charges) and their own random (because of the movement of millions of atoms and their electrons) thermal agitations, should be borne in mind to understand the origins of certain types of internal sources of extremely *tiny* emfs[21] (dipole effects) "noise" (in fact, using sophisticated instrumentation, very tiny "noise" emfs can be observed even across *open-ended* resistors!), which superimpose on the voltages of interest across conductors.

A plot of the potential inside the wires and the resistor is shown in Fig. 1.35 (Chabay & Sherwood, 1995, used by permission)[2] and indicates a large potential drop across the resistor.

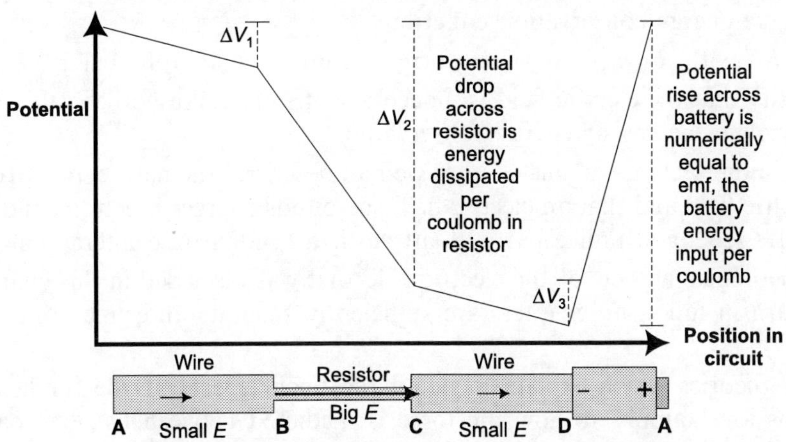

Fig. 1.35 Graph of potential inside the wires and the resistor.

The potential drop is caused by the steep surface charge gradient over the resistor caused by a buildup of positive charges at its positive end and negative charges at it's negative end.

How much current if we double the length of the wire (Fig. 1.33).

If we double the length of the wire:

The amount of charge on the endplate of the battery *stays the same*.

The rate of variation of the surface charge on the wire *decreases;* the progression from "+" to "–" surface charge *spreads out over twice the length*.

Since, $eEL = F_{NC}s$,

Therefore, $E = F_{NC}s/eL$

So if we *double the length L*, we'll get *half as big an electric field E* (since the other quantities *do not change*).

$v = \mu E$ and Electron current is nAv.

With half the field we get half the drift speed and *half the electron current.*

How much current if we double the cross-sectional area of the wire?

If we double the cross-sectional area of the wire:

The amount of charge on the endplate of the battery *stays the same.*

The electric field inside the wire *hardly changes* at all.

The drift speed of individual electrons is *unchanged.*

The free *electron density is unchanged* because the wire is made of the same material

$v = \mu E$ and electron current is nAv.

If n and v are the same, only A matters. Therefore, doubling A doubles the current.

With twice the cross-sectional area, we get *double the electron current.*

In Section 1.15, we stated that the electric field inside a wire 5 meters long and of uniform cross-section connected to a 10 volt battery is 2 V/meter.

Using the energy conservation equation:

$F_{NC}s = \Sigma eEL$ with, $F_{NC}s = 16$ joules corresponding to a 10 volt battery, $L = 5$ meters, we can verify that the electric field is indeed 2 volts/meter. Any other value for the electric field will violate the law of energy conservation.

In the initial sections of this chapter we had described the nature, properties and effects of the electric field and that an electric field surrounds charges, electrons and protons. That like particles repel and unlike particles attract is a fundamental interaction.

The theoretical aspects of the electric field that was discussed in this chapter and which we will learn in following chapters are sufficiently accurate and precise in the analysis of circuits.

Newer theories which explain in detail the finer aspects of the fundamental particle interactions for example, in how the force is mediated between two like particles and of nature of the field will be introduced in Section 4.2 "Theories of Electrodynamics".

I would caution the reader, however, to read through chapters 1, 2 and 3 before reading the description in Section 4.2.

Nature of Currents in Special Situations

It should be noted that current is *not* the *speed* with which the electrons move in wires; current represents the *number* of electrons flowing past a boundary in one second. In the operation of diodes in the *reverse* direction, the current does not *increase* with an increase of the *reverse* voltage (in a certain portion of its characteristic) but is limited by the *number* of mobile charges available to cross the junction of the diode. We will learn about this in Section 9.11. In the theory of operation of bipolar junction transistors, a situation occurs where this aspect of the nature of current becomes more evident. In the active region of operation of the transistor[17,30], the number of electrons injected by an *n*-type emitter to the base of a transistor are *controlled* by the base emitter applied voltage. These are collected by the collector; the current at the junction is limited not by how *fast* the electrons are swept by the electric field at the junction as the reverse-biasing voltage at the collector junction is increased, but rather *how many* have been injected through the base by the emitter, to be swept by the electric field into the collector.[17,30] Therefore, even if the reverse-bias voltage were increased, the current remains fairly constant. The theory and operation of transistors are beyond the scope of this textbook and the reader is referred to Ref. [17,18,30].

1.24 PROCESS OF ENERGY SUPPLY TO MAINTAIN THE CURRENT

As the electrons in the wires and the resistor are being *drawn* from the negative plate of a battery in an operational circuit towards the positive plate there is a *depletion* of charges on the endplates of the battery as electrons from the wire near the positive terminal neutralize the positive charges (deficiency of electrons) on the positive terminal by uniting with them …in reality…..the electrons settle into convenient orbital spaces surrounding the nuclei, producing neutral atoms….(see Section 1.13 on how atoms are neutral) and we call the process, neutralizing or uniting with positive charges.

This in turn *slackens* the strength of the electric field due to the end plate charges *and* the surface charges in the entire circuit momentarily, and the charges on the endplates and the surface charges are immediately replenished (by the process of charge separation) by the motor in the mechanical battery or the chemical action in a commercial battery.

The slackened field is strengthened by the replenished charges and the process is a seemingly continuous though intermittent one occurring at nearly the speed of light… depletion of charges on the plates of the battery and their replenishment by internal charge separation….the kinetic energy of the electrons is lost in the resistive portion of the circuit continuously as they collide and pick themselves up under the influence of the electric field exerting a relentless pull on them.

While a conductor and resistor do not have the capacity to store energy in the form of electricity (we had learned that they dissipate energy rather, in the form of heat), a battery can store energy. A battery is capable of delivering energy, over extended periods of time.

To view an animation of the ionic level reactions in a galvanic (electrochemical) cell visit http://www.chembio.uoguelph.ca/educmat/chm19105/galvanic/galvanic1.htm.

1.25 WHY DOES A CHARGED PLASTIC PEN OR COMB ATTRACT TINY BITS OF PAPER? AND WHY DOES IT LOSE ITS CHARGE QUICKLY? (Chabay & Sherwood, 1995, used by permission)[2]

The comb is made up from large organic molecules and when rubbed in your hair, the organic molecules of the surface strands of your hair break and electrons are transferred from them to the plastic comb. Large organic molecules in the plastic and/or your hair break at their weakest bond in such a way that negative ions (negatively charged fragments) are deposited on the comb and/or positive ions (positively charged fragments) are deposited on your hair. The plastic comb becomes negatively charged.

Molecular breakage or electron transfer provides a rough explanation as to why the plastic comb and the U-tape when pulled off the base tape (see Section 1.14) got charged, but such details as to why the comb rather than your hair becomes negative are the subject of continuing research by physicists and chemists. Part of the complexity of these phenomena is due to the fact that they are surface phenomena and the special nature of intermolecular interactions at the surface of a solid are generally less well understood than those in the interior. The U-tape used in the experiment in Section 1.14 could have charged positive or negative and this would have depended on the type of the tape. When the charged comb is held near tiny bits of paper, induced dipoles (or separated charges formed by the shifting of the electron cloud of the outermost four carbon atoms by a tiny amount, less than 10^{-15} meters) are formed in the paper by their interaction with the electric field of the charged comb. And, since there are millions of atoms, the net force of attraction between the near side positive nuclei and the negatively charged comb is sufficient to lift the paper against the force of gravity. See Appendix A for a calculation of the amount of electron cloud shift of the carbon atoms in the bits of paper attracted by the comb.

Different atoms of different elements will have different strengths of interactions with the applied field. When a small electric field E is applied to an atom or molecule, which is not a permanent dipole like a water molecule, it is found experimentally that for the charge separation s produced, $p = qs$, which we define as the dipole moment, is proportional to the applied field E. It is an excellent approximation to say that the dipole moment $p = qs = \alpha E$, where α is a constant which expresses the "polarizability" of the particular kind of atom or molecule. See the Appendix A for details of the calculation of the electric force of attraction by the pen to lift the bit of paper.

Water molecules are permanently polarized. The water molecule (H_2O) is not symmetrical but has both hydrogen atoms off to one side of the oxygen atoms. Slight shifts of the electron cloud to one side of the molecule leaves that side a bit negative as compared to the other side. So the water molecule is a permanent dipole. When water molecules in the air (when the atmosphere is humid) strike a surface they sometimes become attached to the surface, probably because the charged ends bind to the surface. A film of water builds on all surfaces. Pure water is a poor conductor but does contain small amounts of mobile H^+ and OH^- ions.

More importantly, the water dissolves surface contaminants such as salt, and the impure water provides an effective path for charges to spread onto neighboring objects. After a while a charged surface loses its original charge, which is why the charged pen loses its charge (by spreading over large surface areas), quickly. When grounding is discussed at the end of Chapter 2, this aspect will become clearer. So, experiments with charged objects work better when the humidity is low.

I conclude the chapter, with brief notes on surface charges, the nature of the electric field adopted in this book, and some aids to visualization of atoms and conductors. Wilhelm Eduard Weber (1804-1891) was one of the first scientists (separately from Kirchhoff who arrived at similar results in 1857) who mentioned the distribution of charges on the surface of resistive conductors carrying steady currents (Ref. [37]). He mentions surface charges in his paper of 1852. Weber's detailed analysis were published in 1864.

The descriptions of the electric field as 'an alteration of space' is sufficient to prepare us to understand the operation of circuits using capacitors and inductors in subsequent chapters. The field at a point in space is a concept that enables us to visualize the force that will act on any charge at that point by attaching a local property in a small neighborhood of the point without further inquiry such as "what produced the field". More on these aspects are given in following chapters, and in the Appendices, which the reader may defer to read until after the end of Chapter 6.

I find it useful to visualize the nucleus and electrons (cloud) as if being held together by *tiny rubber bands*. Visualize these as having different stiffness for different atoms. The stiffer they are, the tinier the stretch when influenced by external electric fields; the smaller their stiffness as in conductors, the rubber bands can be stretched more and can eventually be broken and result in producing isolated charges!

In this way, it is not too difficult to visualize the extremely thin bands between the outermost electrons and the lattic ions stretching and snapping to produce the electron sea. At the surface, the sea expands (creating positivity) and contracts (creating negativity) as the circuit warrants!

In Appendix C, a qualitative description of the effects of an electric field on insulators, their surface voltages and field strengths and descriptions of the internal mechanism which enables them to prevent electric shocks from live conducting parts is given.

SUMMARY

In this chapter, we learned that charges (electrons and protons) are sources of electric fields. The electric field may be visualized as an alteration of space, and which has the capacity to cause charged particles to move under its influence. After deriving the fields due to individual charges and dipoles, we learned that their field strengths drop away so rapidly with distance (proportional to $1/d^2$ and $1/d^3$) that the separated charges on the surfaces of the plates of a battery do not have sufficient strength to make the mobile charges or conduction band electrons in circuit wires and resistors, to move and produce a current. The property of Electric fields to penetrate through matter has a deeper impact than probably the notion most among us would have gathered in pre-college curricula that the electric field is "everywhere". This important property combined with the Principle of Superposition gives an accurate tool to compute the electric fields due to charges at points anywhere in space out in the open, or behind objects. The electric field configuration due to charged objects is best depicted using vectors (arrows) of appropriate lengths to indicate the strength of the field and pointing in the direction of the field.

When negative charges (electrons) are separated from nuclei (protons and neutrons) of a stable neutral atom, we learned that the strength of the electric field around the atom grows in strength. In conductors, which comprise millions of atoms, this basic charge separation process will produce an electric field in the surrounding space. We studied in some detail, this simple mechanism of production of an electric field which is crucial to the operation of most electric and electronic circuits.

Conductors, we learned, can be modeled as giant structures of atoms and molecules, held together by the conduction band electrons. A detailed study of the electric forces which various surface charge configurations exert on conduction electrons in uniformly circular conductors such as wires was made. From the study, it was deduced that the real mechanism of conduction is due to a *gradient* of surface charge density which produces a *uniform* electric field everywhere within the conductor producing a current density given by $J = \sigma E_{NET}$. In wires and resistors of closed circuits, uniform fields of appropriate strength develop due to surface charges alongwith excess charges elsewhere in the circuit. The field configuration is produced during the initial transient before the current stabilizes, when the circuit is said to be in the steady-state. An electric field is present in the space around the circuit also. If connecting wires are moved and the geometrical arrangement of an energized circuit was changed, the surface charges will rearrange themselves at nearly the speed of light to maintain the fields within the circuit components.

The fields in the components of circuits such as batteries, wires and resistors produce a current which is kept uniform by an elaborate feedback mechanism resulting in a steady-state of operation. This is current conservation.

We learned that electron current $i = nA\overline{v}$, where n is the mobile electron density, A is the cross-sectional area of wire, and $\overline{v}$ is the average drift speed of electrons. The drift speed $\overline{v} = \mu\vec{E}$, where μ is the mobility of an electron, and $\vec{E}$ is the applied Electric field. Since, circuit theory had already been developed from the viewpoint of current being conventional, a treatment of the characteristics and notations of these two forms viz. electron and conventional currents, with formulae quantifying them, was given.

The discussion up to this point having been opened up to reveal the inner mechanism of how electric fields produce a drift of the electrons and currents in circuits, it was appropriate to provide a microscopic $v = \mu E$, and a macroscopic $J = \sigma E$ viewpoint of conduction. This was followed up by a description of the amazing movement of a mobile conduction band electron during a current and that it continuously makes an astounding 10^{14} collisions per second, and if the movement were to be constrained (its mobility) by the lattice environment, the resulting drift, is extremely small. The movements of the millions of conduction band electrons, as a matter of fact, are so tightly and harmoniously coordinated that the overall neutrality within the conductor is barely disturbed. The slight deviations when there is some charge separation, cause tiny field fluctuations (noise emfs) to superimpose on the applied signal fields, to produce noisy signals.

We also developed from the study of electron movement in conductors, a working definition of current.

The simple circuit we discussed had three components; battery, conductors (wires) and resistor, whose structures and operations were described. As the chapter progressed, we learned how the electrostatic phenomenon of surface charges with their elaborate geometries maintain current conservation by manipulating the strengths of fields in conductors, and an analysis of which led to expressions that could be used to compute the strengths of fields and of currents. We also learned that surface charge configuration changes occur at amazing speeds, of the order of picoseconds (10^{-12}s), enabling field configurations to adjust based on feedback. This is due to the wonderful phenomenon of the separation of charge in situ, where necessary around the circuit, achieved with near zero overhead making use of a very very tiny fraction of the neutral surface atoms of conductors, the supply of atoms being plentiful numbering in the millions and millions.

Resistors, we then learned are conductors whose property makes them oppose the flow of current, and that resistors combine geometry with conductivity of the material they are made from. The property to oppose the flow of current is due to lattice imperfections and the vibration of lattice ions at elevated temperatures.

Then, we learned how to simulate the action of a chemical battery by using a mechanical battery which is fitted with a motor and conveyor belt which separates electrons from their atoms and transports it from the positive terminal to the negative terminal to produce an emf.

We next derived expressions for the energy input to one electron and the work done to transport it around the circuit and showed how the energy is conserved. We also learned to distinguish internal energy mechanisms from external mechanisms in a circuit. Making this distinction is important so when setting up energy balance equations of simple circuits we don't count the same energy twice, one belonging to the external environment and the other internal to the circuit over which we don't have control. Then, we learned that current conservation is a consequence of charge conservation in circuits that prevents charges from accumulating in resistive circuits operating under steady-state conditions.

Next, we learned to use these principles to analyze a simple circuit consisting of a battery connecting wires and a thin wire resistor. The circuit was analyzed by computing the electric field in wires and components and then using resistance values assigned to each circuit component. The analysis using electric fields it was shown conformed to Ohm's Law.

We then studied from the surface charge point of view two cases of the effect on current in the simple circuit when the length of the conducting wire is doubled and when it's diameter is doubled. The study provided a comprehensive technical explanation of the effect on current for both cases.

We then learned that currents in diodes and transistors are controlled not by a change in drift speed but by the number of mobile electrons injected in charge depleted regions.

We next learned that in electrochemical cells, a continuous charge separation and replenishment mechanism maintains the emf across battery terminals. In this we clearly understood at the atomic level the process of charge neutralization when electrons settle into atomic orbital spaces with missing electrons.

In the last section of the chapter we studied in some detail the processes which make a charged plastic pen or comb to lose charge. The process, we learned, is fairly complex since the surface phenomena involved are fairly complex. Yet, we could make a calculation to determine the amount of force that the charged pen will exert to lift small bits of paper.

We then learned that surface charges are really point charges and their presence in circuits to enable conduction were first mentioned by Wilhelm Eduard Weber in a paper in 1852 and separately by Kirchhoff in 1857. Weber published a detailed analysis of surface charges in a paper published in 1864.

Finally, we studied a useful visualization of surface charges as point charges which are conduction band electrons which were attached to their parent nuclei by tiny springs or thin rubber bands which snapped and enabled these electrons to roam freely in the electron sea. In operational circuits, it is these charges that condense or expand on the surface of conducting wires and at their interfaces with circuit components to form surface charge gradients that cause the production of uniform electric fields in them.

The processes described in the chapter, when read with a viewing of the animated power point slides would have given the reader a comprehensive and thorough introduction to the unified treatment of electrostatics and circuits, while opening up the hidden beauty and elegance in simple resistive circuit mechanisms.

Questions

1. Explain what is meant by a "closed system" and how does the conservation of charge principle apply to such a system.
2. What is an electric field? Describe its nature and its penetration through matter and action on charges. What is the resulting velocity when an object is hit once? Why does an electric field produce an acceleration of charges in its influence? Does an electric field penetrate all matter?
3. Describe how there must be an electric field inside the wires of a simple dc circuit that lights a bulb and how the charges on the battery alone cannot be responsible for lighting it because the brightness of the bulb does not change (which should, if we apply Coulomb's law) when brought closer or taken farther away from the battery.
4. What is polarization? Discuss the effect of an external electric field on atoms and molecules. Describe what happens when the external electric field is removed. Using symbols in the "Teacher's Guide" in the CD sketch (a) Coulombic attraction, (b) repulsion, and (c) interaction between an electron and surface charges.
5. What does a battery do? Describe its operation and also explain the conservation of charge principle in the universe.
6. What are permanent dipoles? Describe with the help of examples.
7. Why are the energies of electrons (at different levels) in atoms considered negative? Are these energies potential energies? Explain. Why do the electrons not fall into the nucleus of stable atoms? Explain.
8. What are surface charges and how do they enable the process of conduction? Is electric current a result of pushing of charges as water molecules do when flowing in a pipe? Describe how electric currents are different.
9. Derive an expression for the electric field at a distance x from a ring having a charge q that is uniformly distributed around it.
10. Describe qualitatively what happens when unlike charges are separated to form dipoles.
11. Describe with the help of sketches the surface charge distribution on a round wire which would enable the conduction process. Will surface charge distributions be like rings for conductors of other geometries and when they are crossed?
12. Describe metallic structures and relate to the type of bonding which holds the metal atoms together.
13. Is an atom a dipole? Describe how a stable atom is treated as being neutral and is there zero electric field around it?

14. Describe the atomic structure of metallic substances and how a piece of metal is neutral for practical purposes though it contains thousands of millions of electrons and protons whose potentials range in the thousands of kilovolts.
15. How does the magnitude of the electric force binding the hydrogen atom together compare to the gravitational force exerted by the entire earth on the hydrogen atom? Would you conclude that electric forces are indeed much more powerful than gravitational forces?
16. Perform an experiment using charged (Scotch) tapes prepared by sticking two strips of invisible tapes together and pulling apart one of them. Suspend one of the strips between two books on a table and lower the other tape toward it, its length facing the length of the suspended tape. Describe your observations and calculate how much amount of charge is attached to the surface of one of the strips.
    Write down your observations and make the calculations listing the assumptions you made in your calculations.
    What is the fraction of surface atoms with excess charge?
17. How much (approximately) is one newton of force and one joule in everyday life? Also, how much is the rest energy of an electron in joules? How much force does an ordinary chisel apply when hit by a hammer and explain whether it can prise electrons from a metal.
18. Draw the electric field of a dipole lying centered on the origin along the x-axis consisting of unlike charges $+q$ and $-q$ separated by a distance 's'. Derive an expression for the field at points far away in space along the x-axis and along the y-axis.
19. Describe the structure of metals with reference to metallic bonding and how the structure differs from other structures formed due to ionic and covalent bonding.
20. With the help of neat sketches describe what happens when metals (conductors), insulators (dielectrics) and ionic solutions are subject to an external electric field. Do insulators have a voltage? Explain with the help of neat sketches.
21. When a metal is subject to an external electric field what is the amount of shift of the electron cloud that produces the field necessary to cancel the influence of the external electric field?
22. Do electrons that form currents move all the way around a closed circuit? Is an electric field necessary to establish a current? Describe qualitatively currents in circuit wires and the relation between drift velocity and mobility.
23. A wire is connected to the two terminals of a battery. What is the rule for the very initial charge movement whenever an electric circuit is made? Describe how feedback enables the current to remain constant in a circuit if the incoming and outgoing current was to become unequal.
24. Which 19th century scientists first proposed the presence of surface charges that generate the electric force for a current?
25. What are resistors? Describe qualitatively their lattice structures and defects in them mentioning those which affect resistivity and the effect of temperature on resistivity and currents in them. Is an electric field necessary to establish a current in them?

26. How do conductors and resistors remain neutral in the interior in circuits (not the surface charge) despite a current and the electric field that motivates it? How is this so? Describe the processes that keep it so.
27. When the applied DC voltage to a simple circuit with a resistor is increased (or decreased), surface charges appear due to expansion (or contraction) of the electron sea of the conductor and a new current will be established. What is the source at the end of the charge supply chain (or withdrawal) of surface charges (noting that surface charges have to shift only a tiny amount to produce a significant amount of surface charge) which enables the maintenance of *neutrality* conditions in the interior of the wires and resistor?
28. What is the most important general difference between a system in "steady state" and a system in "equilibrium"?
29. Describe the following attributes of a metal wire in steady state vs. equilibrium:

Metal wire	*In steady-state*	*In equilibrium*
Location of excess charge		
Motion of mobile electrons		
E inside the metal wire		

30. What is electric flux? State Gauss's Law and calculate the net flux surrounding a point charge Q.
31. What is meant by "initial transient" and "steady-state"? Explain how these are applicable to a simple series circuit consisting of a battery, wires and a resistor and describe the conservation of current principle.
What is the speed of surface charge build up in a typical circuit?
32. What are induced dipoles? A dipole is induced of an atom centered at the origin as a result of polarization by a charge +Q located along the x-axis at a distance d which is much greater than 's' the distance of separation of the dipole. Derive an expression for the force F on the point charge +Q.
33. The force of the motor of a mechanical battery is F_{NC}. The terminals of the battery are connected using a wire of length L. Derive an expression for the circuit, when the energy input and output are balanced.
34. What order of velocity does an electron reach in in a current in a metallic conductor? Is it close to the speed of light? or is it non-relativistic? How is it instrumental in the dissipation of heat in a conductor?
35. Describe the Drude model of electron motion inside a conductor. Use it to describe the process of current (flow) in the wires of an operational circuit and then, state a working definition for current.
36. What is Joule heating in an electric circuit? Describe the circuit process which results in the production of joule heat.
37. With reference to the mechanical battery connected to a wire in Question 33, what will the current be in the wire if its length is doubled? Give reasons.

38. With reference to the mechanical battery connected to a wire in Question 33, what will the current be in the wire if its cross-sectional area is doubled? Give reasons.
39. Explain how a charged plastic pen is able to lift tiny bits of uncharged paper from a height of 1 cm and the amount of shift of the electron cloud necessary to achieve this.
40. Explain how heat is produced inside resistors when an electric field from a voltage source is applied to it and a current is established.
41. Using Gauss's Law prove that there is no excess charge in the interior of a current-carrying wire.
42. Using Gauss's law prove that the field and current inside a current-carrying wire are uniform.

Additional practice questions are available in the CD with this book.

2

Capacitors in Circuits

Chapter 2 commences with a description of the role of surface charges on the discharging and charging of capacitors. This is then followed by a description of the behavior of RC (Resistor-Capacitor) circuits to pulse inputs, and the phenomenon of how the voltage across a capacitor cannot change instantaneously, is then treated in detail.

Next, the application of a sinusoidal input to an RC circuit is presented with input and output waveforms, and the fringe field which plays an important role in modulating the current, is described. The way the current leads the voltage in RC circuits is presented next, followed by a description of the difference in the qualitative nature of resistance and reactive capacitance. I felt that it would be appropriate to introduce in this section of Chapter 2, the division of currents in resistive parallel circuits and this was done. The current node rule is then introduced.

The chapter progresses with a description of the concept of voltage, and how it may be visualized and an answer to "where is the voltage" is provided. A brief introduction to the idea of potential energy, potential, conservative and non-conservative fields is also made and is followed by a description of the power and energy in a circuit and the loop rule. A single conductor has a capacitance and this is illustrated by separating a neutral conductor near a charged sphere leading to the calculation of the capacitance of a single conductor, the earth.

This is followed by the all important concept of potential and, in particular, the process which is involved in charging up a single wire connected with one terminal of a battery. In the following sections, concepts of neutrality and equipotential regions in conductors are explained. In the concluding sections, a description of how capacitors share charges and the processes involved before steady state or equilibrium is reached are given. The concept of "grounding" is treated qualitatively next. The aspect that makes a capacitor store more energy when a dielectric is inserted between its plates is then described. The concept of Displacement flux and Displacement density is explained in the following section. Finally, Magnetism and Magnetic interactions are introduced as a prelude to understanding energy flow in circuits described in Chapter 4.

2.1 DISCHARGE OF A CAPACITOR

In this chapter, we will study another important component, the capacitor, and associated with it, the concept of capacitance.

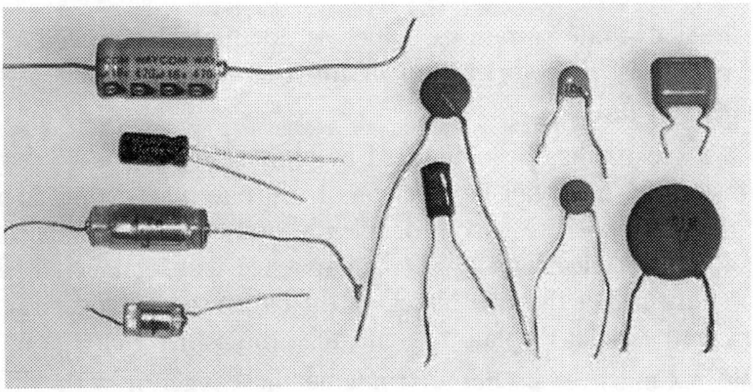

Capacitors of various types and sizes

There are numerous applications for capacitors in circuits, and a few among these are power conditioning, signal coupling, noise decoupling, high pass and low pass filters, snubbers and as start capacitors in single phase motors like the fan motors in homes.

The list goes on, and yet capacitor operation is based on the few processes described in this chapter.

Capacitors—Charging and Discharging

A capacitor is a two-terminal device with each terminal internally connected to one plate of two plates (identical) arranged geometrically in parallel with either air or a dielectric as a separator.

To view the construction of a parallel-plate capacitor, which comprises two conducting plates separated by an insulator (dielectric), view the power point "Regulated_Power_Supply_70min" in the CD accompanying this textbook.

A capacitor in a circuit, charges up by displacing (or separating) an *equal* number of electrons and lattice ions (charges) *always* from its plates by the charging voltage, and a voltage develops across the plates. This process is called charging and we will deal with this in a later section.

A charged capacitor in a circuit discharges, when it allows or forces (by a reversal of the applied voltage) the *equal* number of electrons on one plate and the lattice ions (charges) on the opposite plate to *reunite*, and the voltage across its plates reduces in magnitude. This process is called discharging.

When all the displaced (or separated) charges that is, electrons are reunited with the lattice ions, the capacitor is said to have discharged and the capacitor is uncharged.

A capacitor in a circuit with a net (nonzero) charge that is, with an unequal amount of opposite charge on its plates is not to be found easily. During the transient periods like switch ON, switch OFF, and when responding to sinusoidal voltages, there may be extremely tiny intervals of time of the order of picoseconds, when the plates may hold charges that are not equal. However, we restrict our study of circuit behavior to the steady-state and quasi steady-state situations, which is adequate to understand capacitor charge and discharge processes in the operation of most electronic circuits.

Why does a capacitor discharge?

Consider that a capacitor has been charged by connecting it to a voltage source, such as a battery, and this would have caused electrons from one plate to be displaced to the other plate, leaving one plate positive with a deficiency of electrons and the other plate negative with an excess (local) of electrons. Now, the capacitor is disconnected from the source and a wire used to short-circuit the two plates of the capacitor, to discharge it. This will cause the electrons from the negative plate to neutralize the positive charge on the positive plate.

Why should a capacitor discharge? Imagine the capacitor with all its charges is kept inside a bag and tied at two open ends (open-circuit) as shown in Fig. 2.1.

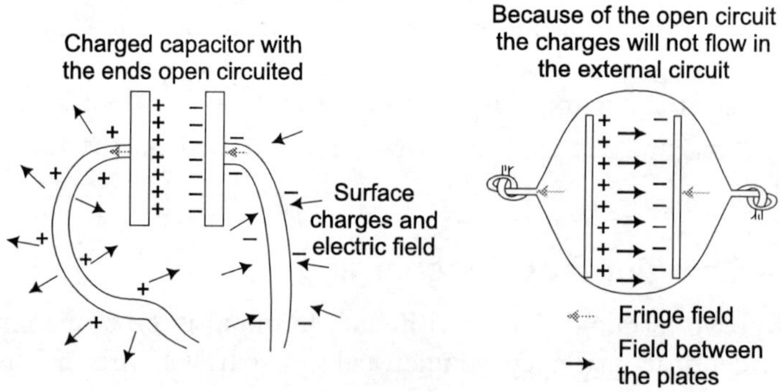

Fig. 2.1 Discharge of a capacitor.

The positive charges on one plate are attracted strongly to the electrons on the other plate, and the electric field is strong and uniform between the opposite charges on the plates. The opposite charges on either plate seem to hold each other tightly. The field is not so strong to cause the electrons to cross the gap and reunite with the positive charges. There is a fringe field at both ends ready to burst the bag, but the open circuit does not allow charge flow in the external circuit. The electric fields of surface charges which move onto the wires from both the plates (We will learn more about this phenomenon in Section 2.21 "How is there a voltage or potential for a single wire connected to one terminal of a battery") and the fringe field will, during the initial transient when the wires are attached, combine to make the resultant electric field in the wires, *zero*. The fringe field is *different* from the field *between* the capacitor plates.

2.2 THE FRINGE FIELD AND ITS ORIGINS

There is nothing mysterious about the origin of the fringe field.

The fringe field (see Fig. 2.2) exists slightly *external* (marked 'x') to the plates and is a result of the uneven effect or contributions of the two powerful fields set up by the opposite charges on the capacitor plates, which are separated by a "*tiny*" distance '*s*'.

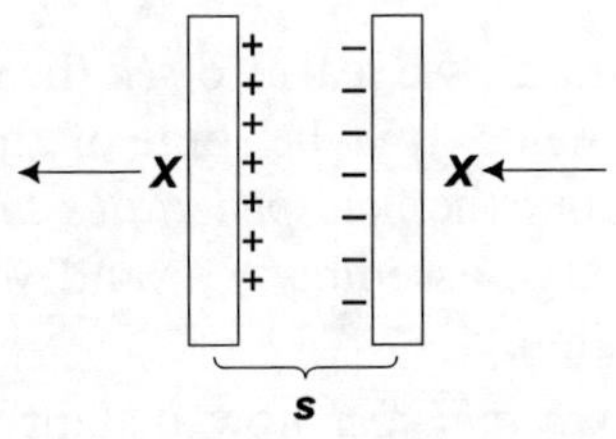

Fig. 2.2(a) Fringe field and its origins.

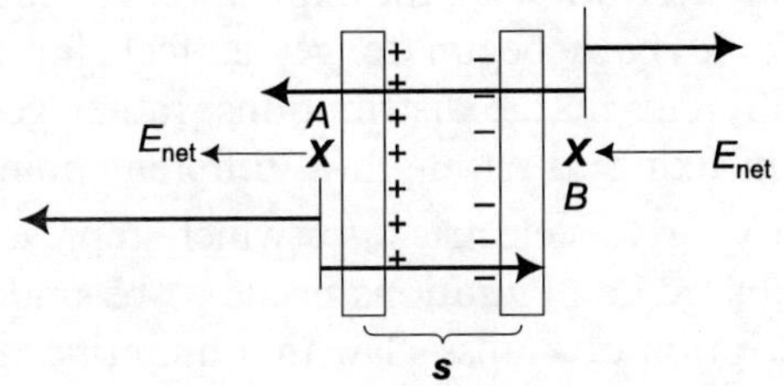

Fig. 2.2(b) Fringe field is the resultant of fields due to charges on the capacitor plates.

At point A (Fig. 2.2(b), the field due to the positive charges is slightly more powerful (indicated by the longer length of the arrow) than that due to the negative charges, because the point A is exactly length s away from the negative plate. The *net* field is the algebraic sum of the two nearly equal though not exactly equal and powerful fields and is *very small.* The reader should note many Physics textbooks indicate that the net (fringe) field *outside* the capacitor is zero, but this is not valid. For disk type parallel plate capacitors with separation s, radius R, area of plates A and carrying a charge Q, it is given by:

$$\frac{Q}{2A\varepsilon_0}\frac{s}{R}$$

See the Appendix A beginning "Electric Field Due to Distributed Charges" through to "Field of Uniformly Charged Disks" for details of calculating the field between two charged plates and the fringe field. I urge the reader to practice the derivations of expressions for the electric field components of a charged thin rod, charged rectangular plate, charged thin circular ring, charged disk and the fringe field of a capacitor. Also see http://www.nickswebpage.com/cacci6_32112/phy-webpages/phy214/214workbook/4-Wkb3ch3c&schap%27s3&4.htm.

A similar argument holds for the point B across and away from the negative plate.

It takes a little getting used to …to *estimate* the e-field at points in space near an assembly of charges; *the Principle of superposition* (see Section 1.10) built-in by nature enables the estimation to be performed easily.

1. Remove all source charges except *one* and estimate the field due to it at the target point.
2. Next, remove all source charges except the *next one* and estimate the field due to it at the target point and so on.
3. Vectorially add the contributions and you have the field due to the complete assembly of charges at the target point.

By making such estimations, we learn to *see* the e-field in space as being due to several contributory source charge fields…maybe some of these are on the other side of plates…or other objects…and so on…but the field *penetrates through matter*…so no bother!....remove one set from the source charge assembly…….and you have the contributory effect of the *e-fields* of the rest……remaining.

In short, we learn to *see and feel* how distant source charges influence the strength of fields at points in space....with or without objects in between the target and the source charges. After reading the derivation of the expression for the fringe field of the capacitor in Appendix A, you would have begun to get a 'feel' for the patterns of electric field produced in space by different charge distributions, rather geometries, and the role of the superposition principle in fixing its' strength at different points in space.

A means of quantifying the electric *flux* which represents the "spread" of electric field lines surrounding charge configurations needs to be understood and this is treated in Section 2.27 with a description of Gauss's law that quantifies the flux surrounding a charge configuration.

2.3 VISUALIZING AN ELECTRIC FIELD AT POINTS IN SPACE

One of my favourite ways of *visualizing* electric fields in space is to imagine a *force or wave* (as in a sea)…filling the space….around and near the target point of interest....

….with a sound …. a gentle…zzzzzz or mmmmmm….

….exerting a *relentless force*….*never tiring* unless destroyed….on some charges at the points of interest in space…..keeping a mental image of the *source charges*…..

...I then imagine the *magnitude and assign a direction* to this force or wave that suits the situation.... that is for the individual parts of the assembly of source charges....

...I imagine the magnitudes in relative terms to the situation....

...I do find it difficult to stick in *absolute* magnitudes.... *relative* magnitudes work fine for me.

Fields do not Travel

Gravitational, electric and magnetic fields do not travel. It is INCORRECT to say, "the electric field goes from the battery to the resistor" and "the electric field comes from the space outside to a negative charge". The field exists around a charge and when a charge shifts, the changes in the electric field that corresponds to the new position of the charge, are effected at the speed of light.

2.4 THE PROCESSES WHEN A CHARGED CAPACITOR DISCHARGES

When the charged capacitor is short-circuited using a wire, the effect is to open the tied ends of the bag and the charges are released.

When the circuit is closed, the effect is as if the closed bag is opened and the charges are released.

Closing the circuit between the two plates... ...has the effect of *opening* the tied ends and the charges are released

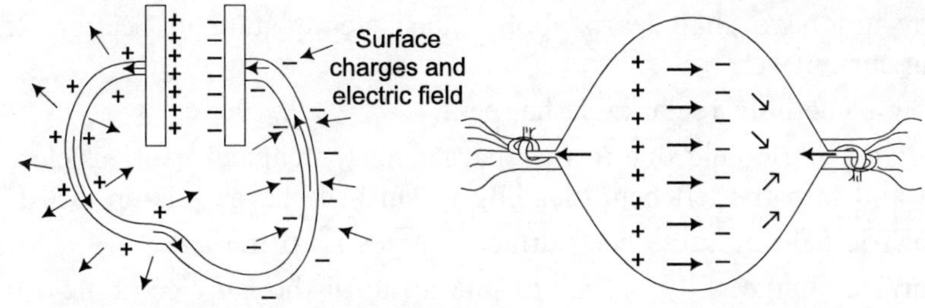

If there were *no fringe field* there would not result a movement of charge. Nature takes care of everything!

→ Flow of electrons

→ Field between the plates

The **fringe field** causes the initial crucial release of charges for further flow of the charges....and development of surface charges that will enable the charges to flow around the wire, and neutralize each other. After all the charges on the plates are neutralized, the circuit attains a state of static equilibrium, a natural relaxed state of equilibrium.

2.5 DISCHARGE OF A CAPACITOR USING A WIRE WITH A BEND

Supposing a charged capacitor is short-circuited with the intention of discharging it using a wire with a bend as shown in Fig. 2.3 after Chabay & Sherwood, 1995.

On closure of the circuit, charges, according to rule (see Section 1.15), move out from the positive plate on the wire and from the negative plate on the wire. The electric field, due to these charges and the charges on the plates of the capacitor, fills the space surrounding the capacitor, and because it penetrates through matter, sets up the field as shown in Fig. 2.3.

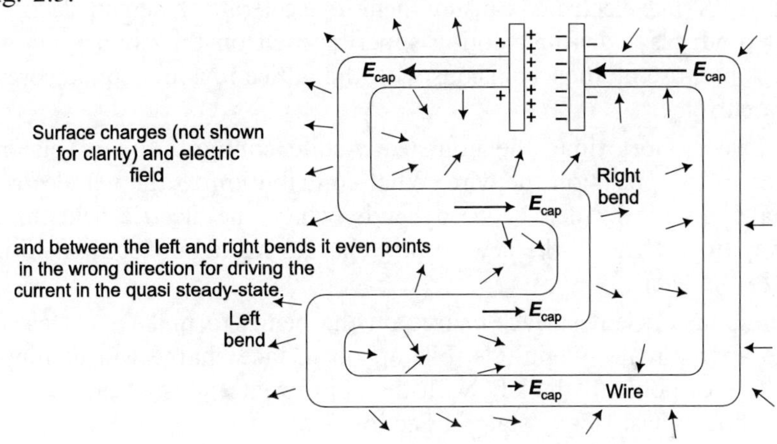

Fig. 2.3 Discharging a capacitor using a wire with a bend.

The capacitor looks approximately like a dipole, so E_{cap} falls off *rapidly* ($\sim 1/r^3$) with distance from the capacitor, as may be seen by the length of the E_{cap} field arrows near and away from the capacitor. The *electric field* E_{cap} due to the charges on the capacitor is *not uniform* in magnitude, and between the left and right bends it even points in the wrong direction for driving the current in the quasi steady-state, that is, having some resemblance to a steady-state. We shall learn about quasi steady-state in Section 2.7. There is no discharging current yet.

But now something remarkable happens.

The *initial* electric field due to the charges on the capacitor drives electrons out of the right bend and into the left bend, leading to surface charge pile-up. See Fig 2.4.

The electric field of these *new* surface charges is to the left.

The surface charge will *continue* to pile up until the *net* electric field is to the left.

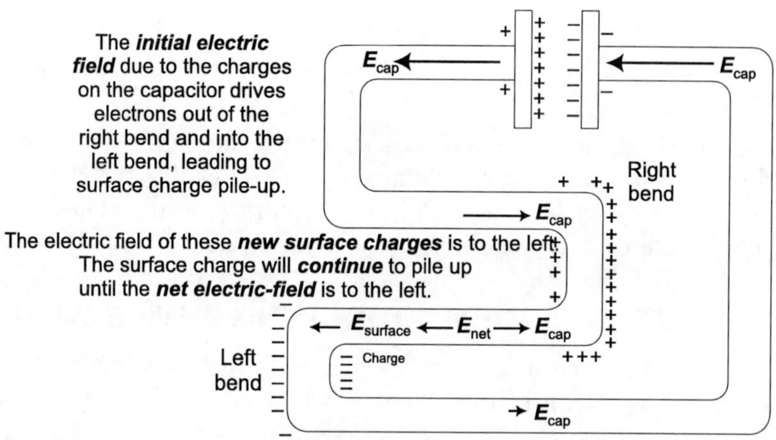

Fig. 2.4 Electric field being reoriented in the bend.

It is useful to review the nature of the electric field as described in Sections 1.4 and 1.11. The electric field is an invisible alteration of space that exerts a force continuously on charged particles (like electrons) causing them to accelerate. When an electric field is applied to wires, it produces a drifting motion superimposed on the rapid random movement of the electrons because of their collisions with the lattice ions or atomic cores, and this drift results in a current.

After a very short time, the non-steady-state configuration of electric field causes surface charges to appear on the wire. What contribution to the net electric field do these charges make? In the branch between bends, where the electric field initially points the "wrong" way (to the right), the new surface charges make an electric field in the "correct" direction (to the left).

If there is insufficient surface charge on the bends to make a net electric field to the left, of the approximate magnitude, pile-up of surface charge will continue until the net electric field does point to the left, with the appropriate quasi steady-state magnitude. We shall learn about quasi steady-state in Section 2.7.

Ultimately, this elegant negative feedback mechanism will result in the production of an arbitrarily complex distribution of surface charge all over the wire such that the electric field due to all the charges (surface charge plus the charge on the capacitor plates) is uniform throughout the wire as seen in Fig. 2.5 and has the magnitude predicted by conservation of energy.

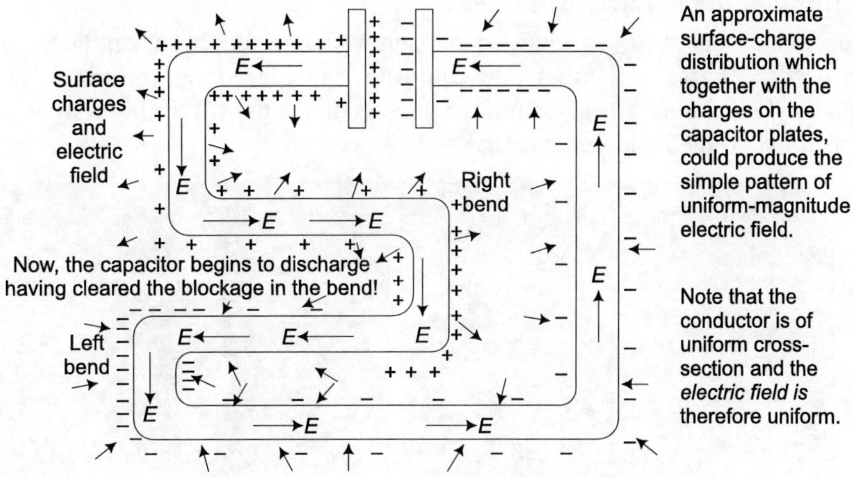

Fig. 2.5 Electric field in the quasi-steady state.

Note that a capacitor cannot *sustain* a flow of charge like a battery because there is no mechanism to supply energy to the circuit once the capacitor plates lose their excess charge to the opposite plate.

The key element which enabled the process of discharge was the *fringe field* of the capacitor outside the capacitor plates.

Electric and Magnetic Interactions, Chabay and Sherwood, Wiley, used by permission.

The fringe field of the capacitor plus the electric field of the charges on the surfaces of the wires, *drives* current (the millions of conduction band electrons) in such a way as to *reduce* the charge on the capacitor plates and to reduce the charges on the surfaces of the wires. As a result of these reduced charges, the electric field inside the wires is reduced, and so the current is reduced.

The process continues with *ever-decreasing current* until all the charges have been neutralized.

The system reaches "static equilibrium".

As a matter of fact, it should be noted that after the capacitor is *fully discharged...*

...no surface charges remain either on the capacitor or on the wires.

In the next section we shall see the results of a quantitative study of how during the initial transient phase of the discharge of a capacitor the fields evolve and establish the quasi steady-state condition. We shall learn about quasi steady-state in Section 2.7.

2.6 THE EVOLUTION OF THE FIELDS AND SURFACE CHARGES OF AN RC CIRCUIT

The evolution of the fields and surface charges of an RC circuit were simulated and studied by Norris W. Preyer, Department of Physics and Astronomy, College of Charleston, Charleston, South Carolina, USA and the results[9] are shown in Figs. 2.6 and 2.7 (reproduced with permission from Norris W. Preyer).

The circuit consists of a charged capacitor connected by a single fat wire of high resistance with charges of equal and opposite sign placed on the plates at $t = 0$. The diagrams in Fig. 2.6(a) are at 0 and 15 picoseconds (ps) after the system is allowed to evolve in time. **Note:** 1 picosecond (ps) = 10^{-12} seconds.

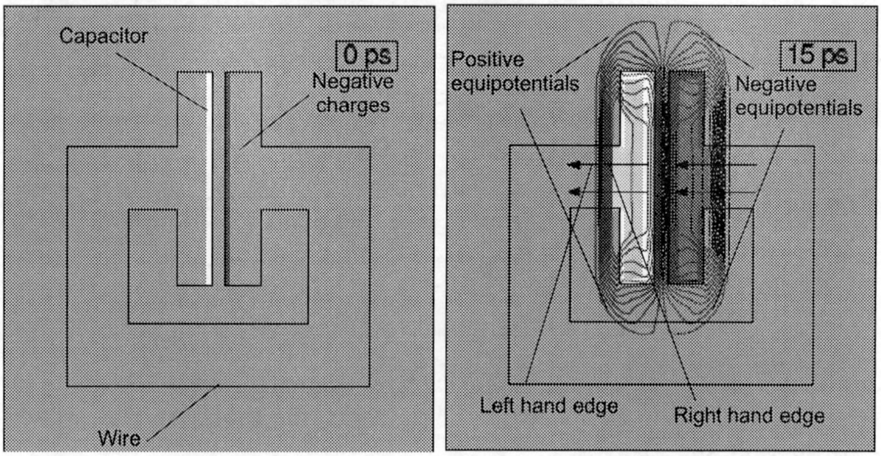

Fig. 2.6(a) Discharge of a capacitor: very intial transient.

The diagrams are a cross-section through the midplane of the circuit and the white shade on the left plate ($t = 0$) indicates the initial positive charge and the dark gray shade on the right plate indicates initial negative charges.

The various gray shades represent the excess charge density, ranging between +/–5000 e (charge of an electron)/mm^3. The charge density on the capacitor plates exceeds this range, which was chosen to illustrate prominently the surface charges on the wires. The arrows plot the square root of the electric field. The large field vectors between the plates have been omitted.

Also shown are the equipotentials from –5 V to +5 V in steps of 0.4 V; positive equipotentials by solid lines and negative equipotentials by dashed lines. As one moves *across* these lines, the potential changes and, this change is what is commonly called "voltage drop". Potential and voltage will be discussed in Section 2.17.

Note the tight bundle of equipotentials to the left of the positive plate when the elapsed time is 15 ps in Fig. 2.6(a). The left-hand edge is caused by the new positive charges, while the right-hand edge is caused by the new negative charges (the two together give a much smaller field than a single plate). The width of this region of very strong electric fields is just about the width of the gap between the capacitor plates.

In Fig. 2.6(b), the situation shown is at 30 ps. The steep potential gradient (closely spaced equipotential lines) and large electric field (shown by the long field vector lines) at the left-hand wire upper are entirely due to the charges on the left-hand positive plate and is the chief cause of rapid removal of charges from the plate.

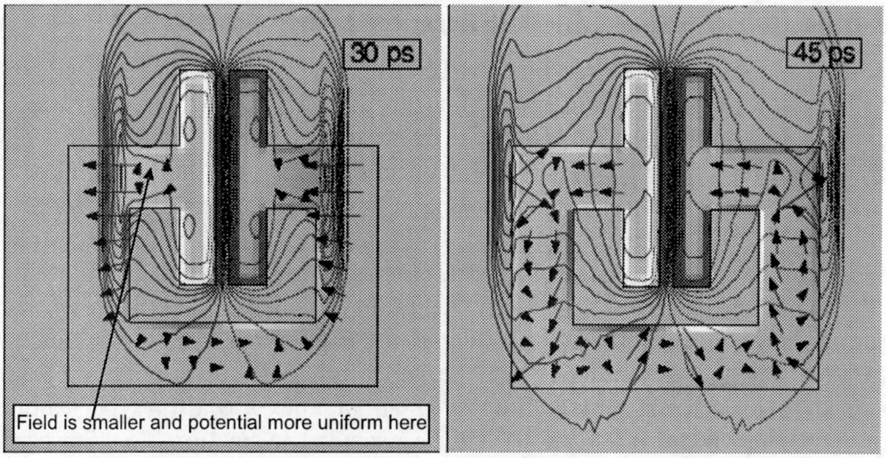

Fig. 2.6(b) Discharge of a capacitor: very intial transient.

At 30 ps the potential is more uniform and the electric field is much smaller to the right of this point due to contributions from both plates. This is the classical weaker fringe field due to charges from both the plates. The pattern extends all around the circuit.

The electric field's vectors are drawn 5 times larger in the last three diagrams (at t = 45, 80 and 160 ps) to display small features more clearly. Notice the surface charge density gradient from the various grey to light shades on the edges of the wires. Each cubic centimeter of the connecting wires (assumed made of copper) contains 8.4830×10^{22} free electrons and the charge density at the surface of the plates and on the connecting wires does not exceed 5×10^{6} charges per cubic centimeter indicating that only a very tiny fraction of the total number of free electron point charges $\frac{1}{6 \times 10^{17}}$ (= $5 \times 10^{6}/8.4830 \times 10^{22}$) move to the surface of the wires and the capacitor plates. In other words, a small number of surface point charges are sufficient to produce a significant potential (a potential variation from –5V to +5V across the plates of the capacitor = voltage of 10 volts) difference. There are plenty of free electrons still in the conducting wires and plates. The fraction of surface charge to free charge in the wires will be tiny even if the wires were assumed to be thinner.

Note that since free electrons move under the influence of an electric field, and since the field vectors in the Figs. 2.6 point in directions not only along the axis of the wire but in several other directions, the electrons move in directions along the axis of the wire and in several other directions as well, of course, opposite to the direction of the field vectors, being electrons. In Sections 3.8 and 5.22 we will learn more about current directions.

At 45 ps (see right diagram in Fig. 2.6(b)), the electric field is seen to reach the first turn in the wires from both the plates, and upto this time there was no significant surface charge to modify the large electric field of the closest plate of charge. It is after this time that we see surface charges building up on the wires as shown in Fig. 2.7, at t = 80 ps.

Note on Equipotential Lines, Surfaces and Regions

Equipotential lines are fictitious lines which are drawn to provide a visual representation of the behavior of the potential at points around electric circuit diagrams and around distributions of charges. An equipotential line usually carries a number that represents the potential (with respect to a point at infinity) on *all* points on the line. Unlike electric field lines which represent a physical quantity namely, the electric field (an alteration of space) that causes a force to act on charged particles, an equipotential line is not a physical entity like a force or a vibration. An equipotential line is simply a graphical representation of the locus of points where there is an electric field and at which points, the *total work done*, (by computing the integral of the dot product $E.dl$ ($\int E.dl$) along infinitesimal distances on the chosen path[13]) to move a unit positive charge from a point at infinity where the potential is assumed to be zero, is equal. The *equipotential* lines are always perpendicular to the electric field lines. More descriptions will be provided in Section 4.5 "Power flow when a capacitor discharges into a resistor" on equipotential lines.

Their usefulness in the study of circuits lies in their meaningful visual representation of potentials at points which lie on *different* equipotential lines and highlight the amount of voltage drop (potential difference) between these points. In Section 2.17 "what is voltage" we will learn that voltage is a way of using numbers to describe an electric field.

Equipotentials can occur on surfaces and in regions. It must be appreciated that in a real conductor or wire (in 3-dimensional space) the equipotentials can be on surfaces (for example, of isolated charged conductors) and in regions (for example, inside a conductor in the interatomic spaces). We will learn more about this when we describe the average field in matter in Section. 2.26.

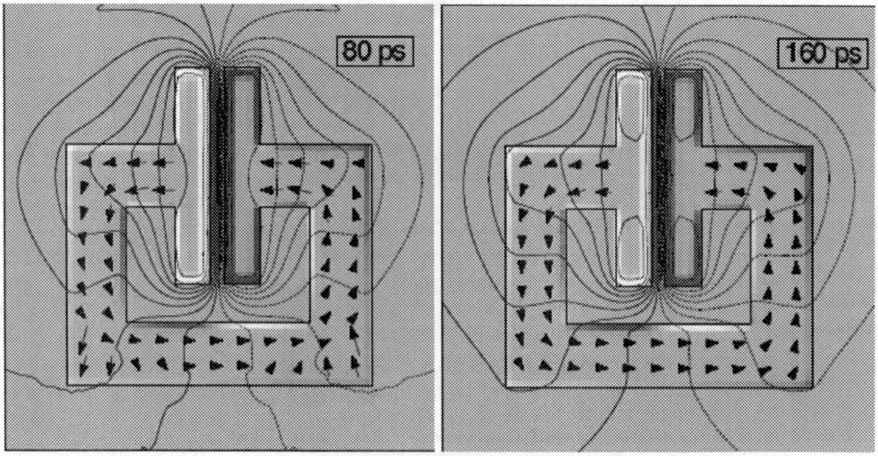

Fig. 2.7 Establishment of quasi-steady state.

It takes a light-crossing time (the time taken for the electric field to cross the length of the circuit), before the electric field vectors begin to point correctly to reach quasi steady-state (Section 2.7).

Note that after 160 ps, the electric field vectors are being tipped towards vertical from being horizontal at the corners of the wires. This is due to the feedback mechanism that begins to take effect when the electric field and surface charge development becomes coordinated. As the charge develops, the field advances. This in turn causes further charge development and the charge is advanced further. There is a further final rearrangement of surface charges and the resulting electric field becomes uniform along the wire (see Fig. 2.7, 160 ps) assumed to be of uniform cross-section.

After this transient period of 160 ps, all the conduction band electrons motivated by the field set-up in the wires, move and begin the discharge cycle. The charges would have been moving during the transient also, but ever so slightly.

While the charges and fields do form a feedback loop beginning around 160 ps, initially there is no feedback. At $t = 0$, charges were placed on the capacitor plates, and the changing electric field from the sudden appearance of the charges on the capacitor plates, races away at the speed of light and a bit after the $t = 45$ ps has moved off the plotted area, never to return (there's no rebound). As this advancing electric field reaches new pieces of neutral conductor, the charges respond by separating. This change in the charge distribution creates a new propagating electric field which in time, modifies all the other charges slightly.

The only thing that has happened at 45 ps is that the first electric fields have finished sweeping across the circuit and separating charges. From here on what we see is feedback; the newly separated charges are not an equilibrium configuration and the charges move (and change the fields) until a quasi steady-state (Section 2.7) equilibrium is reached by 160 ps = roughly 2 or 3 light-crossing times.

For the above computations, six 400 MHz Pentium II computers were run in parallel and each integration step took about 5 minutes.

2.7 CHARGING A CAPACITOR (After Chabay & Sherwood, 1995)

In several figures of this chapter, more than one field vector are drawn in one segment of the circuit to show dominant field components for illustration. In reality, the field is a resultant of dominant field components and those due to source, surface and interface charges of near and distant parts of the circuit. We ignore fields due to charges of nearby circuits.

What happens when an uncharged capacitor is charged?

Shown in Fig. 2.8 is a circuit used to charge a capacitor comprising a battery, connecting wires and a bulb. During the first few nanoseconds, while the surface charges were arranging themselves, it is as though the *capacitor weren't there* – as though there were a continuous wire with no break in it.

There is *no electric field* E_{cap} *and fringe field* because there is *no* initial charge on the capacitor. On closure of the circuit, the *net field* is due just to battery and surface charges.

Note: The voltage across the capacitor is zero at this instant of turn on and therefore the voltage of plate *B* is *equal* to the voltage of the battery positive terminal. The variation in voltage of plate *B* with reference to the battery negative terminal versus time is plotted in the graph of Fig. 2.8.

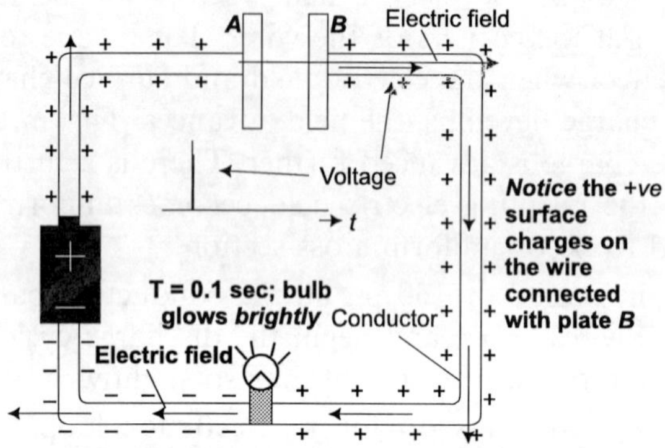

Fig. 2.8 Charging a capacitor.

On closure of the circuit, the *net field E* is due just to battery and surface charges. The field sourced from charges on plate *A* which is in contact with the battery positive terminal, *penetrates* the gap between the plates and then *penetrates* the plate *B*.

Surface charges appear on the plate *B* and further, appear on the wire to the bulb.

The capacitor is not "charged" at this instant, yet.

Then, the free-electron sea *shifts* in plate *B* due to polarization and negative charges *begin* to appear on the inner side of plate *B* as more and more electrons travel along the wire connected to the bulb, and charges will run in the wires, governed by the relation $J = \sigma E_{net}$.

The voltage *across* the capacitor is zero at this instant of turn on and therefore the voltage of plate *B is equal* to the battery voltage.

During the *first few nanoseconds*, while the surface charges are arranging themselves, the *capacitor* makes *no* contribution to the electric field in the wires because there are no charges on its plates yet. It is as though the capacitor weren't there – as though there were *a continuous wire* with no break in it.

The situation is shown in Fig. 2.9a where the plates of the capacitor are shown distorted to afford a clear view. The positive charges appear on plate *B* initially, and these gradually give way to the negative charges as polarization progresses.

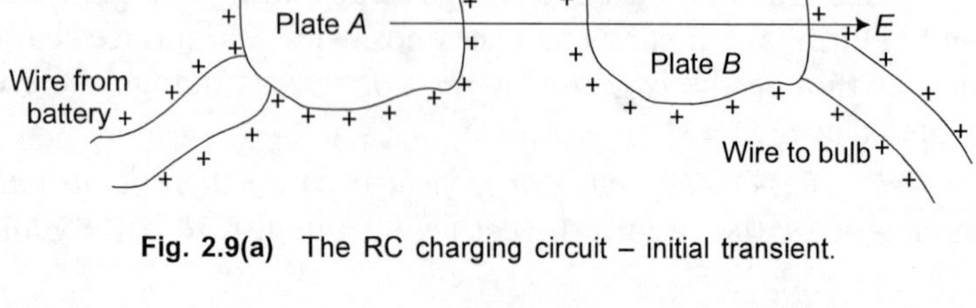

Fig. 2.9(a) The RC charging circuit – initial transient.

Soon, the polarization of plate *B* and the connecting wire begins and negative charges appear on the surface of the plate and the connecting wire, and the resultant field is smaller as shown in Fig 2.9b. The field vectors between the plates of the capacitor are not shown for reasons of clarity.

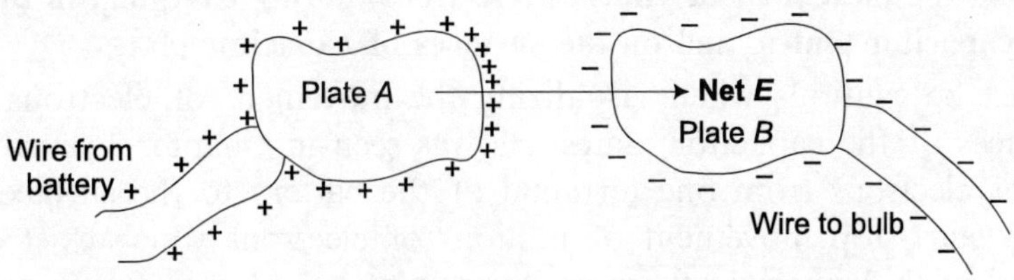

Fig. 2.9(b) The RC charging circuit – polarization of plate ***B***.

If, the initial electric field in the wires is due entirely to the surface charge on the wires and on the battery plates, with *no contribution* from the capacitor, then the initial *current* should be the *same* as it is in a circuit with a bulb and batteries but no capacitor. So, the current is a maximum (as shown by the large block arrow in the wires in Fig. 2.10) at turn ON and the bulb will be bright initially.

That is why at turn ON of the circuit, the electron current, which initiates *polarization* of the plates *A* and *B*, is a maximum.

Note that polarization occurs at *both* plates. And, the electrons *displaced* from plate *A* cause a *corresponding number* to make their presence on plate *B*; the electrons appearing on plate *B* move from the wires and are not the ones that left plate *A*.

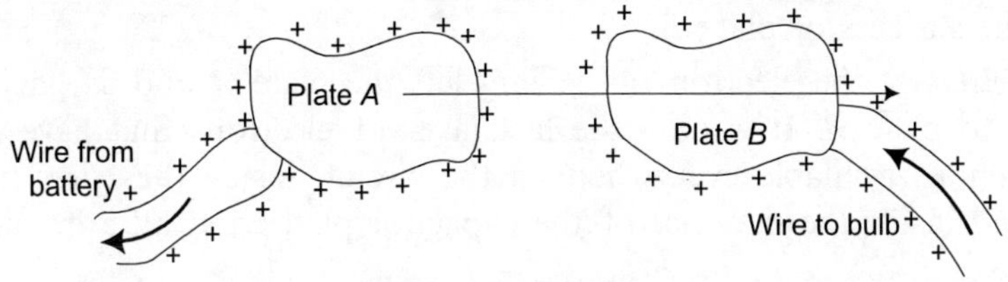

Fig. 2.10 The RC charging circuit – initial transient and maximum current at turn ON.

After the initial transient is over, the charging of the capacitor progresses and now that there is some charge on the capacitor, the *net* field is *smaller* due to the fringe field of the capacitor; and this will result in a smaller current. See Fig. 2.11.

On the wire connecting the bulb and the capacitor note the appearance of a few negative charges, the positive charges having been neutralized. The resultant field in the wire however, keeps the voltage with respect to the battery (-) terminal still positive, as can be noted from the direction of the net electric field in Fig. 2.11.

A graph of voltage vs. time elapsed since turn ON is shown in Fig. 2.11.

Note: While the charging process description makes references to an *applied* field and a *fringe* field and that these are present simultaneously, given the nature of the electric field (end of Sec. 1.6) and the superposition principle (Sec. 1.10) that it obeys, it is neither possible to detect them nor measure their individual strengths. The field in the wire is always a *resultant* of the applied and the fringe fields.

The net field which is a result of the vectorial addition of the applied field and the growing fringe field component gradually begins to reduce in magnitude.

The initial voltage recorded at plate B gradually begins to reduce in magnitude as the capacitor charges and there is a reduction of the net field everywhere in the wires.

Visualizing the movement of charges (electrons) during charging, in the wires, inside the material of capacitor plates, and on the surfaces of capacitor plates

One must be cautious when visualizing the movement of electrons in the wires and on the surfaces of the capacitor plates. As was seen in Chapter 1, current is not a zippy movement of electrons from one terminal of the battery to the surface of the capacitor plate. It is a start-stop movement of millions of electrons which exist everywhere inside the wires and in the material of the capacitor plates, *with a drift superimposed.*

At every instant during charging, it is only the electrons inside the *last layers* of the material of the plate B that shift and rest on its surface. A corresponding number of electrons that leave the *last layer* of the material on the surface of plate A, shift and move into the wire. Do remember that the conductors (wires, capacitor plates) with current in them are everywhere "neutral" inside (of course, leaving the surface charges which determine the voltage), which means that significant dipoles are not formed within them. See Section 1.18 "Neutrality in current carrying conductors".

As the last layer of electrons shift, the resultant electric field everywhere in space due to the charges on the surfaces of the plates, reaches out and around and *slaps* onto the opposite charges on the other plate; the lines of the electric field can be drawn originating on the positive charges of plate A (visualize the positive charges as lattice ions *shining* through the shifted and pulled cloud of orbiting electrons) and terminating on the electrons resting on the surfaces of plate B.

Do not visualize the electrons as having left the plate A and as having moved all the way around to plate B. It is only the last layer of electrons that have moved. After all, there are millions available everywhere in the circuit, inside the wires, inside the battery terminals, and inside the material of the capacitor plates and all are alike!

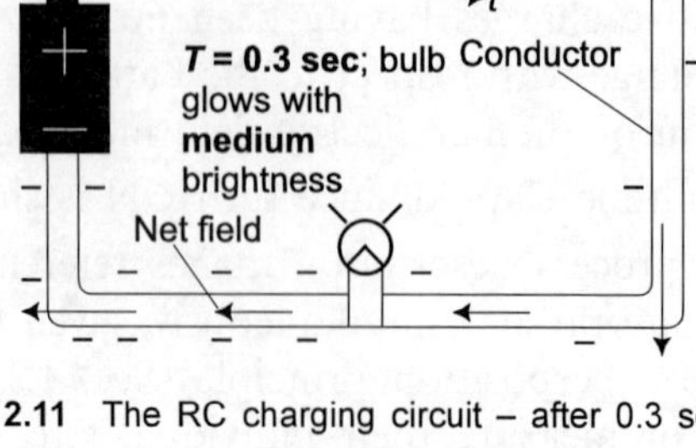

Fig. 2.11 The RC charging circuit – after 0.3 second.

As more and more charges accumulate on the capacitor, net field continuously becomes smaller due to the growing fringe field of capacitor; therefore, smaller current results.

Visualizing the circuit process: One problem most students experience in visualizing circuit processes is the deeply ingrained idea that current flow is sequential. The reader may have by now observed with circuits that one has to think about what happens to *every* element in the circuit all at the *same* time; it normally causes problems if one focuses in on one part and then another. The presence of surface charges should make the visualization comparatively easier. Read the **important note** below Fig. 1.23, Section 1.15.

The applied field (due to the battery) *feels* an opposing force due to the fringe field. The applied field does not reduce in magnitude because the battery holds it as is its duty, but the net field reduces in magnitude because of the opposition.

When the frequency of operation begins to exceed a few hundred megahertz (1 megahertz = 10^6 cycles per second) the visualization described here becomes incorrect due to factors such as the reactance expression (Section 2.10) which will break down about which we shall learn in Section 2.11.

The Action of a Resistor and a Capacitor in Electric Circuits

A resistor *physically* impedes the flow of charge carriers in a circuit. It never opposes the applied voltage (field). That job of opposition to applied voltages (fields) is for capacitors and inductors (which we shall study in Chapter 5).

A resistor *absorbs* all the voltage applied to it. It has a stiffness to charge movement that is uniform throughout its length, if it is geometrically uniform throughout its length, and the entire applied voltage *drops* along its length overcoming this stiffness equally within all its elemental segments. It does not form dipoles inside it, not counting the surface point charges across its ends. A resistor will oppose an externally applied field only if it were to be *isolated* which we learned in Section 1.14 (see Figs. 1.13 and 1.14). In a circuit, the applied field *stokes* a movement of charges *continuously* inside the resistor, which we call *current*. In the relation $J = \sigma E_{net}$, a resistor has *no* influence on E_{net}.

A capacitor does not have the kind of stiffness which a resistor has. A capacitor opposes the applied voltage (field) directly by its fringe field, which is due to the charges that accumulate and form dipoles on its plates. In the relation $J = \sigma E_{net}$, a capacitor *affects* E_{net}.

Going back to the RC charging circuit, there is a slight reduction of positive surface charges on the wire between the capacitor plate A and the battery + terminal. The number of negative charges on the wire connecting plate *B* of the capacitor and the bulb has reduced. The voltage of plate *B* with reference to the battery negative terminal is positive but further reduced after 0.7 seconds as shown in Fig. 2.12.

We have not shown the field configurations around the charging circuit because of the complexity. There is certainty however, in the resultant fields in the wires as depicted. This should be sufficient to graph the voltage of plate *B*. We will learn more about this when we understand conservative fields in Section 2.19 and in Section 2.20 "Visualising voltage". See Fig. 1.10 and Fig. 2.58, to understand the positive voltage of plate *B* with negative charge.

The bulb glows with less than medium brightness due to a reduction in the current as a result of the overall reduction in the net field. This situation is shown in Fig. 2.12.

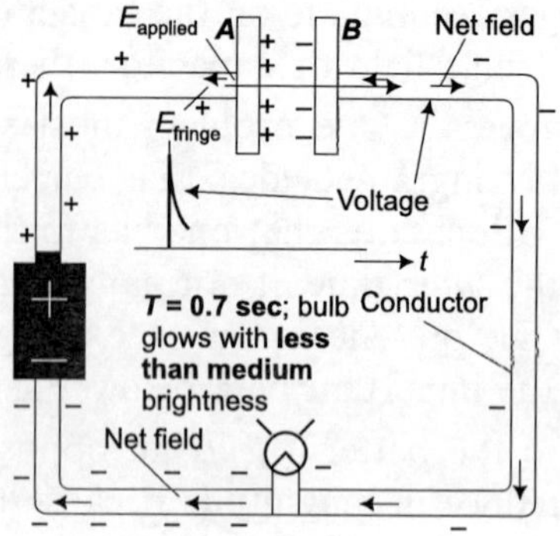

Fig. 2.12 The RC charging circuit – after 0.7 second.

It is instructive to *see and feel* the electric fields.....and to not see them as mere vectors... as described in the Section 2.3 on the fringe field and its origins.

The battery source field *penetrates* the space between the plates ......*assisted* by surface charges on the wires, then surface charges develop on the *plates*....these charges also (never mind if these were induced) ....are surrounded by their own electric field.........all the fields each have their own source ...and strengths....their contributions *add up vectorially* at target points.....in the wires for example........knock off one source charge and the field at the target point is the one due to the *other* source charges. Never forget to visualize the electric field in the resistive bulb filament as well. See Fig. 1.33 which shows the electric field in the resistive portion of a simple circuit.

As charge accumulation continues on the capacitor, *net field* keeps reducing and the current becomes smaller and smaller, and the voltage of plate B is moving closer to the battery negative voltage.

There is a further reduction of the negative charges and of the net field between the bulb and the capacitor and at all points around the circuit. The voltage of plate *B* with respect to the negative terminal of the battery is smaller than that after 0.7 seconds, but remains positive.

The situation 1 second after closing the circuit is shown in Fig. 2.13.

Finally there is so much charge on the capacitor plates that the fringe field "cancels" the fields due to the battery and surface charges.

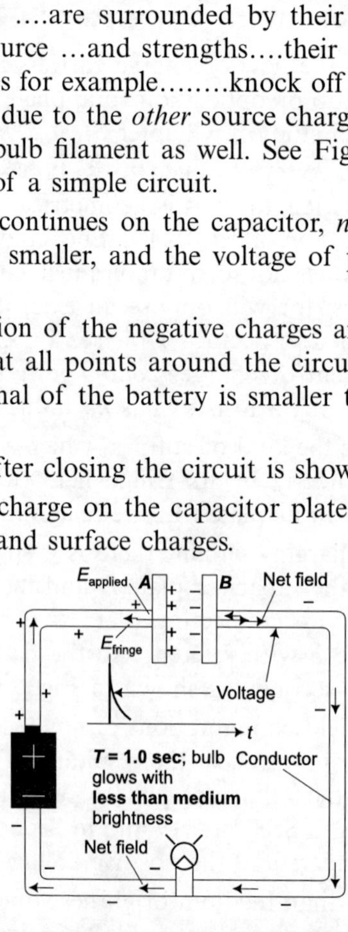

Fig. 2.13 The RC charging circuit – after 1 second.

The situation after 50 seconds is depicted in Fig 2.14. Note that a *dash of negative* surface charges are present on the wire connecting the bulb with the plate *B* of the capacitor.

The voltage of plate *B* is still positive with reference to the battery negative terminal but is nearly 0 volts, and the voltage across the capacitor is nearly equal to the battery voltage.

The bulb *stops* glowing.

To "charge up" a capacitor, electrons are removed from the positive plate and "carried" to the negative plate. (**Note:** In a real charging circuit, it is only the last layer of electrons which are moved). In doing so, the agent charging the capacitor (the battery), has to *fight* against the electric field, which is pulling them back toward the positive plate and pushing them away from the negative plate.

For a calculation of the amount of energy stored when a capacitor is charged, see Appendix A "Energy stored in capacitors".

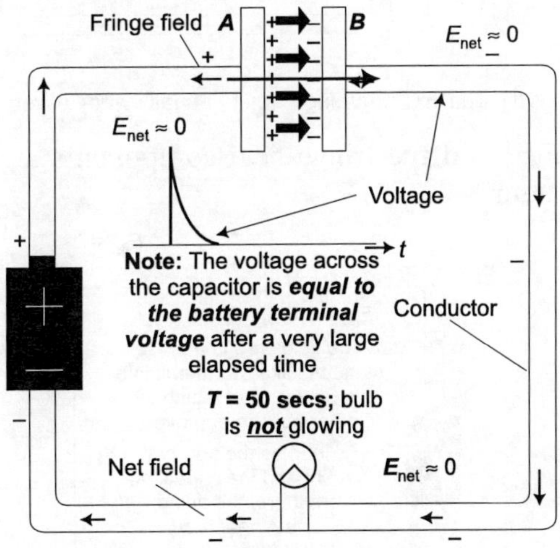

Fig. 2.14 The RC charging circuit – after 50 seconds.

As time further elapses the capacitor is fully charged and there is no net field and no net current.

The potential of the plate *A* will be the same as that of the battery positive terminal and the potential of plate *B* will be the same as that of the battery negative terminal. If they were not, static equilibrium will not have been attained, and a current will result until static equilibrium is achieved.

A very powerful field shown by thick field vectors (see Fig. 2.14) has come into existence between the plates of the capacitor. This is due to charges of opposite sign tightly packed on the inner surface of each capacitor plate. One of the specialities of this field is that it is *uniform between the plates* and does not reduce with distance moving from one plate to the other in the gap.

Expressions for the field are given in Appendix A, and I urge the reader to derive the expressions for the field of two uniformly charged disks, beginning with the electric field of a dipole.

The final situation is depicted in Fig. 2.15.

However, let us not forget that the field components from the battery and the surface charges are still *real and existing* and so are the fringe fields of the capacitor that are *alive and existing as well.*

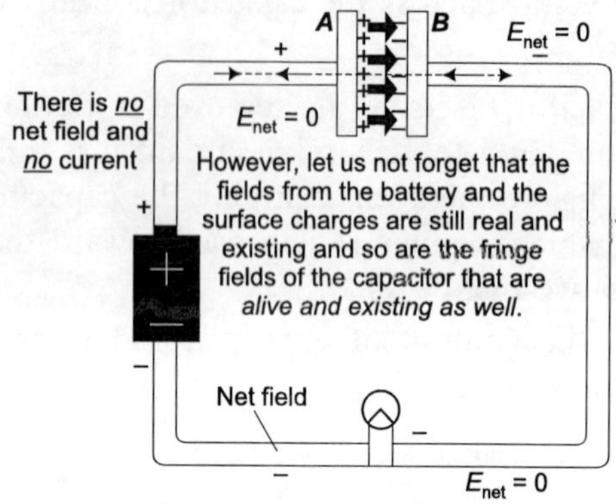

Fig. 2.15 The RC charging circuit – finally after a very long time.

The fields, the applied and the fringe field components are in a tense standoff in the wires as shown in the figure.

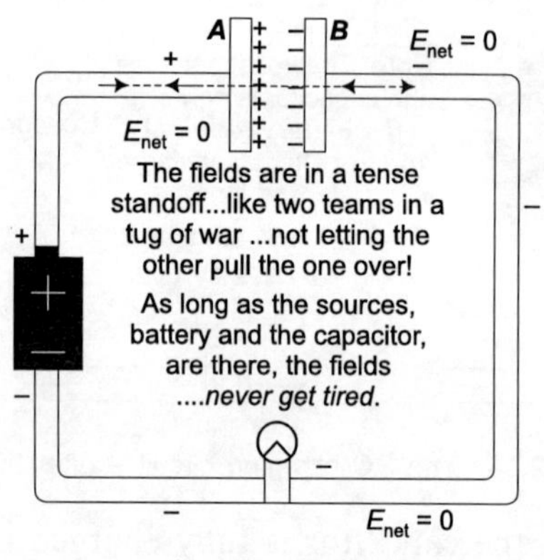

The field components are in a tense standoff

Note that the charges on the plate *B* were sourced from the plate *A* of the capacitor. A current is associated between its plates (charges do not crossover between the plates to the other plate), and is termed *displacement current* associated with a changing electric field, which will be described in Section 4.7.

After the capacitor has been *charged, some* surface charges exist on the wires connecting the capacitor and these are responsible for *holding* off the capacitor from discharging.

Here is One Way to Visualize the Charging of a Capacitor

Whenever we "charge" a capacitor, the path for the current is *seemingly* through the capacitor and back out again. Of course, between the plates no charges ever cross over; but an ammeter connected to either plate of the capacitor will indicate the *same* current.

Why? How?

A certain amount of charge has been forced to flow momentarily during charging, and a *rising* potential difference has appeared. *But the* current is directed *through* (only apparently, not really) the capacitor and the *incoming* electrons on one plate force *other electrons to leave* at the same time.

Every bit of charge that is injected into one terminal must be *forced* out of other terminal at the same time.

The amount of charge *inside* the capacitor *never changes.*

The net charge on each plate is *cancelled* by the opposite charge on the other plate. When charging a capacitor, a *momentary current* causes the voltage to rise. Volts times the electron-flow equals *energy-flow* ($V \times I = P$). Therefore, during a momentary current through a capacitor, there is a joules-per-second *transfer of energy* from the power supply to the capacitor.

During the *charging* process, *energy* is placed in the capacitor.

The energy is *transferred* from the battery to the capacitor, at the moments when the charge, driven by the applied field, overcomes the fringe field and accumulates on the plates of the capacitor.

Capacitors store *energy*, not charge. When we "charge" a capacitor, we *give it a charge of energy!*

Charging a capacitor means *injecting* electrical *energy into* the device. In Section 2.27, we will learn how electric fields store energy. When there is enough charge on the capacitor plates to produce a large enough fringe field to counteract the field made by all the other charges, the current stops completely.

The initial establishment of surface charges takes a few nanoseconds, while it takes several seconds to reach static equilibrium. In Section 5.22 "Transmission line model of a lumped capacitor", a model is referred to which shows the exact process of growth of the charge on a capacitor plate as it charges.

The Quasi Steady-State

During the several seconds while the bulb is glowing, the *current* is not constant, it *changes so slowly* that it makes sense to describe the situation as a **quasi steady-state**; *quasi* means "having some resemblance usually by possession of certain attributes".

During every tiny interval of time, when the voltage across the capacitor, (and the current in the circuit) is changing, the surface charges rearrange themselves to maintain appropriate electric field conditions around the circuit and to uphold the conservation of current law, in a sort of steady-state during each of those tiny intervals, or quasi steady-state. It may be said that every interval of a quasi steady-state includes a small percentage transient.

In some literature and textbooks, the quasi steady-state is referred to as "quasistatic".

Now for some Mathematics

If the instantaneous current through a capacitor is i, then the change in voltage across the capacitor in time t_1 is $\frac{1}{C}\int_0^{t_1} i \cdot dt$. Practically, zero resistance is never achieved, therefore, the current is always of finite magnitude, then the above integral approaches zero as $t_1 \to 0$.

Hence, it follows that the voltage across a capacitor cannot change instantaneously, provided that *the current remains finite;* charges cannot move with the speed of light (because no object or particle is known to move faster than the speed of light) to charge up the capacitor to the applied voltage, which we reiterate is always so in practical circuits because zero resistance is never achieved in electrical and electronic circuits.

The above reasoning is in line with what we observed in the action of charging a capacitor and the descriptions given with reference to Figs. 2.9 and Fig. 2.10. The output voltage and the current is given by the expressions $Ve^{\frac{-t}{RC}}$ and $\frac{V}{R}e^{\frac{-t}{RC}}$ respectively, where V is the battery voltage and R is the resistance of the bulb. These may be easily derived from $V - v_c = RC \cdot dv_c/dt$ written around the loop beginning with $i = C \cdot dv_c/dt$. Refer to "Electrical Technology" by Edward Hughes for this derivation.

The waveform of the output voltage is shown in the left figure.

Voltage across the resistor

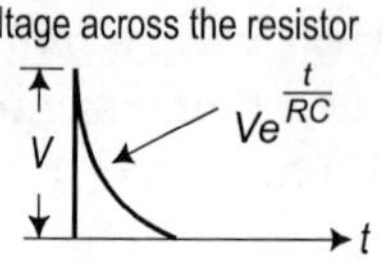

Voltage across the capacitor

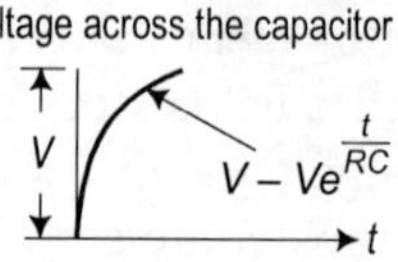

As charges accumulate on the plates of the capacitor, the voltage across it grows. Its voltage is plotted in the right figure above and is given by

Voltage across capacitor v_c = Battery voltage – the output voltage across $R = V - Ve^{\frac{-t}{RC}}$

The output voltage decays exponentially with respect to time and note the dependence of the rate of decay on the product RC. The greater the product, slower is the decay, and the smaller the product, faster is the decay.

Why does the voltage decay have to be "exponential"?

'e' is the base amount of growth (or decay) shared by all continually growing (or decaying) processes just as Pi or 'π' is the ratio between circumference and diameter shared by all circles. The constant 'e' allows you to extend a simple growth rate such as the end of year interest added to a savings bank account and produce a compound, continuous growth at much faster rates down to fractions of a second. 'e' shows up whenever systems grow continuously, and is used in the field of many natural growing processes: population, radioactive decay and so on. Even jagged systems that don't grow smoothly can be *approximated* by e.

Pi or 'π' is a fundamental ratio inherent in *all* circles and therefore impacts any calculation of circumference, area, volume, and surface area for circles, spheres, cylinders, and so on. Pi shows that all circles belong to one family, not to mention the trigonometric functions derived from circles (sin, cos, tan). Similarly, '*e*' shows that all growth and decay belong to *one family,* that of '*e*'.

Just like every number can be considered a *scaled* version of 1 (the base unit), every circle can be considered a *scaled* version of the unit circle (radius 1), and every rate of growth (or decay) can be considered a *scaled* version of '*e*' (the *unit* rate of growth or decay).

To gain an intuitive understanding of the constant '*e*', and why it has the value '2.718', read the article "What_is_e_in_diffl_equation_solutions" in the accompanying CD which includes a simple Excel program that demonstrates the convergence of the compound growth formula to '*e*' in simple and easy steps. Also visit http://betterexplained.com/articles/an-intuitive-guide-to-exponential-functions-e/ for an intuitive explanation of this amazing constant.

2.8 APPLYING A VOLTAGE PULSE TO A CAPACITOR-RESISTOR CIRCUIT

Assume we apply a voltage pulse to a capacitor-resistor circuit.

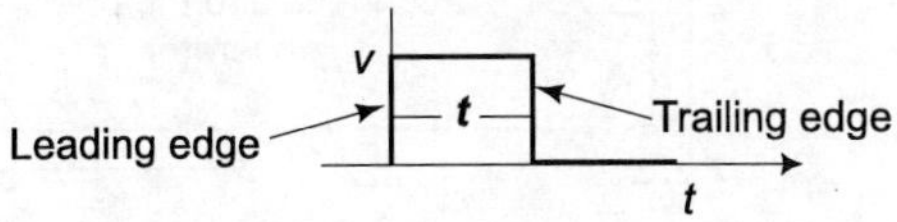

Applying such an input is like connecting a battery of voltage '*v*' for a time '*t*' and then "*shorting*' the capacitor resistor series combination subsequently...while disconnecting the battery from the circuit.

A capacitor-resistor circuit is also called a "high-pass RC circuit" in a class of linear wave-shaping circuits.[34] Such a circuit is shown in Fig. 2.16.

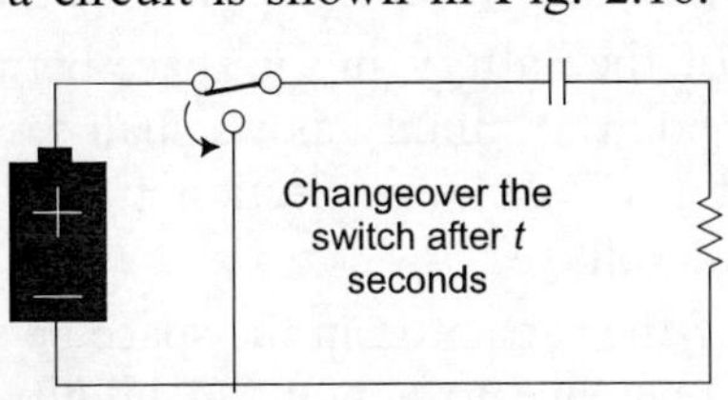

Fig. 2.16 Applying a voltage pulse input to an RC combination.

If the product of the capacitance and the resistance is *large*...***(RC >> t),*** that is, either one of the two is *large* or maybe even *both...are large,* the capacitor will take a *long time* to *charge up*; because the *resistor restricts* the charging current..., if it were to be large... else if, the capacitance were to be large....the capacitor will *thirst* for a *huge* amount of charge, Q being = CV.

In either case, the output voltage will initially *jump to* '*v*' as described earlier and then will reduce *slowly* towards the negative voltage of the battery. This is shown in Fig. 2.17.

Due to the slow passage of current and resulting *slow* charging of the capacitor, the *strength* of the fringe field grows *slowly*.

The output voltage *lingers* and reduces slowly from its initial jump to the input voltage '*v*'. The electric fields inside the wire are shown only near the capacitor for clarity though the fields exist everywhere in the connecting wires. Fewer negative surface charges on the wire between plate *B* and resistor than on the wire between battery and resistor produce a positive output voltage (also see Fig. 1.10 and Fig. 2.58).

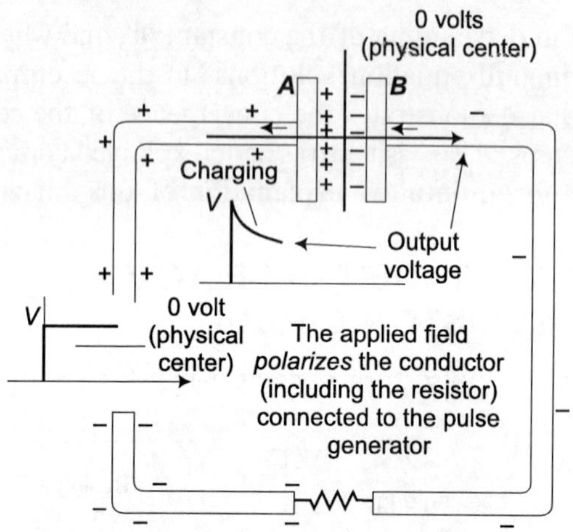

Fig. 2.17 Applying a pulse input to a CR combination.

The electrons in the wires connecting plate *B* slowly *neutralize* the positive surface charges and make the wire less positive....polarization progresses as seen in Fig. 2.17.

The applied field is *driving* charge (sourced from opposite plates) inside the wires onto the plates; negative on *B* and positive on *A* despite the growing fringe field.

At the physical center of the battery, in the space between the terminals, the potential is zero with reference to a point at infinity as we shall see in Section 2.21 "Potential of a point charge relative to infinity". Also see Section 1.8 "Conservation of charge" and the example of battery terminal voltages.

At the physical center of the capacitor, in the space between the plates of the capacitor, the potential is zero with reference to a point at infinity. These are shown in Fig. 2.17 above.

These potentials are a natural consequence of separating unlike charges, a speciality in electricity.

All of a sudden, the *applied voltage is made zero...from 'v'*.

It is like shorting out the CR series combination, and causing the capacitor to discharge through the resistor.

The situation is shown in Fig. 2.18.

Note: The voltage of plate *B* is still short of becoming *equal* to the pulse generator's negative terminal voltage. Since unlike charges neutralize in a flash, the positive and negative charges cannot be shown across the short in Fig. 2.18 and are shown only for visualization.

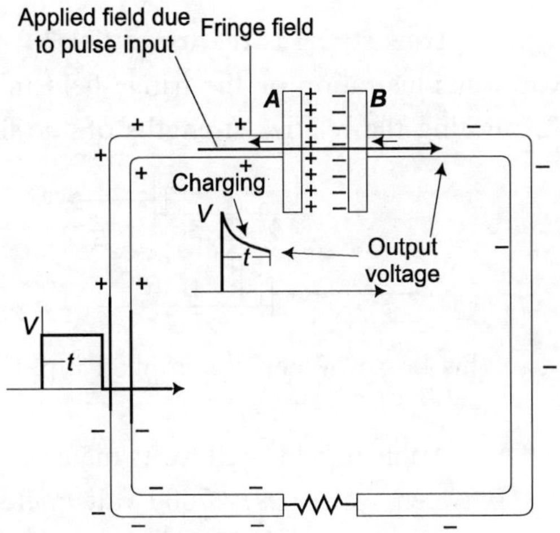

Fig. 2.18 Pulse input – input voltage is made zero.

How does the fringe field *oppose* the electric field due to the applied input voltage?

After all, we had stated that the fringe field is the vectorial sum of slightly unequal though large capacitor plate fields?

As a consequence, the fringe field cannot be very large but only a *very tiny fraction* of the individual fields on the plate.

How strong is the fringe field?

Let us put in a few numbers to check what the *real value* of the strength of the fringe field is.

In Chapter 1, we had seen that the applied field to a resistive circuit for a typical 1.5 volt battery is 2.1 volts/meter in the wires…and for a 10 volt supply it will be, say 21 volts/meter.

Now, consider that a 10 V_{rms}, 1 kHz wave is applied to a capacitor of value...0.01 microfarad...through a 15 kiloohm resistor.

The voltage will be 7 volts across the capacitor, and since the capacitor is 0.01 microfarad, then assuming a typical air insulated disk plate construction…with area A = 110 cm^2, the airgap is 1 mm ("s") and radius is about 6 cm.

The resulting field strength is 7 V/1 mm = **7 × 10^3 volts/meter** *between* the plates or 7000 volts/meter *between* the plates, and the fringe field at a point a little distance away from one plate, outside the capacitor is: (see Appendix A for the expression for the fringe field).

The fringe field is …(s/2R) times the "field inside" = (1 mm/2 × 60 mm or 0.008) × 7 × 10^3 volts/meter = **56 V/meter.**

The fringe field is *comparable* with the field in the wires....which was 21 V/meter.

And it is *in the wires* that the fringe field is called upon to *oppose* the applied field.

We have indicated only sample values for illustrative purposes.

How strong is the fringe field?

Recap our illustration of the fringe field in Section 2.2…noting the relative strengths of the fields…

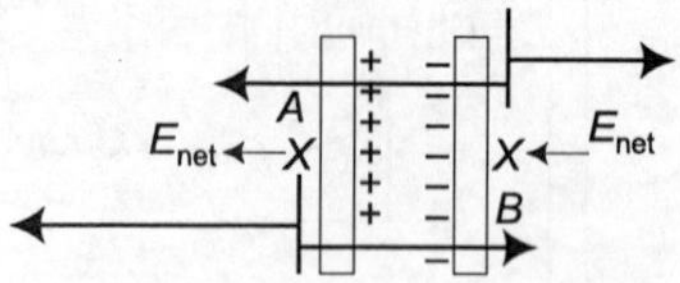

How can this be so? When the applied input field is so *"weak"*?

Applied field = 21 volts/meter

Between the plates = 7000 volts/meter

Fringe field = 56 volts/meter

Are we to believe this?

Yes....We must be careful not to forget that the charges on the plates of a capacitor are extremely tightly packed together….therefore, the **"*strong* 7000 V/m"** field strength! Giving us our tiny fringe field to *oppose* the small input field.

The voltage across the capacitor in our example maybe a *mere 7 volts* but the field is very powerful...

So, when you visualize the charging of a capacitor…

......bear in mind

…….to visualize the *small* applied field…with a *tiny* sound of zzzz… or mmmm ….. pushing some charge on the plate…source charges for the capacitor field…

…the comparable *small* fringe field outside the plate with a *tiny* sound of zzz… or mmmm…. which is *due*…..

…to the powerful fields *between* the charged capacitor plates...with a loud sound of zzzz… or mmmmm…

…and the charging process with the applied field…

….pushing charges *against* the fringe field

Charging and Discharging a Capacitor–Applying a Pulse Input-contd.

Refer to Fig. 2.18. When the pulse input goes to zero volt, the effect is to *remove the* applied field, of applied voltage '*v*', by which it was ***holding up*** the voltage of plate *B* and its connecting wire.

What *remains* is the fringe field and the polarization of the plate *stops* and the incoming electrons are *reversed* in direction. This situation is depicted in Fig. 2.19.

The fringe field takes charge of the tug of war !...and electrons from plate *B* makes the plate *B* voltage *negative*, down instantaneously by the amount of voltage '*v*' by which they were held up by the now "removed" applied field.

In Section 2.26 "Applying a pulse to an RC circuit: Another close look", this aspect will be discussed in some more detail. Also see the topic "Superposition of Potential" which is a consequence of the "Superposition of electric field" principle in Section 3.1.

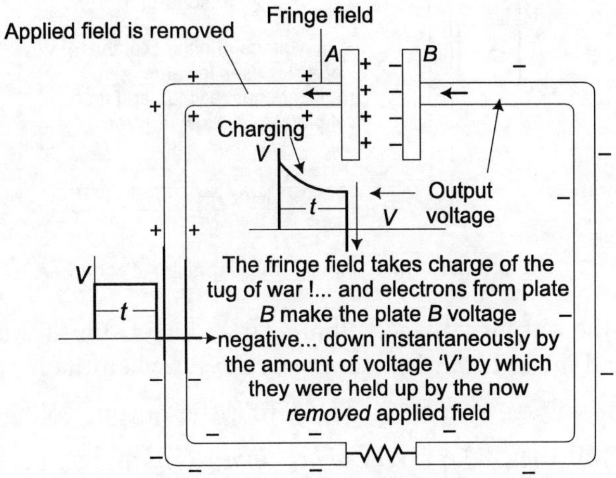

Fig. 2.19 Pulse input – input voltage is made zero.

Note: Since unlike charges neutralize in a flash, the positive and negative charges cannot be shown across the short in Fig. 2.19 and are shown only for visualization.

Gradually, more and more *excess* (local) electrons from plate *B* move towards plate *A* as the capacitor begins to *discharge*.

This is depicted in Fig. 2.20.

The voltage of plate *B* of the capacitor *rises* towards "zero" from its negative value and that of plate *A falls* towards zero.

The surface charges simply *disappear* and so do the charges on the plates of the capacitor and finally, the fringe fields disappear.

Energy is simply radiated away in the resistor, not to return. A resistor does not *store* energy. It takes it away from a circuit.

If the product of the capacitance and the resistance is *small* and $RC << t$, that is, either one of the two is *small* or maybe even *both...are small*: the capacitor will take a *very* ***short time*** to *charge up*, because the *resistor allows a large* charging current....if it were to be small else if, the capacitance were to be small, it develops a voltage with small amounts of charge.

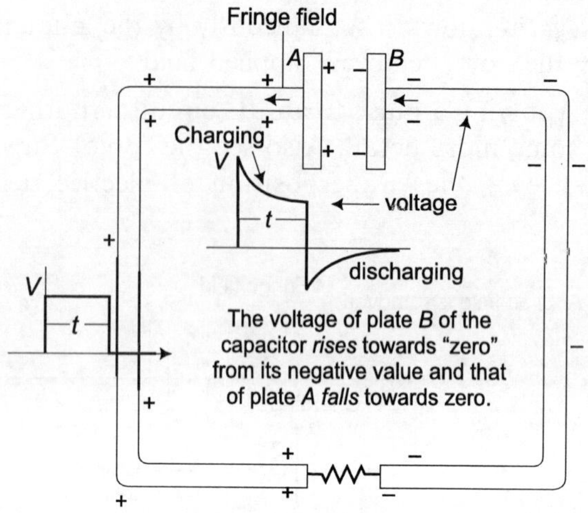

Fig. 2.20 Pulse input – the capacitor discharges.

In either case, the output voltage will *quickly* reduce from the value '*v*' to which it had jumped initially...and move towards zero *much before t* when the input voltage goes to zero.

When the input voltage is reduced to "zero", the output voltage will jump to "–*v*".

The capacitor will then take a ***very short time*** to discharge, because the ***resistor allows a large*** discharging current.

In either case, the output voltage will *quickly* rise from –*v* to "zero" as the capacitor is *fully discharged.*

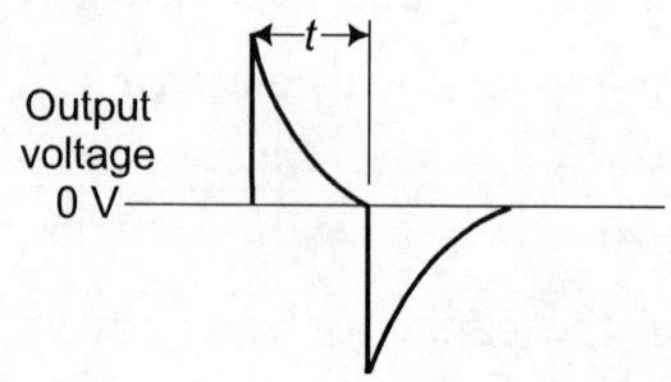

Summarizing, for a pulse input of width '*t*' to a CR circuit, if the *product* of the capacitance and the resistance......

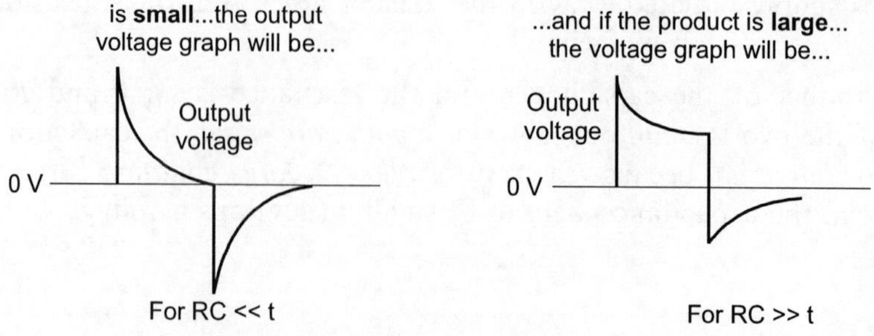

Voltage Changes Across Capacitors

Provided that the current remains finite, *the voltage across a capacitor cannot change instantaneously*. What this means is, as seen in Fig. 2.20, at $t = 0$, when the input voltage changes discontinuously or instantaneously, the capacitor does not charge instantaneously and instead seemingly "passes" the voltage to appear across the resistor. In reality, the electric field *penetrates* the gap between its plates. The capacitor begins to charge only *after* this instant.

The same comments are applicable at the trailing edge of the pulse input, when the output voltage becomes negative and *then* the capacitor begins to discharge.

It should be noted that a sinusoidal voltage does not change instantaneously and has instantaneous-like properties only at very high frequencies.

Important: Energy considerations when charging a capacitor using a battery (application of a step voltage).

When a capacitor of capacitance C is charged using a battery of voltage V and a switch, the battery expends an amount of energy QV while the energy stored in the capacitor (after charging) is $QV/2$ where Q represents the amount of charge transferred when the capacitor voltage is V. Note that with a potential difference V, a capacitor C will store charge Q, but the energy stored is *not* $Q \times V$. Why not? Because, as the capacitor charges, both *Q and V* increase, so we have not moved *all* the charge with a potential difference of V across the capacitor (fighting against the growing fringe field to transfer energy from the battery), which the battery did. See http://230nsc1.phy-astr.gsu.edu/hbase/electric/capeng.html.

Where is the missing energy? $(QV)/2$ from the battery?

Directly connecting the battery to the capacitor (application of a step voltage) causes an instantaneous energy *input* of QV to the capacitor, which is capable of *receiving only an amount* $(QV)/2$. The circuit shakes away excess energy by: heating up the parasitic resistance such as the capacitor equivalent series resistance or wiring resistance; arcing across switch contacts; or, by electromagnetic radiation (in digital circuits this disturbs operation of nearby circuits). This method of charging is analogous to standing on a table and lifting a weight from the floor by jerking a rope tied to the weight with the intention of placing it on the table. The jerk causes the weight to accelerate, and make it land on the table with a 'thud', dissipating "excess" kinetic energy in the form of frictional heat. If however, the weight were to be pulled with a slow and steady speed, it can be placed gently on the table surface without loss of excessive energy! Likewise, if the capacitor is charged using a *ramp* voltage with a uniform rate of rise, energy losses could be kept to a minimum. See also Section 2.23 "The process of capacitors sharing charges".

2.9 APPLICATION OF A SINUSOIDAL VOLTAGE (A.C.) TO A CAPACITOR *C*

A sinewave depicts movements in the form of waves. It has a datum, rapid, slowing and steady growth in one direction for a quarter cycle with reference to a datum (a reference), and then slow and faster decay in one direction for the next quarter cycle, and all these elements again in the opposite direction (reversal) of the forward movement for the next half-cycle. It has peaks and valleys. In essence, the sinewave is a perfect embodiment of oscillatory movements like springs and quantities, like a voltage. It retains its waveshape when added to another sinewave of the same frequency and arbitrary phase and is the only periodic waveform which has this property. When a sinusoidal (a.c.) input voltage $v(t)$ is applied to a capacitor, then during the positive half-cycle, notice that in the *absence of a resistance* the capacitor *instantaneously* attains the voltage of the input.

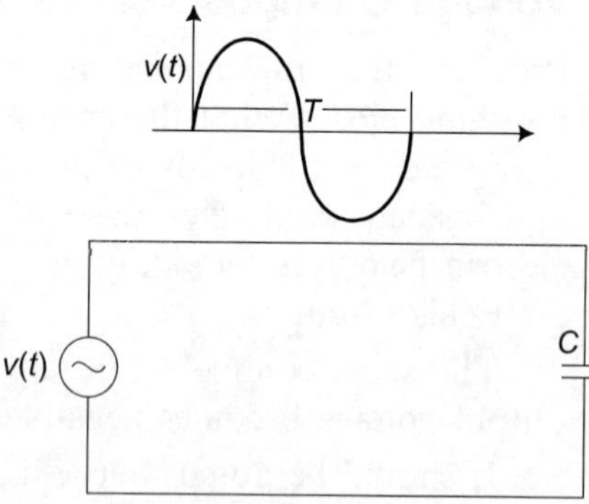

Application of a sinusoidal voltage to a capacitor

In short, the voltage *across the capacitor* changes by *charging* and then *discharging* as a half-sinusoid *following* the input voltage *at every instant.*

A sinusoidal voltage generator operates by separating charges for the positive half-cycle in one direction initially rapidly and slowing down and momentarily halting when its voltage is at the positive peak, and then gradually reuniting the separated charges to make the separation zero; when all charges reunite, the generator voltage is zero.

The negative half-cycle now commences, when the generator operates by separating charges in the *opposite* direction initially rapidly and slowing down and momentarily halting when its voltage is at the negative peak, and then gradually reuniting the charges to make their separation zero; when all charges reunite, the generator voltage is zero. It then begins the next positive half-cycle followed by the negative half-cycle and the process repeats.

If connected to a resistor or a capacitor, then charges (surface) move onto the wires and their number and orientation changes at *every* instant of time. Remembering that surface charges are point charges (produced by a shift in the sea of electrons), it may be noted that the surface charge changes that occur at nearly the speed of light, produce the varying electric field inside the wires. The voltage at the source and the load terminals if plotted, produces a sinusoidal waveform.

The situation when the positive half of the sinusoidal voltage is applied is shown in Fig. 2.21.

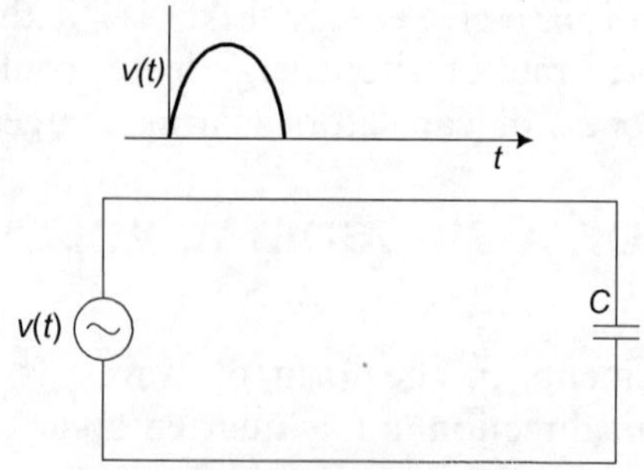

Fig. 2.21 Positive half sinusoid applied to a capacitor.

Note the subtlety in this, the capacitor can charge instantaneously following the input because *there is no (or rather near zero) resistance* and develop the *fringe* field almost instantaneously to closely *match* the e-field of the applied input voltage…of course due to surface charges. See Fig. 2.22.

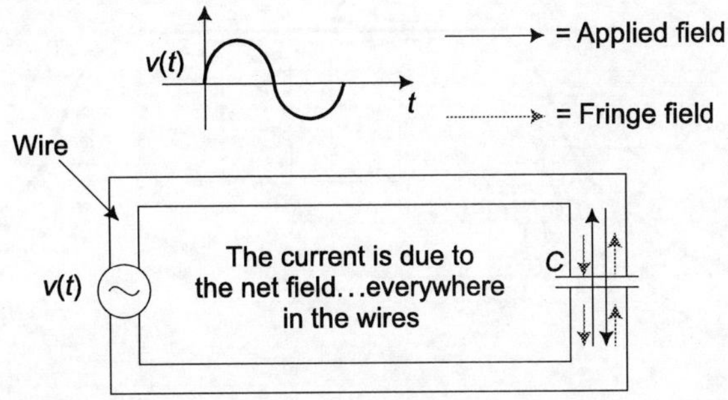

Fig. 2.22 Capacitor charges instantly following the input voltage.

The input voltage rises…the e-field *slams* into the capacitor plate…and through it.

There is *no* fringe field… and the resistance being near "*zero*"…the current zooms instantly… governed by the relation $J = \sigma E$, and *positive charge* is *sprayed* on to the capacitor plate…(shown is one of the pairs of the capacitor plates)……the current is large…!

The situation is depicted in Fig. 2.23.

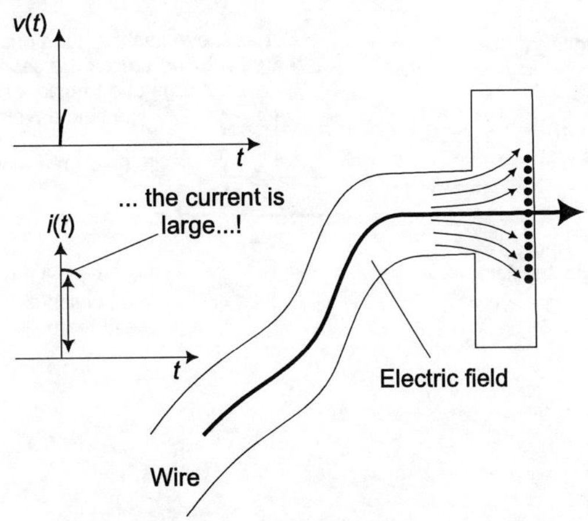

Fig. 2.23 Charge movement when a sinusoid is applied to a capacitor.

From the movement of charge depicted in Fig. 2.23 it is easy to understand the e-field *due* to the applied input. It is instructive to see how the *fringe field* develops which is described next.

In Fig. 2.24 the action at both the capacitor plates is depicted.

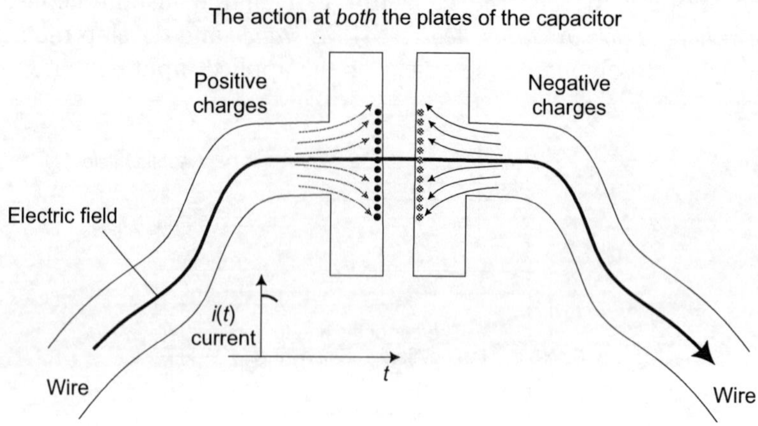

Fig. 2.24 Charge "spray" on both the plates of a capacitor.

The electric field, remember, always "fills" the wire and should not be visualised as being present in a narrow central portion of the wire.

At the initial instant, positive charge is sprayed on to the left capacitor plate and even as the charges are being sprayed on the capacitor plate, a fringe field develops and this is in opposition to the applied electric field due to the sinusoidal signal generator.

This is shown in Fig. 2.25.

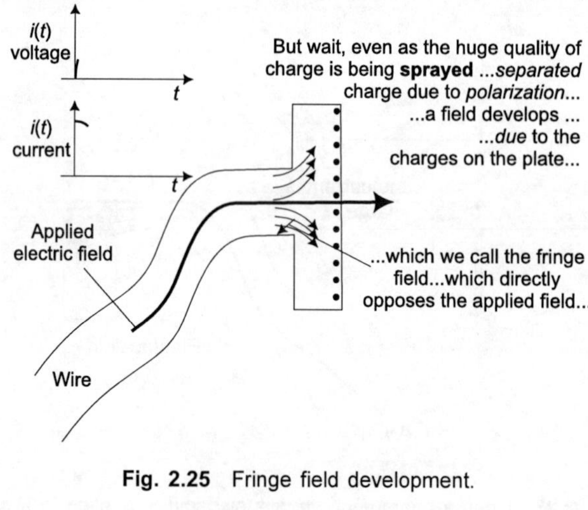

Fig. 2.25 Fringe field development.

The opposition force of the fringe field to the force of the applied field may be likened to the opposing force felt when compressing a ballpoint pen spring.

As time advances, the input voltage increases but, with a smaller rate of growth. Accordingly, the electric field (applied) also increases but with a smaller rate of growth. The force on the electrons in the wires increases but with a smaller rate of growth.

The charge *spray* and therefore, the **current** is reduced in magnitude due to the slow growth rate of the motivating applied field of the input voltage giving more time for charges to accumulate on the capacitor plates and so allowing a stronger growing fringe field to *oppose* the applied field.

Remember that the current in the wires is a start-stop motion of millions and millions of electrons everywhere within with a drift superimposed.

Now, as the applied voltage increases but with a *smaller* growth rate (see the graph of $v(t)$ in Fig. 2.26), the current *reduces…to zero.*

This is shown in Fig. 2.26.

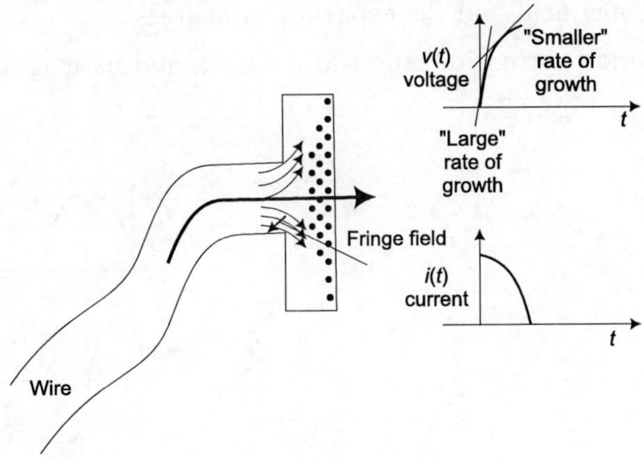

Fig. 2.26 Reduction in current due to the fringe field.

In Fig. 2.27, is shown how the fringe field *brakes* the oncoming charges during the positive half of the sinusoid.

Note the production of current...the charge "spray" or current may be huge...yet cannot be infinite, because the fringe field the spray (on both the plates) develops, is in *direct opposition* to the applied field, matching the applied field at every instant...

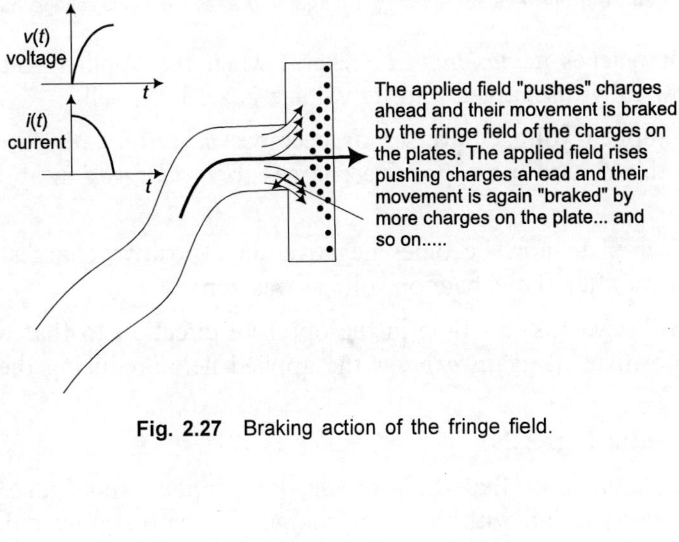

Fig. 2.27 Braking action of the fringe field.

The actions are described considering positive charge movement for simplicity and clarity in understanding the circuit processes. This does not in any way detract from the accuracy of description of the process or its mechanism being considered as opposite to the movement of electrons.

As time advances further...and the input voltage *reduces*...as a sinusoid the applied voltage causes the applied field to *reduce* in magnitude and...the positive charges on the capacitor plate are *dragged* away from the plate and move into the wire under the influence of the *dominant* fringe field and the capacitor discharges.

This will cause the current direction to *reverse*....and its magnitude *grows*, in reverse.

This is shown in Fig. 2.28.

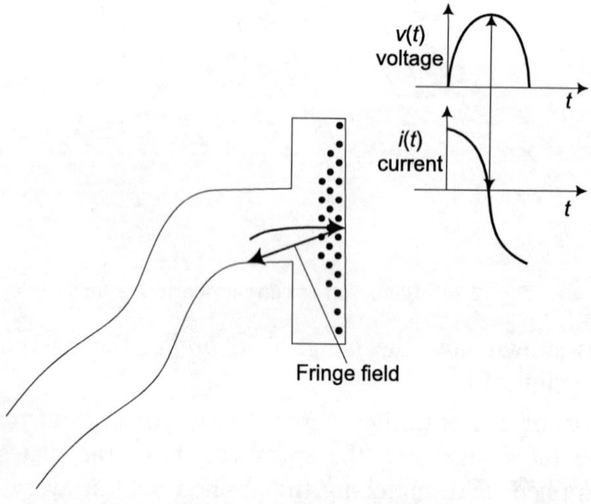

Fig. 2.28 Current reversal led by the fringe field after the input voltage has peaked.

The current reaches a *maximum* in reverse when the applied voltage is "zero". The capacitor is now fully discharged and its voltage is zero, as well.

When the applied voltage reverses or becomes *negative*, ...the applied field changes direction...and this causes more positive charges to be *dragged* away from the capacitor plate.

The capacitor plate now becomes negative since positive charges are being dragged away into the wire after the capacitor voltage was zero.

A fringe field develops this time in the opposite direction to that when the charge on the plate was positive....it again ***opposes*** the applied field...reducing the magnitude of the current.

This is shown in Fig. 2.29.

The reader should note that the processes (the applied and fringe field development) and effects (the deposit and withdrawal of charges on the plate) were described in relation to one plate, but identical though reversed processes occur and inverse effects produced simultaneously at the other plate as well (see Fig 2.24). This is how the charges on one plate are sourced by the other plate.

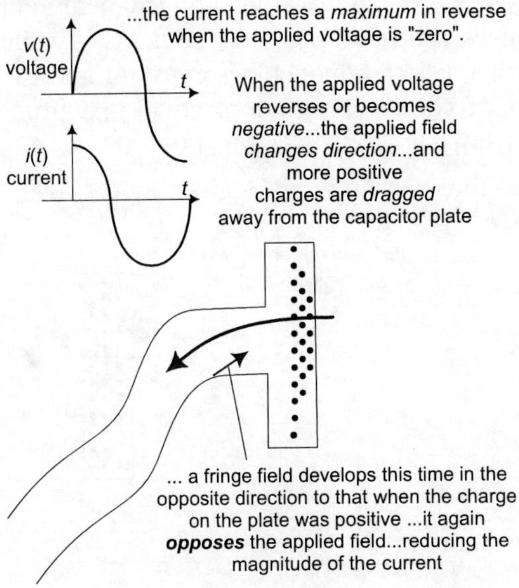

Fig. 2.29 Current reduction after reversal.

Note that the current is a *maximum* and *reducing* even as the voltage is increasing (or decreasing through "zero"). This is due to the *slow* growth rate of the applied field as time progresses...allowing time for the fringe field to develop...and *oppose* the applied field more.

The effect is that the current is large when the input voltage is small and rising and the current is zero when the input voltage peaks.

We say that the current *leads* the voltage by 90 degrees for a sinusoidal input.

This is shown in Fig. 2.30.

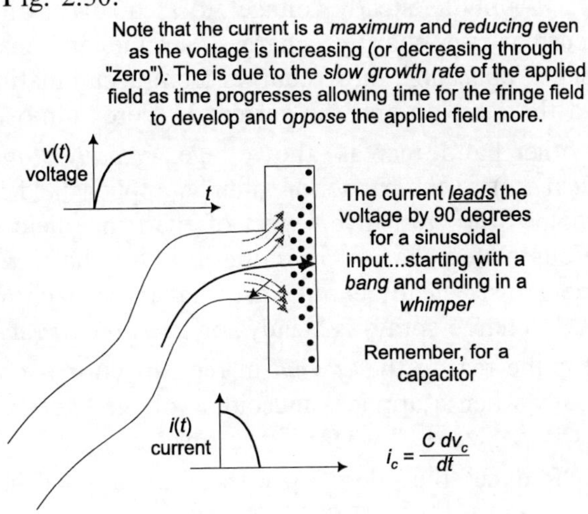

Fig. 2.30 Current "leads" the voltage for a capacitor.

That the current formula for a capacitor is correct may be observed when a ramp voltage is applied to a capacitor.

The *rate* of growth of the motivating applied field remains *constant* (unlike in a sinusoid where it changes continuously) and the *net* field which is the difference between the applied and the fringe field components is constant and the charge spray is *continuous*, and successive batches of charge spray are of equal magnitude.

The applied field is ahead of the fringe field.

The current is therefore *constant* and is given by $J = \sigma E_{net}$ as shown in Fig. 2.31.

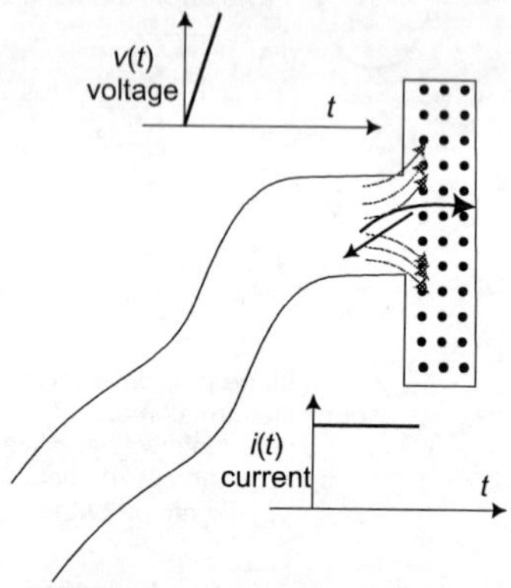

Fig. 2.31 Current remains constant for a ramp input voltage.

Visualizing the Charging of the Capacitor

One very powerful visualization of "charging" a capacitor...
...*one of my favourites* is this.

For a "steady" current in a resistor, one has to apply a
steady voltage... DC

Apply a "steady voltage" to a capacitor and
the current rises sharply....charges "accumulate" instantly on
the plate "*opposing*" more charges from accumulating...due to
the e-field these set up and the current is "zero" almost immediately.
On the other hand, increase the voltage "*steadily*" (*and "continue" to increase*) to the capacitor and the applied field is able to
"overcome" the "resisitive" effect of the fringe field due to the
accumulated charges, resulting in a "net field"...whose
value is just "steady enough" so that the capacitor current
(charge spray) is steady...or ***constant current.***

That is the reason the ***current*** in the capacitor is constant or
steady when a applied sinusoidal voltage "goes through
zero..." where it resembles a ramp.

"Reduce" the rate of growth of the applied field...
...and the "*net field*" (i.e applied minus the fringe field)

"rate of growth" is *reduced...*

…never forgetting that the accumulating charges introduce
a field that is ever in "*opposition...*"

...the current is "reduced..." in magnitude

...make the applied voltage "*steady*" or constant and quickly
the accumulated charges develop a fringe field that directly, exactly
and continuously "opposes" the applied field... the net field is "zero"....

...and the current is "*reduced to zero.*"

Why does the current remain steady and maximum when the applied sinusoidal voltage crosses zero when swinging from a positive to a negative magnitude or the reverse?

The rate of change of voltage is quite similar to a *ramp* voltage at this instant of time and applying the reasoning for determining the current in a capacitor subjected to a ramp voltage described above (see Fig. 2.31), the current is seen to remain steady and at a maximum at this instant.

This is depicted in Fig. 2.32.

This is quite unlike the magnitude of current if this voltage were to be applied to a resistor, where there is no fringe field to oppose due to accumulated charges and the current will be zero when the voltage is zero.

VISUALIZING THE CHARGING OF THE CAPACITOR...ONE MORE AID!

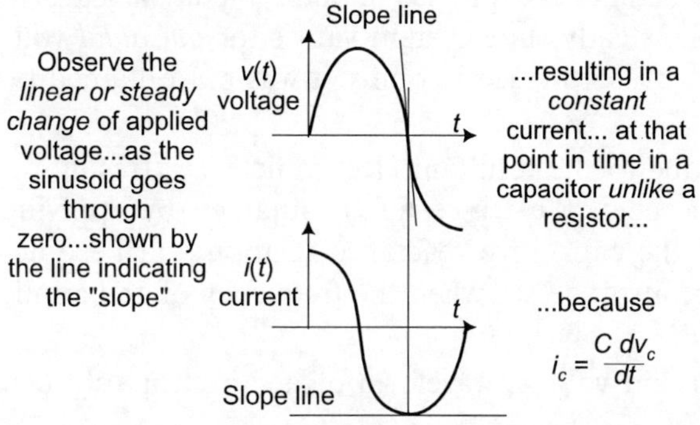

Fig. 2.32 Current is constant at the zero crossing instant of applied voltage.

Note that it is at the *zero crossing point* in a sinusoid that the rate of change of the input voltage is a *maximum... Cdv_c/dt* = $C\cos\omega t$ = $C\cos 0°$ = C and so the current is also a maximum and naturally also *steady*.

2.10 CURRENT IN A CAPACITOR CIRCUIT – SINUSOIDAL INPUT

By a little stretch of imagination, it is possible to see that the *current* in the circuit is a sinusoid and *leads* the sinusoidal voltage *across* the capacitor ….which is also the applied voltage …..by 90 degrees as shown in Fig. 2.33.

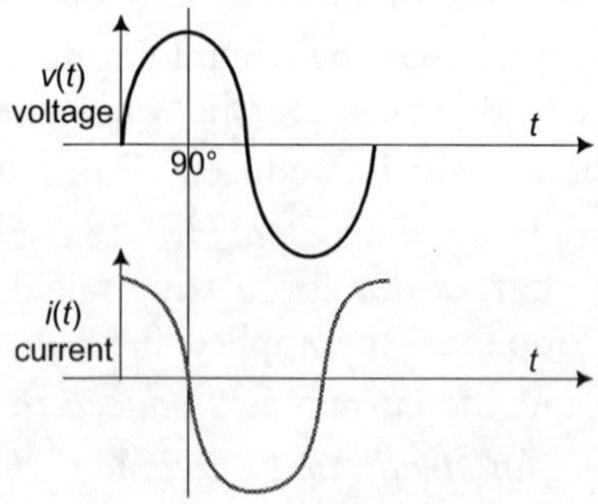

Fig. 2.33 Current leads the voltage for a sinusoidal input.

The existence of a *sinusoidal* current resulting from the application of a sinusoidal voltage to a capacitor is a characteristic of fringe field development due to charges on the plates of the capacitor.

The relation between the "RMS" (square root of the mean of the squares) current and voltage of a capacitor is given by:

$$\frac{V}{I} = \frac{1}{2\pi fC} = \text{capacitive reactance.}$$

where f = frequency of the sinusoidal voltage and C is the capacitance of the capacitor.

The capacitive reactance is expressed in ohms and as we learned in Section 1.2 "current", is useful to determine steady-state current values to *sinusoidal* voltage inputs. Although the value of the reactance is expressed in ohms, it was not obtained by computing a resistance, as of resistors.

The current is due to the resultant electric field E_{net} (resultant of the applied field and an opposing electric field, the fringe field), obtained by applying the relation $J = \sigma E_{net}$. The implication of the capacitive reactance relation will become clear when we study in Section 2.11 the circuit processes when the frequency of the applied voltage to a capacitor is varied.

The full current and voltage waveform of a capacitor subjected to a sinusoidal voltage is shown in Fig. 2.34.

Alternately during every quarter cycle, the voltage source overcomes the fringe field to separate charges thus polarizing the capacitor and establishing an electric field between its plates. The capacitor, during every other quarter cycle overcomes the applied field returning the energy of its electric field (by dismantling the separated charges on its plates) to the voltage source as shown in Figs. 2.27 and 2.28. This amounts to electrical power passing rhythmically and sinusoidally between the two. More on this is described in Sec. 4.8 "Reactive Power in a Capacitor Circuit".

What is meant by steady-state in a capacitive circuit subject to a sinusoidal voltage input?

Imagine that you and your friend are playing a game of "swing". When your friend sits on a stationary swing and you begin pushing it, it will take a few hard pushes initially to overcome inertia when the swing moves with a small displacement. You then synchronize your pushes by progressively moving slightly away from the stationary position of the swing, initially, pushing gently, and then pushing harder as you move away from the central stationary position of the swing.

The point of pushing the swing is usually at the top of the swing cycle at one end. It takes a while of pushing before you are able to establish a rhythmic swing.

The period before the rhythmic swing is established is the "transient", and the rhythmic swing that is established after the few transient cycles elapse is the "steady-state".

This is analogous to establishing the steady-state in a capacitive circuit subject to a sinusoidal input. The capacitor being initially uncharged, will cause the current during the transient period to assume a value that will be quite different from that at the same voltage angle after steady-state is established. It will take a few cycles before steady-state is established, and the current will then begin to follow a pattern that is rhythmic. There are some differences with the swing analogy though. The person pushing the swing can adjust the "push" rhythm initially so that the swings increase in amplitude gradually, while a sinusoidal applied input voltage is normally of fixed amplitude and there is no adjustment possible once the circuit is switched ON.

The descriptions of circuit behavior given in these and subsequent sections assume that *the steady-state has* been established.

These remarks on transient and steady-state are applicable to inductive circuits that will be described in Section 5.17 "Current in an inductive circuit-sinusoidal input".

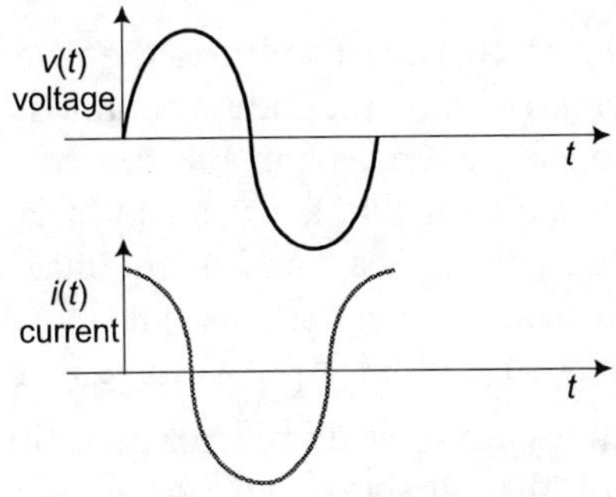

Fig. 2.34 Voltage and current waveform of a capacitor – sinusoidal input.

I urge the reader to visualize all the circuit processes which produce a sinusoidal current when a sinusoidal voltage is applied to a capacitor....the *changing* rates of change of the applied voltage ....the surface charge set up changing at every instant....the applied field changing in the wires ....and the current at each and every instant in time...the charges packing on the plates forming a *dipole*....the fringe field development....dismantling of the charges on the plates as the applied field reduces ...dominance of the fringe field when the applied field reduces....and so on. One critical aspect of the visualization, the one that many among us miss...when we plot the current at the zero-crossing point is its *reversal*.... the charges that were moving rapidly..in one direction slow down..as if braking, ...then come to a standstill...when the current is zero and then reverse..to move in the opposite direction as if at the end of a pendulum's swing.

A resistor obeys Ohm's law: $V = IR$ and for a sinusoidal applied voltage, the current is always in phase with the applied voltage. The interior of a resistor carrying a current is always neutral despite the presence of the applied field and the surface charges; no dipoles are formed in the interior, therefore there is no opposing field and the current simply follows the voltage in phase. See Section 1.23.

A capacitor is different. At low frequencies the capacitor "fills up", but at high frequencies it begins discharging before "filling up" as we shall see in Section 2.11.

The current *leads* the voltage; current flows *before* the charges/potential difference on the capacitor appear.

If the capacitance of the capacitor C is made large, then the fringe field does not build as fast as it would have if C were to be smaller. With a large C, the charge sprays on the plates do not result in developing a large voltage in a given interval of time as evident from the capacitor voltage-charge relation $Q = CV$. The fringe field is smaller and the net field consequently is greater. Therefore, at a fixed frequency, the current increases ($J = \sigma E_{net}$ in the wires) as the size of the capacitor is increased and this is encapsulated in the capacitive reactance expression, $\frac{1}{2\pi fC}$.

2.11 VARYING THE FREQUENCY OF INPUT SINUSOIDAL VOLTAGE TO A CAPACITOR

If we increase the *frequency* of the input voltage, the *rate* of change of the input voltage and the applied field is *greater than* the rates obtained with applied voltages at lower frequencies. At high frequencies, energy from the source is transferred at quicker rates to the circuit as if being hurled into it like a javelin by a javelin thrower! As the frequency increases, the applied electric field is near, and at the maximum and minimum for smaller lengths of time, as may be seen by the steep ascent and descent near the wave crest and trough of the high frequency sinusoidal input voltage in Fig. 2.35.

At high frequencies charges are moved in larger quantities within small intervals of time in the wires to the plates of the capacitor but do not give enough time for the plates to accumulate charges in large numbers (as would have if the frequency was smaller) due to the rapid forward and reverse movements of charges. Therefore, a smaller fringe field builds in opposition to the applied voltage (field). E_{net} therefore, is greater at higher frequencies in the wires.

The result: The current is *greater than* that at lower frequencies of input voltage as seen in Fig. 2.35.

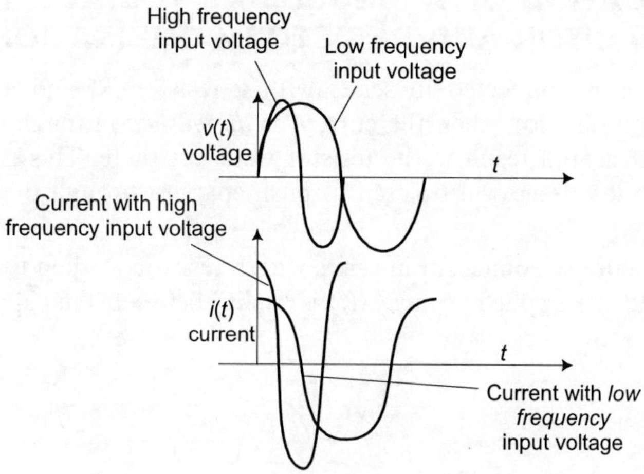

Fig. 2.35 Current when the sinusoidal input voltage frequency is increased.

$J = \sigma E_{net}$ *and* $\dfrac{V}{I} = \dfrac{1}{2\pi fC}$, which one among the two relations is satisfied in a capacitor circuit?

It may appear that there are two relations (see the previous section) that the current is required to satisfy: J = σE_{net} and $\dfrac{V}{I} = \dfrac{1}{2\pi fC}$. This is always a source of confusion in most students. The answer to this is: both relations are satisfied. Between two circuits whose resistances are insignificantly small (one cannot realize "zero resistance" in any practical circuit) yet different, the E_{net} change in the wires of the two circuits is so *tiny* that the capacitive reactance relation will appear not to have been violated, and becomes an ideal approximation (Refer "Ideal Approximation" in "Electronic Principles" by Paul Malvino[20]) while noting that E_{net} regulates the total charge transfer and withdrawal from the capacitor plates, and is not easily measured nor computed in a practical circuit. If the resistance were to assume larger values, then the capacitive reactance relation does not apply to the circuit as a whole and the circuit becomes a capacitor-resistor circuit which is described in Section 2.12.

When the frequency of the applied signal voltage approaches a few hundreds of megahertz (1 MHz = 10^6 cycles per second), the relaxation time (Section 1.16) begins to affect the transient processes which maintain the quasi steady-state (quasistatic) conditions, and the reactance expression ($1/2\pi fC$) will no longer hold. We will learn more about this aspect in Section 5.22.

In the solutions to problems of circuits subject to sinusoidal voltages in the steady state, Euler's identity $e^{j\theta} = \cos(\theta) + j\sin(\theta)$ which is the constant '*e*' with an imaginary exponent, is frequently used. I urge the reader to visit https://ccrma.stanford.edu/~jos/mdft/Proof_Euler_s_Identity.html for a description of how imaginary exponents came about and for a proof of this important identity.

2.12 APPLICATION OF A SINUSOIDAL VOLTAGE TO A CAPACITOR AND RESISTOR COMBINATION

When a capacitor is connected in series with a resistor R, and a sinusoidal voltage is applied to the combination, then the current is *diminished* in magnitude in comparison to the current which would result if the resistor were not there. This is to be expected, since a part of the input voltage will be used to push charges through the resistor and is shown in Fig. 2.36.

When a capacitor is connected in series with a resistor R, then the current is *diminished* in magnitude and the applied voltage $v(t)$ is "**split**" between the capacitor and the resistor in a slightly more complex way.

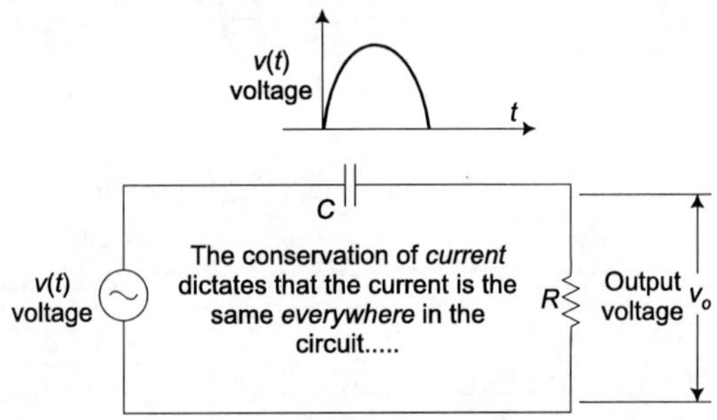

Fig. 2.36 Current when a sinusoidal voltage is applied to a capacitor resistor combination.

Surface charges appear and *bunch up* at the junction of the capacitor and the resistor.

The ***initial*** current is not as large as that in the capacitor circuit. It is restricted by the resistor in series.

All the processes that lead to the current *leading* the voltage *across* the capacitor ….mind …..despite there being a *resistor in series*….by 90 degrees….hold.

But, due to the resistor, the voltage *across* the capacitor is not the same as the input voltage….it is smaller….overall, smaller in magnitude.

This is shown in Fig. 2.37.

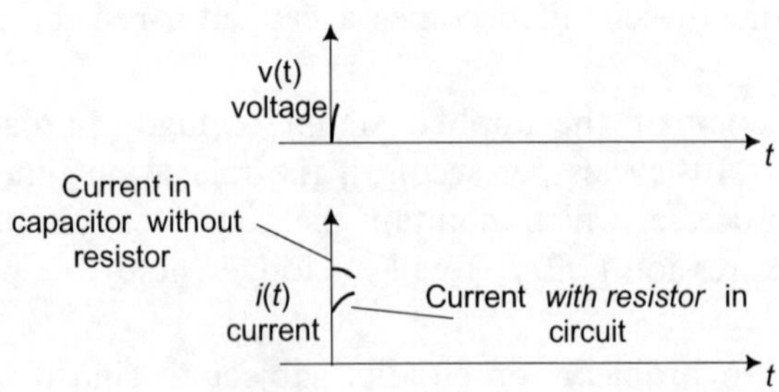

Fig. 2.37 Comparison of currents in the capacitor circuit (Fig. 2.22) and in a capacitor-resistor combination circuit subjected to a sinusoidal voltage.

The ***initial*** current is not as large as that in the capacitor circuit....it is *restricted* by the resistor in series.

Note that we are considering the situation when steady state is reached and in its previous cycle the voltage across the capacitor was negative....and the input voltage is now rising from zero. See Fig. 2.41.

The ***initial*** current grows with the applied input voltage ...and the *capacitor voltage* rises in accordance... in the *positive* direction.

The current levels off and remains constant as the voltage across the capacitor reaches zero and is just becoming positive. The instant this occurs is before the applied voltage *peaks*.

Remember, the current is a constant when the *rate* of change of the voltage is constant across a capacitor.

The situation is shown in Fig. 2.38.

At this "*instant*", the rate of growth of the input voltage and the rate of growth of the capacitor voltage result in a "*constant*" *net* electric field in the wires which maintains the current *constant before* the input voltage has reached its peak.

Also note that *at this instant*, the voltage across the resistor which has the same phase as the current is also "constant though for a very brief period..." (see Fig. 2.41). Why does the current peak and remain "*constant*" briefly "*before*" the applied voltage has "peaked"? While this happens in the capacitor circuit (see Fig. 2.32) when the applied voltage was crossing zero?

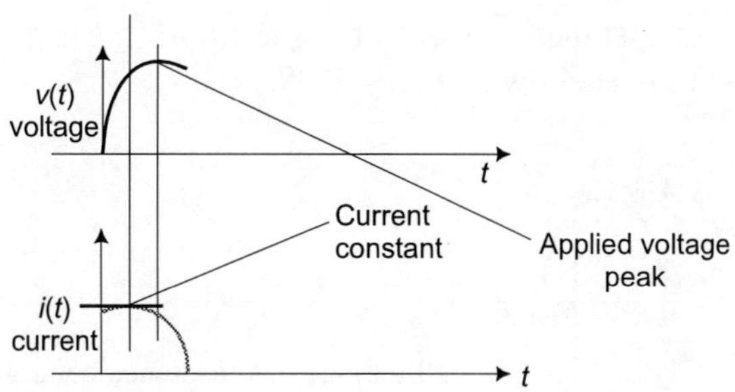

Fig. 2.38 Constant current in a capacitor-resistor combination circuit subjected to a sinusoidal voltage.

We learned (Section 2.9) how the current in a simple capacitor circuit is constant when a ramp voltage is applied, and when subject to a sinusoidal voltage, which is nearly linear when crossing zero, the current peaks and remains constant.

When there is a resistor, the capacitor has a smaller voltage across it...because the current is smaller, fewer charges are transferred in a given time interval...and so, the rate of change of its voltage is also smaller, (draw a couple of sine waves of the same frequency one with a "large" amplitude and the other with a *smaller* amplitude both starting at $t = 0$ and compare the slopes at the same instant of time, any time).

Note that the capacitor voltage's *zero crossing* rate of change will *exactly match* the rate of change of the input voltage at an instant of time *later* than when the input crosses zero...(slide around the smaller amplitude waveform on the time axis to convince yourself of this point). Closer observation reveals that the rate of change of the input voltage and the rate of change of the voltage across the capacitor is the *same* at *this instant* of time. See sketch below and Fig. 2.41.

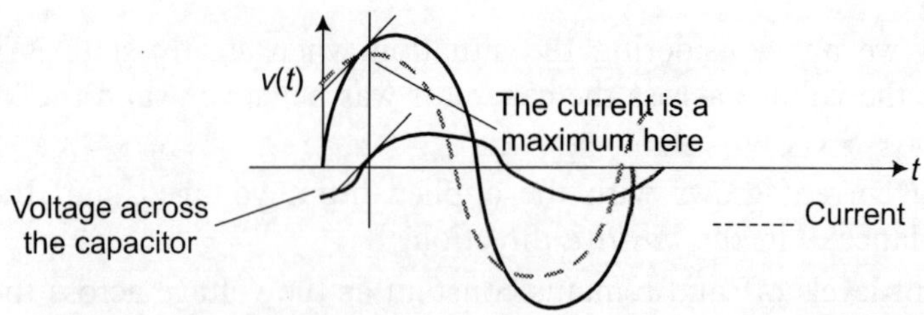

In other words, the rate of growth of the applied input field and the rate of growth of the fringe field (at different magnitudes) is the "*same*" at this instant of time, as we shall soon see, resulting in a *constant* net field, therefore, the current remains "*steady*" since $J = \sigma E_{net}$ = a constant.

In the capacitor circuit, this instant occurs when the input voltage and so the capacitor voltage (being the same as the input voltage) is going through "zero"...and their rates of change match at every instant.

As the input voltage reduces in magnitude, the current crosses "zero" and reverses under the influence of the *dominant* fringe field. At this time, energy will be returned to the source. (The transport of energy and the processes involved are described in Chapter 4).

The fringe field has assumed control at some point "*ahead*" of 180 degrees when the input voltage is zero. This is shown in Fig. 2.39.

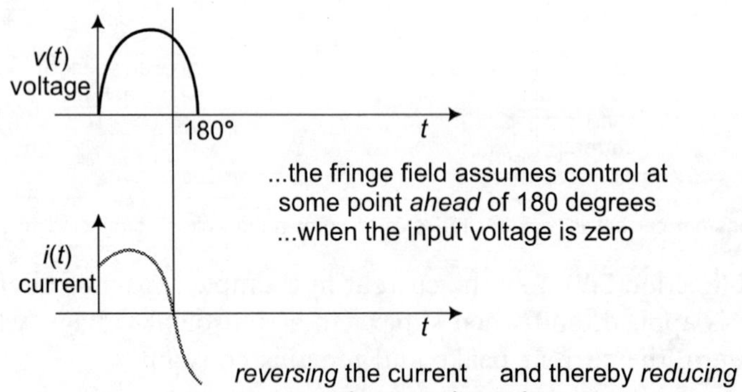

Fig. 2.39 The current in the capacitor resistor combination circuit subjected to a sinusoidal voltage.

The current in a circuit comprising a capacitor with a resistor is smaller and *leads* the input voltage by an angle "*less than*" 90 degrees.

This is shown in Fig. 2.40.

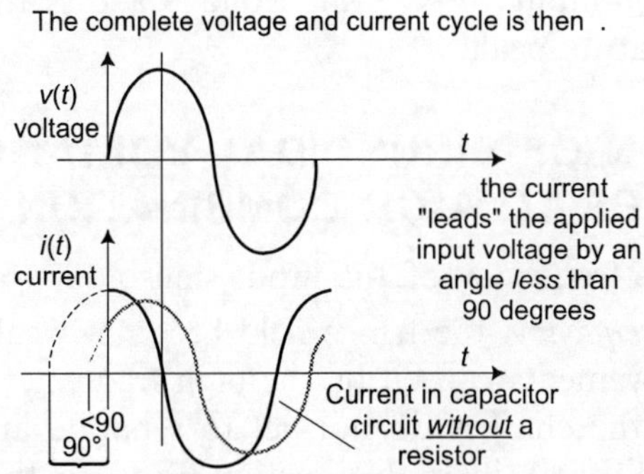

Fig. 2.40 Comparison of the current in the capacitor resistor combination circuit with that in the capacitor circuit (Fig. 2.22) subjected to a sinusoidal voltage.

The output voltage V_o is measured across the resistor.

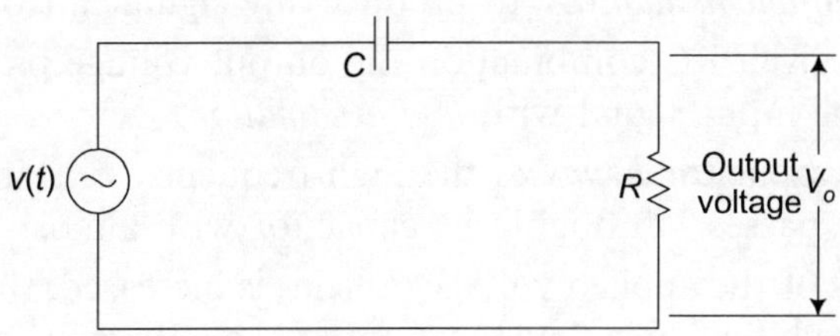

The output voltage across the resistor, *leads* the input voltage by the same amount as the current, because they are in phase for a resistor.

The current which has to be the same everywhere in the circuit leads the voltage *across* the capacitor by 90 degrees; a rule which is *never violated.*

Note that the *slope* of the capacitor voltage change *at the zero crossing point* is the *same* as the slope of the input voltage change before the input peaks, where the current is constant.

Observe that the output voltage is *smaller* in magnitude than the input voltage, as it should be.

The waveforms of input and output voltages and the voltage across the capacitor are shown in Fig. 2.41.

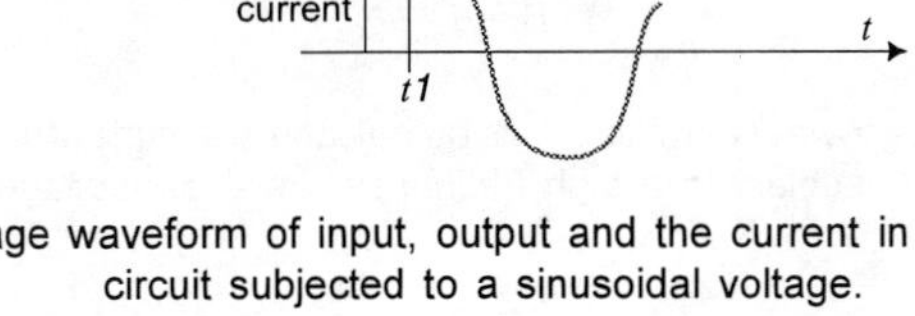

Fig. 2.41 Voltage waveform of input, output and the current in a CR combination circuit subjected to a sinusoidal voltage.

Note equal slopes of input voltage and voltage across the capacitor at time $t1$ when the current is "steady" at its peak.

2.13 APPLICATION OF SINUSOIDAL VOLTAGES AT VARIOUS FREQUENCIES TO A CR COMBINATION

What happens when the frequency of the input sinusoidal voltage is increased?

If we *increase the frequency*, the fringe field becomes weaker. Due to the rapid charge forward and reverse movements caused by the high frequency positive and negative cycles of the applied voltage, the charges do not get to separate and stick to the plates of the capacitor...these being buffeted in both directions, and so the fringe field is weaker. The current therefore increases and as the frequency is further increased is limited entirely merely by the resistor…and the input voltage drops almost entirely across the resistor.

This resistor limited current is not enough to *dump charge fast enough at such high frequencies and of sufficient quantity* to produce any significant opposing fringe field.

Therefore, for a given RC combination the output voltage picked across the resistor is able to reproduce the input signal with *less attenuation.*

We say that the capacitor *bypasses* the high frequencies .....in reality, the electric field of the input voltage passes "through" the capacitor with almost no opposition.

As the frequency of the applied voltage (e-field) is increased, the applied voltage (e-field) is not opposed as much by fringe fields that are weaker (thereby enabling the applied e-field an easy penetration through the capacitor), and the RC circuit *impedance* (see Section 1.2 "Current") will be *small* ($1/2\ \pi fC$).

The output voltage is almost *in phase*…with the input voltage.

The waveforms for the circuit are shown in Fig. 2.42. Also depicted are the voltage and current waveforms for input voltages of a lower frequency.

CR COMBINATION—SINUSOIDAL INPUT—HIGH FREQUENCY

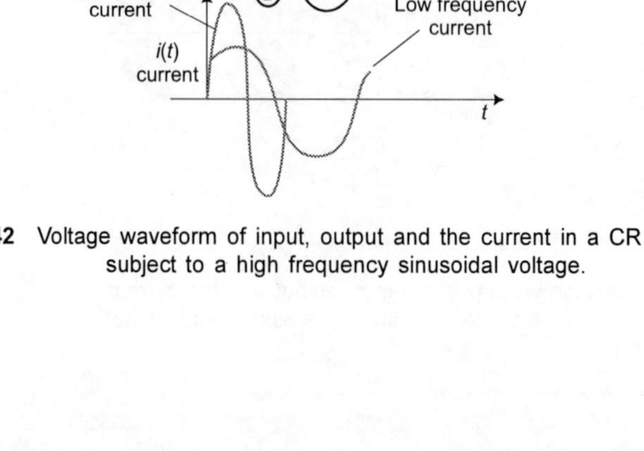

Fig. 2.42 Voltage waveform of input, output and the current in a CR combination subject to a high frequency sinusoidal voltage.

The Capacitor and Resistance Combination – More on Fringe Field

The fringe field is *due* to the charges on the capacitor plates and the charges *get there* by the current. Read the important note below Fig. 1.23.

The current is limited by the resistor. The oncoming charges cannot move at the speed of light and are *opposed* by the fields of charges which had earlier *accumulated* on the plates.

And that is the reason why at high frequencies the fringe field is unable to build fast enough to *oppose* the applied field.

Therefore, we say the capacitor acts as a "short circuit" at high frequencies.

What happens when the frequency of the input sinusoidal voltage is decreased?

If we *decrease the frequency* of the input voltage there is sufficient time for the fringe field to develop, because charges have *ample time* to accumulate on the plates in relation to the voltage change and the current is *small*. See Fig. 2.43.

The output voltage is unable to follow the input signal being increasingly blocked by the capacitor's fringe field and therefore, the *attenuation* is also high.

CR COMBINATION—SINUSOIDAL INPUT—LOW FREQUENCY

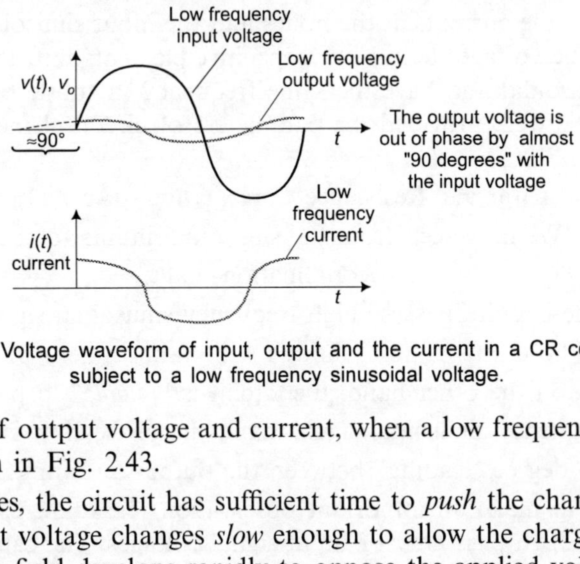

Fig. 2.43 Voltage waveform of input, output and the current in a CR combination subject to a low frequency sinusoidal voltage.

The waveforms of output voltage and current, when a low frequency sinusoidal voltage is applied, are shown in Fig. 2.43.

At low frequencies, the circuit has sufficient time to *push* the charges on the capacitor plates, since the input voltage changes *slow* enough to allow the charges to accumulate on the plates. The fringe field develops rapidly to oppose the applied voltage.

The potential difference across the capacitor is frequency dependent and current flows *before* the charges/potential difference on the capacitor appear; therefore, the current *leads* the applied voltage by a phase angle greater than 0° but less than 90°.

So, we say, the capacitor acts as an *open circuit to low frequency input voltages.*

As the frequency of the applied voltage (e-field) is reduced, the applied voltage (e-field) is opposed by fringe fields which are *stronger* (thereby making it difficult for the e-field to penetrate through the capacitor), and the RC circuit *impedance* (see Section 1.2 "Current") will be *high*.

Sinusoidal Steady-State

Note that the current in the circuit of Fig. 2.36 and the voltage across the resistor and the capacitor are sinusoidal. The current is phase-shifted with respect to the input voltage (Fig. 2.40). The quasi steady-state is applicable to the circuit as a whole since the input voltage is varying sinusoidally with time.

From switch-ON, the circuit will continue to develop these voltages and currents across and in the elements respectively, as long as the input voltage is sinusoidal. Assuming the input voltage is a pure sinusoid, there will be a transient period lasting a few cycles before a quasi steady-state, when the voltages and currents become purely sinusoidal, is established (Section 2.10).

Circuit analysts prefer to call this a "sinusoidal steady-state", which in a sense is a special case of a quasi steady-state.

In general, it can be shown that any circuit or network made from ideal resistors and capacitors (by "ideal" is meant that their resistance and capacitance do not change with time) and excited by a sinusoidal voltage source at a fixed frequency, will produce voltages across the elements and currents at the nodes where they are connected together, and that these currents and voltages measured in and across the elements respectively will also be sinusoidal, after the transients have died down. There may be amplitude and phase differences between the currents at the nodes and the input sinusoidal voltage and between the currents and the voltages across the capacitive elements but, the voltages and currents will be purely sinusoidal and have the same frequency (a subtle but noteworthy point) as the input voltage as well, provided the parameters of circuit elements do not change with time.

The Capacitor and Resistance Combination – As a High-Pass Filter

We have seen from the sinusoidal inputs to the CR combination that...

...the circuit "passes" high frequency sinusoidal signals easily with "little attenuation" though with "almost zero phase shift..."

...on the other hand, the circuit is "*unable*" to pass low frequency sinusoidal signals...and the phase shift is "almost 90 degrees leading" between the output and input voltage.

This leads us to make an important observation regarding CR circuits: A capacitor-resistor circuit is a high pass *filter* of input signal voltages. The capacitor effectively blocks low frequency and dc signals and transmits the high frequency content of signals which maybe a combination of several frequencies, to the resistor.

High pass filters are built using CR circuits in simple applications while in applications with stringent requirements, filters are built using electronic components such as, transistors and operational amplifiers, which provide a sharp demarcation between the low frequencies which are blocked and high frequencies, which are passed. To view the output waveforms generated by computer simulation programs of high pass filters to signals which are a combination of several frequencies, the reader is referred to the "Links_to_animations_articles" file in the CD accompanying this book.

If one is interested in the low frequency content of the signal, this can be picked across the capacitor.

The Capacitor and Resistance Combination – As A Low-Pass Filter

Referring to Fig. 2.41, we observe that the voltage across the capacitor has a very small rate of rise and fall and if the capacitance is made large enough, then the voltage across the capacitor will indeed vary very slowly and may be used as a filter to suppress "ripples" in rectified ac voltages to produce dc voltages for power supplies in electronic circuits.

One such filter circuit using a capacitor is shown below in which the output voltage is picked across the capacitor.

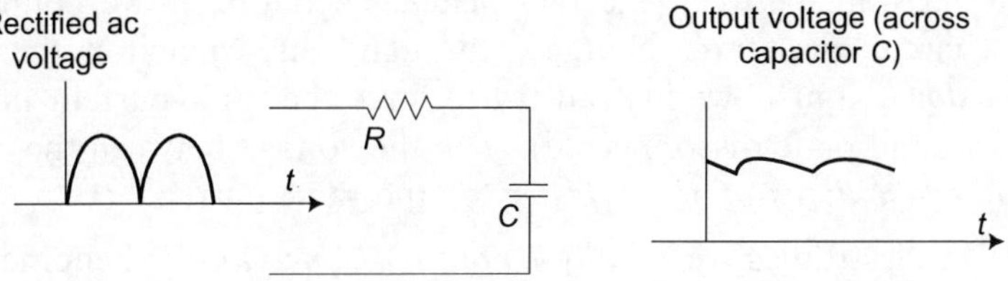

There are some interesting observations we can make to understand and get a feel for the mechanism of capacitive circuits and capacitive resistive circuit operation.

2.14 FOR CHARGE TO FLOW IN A CIRCUIT – THERE HAS TO BE A CLOSED PATH – IS THIS NECESSARY?

From: http://amasci.com/emotor/cap1.html

No!

There must be a closed path ONLY if the charges must *flow continuously* in the circuit forever (dc).

But if the charges flow briefly, or if they flow back and forth (ac), then an *open path* is "acceptable".

For example, with a single metal rod 1 meter in length, it is possible to create a standing wave of charges oscillating lengthwise in the rod with frequency 150 MHz. Inside the rod, *the charges move back and forth* (this resembles the motions of the compressible air inside the pipe of a church organ).

The charges within the rod are acting like a *"compressible"* fluid.

At lower frequencies, the electric current can still be large if the ends of an open-circuit are joined to the plates of a capacitor.

On the metal surfaces of the plates of a capacitor, the quantities of charge carriers behave as a "*compressible*" fluid, while the *charges within a wire* behave as an "*incompressible*" fluid.

Note on incompressible and compressible fluids.

An incompressible fluid flows so that the net flux of fluid into or out of a given volume within the fluid is zero.

If the fluid is compressible, however, and its density $\rho(r)$ varies with position because of pressure or temperature variations, the net outward flux of mass from some small element is determined by divergence (ρv), and this must be related to the rate at which the density of the fluid within is changing.

Suppose we *force* (use a Non-coulomb force!) some charge ***into one plate***, but we ***do not connect the other plate to anything.***

In this case, the **lines** of electric force extend outwards from both plates to the earth. Important question: what is the capacitance of the "capacitor" formed by one plate and the distant earth? Very small.

VERY small, possibly millions or even billions of times smaller than the number of microfarads existing between the two plates. Therefore, if we connect both capacitor terminals to a circuit and force charges "through" this capacitor, the voltage across the plates ***will rise slowly***, but if we instead try to force the same current into only ***ONE plate*** (and leave the second plate disconnected), then the voltage between the capacitor and earth will ***rise millions or billions of times faster*** for the **same** current.

In electronics we *usually ignore the "one-plate"* phenomenon entirely, and imagine that capacitors are a ***strange kind*** of two-terminal conductor.

2.15 WHAT ARE SURFACE CHARGES REALLY?

Surface charges are point charges produced by the shifting of free electrons (see Section 1.15 and Section 1.16) as if compacted or dispersed, and arranged in the form of rings in long wires of circular cross-section. They guide the motion of charges and create the appropriate electric potentials around the circuit, and hence are an important factor in the study of circuit theory.

Charges on charged capacitors are *excess* charges local to that capacitor plate and that are sourced from the other capacitor plate (the two plates form a closed system, refer section 1.8 "Conservation of charge").

Surface charges perform three roles in circuits

They appear upon closure of electric circuits

1. To *maintain the potential* around the circuit
2. To *provide the electric field* in the space around the circuit
3. To assure the *confined* flow of current

In Fig. 2.44 is shown the surface charges that are present in a simple series circuit.

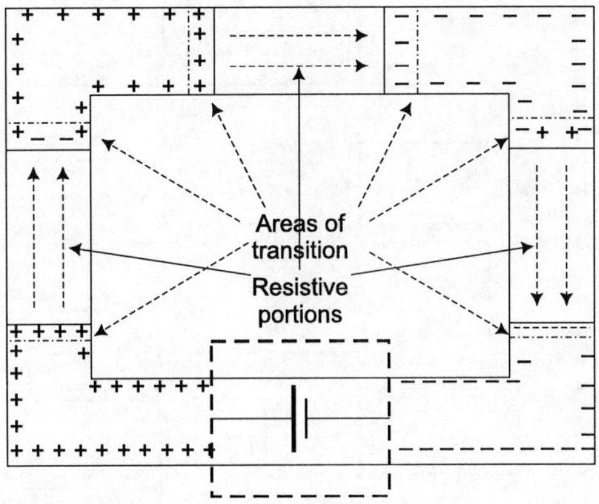

Fig. 2.44 Surface charge distribution in a series circuit.

The figure is not to scale and the gradient of surface charge density along the wires is not shown, because it is quite small.

The presence of the interface charges where the resistive portions join the wires provides a steep surface charge gradient, and a resulting powerful field in the resistive portions.

Figure 2.45 shows the interface charges which are present at the junction of two materials that carry a current. (From: Massachusetts Institute of Technology lecture notes).

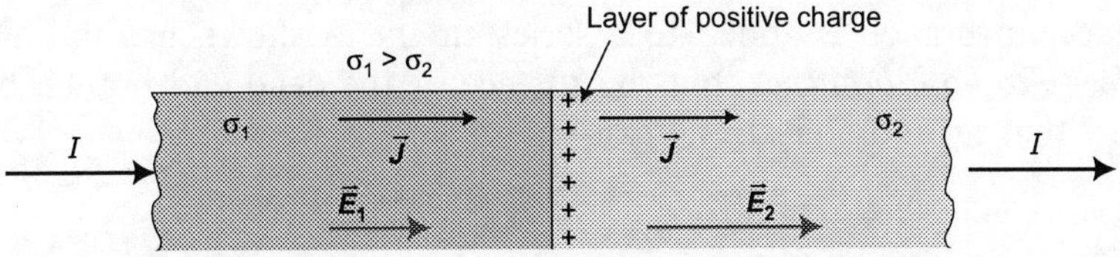

Fig. 2.45 Interface charges at the junction of two materials.

E_1 is smaller than E_2 and the reduction in E_1 is effected by the layer of positive (interface) charge in the material with smaller conductivity.

During a steady-state of current flow, in the circuit of Fig. 2.45, the normal component of current density J must be the same on both sides of the junction. Since the current is the same because the area of cross-section is the same, *the layer of charge develops a field* that is "different" in the two resistor materials of different conductivities.

Note: $J = nqv = nq\mu E = \sigma E$, where n = number of free electrons per unit volume, v = drift velocity, μ = the mobility for the material, and σ = conductivity of the material, and $\sigma = nq\mu$.

In Fig. 2.46 is shown the surface charge distribution in a parallel circuit comprising two resistances in parallel.

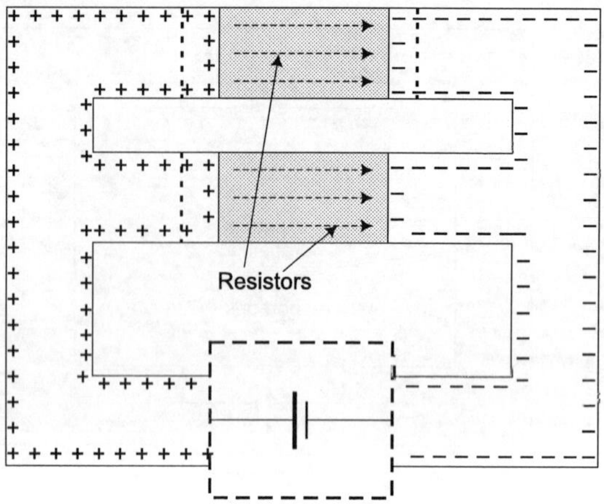

Fig. 2.46 Surface charge distribution in a parallel circuit.

We will now study the mechanism by which current is appropriately steered down a parallel path that is introduced in a simple series circuit as an answer to the question: "How does the current know how to split when there are parallel branches?"

2.16 CURRENT SPLITTING IN PARALLEL BRANCHES

(Chabay & Sherwood, 1995, used by permission)[2]

When a dead end branch is added to a series circuit as shown in Fig. 2.47, then initially electrons *flow into both branches*, but the surface of the dead-end branch becomes charged so negatively that no more electrons can enter.

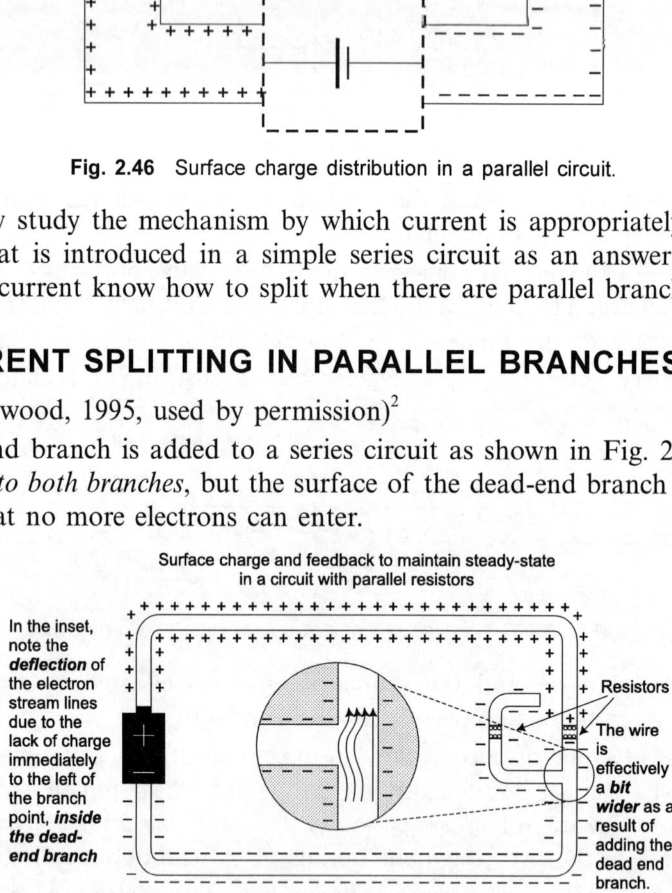

Fig. 2.47 Parallel circuit—dead-end branch.

Also the wire is *effectively* a bit wider as a result of the dead-end branch where conduction electrons make an attempt to enter the dead end branch but are deflected away by the highly negatively charged branch.

Now *complete* the parallel connection, which leads to a rearrangement of surface and interface charges.

The new arrangement is shown in Fig. 2.48.

During the initial transient phase, current may run equally down two parallel branches, but different resistance in the two branches leads to different surface-charge build-up along the two branches, which in the steady state will ***steer*** appropriate amounts of current down each branch.

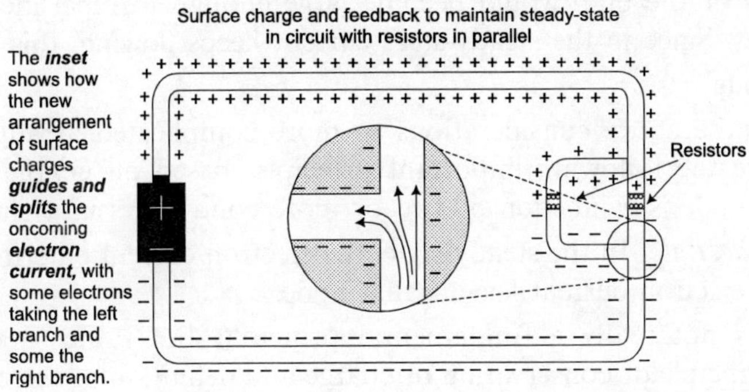

Fig. 2.48 Parallel circuit-closed.

Current in Different Parts of a Circuit

We can measure the current flowing in different parts of a circuit in the steady-state (Section 1.15) using ammeters or current meters. We can also measure the current using compass deflections as an indicator of current, since the magnetic field (about which we will learn in Section 2.26) made by a wire is proportional to the current in the wire.

Consider a circuit consisting of two batteries and a bulb as shown in the figure.

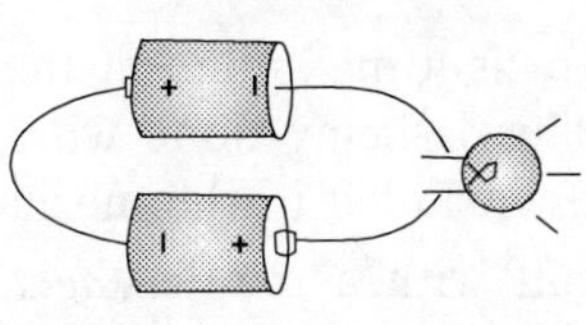

A series circuit with two batteries and a bulb

Electron current flows from the negative terminal of the battery towards the bulb.

The Steady State

Once the circuit has been assembled, current keeps flowing at about the same rate for quite some time. The brightness of the light bulb or the current measured by the ammeter (or the amount of compass deflection) does not change noticeably over several minutes. In fact, it takes quite a while for the batteries to run down under these circumstances, the current remains about the same for several hours. We say that the circuit is in a "steady state."

Can current be used up in the bulb?

It is reasonable to expect that something will be used up in the bulb in order to produce heat and light.

Electrons cannot be destroyed, because this would violate the fundamental principle of conservation of charge.

Could electrons just accumulate in the bulb, so that the current in the wire attached to the positive terminal is less than the current in the wire attached to the negative terminal?

If electrons accumulated in the bulb, the bulb would become negatively charged. The negative charge of the bulb would become large enough to repel incoming electrons and stop the current. Since in the steady-state current keeps flowing, this can't be happening.

Current at a node

We can extend the above considerations to more complicated arrangements of wires. In general we have the following important principle, based on charge conservation in the steady state (a node is a junction of two or more conductors in a circuit):

Current Node Rule: In the steady-state, the electron current entering a node in a circuit is equal to the electron current leaving that node.

This rule is not really a fundamental principle, but rather a consequence of the fundamental principle of conservation of charge and definition of the steady-state. It is also called the "the Kirchhoff node rule". According to the current node rule, in the portion of the circuit shown in the figure, $i_1 = i_2$, and $i_2 = i_3 + i_4$.

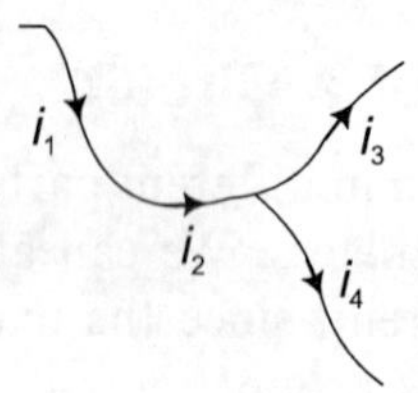

Electron current at two different locations in part of a circuit: $i_1 = i_2$, and $i_2 = i_3 + i_4$

Recapitulate why the current is a maximum at the instant of closure of the switch of a capacitive resistor circuit. This is being done with a view to get a feel for the circuit processes with the electric fields included in the mechanism.

When the switch in a circuit with a parallel *dead-end branch* was closed, the circuit initially was not "aware" of the presence of the dead-end and caused the charges to flow into the dead-end which then accumulated there and *prevented further movement* of electrons in the dead-end branch.

There must be *positive and negative charge* on the surface of the wires near the "+" and "–" plates of the battery, and in the steady state the surface charge must *vary smoothly* from positive to negative around the circuit, possibly with extra charge on the bends in the circuit.

When the switch in a charging circuit of an uncharged capacitor is closed, similarly, the surface charges, developed by the e-field which is produced by the charges from the battery, initially appear around the circuit. The electric field produced by the surface charges on the wires, is ***unopposed*** by any fringe field on the uncharged capacitor and the charges also cause an electric field to appear in the space surrounding the circuit, in, ***"through"*** the capacitor, and around the wires in the circuit as if ***"unaware"*** of the presence of the capacitor.

Remember: *An electric field penetrates through matter.*

In the circuit used to charge a capacitor from a battery with a bulb, at the instant of turn ON, surface charges appear causing an electric field that penetrates through the plates of the capacitor and appears in the connecting wires.

The field also fills the space surrounding the capacitor and the wires. The situation is shown in Fig. 2.49.

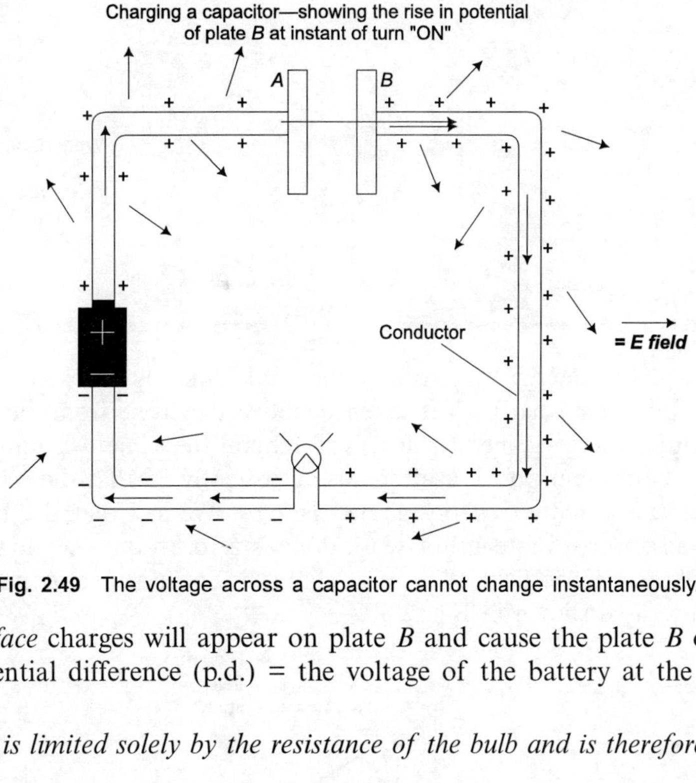

Fig. 2.49 The voltage across a capacitor cannot change instantaneously.

Positive surface charges will appear on plate *B* and cause the plate *B* of the capacitor to attain a potential difference (p.d.) = the voltage of the battery at the instant of turn ON.

The current is limited solely by the resistance of the bulb and is therefore a maximum at that instant.

As the *fringe field* develops and grows in magnitude on the capacitor plates, the capacitor charges and stores energy. The fringe field *opposes* further flow of charges and the current gradually decreases in magnitude.

2.17 WHAT IS VOLTAGE?

We now turn our attention to answering qualitatively the question: "What is voltage?"

This is an important concept that should be studied carefully, because voltage is intimately linked to electric fields. The transport of energy in circuits is through the electric and magnetic fields and therefore a qualitative understanding of these is very important.

A working definition of voltage: Voltage is a way of using numbers to *describe an electric* field. An electric field exerts a force on a charge placed in it.

Electric fields or "E-fields" are measured in volts over a distance; volts per centimeter for example.

"Voltage" and "e-fields" are two ways to describe the *same basic concept.*

Consider the potential energy of objects in a gravitational field.

A stone on the ground has a lower potential energy (PE) in comparison to the PE it possesses when raised some distance above the ground.

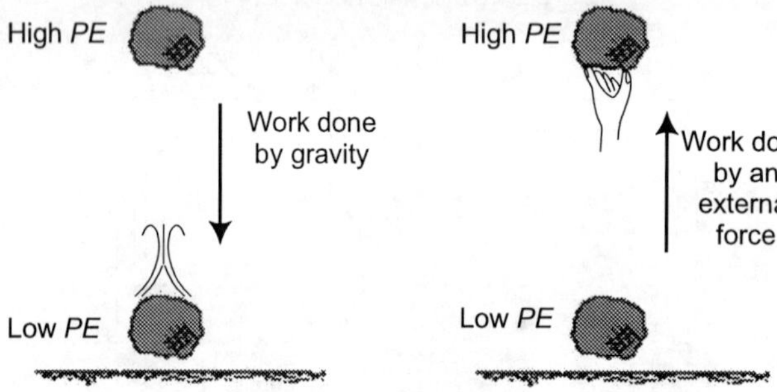

It has gained this energy by virtue of the work done by an external agency in raising it. Note that no matter what path is taken to move the stone from the ground to its final position, potential energy gained (or lost) will remain the same. Also note that if the stone is returned to its original position after raising it gently so that there is no conversion to kinetic energy, the potential energy gained is zero. We say that the field is *conservative* because a round trip path integral of work done is zero. In quite analogous fashion, it can be shown that the electric field of a point charge is conservative. A positive point charge has a field pattern as shown in Fig. 2.50.

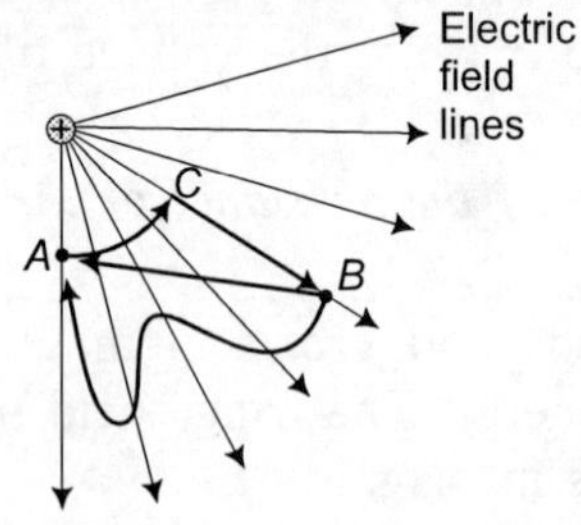

Fig. 2.50 The concept of electric potential being independent of path.

The potential difference between points **A** and **C** is given by ΔPE/charge where ΔPE is the change in potential energy when moving the charge from location **A** to location **C**. The field is said to be "conservative", because the work done to move a test charge from **A** to **C** to **B** and back to **A** is "zero".

In other words, the round-trip path integral of the electric field is "zero" *no matter which path is taken.* Intuitively, this may be understood by considering that the potential energy gained when moving a test positive charge *against* the field is lost when moving *with* the field measured in small incremental movements with direction taken into account. The analogy with the stone moved against the gravitational field and gently lowered (so as not to convert into kinetic energy) is fairly obvious.

Note: Care should be exercised to ensure that the force applied to move the test charge towards the point charge is *equal to qE and not greater* because if it is, then the charge will *accelerate* and the work done to push will be converted into kinetic energy.

Hence, the potential difference between two locations near a point charge is *independent of the path.*

We say that a property of the electric potential due to a point charge is *path independence* and the electric field due to point charges is *conservative.*

In Section 2.19, we shall see that voltages are associated with electric fields due to several charges that are separated. This is due to the production of resultant electric fields in the space surrounding the separated charges as was described in Section 1.10 "Separation of charges".

2.18 WHAT IS ELECTRICAL ENERGY?

Electrical *kinetic energy* (KE) appears whenever electrons ***flow*** through the lattice of atomic cores in conductors. We call this *"electric current",* and it causes *magnetism.*

Electrical potential energy appears when positive charges are ***yanked*** away to a distance from their corresponding negative charges (separation of charges). We call this "net electrostatic charge", and it causes *voltage.*

Electrical kinetic energy is associated with current, and electrical potential energy is associated with voltage.

This will be discussed in more detail in Chapter 4.

2.19 CONSERVATIVE AND NON-CONSERVATIVE ELECTRIC FIELDS

The working definition of voltage is simple but the details seem a little complex. The field being invisible, 19th century scientists gave multiple definitions of potential. While Volta thought it as electric tension and potential as the difference in tension between the ends of conductors, Poisson defined the field as a potential function as charge divided by distance to a point. The potential function was introduced to electrostatics as a mathematical artifact only. Ohm described potential as gradient of electrical tension until, Kirchhoff combined Helmholtz's potential energy concept with Poisson's potential function and said that current is driven by the electric field in a conductor, and voltage is related to the energy supplied.

Since the time of Kirchhoff's definitions there has not been any noticeable change, but engineers and physicists do distinguish four related concepts namely potential, e.m.f., circuit voltage, and electrical potential energy about which we learned in a few sections alongwith the property of fields due to any configuration of charged objects; the round trip path integral must be zero. That field is conservative.

Visualize an electric circuit as an area in space in which the battery provides the + and the – between which electric field vectors are spread out around the space and the *round trip integral* around this of the electric potential is *"zero"*. We will shortly explain why.

Before we do that, let us invert the argument and study whether it is possible to produce an electric field pattern that does *not* have the divergent spread pattern associated with point charges.

In the light of path independence of electric potential, "*Is it possible* to set up a *configuration of charged objects in such a way that the electric field along a circular path* looks like this?"

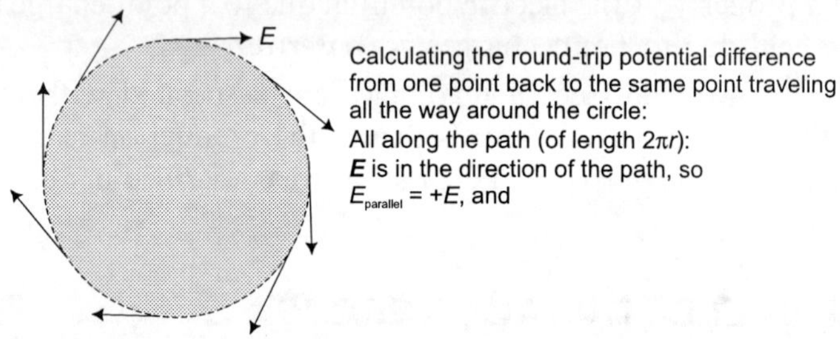

$$\Delta V_{round\cdot trip} = -\int_{initial}^{final} E \cdot dl = -E{\cdot}dl = -E(2\pi r) \neq 0$$

We state one important attribute of charged objects: For any collection of point charges, $\Delta V_{\text{round·trip}}$ of a round trip ***must be zero***.

So, this curly field cannot be produced with charged objects, since $\Delta V_{round\text{-}trip} \neq 0$.

That is the reason we say, that the *curl* (wherein the electric field vectors will close on themselves) of any static charge distribution (including surface charges in circuits) as seen in Figs. 1.24, 2.5 and 2.50 is always "zero". The electric field vectors *diverge* but do not close on themselves. See Section 5.11 "Conservative fields" also.

Allowable Patterns of Electric Fields Using Charged Objects

It ***might seem*** that an electric circuit has a curly pattern of electric field as if the field vectors close on themselves. But examine the diagram below ***carefully***.

Here you will see that ***inside the battery* the electric field is in the *opposite direction to* the electric field in the wire** and the round trip potential-difference is therefore "zero".

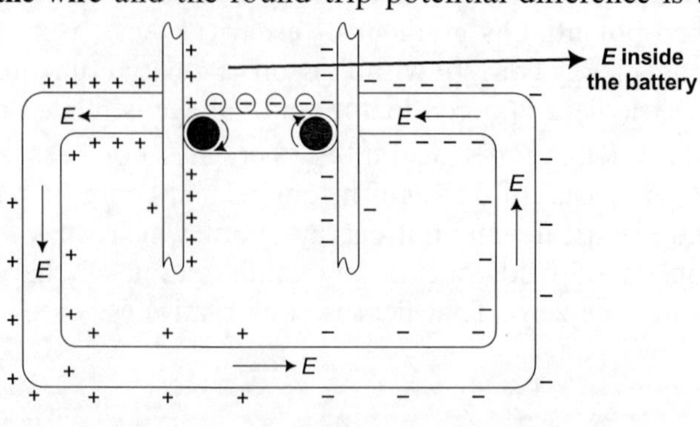

This is also indicated in the voltage "drop" waveform plotted in the circuit shown below.

The electric field in the battery is labeled E_{NC} to indicate a non-Coulomb electric field which is produced by the separation charges owing to a chemical reaction.

Round-trip path integral in circuits

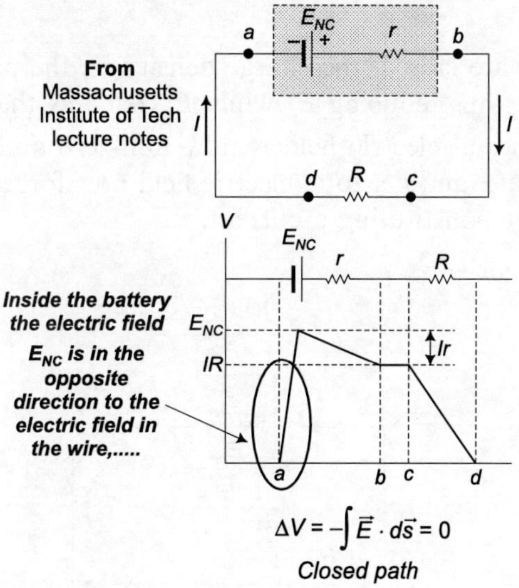

The following description of the non-conservative electric field in a circuit containing a battery is from a paper titled "Charge distribution on dc circuits and Kirchhoff's laws" by W R Moreau, Dept. of Physics, University of Canterbury, New Zealand in *European Journal of Physics*, Volume 10, 1989.

The electric field in the vicinity of a conducting dc circuit can be expressed as:

$$E = E_c + E_{nc}$$

where, E_c is a conservative field with zero curl and E_{nc} is a non-conservative field with zero divergence (and non-zero curl).

Mathematicians call E_c irrotational and E_{nc} solenoidal, but we prefer to call them *conservative and non-conservative* to emphasize the fact *that only the conservative field is related to electric potential.*

Consider the circuit shown in Fig. 2.51.

With the switches open $E_C = E_{NC}$ and so the net field $E = 0$ and there is no current.

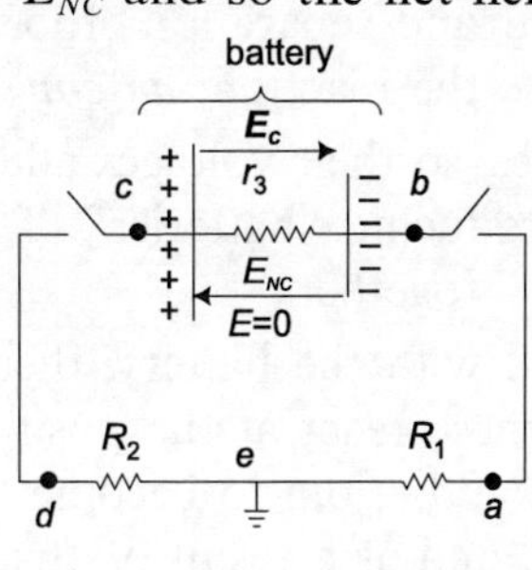

Fig. 2.51 Conservative and non-conservative fields (switch open).

The *non-conservative* part may be generated by chemical action inside a battery or a time-varying magnetic field with rectification inside a dc power supply and is usually non-zero only inside these devices.

We associate the conservative part of the electric field with a charge distribution on the circuit.

When the switches are closed, the charge density on the plates has *decreased* from the open-circuit condition, thus reducing E_C while E_{nc} remains the same.

In the battery the total electric field is now non-zero and directed from the negative plate to the positive plate, and net total electric field now forces charge *through the internal resistance of the battery* constituting a current.

This is shown in Fig. 2.52.

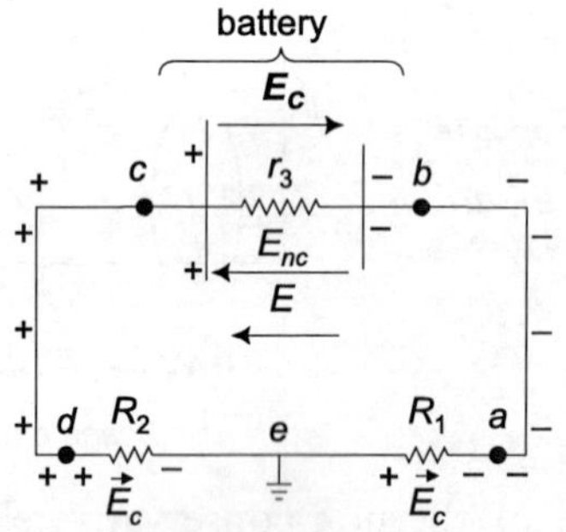

Fig. 2.52 Conservative and non-conservative fields (switch closed).

Effects inside a source of emf - battery

What is the "*direction*" of electron flow?

"*Inside*" a battery?

Source: http://ocw.mit.edu/high-school/physics/electric-circuits/steady-state-direct-current-circuits-batteries-resistors/8_022_fall_2004_lecture8.pdf and www.amasci.com

When a battery is assembled, charge *"pumps"* are located on the *surface* of the battery plates…one on each plate. These are called the "half-cell reaction sites".

Electrons and positive ions are provided by the *negative* plate reacting with the electrolyte. When a number of these are produced, the electric field they give rise to, increases in magnitude, large enough to *"block the migration"* of positive ions to the positive plate. An equilibrium is reached and separation stops when an electric field of sufficient strength is established to oppose the reaction; *an emf has developed between the plates.*

These cells are wired in series, so their voltages add together. The *surfaces* of the metal plates act as the true energy-producing "batteries". The conductive liquid acts as a wire. It connects these two "batteries" together.

When a circuit is completed with the battery, then the charges on the *surfaces* of its plates are *"depleted"*, because these react at the positive plate in the presence of positive ions provided by the electrolyte and rejoin to become "neutral". The electric field between the plates *reduces* in magnitude, and as a result of this, the *chemical reaction restarts,* with negative and positive ions reappearing at the terminals of the battery.

Separation of charges *is resumed* and the depleted charges are *continuously replenished.* The direction of electron flow is *towards* the negative plate inside the battery while the direction of electron flow is *away* from the negative plate in the external circuit.

When the battery is delivering energy to the circuit, the separation of charges is a continuous process and the depleted charges are continuously replenished, restoring the electric field and the "emf". All these effects take place at nearly the speed of light. All sources of emf behave in this fashion. We will be seeing another example when we discuss generators.

2.20 VISUALIZING VOLTAGE

How can we visualize voltage, which is a number which describes the electric field? Considering the simple series circuit in Fig. 2.53(a) consisting of a battery, connecting wires and a resistor, how can there exist a *potential difference* between *A* and *B*, and practically *no potential difference* between *B* and *C*?

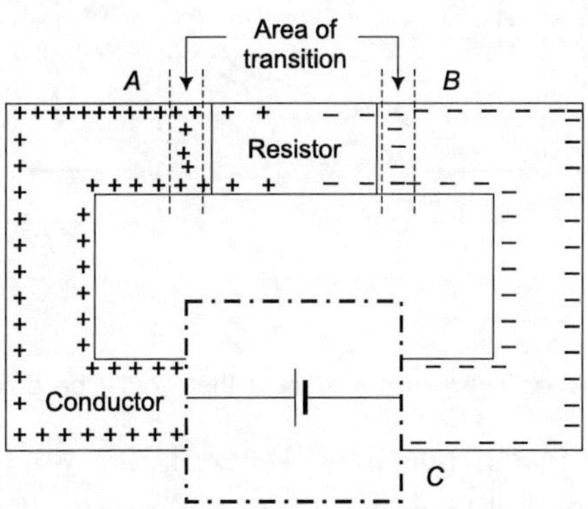

Fig. 2.53(a) Simple series circuit and the existence of a potential difference.

The three cross-sections at *A*, *B* and *C* seem to be identical on a microscopic scale. Also, the same amounts of electrons per second are passing through as follows from Kirchchoff's first law.

Most textbooks explain that there is a potential difference between the points *A* and *B*. But what corresponds to this difference in microscopic dimensions?

There is an electric field, alright. It is constant within the resistor. But what causes this field?

The problem is, of course, not one of the physical theory.

In the case of flow of fluid in pipes there is a pressure difference that can be directly measured at different points but inside a conductor or resistor nothing seems to change over its length. There is a condition of conservation of charge resulting in charge neutrality in every elemental volume within the conductor.

Voltage is intimately connected with electrical energy. So is magnetism. We can say that electrical energy is the fundamental object, while voltage and magnetism are the two faces it displays to the outside world.

Another analogy: in mechanical physics, both the kinetic energy (KE) and the potential energy (PE) are part of matter: relative motion of an object has kinetic energy, and stretched or compressed objects (for example, springs or rubber bands) have potential energy.

In a similar way, electrical kinetic energy appears whenever electrons flow through (really, around) the atomic cores. We call this "electric current," and it causes magnetism. Electrical potential energy appears when positive charges are yanked away to a distance from their corresponding negative charges. See Section 1.15 "The nature of surface charges in circuits". We call this "net electrostatic charge," and it causes voltage.

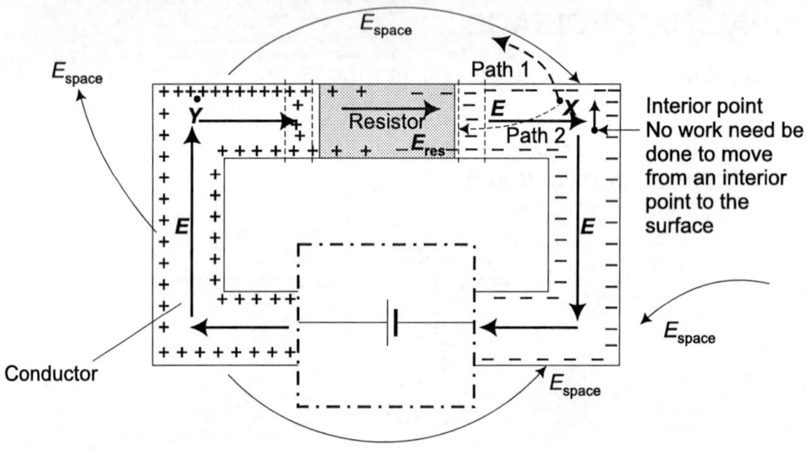

Fig. 2.53(b) When charges are present there must be e-fields and $V = -E\int dx$.

Therefore, the answer to our question "Where is the voltage" is:

The voltage or potential difference is *entirely due to the existence of surface charges.* Surface charges accompany any current flow within a conductor as in Fig. 2.53(b). The work done to move a unit charge from one point X in the circuit to another point Y, whether via the interior of a field-filled connecting wire (path 2), or via the field-filled space (path 1) will be the *same* and thus, the *potential difference* between these two points will also be the same whichever path one takes.

Consider the circuit of the capacitor and the lamp in Figs. 2.11 to 2.15. We compute the voltage between any two points as a path integral of the field between the points say, the capacitor plate B and the battery negative terminal moving either, along the field in the wire and the lamp in which the field will be uniform at a given instant, or along another path say, through the surrounding space in which the field will not be uniform (will be complex) at any given instant, or for that matter along any path. The voltage so computed will be the same irrespective of the path we choose *and* whether there are positive *or* negative surface charges at that instant on the wire between the capacitor plate B and the lamp (see Fig. 1.10 and Fig. 2.58 also). This is because the fields in the wires and in the surrounding space which are due to charges on the battery and the surface are conservative.

See in Section 1.16, the three roles of surface charges. For voltage measurements in circuits, we *always* consider that the points for computing potential differences are located *outside* the surface charges with their small electric fields.

See the last paragraphs of this section and also "Comparison of electric and gravitational fields" in Appendix B on the force exerted by a voltmeter probe on a wire surface and also see Appendix D on the forces exerted when objects touch. If considering points lying *interior* (of wires and resistors), these are always *outside* atoms since dipoles do not exist there (neutrality conditions prevail despite the current and the driving field) and these points are equipotential with the nearest surface.

Therefore, for the path of a test unit charge there is no work done (and so, there is no voltage drop to consider) to move from an interior point to the nearest surface or move from the nearest surface to the interior point.

The reason why most textbooks ignore this simple fact of the existence of surface charges could be that the surface charges are rather small and normally not detectable (the electric field in space due to them in simple low voltage circuits will not make the hair on your hand stand as a charged plastic comb can as described in Section 1.3!) unlike the fluids in pipes where most parameters associated with its flow such as pressure, velocity can be easily measured using suitable instruments and meters. In the case of fluids, the energy associated is carried along in the fluid. This is not the case with the flow of charge in a wire.

Electrical KE is associated with current, and electrical PE is associated with voltage. If electrical energy is the same as electromagnetism, then maybe we should be more sensible and name it "VoltageCurrent-ism." Is voltage a type of potential energy? Think of it like this. If you roll a big boulder to the top of a hill, you have stored some potential energy. But after the boulder has rolled back down, the hill is still there. The hill is like voltage: the height of the hill has "Gravitational Potential." But the hill is not made of Potential Energy, since we need both the hill and the boulder before we can create potential energy. The situation with voltage is similar. Before we can store any electrical potential energy, we need some charges, but we also need some voltage-field through which to push the charges.

A charged metal object is like a flat-topped mountain, where the altitude of the mountain top is like potential of the charged object, and the mountain slopes are like e-fields in the space around the object. On one side of the mountain, the slope might be very gentle. And, on some other side, steep. Yet no matter how gentle (or steep), we arrive at the same altitude when we walk to the top.

If a charge lands on the mountain top where the potential is positive, after a long and arduous climb along its slopes, it is well received and given a free pass to roam around the top without losing energy! The "equipotential" inside a conductor (of any shape) is like the mountain top.

The free electron sea shifting inside the conductor, in every elemental volume, keeps the entire conductor an equipotential "pushing" any excess charge all around to its surface which will maintain its voltage condition and darting around and stirring up the charges within to maintain equipotential conditions inside yet maintaining neutrality in the bulk of the conductor. Neutrality is maintained because the interaction between the electrons and the enforced increase of the velocity of the electrons is so very small.

If a conductor has a potential which is not zero, there must always be an e-field surrounding that conductor and extending to infinity, even if that e-field is very weak.

No matter how weak, it must still add up to the same potential at the object surface. And, since e-field flux lines must terminate upon charges, therefore the presence of the e-field requires that surface charges must be present on that conductor. Field configurations around circuits (outside conductors) can be complex as we saw in the Figs. 2.11 to 2.15 in connection with the charging of capacitors. But the certainty in the fields in the wires enabled us to obtain the voltage or potential difference between points on the circuit. See also the article "Potential around a pair of charges" in the accompanying CD.

Recollect from Section 1.5 that the potential energy (stored energy) is proportional to the number of separated charges and the distance separating them. The same remark applies to voltage when we learned in Section 2.7 "Charging a capacitor" that the voltage across the capacitor grew in magnitude as the number of separated charges closely *packed* on its plates increased as time elapsed, though the distance they were separated remaining fixed, being the distance between its plates.

Power and Energy in a Circuit and the Loop Rule

(Chabay and Sherwood, 2010, used by Permission)

Consider the simple circuit involving batteries (D cell of 1.5 volt each fitted in a battery holder), hookup wire and a flashlight bulb (The bulb resistance is assumed to be 10 ohms) as shown in the figure.

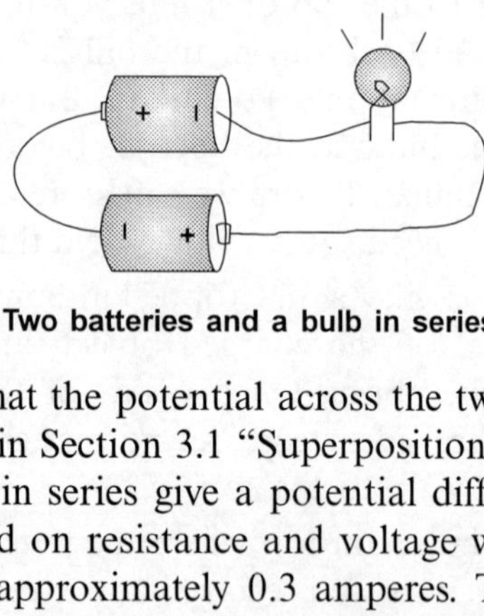

Two batteries and a bulb in series

Assume for the moment that the potential across the two batteries in series will be 2 × 1.5 V = 3 Volts. We will learn in Section 3.1 "Superposition of Potential" and "Batteries in Series" how the two batteries in series give a potential difference = 3 volts. We can show by using the calculations based on resistance and voltage which we carried out in Section 1.22 that the current will be approximately 0.3 amperes. To use the energy conservation equation we would have had to know the electric field strength in the filament of the bulb.

Charge conservation (Section 1.8) plus the definition of a steady state led to the current node rule, which tells us that the current in a simple battery-and-bulb circuit must be the same everywhere in the circuit.

Consider a circuit containing various circuit elements such as resistors and batteries. What determines the magnitude of the current? The Energy Principle provides the answer, plus the properties of batteries.

We could discuss energy conservation in the circuit by considering one electron as it makes a complete trip around the circuit. The energy gained by the electron as it is transported across the mechanical battery is dissipated in collisions with atomic cores as it travels through the wire. Alternatively, we can analyze the energy per unit charge gained and lost in a trip around the circuit. We know that over any path the round-trip potential difference must be zero, since the electric field penetrates and exists in circuit wires and resistors and components as it exists in space, the application of the round-trip path integral (Section 2.17) to circuits follows logically. And, if we follow a round-trip path through a circuit, through circuit element 1, circuit element 2, and so on, we find the following:

ENERGY CONSERVATION (THE LOOP RULE)

$\Delta V_1 + \Delta V_2 + \ldots\ldots = 0$ along any closed path in a circuit. This is essentially the Energy Principle, but on a per-coulomb basis. The electric potential is defined as potential energy per unit charge: $V = U/q$. See 'Electric Potential at One Location and Potential Energy' at the end of Appendix A. The loop rule says that you can't gain energy in a round trip. In circuits this energy principle is often called the "Kirchhoff loop rule".

The current node rule

In the steady state, for many electrons flowing into and out of a node,

Electron current: net i_{in} = net i_{out}, where $i = nA\mu E$

Convention current: net I_{in} = net I_{out}, where $I = |q|nA\mu E$ and where, for the component connected to the node, n is the number of electrons per unit volume (/m^3), A is the cross-sectional area (m^2) of the component, μ its mobility in (m/s)/(N/C).

Energy Conservation (The loop rule)

In the steady state, for any round-trip path,

$\Delta V_1 + \Delta V_2 + \ldots = 0$, where ΔV is the potential difference between ends of circuit elements.

The current node rule refers to many electrons moving together, whereas the energy equation is on a per-unit-charge basis.

Power in a circuit

If you move a small amount of charge Δq from one location to another, the small amount of electric potential energy change ΔU_e is equal to the charge Δq times the potential difference ΔV, $\Delta U_e = (\Delta q)(\Delta V)$, since electric potential is electric potential energy per unit charge (joules per coulomb, or volts). If this work is done in a time Δt, we can calculate the power, which is energy per unit time:

$$\text{Power} = \left|\frac{\Delta U_e}{\Delta t}\right| = \left|\frac{(\Delta q)(\Delta V)}{\Delta t}\right|$$

However, $\Delta q/\Delta t$ is the amount of charge moved per unit time, which is the current I. Since $\Delta q/\Delta t = I$, we have the following important equation:

POWER FOR ANY KIND OF CIRCUIT COMPONENT

$$\text{Power} = |I\Delta V|$$

Application of the two rules

We had seen two examples of the application of these rules in Section 1.23 when we studied the effects of doubling the length of the wire and when the cross-sectional area of the wire is doubled.

We will consider the case here when two bulbs of different wattage (and filament resistance) are connected in series as in the figure.

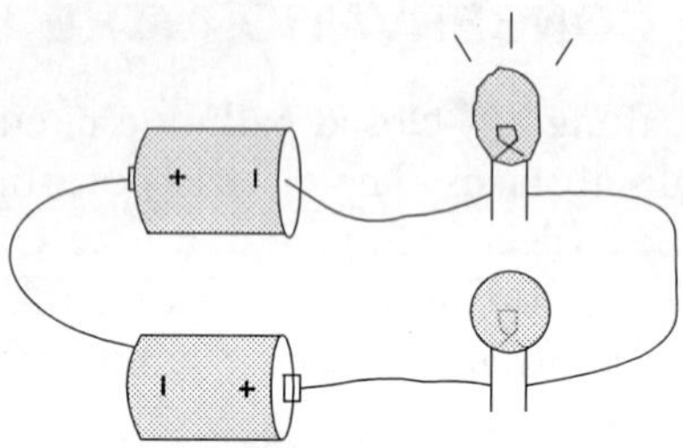

A thin-filament bulb (long) and a thick-filament bulb (round) in series—why doesn't the thick-filament bulb light up?

If you put a thin filament long (elongated) bulb (0.12W) in series with the thick-filament round bulb (1W) bulb, the thin-filament long bulb lights up but the round thick-filament bulb doesn't light up!

What happened to the current that ran through the thin filament bulb?

Consider that

Both bulb filaments are made of tungsten.

The round thick-filament bulb has large cross-sectional area A_r, length L_r, electric field E_r, and drift speed $\bar{v}_r$.

The long thin-filament bulb has small cross-sectional area A_l, length L_l, electric field E_l, and drift speed $\bar{v}_l$.

Assume for the moment that the electron mobility is the same in both bulbs, even though in fact the glowing thin-filament bulb is very hot (which should mean low mobility) and the thick-filament bulb is apparently cool (since it isn't glowing, which should mean high mobility).

Using the current node rule, write an equation relating the drift speeds $\bar{v}_r$ and $\bar{v}_l$.

In the steady state, the number of electrons per second passing through each bulb must be the same:

$$nA_r\ \bar{v}_r = nA_l\ \bar{v}_l$$

Based on the equation you just wrote, write an equation relating the electric fields E_r and E_l, and state which is larger, E_r or E_l.

The electric field E_r will be uniform in the filament of the round bulb due to surface charges and so will the electric field E_l be uniform in the filament of the long bulb due to surface charges.

Assuming the same mobility in both filaments, since

$$\bar{v} = \mu E$$

we know that $nA_r\mu E_r = nA_l\mu E_l$, so

$$E_l = \frac{A_r E_r}{A_l}$$

since $A_r > A_l$, $E_l > E_r$.

Applying the loop rule, we find that 2emf = $E_r L_r + E_l L_l$, which is approximately equal just to $E_l L_l$, since E_r is very small. Explain why the thick-filament bulb doesn't glow.

If the thick-filament bulb were alone in a circuit, it would be bright, and the electric field in the filament would be:

$$2\text{emf} = E_{bright}L$$

so

$$E_{bright} = \frac{2\text{emf}}{L}$$

In the two-bulb circuit (if the length of the filaments are the same) we have

$$2\text{emf} = E_r L + E_l L = L\left(E_r + \frac{A_r E_r}{A_l}\right)$$

$$= LE_r\,(1 + A_r/A_l)$$

so

$$E_r = \frac{2\text{emf}}{L(1 + A_r / A_l)} < \frac{2\text{emf}}{L}$$

Thus the electric field in the thick-filament bulb is significantly smaller than it is in the single-bulb circuit. We can estimate how much smaller by noting that one typically measures about ¼ as much current in a circuit with a single thin-filament bulb as in a circuit with a single thick-filament bulb. This suggests that $A_r \approx 4A_l$, so the electric field in the thick-filament bulb has been reduced by a factor of 5. Evidently this field does not produce sufficient current through the filament to make it glow.

What determines Bulb Brightness?

For each electron that moves through a wire of length L due to an electric field E, the electric potential energy of the system (electron plus wire, remembering that potential energy refers to a system of objects) decreases by an amount $e\Delta V = eEL$.

But how is this equation $e\Delta V = eEL$ applicable since we know that the electron drifts so slowly that it hardly, if ever, will traverse the length of a physical circuit?

This is an important question. Recollect from Section 1.18 that there are millions and millions of electrons everywhere in the wire and in the filament of the bulb which are participating in the current. These electrons are moving everywhere within the wires and in the filament at every cross-section. When we derived the expression for the electron current in Section 1.17, we drew a section of the disk shaped portion of the electron sea, but the disk shaped sections exist throughout the length L of the wire, and when one electron in the disk shaped section to the left of the wire has moved right another disk shaped section to the right of the wire has moved out of the right end of the wire as shown in the exaggerated view of the piece of wire between the battery negative terminal and the long bulb in the figure.

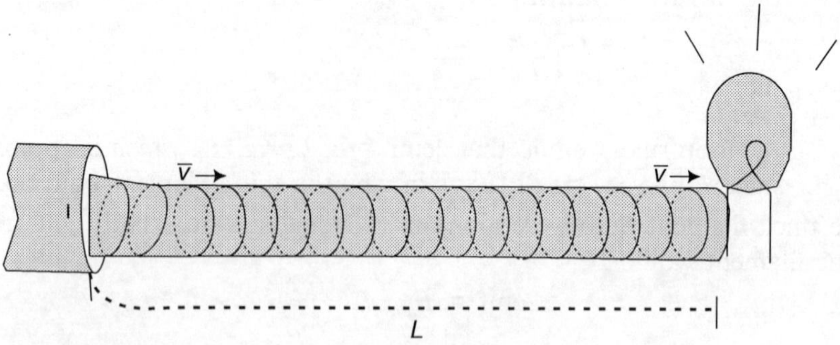

There is a disk shaped section of the electron sea that has moved in every section of the wire of length L

That is true for every section of the wire and that effectively amounts to having one electron move through a wire of length L.

The formula is applicable over any interval of time including Δt.

This loss of electric potential energy is equal to the gain in kinetic energy of the electron, which is dissipated in collisions with the atoms in the wire, with consequent increase in the thermal energy and temperature of the wire. As the wire's temperature increases, the rate of energy transfer to the surroundings increases in the form of emission of light. The temperature of the wire stops rising once the temperature is high enough that the rate of emitted energy (as light) has become equal to the rate of energy dissipation in the wire.

In a time Δt the number of electrons entering (and leaving) the wire is $nA\overline{v}\Delta t = nA\mu E\Delta t$, where n is the number of electrons per unit volume of the material. So, the energy dissipation per second ("power" in joules per second, or watts, see "Power in a circuit" above) is $(eEL)(nA\mu E) = enA\mu LE^2$.

Why does the thick-filament bulb not glow brightly in the series circuit with the thin-filament bulb?

We see that the rate of energy dissipation is proportional to the square of E, or alternatively to the square of the current $I = enA\mu E$. In the example above, the field E in the thick-filament bulb was reduced by a factor of 5, so the rate of energy dissipation in the wire was reduced by a factor of 25. This in turn means that the rate of energy emitted as light was reduced by a factor of 25.

The bulb is so dim that you don't see it glow.

Parallel circuits

The circuit of Fig. 2.46 is redrawn and shows two different paths with the resistors replaced by thick and thin wires to make it easy to visualize the presence of the electric field in the resistive portions. There will be electric fields in the resistors in the circuit of Fig. 2.46 as well but this may be hard to visualize since real resistors are only a centimeter or two long.

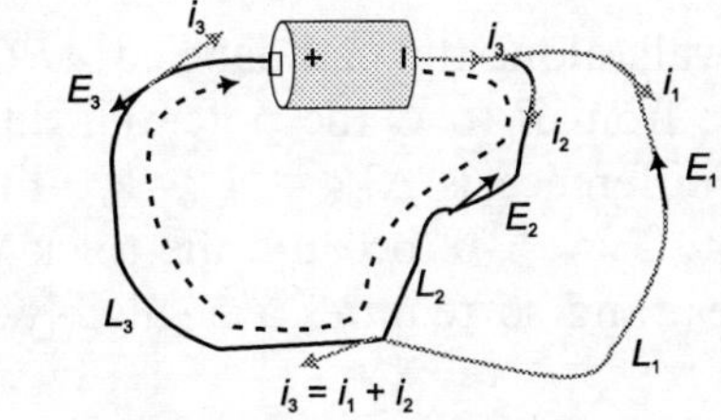

Follow the dashed path through wires L_2 and L_3

There is more than one path for current to follow. Start at the negative end of the battery and follow the dashed path clockwise through the wires L_2 and L_3, and then through the battery to the starting location. There is a potential rise (see Figs. 1.34 and 1.35) + $E_2\ L_2$ along the part of the path of length L_2, a potential rise $+E_3\ L_3$ along the rest of the path of length L_3, then a potential drop through the battery of amount -emf, yielding $E_2\ L_2 + E_3\ L_3 + (-\text{emf}) = 0$ for the round-trip potential difference. If we instead follow the path through L_1 and L_3, we find a round-trip potential difference $E_1\ L_1 + E_3\ L_3 + (-\text{emf}) = 0$.

These two equations imply (correctly) that $E_1L_1 = E_2L_2$, which makes sense, since with the same starting and ending points the two wires have the same potential differences. Also notice in the figure that $i_3 = i_1 + i_2$, due to the current node rule (charge conservation in the steady-state).

The loop rule can be applied along any path through an arbitrarily complicated circuit. Consider a circuit made up of five circuit elements connected together by thick copper wires that have high electron mobility (see figure below).

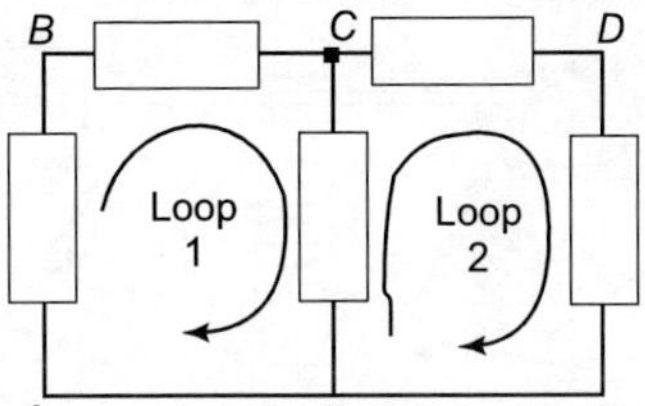

A circuit with two loops

The circuit elements could be resistors, batteries, capacitors, or even complicated devices such as semiconductor diodes or even refrigerators!

In the figure we show the left portion of the complete circuit (loop 1).

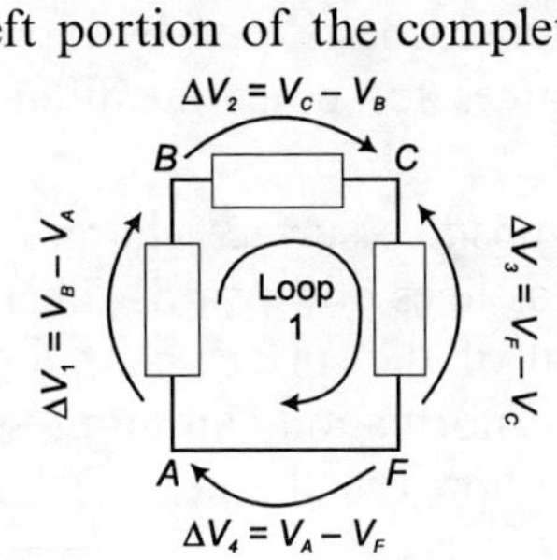

Walk around loop 1, following the path *ABCFA*

Start at location *A* and walk along the path labeled *ABCFA*. From *A* to *B* the potential difference is $\Delta V_1 = V_B - V_A$; from *B* to *C* the potential difference is $\Delta V_2 = V_C - V_B$; and from *C* to *F* the potential difference is $\Delta V_3 = V_F - V_C$. From *F* to *A* there is negligible potential difference $\Delta V_4 = V_A - V_F \approx 0$, because the thick wire has high electron mobility and large cross-sectional area, and so requires a negligibly small electric field to drive the current.

Any round-trip potential difference must be zero, and the loop rule gives this:

$$\Delta V_1 + \Delta V_2 + \Delta V_3 + \Delta V_4 = 0$$

$$(V_B - V_A) + (V_C - V_B) + (V_F - V_C) + (V_A - V_F) = 0$$

We could write a similar equation corresponding to walking along loop 2 (path *FCDEF*). The idea is that you start anywhere you like on the circuit diagram and walk around a loop that brings you back to your starting location, adding up potential drops and rises as you go.

You can even walk along a round-trip path through the air, outside the circuit elements (as we had done along the two paths in Fig. 2.53(b)), but unless you know something about the electric field and potential differences outside the devices and connecting wires, this usually isn't very useful.

A complete circuit with an indication of surface charges could be represented as depicted for the circuit on the left of Fig. 2.54 by Hermann Härtel, in his paper "**Constant Electric Current and the distribution of Surface Charges1**". (Reproduced with permission from Hermann Härtel).

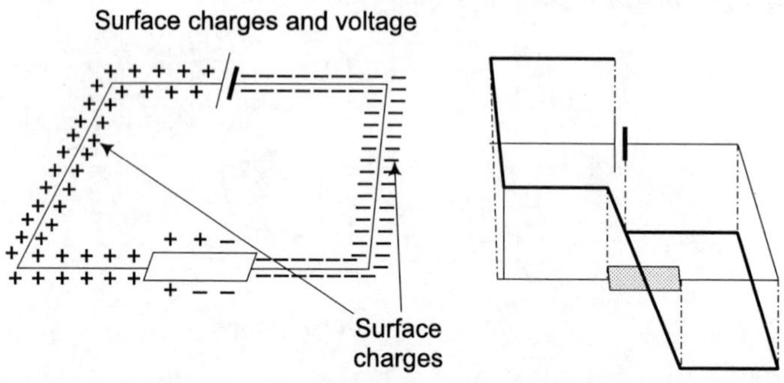

Fig. 2.54 Different possible respresentations of voltage.

A different representation could be chosen by representing different values of surface charges, or better: different values for potential differences by vertical lines of different length.

In Fig. 2.54 on the right, positive surface charges are indicated as lines in the upper direction and negative charges as lines downwards. This representation is more abstract but better to read, and the gradient of the surface charge distribution (potential difference) is immediately visible. This same information can be presented by choosing one side as zero (same height as the circuit) as shown in the next Fig. 2.55.

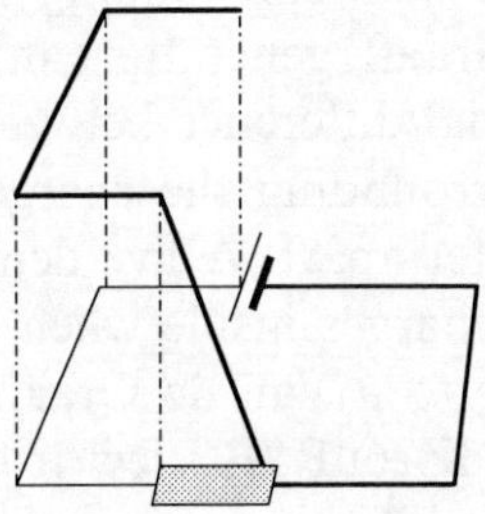

Fig. 2.55 Representation of voltage (change of reference point).

It is not the absolute value of the surface charges (potential) but *only the difference* in density between two points (potential difference) that causes a net driving force on the mobile electrons within the conductor and the resistor.

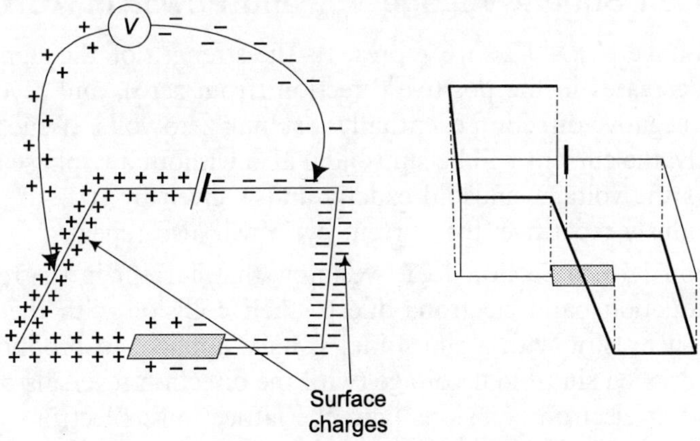

Fig. 2.56 Measurement of voltage.

When probes to measure voltage using a voltmeter are touched (see Fig. 2.56) at two different points in a circuit, then surface charges appear on the probe wires. The surface charges enable a tiny current in the wires, and this current deflects the coil in the meter and the pointer attached to the coil swings to indicate the voltage. Note that a voltmeter probe merely comes *very close* (where the e-field is quite strong) to the surface charges when it touches a terminal and it *never* really touches neither the valence electron nor gets inside a weak atomic dipole. Refer Appendix B. In electric and electronic circuits the electric field *drops away so steeply* with distance from the wires into space that no voltage is recorded in voltmeters if its probes are held even micrometers away from the wires.

The experiment and the explanation above of how a meter measures voltage raises another interesting question, as discussed in the next Section 2.21.

Since a pair of voltmeter probes should touch the terminals on a circuit board to measure voltage across them, does that mean that it will not be possible to detect a current in the space surrounding a thick bunch of charges, which produce a powerful electric field around them?

It is possible for the electric field surrounding an isolated charge to set up a current which is detectable. A Van de Graaff generator can generate thousands of volts potential and hundreds of volts of potential difference between two points in space (obviously, with an electric field in the space surrounding the charge on the collecting sphere of the Van de Graaff generator) upto a meter apart. A live demonstration of a fluorescent lamp that glows due to a flow of ionized charges inside when the lamp is placed in the electric field (without wired connections) due to a Van de Graaff Generator may be seen here https://www.youtube.com/watch?v=5aQdZ26uR7E View from the 41st minute.

Most electric control and electronic circuits however are designed to operate at lesser than a few hundred volts and this is insufficient to trigger and sustain currents in voltmeters when the probes are brought near (without touching) a pair of circuit terminals between which the voltage is to be measured.

The Root Mean Square Voltage V_{rms} and Power in AC Circuits

A sinusoidal voltage $v(t) = V_m \sin\theta$ represents the strength of the signal as it progressively increases and decreases in the positive direction (from zero), and reverses to increase and decrease in the negative direction eventually reaching zero volts. If such a voltage is applied to a resistor only, the current will be sinusoidal also without any phase shift, i.e., the current sinusoid follows the voltage sinusoid exactly and is given by $i(t) = I_m \sin\theta$. If the voltage cycle is continuously repetitive, the current cycle will also repeat.

From Section 1.18 to Section 1.21, we know that current in the resistor is a start-stop motion of conduction band electrons due to their collision with the rocking lattice ions, and this causes a resistor with a sinusoidal voltage applied, to produce heat. The polarity reversals of an applied sinusoidal voltage (with the direction reversals of the applied electric field) do not affect electron collisions with the lattice ions. Electrons colliding with lattice ions from either direction will continue to produce heat and there is no cancellation of the heat developed!

Mathematically, the average value of a sinusoid is zero, and so, the average value of the current will also be zero. The average values cannot therefore represent the heat developed in a resistor with a sinusoidal current, as shown in the figure.

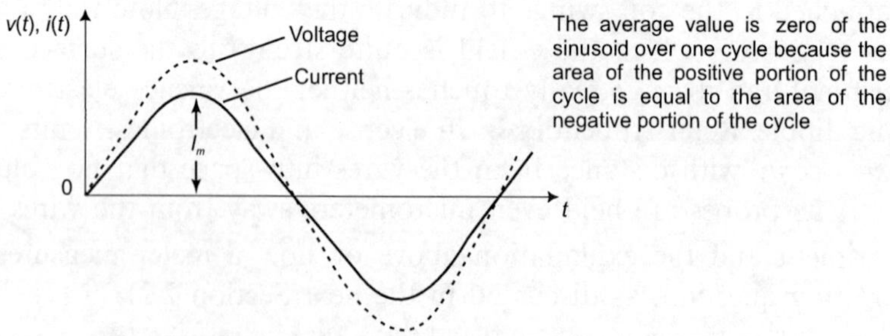

Instantaneous voltage and current in a resistor with a sinusoidal voltage applied showing their average values are 'zero'

A resistor cannot develop heat due to a current in one half-cycle and then cool itself by a like amount of heat during the next half-cycle! It develops heat either way whether the current is positive or negative. The lattice ions vibrate from collisions irrespective of the directions in which the electrons collide with them. Therefore, since the average value is zero, it necessitates the use of the root-mean-square values of the voltage and current to compute the power, which is a statistical measure of the *magnitude* of a varying quantity and is the square root of the arithmetic mean of the square of the sinusoidal function.

Therefore, since the positive and negative cycles of current *will* produce a heating effect, the average *power* delivered (P) will be given by $P = V^2/R = I^2R$ where $V(I)$ = the *root-mean-square* value of the voltage (current) and *not* the average value of the voltage (current). The instantaneous value of power is $v(t)\ i(t)$. The waveforms of instantaneous power and the average power are shown in the figure.

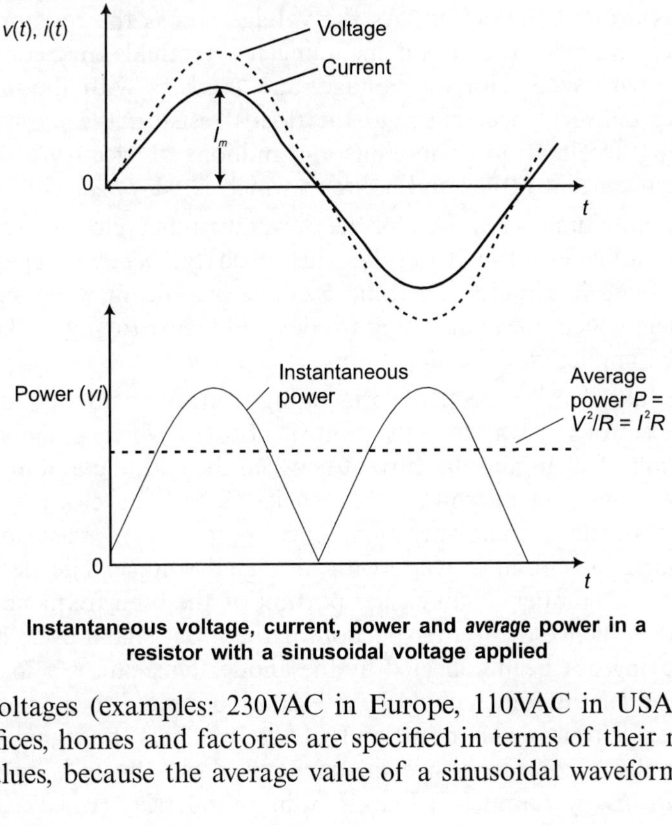

Instantaneous voltage, current, power and *average* power in a resistor with a sinusoidal voltage applied

The mains voltages (examples: 230VAC in Europe, 110VAC in USA and Japan) and ac currents in offices, homes and factories are specified in terms of their root-mean-square (V_{rms} and I_{rms}) values, because the average value of a sinusoidal waveform for one cycle is zero.

Digital voltmeters used to measure ac voltages and currents display their RMS values.

Transit Time

While the following discussion of transit time should rightly be read after an advanced course in the operation of *active* devices such as vacuum tubes and transistors, it is recommended that a cursory reading now, will give the reader insights into how deviations from the ideally approximated quasi-steady state operations affect circuit dynamics and consequently the computation of ac power of active devices.

In writing the expression of power (instantaneous power = $v(t)i(t)$) in AC circuits, we had implicitly assumed that the current is *in-phase* with the voltage across a resistor or a filament bulb. Since in the ideal approximation, a resistance does not change its value with the frequency of the voltage across it, the formulae given above for instantaneous and average power are valid and the surface charges in the quasi-steady state (set up in picoseconds) ensure that the current (the circuit being filled with mobile charges everywhere around the circuit) indeed is in phase with the voltage.

Recollect that after the very initial transient, *all* the electrons in the circuit feel the force of the electric field at the same time. The current *everywhere* in the circuit and at the resistor end terminals varies, in response to the voltage across the resistor at the *same time*. It is assumed, in the ideal approximation that there is neither stoppage nor slowing down of the electrons around the circuit. As the voltage across the resistor varies sinusoidally, the current everywhere in the circuit including its terminals and connecting wires varies sinusoidally, synchronized with the voltage, and *in-phase* with the voltage. This ensures a perfect power delivery mechanism to the (load) resistor; the *push* by the electric field and the resulting acceleration of millions and millions of electrons, all at the same time, producing a drift current all around a circuit which is *pre-filled* with electrons.

However, a finite time is required for the slowly drifting (velocities smaller than 1/200 the speed of light) mobile electrons to bridge narrow charge *depleted* regions *within* the active devices incorporated in circuits. The time becomes significant when signal frequencies are high in amplifiers which are made using triodes and transistors, and the power expressions above should be applied with care.

In vacuum tube triodes, electrons originating from the cathode (vacuum tubes are not prefilled with electrons !) travel to the control grid (to which a signal voltage is applied with the intention of changing the current between the cathode and anode) in a short time called the *transit time*. At microwave frequencies (2×10^9 cycles per second and above), the transit time affects normal operation, since the current *change* does not arrive at the anode and around the circuit *in phase* with the signal voltage. The electrons near the grid, being modulated in number by an earlier portion of the high frequency input signal cycle, come under the influence of a latter portion of the input signal cycle because of the finite transit time; having not been collected by the anode. This causes a loss of high-frequency power in the load circuit, which could be a resistor connected to the anode (imperfect high-frequency power delivery mechanism) and adversely affects the signal source connected to the triode as well. A comprehensive treatment of these effects is available in *Microwave Devices and Circuits* by, Samuel Y. Liao, Dorling Kindersley (India), 2008.

In transistor amplifiers, the electrons take a finite time, also called the *transit time,* to travel the distance across the base, which is bordered by two regions *depleted* of charges where, the input signal modulates the current in the path between the emitter to the collector. The transit time is not the same for all carriers (say, electrons of an *n-p-n* transistor) injected into the base. Had the transit time been the same for all carriers, the effect of an identical transit time for all electrons would be a simple delay in the output compared to the input signal. However, the injected carriers do not all arrive at the collector at the same time because of the different path each electron may take in the base region. The resulting difference is very small and is of no consequence in the frequency below about 100 KHz. At the higher frequencies, however, this difference becomes a measurable part of the operating cycle and the current is therefore, diminished in magnitude at the collector. Read the **important note** below Fig. 1.23, Section 1.15.

Besides, there are junction capacitances which need to be charged and discharged creating the voltage terminal conditions. During every tiny transient segment of the quasi-steady state of an input sinusoidal voltage cycle (see "The Quasi Steady-State" in Section 2.7), charges accumulate and disperse extremely rapidly in the base (emitter-base capacitance). The rates of charge and discharge were not significant at the low frequencies and the output voltage, current and input signal variations were in a sense, synchronized.

At high frequencies (above a few 100×10^6 cycles per second), due to the transit time effect (note that current is rate of flow of charge, or charge/time) the delivery (or removal) of sufficient amount of charge is not fast enough to (and from) the base capacitance and the output voltage (charge and discharge of the collector junction capacitance), output current and input signal variations, in a sense, are no longer synchronized. This causes a loss of power in the output circuit (imperfect high-frequency power delivery mechanism) and in amplification. There are other mechanisms as well, such as the interelectrode capacitances which limit high frequency output power delivery which will be discussed in Section 3.8.

A comprehensive treatment of these and other effects is available in Ref. [30].

2.21 HOW IS THERE A *VOLTAGE* (OR POTENTIAL) FOR A *SINGLE* WIRE CONNECTED TO ONE TERMINAL OF A BATTERY?

Before we answer this question, it is necessary to understand how charges spread when placed on an insulated conductor and the concept of capacitance applicable to a single conductor. When an excess charge is placed at random on an insulated conductor (Fig. 2.57a), it will set up electric fields inside the conductor.[13] These fields act on the charge carriers of the conductor and cause them to move, that is, they set up internal currents. These currents redistribute the charge in such a way that the internal electric fields are automatically reduced in magnitude. Eventually the electric fields inside the conductor become zero everywhere, the currents automatically stop, and electrostatic conditions prevail as shown in Fig. 2.57b. The potential energy after the charges redistribute themselves will be *less* than that when the charges were placed; the difference will appear as heat in the sphere or radiated away.

Fig. 2.57(a) Excess charges placed on a conductor **(b)** the charges redistribute themselves.

There were electric fields and tiny currents inside the conductor for a short duration (we will learn in Section 5.22 that this duration is called the "relaxation time") which redistributed the excess charges even though there was no voltage maintained as by a battery between any two points on the conductor. The fields of the excess charges exist, but their resultant or net inside the conductor is zero.

Next, we consider that there are two *uncharged* conducting spheres and these are separated by pulling one sphere far away to infinity. See Fig. 2.57c.

The potential of the conductive sphere at infinity may be assumed to be zero. That is a convention. Thus, we can introduce a notion of the concept of capacitance of a single conductor with reference to a spherical conductive shell at infinity.

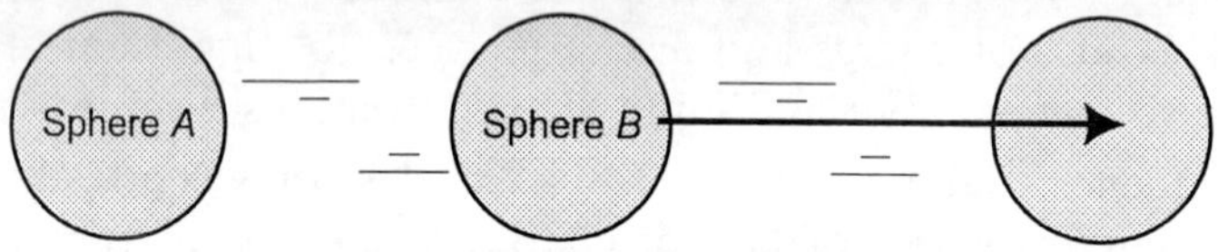

Fig. 2.57(c) Two conductive spheres near to each other are separated and one is taken away to infinity.

It is reasonable to assume that if some positive charge is placed on the conductor nearby then electric field flux lines will originate on the surface of the conductor and terminate on "some induced negative charge" on the conductive shell placed at infinity and that is how we would normally depict the electric field flux lines of isolated positive charged spheres: lines extending in all directions towards infinity. See Fig. 2.57d. It is as if the positive charge is in search of a companion negative charge wherever possible.... infinity. Thus we may consider and compute the capacitance of a single conductor with reference to a spherical conductive shell at infinity.

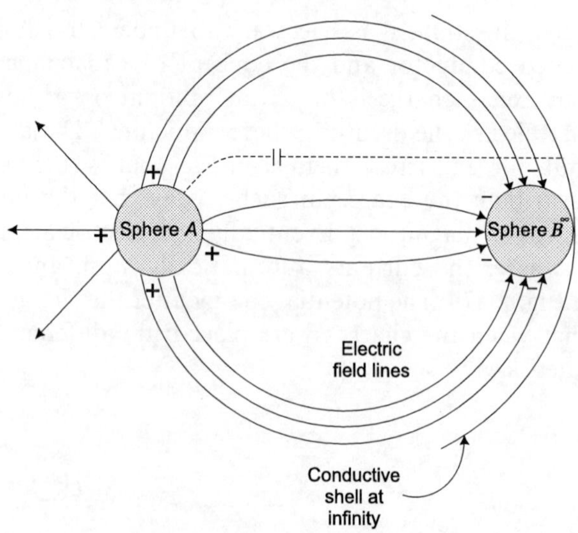

Fig. 2.57(d) Positive charges placed on the conductive sphere nearby will induce negative charges on the sphere at infinity.

All matter in the universe may be considered to be neutral. This is the natural and neutral world we are familiar with and used to in our daily life.

Whenever an electric charge is isolated, it is always in search of a companion opposite charge and an isolated charge searches for its companion opposite charge because its electric field extends in all directions in space. If it finds one, it terminates its electric field on it.

The field lines between the pair of opposite charges show the direction and strength of the electric field which is a resultant computed from the fields of the opposite charges, whose fields *always* exist in all directions in space. The "termination" of lines of a positive charge on a negative charge, is simply an "illusion" created by the *resultant field* of the two charges.

If the isolated charge were free to move, it will move towards that opposite charge and produce a neutral atom. It seems natural that this should happen, and all the observations in electrostatic experiments confirm it.

Therefore, it seems reasonable to define a capacitance for single conductors that will tell us what potential that conductor will (which it does, practically) attain if some charge were to be pumped on to it with reference to another conductor placed at infinity (of course, which we cannot). This is the next topic of this section.

Potential of a Point Charge Relative to Infinity

It is often useful to *choose one reference position* (Chabay & Sherwood, 1995, used by permission),[2] and measure the potential at all other locations relative to that reference position.

When dealing with individual point charges, it turns out to be particularly convenient to *select infinity as the reference* position, infinitely far away from all charged particles. The rationale for this is that all points at infinity (that is, very far from all charges) have the same potential.

If we go from one distant point to another distant point along a distant path, along which the electric field is negligible (proportional to $1/d^2$ and because we are *faaaa….r* from the electric fields due to dipoles whether permanent or induced and multipoles of manmade capacitors, batteries, generators and all the not so perfectly neutral atoms of all matter with their lattice ions consisting of protons and inner electrons and outer electrons of, the earth, atmosphere, planets, stars, galaxies, asteroids, comets and so on), the potential difference between two different *distant* points is "zero". See Note in Fig. 2.58 and "Is an atom a dipole?" in Appendix A.

So all locations at infinity, in all directions, are at the same potential namely, *"zero"*, and is an ideal approximation ("Ideal Approximation" in "Electronic Principles" by Paul Malvino[20]). So is the case of the isolated point charge; it is just that, an ideally approximated *isolated* point charge.

The potential due to the point charge can be computed as shown in Fig. 2.58. A test charge q_0 is moved by an external agent from point A to point B in the field set up by a point charge q. The potential difference ΔV (See end of Appendix A) between A and B is

$$\Delta V = V_B - V_A = -\int_A^B E.dl = -\int_A^B Edr$$

The field E is given by

$$E = \frac{1}{4\pi\varepsilon_0}\frac{q}{r^2}$$

Combining the above two equations, the potential difference between A and B:

$$V_B - V_A = -\frac{q}{4\pi\varepsilon_0}\int_{r_A}^{r_B}\frac{dr}{r^2}$$

$$= \frac{-q}{4\pi\varepsilon_0}\left(\frac{1}{r_B}-\frac{1}{r_A}\right)$$

Choosing reference position A at *infinity* and choosing $V_A = 0$ ***at this position*** and dropping the subscript B leads to

$$V = \frac{1}{4\pi\varepsilon_0}\frac{q}{r}$$

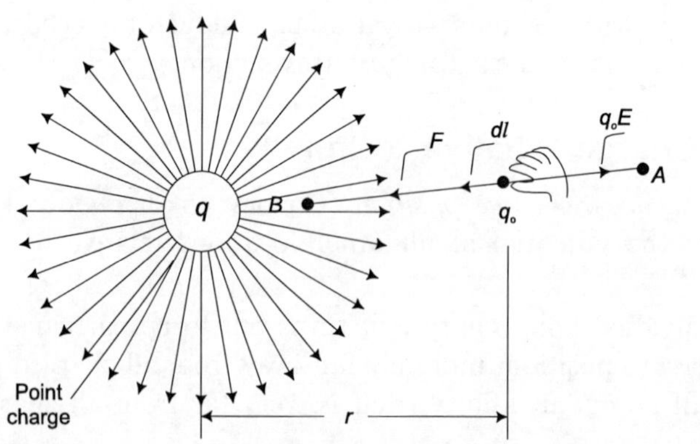

Note: The expression for $V_B - V_A$ also bears out the ideal approximation that the potential difference between two different distant points at infinity is zero because as both r_B and r_A tend towards infinity, we can replace these two variables by a single variable r and so $\left[\frac{1}{r}-\frac{1}{r}\right]$ becomes '0'.

Fig. 2.58 Potential due to a point charge.

From the above computation we see that the potential at a location r away from a point charge q, with reference to a point at infinity is given by

$$V = \frac{1}{4\pi\varepsilon_0}\frac{q}{r}$$

Using the above formula it is possible to calculate the capacitance of the earth. Refer to the figure.

Assume that the earth is a conductive sphere of radius R (=6400 km) and that it carries a charge Q. Then its potential (at the surface or inside the sphere) is given by

$$V = \frac{1}{4\pi\varepsilon_0}\frac{Q}{r}$$

The expression for potential of the earth carrying a charge Q is similar to that of the potential due to a point charge. This is because the electric field due to a sphere (the earth is assumed to be a sphere carrying a charge Q uniformly distributed over its' surface) is similar to that of one point charge Q at the center of the sphere. How this equivalence may be made is given in "A Spherical Shell of Charge" in Appendix A.

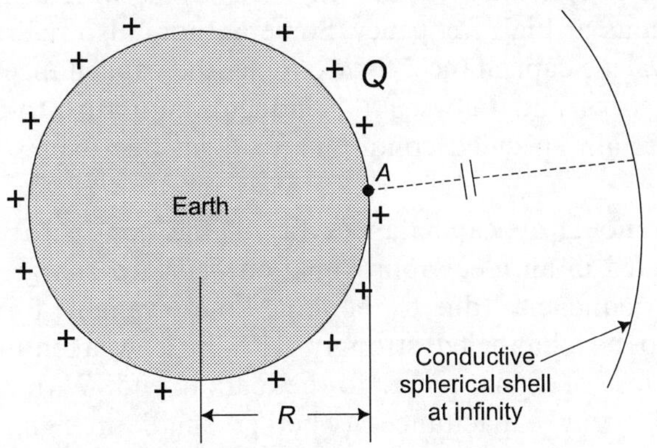

Capacitance of the earth

We now assume that a *second* conductive shell is located *at infinity and carries an equal and opposite charge Q*, therefore the capacitance of the earth is given by rearranging the above equation to give $C = Q/V$.

$$C = \frac{Q}{V} = \frac{6.4 \times 10^6 \text{ meter}}{9 \times 10^9 \text{ Newton - meter}^2 / \text{coulomb}^2} = 711 \times 10^{-6}\,\text{F} = 711\ \mu\text{F}.$$

In electromagnetism and electronics, capacitance is the ability of a body to hold an electrical charge. There are therefore, types of capacitance other than the simple capacitor consisting of two conducting plates separated by a distance, as we saw in the above example.

Self-capacitance is the amount of electrical charge that must be added to an isolated conductor to raise its *electrical potential* by one volt; the reference point for this potential is a theoretical hollow conducting sphere, of infinite radius, centered on the conductor. A few authors designate the capacitance between the plates of capacitors, a *mutual* capacitance. The term mutual capacitance also finds use in the analysis of discharge processes of orbiting satellites that are charged and high-voltage discharges from power-transmission lines. In such analyses, engineers usually designate self-capacitance as that due to charge accumulation on the surface of structural components and mutual-capacitance is attributed to charges on one structural component due to another structural component which may not be of the same shape, nor surface contour and the two components may not be aligned parallel to each other.

In Chapter 2, Fig. 2.63 the field lines of an uncharged conductor brought near a positive charge is shown. The uncharged conductor has a self-capacitance and a mutual-capacitance. For a comprehensive mathematical treatment of self and mutual capacitances of systems of conductors, or rather their geometrical shape and arrangement in space, because a conductor has a plentiful supply of mobile electrons, the reader may refer *Static and Dynamic Electricity* by William R. Smythe.

Any two adjacent conductors can be considered as a capacitor, although the capacitance will be small unless the conductors are close together for long. This (unwanted) effect is termed *stray capacitance*. Stray capacitance can allow signals to leak between otherwise isolated circuits (an effect called crosstalk), and may become a limiting factor for proper functioning of circuits at high frequency. Some authors also refer to the stray-capacitance to denote an equivalent capacitance; for example, the capacitance of an inductor between its ends, between turns and between the inductor and turn-to-core capacitance are all combined under certain specified conditions to form one expression that is taken as the stray-capacitance.

With small or no stray capacitances (as if the conductor is *very far* from other conductors or isolated in an ideal approximation), surface charges can accumulate rapidly at low to medium frequencies (due to the large times available for setting up the charges) enabling signals to be channeled strongly with little weakening of electric fields and consequently little loss of signal power to the loading end. With higher signal frequencies, in the presence of stray capacitances (whether small or significant) however, surface charges do not accumulate rapidly enough (due to the very small times available for setting up the charges) causing weak signal voltage and losses in the power transmitted due to losses in the stray capacitance charging processes. And, if the signal or applied voltage is varying, the charges of these capacitances (both self- and stray-) have to readjust their configurations to reflect the varying voltage at those points on the surfaces on wires, conductors, copper lines or whatever at the expense of time and energy with magnitudes appropriate to voltages of high-power electric circuits or low-power microelectronic circuits. We will study the effects of stray capacitances on the operation of amplifiers in the Section 3.8. See also "Links to animations articles" in the CD accompanying this book for a link to an article on the use of low k dielectrics in integrated circuits.

The concept of the capacitance of a single conductor may be extended to explain the potential of a single wire connected to a battery.

I owe gratitude to William Beaty of the University of Washington for providing the answer and which I have explained below with figures.

"Is there a voltage (or potential) for a single wire connected to one terminal of a battery?"

If the battery's net charge is zero, then for a 12 V battery, one terminal would be at a potential of +6 V with respect to a conductor at infinity, while the other terminal would be at –6 V. At the physical center in the space between the two terminals of the battery, the voltage is zero.

If we then connect a small wire to, say, the positive terminal, a few excess point charges will move onto the wire, and that wire will now be at the same potential as the battery terminal: +6 V. The wire will be surrounded by an e-field, and since the lines of electrostatic flux must always terminate upon charges, the wires *must possess an excess surface charge.*

Actually the wire upsets the picofarad values of each battery plate.

This is shown in Fig. 2.59.

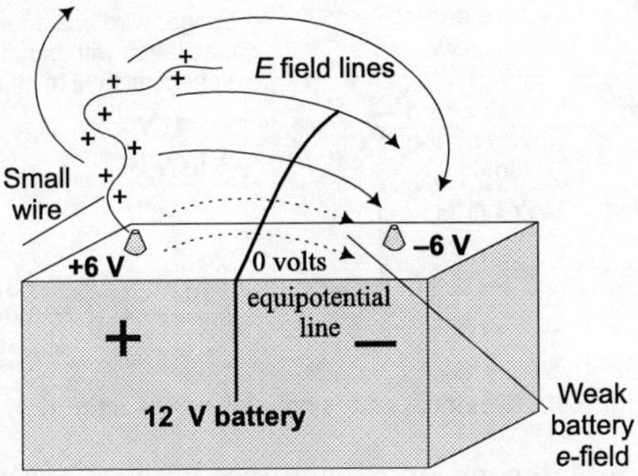

Fig. 2.59 Potential of a single wire connected to a battery.

Case 1: Treating the battery as an inert object with no net charge

We need to know if the net charge of the battery as a whole is zero. If we treat the battery as an inert conductive object hanging in space, it will have a capacitance of perhaps a few picofarads.

For such a single conductor, we assume that there exists a conductive shell surrounding it at an infinite distance as described with reference to Fig. 2.57d.

In other words, does the battery have an e-field which reaches out to infinity? Or is the battery a neutral object, where the only e-field is the weak one produced by the potential across the battery terminals?

Actually, the small wire would upset the picofarad capacitance values of each battery plate, so the voltage at the wire might only be +5.99 V, while the other battery terminal would be –6.01 V.

Effect of Picofarad Capacitance on the Potential with Respect to a Conductor at Infinity

If we connect a very long wire to the battery positive terminal, the positive battery terminal must supply significant excess charge in order to electrify the wire (electrons move from the wire into the terminal). The battery then pumps charge through itself in order to maintain the 12 V difference. When things settle down, perhaps the wire is only charged to +1 V with respect to infinity, while the battery's negative terminal is now at –11 V.

This is shown in Fig. 2.60 with different values of the potential with respect to infinity: +1V, +5.99 V and 6 V for different picofarad values of the wire.

But the battery always maintains a 12 V difference across its plates by charge separation.

Potential difference between the plates is always 12 V.

+1 V – (–11 V) = +6 V – (–6 V) = 12 V = +5.99 – (–6.01)

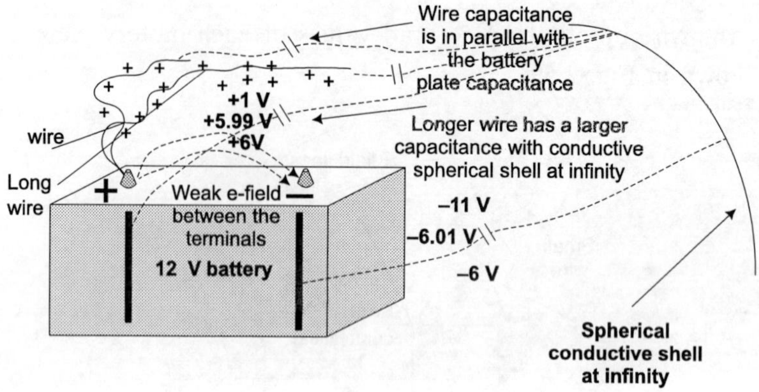

Fig. 2.60 Potential of a single wire connected to a battery.

Thus, the values would depend on capacitance with respect to earth, so it would depend upon the size of the wire, the size of the battery, and upon the proximity to any other objects such as the earth's surface.

Case 2: Now supposing we connect a long wire to the battery with a "net charge"

We assume the net charge of the battery was not initially zero. Perhaps someone with rubber-sole shoes carried the battery while walking on a wool carpet or perhaps the battery was charged to –5000 V, and would deliver a tiny spark if touched. See Fig. 2.61. (In that case one terminal would still be 12 V different than the other terminal, perhaps –5000 V on the positive terminal and –5012 V on the negative).

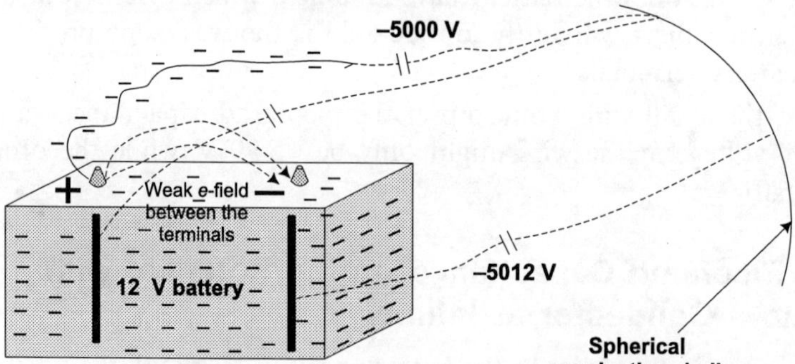

Fig. 2.61 Potential of a single wire connected to a battery having a net charge.

Or, perhaps the battery had a much smaller net charge. If we wish, we can deposit just enough charge onto both terminals, onto the battery as a whole, so that one terminal will have a potential of zero with respect to infinity (or with respect to the earth) while the other terminal is 12 V different as shown in Fig. 2.62.

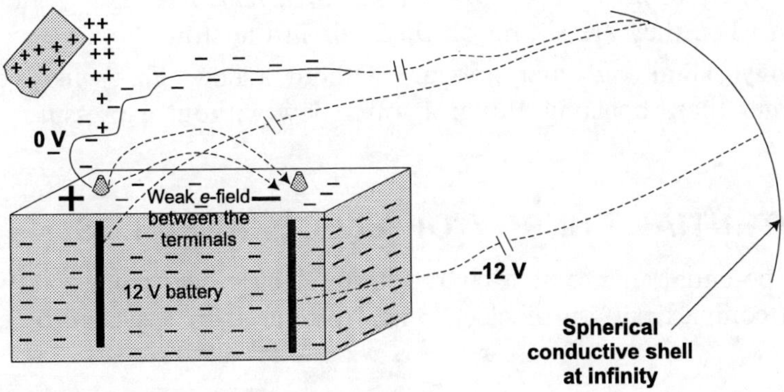

Note: Potential difference between the plates is always 12 V.
0 V–(–12V) = 12 V

Fig. 2.62 Cancelling the excess charge of the battery.

By depositing a *net* charge upon the battery as a whole, we can choose to place one battery terminal at 0V with respect to infinity. If this terminal happens to be connected to a long wire, then there will be no e-field surrounding most of that wire, and most of the wire will have zero surface charge excess.

Thus, we see that the *voltage* of a single wire connected to a positive (say) terminal of the battery will be that of the terminal, and this is due to excess charges on the wire surface that the battery will pump (in reality, electrons enter the terminal from the wire to the positive terminal and neutralize positive charges there, exposing positive lattice ions in the wire and causing the battery to pump up the depleted charges on its plates), in order to maintain its difference of potential with respect to the other (negative) terminal.

The *potential* of the wire will depend on the *state* of the battery potential (taken as a whole with respect to a point at infinity).

However, the difference of potential between the terminals and between the connected wire and the other terminal will never be different from the emf of the battery, which in the example above we have assumed to be 12 volts.

The potential of the wire itself (and therefore of the terminal to which it is connected) depends on its capacitance and the capacitance of the terminal and the amount of excess charge that is already present.

What are Currents and What is Voltage?

It is wrong to associate currents with voltage.

Currents don't have Voltage.

Voltage is *not a characteristic* of electric current. It's a common mistake to believe that a current "has a voltage" (and this mistake is probably associated with the 'current electricity' misconception, where people believe that 'current' is a kind of substance that flows).

Voltage and current are *two independent things*. It is easy to *create a current that lacks a voltage*: just short out an electromagnet coil.

It is also easy to create a *voltage without a current*: flashlight batteries maintain their voltage even when they are sitting on the shelf in the store.

Water analogy: Think of water pressure without a flow. That's like voltage alone. Now think of water that's coasting along; a water flow without a pressure. That's *like electric current alone*.

2.22 POTENTIAL ENERGY OF ELECTRIC FIELDS

How is the potential energy of a charge in an electric field gained and how is it lost? Of course, in a complex pattern of electric field produced by a system of charges.

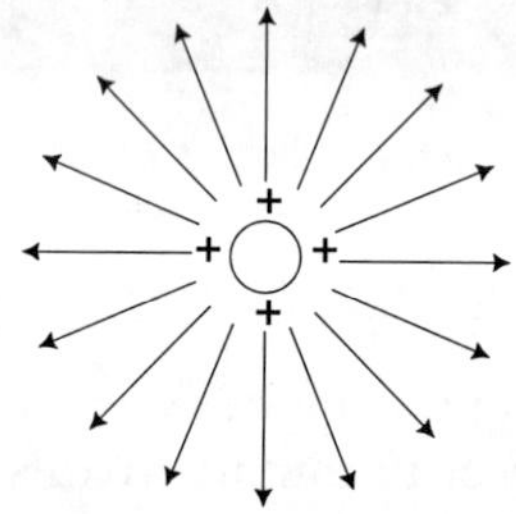

An *electric field* surrounds an isolated positively charged sphere

An *uncharged* conductor is brought near the positively charged sphere. Charges will be *induced* on the surface of the conductor and there will be a net attraction between the sphere and the conductor. See Fig. 2.63. The system did some work to cause the *separation* of charges when the conductor was brought near. The energy of the system has gone down in comparison to the energy in the field "before" the conductor was brought near the positively charged sphere. Look at it this way: If you were to stand very far away the net charge *will appear* like a *smaller* point charge. In the discussion in the topic "Induced Charge configurations" in Appendix C, it is seen that the induced charge configuration depends upon the shape of the conductor and is quite complex.

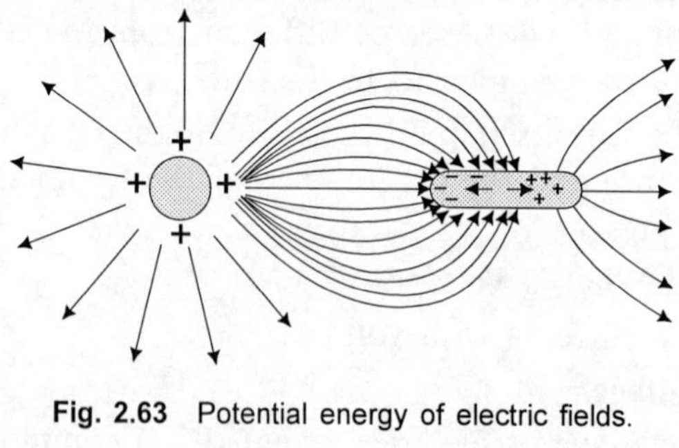

Fig. 2.63 Potential energy of electric fields.

There is a *slight loss in energy* in the conductor *during* the charge separation.

This is *carried away* in the form of heat (kinetic energy of atomic vibrations) and electromagnetic (EM wave) radiations.

Of course, the *bulk* of work done by the positively charged sphere is *stored* as electric potential energy in the *separated charges* of the conductor.

A graph of the potential of the system is plotted in Fig. 2.64a. The potential of the system *before* the conductor is brought near is shown with a dashed line. The altered potential *after* the conductor is brought near is shown with a full line. Far away from the system, the *electric field* strength will become smaller due to the polarization of the conductor. Between the conductors the electric field has *increased* due to the presence of the image charges and so the potential *decreases* more rapidly as compared to the isolated sphere.

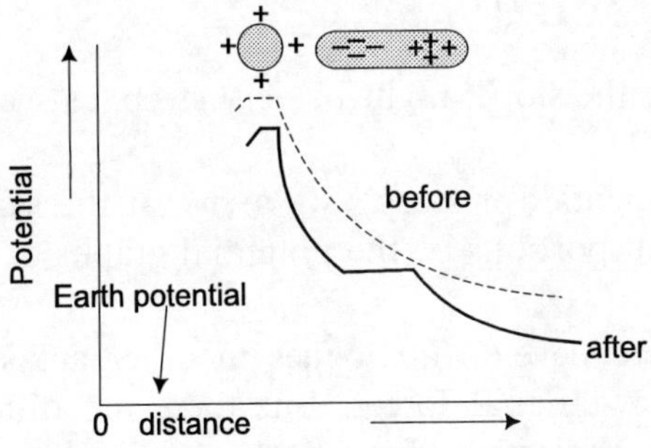

Fig. 2.64a Graph of the potential of the system.

If we move the conductor nearer the charged sphere, the potential graph will shift lower and lower and the potential near the left end of the conductor becomes sharply negative as seen in Fig. 2.64b.

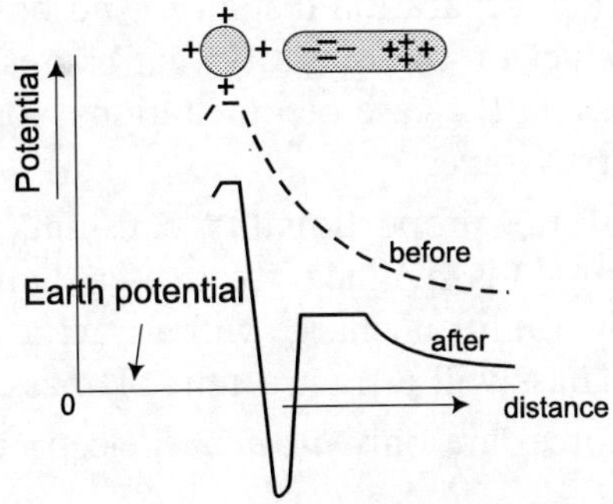

Fig. 2.64b Graph of the potential of the system.

The figure shows the electric field flux lines of the system of Figs. 2.64 comprising a positively charged sphere and the uncharged conductor brought near it. A charged metal object is like a flat-topped mountain, where the altitude of the mountain top is like the potential of the charged object, and the mountain slopes are like e-fields in the space around the object. On one side of the mountain the slope might be very gentle as the fields we see surrounding the positively charged sphere to the left, top, below, and to the right of the conductor.

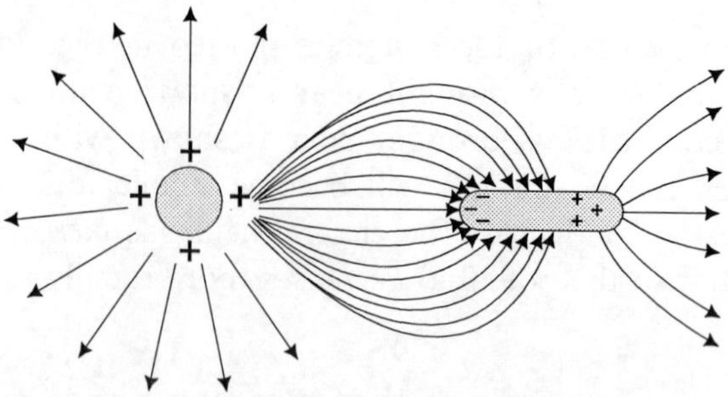

On some other side, the slope might be very steep, as seen *between* the charged sphere and the conductor.

Yet no matter how gentle (or steep), we arrive at the same altitude when we walk to the top as seen in the flat portions of the potential graphs of Figs. 2.64 for the sphere and the conductor.

Many students try to develop analogues in other areas such as mechanics between electrical forces and gravitational forces. But there are differences that complicate this. Coulomb's law equation for electrical force bears a strong resemblance to Newton's equation for universal gravitation.

$$F_{elect} = \frac{k * Q_1 * Q_2}{d^2} \quad \text{and} \quad F_{grav} = \frac{G * m_1 * m_2}{d^2}$$

$$(k = 9.0 \times 10^9 \text{ N-m}^2/\text{C}^2) \quad \text{and} \quad (G = 6.67 \times 10^{-11} \text{ N-m}^2/\text{kg}^2)$$

The two equations have a very similar form. Both equations show an inverse square relationship between force and separation distance. And both equations show that the force is proportional to the product of the quantity that causes the force – charge, in the case of electrical force, and mass, in the case of gravitational force. Yet, there are some striking differences between these two forces.

First, a comparison of the proportionality constants — k versus G — reveals that the Coulomb's law constant (k) is significantly greater than Newton's universal gravitation constant (G). Consequently a unit of charge will attract a unit of charge *with significantly more force* than a unit of mass will attract a unit of mass.

Second, gravitational forces are only ***attractive***; electrical forces can be ***either attractive or repulsive.***

2.23 THE PROCESS OF CAPACITORS SHARING CHARGES

In this section, we study the processes using the concepts that we have learned to understand how capacitors connected in parallel, share charges.

When capacitors connected in parallel share charges, the actual processes involved *before* steady-state occurs are obscured, and it is only when the *energy balance equations* are written do these *become more apparent*.

A capacitor C_1 is charged by a battery, and the battery is disconnected and the capacitor then connected to another *uncharged* capacitor C_2. (Refer to "Physics" by Resnick and Halliday).

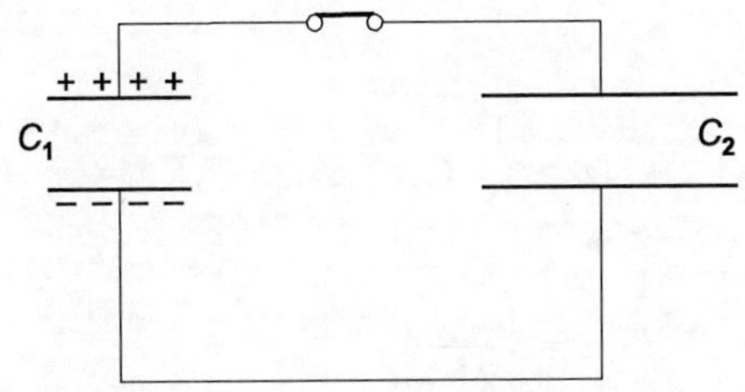

The charges *migrate* from capacitor C_1 to the plates of capacitor C_2.

How long should they migrate?

They will migrate until the *potential difference* between the plates of the respective capacitor are the *same*...when further movement stops, and that is *the rule.*

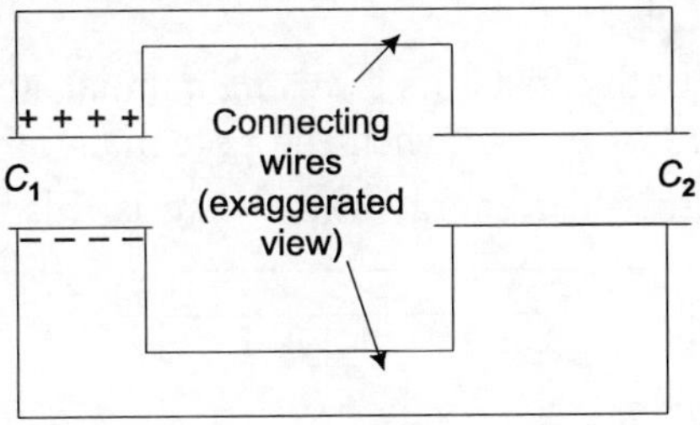

Consider the following example:

A capacitor C_1 is charged to a potential difference V_O. The charging battery is then removed and the capacitor is connected to an uncharged capacitor C_2.

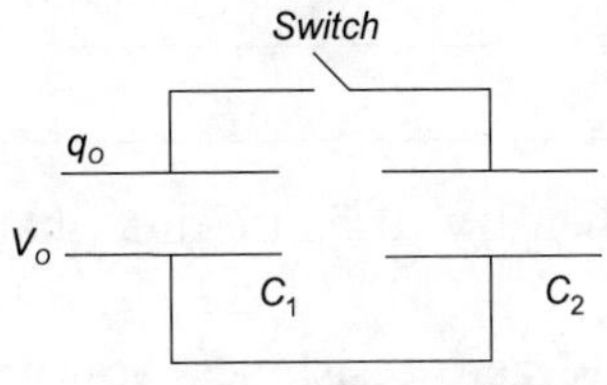

What is the potential difference V across the combination?

What is the stored energy before and after the capacitor C_1 is connected to capacitor C_2?

The original amount of charge on capacitor C_1 is q_o. After the two capacitors are connected the charge q_o is shared by the two capacitors. Thus,

$$q_o = q_1 + q_2$$

Applying the relation $q = CV$ to each of these terms yields

$$C_1V_o = C_1V + C_2V$$

or

$$V = V_o \frac{C_1}{C_1 + C_2}$$

The initial stored energy is $U_o = \frac{1}{2} C_1 V_o^2$ (see Appendix A "Energy stored in capacitors").

The final stored energy is

$$U = \frac{1}{2}\ C_1V^2 + \frac{1}{2}\ C_2V^2 = \frac{1}{2}\ (C_1 + C_2)\left(\frac{V_oC_1}{C_1+C_2}\right)^2$$

$$= \left(\frac{C_1}{C_1+C_2}\right)U_o$$

Final stored energy (in the two capacitors) = initial stored energy in C_1 multiplied by $\left(\frac{C_1}{C_1+C_2}\right)$.

Thus, *U* is less than U_o!

Final stored energy in both capacitors is ***less than*** initial stored energy of capacitor C_1. This means that there was a loss of energy when the potentials of both capacitors became equal.

What is the form of this lost energy and where is this energy?

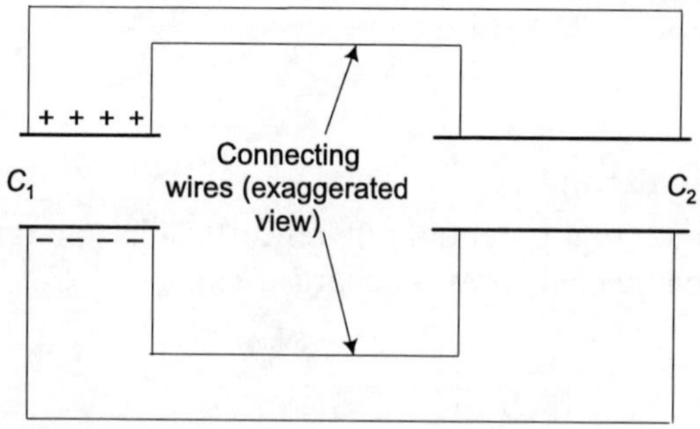

Intuitively, many of us then try to figure out what will happen if the resistance of the wire were to be *made zero.*

Assuming you can achieve zero resistance, you will find however that there is *still an impedance* due to the self inductance of the wire leads. We will discuss the phenomenon of inductance and self-inductance in Chapter 5.

Even if this self inductance is infinitesimally small it will result in the capacitor actually "overcharging" to a voltage greater than that of the battery. Think of the inductance as an *"effective inertia"* in the flow of the current.

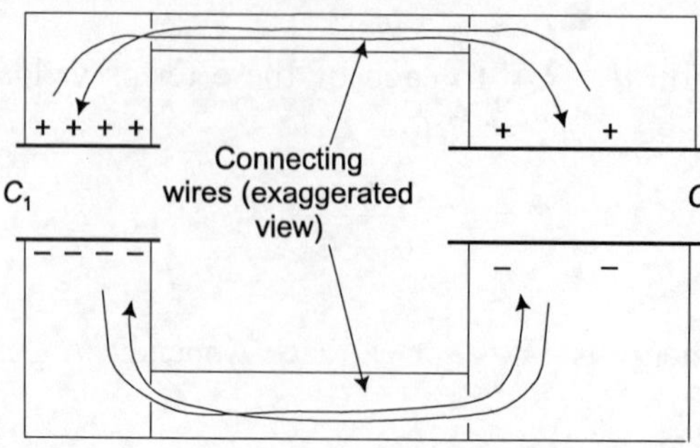

The circuit will then oscillate at a frequency √{C/L} between the initial zero charge on the capacitor and a maximum greater than the voltage of the battery.

If the circuit is not shielded the difference in energy will ***radiate away*** as electromagnetic (em) waves and the voltage of the capacitor will be a decaying oscillation centered around the battery voltage. The phenomenon of electromagnetic radiation is treated in Chapter 10.

Eventually it will settle down to the static case and you'll find the *missing energy is exactly the energy of the radiated waves.*

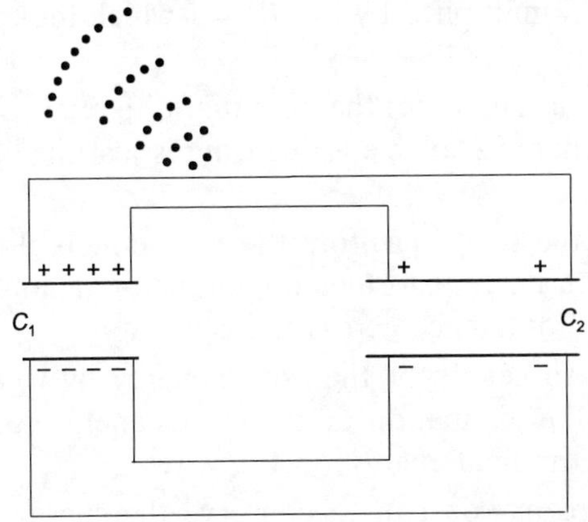

The amount of missing energy dissipated is independent of the impedance value and so is the same in the limit of zero impedance. The only variable is the *complex phase* of the impedance (how much is real = resistive and how much is imaginary = inductive).... and this tells you how the energy is dissipated, either as heat or as E-M radiation.

If the resistance is large, the missing energy is ***dissipated*** as heat in the wires!

The *em-waves/photons* of the missing energy result from accelerating charges. This is a fundamental property of a charged particle; it radiates energy when accelerated. The collisions essentially result in acceleration of the charges, and some of the energy is *lost directly* as photons or electromagnetic wave-packets.

Some of it results in energy being directed to the lattice. However, since the lattice points are also charges, they too undergo *acceleration*. There may also be collective mode of *charges sloshing* back and forth in the solid (also called *plasmons*), or lattice vibrations (called *phonons*). But the essential picture remains the same... charges are jostling around, and therefore *coupling energy out of the system* in the form of photons.

The exact spectrum of energy loss will of course depend on the geometry, material and other details of the system under consideration.

The result of all these charges wiggling around results in a spectrum that is predominantly in the infrared, and therefore they show up as *'heat'* as we commonly understand it. When you touch the body, the random kinetic energy or vibrations of all its ions is what you feel as heat (transferred to you) and if you direct an infrared detector at it, you will detect the energy of the photons in the infrared band.

I thank Dr. Dipankar, Treelabs, Mumbai (www.treelabs.org) for providing the process details.

Let us try to understand all that has happened when the *charged capacitor* C_1 was connected to the *uncharged capacitor* C_2.

In the act of connecting a capacitor charged initially with an amount of charge Q with another capacitor, we have ***created a system*** whose steady-state (final) total energy has to necessarily be *less than* the original energy we started with in the singly charged capacitor.

The formula cited earlier proved this...Final stored energy (in the two capacitors) = initial stored energy in C_1 multiplied by $\left(\dfrac{C_1}{C_1 + C_2}\right)$ which took into account the constraints: in the steady-state (after a long time) the potential difference of both the capacitors should be same, and the total charge available for sharing is just the initial charge of the capacitor C_1.

When you connect the two capacitors the **final** potential difference can only ***go down*** and ***not up*** since there is no transformer nor inductor or any other voltage source other than the initial potential difference across the capacitor C_1.

The system will try to get rid of the excess energy by whatever means available at its disposal and therefore, the dissipation of the excess energy (one may call it the difference between the initial and the final energy) in the wires.

You may try as hard as you can to conserve the energy and prevent the loss.....but there is a *poor chance* that you will succeed.

In the act of creating the system of the uncharged capacitor connected to the charged capacitor the *total final energy* possible to store in the capacitors, given a fixed amount of initial charge on one of them, *is fixed.*

The system will "shake away" the excess energy either in the form of *heat* or in the form of *em radiation* .....depending on the phase of the impedance.

This will continue until charge migration ***stops*** and the dissipation stops and *steady-state* is reached.

In the rare and impossible event that you prevent both, heat dissipation and em radiation by some means from occurring, then you will *never get* the assumed *rest state.*

The table below shows the final p.d. and the final energies for different sets of capacitor values C_1 and C_2.

Cap C_1	*Cap C_2*	*Voltage*		*Energy*	
		Initial	*Final*	*Initial*	*Final*
10 pF	10 pF	10 V	5 V	500 pJ	250 pJ
10 pF	10 nF	10 V	0.009 V	500 pJ	0.5 pJ
10 pF	10 μF	10 V	0.01 mV	500 pJ	0.0005 pJ

As the (uncharged) capacitor $\mathbf{C_2}$ is made larger and larger in comparison with the capacitor $\mathbf{C_1}$ the final p.d. becomes smaller and smaller and most of the energy is dissipated away.

When *both* the capacitors are 10 picofarad, the final p.d. is **5 volts** or *half* the initial voltage on C_1 and the energy dissipated is half the initial stored energy.

When the capacitors are **10 pf** (C_1) and **10 μF** (C_2), however, the *final p.d., is a tiny* **0.01 mV** and the energy dissipated is **99.99 percent.**

As we shall see, this is the concept behind providing **safety grounds** in electrical and electronic equipment.

2.24 STEADY-STATE AND DYNAMIC EQUILIBRIUM AND TRANSIENT STATE

The situation when the circuit was shaking away the excess energy in the form of heat or *em* waves is a *transient* state.

In many systems, steady-state is not achieved until some time has elapsed after the system is started or initiated. This initial situation is often identified as a transient state, start-up or warm-up period.

In electrical and electronic circuits these *sudden changes* are mostly found as the result of the operation of switching devices. Circuit designers use voltage regulators and surge protectors to prevent transients in electricity from affecting delicate equipment.

Steady-state is a more *general* situation than dynamic equilibrium.

While a **dynamic equilibrium** occurs when two or more **reversible processes** occur at the **same rate**, and such a system can be said to be in steady-state.

(There are many physical processes that attain a state of dynamic equilibrium. For example, when a substance boils, the vaporization process and the condensation process are occurring at equal rates. This occurs at a specific fixed temperature called the boiling point of the substance.)

A system that is in steady-state may not necessarily be in a state of dynamic equilibrium, because some of the processes involved are not reversible.

For Example: The flow of material through a system, or electricity through a network, could be in a **steady-state** because there is a constant flow of material, or electricity.

Conversely, the filling or draining of a tank with material would be an example of a **transient** state.

2.25 WHAT IS GROUNDING?

If you consider a piece of electrical or electronic equipment with a capacitance of a few picofarad has some excess charge and its potential is say, 415 V with respect to some point at infinity....and if the equipment were to be grounded by making a connection to the *earth*....whose capacitance we learned is a robust *700 odd microfarad*, the system will immediately reach a safe value of *6 microvolts.*

The process? Simple mutual repulsion of like charges and a desire to reach static equilibrium with minimum energy at rest (Sec. 2.23) caused by charge spreading over a large area using electric fields; $E_{net} = 0$ after equilibrium is reached, all in a flash!

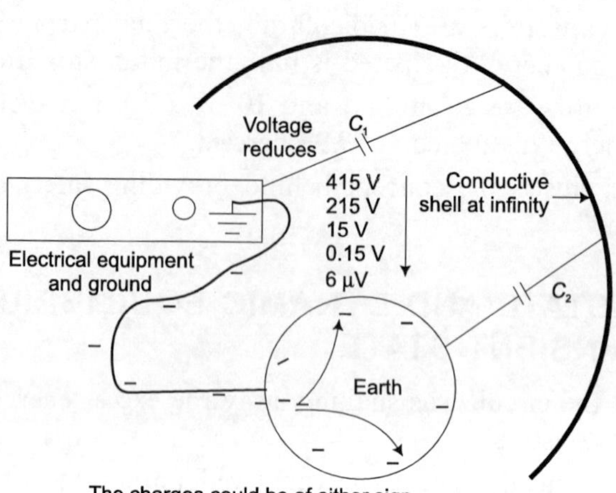

2.26 APPLYING A PULSE TO A CR CIRCUIT—ANOTHER CLOSE LOOK

Shown in Fig. 2.65 is the CR circuit of Fig. 2.15 with a "signal ground" added which will be explained shortly.

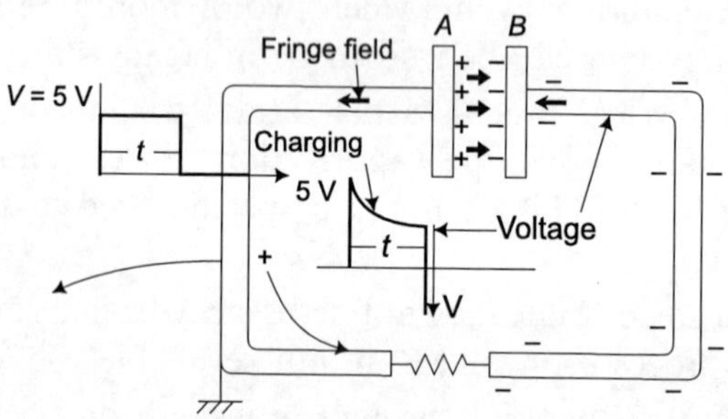

Fig. 2.65 Pulse input – voltage input is made zero.

At the trailing edge of the pulse, shorting the input voltage source (stopping the charge separator machine) makes the few positive (surface) charges on the wire from plate *A* to neutralize the few negative (surface) charges on the resistor end....zero volts or close to zero volts.

Note carefully that the electric field flux lines on this section of the wire are *thinly* spaced. This means that the field strength is very tiny here. And at one end the positive charges on the inner surface of plate *A* are tightly bound with the negative charges on plate *B*, shown by tightly spaced thick arrows indicating a powerful field strength. But that does *not change* the voltage of plate *A,* at the instant of the trailing edge transition, with respect to plate *B*.

How is Zero Volts Possible with Positive Charges on the Plate *A*?

The truth of the matter is, there was never really "zero" volts, nor a true zero potential either.

All that we began with, when we constructed the circuit was a reference voltage that we can label in the diagram with this symbol ⏚ or ⏚ and as '0' for the origin point in voltage vs. time plots. Physically it is a piece of wire, of conductive material whose bulk is neutral and as per convention we should label it "signal ground" or "circuit zero". The circuit zero is marked by a circuit analyst as the node or terminal or line in reference to which all other node voltages of the circuit are indicated or measured. The symbol is *not meant* to indicate that the point where it is shown *is connected physically to the earth*. If it is earthed, then this should be explicitly stated in the diagram.

Many students assume there is no current in the wire connected to the terminal designated circuit zero when the circuit is operational. That notion is wrong. Recollect from Section 1.18 that the electron current *in* a battery closes the circuit. It doesn't matter if the negative of a battery powering a circuit is assigned 'circuit zero' as is normally the practice; there will still be an electron current in the battery which flows from the wires connected to the battery negative terminal. Therefore, it is wrong to assume that the current terminates in the circuit zero.

It is meaningless to imagine that we can, achieve a 'true' 'zero' volts in the circuits that we build and to make voltage measurements with reference to 'true' 'zero'.

Why? Because, recollect that 'zero' potential (see Section 2.21 "Potential of a point charge relative to infinity") is defined only at a point far away from all charges where there would be no electric field and the nearest point which would qualify to be at zero potential would be at "infinity". I have not seen any engineer or scientist with one positive probe of a voltmeter fixed at a point to a circuit and a negative probe of the voltmeter with a one-way ticket boarding an aeroplane to "infinity" to fix the negative probe; and I, for one, have no intention to become the first one!

Those were in a lighter vein. But do not scoff at those who set the zero potential at a point at infinity, because scientists have devised the convention to make it easier to compute the energy stored in various charge configurations.

So, our neutral piece of wire with its millions and millions of neutral atoms is good enough to be labelled as a 'reference' and also as '0' volts. All 'other' voltages in the circuit, are measured with reference to this. And if the wire were connected to the earth (see figure below), the earth being the best charge "spreader" to all its far corners (!) as compared to a simple wire, the potential (compared to a point at infinity) of the wire when connected to it will be even closer to zero volts potential. If allowed by a large enough pulse duration time '*t*' to become fully charged, the plate *A* of the capacitor will be at the *potential* of the positive terminal of the source i.e +V volts, and the physical center between the plates of the capacitor will be at +V/2 volts (*potential*).

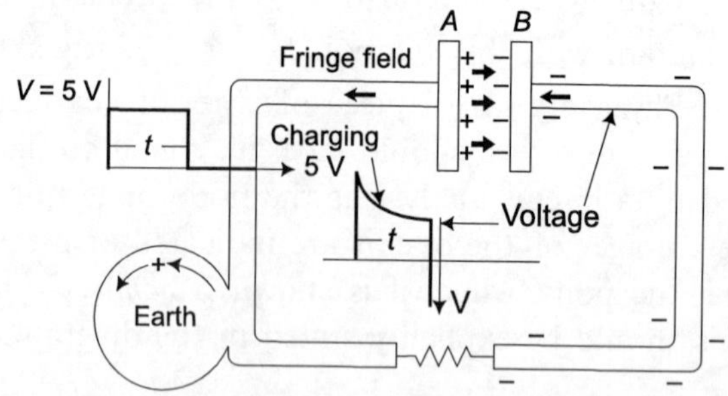

Either way, whether the source uses the earth as its reference or not, the charges on plate *B* haven't been neutralized instantly at the trailing edge of the pulse input, prevented by the resistance, and therefore, the negative value for the *voltage* at *B* (with respect to the signal ground or circuit zero).

Why does the Wire not Qualify to be labeled as 'true' Zero Potential, Since after all, all its Atoms are Neutral?

There are several reasons for the disqualification of the piece of wire connecting plate *A* to the resistor from being labeled 'true' zero volts; the foremost being the presence of tiny electric fields which are present to move the electrons when discharging. See the Fig. 2.65 and note the presence of the fringe field even as the capacitor begins to discharge. This is necessary and there we have a field in the "neutral" conductor and the attendant surface charges to enable the conduction process and the neutrality is certainly disturbed on the surfaces, and the field in the wire and the surfaces though tiny, are there, all lined up against qualifying the wire to be at 'true' zero volts.

We can make the wire thicker and reduce the potential difference across its ends but that is the best we can do. In fact, in printed circuit boards the reference lines are made using broad traces for the very same purpose.

Among other reasons, there could be stray charges produced by interfering electric and magnetic fields coupling into the circuit. This is shown by a lone stray positive charge, it could have been negative as well, in the "reference" wire.

The potential of any charged conductor is uniform throughout its interior and on its surface. If it is at 100 volts, the potential of any point on its surface and interior will be 100 volts. If it is at 0 volts, the potential at any point on its surface and interior will be 0 volts.

We now turn our attention to how the voltage of the source is made zero when the pulse input reduces to zero and the sudden transitions in voltage which occur in such circuits. Students find it difficult to visualize the rapid voltage transitions occurring at different points in the circuit especially when circuits are subjected to suddenly transiting voltage inputs like the pulse input, square wave input and so on.

Consider the circuit of Fig. 2.65. When the pulse height was *V*, the voltage of plate *A* was at *V*, say, $V = 5$ volts and when the input was shorted, plate *A* is at 0 volts, the reference terminal of the voltage source. This is a sudden transition. Before the transition, there were positive surface charges (positive) on the wires between the pulse generator terminal and the plate *A*, and negative surface charges on the wire between the resistor and the plate *B* of the capacitor. See Fig. 2.66.

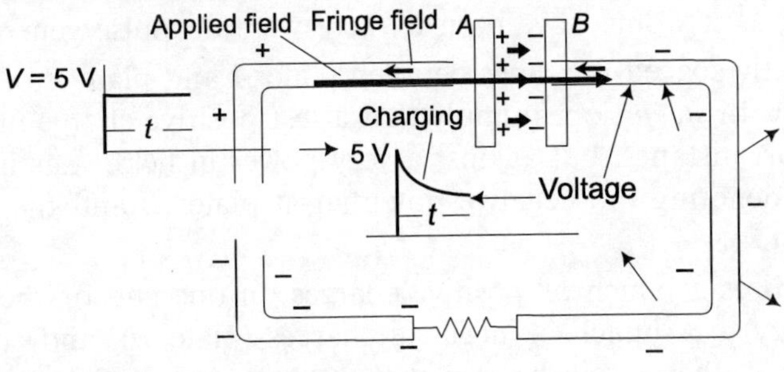

Fig. 2.66 At the instant before the input voltage is made zero.

During polarization, there were negative charges on the plate *B* and these were tightly bound to the positive charges on plate *A* and the negative charges did not allow the latter (positive charges) to move away. The field vectors will be pointing inward at plate *B*; the work done to bring in a positive charge from plate *A* would have been negative, but the charge would have landed on a potential which is still positive at a time '*t*' much less than RC.

When the pulse generator voltage became zero, the surface charges that had kept the (source) mountain flat top at voltage *V* simply disappeared, and with them the electric field (the field represents the effort the generator used in driving electrons onto plate *B* by overcoming the fringe field of the capacitor and the obstacles in the resistor), is turned off. This is another speciality of electricity; it is not possible to switch off a gravitational field easily.

As the pulse generator voltage became zero, it left behind in the circuit a mini flat top mountain which it had constructed by charge separation, of the size determined by the voltage between plates *A* and *B* with its flat top at voltage *V*. While the source mountain crumbled when its voltage became zero, it brought down the flat top of the mini mountain to zero volts (input short) with its "stiff" wedge (the resistor in the circuit) between the separated charges causing the voltage of plate *B* to go down below the reference zero volts by an amount equal to *V* (the driving field cut-off everywhere) from whatever value the plate *B* voltage was at as if a spade was used to dig out a chasm at the source mountain's base! see the Fig. 2.65.

There is a reduction in the strength of the field emanating from this section of the circuit (the wire from the pulse generator to the plate *A* of the capacitor) because the driving field is off.

The driving field did not become zero merely in the wire from the generator to the plate *A* of the capacitor; it became zero everywhere in the circuit including the wire from the generator to the resistor and the wire between the plate *B* and the resistor, with which it was driving and holding up the charges on plate *B* as well, with a strength of *V*.

So, when the pulse generator voltage became zero, what the thin electric field flux lines mean (see Fig. 2.65; like the long gentle slopes of a mountain) is that the work done by a test positive charge to move from far away towards this section of the wire is "zero"... or close to zero, as it would have been if there were no pulse generator at all.

And the tightly spaced flux lines between plate A and plate B of the capacitor (on the same mountain with *steep* slopes) imply that a test positive charge moving from plate B to A will rise a short distance, but against a very powerful field,....as if from a chasm at the mountain base, denoting the negative potential at plate B, and the test charge will reach zero volts at plate A.

If one visualizes a bunch of positive charges on one end of the wire, that is, plate A in close proximity to a bunch of negative charges (plate B), and computes the potential around the plate A, then it will be seen that the potentials will be overall much lowered at fixed points due to a thinning of electric field flux lines elsewhere and a thickening at points on field flux lines directly joining the unlike charges in comparison to the potentials which would occur at these points without the presence of the negative charges. And, when the voltage of plate A became zero, the voltage of plate B will be *negative, with reference to this "zero" of plate A* which is at the other end of a resistor that maintains the difference between the two below say, a voltage V = 2.5 volts as depicted in Fig. 2.66. If a piece of wire were in place of the resistor, there would have been a dead short across the capacitor and you would have zero volts everywhere!

Therefore, it would seem all that the pulse generator did was to separate charges (electrons from lattice ions) which were held together as if by tiny springs, by injecting energy into them against the restriction of the resistor and when the generator packed up its bags and left, the capacitor had to rely on its fringe field to begin reuniting with the opposite charges against the resistor!

Let us assume, for a moment that the capacitor was 0.1 μF and was charged to 2.5 V. Now that would mean it had a charge of $Q = 0.1 \times 10^{-6} \times 2.5$ coulombs on it. If we had the same amount of charge kept on a sphere with a radius of 1 meter, its potential (with respect to a point at infinity) will be: $(9.0 \times 10^{9} \times 0.1 \times 10^{-6} \times 2.5)/1 =$ **2.25 kV.** It is the presence of the charges of opposite sign on plate B that kept the voltage between the plates at 2.5 V.

Many of us attempt to visualize point charges as single charges and dipoles from electrostatics and the electric fields and potentials around them. There is nothing wrong with that. The potentials of single charges and dipoles is of the order of kilovolts at points very close to the charges (for example, potential is $q/4\pi\varepsilon_o r$ for point charges). A plot of potential lines is shown in Fig. 2.67 of a dipole in isolation. But note that there is a line of zero potential which means the work done to move a test positive charge from infinity along the line will be zero, and work needs to be done to move the test charge to the positive charge, and if moving the test charge towards the negative charge, the electric field does the work (you can simply let it go).

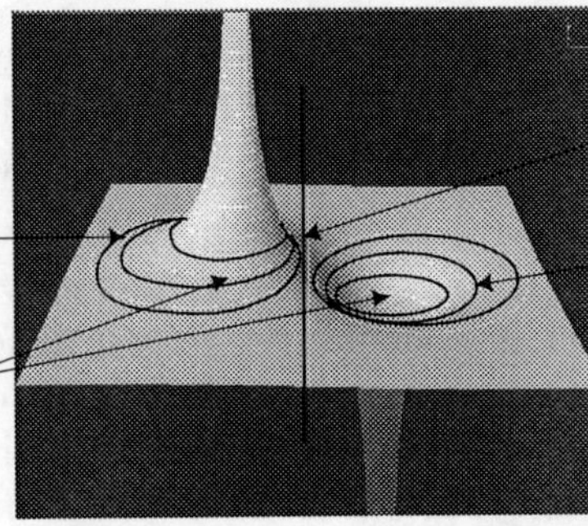

Fig. 2.67 Potential and equipotential lines of a dipole (positive charge on the left).

If the dipole charges come nearer as in Fig. 2.68, the potentials decrease at points surrounding the charges (not in between) due to cancellation of fields of unlike charges. And the voltages are at levels of tenths or multiples of volts at the same locations where they were 100's or 1000's of volts when the charges were farther apart.

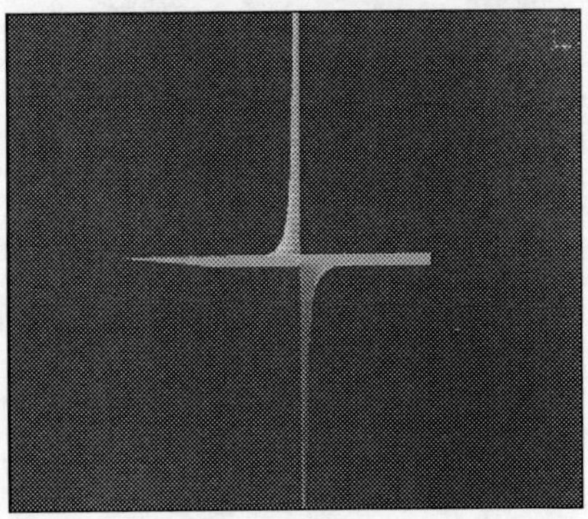

Fig. 2.68 Potential and equipotential lines of a dipole (charges close to each other).

The zero volt level midway between the dipole exists despite the presence of an electric field and does not qualify for a "true" zero potential definition, which is best defined at infinity, where we would believe there is no electric field at *all,* being *far away* from *all* charges.

Similarly, there is a *line of zero volts* midway between the plates of a parallel plate capacitor. And that is why, the capacitor may be seen as a dipole from far away!

We digress here to consider an average field inside matter, which consists of millions and millions of atoms and molecules.

The Average Field Inside Matter

According to David Griffiths'[8] descriptions in "Introduction to electrodynamics" of the fields inside matter: The electric field inside matter is quite complex at the microscopic level. Very near an electron, the electric field is gigantic, whereas a short distance away it may be small or point in a totally different direction even, due to the interactions with neighboring atoms and molecules. Moreover, an instant later, as the atoms move about, the field will have altered entirely, due to the rocking ions and darting electrons.

This true microscopic field would be impossible to calculate; nor would it be of interest to calculate. Just as, for macroscopic purposes, we regard water as a continuous fluid, ignoring its molecular structure, so also we can ignore the microscopic bumps and wrinkles in the electric field inside matter, and concentrate on the macroscopic field. This is defined as the *average field* over regions large enough to contain many thousands of atoms so that the uninteresting microscopic fluctuations near the lattice ions and close to the conduction band electrons are smoothed over. The *macroscopic* field is what people mean when they speak of *the* field inside matter, and is also sometimes called an "equipotential" region. In an aluminium metal crystal the regions of the macroscopic field would refer to the volume shown grey in Fig. 2.69.

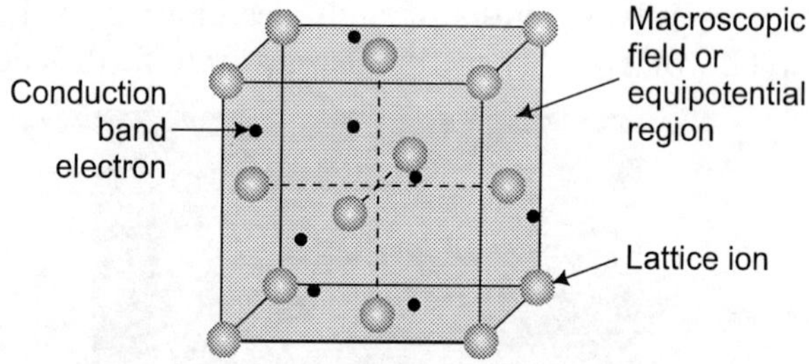

Fig. 2.69 Aluminium metal crystal structure.

While one is tempted to quickly apply the potential profiles of Figs 2.67 and 2.68 to excess charges on the plates of charged capacitors, one should remember that *capacitor plates and wires are conductors and excess charges cannot exist in them as they do in space, and* the close spacings and interactions with neighboring atoms of the excess charges cause the potentials of the conductors (on the surfaces of which we measure potentials and potential differences) to assume more moderate values.

All the ions and charges within conductors (seemingly) define a *no-man's land* that is out-of-bounds for the voltages and accompanying fields that are applied in common electric and electronic circuits; one cannot get these voltages to rip off the proton or cause the electrons (or protons) to fuse into each other. In case of metallic conductors one can visualize the conduction band electrons with their incessant movement smoothening out the wrinkles in the electric field and causing the average electric field to be zero or near zero inside, (in static equilibrium) whether *electrified or neutral* and make the interior to be equipotential with the surface.

A simple calculation will show that the potential (relative to infinity) at a location 10^{-10} meter from the centre of a singly charged ion treated as a point particle is a mere 14.4 volts and much less as one moves farther away. The potentials still rise sharply very near the charges. Besides, one is interested in the potential differences of groups of charges in conductors at far greater distances than those between say, electrons and protons, evened out by the properties of conductors which maintain a net *nearly* zero e-field in static equilibrium (near perfect field cancellation between the orbiting electrons and rocking protons noting that the electric fields of the protons and electrons exist all the time). Therefore, we speak of a "net" field that is very tiny with no significant dipoles, as though the two fields are engaged in a tug-of-war competition (both teams alive and existing though the rope barely moves!) and that cause potentials in a conductor to be the same throughout their volumes, and not the individual potentials of the electrons and lattice ions themselves.

How is it that charged conductors (isolated) and current carrying wires and resistors have a potential on the surface as well as the interior if there are no excess charges (being neutral) in the interior but only on their surfaces?

Imagine you wish to lift a rubber ball from the ground. When you grip the ball between your fingers and thumb, a force that is normal to the surface is applied, derived from the fundamental electromagnetic force of repulsion between the electrons of the surface atoms of the skin and the electrons of the surface atoms of the ball. This force squeezes the ball penetrating through the rigidity of all the atoms of the ball. The squeezing force cannot lift the ball because it acts perpendicular to the force of gravity. However, *frictional* forces (again derived from the fundamental electromagnetic force....see Appendix D for the origin of such forces) between the surfaces, then arise, and oppose the weight of the ball, the gravitational force that tends to pull it to the ground. The gravitational force is balanced by the frictional force which is a fraction (called the coefficient of friction) times the normal force due to squeezing. Finally, this balance of forces permits you to lift the object above the ground.

The absence of friction and accompanying frictional forces is what makes it difficult to pick up smooth and hard objects such as ice cubes.

The ball you wish to lift from the ground could have been made from metal and none of the processes would have changed excepting that the forces acting and balancing would be more powerful. After lifting the ball, you would not say, you have lifted only the surface contact atoms of the ball above the ground and that only these have gained in potential energy of the system (of ball and ground)! The (entire) ball (its atoms held together by its rigidity) and ground system gained potential energy.

Analogously, when a conductor is electrified by surface charges, it is the equipotential regions of the conductor with its millions of electrons that gives the whole conductor (interior *and* on the surface) the quality of potential. You would not say "only the surface" of the conductor has attained a potential.

This potential is manifested in the electric field surrounding the conductor, while the *resultant* electric field in the interior is *zero* due to the geometry of the arrangement of the surface charges that moved, which is a *natural process over which we can exercise little control.* In the interior the potential is the same as on the surface because the region is equipotential with zero resultant field; a test charge if "pushed" from the outside on the surface of the conductor needs no force to move around in the interior because there is no resultant field there, maintained by the millions of free electrons moving inside.

The electric field around the conductor extends its reach to the entire universe, seeking to induce and *pull opposite* charges towards it and *push like* charges away from it so that it may reach levels of potential energy tending towards zero, whenever possible.

If the charged conductor is isolated, the entire volume of the conductor is equipotential and if it is a current-carrying wire or a resistor in a circuit with a potential difference (see Figs. 2.7, 2.53(b) and 4.16(b)), it has *planes* (*or slices*) of equipotential regions with the potential varying along its length. The slices of equipotential regions are maintained by the millions and millions of electrons within the conductor where they are in abundant supply. The uniform applied field the surface charges impose (which are few in number in comparison with the number of free electrons in the interior), produces the varying potential because voltage (along the length) is E times distance. Significant dipoles are not formed in the *interior* of a conductor; whether it is isolated or steady-state current-carrying, due to an applied electric field. In current-carrying conductors, the maintenance of the equipotential regions is due to the continuous feedback mechanism between the current and the surface charges, as we learned in Section 1.15, Section 1.18 "Neutrality in current carrying conductors" and in Section 1.23 "Field strength inside the wires and the conservation of current".

The fields which are applied at voltages in most low voltage and electronic circuits shift the outermost electrons and hardly influence the inner ones. This behavior makes capacitors store energy*; packing* all displaced electrons on one of its plates. For a comparison of the *binding* energies indicative of the strength of force of attraction to the nucleus of innermost and outermost electrons of Copper see Appendix B "Inner electrons and the binding energies of electrons".

Returning to the RC circuit to which a voltage pulse is applied, if the input is held at V volts for a long time, then treating the circuit as a closed system (see Section 1.8 "Conservation of charge"), plate B attains zero volts with respect to the circuit ground (assumed kept *unearthed*) *and* potential –V/2 volts with reference to a point at infinity (called "potential" by convention when referred to points at infinity) because it is holding a huge bunch of negative charges. See Fig. 2.14. If the circuit ground were perfectly earthed, the plate B will attain "zero" potential if the input is held at V volts for a long time.

There will be very few or no surface charges on the wire between the resistor and the capacitor plate B, because the presence of charges on plate B has effectively canceled the driving field of the pulse generator; the net field is zero. Compare the charges and field on this section of the wire with Fig. 2.66 (during charging) and Fig. 2.15 (after charging). Electric field flux lines that terminate on the negative charges on plate B originate on the positive charges on plate A.

The charge configurations (excess surface and the bound) everywhere in the circuit determine the potentials near and on conductors and the electric field orientations near the conductors. In conductors and in wires, the potentials have to be same everywhere within; the potential will be different across low conductivity regions (resistive portions) when the charges are moving; the electric field flux lines emanating from conductive portions may be inward, outward, thin or thick at different points in one conductor depending on the charge configurations at a particular instant of time.

Putting all the above fragments together, I find the following a suitable way to visualize the potentials, their formation and destruction for the circuit of Fig. 2.15: When the pulse voltage is high, treating the (physical) center of the pulse generator (with the circuit ground *unearthed*) and as a "very close" to zero potential, imagine two mountains of concrete (representing charges separated by the pulse generator) stacked one above the other are constructed; $V/2$ volts above zero and $V/2$ volts below zero. This is shown in Fig. 2.70(a). Since the negative terminal of the pulse generator is considered to be the reference or circuit zero volts; the generator mountain is shifted above circuit zero as shown in Fig. 2.70(b).

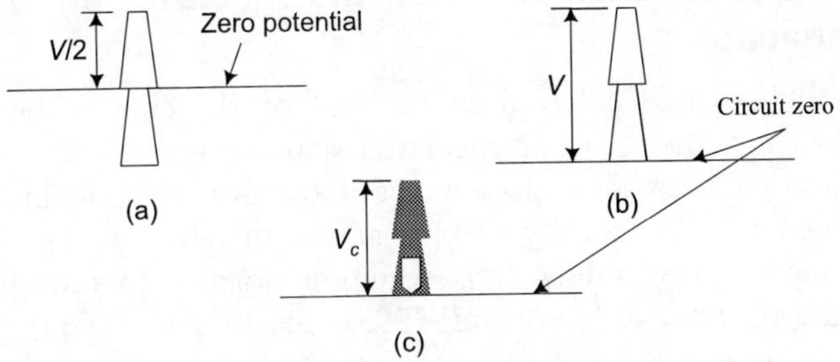

Fig. 2.70 Pulse and capacitor voltage representations for the circuit (Fig. 2.15).

While charging the capacitor, the pulse generator has created two mini-mountains stacked one above the other; one of height $V_c/2$ above zero potential (if the circuit is a neutral object as in Fig. 2.59) and one of height $V_c/2$ below zero potential upto the time t (see Fig. 2.66) to represent the voltage across the capacitor. If one were referring to the figures below Fig. 2.20 with $RC << t$, then $V_c = V$.

The mini mountain representing the voltage across the capacitor may be visualized as made from a mixture of sand and water with a spade-like base; this is indicative of the fragility in its construction so that it may be easily dissolved by a small wave, as on a beach (because it has no internal source of energy that would maintain its rigidity and an ever-present fringe field ready to discharge it, should it find a path) and the ability to use the spade to sink below zero volts as shown in Fig. 2.70(c). Again, since the negative terminal of the generator is considered a circuit zero volts the base of the lower mountain of the pair is at circuit zero as shown in Fig. 2.70(c).

The visualization using mountains is useful, though not perfect and the reader should refrain from developing analogies in other areas because negative particles (and negative potential energies of electrons near positive ions) are observed only in electricity. See Section 1.13 "Structure of electrical conductors" and Section 2.22 "Potential energy of electric fields".

Circuit references if earthed in addition to metallic chassis to meet safety regulations, could become critical in industrial environments where several electronic equipment are in operation with long signal lines carrying analog and digital data.

Earthing can result in shifting voltage reference levels caused when currents flow in the finite resistance earth wires when there is a fault (such as insulation failure between live and neutral which is earthed at the power-substation transformer), or due to other unknown causes.

These shifting reference levels not only provide inaccurate representations of measured parameters in analog/digital signal forms, also, if operations are controlled using these signal forms, unintended behavior of equipment may result in threats to safety of personnel.

Sometimes, the control equipment earth and power earth are separated to prevent strong earth fault currents from interfering with signal integrity. It is always advised to follow equipment manufacturers' recommendations regarding earthing of equipment.

Inserting a Dielectric Slab between the Plates of an Air-Filled Capacitor

This topic was deliberately deferred to the end of the chapter because to understand what happens when a dielectric or insulator slab is inserted between the plates of an air-filled capacitor requires some notion about the average field inside matter described previously. To increase the quantity of charge and therefore energy stored in a capacitor without increasing its size unduly, it is common practice to introduce a dielectric slab between its plates. (Chabay & Sherwood, 1995, used by permission).[2]

We consider a plastic insulating slab inserted between the metallic plates of a charged air-filled capacitor and connected to a voltage source. The field applied by the external charges polarizes the molecules in the plastic. The resulting molecular dipoles produce an average field inside the insulator that is in the opposite direction to the applied field as shown in the figure. The net field inside the plastic (insulator) is weakened.

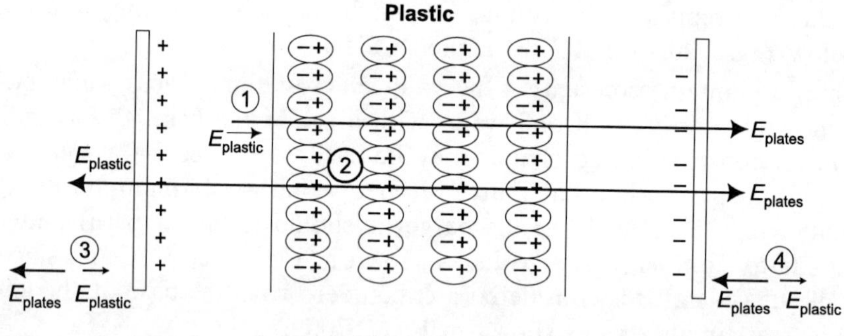

An initial glance would seem to indicate that the field at the location 1 should be more powerful than without the insulator and this is rightly so, due to the presence of the field of the polarized plastic as indicated in the figure. It is not increased much though, because the field due to the plastic is small when compared to E_{plates}. At points 3 and 4, the net field is significantly weakened by the plastic, as can be seen from the field vectors drawn at these points due to the plates and due to the plastic.

At point 2, at the molecular level, the electric field inside the plastic varies a lot in magnitude and direction, because the electric field of a dipole is complicated. So what is shown is an average field, averaged over many (millions) molecules, which is the appropriate field to use in calculating the potential difference from one side of the plastic to the other.

The direction of the field at point 2 which is *inside* the plastic, is shown pointing to the left. How do we know this? The electric field pattern of each molecular dipole is quite complicated, and it is not obvious how to take the right average over a large number of dipoles. To determine this, we use a method of counterargument.

We do know the field at some points due to the polarized plastic as shown in the figure.

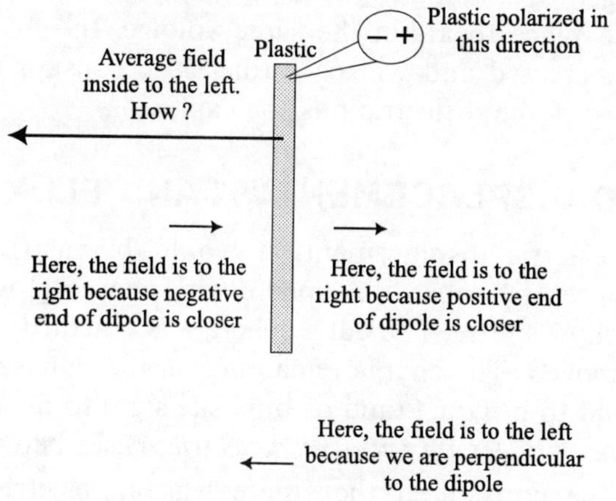

Consider the electric field *just due to the polarized plastic* along this oval path.

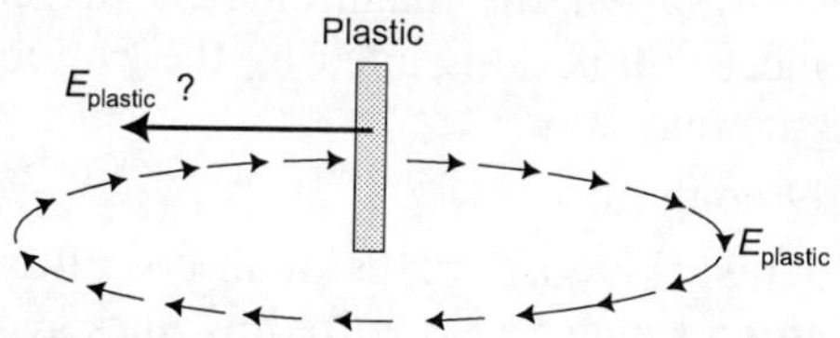

Since the molecular dipoles can be considered to be point charges, the path integral of the electric field due to the molecular dipoles *must* sum to zero. If the average electric field E pointed to the right inside the plastic, then $-\int E \cdot dl \neq 0$ along the round trip path since there are no negative contributions. This is because point charges cannot produce this pattern of electric field.

Therefore, E must have a component opposite to dl somewhere along the path for the integral to evaluate to zero. This is possible only if the average field direction inside the plastic is to the left.

If a plastic slab were to be introduced into a charged air-filled capacitor that has been disconnected from the voltage source, the energy after the plastic slab is introduced is less by a factor $1/k$ where k is the dielectric constant of the plastic. The voltage across the capacitor goes down. The *missing* energy would be felt by the person who inserted the slab, who would feel a *tug* and would have had to restrain it if he had to insert it without acceleration.

The force of attraction arises due to the free charge on the plates and the induced surface charges that appear on the slab. The charged air-filled capacitor expended some energy, the *missing* energy, to produce the charges on the slab by charge separation or induction.

See also "Links_to_animations_articles" in the CD accompanying this book for a link to an article on the use of low k dielectrics in integrated circuits.

Evidently, if an uncharged air-filled capacitor had a plastic or dielectric slab introduced and were to be charged by a voltage source, the source will need to pump up more charge with the dielectric in place to attain the same voltage. In effect, the capacitance of the capacitor has been increased and will be k (dielectric constant) times the capacitance of the air-filled capacitor if the dielectric fills the capacitor.

2.27 ELECTRIC DISPLACEMENT 'Ψ' AND FLUX DENSITY 'D'

Before we discuss electric displacement, it would be instructive to recall Faraday's experiments with concentric spheres. A sphere with charge Q was placed within, but not touching, a larger hollow sphere. The outer sphere was "earthed" momentarily, and then the inner sphere was removed. The charge remaining on the outer sphere was then measured. This charge was found to be equal (and of opposite sign) to the charge on the inner sphere *for all sizes of spheres* and *for all types of dielectric media* between the spheres.

Thus, it could be considered that there was an electric displacement from the charge on the inner sphere through the medium to the outer sphere, the amount of this displacement depending only upon the magnitude of the charge Q. In MKS units, the electric displacement or electric flux designated by the symbol Ψ is equal in magnitude to the charge that produces it, that is

$$\Psi = Q \text{ coulombs.}$$

In Faraday's experiments, there are paths of electric flux Ψ extending from the inner sphere to the outer sphere and may be visualized as lines symmetrically distributed drawn radially between the two spheres.

At the surface of the inner sphere, Ψ coulombs of flux are produced by the charge Q ($= \Psi$) coulombs and this is distributed uniformly over the spherical surface area $4\pi d^2$ m^2 where 'd' is the radius of the inner sphere. 'Ψ' is called the displacement flux.

The **flux density**, designated by the symbol $\vec{D}$, is measured in coulombs per square meter. The displacement density $\vec{D}$ may be written

$$\vec{D} = \frac{Q}{4\pi d^2}$$

Comparing this with the electric field intensity $\vec{E}$, due to a point charge of value Q in free space

$$\vec{E} = \frac{1}{4\pi\varepsilon_0}\frac{Q}{d^2}$$

we obtain

$$\vec{D} = \varepsilon_0 \vec{E}$$

The above is a simple description of the electric displacement flux Ψ and the flux density $\vec{D}$ for free space and media which does not take into account the effects of polarization of intervening media due to the free charge.

The flux density $\vec{D}$ may be viewed as a "stress" that produces the electric field and is applicable to charges in free space.

In the case where the charge Q is surrounded by media, it is necessary to include the effect of polarization of media due to the charge Q. This is made by introducing a polarization $\vec{P}$ (in coulombs/meter square) as the dipole moment per unit volume of the dielectric (for dipole moment, see Appendix A "Electric field due to a dipole").

Then, the electric displacement density $\vec{D}$ is given by

$$\vec{D} = \varepsilon_0 \vec{E} + \vec{P}$$

where $\vec{P}$ is the Polarization.

It should be noted that the units for $\vec{P}$ and $\vec{D}$ are coulombs/meters and are vector fields represented by lines while $\vec{E}$ has units of Force/charge (Newtons/Coulomb), suggesting that it is $\vec{E}$ which exerts a force on charges and $\vec{E}$ is represented by "lines of force".

Because of the difficulty in locating and measuring the millions of polarization dipoles when studying polarization effects in a dielectric, engineers and physicists use an equivalent volume charge density ρ_{pv} as if formed throughout the dielectric and an equivalent surface charge density ρ_{ps} as if formed over the surface of the dielectric. The subscript p denotes charges due to polarization. These are referred to as bound surface charge (ρ_{ps}) and volume charge (ρ_{pv}) densities. Bound charges are those that are not free to move within the dielectric material; they are caused by the displacement which occurs (forming dipoles) on a molecular scale due to polarization. See Section 1.14 "Electric Fields in Matter".

If a dielectric slab is introduced between the plates of a charged parallel-plate capacitor with an air gap on either side between the slab and each plate, then $\vec{D}$ has the *same value* in the dielectric and in the gap. $\vec{P}$ vanishes outside the dielectric and $\vec{E}$ has different values in the dielectric and in the gap.[13]

The reader is referred to topics on electric displacement flux and flux density $\vec{D}$ in Ref [13] and [16].

Gauss's law which relates the flux on a closed surface around a charge is explained in several textbooks in terms of the flux density $\vec{D}$ rather than the electric field $\vec{E}$. The statement and a geometry-based proof of Gauss's law is given in Appendix A "Gauss's law and its application".

Spherical geometries and the electric fields they produce have a few surprises in store different from even circular disks. The fields of spherical distributions of charge are easily determined by the application of Gauss's law. The reader is advised at this point to review Gauss's law and its application in determining the electric field of an infinite plane sheet of charge, symmetrically spherical distribution of charge and a hollow sphere of charge in "Application of Gauss's law to charge distributions and energy stored in fields" in Appendix A.

The reading will afford rich glimpses of the imagery of spherical distributions of charge, knowledge of which will be useful in deriving the expressions which represent the energy stored in electric fields.

While in the computation of the energy stored in electrostatic systems in vacuum, the energy stored in the electric field is given by $U = \frac{\varepsilon_o}{2} \int_{\substack{Entire \\ Field}} E^2 \, dv$ where, dv is an infinitesimally small volume in space around the charge configuration, (see "Energy stored in the fields", Appendix A), for the energy stored in *dielectrics*, the expression for the energy stored should be modified to $U = \frac{1}{2} \int_{\substack{Entire \\ Field}} \vec{D} \cdot \vec{E} \, dv$. This modification is necessary because of the polarization of the dielectric molecules and the flux density $\vec{D}$ becomes an indispensable tool for the computation of the total energy stored. For a description of the physical processes and computation of total energy see Appendix C, "Energy in Dielectric Systems".

Energy Stored in Electric Fields

In Section 2.7, we had indicated that capacitors store energy and not charge. The presence of the charge on the plates of a charged capacitor causes a strong electric field between its plates where there was no field when the capacitor is uncharged. The battery of Fig. 2.15 caused a conversion of chemical energy (by charge separation) into electric energy and this energy is stored in the electric field between the charged capacitor plates. The energy grew in strength as the field between the plates of the capacitor grew in strength when the capacitor was charging.

To gain intuitive insights into how the electric field *stores* energy, refer to "Force on a layer of charge" and "energy stored in fields" in Appendix A.

I end the section with a picture of a charged capacitor showing its fields between the plates and the fringe field (courtesy www.falstad.com) and the center zero potential line. The top plate is positively charged with respect to the bottom plate (see picture below).

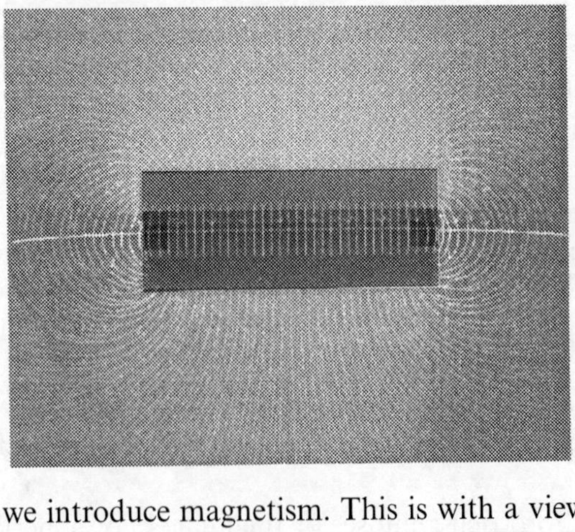

In the next section, we introduce magnetism. This is with a view to explain the concept of electromagnetic energy and how it is transported that will be treated in Chapter 4.

2.28 INTRODUCTION TO MAGNETISM

Ampere's Law

Careful experimentation shows that a *moving point charge* not only makes an electric field *E* but also makes a magnetic field *B*.

The magnetic field of a single moving point charge is not static in space.

We see that a region of space is *altered* invisibly when an electric charge is introduced, which we call an electric field.

If the charge is *moving*, the region of space surrounding the charge is *further altered invisibly*, and we call this a *magnetic field* as seen in the figure.

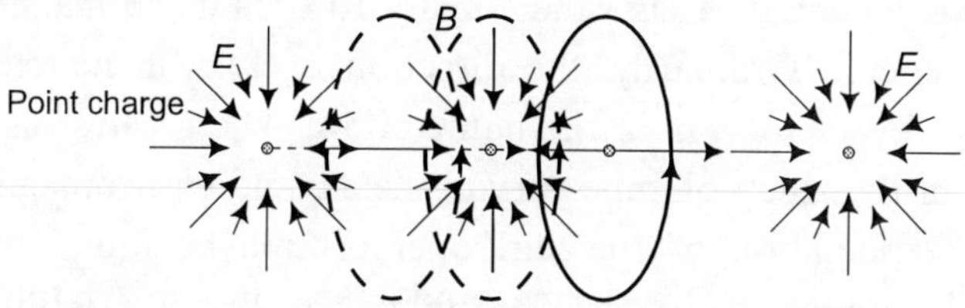

The magnetic field of a *single* moving point charge is not *static* in space.

A mathematical expression of this phenomenon is:

$\frac{1}{\mu}(\nabla \times \vec{B}) = \vec{J}$ where μ is a constant called the permeability.

This is the statement of Ampere's Law. The law is named after Ampere, though Ampere never worked with the concept of a "magnetic field". However, it is possible to *express* Ampere's formulas of the force between current elements in terms of this magnetic field (Refer Section 10.1.1, Ref. [38] and Section F.1, Appendix F]). Stated simply, a steady current density $\vec{J}$ produces a magnetic field $\vec{B}$ which has a curly pattern as can be seen in the figure.

We have introduced in the expression two new mathematical symbols; '∇' and '×'. These look understandably quite daunting. Faced with the task of expressing physical patterns of say, the kind produced on the surface of water in a pail when the water is stirred, mathematicians introduced this symbol to express the velocity ($\overrightarrow{\text{Vel.}}$) at points on the surface. The curl operator which is a combination '$\nabla \times \overrightarrow{\text{Vel.}}$' comes to the rescue and expresses such velocity patterns as a *vector sum* of the three (dimensional) velocity vector components at each and every point on the surface of the swirling water in the pail.

It is not possible to express for example, the magnetic field pattern (shown circular for simplicity) produced by the single moving point charge using the known simple algebraic (linear or quadratic) and trigonometric functions. The magnetic field does not swirl around as seemingly suggested in the figure but simply comes into *existence* in a curly pattern associated with the current $\vec{J}$. A more accurate representation using vectors would be this:

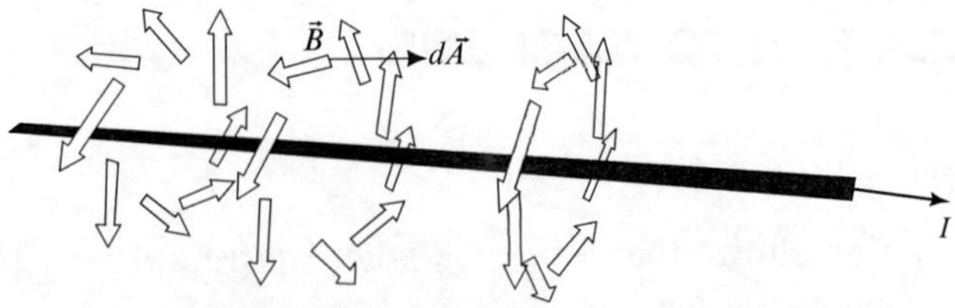

Depicting a magnetic field using vectors

In Ampere's law, $\nabla \times \vec{B}$ is the *curl* of the magnetic field, and *not* the vector field $\vec{B}$. To get the field, the vector potential $\vec{A}$ is used, defined by $\vec{B} = \nabla \times \vec{A}$. Ref. [22] gives an excellent introduction to the vector potential and how the Biot-Savart Law (Chapter 5), without the vector potential $\vec{A}$, is valid for use to obtain the magnetic field $\vec{B}$, correctly over a complete circuit. Ordinarily, the direction of $\vec{A}$ will match the direction of the current[8], as shown by vector $d\vec{A}$ in the figure above. The greatest usefulness of the vector potential $\vec{A}$ lies in the study of time-varying fields, and electromagnetic radiation.

An intuitive explanation of the curl operator maybe found in the topic "Physical Interpretation of Gradient, Divergence, and Curl" in the Chapter "Fundamentals of Electromagnetic Analysis" Ref. [16]. Edward Purcell in Chapter 2, Ref. [22] provides lucid descriptions with illustrative figures of the three operators, a reading of which will be instructive.

Magnetic Interactions

Magnetic effects are associated solely with moving charges and currents in circuits, whereas Electric effects are evident with *both* stationary and moving charges. *Magnetic fields penetrate through matter*.

Magnetic fields ***affect*** moving charges.

The magnetic force in a charge $\boldsymbol{q}$, moving with velocity $\boldsymbol{v}$ in a magnetic field $\boldsymbol{B}$, is

$$\vec{F}_{mag} = q(\vec{v} \times \vec{B})$$

This is known as the "*Lorentz force*" law.

In the presence of both electric and magnetic fields, the net force on q would be

$$\vec{F}_{mag} = q[\vec{E} + (\vec{v} \times \vec{B})]$$

SUMMARY

In this chapter, we learned an important capacitor process; that of the development of a fringe field when charges are displaced from the plates of capacitors by voltage sources. The fringe field triggers capacitor discharge and regulates the amounts of surface charges to produce a resultant field inside connecting wires. The field drops in strength as the charges reunite on the plates and therefore, the current is ever decreasing during the discharge. Electric fields are present around the circuit as well, whether a charged capacitor is open-circuited, or when discharging due to surface charges which rearrange their configurations, in what was described as a quasi steady-state.

A strong current flows when a voltage is first applied to an uncharged capacitor, and we learned that this is due to the electric field inside the wires, which penetrates the capacitor plates due to charges from the source and surface charges. This lasts very briefly. Immediately, the first few displaced charges are sufficient to begin the production of the fringe field which opposes the applied field, and alters the surface charge configuration, thus bringing down the resultant field in the wires connected to the capacitor. The action regulates the current in a quasi steady-state condition until, the capacitor is fully charged (the voltage across its plates equals the applied voltage), and the surface charges remain, to produce a resultant zero electric field in the wires thus making the current zero. There is a powerful electric field between its plates.

Beginning with the electric field of a charged thin rod, we used the principle of superposition to derive expressions for the field due to a charged rectangular plate, a charged thin circular ring, a charged disk and finally, *the fringe field of a capacitor*. The fringe field, we understood, regulates the flow of charging and discharging current in the wires of a capacitor circuit. When a pulse voltage is applied to a CR circuit, the output voltage rises sharply at first when there is no charge on the capacitor plates and then there is a fall in output voltage when the capacitor begins to charge. This occurs, we learned, because the applied electric field for a short duration before charging commences is unopposed and appears across the output resistor. Similarly, at the trailing edge of the pulse, there is sharp fall in voltage across the resistor. This occurs, because when the applied voltage goes to zero, the applied field is suddenly reduced to zero and there is a bunch of negative charge on the plate connected to the end of the resistor which makes the voltage dip sharply.

We then learned that a sinusoidal voltage applied to a capacitor causes both charging and discharging processes which makes the capacitor follow the applied sinusoidal voltage. The current in the circuit and the voltage across the capacitor is sinusoidal, and this is the sinusoidal steady-state. We also learned that if the frequency of the sinusoidal voltage is increased, the current is larger, because the charges being pushed and dragged away rapidly due to the applied field frequency variation, do not accumulate on the plates thereby reducing the magnitude of fringe field development. The end result is that the net field in the wires is stronger, therefore a larger current with increasing frequency of applied voltage. If the frequency is decreased, charge accumulation on the plates is easier and the fringe field is stronger, which in turn reduces the net field in the wires, thereby reducing the current.

We also studied how the current leads the voltage across the capacitor by 90 degrees, expressed by the reactance of a capacitor, $\frac{1}{2\pi fC}$. The reactance expression breaks down above a few hundred megahertz due to the finite relaxation time of charges.

We learned to clearly distinguish the nature of the opposition made by resistors and capacitors in circuits. While a resistor opposes the flow of mobile charges either when there are more lattice imperfections or across boundaries where there are fewer charges available, a capacitor opposes the applied voltage (field) directly by its fringe field, which is due to the charges that accumulate and form dipoles on its plates. In the relation $J = \sigma E_{net}$, a resistor affects $\sigma = qn\mu$, while a capacitor affects E_{net}. In the sinusoidal steady-state, it is the regulating effect of E_{net} that led to the expression for reactance $\frac{1}{2\pi fC}$.

With a resistor in the capacitor circuit, the current being restricted, we learned how the fringe field development process makes the current lead the voltage (across the capacitor), by an angle which is smaller than 90 degrees.

We then made the observation that it is this process that makes the CR circuit, a filter. Since slowly varying or low frequency sinusoidal voltages are slow enough to cause charge accumulation across the capacitor (associated with large fringe fields), the low frequency voltage produces voltage "drops" across it. The current in the circuit is small, and the voltage across the resistor is smaller than the signal voltage and we say, the signal across the resistor is *attenuated.*

With rapidly varying or high frequency sinusoidal voltages, rapid enough to cause little charge accumulation across the capacitor (associated with smaller fringe fields), the high frequency voltage produces large voltages "drops" across the resistor. So, the current in the circuit is large, the voltage across the capacitor is small and we say, the signal across the capacitor is *attenuated.*

When signal voltages comprising a combination of several frequencies are applied to the CR circuit, we can "tap" the low-frequency content of the signal across the capacitor even as the rapidly varying portion of the sinusoidal voltage is busy producing surface charges across the resistor causing a large voltage drop across the resistor. Therefore, the high-frequency content of the signal can be "tapped" across the resistor. These effects of CR circuits are the basic processes that make for simple low-pass and high-pass filter designs.

After we had understood the processes which enable a capacitor to charge and then discharge, which is a confirmation that electrostatics and circuits are indeed united, we applied the principles to understand how current splits in parallel branches; their strengths in the branches adjusting to magnitudes such that the charge and current conservation rules are not violated.

From these observations on the conservation of charge and of current, and the definition of steady-state, the Kirchhoff current node rule was logically derived which states "the electron current entering a node in a circuit is equal to the electron current leaving that node".

After a study of potential, potential energy of masses and charged objects, we then were introduced to a working definition of voltage "Voltage is a way of using numbers to describe the electric field".

We learned that between any two points in a circuit, the voltage is a cumulation of energy spent or absorbed as a unit charge is pushed in the field which is, of course, due to surface charges. The charges keep the cross sections of the conductor at the same potential as on the nearest point of the surface from that interior point. The field inside the conductor under consideration is always the average field (in the interatomic spaces) and we do not consider the strong electric fields near the charges themselves. Because the field due to point charges (surface charges) is conservative, we can measure the voltage between any two points of a circuit moving a test charge in *any* path between the points and the voltage will be the same regardless of the path chosen.

We then were given a brief introduction to the idea of a non-conservative field in which the round trip potential difference is not zero; these fields come into existence with changing magnetic fields. Circuit analysts and mathematicians also treat the field between the plates of a battery as non-conservative, we learned.

After studying the field configuration of a simple circuit, we learned that the potential difference computed by applying the formula $V = -\int \vec{E} \cdot dx$ is always the same irrespective of the path chosen.

We then learned an application of the loop rule which states "any round-trip potential difference must be zero", to show that the power and energy in a circuit can be estimated from a measurement of current and voltages across components. The loop rule was illustrated with circuits constructed using batteries and low and high watt bulbs. We learned to visualize the electron sea shaped in the form of tiny disks strung together that shift in the wires between the battery and the bulbs. We learned that the loss of potential energy of an electron is equal to the gain in kinetic energy, which is dissipated in collisions with the atoms (lattice ions) and the temperature of the filament wire of the bulb stops rising when the rate of energy emission (light) equals the rate of energy dissipation in the filament wire. In the formula $e\Delta V = eEL$ applied to an electron drifting in a circuit we made an important observation that while in reality an electron may not traverse the entire length L of the circuit in one operational duration of the circuit, since for every electron entering the circuit from the source (negative) a corresponding electron reunites with a lattice ion of the source (positive), it can be considered that each electron traversed the entire length L of the circuit.

We next studied several possible representations of voltage propounded by Hermann Härtel, and observed that voltage measurement is possible entirely due to surface charges. The voltmeter probes when touching circuit surfaces do not touch any charge, be it an electron or a lattice ion, and one can be certain that it *senses* the average electric field in the interatomic spaces. Next, we studied the definition and meaning of the root-mean-square value (RMS) of a sinusoidal voltage and that it signifies the heating effect produced by the sinusoidal voltage.

Before we studied the concept of capacitance of a single conductor (with respect to a point at infinity), we learned the process that makes charges move on to a single wire connected to one terminal of a battery and equalizes the potential of the wire with that of the terminal.

The concept of capacitance of a single conductor was illustrated by considering a charged sphere near a spherical neutral object on which charges are induced. The neutral sphere was then moved to a point at infinity, indicating how the electric field lines which surround charged objects in isolation, seemingly extend to infinity. This enabled us to treat the system as two charged objects separated by an infinite distance, and thus compute the capacitance. We then computed the capacitance of the earth and found it to be around 711 microfarads.

The processes which make the potential of charged surfaces lower their potential when connected with conductors with a large surface area, is called "grounding", and this is best explained by studying the manner in which a pair of capacitors share charges. Here, we learned that the energy of the system, after sharing is completed, is smaller than the energy in the beginning; the difference having been radiated or dissipated if the connecting wires had some resistance.

We then gave special attention to the aspect of assigning a circuit zero or circuit reference to a circuit for the purpose of simplifying, measuring or computing potential differences across different points and components of a circuit. We then understood clearly the differences between grounding and the circuit zero.

That the potential everywhere in the interior of a charged *isolated* conductor is the same as that on the surface is a property, which is linked with the field inside matter kept average by the busy conduction band electrons which make the field in the interior zero. We learned that it is the field outside the conductor (that is in a direction normal to the surface) which gives it the potential, and when we compute it in the interior, we are carrying forward the same potential which was at the surface. Therefore, the entire isolated conductor is an equipotential region comprising, the surface and interior.

The capacitance of a capacitor is dependent on the area of its plates the distance separating them (geometry), and the dielectric constant of the dielectric material between the plates. The molecules of the dielectric polarize under the influence of an electric field between the plates, and the polarized charges make a field that is directed opposite to the applied field. The applied voltage in turn, displaces a larger amount of charge than when the dielectric was not present, and the capacitor stores the additional energy supplied. Because of this process, we learned that capacitors with dielectrics between the plates instead of air, store more energy.

Next, we studied the energy exchange and dissipation process when a charged capacitor is connected to an uncharged capacitor and learned that in practical circuits there is always some loss of energy when equilibrium is reached.

Faraday's experiments were then described with charged spheres that led to the introduction of the quantities electric displacement and flux density. The ideas of displacement flux Ψ and the flux density $\vec{D}$ for free space and media were introduced which are more fundamental and apparent without the effects of polarization of intervening media due to *free* charge. The expression for flux density $\vec{D} = \varepsilon_o \vec{E}$ relates $\vec{D}$ to the more familiar and measurable electric field $\vec{E}$. Some engineers prefer to use the displacement density $\vec{D}$ when analyzing electrostatic configurations. The same results can be obtained by using $\vec{E}$ and the electric flux $\vec{E}.\hat{n}\Delta A$ as we learned in Appendix A "Electric flux".

At the end of the chapter, we learned that Ampere's law associates a current in a wire with a curly-patterned magnetic field surrounding it. We also learned that magnetic fields affect moving charges, and not stationary charges. The magnetic fields cause a moving charge to deflect by a force which is expressed by the Lorentz force law $\vec{F} = q(\vec{v} \times \vec{B})$.

Questions

1. Derive the expression for the force on a point charge due to a permanent dipole.
2. Derive expressions for the field components at a location x meters from the midpoint of a charged rod. Using the expressions, write a computer program and obtain the electric field components at a distance of 18 mm from the center of a clear plastic pen 12 cm long which is rubbed all over with wool. The pen acquires a negative charge of –2 nanocoulombs(nC).
3. Derive expressions for the electric field components at a location x meters from the midpoint of a charged rectangular plate. Using the expressions, write a computer program and obtain the electric field components at a distance of 3 cm from a strip of invisible tape 12 cm by 2 cm which is charged uniformly and has acquired a net positive charge of 10 nanocoulombs (nC).
4. Derive expressions for the electric field components at a location x meters from the center of a thin charged ring.
5. Beginning with the expression for the electric field of a charged ring, derive expressions for the field components of a charged circular disk and explain briefly how knowing the electric field of a ring helps in calculating the field of a disk.
6. Beginning with the expression for the electric field of a charged disk, derive expressions for the electric field of two uniformly and oppositely charged disks a small distance apart.
7. How does a capacitor discharge? Describe with the help of neat sketches.
8. A capacitor is discharged using a wire with a bend in it. Describe the processes which enable the capacitor to discharge and attain static equilibrium with the help of neat sketches.
9. What is meant by "charging" a capacitor? A capacitor is charged with the help of a battery and a resistor and a switch which is closed at a time t = 0. Describe the processes that enable charging until it is fully charged and sketch the voltage across the resistor at suitable intervals. Explain the reason for the voltage spike at the time $t = 0$.
10. What is meant by "quasi steady-state" when charging a capacitor? Explain with the help of a simple capacitor charging circuit.
11. What is the fringe field of a capacitor? How does it develop? Derive an expression for the fringe field outside a capacitor (near one of its connecting leads) constructed of two disks of radius R separated by a distance 's' with respect to the field between its plates.

Indicate typical values of fringe field common in electrical and electronic circuits and compare this with the applied voltage fields again in common electrical and electronic circuits.

12. When a capacitor is charging (or discharging) during the quasi steady-state, surface charge changes are due to a contraction (or expansion) of the electron sea of conducting wires. What is the source of the charges (current) which maintain neutrality in the interior of the conductors?
13. The expression $J = \sigma E_{NET}$ where J is the current density, σ is the conductivity and E_{net} is the electric field is a very fundamental relation. Derive the expression from fundamental principles.
14. When a capacitor of capacitance C is charged using a battery of voltage V and a switch, the battery expends an amount of energy QV while the energy stored in the capacitor (after charging) is QV/2. Explain where the missing energy is.
15. Assume we apply a voltage pulse input to a capacitor-resistor circuit and that the capacitor is reasonably large. With the help of neat sketches of the input and output voltage (across the resistor), describe the processes with emphasis on the role of the fringe field which gave the output voltage waveform the shape you have drawn.
16. The current in a capacitor circuit leads the input by 90 degrees. Describe with the help of neat sketches the processes which maintain this lead angle in the steady-state. Is it possible to achieve 90 degrees lead angle between the current and the voltage accurately in a typical capacitor circuit. Explain giving reasons.
17. Explain briefly the "sinusoidal steady state" of a circuit with a capacitor.
18. Describe the effects of varying the frequency of a sinusoidal voltage to a capacitor with respect to the current and its phase. Around what frequency does the reactance expression breakdown? Give reasons.
19. A capacitor resistor combination is subject to a sinusoidal voltage. With neat sketches explain how the current *in the circuit* leads the voltage by an angle which is less than 90 degrees. If the frequency is varied, what effect does this have on the current and the phase with respect to the voltage?
20. What are surface charges really? How do these enable a circuit to split the current in parallel branches? Explain with the help of neat sketches.
21. State and explain the current node rule (or Kirchhoff node rule). Sketch a portion of a circuit with a node and branches and label it with currents. Write expressions of the relationship between the currents in the steady state.
22. What are conservative fields and non-conservative fields? Explain with reference to their curl and divergence. Can a non-conservative field be produced using charged objects?
23. With the help of a simple circuit consisting of a battery, resistor and wires, explain what the voltage is due to and state where it exists. What is the nature and strength of the electric field inside the wires and in the space surrounding the wires? Are they uniform?

24. State the *loop rule*. Sketch a simple two-loop circuit with resistive elements and explain how the rule can be applied to determine the electric field strength inside wires and components. What data will you need about circuit elements to be able to determine these fields?
25. How much force is normally applied by a hammer and chisel? How much force is normally applied when we lean against a wall? Do the atoms of voltmeter probes touch the atoms of wires on which they are pressed to measure the voltage? Explain mentioning the values of nominal force/stress involved at the atomic level.
26. Can a capacitance be associated with a single conductor? Explain with sketches and obtain the capacitance of the earth assuming that the radius of the earth is 6400 km.
27. Can there be a voltage or potential for a single wire connected to a battery? Explain assuming 3 cases; when the battery has no net charge and when the battery has a net charge and when the excess charge on the battery has been neutralized.
28. Explain with a neat sketch how *large regions* within a metal crystal which the conduction band electrons occupy are maintained equipotential despite the huge potentials near individual protons and electrons. What is the "average field" within conductors and where is this field?
29. Explain how the potential is the same as the surface in the interiors of charged isolated conductors and current-carrying wires and resistors with a potential difference.
30. Explain what is meant by "Grounding" and the difference between "signal ground" or circuit zero and a circuit ground which is earthed.
31. Assume we apply a voltage pulse input to a capacitor-resistor circuit (of a small RC time constant) and that the resistor connected to the generator ground is circuit ground. The output voltage is picked across the resistor. Explain how a negative spike voltage is observed at the output corresponding to the trailing edge of the pulse input.
32. Explain how the electric field inside matter may be considered to be an average field. How does the average field of insulating slabs inserted between the plates of an air filled capacitor increase its energy storage capacity of capacitors? Explain with the help of neat sketches.
33. Will a capacitor whose leads are brought out from the inner surfaces of its plates, charge when connected with a current limiting resistor to a dc voltage, despite the fact that a strong field that penetrates through matter will come into existence as the capacitor charges? Describe the process and compare with the process of charging a capacitor whose leads are brought out from the external surfaces of its plates.

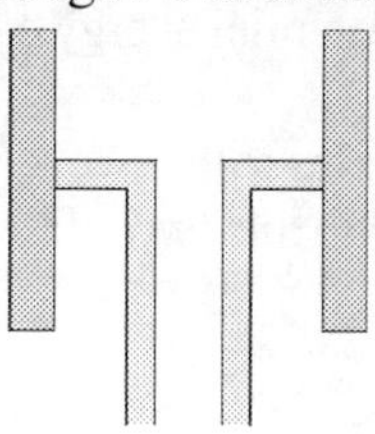

Hint: The round trip path integral of a conservative electric field (due to a distribution of point charges) should always be "zero".

34. With reference to Fig. 2.65 explain what is meant by circuit zero. If it is necessary to "earth" the circuit zero how should this be indicated in a diagram?
35. Is it necessary that the circuit zero of every circuit be earthed? Explain briefly how and in what circumstances this may cause undesirable effects.
36. What do safety regulations stipulate with reference to providing an earth terminal? Which part of the equipment enclosing the circuit is earthed? Explain the reasons.
37. Sketch a simple two-loop circuit with a voltage source and mark the circuit zero in it. Do the branch currents terminate in the circuit zero? Explain what happens to all the currents flowing into the circuit zero.
38. Why is the root-mean-square voltage and current used to indicate the voltage and current in AC circuits? Explain.
39. Describe qualitatively using waveforms the three ac representations of a sinusoidally varying voltage i) instantaneous, ii) average, and iii) root-mean-square (rms).
40. Explain why the transit time of electrons does not affect the power delivery mechanism at high frequencies in resistors.
41. In what devices is the transit time for electrons observed? In which sections of these devices do electrons take a transit time and why do they take a transit time to traverse these sections?
42. At what frequencies and how does the transit time affect the power delivery mechanism in i) triodes, and ii) transistors.
43. State Gauss's law and write two equations of it, one in terms of the flux ψ, and the other in terms of the electric field $\vec{E}$.
44. Explain briefly how a uniformly charged spherical shell carrying a total charge Q is equivalent to placing one point charge Q at the center of the sphere.
45. Using Gauss's law show that the field of infinite plane sheet of charge is $\frac{\sigma}{2\varepsilon_0}$ where σ is the surface charge density in coulombs/cm^2.
46. Show that the field inside a "hollow" spherical charge distribution with a surface charge density σ coulombs/cm^2, is zero.
47. If the fields on either side of a charged layer with a surface density σ coulombs/ cm^2 are E_1 and E_2, then show that the force on the layer of charge is $\frac{\sigma}{2}$ $(E_1 + E_2)$. What will be the force on the surface of a sphere with uniform surface charge density σ? What is the direction of the force?
48. Derive the expression for the potential energy of a system of charges $U = \frac{\varepsilon_0}{2}\int_{Field}^{Entire} E^2\, dv$
49. What are lossy dielectrics? Describe two effects that can cause a loss of energy in dielectrics. Which one among the two effects causes energy loss by storing energy? Explain.

50. How does one feel an electrical shock? What makes an electric shock fatal in humans? Is it voltage or current? Explain giving range of voltage and current.
51. Why does a charged insulator such as a balloon rubbed in one's hair not give a shock? Explain.
52. The inner shell of radius a of two concentric spherical shells carries a charge $+q$ and the outer one of radius b carries a charge $-q$. Assuming that the charges are distributed uniformly over their surfaces, show from i) the field of the total assembly, and ii) using the fields of individual shells that the total energy contained is $\dfrac{q^2}{8\pi\varepsilon_0}\left(\dfrac{1}{a}-\dfrac{1}{b}\right)$.
53. State Ampere's law and describe qualitatively the relation between current and the magnetic field of a current-carrying wire.
54. Describe the curl operator in mathematics and explain its physical significance.
55. How are magnetic effects associated with charges? Explain.
56. How do magnetic fields affect charges? Explain.
57. The electric field around a charge configuration can be determined if the potential (a scalar function of position) around the configuration is known from the formula $\vec{E} = -\nabla V$, i.e the electric field is given by the (negative) gradient of the potential V at a point around the charge configuration. i) Explain the reason for the minus sign in the formula ii) What is the direction of $\vec{E}$? Hint: See Sections 2.3 "Gradient of a scalar function" and 2.4 "Derivation of the field from the potential" in Ref. [22].
58. In Fig. 2.53(b), assuming magnitudes and assigning directions for the field set up by surface charges near and at distant points (such as the charges on the bends and on the battery) at the point X, explain qualitatively how the formula $\vec{E} = -\nabla V$ for the magnitude and definition of direction of $\vec{E}$ is consistent with the direction of the electric field depicted at point X. Hint: See Figs. 1.5 to 1.11.

Additional practice questions are available in the CD with this book.

3

The Principle of Superposition in Circuit Processes

Chapter 3 was developed with a view to describe how superposition of ac signals on dc voltages is achieved. In this, I have brought in the role that surface charges and fields play in developing the output voltage of a capacitive coupling network. Beginning with a description of superposition of dc voltages in simple capacitive circuits, the chapter progresses to describe how ac signals are coupled to transistor amplifier inputs.

The graphical analysis of circuits using *VI* characteristics and network load lines is treated in the next section.

Circuit analysts, routinely apply the principle of superposition, to analyze circuits and networks and this is treated in one section, which also deals with aspects of ideal voltage sources that simplify the analysis.

Transistor amplifiers, which strengthen weak signals, use a biasing network which sets the operating point of the transistor with an emitter resistor and a bypass capacitor across it, to enhance stability. One section in this chapter, provides a qualitative description of the function and operation of the bypass capacitor, from the electric field viewpoint.

The next two sections deal with, the application of the principle of superposition, to explain the simplification of complex transistorized circuits into ac small-signal equivalents, and to explain the effects of stray capacitances which limit transistor frequency of operation. The final section, deals with the use of the principle of Superposition in the operation of voltage shunt feedback amplifiers, a knowledge of which is important to understand the principle of operation of a class of linear integrated circuits namely, operational amplifiers.

3.1 SINUSOIDAL VOLTAGE APPLIED TO A CHARGED CAPACITOR AND RESISTOR

When designing amplifier circuits, circuits that require the superposition of a small sinusoidal voltage on a steady dc voltage (which biases transistors in the active region), are used.

How is this Achieved?

Consider a capacitor-resistor circuit in which the capacitor was previously **charged** to a voltage of say, 2 V dc, by an external agent.

A sinusoidal voltage $v(t) = 0.5\sin(\omega t)$ with a peak $V_m = 0.5$ volts, is applied to the CR combination, as shown in Fig. 3.1.

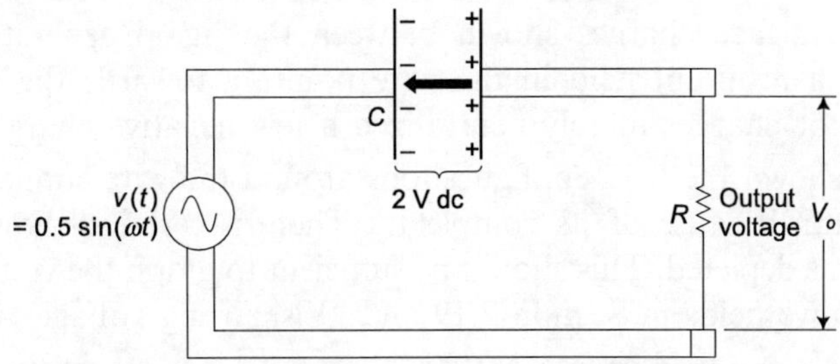

Fig. 3.1 Sinusoidal voltage applied to a charged capacitor and resistor.

Before the sinusoid voltage generator is connected, surface charges appear, positive charges on the wire connected to the resistor R, and negative charges on the wire connected to the negative plate of the capacitor. The charges form a tight net, all over the circuit and there is static equilibrium and *no current*, because there is no signal generator as shown in Fig. 3.2.

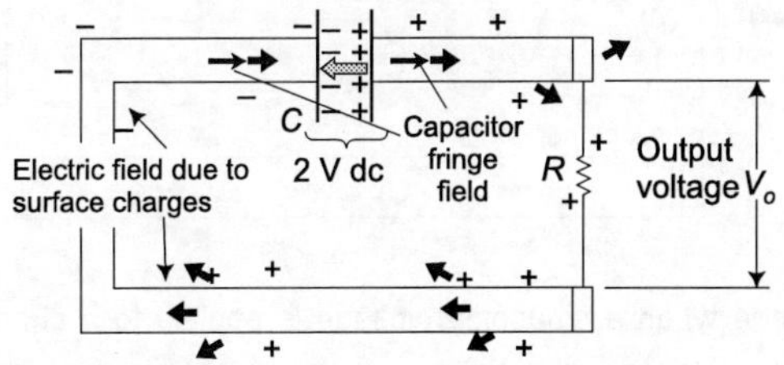

Fig. 3.2 A charged capacitor and resistor (before a sinusoidal voltage is applied).

Now, when the positive half of the sinusoid is applied, positive surface charges move on the wire to the negative plate of the capacitor, to neutralize the charges, and negative surface charges move to the wire connected to the resistor, resulting in a new configuration of the electric field, as shown in Fig 3.3.

The electric field (which penetrates matter) due to the signal penetrates through the capacitor plates and adds to the fringe field (remember this is s/2R times the field between the plates as was described in Section 2.2).

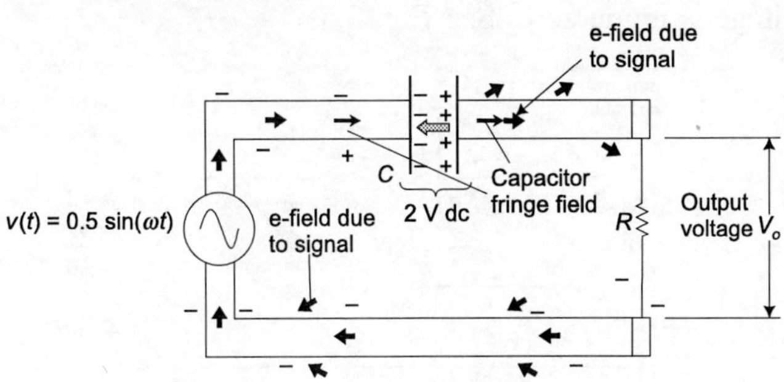

Fig. 3.3 Sinusoidal voltage applied to a charged capacitor and resistor.

The net field in the wire from the capacitor to the resistor *R*, is now, signal e-field + fringe field.

The negative surface charges shown between the signal generator and the capacitor plate C, produce a resultant field in the wire pointing towards the capacitor plate, since the signal generator charges merely neutralized a few negative charges on the wire.

We have not shown the field configurations around the wire connecting the capacitor to the signal generator, because of its complexity. There is certainty however, in the resultant fields in the wires as depicted. This should be sufficient to graph the voltage across the resistor *R*. Refer conservative fields in Section 2.19 and "Visualizing voltage" in Section 2.20.

The resulting large e-field causes, surface charges to increase in the wire connected to *R*.

The output voltage is shown on the right-hand side in Fig. 3.4.

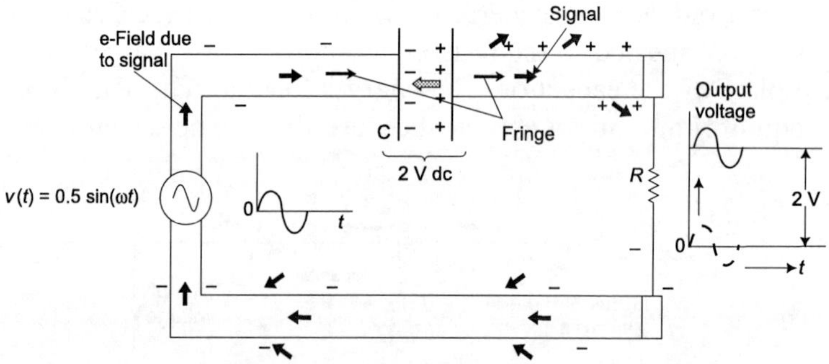

Fig. 3.4 Output voltage when a sinusoidal voltage is applied to a charged capacitor and resistor.

What is the Circuit Attempting?

To understand the behavior of the circuit to a complete sinusoid, let us for a moment assume that the signal is very tiny, or better, close to zero or, even better, at zero voltage. But before we do, let us remove the source for a moment, and assume the capacitor charged to 2 V with polarity as in Fig. 3.2.

Charges will move, from the positive plate onto the wire and negative charges from the negative plate onto the wire **before** the voltage is made zero.

Now, the voltage is made zero. See Fig. 3.5a.

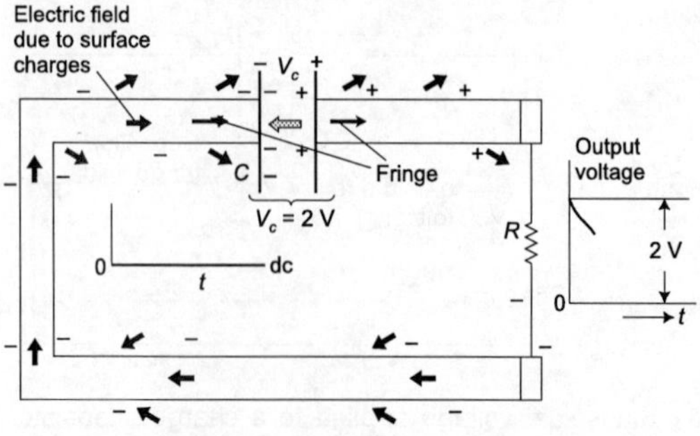

Fig. 3.5a The discharge of a charged capacitor.

Observe that the correct way to depict zero voltage is NOT to open-circuit but to the signal source with a short-circuit.

The surface charges and electric field the capacitor and resistor set up, cause a discharging current *inside* the wires and resistor. The output voltage decreases exponentially towards zero, as the capacitor discharges, and the resistor heats up. Finally, all the charges in the capacitor will be neutralized and the output voltage reaches zero, and the fields reduced to zero everywhere, as shown in Fig. 3.5b.

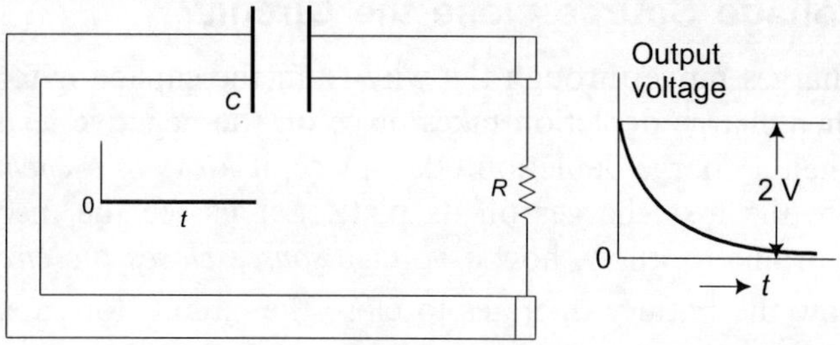

Fig. 3.5b The discharge of a charged capacitor—after a long elapsed time.

Next, we consider the application of a tiny 0.2 V to the capacitor charged to 2 V and let us make it steady, say, a dc voltage.

Positive surface charges move out of the positive terminal of the voltage source and negative surface charges move out of the negative terminal, according to rule. (See Section 1.15)

The net electric field is raised sharply everywhere in the wires and the resistor, and the voltage across the resistor rises, not to 2 volts, but to 2.2 volts—the capacitor has not begun discharging yet. See Fig. 3.6.

The effect of the addition of the source voltage (0.2 V) is felt around the circuit by a zip in the electric field (not forgetting that an electric field is an invisible *alteration* of space). The capacitor will have to discharge first, before it assumes the voltage of the source and reach static equilibrium. This is what the circuit is attempting.

A dc voltage applied to a charged capacitor and a resistor

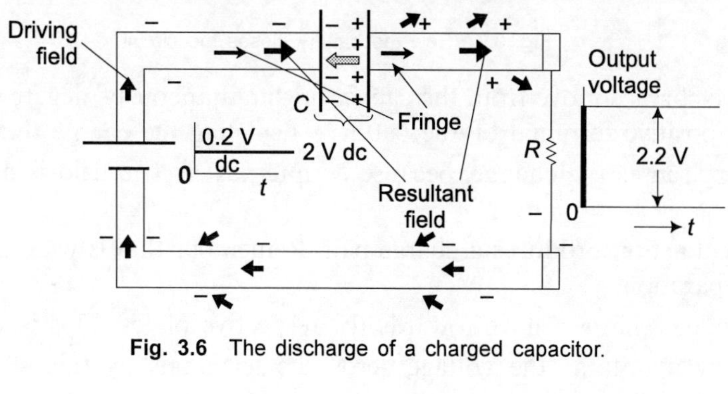

Fig. 3.6 The discharge of a charged capacitor.

What is the Discharge Path of the Capacitor?

The discharge path is *through the resistance and through the voltage source*!

The voltage source maybe a battery, which is two conducting plates separated in a container containing an electrolyte—and the chemical action had separated charges—negative charges on the negative plate and positive charges on the positive plate.

How can a Voltage Source close the Circuit?

When positive charges move through the wire from the capacitor to the negative terminal of the battery, then charge depletion takes place on the negative terminal. The *duty of the battery* is that when a charge depletion takes place, it *kicks in a chemical action*, inside the battery, to restore the lost charges on its plate. Let us see the mechanism closely, once, because it is important to know, how a voltage *source closes the circuit,* any circuit.

Let us see how the battery operates to close the circuit, for capacitor discharge.

For each and every charge that leaves the capacitor, a corresponding one enters the battery terminal, and that is the real movement. But, that does not diminish in any way our depiction of the movement. The Fig. 3.7 exaggerates the movement of positive charge from the capacitor plate towards the negative terminal. In reality, it is not the same positive charge, which left the capacitor plate, but another positive charge in the wire close to the battery negative terminal.

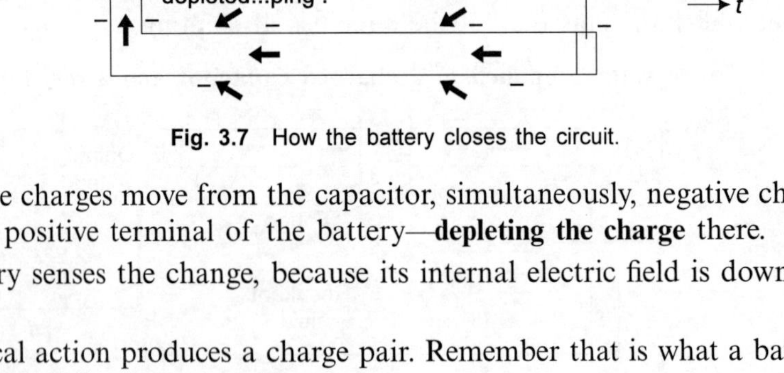

Fig. 3.7 How the battery closes the circuit.

As positive charges move from the capacitor, simultaneously, negative charges will have moved to the positive terminal of the battery—**depleting the charge** there.

The battery senses the change, because its internal electric field is down, and operates now.

Its chemical action produces a charge pair. Remember that is what a battery is *capable* of—charge separation.

The individual charges move towards the respective plates. This is shown in Fig. 3.8.

The battery maintains the voltage across its terminals by this chemical action that replenishes depleted charge.

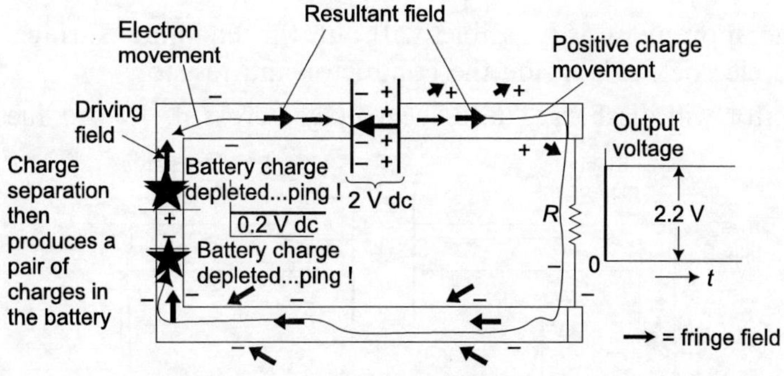

Fig. 3.8 How the battery closes the circuit (contd...).

Let us look at the movement closely, with all the frills removed—the surface charges, the fields. We show the resultant field (which is a combination of the driving field and the fringe field components) only. Read the text below Fig. 2.10 about the net field as a result of the vectorial addition of fields.

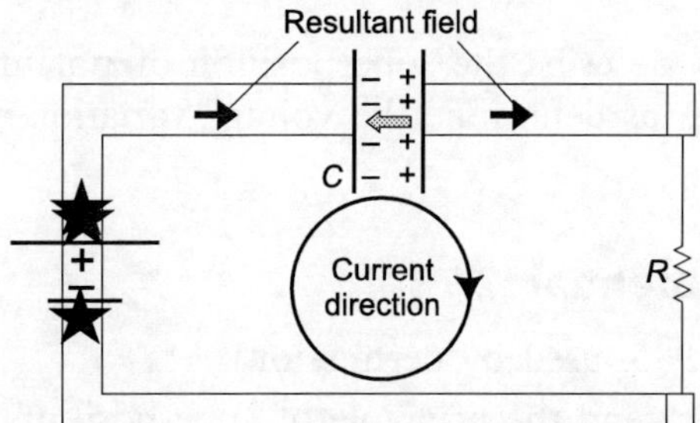

Looking closely, what we see is an upward movement of positive charge, and a corresponding downward movement of negative charge inside the battery—both of which correspond to a *conventional current* which *is upward*, and that is how the *circuit is completed.*

Of course, there cannot be any movement of charges in between the plates of the capacitor, but for the sake of completeness, we say that there is a *displacement current*, all fictitious, but good for bookkeeping.

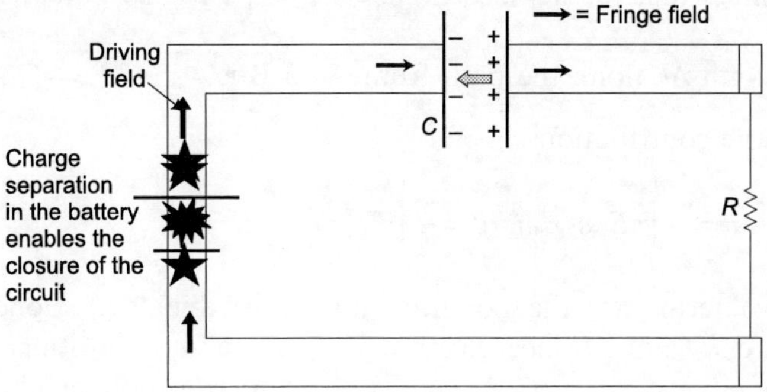

No charge movement is possible without the enabling surface charges with the accompanying electric fields inside the conductor and resistor.

The capacitor will discharge, until the voltage across it, V_C becomes 0 volts.

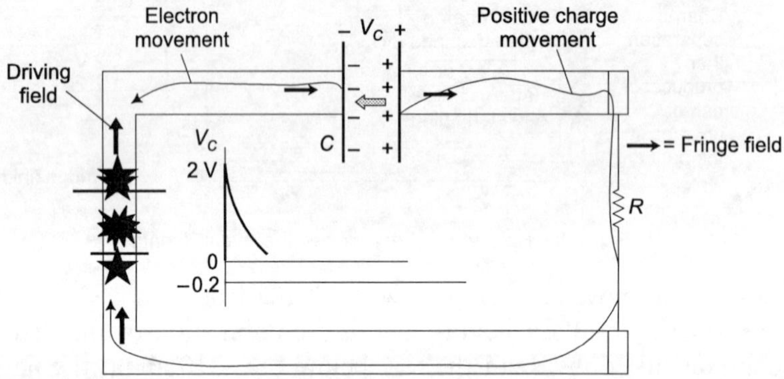

It will then *charge*, until the voltage across it reaches 0.2 volts viz. the voltage of the source.

We briefly digress to describe the superposition of potential principle at work behind the scenes, which enables us determine the voltage variation with time, at different points around the circuit.

Superposition of Potential

(Chabay & Sherwood, 1995, used by permission)

It is not difficult to extend the principle of superposition of electric field to prove the superposition of potential. There are two charges $+q_1$ and $-q_2$ located a distance apart, and we have to compute the potential difference between two points A and B.

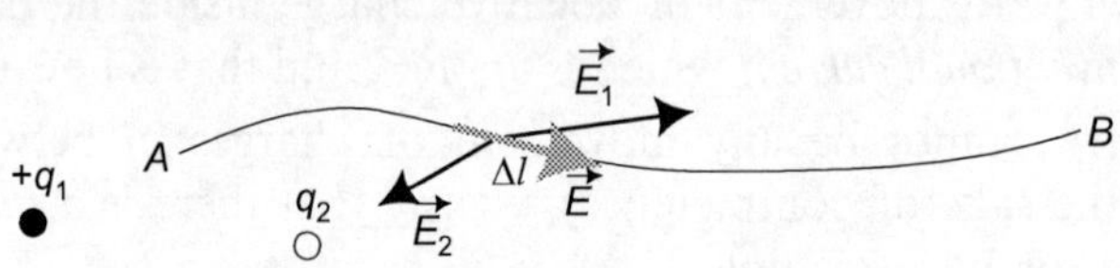

The key point is that the net electric field $\vec{E} = \vec{E}_1 + \vec{E}_2$, so potential difference for each small step of length Δl along the path from A to B is $-\vec{E}\cdot\overrightarrow{\Delta l} = -(\vec{E}_1 + \vec{E}_2)\cdot\overrightarrow{\Delta l}$

Adding up the contributions

$$\Delta V = -\int_A^B \vec{E}\cdot dl = -\int_A^B (\vec{E}_1 + \vec{E}_2)\cdot dl = -\int_A^B \vec{E}_1\cdot dl - \int_A^B \vec{E}_2\cdot dl.$$

The last two terms are the potential difference due to q_1 alone, and the potential difference due to q_2 alone. Since electric field is the superposition of the electric field contributions of all the charges (Appendix A "Superposition of electric fields"), we see that potential difference is also the superposition of the potential difference contributions of all the charges. It is easier to add potentials than it is to add fields, because potential is a scalar, and so is potential difference, while electric field is a vector, which requires vector addition with the use of vector components.

Batteries in Series

Two batteries in series can drive more current through a wire than one battery can. The potential difference across two batteries is 2 emf, and this doubles the electric field everywhere in the circuit. Double the electric field, double the drift speed, double the current (if the mobility doesn't decrease due to a higher temperature). By a simple addition of the emf (potential) of two batteries, we obtain the emf of both the batteries when connected in series.

The principle of superposition of electric fields leads to the Principle of superposition of potential differences or voltages which finds wide application in the circuit analysis of complex electric and electronic circuits. We will learn more about these techniques in Section 3.7.

In the circuit comprising the battery, capacitor and the resistor, we have been using the fields due to the charges on the battery, surface charges and the fields due to the charges on the capacitor plates, implicitly when applying the superposition of potential principle to graph the voltage change with time across the capacitor and resistor.

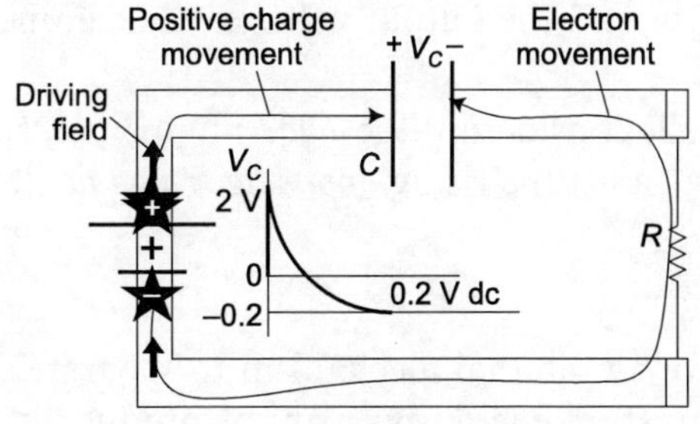

Returning to the circuit, the voltage of the capacitor 0.2 V is shown below the zero line in the graph of the capacitor voltage, because we have defined the polarity of V_C that way (see the polarity assumed for V_C in Fig. 3.5a). Also, note the capacitor initial voltage is 2 V, and not 2.2 V, which is the initial voltage of the output. The **output voltage** decreased **exponentially** from its initial 2.2 volts. The action is shown again in the figures below with the output voltage also plotted alongside for the sake of clarity.

As the voltage across the capacitor reaches zero volts, the voltage across the resistor will be 0.2 volts, under the influence of the electric field due to the voltage source.

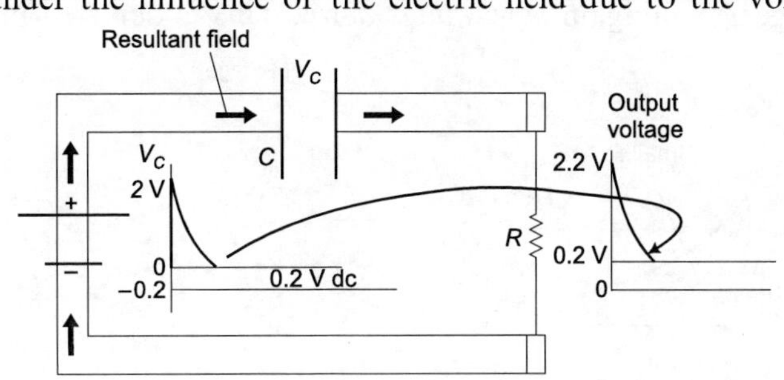

The voltage source now begins to charge the capacitor and V_C therefore, becomes negative.

The voltage source will charge it up until, no further movement of charge is possible....a state of *static equilibrium*.

The voltage across the resistance *R* becomes zero; static equilibrium.

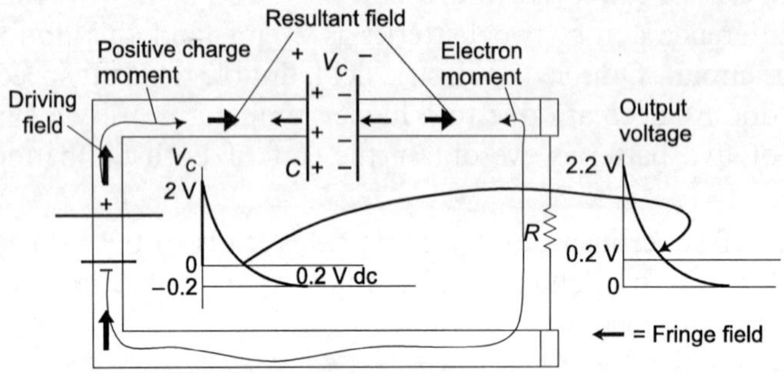

Did you notice that the **current never reversed** during the entire cycle! yet, discharge and charge were built into the unidirection movement.

In all the processes that we saw, only the change in the net field in the wires occurs at nearly the speed of light, and the charge transport (drifting velocity) and the *voltage change* across the capacitor and the output voltage occur in measurable times ranging from nanoseconds to seconds.

That is to be expected, considering the ***huge*** number of charges, millions and millions which have to be moved against the rising *opposing fringe fields* due to charge accumulation.

A Reminder

A capacitor does not have an internal mechanism to separate charges, it is just that simply, a passive device that may store energy but cannot produce it.

This is unlike a battery or a voltage source which have internal mechanisms to effect charge separation and maintain the voltage across them.

This also means that the *sizes* of the capacitor and the resistance will influence the voltage *decreasing times* across the capacitor and the output voltage. Because $Q = CV$, then for a fixed V, a small capacitor will have less charge than a large capacitor.

If RC is large, then it may be expected that huge amounts of charge have to be removed—so the time is prolonged—we say the time constant is large. Even if C were small, then a large R will again *slow things down*. This is depicted in Fig. 3.9.

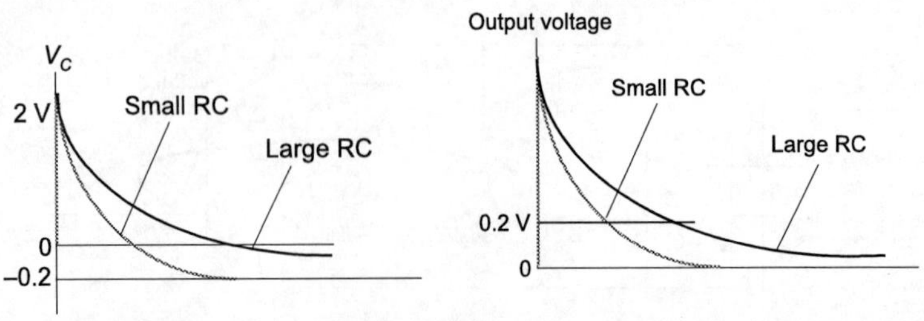

Fig. 3.9 How the sizes of the capacitor and the resistor influence the charging and discharging times.

We depicted the voltage source in the circuits with a shorting line across the supply terminals. This was done merely to emphasize the fact that the circuit has a closed path.

In case the source is the winding of a transformer the winding may be considered a closed path as well, and the induction process causes surface charges to appear to develop the so called induced emf or voltage.

In either case, the *depletion* of charges at their terminals *causes* the charge separation mechanism to kick in; chemical action in the case of batteries, and induced emf action in the case of transformer windings—and the circuit is completed.

3.2 APPLICATION OF A SMALL SINUSOIDAL VOLTAGE TO A CHARGED CAPACITOR

From the viewpoint of what was seen with the application of a small dc voltage to the CR combination, let us take a look at what happens when we apply a sinusoidal voltage of amplitude 0.5 V with the capacitor charged externally (not shown to 2 V dc). This is shown in Fig. 3.10.

To understand intuitively the operation of the circuit, we depart from convention and show positive excursions of the voltage with a battery whose voltage varies sinusoidally, and negative excursions with the battery whose voltage varies sinusoidally, connected in reverse; the e-fields and the surface charges are not shown for the sake of clarity now, but they are present in the circuit to enable the processes.

Positive Half of the Input Sinusoidal Voltage

The movement of charges (discharging cycle) is clockwise as per convention during the positive cycle of the sinusoid. Also refer to Fig. 3.6 for a confirmation of the direction of charge movement where a small dc voltage of 0.2 V dc was applied to the circuit.

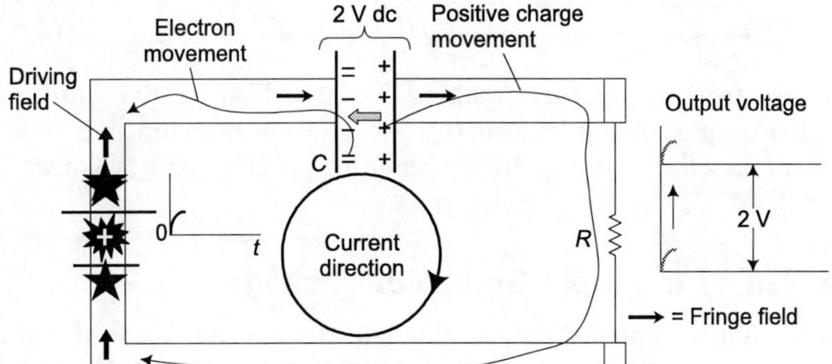

Fig. 3.10 The discharging cycle with a charged capacitor.

We make one assumption though that the CR product is large, specifically, we make the capacitor large.

We could have as well made the resistance also large ….and that will not affect the subsequent operations, as we shall explain now.

Because CR is large, *not enough charges* have moved away from the capacitor plates during each cycle of the sinusoid which could have caused the voltage across the capacitor to reduce to zero. At the end of the positive cycle therefore, there are still charges, quite large, in fact, left on the capacitor plates.

Note that the capacitor is still charged with a positive voltage. It lost only a few charges while discharging during the positive cycle of the input. It is therefore, at a voltage less than 2 V dc but still a little close to 2 V dc, at the end of the first positive cycle. The large circle in the middle of the diagram, is drawn to indicate that conventional current is clockwise in the circuit.

Accordingly, the output voltage is a sinusoid riding on the 2 V dc (nearly) as shown in Figs 3.10 and 3.11.

This will happen if the sinusoid is fast enough, or technically speaking RC>>T, the period of the input....and further, the amplitude of the signal 0.5 volts is small, in comparison with the initial voltage of the capacitor 2V dc, so its influence is small.

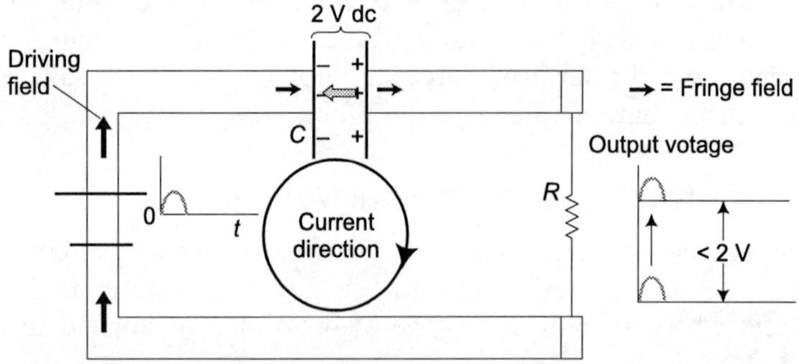

Fig. 3.11 The discharging cycle with the positive half sinusoid applied.

One can visualize the process of making CR large, by *imagining a tall and large diameter tank* connected from a tiny outlet near the bottom to a small pipeline, wherein the pipeline may have *small* flow fluctuations, but the liquid in the large tank will always drain into the pipeline slowly.

Negative Half of the Input Sinusoidal Voltage

We reverse the battery connections, to simulate the negative cycle of the input as shown in Fig. 3.12 and remember this is the first cycle, and also that the input peak voltage is a mere 0.5 volts.

The driving electric field reverses direction. Notice that there is *no change* in the directions of the movement of the charges from the capacitor plates; the positive charges on the capacitor (which is less than 2 V and for simplicity we assume 1.8 V) but close to 2 V, move towards the weaker positive battery terminal. And similarly, the negative charges (electrons) move towards the negative terminal.

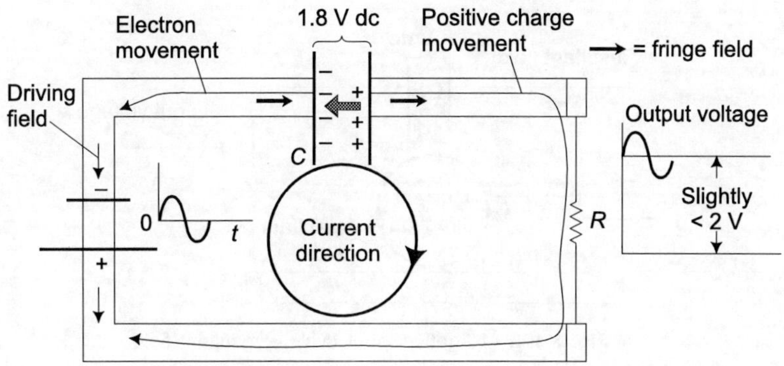

Fig. 3.12a The negative input cycle.

We now enter into one of the most *crucial phases* of the operation of the circuit, which is the current direction during the negative cycle.

During the negative cycle, note that the current in the battery continues in the *same direction*. One way to visualize, without cluttering the mind with visions of the chemical processes involved, is to open a separate line inside the battery providing a path for the charges in motion.

The charges continue through the battery. The battery acts as a *policeman* keeping a watch on its terminals and *separating or combining charges* as and when necessary to *maintain its emf* according to its nature.

During the negative cycle, there *will hardly* be any replenishment by the battery, because the charges moving in from the capacitor plates have a sign that is the same as the sign of the charges on its (battery) terminals.

Most of the charges will use the bypass route, and the battery's *replenishing mechanism* is not called upon to do much.

But should there be movement at all from the capacitor plates to the battery terminals? Because it *appears* as if, negative charges are moving from the negative plate of the capacitor towards the negative terminal of the battery, and similarly positive charges are moving from the positive plate of the capacitor towards the positive terminal of the battery.

And, the answer

There will be a movement of charges, and these are moving no doubt from a positive to a positive potential and from a negative to a negative potential. But, note that positive charges are moving from a larger positive potential (capacitor) towards a *smaller* positive potential (battery) and negative charges (electrons) from a *more* negative (capacitor) potential towards a *smaller negative potential* (battery). See the resultant field and its direction in the wires, as shown in Fig. 3.12b.

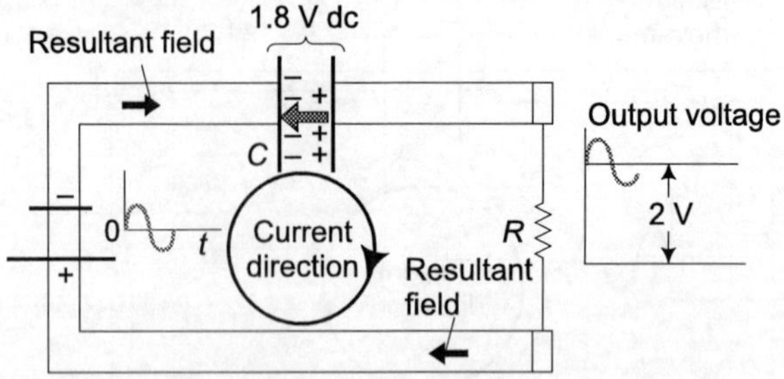

Fig. 3.12b Negative input cycle – resultant field.

Let us study the action one more time during the negative cycle of the input sinusoid. Note how the *unlike* charges cancel each other inside the battery *and register as a current.*

The capacitor voltage is shown reduced, say, 1.8 V dc that reflects the small amount of charge that was lost during the positive cycle. This is shown in Fig. 3.13.

When necessary, a few charges on the battery terminals may cancel each other as well, to provide the sinusoidal input voltage condition. This is a valid visualization after all, because we are using a battery to simulate the sinusoidal voltage.

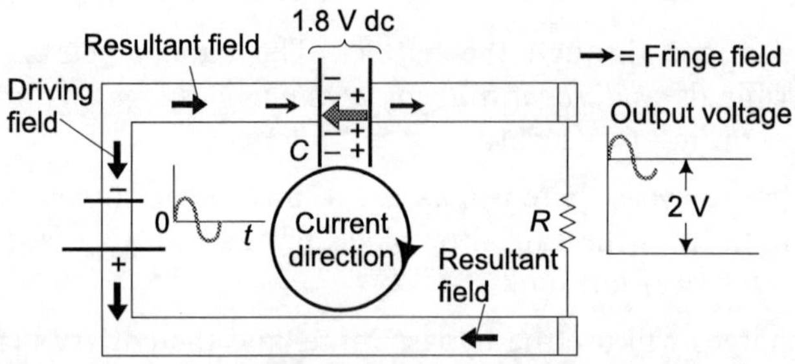

Fig. 3.13 A second look at the negative input cycle.

Note that the current did not change direction during the negative cycle; it maybe said, it was pulsating. This was on account of the small signal input voltage that was assumed, a valid one, because in most amplifiers where such circuits are employed, the sinusoidal input voltage is rarely greater than the biasing voltage. During the positive half-cycle the strength of the current in one direction increased, then in the negative half-cycle, the strength of the current reduced in the same direction; the current *pulsates* in strength without changing direction, until the capacitor discharges after a few cycles.

There is yet one more **nagging** question that needs to be answered.

How does the voltage vary at the output? How is the signal voltage *conveyed across* the capacitor? Is not the capacitor voltage influencing the voltage on the resistor?

Yes, and as the input voltage varies, the *electric field varies* in the wires connected to the capacitor plates.

These in turn, cause a change in the surface charge conditions on the wires and the *potential varies* accordingly.

We will **combine** these actions to understand the operation—first few cycles.

Note that it is the electric fields and surface charges combination that enabled the voltage at the output (across the resistor) to vary.

The actions are shown in Figs. 3.14.

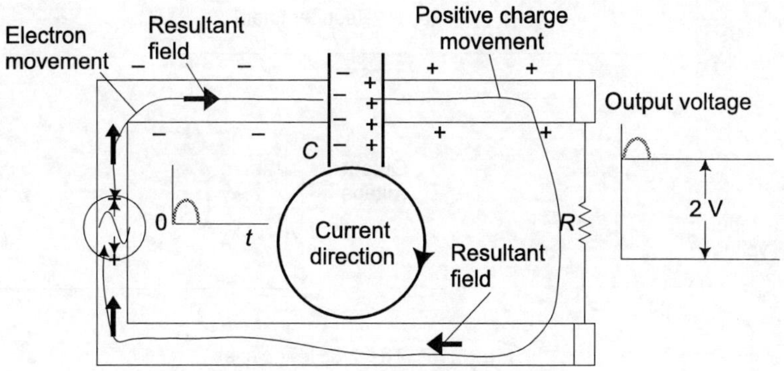

Fig. 3.14a The positive input cycle.

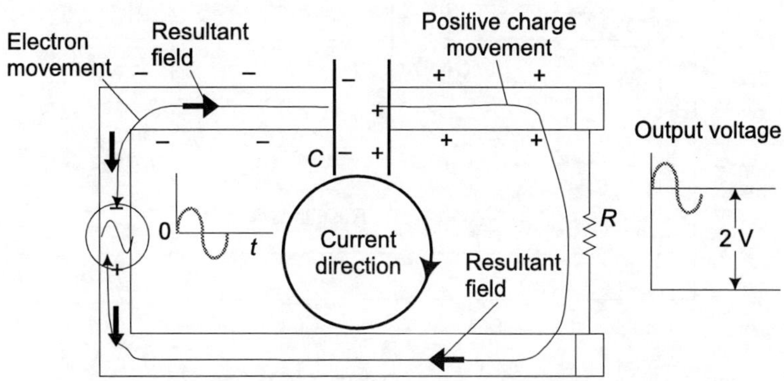

Fig. 3.14b The negative input cycle.

The output is a sinusoid *superimposed* on the 2 V dc steady voltage which *progressively decays* as more number of cycles pass. The resultant electric field in the resistor *R* is not shown for clarity, but a resultant field which maybe of a different overall strength than that in the wires exists inside the resistor *R*. See Fig. 1.33 which shows an electric field in the thin resistive portion.

The capacitor will be *discharged* in a *few cycles* and the few charges that may be present on it in every cycle cancel out each other during the *alternating* positive and negative cycles.

This is shown in the Figs. 3.15 and 3.16.

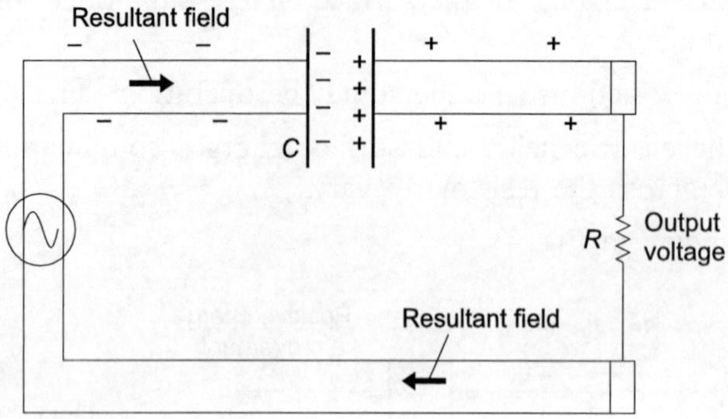

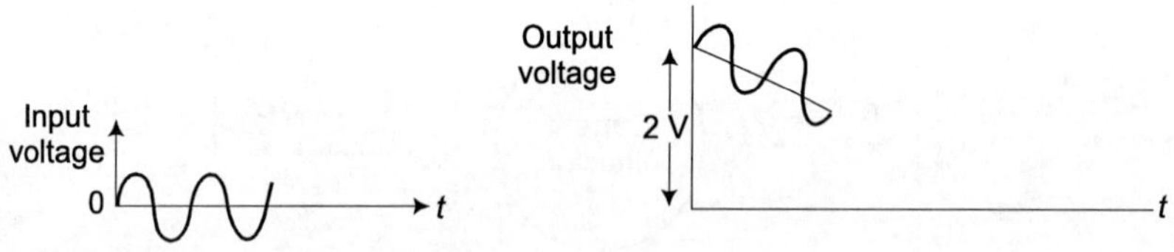

Fig. 3.15 The first few cycles.

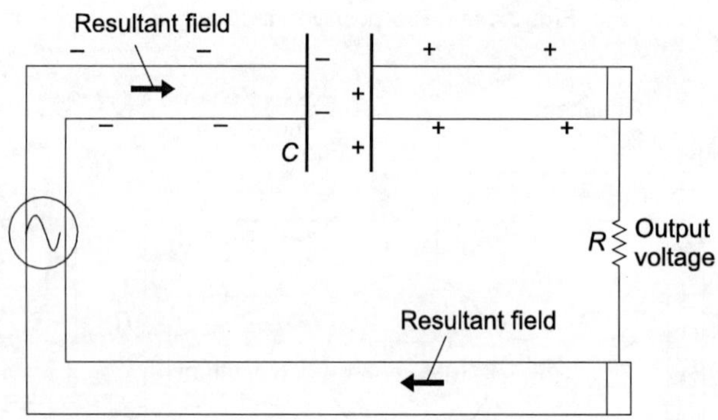

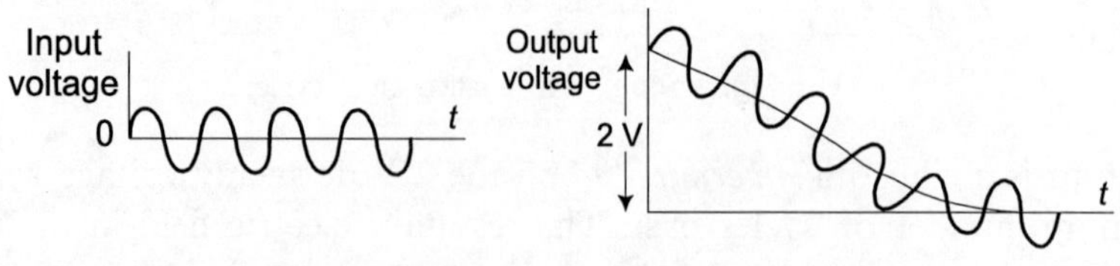

Fig. 3.16 The first few cycles. (Contd.).

After several cycles have elapsed, the capacitor is *discharged*, and the current will be clockwise during the positive and counterclockwise during the negative half of every cycle.

The output voltage is now a faithful *reproduction* of the input under the conditions that the RC time constant is *large* compared with the period T of the sinusoidal input. This is shown in Fig. 3.17.

The first few cycles are shown compressed in the figure for the sake of clarity.

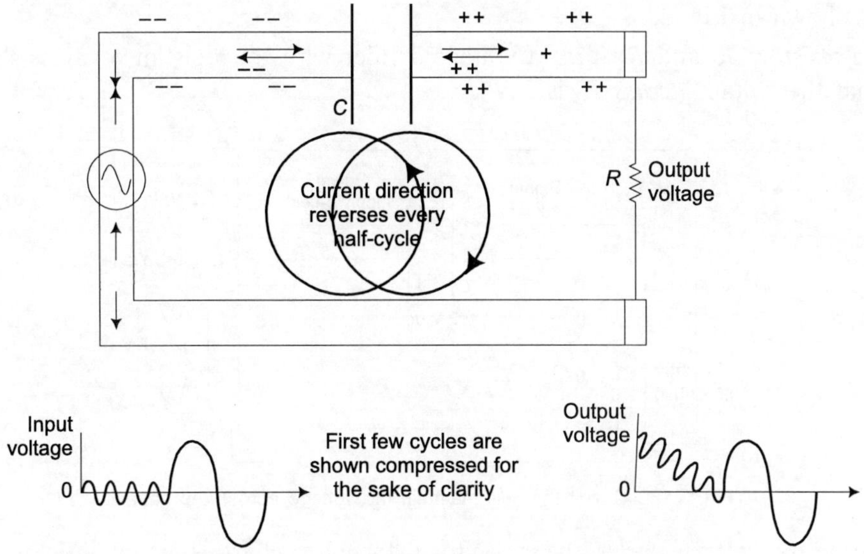

Fig. 3.17 After several cycles have elapsed.

It must be borne in mind that the current is now (after several cycles have elapsed and the capacitor is discharged), *forward* during the positive cycle, and *reverse* during the negative cycle.

A screenshot of the waveforms generated by a simulated run of a CR circuit with a charged capacitor and a sinusoidal generator using TINA TI V9 software (special complementary basic edition) is included in the power point slides of Chapter 3 in the CD with this book. I urge the reader to follow the detailed set of instructions available in the pdf file "Simulation_super_ac_sig_on_dc_volt" in the pdf files folder in the CD with this book, to compile a simulation program using TINA software and then run it, which will be instructive.

3.3 SINUSOIDAL VOLTAGE SUPERIMPOSED ON A DC VOLTAGE

We are now ready to develop a circuit which will superimpose a sinusoid on a dc voltage.

Such a circuit, will *maintain the operating point* of a transistor in the *active region* permitting the input, which is a sinusoid to be *superimposed* on a dc voltage. The circuit amplifies and provides power gain to drive loads, such as, loudspeakers.

The matter rests on being able to somehow *maintain the voltage* across the resistor *R* of the CR combination.

It is really very simple.

We simply incorporate the resistor in a *divider* arrangement from an available regulated power supply or a battery, using one more resistor.

We maintain the requirements of a $0.5\sin\omega t$ sinusoid to be superimposed on a 2 V dc steady voltage.

The circuit comprising two resistors R_1 and R_2 and a voltage source (a battery) are used as shown in Fig. 3.18.

The resistors R_1 and R_2 form a voltage divider with the resistance values so computed such that the voltage *across* R_2 is 2 V dc.

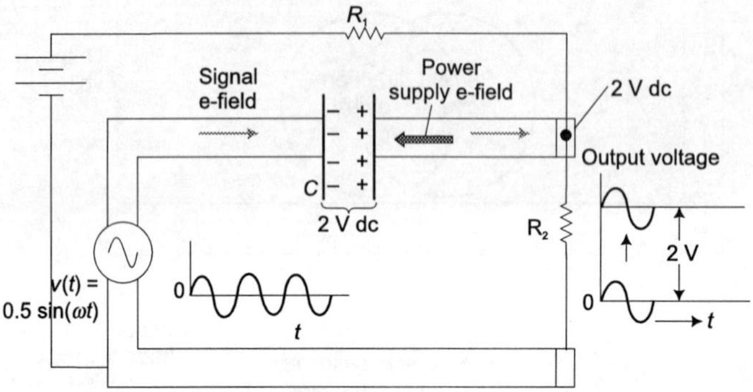

Fig. 3.18 Circuit that will maintain the voltage across the capacitor.

During the *first few cycles*, the capacitor is forced to charge up to 2 V dc, to the voltage across R_2 which carries the input sinusoidal voltage along with it. This is shown in Fig. 3.19.

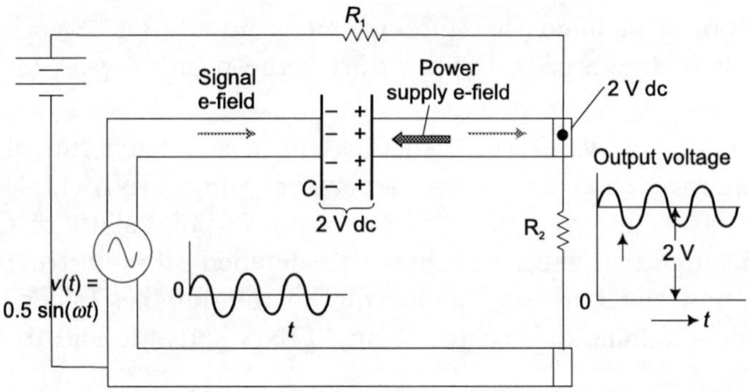

Fig. 3.19 Circuit that will maintain the voltage across the capacitor.

Supposing, a *deficit* of charge occurs on the capacitor (should be equal to 2 V dc), as a result of a *failing* fringe field. This is shown in Fig. 3.20.

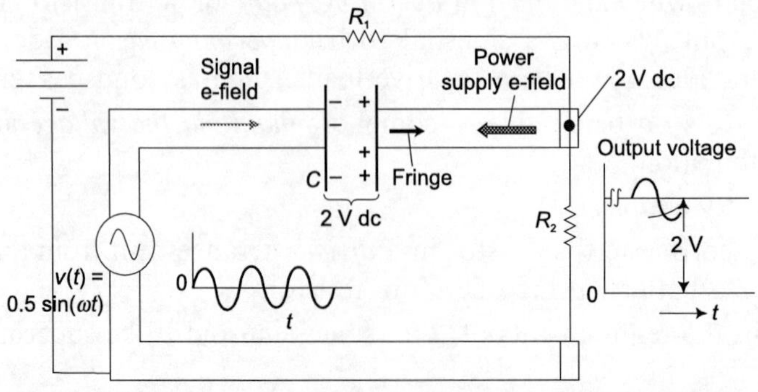

Fig. 3.20 Deficit of charge on the capacitor and falling fringe field.

This will be immediately *replenished* by charges pushed by the power supply e-field, and the dc voltage restored. This is shown in Fig. 3.21.

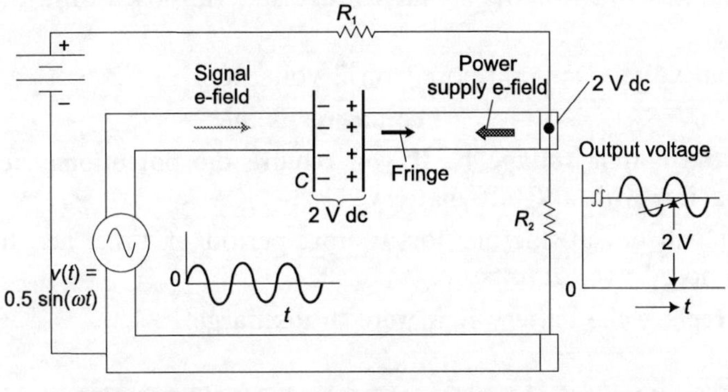

Fig. 3.21 Deficit charge on the capacitor being replenished.

This is how the voltage divider circuit with a capacitor provides a means to *superimpose* a continuously varying voltage on a steady dc voltage.

Using this circuit technique, ac signal voltages are coupled to transistor amplifiers as shown in Fig. 3.22, for the self-biasing arrangement of a common emitter amplifier.

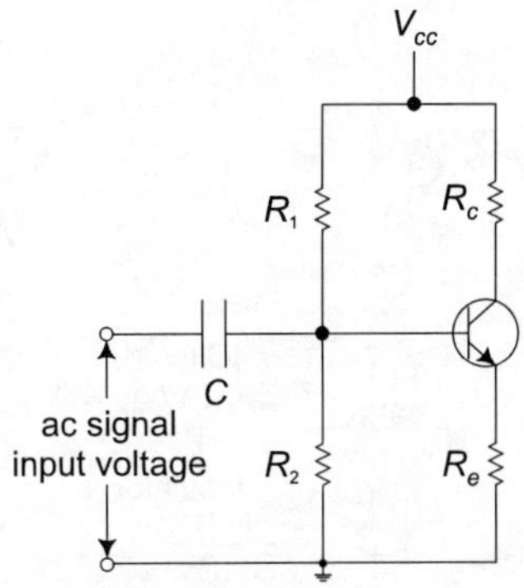

Fig. 3.22 A transistor self-biasing circuit with ac signal input voltage.

A screenshot of the waveforms generated by a simulated run of a circuit with a capacitor, a resistor divider network, a dc source and a sinusoidal generator with TINA TI V9 software (special complementary basic edition) is included in the power point slides of Chapter 3 in the CD with this book.

The following questions are with reference to Fig. 3.21.

What will the output voltage be if the capacitor is replaced with a short circuit?

(a) Zero

(b) Two volts dc

(c) The input signal voltage

The answer is 'c'.

What will the output voltage be if you replace the capacitor with another resistor?

(a) The signal voltage superimposed on a dc voltage less than 2 volts

(b) An attenuated form of the signal voltage superimposed on a dc voltage less than 2 volts

(c) The signal voltage superimposed on 2 volts dc

The answer is 'b'.

What will the output voltage be if you remove the potentiometric divider and simply replace the capacitor with a 2 volts battery?

The output will be satisfactory for a short period of time, i.e., the ac signal voltage will be superimposed over 2 volts dc.

But, who is to replace the battery if it were to discharge!

3.4 ANALYZING ELECTRIC CIRCUITS GRAPHICALLY USING *VI* CHARACTERISTICS AND NETWORK LOADLINES

Before we discuss graphical techniques in electric circuit analysis, it would be useful to review how equilibrium is attained when a single wire is connected to a battery described in Section 2.21. If we connect a single wire to the 12 V automobile battery shown in Fig. 3.23(a), surface charges appear on the wire and make its potential 12 V within a few pico-seconds.

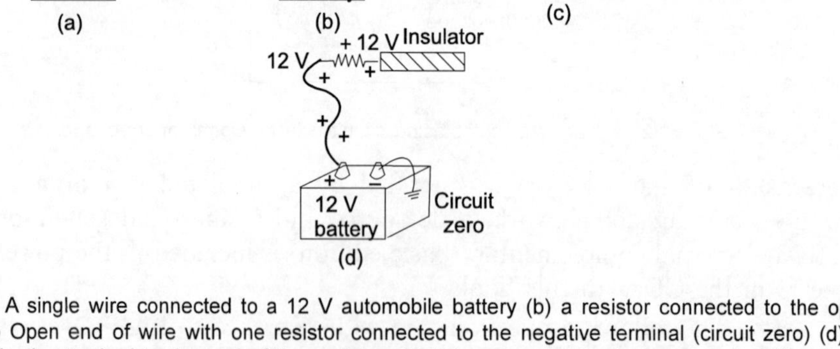

Fig. 3.23 A single wire connected to a 12 V automobile battery (b) a resistor connected to the open end of the wire (c) Open end of wire with one resistor connected to the negative terminal (circuit zero) (d) A piece of plastic attached to the end of the resistor which is connected to the open end of the wire.

If a resistor is tied to the open end of the wire, then since a resistor is a conductor (Section 1.19), surface charges appear on the resistor as well (Fig. 3.23(b)) and tiny currents flow for a few pico-seconds before the potential of the resistor at every point on and in it attains 12 Volts, the potential of the wire. This may be easily verified by measuring the voltage at the open end of the resistor with reference to the negative terminal of the battery designated circuit zero or "ground", using a high impedance voltmeter. See Fig. 2.56.

Now, if the open end of the resistor is returned to circuit zero or "ground" (Section 2.26), a steady current (I_1) will flow as shown in Fig. 3.23(c). The current is the result of surface charges, which would have moved from the negative terminal of the battery towards the open end of the resistor, which by establishing a surface charge gradient across the resistor, produce an electric field E in it, as shown in Fig. 3.23(c).

The potential at the end of the resistor is now 0 V, which is the value of the potential of the negative terminal, designated circuit zero or "ground".

It is important to note that the potential difference (p.d.) of the open end with the circuit open, is 12 volts with reference to circuit zero but, when the circuit is closed and a flow of steady current established, *the voltage dropped from 12 volts to 0 volts* (Fig 3.23(c)).

We can connect one more resistor (of resistance, say, the same value as the first resistor) to the open end of the first resistor (Fig. 3.24(a)), and again, tiny currents flow for a few pico-seconds and surface charges appear on the second resistor. The potential of the wire and resistors at every point on, and in them, attains 12 volts which can be verified, by making voltage measurements between the junction of the resistors and the circuit zero, and between the open end of the second resistor and the circuit zero.

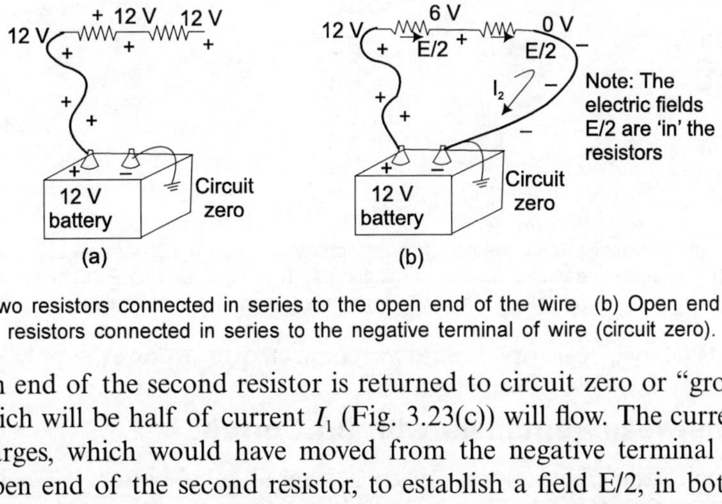

Fig. 3.24 (a) Two resistors connected in series to the open end of the wire (b) Open end of wire with two resistors connected in series to the negative terminal of wire (circuit zero).

If the open end of the second resistor is returned to circuit zero or "ground", a steady current (I_2) which will be half of current I_1 (Fig. 3.23(c)) will flow. The current is the result of surface charges, which would have moved from the negative terminal of the battery towards the open end of the second resistor, to establish a field E/2, in both the resistors, by surface charge gradients across the resistors, as shown in Fig. 3.24(b).

The potential at the junction of the two resistors is now 6 V, half of the voltage (12 Volts). This may be expected, since the electric field strength is E/2 in both the resistors, causing the potential difference across their ends to be half of that when the field established had a strength E as in Fig. 3.23(c). Also refer "How much current if we double the length of the wire" in Section 1.2.

It is important to note that the voltage at the junction of the two resistors with the circuit open was 12 volts but, with the circuit closed and with a flow of steady current established, has *dropped from 12 volts to 6 volts* (Fig. 3.24(b)).

Similarly,with the circuit open, the open end of the second resistor was 12 volts but, with the circuit closed and with a flow of steady current established, *has dropped from 12 volts to 0 volts.*

If a piece of insulator (say a plastic stub), were to be attached to the end of the resistor as shown in Fig. 3.23(d), it is not possible to predict the voltage on the insulator. Since an insulator does not contain mobile charges, there is no mechanism for the potentials to equalize. See the last few paragraphs, at the end of this section also, on what the voltages will be if the battery were to carry a net charge and the opening paragraph of Appendix C.

Operating Point of Devices in Networks

Often, in the analysis of circuits that make use of resistors, diodes and transistors, it is necessary to determine the operating point (voltage and current) of the device connected in a network, which is made from devices whose parameters are known.

We will first analyze the simple case of a network of resistors and the unknown device is also a resistor.

Consider for example the network of Fig. 3.25(a).

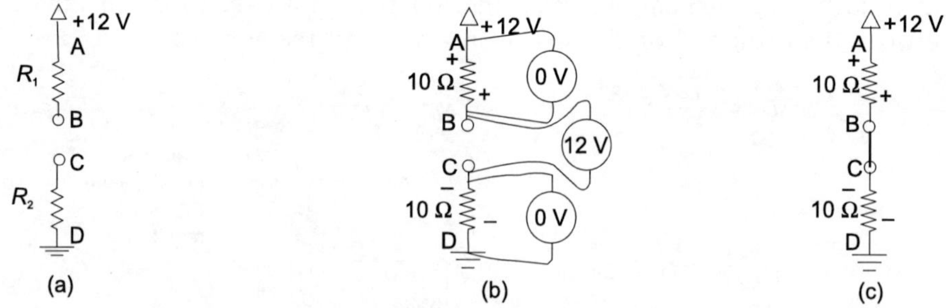

Fig. 3.25 (a) Two resistors connected in a network with a power supply (say, 12 V dc) - open-circuit (b) Voltages measured across terminals i) A, B ii) C, D and iii) B, C of the network (c) the network with terminals B and C short-circuited.

It consists of two resistors R_1 and R_2 connected to a power supply of 12 V dc.

Voltages between Terminals with no Current

We could depict the power supply with a symbol of a battery (⊣⊢), but because most complex circuits operating in electronic laboratories or fixed equipment in industries, use a regulated power supply, it is preferred to depict the power supply using the symbols (⍋) and (⊥) for the positive and negative terminals respectively, where, we assume that the negative terminal is the circuit zero. The '+' symbol is used to indicate the positive terminal with value of its potential difference from the negative terminal. The minus symbol '–' is usually dropped from the circuit zero symbol.

Suppose that the resistors R_1 and R_2 are 10 ohms each, as in Fig. 3.25(b). What will a voltmeter read if it is connected between terminals A and B?

When the network is made, electrons will enter the battery positive terminal and the potential of the resistor R_1 would have been raised. The battery positive terminal, would drop in potential due to the neutralising action of the few electrons, but then its chemical action will quickly restore its potential difference (p.d.) to 12 volts. The migration of electrons will stop, when the potential of the resistor equals the battery positive potential. And all this happens within a few picoseconds. There will be a few surface charges on the resistor R_1, and in Fig. 3.25(b) these indicate its new status of potential, in the same manner as was indicated in Fig. 3.23(b) of a resistor connected to a battery.

Returning to our question, it should be clear that a voltmeter when connected between terminals A and B of circuit (Fig. 3.25(b)) will read '0' volts, because the potential of the resistor and the battery positive, will be the same, namely, 12 volts.

In the case of the resistor R_2 connected to circuit zero, a few electrons would have entered the wire and the resistor R_2 from the negative terminal of the power supply which is circuit zero, making the potential of the resistor R_2 the same as circuit zero. There will be a few surface charges on the resistor R_2 and in Fig. 3.25(b), the minus symbols indicate its new status of potential. A voltmeter when connected between terminals C and D of circuit (Fig. 3.25(b)) will read '0' volts, because the potential of the resistor R_2 and the battery negative will be the same, namely '0' volts.

The current in the circuit of Fig. 3.25(b) will be zero because of the open circuit.

What will be the reading of a voltmeter connected between the terminals B and C?

Since the potential of terminal B is the same as that of A, its potential is 12 volts. Since the potential of terminal C is that of the battery negative or circuit zero (terminal D), it is '0' volts. Therefore, a voltmeter connected between the terminals B and C will read the same p.d. as that between the terminals A and D namely, 12 volts.

We can allow a current to flow in the circuit by simply connecting a wire between terminals B and C thus, short-circuiting them as shown in Fig. 3.26(a).

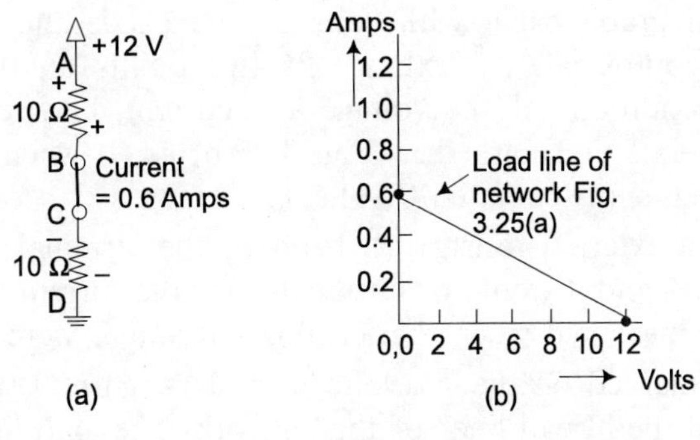

Fig. 3.26 (a) Terminals B and C short-circuited (b) Load line of network of Fig. 3.25(a), and (c) *VI* characteristic of a resistor.

The current in that case will be $12\,V/(10\ \Omega + 10\ \Omega) = 12\ V/20\ \Omega = 0.6$ amps.

A Word of Caution

Many students assume that the current flows from the positive terminal and *terminates* on the circuit zero terminal. This is wrong ! The current continues *in* the circuit zero terminal and around through the source, which could be a battery or a bench top adjustable regulated power supply, to the positive terminal of the source, of which the zero terminal is an associated part by the mechanism described in Section 1.18, to complete the circuit. The reader should always remember that the network of Fig. 3.26(a) is a convenient method of representing the circuit, and this will become apparent, when analyzing complex circuits with several parallel paths, in later sections of this chapter.

Every network's voltage source operates by separating charges (Section 1.8), and there will be +V/2 volts on the positive terminal and –V/2 volts on the negative terminal. The computation of network currents becomes difficult and complex to work with these voltages. Hence, in network analysis it is customary to mark the negative terminal the circuit reference or circuit zero, or what is commonly called *ground* and is assigned '0' volts. This automatically qualifies the positive terminal voltage to be +V volts with reference to the circuit zero. This simplifies the network current computations. While the assignment of the circuit zero simplifies the computations, it does not affect the physics of current flow in the network. See also Section 2.29 "Grounding".

Also, since conventional current flows out of the positive terminal of the source and returns to the negative terminal, the current is termed *differential*. We will learn more about differential-mode currents in Chapter 5, Section 5.22.

The Operating Point (Voltage and Current) of the Network

We can plot a graph of the voltage and current between the terminals B and C on linear graph paper, as depicted in Fig. 3.26(b) for the two conditions of the network (i) terminal B and C open circuit, and (ii) terminal B and C short circuit. The network comprises a power supply of 12 volts, and two resistors of 10 Ω each, connected between the positive and negative terminals of the power supply.

We label a linear graph paper with volts on the *x*-axis and amps on the *y*-axis and scale the two axes appropriately. Next, we fix two points on the graph, corresponding to the two known conditions of the voltage and current, between terminals B and C in the network. Two points are marked in Fig. 3.26(b) as (i) when open-circuited 12 volts, 0 Amps, and (ii) when short-circuited 0 volts, 0.6 Amps.

If we connect an external resistor R_3 between the terminals B and C, there will be current in the network, and it would be impossible for the current to have a value exceeding 0.6 amps which was measured when the terminals B and C were short-circuited.

The two points marked on the *x*- (voltage) and *y*- (current) axes are joined together, and this line is called the "Load line" of the network. The load line is the locus of points, or pairs of voltage and current values for the voltage between terminals B and C and the current in the circuit, when the terminals B and C are bridged by a resistance. The load line is fixed for a fixed power supply voltage and fixed values of resistors R_1 and R_2.

In electronics, most devices are characterised by *VI* characteristics, or a sketch of the values of Current vs. Voltage of the device.

For example, the *VI* characteristic of a resistor will be a straight line, as shown in Fig. 3.27(a). From the slope of the graph which is a straight line, we note that the resistor has a value of, 12V/1 amp = 12 ohms.

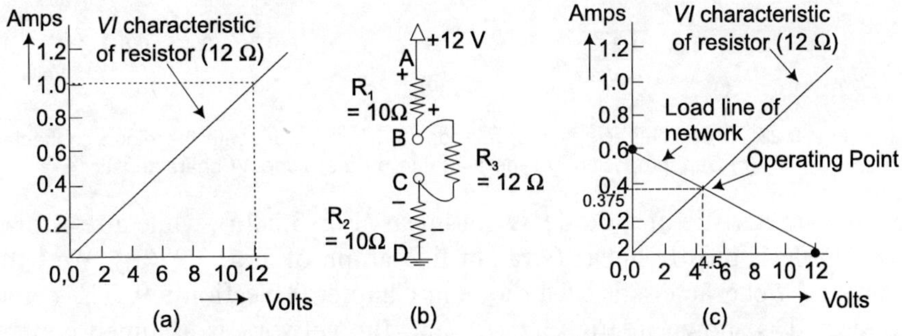

Fig. 3.27 (a) *VI* characteristic of resistor (12?) (b) A network of two resistors and an external resistor of 12 ohms (c) the load line of the network superimposed on *VI* characteristic curve of the device (straight line for a resistor).

If a resistor R_3 of 12 ohms, is connected between the terminals B and C, as shown in Fig.3.27(b), we will find the current in the circuit should be 12V/(10 Ω + 10 Ω +12 Ω) = 12 V/32 Ω = 0.375 amps, and the corresponding p.d. between terminals B and C is 4.5 volts.

This can be easily verified by *superimposing* the load line of the network on the graph of the resistor, and reading off the values of voltage and current, as shown in Fig. 3.27(c) from the point of intersection of the load line and the device characteristic.

Their point of intersection is called the *operating point* of the device, since the values of current in the network and voltage across the terminals B and C, will be those of the device, when the network is energised.

It is now a simple matter to read off the values of network current and voltage between terminals B and C (i.e. across the external device) at the operating point: 0.375 amps and 4.5 volts, as shown in Fig. 3.27(c).

The load line characterises the network comprising, the power supply (+12 V dc), the resistors R_1 and R_2 (10 ohms each), in which the resistor R_3 was the external device. If any parameter of the network was to change, the load line will need to be redrawn. For example, if the power supply were to be changed to say, 8 volts, then the new load line will be a straight line joining 8 volts on the *x*-axis to 0.4 amps on the *y*-axis.

If the external device were changed to a different value, then the *VI* characteristic will be different; for example, if R_3 was a 5 ohms resistor, then because 12 V/5 Ω = 2.4 amps, the *VI* characteristic would be a straight line joining the origin to the point (12 V, 2.4 amps).

If the external device is a diode, we draw the *VI* characteristic of the diode (Chapter 9 describes the operation of a p-n junction in a diode), and then superimpose the load line of the network on the *VI* characteristic, to obtain the operating point. For example, we have plotted the device characteristic of a 1N4007 diode with the *x*- and *y*-axes appropriately scaled, as shown in Fig. 3.28(a).

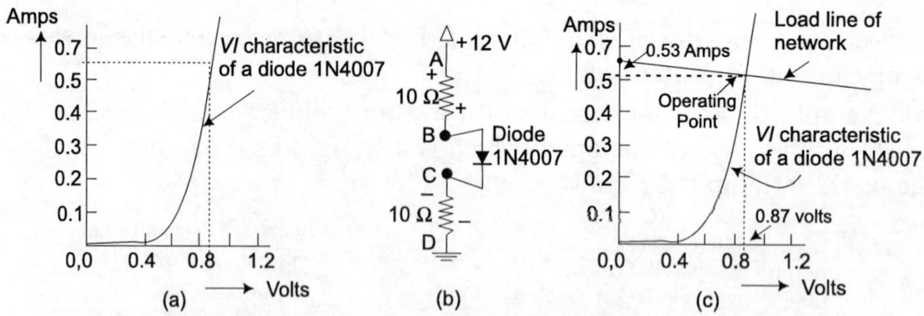

Fig. 3.28 (a) *VI* characteristic of a diode (b) The network with the diode connected (c) Superimposition of the load line on the diode *VI* characteristic.

The *VI* characteristic of a diode, is not linear (Fig. 3.28(a), while a resistor in its normal operating region, is mostly linear (straight line graph of Fig. 3.27(a)). We will learn about the non-linear *VI* characteristic of a diode in Chapter 9, Section 9.9. The network with the diode connected is shown in Fig. 3.28(b), and the network is assumed connected to a 12 V dc supply. The diode is shown in the forward-bias connection with its anode connected to terminal B and the cathode to terminal C.

In Fig. 3.28(c), we have drawn the segment of the load line of the network of Fig. 3.25(a) which is visible in the 0 to 1.2 volts range, because the zero-current voltage intersection point on the *x*-axis is out of the range of the graph. The load line should intersect at (12 volts, 0 Amps) (when terminals B and C are open) as in Figs. 3.27(c) and 3.26(b).

We can read off the current and voltage of the diode in the network of Fig. 3.28(b) from the operating point in the graph, as 0.53 amps and 0.87 volts respectively.

Also note that conventional current (Section 1.17) flows out of the positive terminal of the source and returns to the negative terminal and so, the current is termed *differential* (Section 5.22).

If the external device is a transistor used as an amplifier in the Common-Emitter (CE) configuration, then we draw the output *VI* characteristics of the transistor in the CE configuration. Next, superimpose the load line of the network on the output *VI* characteristic, to obtain the operating point.

For the example, we have plotted the output characteristic (Collector current I_C vs. Collector-emitter voltage V_{CE}) of a BC 107 transistor, with the *x*- and *y*-axes appropriately scaled, as shown in Fig. 3.29(a).

The family of output characteristics are different at various base currents. For clarity, only the individual characteristics for base currents of I_{B1} = 20 microamps and I_{B2} = 40 microamps are indicated in Fig. 3.29(a).

The network with the transistor connected is shown in Fig. 3.29(b). Note that the network is now connected to a power supply of 12 volts and the resistors R_1 and R_2 (Fig. 3.25(a)) are 1.2 kΩ and 400 Ω respectively. The maximum current which can flow in the network with the terminals B and C short-circuited is therefore, 12 volts/1.6 kilo ohms

= 7.5 milliamps. The load line points are: 12 volts on the x-axis (Collector to Emitter voltage V_{CE}) and 7.5 milliamps on the y-axis (Collector current I_C).

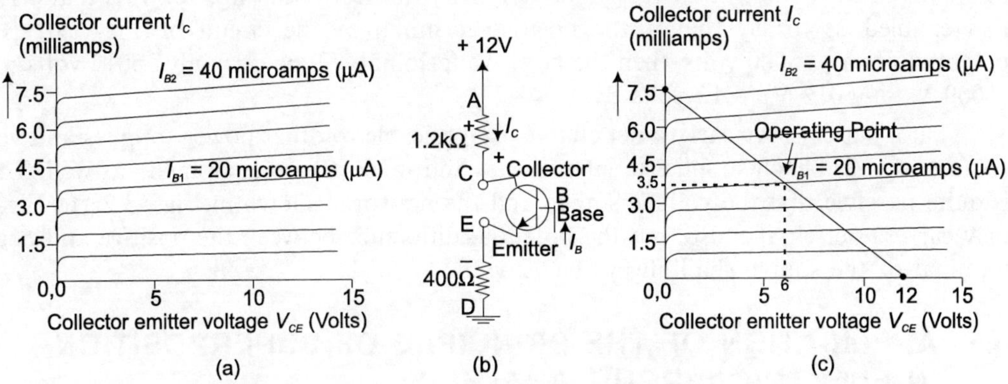

Fig. 3.29 (a) Output characteristic of a transistor in common-emitter configuration (b) The network with the transistor connected (c) Superimposition of the load line on the diode *VI* characteristic.

In Fig. 3.29(c), we have drawn the load line of the network of Fig. 3.29(b) superimposed on the transistor's output characteristics.

We assume that the base current I_{B1} of the transistor is 20 microamps (μA). We can read off the collector current I_C in the network and the collector to emitter voltage V_{CE} at the operating point from Fig. 3.29(c); 3.5 milliamps and 6 volts respectively.

Also note that conventional current flows out of the positive terminal of the source and returns to the negative terminal and so, the current is termed *differential* (Section 5.22).

The voltages (or more correctly potential differences) in the circuits shown are with reference to the negative terminal of the battery assumed to be the circuit zero. If the battery were to be carrying a net charge, as was described in Section 2.21 (see Fig. 2.61), then it is likely that the potential of the battery as a whole, including it's terminals, will be offset by some potential say, –5000 volts on the positive terminal, as was assumed in the battery of Fig. 2.61. In that case, if we used the battery with such a net charge in Fig. 3.23 and Fig. 3.24, the potential of the resistor (Fig. 3.23(b)) and of the resistors (Fig. 3.24(a)) will be –5000 volts too and the circuit zero would be at –5012 volts (absolute potential). It is not possible to predict the amount of charge nor the voltage on the insulator attached to the end of the resistor as shown in Fig. 3.23(d) since the insulator has no mobile charge. In Appendix C we learned that insulators do not have a voltage, though they may carry a net charge.

There is a very high degree of uncertainty in the amount and distribution of excess charge (complex geometries make analysis difficult) on both conductors and insulators in electrical and electronic circuits and assemblies. The processes which cause the accumulation of such charges and their dispersion are conduction (say, by the touching of charged objects) and induction (presence of accumulated charges on nearby objects), and the complexity of these are compounded by the highly randomised nature of the environment surrounding the equipment and assemblies. The variables may be moisture, contaminants, high energy particle bombardment from space producing ionization and the list is endless.

There is however, certainty in the charge separation process of the voltage source (dc as of a battery or ac if a sinusoidal source) which maintains the voltage across the terminals of the source at a fixed value (dc) or rhythmically varying (ac). It is that certainty that enabled us to say that if the positive terminal of the circuit of Fig. 2.61 is at a potential of say, –5000 volts, then the negative terminal will be certainly –5012 volts to give –5000 V – (–5012 V) = 12 volts dc.

That a battery *may* carry a net charge, is applicable to other power supplies and voltage sources used as bench supplies in laboratories and in electronic equipment, as well. Thus, it is quite possible that the voltage source and the networks shown in Figs. 3.25 to Fig. 3.29 may carry a net charge also, but the potential difference between the positive and negative terminal of the source shall always be 12 volts.

3.5 APPLICATION OF THE PRINCIPLE OF SUPERPOSITION IN ELECTRIC CIRCUIT ANALYSIS

The method of superposition is frequently used to determine unknown currents in complex electrical circuits, which contain two or more voltage sources. Consider the circuit of Fig. 3.30, for which it is required to determine the unknown current I_1 in the 10 Ω resistor.

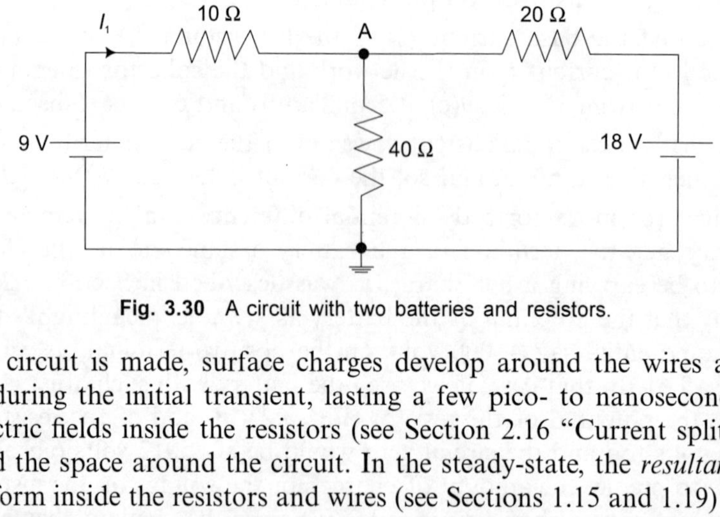

Fig. 3.30 A circuit with two batteries and resistors.

When the circuit is made, surface charges develop around the wires and accumulate at the nodes during the initial transient, lasting a few pico- to nanoseconds at the most, producing electric fields inside the resistors (see Section 2.16 "Current splitting in parallel branches), and the space around the circuit. In the steady-state, the *resultant* electric fields which are uniform inside the resistors and wires (see Sections 1.15 and 1.19) and the field in the batteries, drive the mobile charges (free electrons) to produce currents. The distribution of surface charges and patterns of electric fields, which are present in and around the circuit of Fig. 3.30, will be complex and are omitted from the figure, for the sake of clarity.

The principle of *superposition* of electric fields, is implied when engineers apply the *superposition* method of *potential* (see Sec. 3.1), to solve for voltages and currents in an electric circuit.

The superposition principle (some authors treat the topic as superposition method or theorem) states: the output of a circuit is determined by summing the responses to each independent voltage source acting alone. The method of superposition is used to first obtain the response to the 9 V battery, and then to obtain the response to the 18 V battery. The resultant response is the sum of the two responses.

The response of the circuit to the 9 V battery may be obtained, by replacing the 18 V battery with a short circuit, as shown in Fig. 3.31(a). Strictly speaking, the battery should be replaced with its internal resistance. See "How a real battery operates" below.

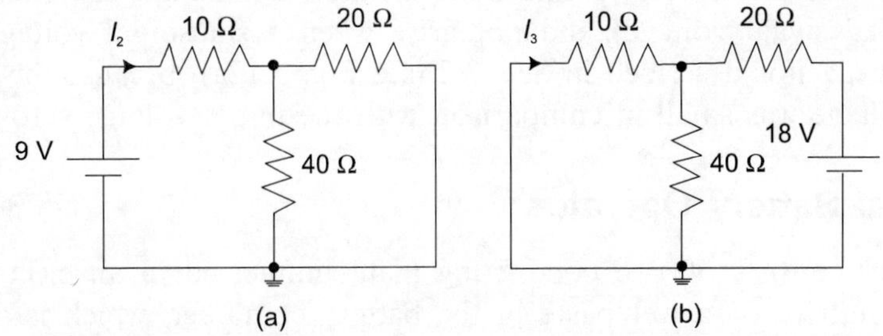

Fig. 3.31 The circuit of Fig. 3.30 solved by the method of superposition.

The resulting parallel 20 and 40 ohm resistances is next replaced, by a single equivalent resistor, and the current I_2 solved due to the action of the 9 V battery alone.

In a similar manner, the response of the circuit to produce the current say, I_3 due to the 18 V battery acting alone, can be determined by replacing the 9 V battery with a short-circuit, as shown in Fig. 3.31(b). The current I_1 with both the batteries in circuit, will be given by the method of superposition as, the sum of the individual responses namely, I_2 and I_3. Hence, $I_1 = I_2 + I_3 = -0.126$ Amperes.

It is found that the current I_1 is *negative* which means that current flows, *into* the 9 V battery.

Analysis of the circuit, using Thevenin's theorem (Ref. [15]) to which the circuit can be converted, using the method of superposition, will yield the same current. Refer http://ocw.mit.edu/courses/electrical-engineering-and-computer-science/6-002-circuits-and-electronics-spring-2007/video-lectures/lecture-3/.

If the circuit was assembled using real batteries, such as non-rechargeable dry cells (several of them may be connected in series to provide 9 V and 18 V), the current I_1 would quite certainly not be –0.126 Amps, because the current through the 9 V battery would be reversed, and would depend on among several other factors, the type of battery used. It is quite likely, the voltages at the node A would also not be as calculated.

In a desktop setup using one 9 volt dry cell and an 18 volt dc supply from a laboratory adjustable voltage regulator of 1 ampere capacity (to handle the large current of the order of 500 mA), it was found that the current I_1 initially shot up and then went down to around 180 mA. After a minute or so, the current again shot up, and the 9 volt cell began to warm up.

The reason for the variance is, the Superposition Theorem and Thevenin's Theorem assumes, in an ideal approximation, the "linearity of all circuit elements, including the *reversibility* of currents through batteries".[22]

It is for this reason that circuit analysts prefer to model circuits, using *ideal* voltage sources. An ideal voltage source (impossible to produce practically), maintains the same voltage across its terminals, no matter, what current is drawn *from* its terminals, or what current flows *into* its terminals.

Note that a situation of current flowing into the terminals of the battery (a small battery in forward and reverse connection was used to simulate the strength and polarity of a sinusoidal signal voltage), did not arise when a sinusoidal voltage was applied to the charged capacitor described in Sec. 3.2 (see Fig. 3.12a), because the amplitude of the sinusoidal voltage was small in comparison with the initial voltage across the capacitor.

How a Real Battery Operates

A real battery consists of two conducting plates immersed in an electrically conductive electrolytic solution, or a wet paste in the battery container, which is a soup of mobile positive and negative ions in an aqueous solution (neutral molecules). Note that *all* the components (excepting the plate separators and the housing) in a battery system consist of conductors (whether metallic or ionic), with some amount of separated charges on the battery terminals giving rise to an emf. There are plenty of mobile electrolytic ions in the battery container.

During a discharge of electricity (in a circuit), the chemical on the anode releases electrons to the negative terminal, and ions in the electrolyte through an oxidation reaction. Meanwhile, at the positive terminal, the cathode accepts electrons, completing the circuit for the flow of electrons. The electrolyte is there to put the different chemicals of the anode and cathode into contact with one another, in a way that the chemical potential can equilibrate from one terminal to the other, converting stored chemical energy into useful electrical energy. These two reactions happen simultaneously, and the ions transport current through the electrolyte, while the electrons flow in the external circuit, generating an electric current. Refer http://engineering.mit.edu/ask/how-does-battery-work and http://www.ee.sc.edu/personal/faculty/simin/ELCT102/13%20Voltage%20and%20Current%20sources.pdf.

The current flowing in any circuit connected to the battery also flows THROUGH the battery (through the electrolyte and the electrodes forming the battery). Both electrolyte and electrodes have resistance. This is called "internal" resistance.

Therefore, in one sense, when applying the superposition method to analyze circuits with batteries, replacing the battery with its internal resistance, amounts to removing the plates (with its separated charges) while leaving the resistance due to the (conductive) electrolytic aqueous solution, with the tacit understanding that the solution has a capability to conduct in *both* directions, making Thevenin's and Norton's theorems (Ref. [15]) also applicable. However, conduction in both directions may not be possible with several types of *real* batteries.

In order to avoid confusing an ideal voltage source with a real battery, analysts use a different symbol for an ideal voltage source (a circle with polarity indicated) as shown in Fig. 3.32, which is the circuit of Fig. 3.30 redrawn, with ideal voltage sources replacing the batteries.

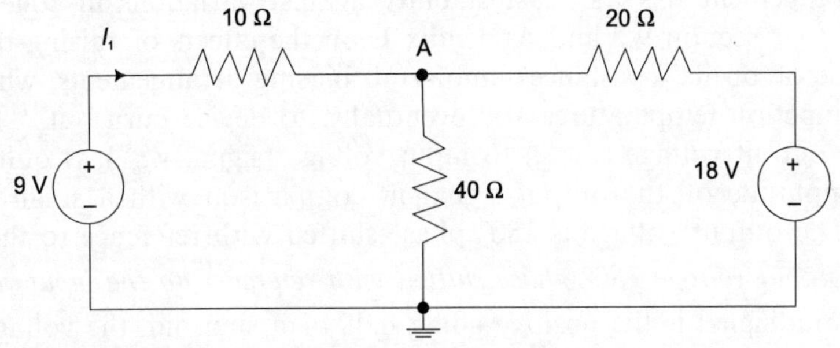

Fig. 3.32 A circuit with two ideal voltage sources and resistors.

The ideal voltage source, is assumed to deliver a steady voltage forever, and to have zero internal resistance. The ideal voltage source, no matter what you do to it, like drawing a lot of current (remembering that in a circuit, the current whether small or large also flows through it), or pushing current through it, as in the example circuit shown in Fig. 3.30, therefore, can be replaced by short-circuits for analysing circuits to determine node voltages and currents.

I urge the reader to visualize the circuit processes viz. superposition of fields and potential, and the transient and steady-state made by the development of surface charges which simplified so elegantly the voltage and current analysis of the circuit of Fig. 3.30 using the principle of superposition.

3.6 APPLICATION OF THE PRINCIPLE OF SUPERPOSITION TO STUDY THE OPERATION OF A BYPASS CAPACITOR IN A TRANSISTOR FEEDBACK AMPLIFIER

The reader may defer to read this section and the following section, until after learning the theory and operation of transistors[17] and their application in amplifiers. The theory of transistor operation and its application is beyond the scope of this book.

We have shown how a capacitor bypasses high frequency signals and how it attenuates the low frequency signals in Section 2.13. One application of a capacitor to bypass signals is in Common-Emitter amplifier circuits. Shown in Fig. 3.33 is a Fixed-Bias arrangement for a common-emitter bipolar junction amplifier.

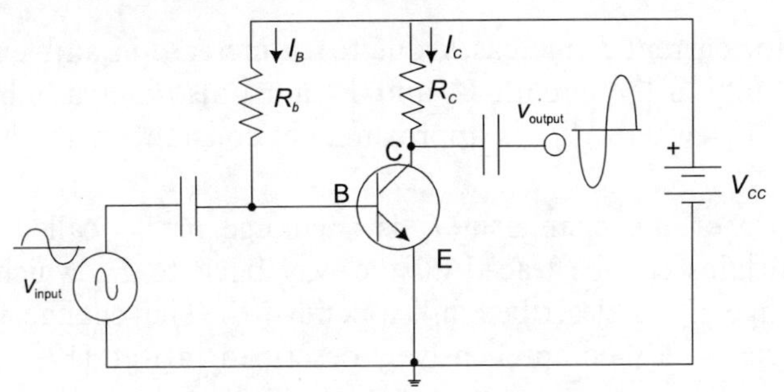

Fig. 3.33 A fixed-bias circuit of a common-emitter amplifier.

This arrangement has the least stability against variations in collector current with temperature (see Section 9.2 and Appendix E on the effects of raising the temperature of a semiconductor on its' resistance) among all biasing arrangements, which could lead to increase in junction temperatures and eventually, to device burn out.[17] The gain which is the ratio of output voltage (v_{output}) to input voltage (signal v_{input}), is quite large as is seen from the amplitude of the output signal in comparison with a small signal input v_{input} amplitude. The output voltage is 180° phase-shifted with reference to the input voltage. *Why is the output voltage 180° phase-shifted with reference to the input voltage?*

If the input signal is the positive going half of a sinusoid, the voltage at the collector reduces and is negative going, because the positive going input signal establishes a current of increasing strength flowing from the collector to emitter of the transistor. The collector and emitter junction depletion regions are bridged by mobile carriers in increasing numbers, increasing the conductivity of these regions and of the transistor from collector to emitter. To gain an intuitive understanding of the negative going output voltage as the collector current in the transistor increases, refer to Fig. 3.24, Section 3.4.

The introduction of a resistor R_e in the emitter leg of the transistor, as shown in Fig. 3.34 introduces negative feedback, which improves the stability.

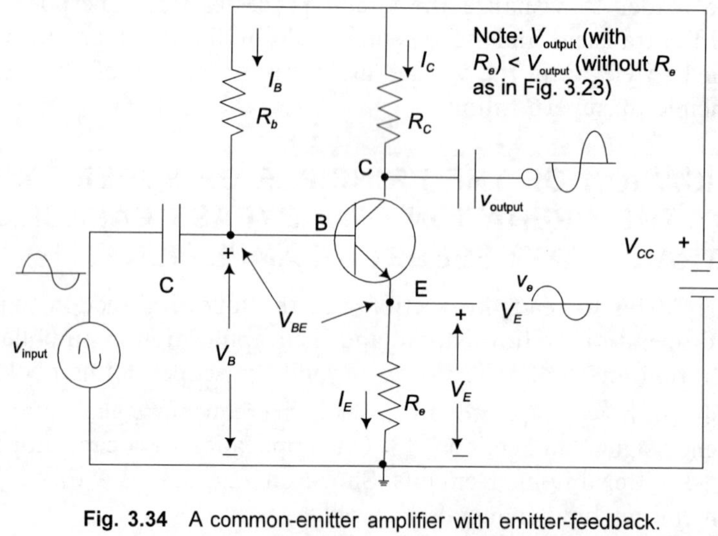

Fig. 3.34 A common-emitter amplifier with emitter-feedback.

If the collector current I_C increases, due to an increase in ambient temperature, leading to thermal instability in the circuit, V_E and V_B must also increase, because $V_B = V_E + V_{BE} = (I_C + I_B) R_e + V_{BE}$ where, V_{BE} is approximately constant in the active region and $V_E = I_E R_e = (I_C + I_B) R_e$.[17]

The real cause of the change in I_C is a change in I_{CO} called the reverse saturation current, whose origins can be traced all the way back to I_O, which is highly sensitive to temperature change. I_O is described in Appendix E. The change in I_C due to I_{CO} from temperature effects is very comprehensively described in Ref. [17].

From $I_B = \dfrac{V_{CC} - V_B}{R_b}$, it may be observed that the increase in V_B must cause a decrease in the base current I_B, which in turn, reduces V_B, the *biasing* dc voltage. In other words, the decrease in I_B *opposes* the original increase in I_C (negative feedback). The current and voltage adjustments occur extremely rapidly since surface charge changes are effected (Section 1.16) within a few pico- to nano-seconds. Thus, the emitter-feedback bias circuit due to the negative-feedback effect, has a better thermal stability than the fixed bias circuit of Fig. 3.33. The operating point of the common-emitter amplifier of Fig. 3.34 will be shifted less, due to changes in I_C with temperature, thereby improving stability.

The amplitude of the emitter signal (v_e) is about the same as the input signal, and the signal varies about the dc voltage V_E as shown in Fig. 3.34.

However, the circuit of Fig. 3.34 introduces signal degeneration and loss of gain due to *signal* voltages, because of the feedback caused by R_e. The transistor is not capable of distinguishing signal voltages from the biasing dc voltage, and this causes a reduction in change in I_B due to changes of signal (say, from a microphone) voltages and loss of gain. This may be seen by the smaller amplitude of the output signal in Fig. 3.34, in comparison with the amplitude of the output signal of the amplifier in Fig. 3.33, which does not incorporate an emitter resistor.

To avoid the loss of gain due to signal voltages, a large capacitor C_E is tied across the emitter resistor R_e, to bypass the varying signal voltage, as shown in Fig. 3.35. The reactance of C_E at the frequencies under consideration (say, audio frequencies), is very small.

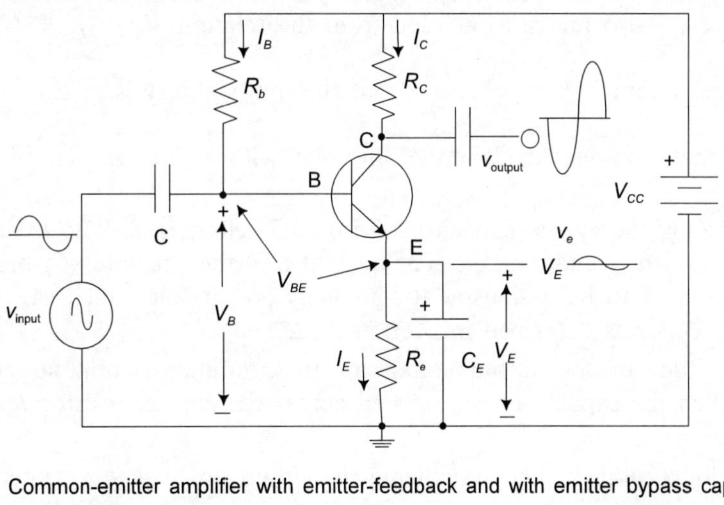

Fig. 3.35 Common-emitter amplifier with emitter-feedback and with emitter bypass capacitor C_E.

The capacitor C_E ensures that there is no loss of gain at the signal frequencies due to feedback which causes the base current (dc) to be limited. The *varying* voltage v_e across R_e is now much smaller in magnitude, as compared to the situation without the capacitor tied across the emitter resistor in Fig. 3.34. The signal at the emitter terminal varies (v_e)

around the dc voltage level V_E, to which the capacitor would have charged up, within a fraction of a second of switching ON the amplifier circuit.

The current *variations* of I_E are *bypassed* or absorbed by the capacitor C_E assumed large (See Section 2.11) with its reactance in relation to R_e being small. The current variations occur only in the bypass capacitor arm of the parallel $R_e \parallel C_E$ combination. The input signal variations, even at audio frequencies, maybe considered fast enough to cause currents in the capacitor arm, but little charge is displaced across C_E. The quiescent dc voltage V_E (the voltage condition across R_e without a signal) remains, since it is a dc voltage. This explains the tiny amplitude of the signal v_e across $R_e \parallel C_E$ as shown in Fig. 3.35.

In Sections 2.13 and 3.2 we learned that this is the effect of the resultant electric field variations in the connections to the capacitor C_E due to time-varying signals. These signals are not slow enough at signal frequencies to accumulate sufficient charge on the large capacitor to cause the voltage across it to follow the variations in voltage. The fringe field it develops will be small. The potential variations of the emitter are now no longer that across the resistor R_e used singly, as in the circuit of Fig. 3.34, but that of the potential of the combined $R_e \parallel C_E$. No potential change can be observed across $R_e \parallel C_E$, unless the plates of C_E hold displaced charges, corresponding to the amount of potential change.

There *is* a varying current component in the resistor R_e no doubt, but it is so small that there is not a significant voltage developed across the R_e and C_E parallel combination. Therefore, there is little to no opposing voltage effect to the base current variations due to the input signal and little, to no loss of gain.

But with temperature changes, the increase in I_C which is a very slow change, can cause charges to accumulate on the plates of the capacitor C_E, and this increases V_E. Due to the increase in I_C, V_B also increases as seen from the relation $V_B = (I_C + I_B)\ R_e + V_{BE}$. I_B, in turn, decreases because $I_B = \dfrac{V_{CC} - V_B}{R_b}$, and thermal stability is retained almost exactly in the manner that was described in the amplifier without the emitter bypass capacitor of Fig. 3.34.

Summarizing, the bypass capacitor of Fig. 3.35 charges itself to the dc voltage V_E but bypasses the varying signal voltage. The current i_e in the capacitor C_E branch will lead the voltage v_e assumed to be a sinusoidal signal, by 90°, a rule which can never be violated for a capacitor as was described in Section 2.12.

The amplitude of the output voltage of the amplifier, is now larger, as seen in Fig. 3.35 than, when the capacitor was not tied across the emitter resistor R_e as seen from the Fig. 3.34.

Pausing for a moment and visualizing the circuit processes viz. superposition of fields and potential, the transient and sinusoidal steady-state (Section 2.13) phenomenon by the development of surface charges at every point in the circuit, and the harmonious blending of all of these will, provide deeper insights on how the weak fringe field development in the bypass capacitor C_E enabled it to restore the amplifier gain which was reduced by the emitter resistor.

3.7 APPLICATION OF THE PRINCIPLE OF SUPERPOSITION IN THE CIRCUIT ANALYSIS OF TRANSISTOR AMPLIFIERS

It was seen in Sec. 3.1 that the principle of superposition of electric fields leads to the principle of superposition of potential differences or voltages. This also finds wide application in the analysis of voltages and currents in complex electric and electronic circuits.

One classic application, is in the analysis of transistor amplifier circuits which engineers routinely perform. An amplifier circuit is used to strengthen weak signals from sensors (for example, temperature sensors) and transducers (microphones) and consists, usually of Bipolar Junction or Field Effect transistors connected in a dc source powered biasing network. Shown in Fig. 3.36 is a biasing network NT1 for a Bipolar junction transistor Q_1 whose three terminals are (i) emitter E, (ii) base B and (iii) collector C.

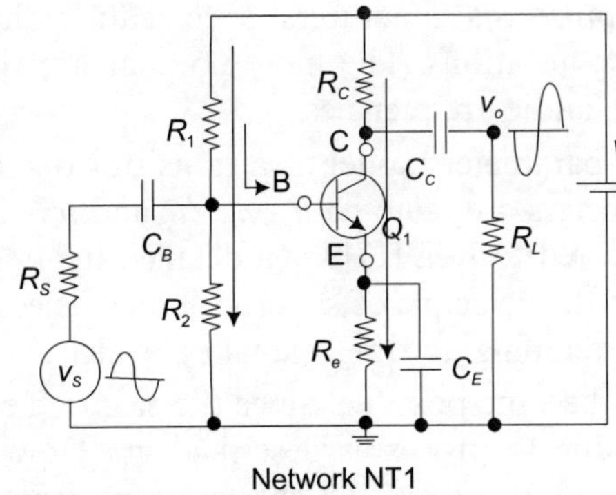

Fig. 3.36 Principle of superposition of potential applied to the analysis of a transistor amplifier.

The biasing network comprises resistors R_1, R_2, R_c and R_e connected to the transistor in a manner which enables a steady (dc) current to flow in the transistor, with a steady voltage across its collector and emitter terminals.

When the circuit is made with no input signal v_s (assume replaced with a short-circuit, not short-circuited because that would short the signal source !), surface charges develop around the wires spreading around the components, while a few accumulate at the nodes during the initial transient lasting a few pico- to nano-seconds at the most, producing electric fields inside the resistors, the transistor, and the space around the circuit. In the steady-state, the *resultant* electric fields inside the resistors, the transistor, the wires and the battery drive the mobile charges (electrons and holes) to produce currents, dc currents.

These currents (shown by arrows) and voltages set by surface charge gradients and interface charges, establish establish a quiescent operating point of a collector current, and collector to emitter voltage, in a linear region of the volt-ampere characteristics of transistor Q_1. The capacitors C_B, C_C and C_E charge up and do not pass dc currents in the steady-state.

When a small sinusoidal input signal v_s is applied to the circuit, the variation of input signal voltage v_s causes a sinusoidally varying electric field to be superimposed on the dc voltage, at the base of the transistor (Sections 3.2 and 3.3). A sinusoidally varying electric field is superimposed, over the electric field of the quiescent dc voltage, across the collector resistor R_C. We say, the transistor "amplifies" the input signal and the output voltage v_o is seen, to be larger in amplitude than the input v_s. A corresponding amplified voltage output v_o is seen, to develop across the load resistor R_L. The circuit reaches a sinusoidal steady-state (Section 2.13) condition.

The ratio v_o/v_s is called the gain of the amplifier.

Since the voltage gain, current gain, power gain, and input and output impedances of the amplifier to alternating currents and voltages (ac) are of chief interest to circuit designers, they prefer to develop an *ac model* from which these parameters may be easily calculated. The ac model does not incorporate the dc quiescent currents flowing, and dc quiescent voltage drops across the various components. This allows the designer to focus on studying the response of the circuit to small ac signal voltages.

To do this, they scrutinise manufacturers' specification sheets and pick from among several styles of parameter notations. The most popular are the *h*- or hybrid parameters, which are real numbers at audio frequencies.

A description of the *h*-parameter model of a transistor operated in the common-emitter configuration whose parameters are denoted h_{ie}, h_{re}, h_{fe} and h_{oe}, are beyond the scope of this book and the reader is referred to Ref. [17] for a detailed treatment. The aim of this section is to describe the aspects of the principle of superposition of electric fields and voltages which assist circuit analysts in developing ac small-signal equivalent circuits of complex circuits.

We will study how this became possible. Since the electric fields of the ac signals pulsate in the dc voltage source, due to processes described in "How can a voltage source close the circuit?" (see above Fig. 3.7), across the charged capacitors C_B of the biasing network and the output capacitor C_C, setting up pulsating components of signal currents in the dc voltage source and in the leads and plates of the capacitors, these can be replaced by short-circuits. Therefore, in the ac small-signal equivalent, each dc voltage source is replaced by its internal resistance, and if the source is ideal, by a short-circuit (Sec. 3.4). We should not forget that the principle of superposition of electric potential (signal and dc) that stemmed from the principle of superposition of electric fields (Section 3.1), allows us to do this. Thus, the dc biasing supply voltage V_{CC} is replaced by a short circuit, and the redrawn network NT1 in Fig. 3.37 shows, a short circuit across the supply voltage terminals.

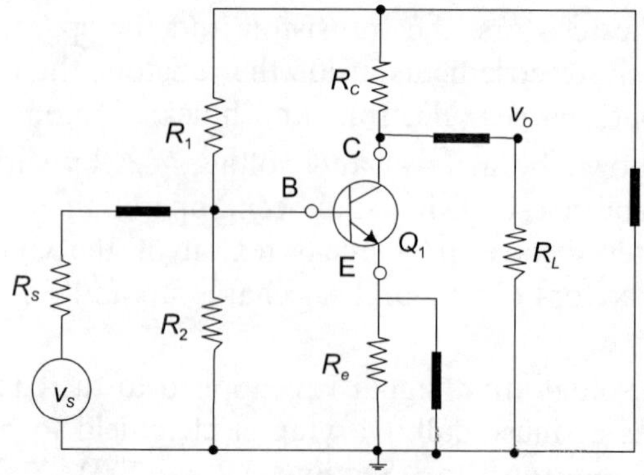

Fig. 3.37 Network NT1 redrawn with the capacitors and power supply replaced by short-circuits.

The circuit zero or "circuit ground" terminal of network NT1, is effectively the emitter E of the transistor, due to the following analysis: (i) the capacitor C_E passes the ac signal currents easily, and (ii) these ac currents produce a negligible voltage drop across R_e, and so the resistance R_e can be omitted from the ac equivalent network.

Since the capacitor C_E passes all signal currents to circuit zero or "ground", we are justified in considering the emitter E as circuit zero or "ground" in the small-signal ac equivalent network. The capacitors C_B, C_C and C_E in the network NT1 also easily pass ac signal currents to ground and are replaced by short-circuits in the redrawn network of Fig. 3.37.

The network NT2 is drawn, as if the signal voltages and currents through the biasing, emitter and collector and load resistors and transistor, are *extracted* from the dc (biasing) voltages and currents through these components, for the purpose of making the ac equivalent network and the principle of superposition of fields/potential allows us to do this.

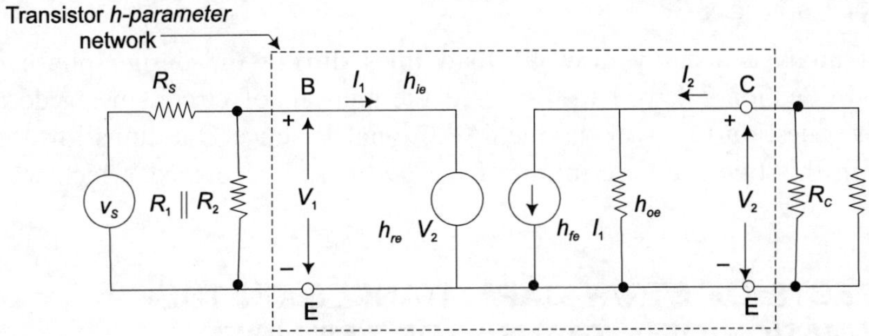

Fig. 3.38 ac small-signal equivalent Network NT2.

With the dc power supply replaced by a short-circuit (assume replaced with a short-circuit, not short-circuited because that would short the power supply!), the top of the resistance R_C which was connected to the positive terminal of V_{CC} is effectively connected to the E terminal, which is now the circuit zero or "ground" terminal.

Circuit analysts then incorporate the h-parameter model of the transistor of network NT1 into the network of Fig. 3.37 to produce the *ac small-signal* (since signal voltages typically are a fraction of the power supply voltage and the transistor will amplify such small signals linearly with little distortion) *equivalent* network NT2, of Fig. 3.38.

The *h-parameter* model is shown in the box of dashed lines in the ac small-signal equivalent network NT2 of Fig. 3.38.

Note that the source and load resistance connection values from the biasing network NT1 are incorporated into the network NT2. The collector and load resistors form a parallel network between the collector and the emitter "ground". Also, the resistances R_1 and R_2 are paralleled from the base terminal B, because V_{CC} is replaced by a short-circuit. Thus, the signal source "sees" a parallel combination (shown by the symbol $R_1 \| R_2$ in Fig. 3.38) connected to circuit zero or "ground" viz. the emitter E of the transistor.

When a weak time-varying input signal (a small signal) v_s from a temperature sensor or a microphone, is applied to the network, it produces time-varying currents, which superimpose over the dc currents in the elements (capacitors and resistors). An amplified version v_o of the input signal v_s, is obtained across the load resistor R_L, the dc component of the amplified signal having been blocked by the capacitor C_C.

The analysis for setting up the quiescent operating point by a dc voltage source is performed separately from the time-varying voltages and currents.

Once again, the reader should pause for a moment and visualize the rich blending of circuit processes viz. superposition of fields and potential, the transient, sinusoidal steady-state (Section 2.13) and steady-state phenomenon made by the development of surface charges, not forgetting the weak fringe field development of the capacitors C_B, C_E and C_C, which enables the dc (quiescent) and ac voltage (signal) and current analysis of the circuit of Fig. 3.36 to be performed easily and leads to the simplified small signal ac equivalent network NT2 of Fig. 3.38.

Circuit analysts usually draw ac load lines during the design phase of transistor amplifiers. In Section 3.4 we had discussed the analysis of circuits using dc network load lines. In the question Q.3.1 in the file "Additional Practice Questions for all chapters" in the CD with this book, a description of ac *load lines* is provided, which would be useful to read.

3.8 EFFECTS OF STRAY CAPACITANCES ON THE OPERATION OF A TRANSISTOR AMPLIFIER

In Section 2.21 we had studied the presence of stray capacitances between conductors. Stray capacitances limit the frequency range of operation of transistor amplifiers which strengthen weak signals. The theory and operation of transistor amplifiers is described comprehensively in Ref. [17]. The reader may defer to read this section until after becoming familiar with the theory and operation of transistor amplifiers.

The stray capacitances (Section 2.21), also called parasitic capacitances, described here, are between the base, emitter and collector leads and connecting wires of the transistor. These capacitances are *not* the *internal* capacitances of the transistor, which are known as transition and diffusion capacitance.

Knowing the circuit processes of the stray and parasitic capacitances of a transistor amplifier and their effects are useful to understand the effects that *internal* capacitances (transition and diffusion), have on transistor amplifier operation. The reader is referred to Refs. [17, 30] for a treatment of internal capacitances or interelectrode capacitances, which are beyond the scope of this book.

To understand stray capacitances, a visualization of capacitors connected in parallel is useful, as seen in Fig. 3.39.

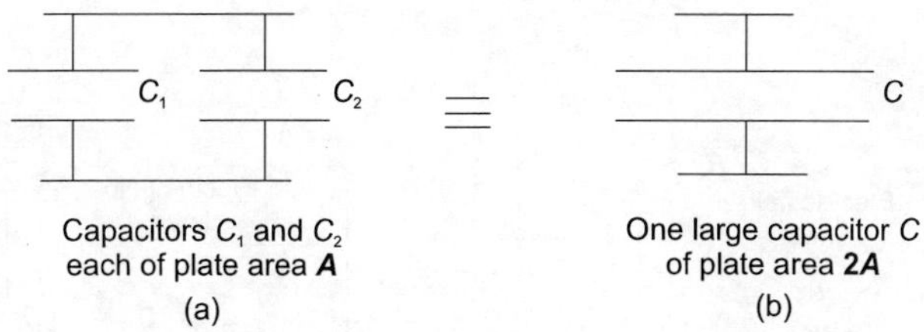

Fig. 3.39 (a) Capacitors connected in parallel (b) Equivalence to a single capacitor with large plates.

If the capacitors are constructed using plates of area A, then when two (or n of them) are connected in parallel (Fig. 3.39(a)), it is as if the plates of one of the capacitors, is extended to form one large capacitor of area $2\times A$ ($n\times A$) shown in Fig.3.39(b).

The other aspect, is the fact that the plates and the lead wires of the capacitor have a finite resistance. Remember that there is no ideal conductor with zero resistance. A capacitor connected to a voltage source exhibiting a leading current of 90° *exactly* with voltage, is an ideal approximation, and cannot be realized in practice. One other important aspect, is that the appearance of even a few charges on the plates, creates a fringe field, and this causes the current (charging and discharging current) to reduce, because $\vec{J} = \sigma \vec{E}_{\text{net}}$, where $\vec{E}_{\text{net}}$ is a resultant of the applied field and the fringe field. At high frequencies, the fringe field is weaker and the current density $\vec{J}$ is large, because the resultant field $\vec{E}_{\text{net}}$ will then be large. We have shown the current density as a vector $\vec{J}$ to indicate that the currents due to the stray capacitances are charging and discharging currents, and their directions driven by complex field configurations are from the interiors of the wires *towards* the surface, and this may not necessarily be entirely perpendicular to the axis of the wire. The direction of the current will depend on the geometry of the layout of the leads of the transistor and the connecting wires, a complex geometry, which produces a capacitance. We did get a hint of currents in a direction, other than along the axis of wires in Section 2.6, but since the focus was more on the charging and discharging processes and other aspects we did not care to dwell upon this aspect in that section. As a matter of fact, currents during the initial transient at circuit switch ON, and capacitor charging and discharging displacement currents are in several directions other than along the axis of wires. The induced charges on the surfaces due to the currents will have a complex configuration depending on the shape of the conducting leads, wires and resistors (See "Induced charge configurations" in Appendix C, self-, mutual- and stray-capacitance in Section 2.21).

We will study another example of different current directions other than along the axis of wires, in guided waves and transmission lines in Section 5.22.

Shown in Fig. 3.40 is an exaggerated view of the base, collector and emitter leads and wires, and the collector and emitter resistors of the input section of the transistor amplifier of Fig. 3.36, without the biasing resistors. The stray capacitances may be visualized as opposite plates of multiple capacitors, with one large plate being the base lead and wire, and the second plate(s) as the emitter and collector leads and wires.

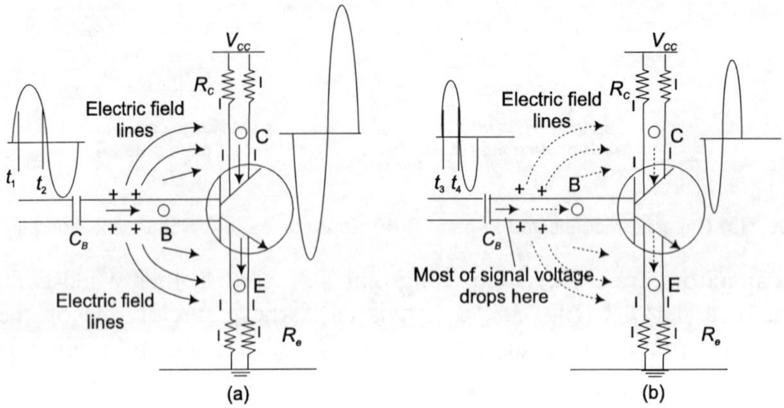

Fig. 3.40 Input section of a transistor amplifier with exaggerated views of transistor terminal leads and wires (a) with a low frequency input and (b) with a high frequency input.

The transistor leads and wires may be treated as long two-wire lines (base and collector pair and base and emitter pair) having a capacitance and a finite though very small resistance. This is sufficient to understand the effects which are described in this section.

When the signal frequency is *low*, when sinusoidal steady-state applies (Section 2.13), the surface charges (see positive and negative charges shown in Fig. 3.40(a)) set up by the signal, quickly allow the input base wire and lead to attain signal potential. The stray capacitances between the base lead and wire with their tiny resistances, and the collector lead and wire with their tiny resistances, charge (or discharge) easily, following signal excursions, because the time interval of signal changes (charge accumulation and dispersion) as compared to the time for charge accumulation and dispersion, is large. Note the large time interval between instants t_1 and t_2 shown by the low frequency input signal waveform in Fig. 3.40(a). Similarly, the stray capacitances between the base lead and wire with their tiny resistances, and the emitter lead and wire with their tiny resistances, charge (discharge) easily, following signal excursions.

The surface charges are set up by tiny pulsating currents which flow and permit surface charge accumulation, as well as dispersion, from within the base (*B*) lead, emitter (*E*) lead and collector (*C*) lead to the surface and on the surface, as well. See the source of the tiny pulsating currents in the **important note** below Fig. 1.23, Section 1.15. Based on current continuity conditions, the tiny pulsating current paths, can be identified through the power supply V_{CC}, the signal source v_s (see Figs. 3.36 and 3.38), the resistors R_c and R_e, the transistor base (*B*), emitter (*E*) and collector (*C*) leads and their bulk and junction crystal regions. The paths will form a complex web, but for the sake of clarity, focus is made on the transistor's base (*B*), emitter (*E*) collector (*C*) leads and the resistors.

The tiny pulsating currents, in all the components of the circuit, are displacement currents of the tiny capacitive reactances, superimposed on the normal ac and dc currents of operation. These tiny pulsating displacement currents, are small and slow enough, to be maintained by the circuit processes with their rapid responses, and cause the tiny (varying) signal electric field lines in the terminal leads, and deliver the surface charges which maintain the potentials (voltages) on the components (Section 1.16). The development of the fringe field by the tiny stray capacitances is very rapid. The electric field lines, seemingly originate on the positive charges, and seemingly terminate on the negative charges as shown in Fig. 3.40(a).

There are strong signal field lines (superimposed on the steady field lines of the dc current that set the operating point) in the resistor R_c. This is a desired state for operation of the amplifier; the base terminal and the base region in the semiconductor everywhere have attained the voltage of the signal, which allows the emitter to inject electrons into the base and they are collected by the collector following the signal potential, at each and every instant of time. There is almost no loss in amplifier gain, as seen in the voltage increase at the output at the collector C terminal (Fig. 3.40(a)), over the signal input voltage. The amplifier thus strengthens the weak input signal of low frequency.

However, when the signal frequency is *high*, assuming it's amplitude remains unchanged, the surface charges cannot accumulate and disperse fast enough, in response to signal variations, to charge and discharge the stray capacitance between the base lead and wire with their tiny resistances, and the collector lead and wire with their tiny resistances, within say, the very short time interval between instants t_3 and t_4 shown on the input signal waveform in Fig. 3.40(b). There is no fringe field development of significance; no surface charges that can maintain the potentials (voltages) on the components. The stray capacitive reactance (1/jωC) reduces, and the output voltage at the collector which is a 180° phase shifted amplified version of the input signal, as we will see in Section 3.9, is being directly negatively fedback at such frequencies to the base terminal which is carrying the input signal. This causes the input signal at the base (not the signal source) to reduce in magnitude, thereby reducing the output voltage.

The stray capacitance between the base lead and wire with their tiny resistances, and the emitter lead and wire with their tiny resistances, *prevents* surface charge accumulation and dispersion easily. The associated currents from within the conductor to the surface and on the surface are *stronger*, than if the signal were of a smaller frequency, because as described in Section 2.11, charges do not accumulate easily on the surface whose electric field (fringe field) can oppose the applied signal field.

The strong current changes so quickly in time and direction that the signal's voltage drops in the finite resistances of the base lead and the source resistance, noting that a strong current multiplying a tiny resistance, is a significant voltage drop (see Fig.3.40(b)). Coupled with the fact that insufficient surface charges cannot sustain the electric field *axially* in the wire near the base lead, the electric field is weak as shown by the dashed arrows in the base, emitter and collector wires near their leads in 3.40(b), so a weakened or attenuated signal level is on the base lead. See the effect of increasing the sinusoidal voltage frequency on capacitor current in Section 2.11 and Fig. 2.35. Another effect is that since the input current is bypassed through the parasitic or stray base to emitter capacitance to the emitter lead, this current does not register in the transistor as an input current. Thus, there is a further reduction of the amplifier output voltage, because the injected emitter current into the base, which produces amplification, has been reduced.

Thus, there are weak signal field lines (superimposed on the steady field lines of the dc current that set the operating point) in the resistor R_c.

Due to the low amplitude base voltage, the emitter injects fewer electrons into the base and the collected current by the collector is small. As a consequence, the amplifier gain drops and the output, at the collector of the transistor amplifier, will also be smaller (Fig. 3.40(b)) than the output of a signal at a lower frequency (Fig. 3.40(a)).

The circuit becomes sluggish in response, and is unable to keep pace with high frequency input signals. This is not a desired state for operation of the amplifier. There appear to be strong currents all around in the circuit yet, they are not in the places where they should be, or rather where we would like them to be, in the base, the emitter with its resistor R_e (axially to ensure the confined flow of current, not sloshing perpendicularly, Section 1.16) and the collector with its resistor R_c (axially to ensure the confined flow of current, not sloshing perpendicularly, Section 1.16).

The above descriptions are provided to develop an intuitive understanding of how the circuit processes involving the stray capacitances between a transistor amplifier's input leads and wires, limit the voltage gain at high frequencies. Input and output wire connections can be analyzed by modeling them as two-wire transmission lines and the reader may refer to Section 5.22 on Transmission Lines for more descriptions, where the finite inductance (that will be discussed in Chapter 5) of the lead and wire also begin to assume significance, as the frequency of the input signal is increased further.

3.9 APPLICATION OF THE PRINCIPLE OF SUPERPOSITION TO FEEDBACK AMPLIFIERS

The reader may defer to read this section until after learning the theory and operation of transistors[17] and their application in amplifiers. The theory of amplifier operation is beyond the scope of this book. The discussion is aimed to provide insights into the circuit processes which enable feedback in amplifiers. The section begins with the benefits of providing feedback in systems.

The Feedback Mechanism

Many tasks which we perform in our daily lives such as reaching out to pick up a glass of water are complex to study. Nature has built-in an elegant feedback mechanism involving sensors (touch and vision) and actuators (muscles in the hands and fingers) that accomplish the task. These components, combine in a way that makes the process seemingly simple, but are really complex.

One of the key points of the mechanism, is the steadiness brought about by the feedback built into the process of picking a glass of water, which continuously adjusts the movement of the hand (its velocity and acceleration) to reach the target. The signal to pick up the glass is initiated by the thirst we feel for water, and feedback is the moving image captured by our eyes of our hands and fingers. The images are interpreted by the brain, which sends signals to the nerves, in the hand and fingers, which constantly adjust triggers to the muscles. The muscles adjust hand and finger velocity and deceleration till the fingers curl around the glass. If there was no visual feedback, the movements of the hands and fingers would be distressingly clumsy and slow, or at the other extreme, grossly rash with disastrous consequences.

Designers frequently incorporate a negative (degenerative) feedback in amplifiers to modify its characteristics, by combining a portion of the output signal with the input signal to generate a correction signal, which is input to the amplifier. While negative or degenerative feedback does reduce the overall gain, the benefits of providing such a feedback far outweigh the loss of gain.

Some of the major benefits are reduction of (i) frequency and phase distortion, (ii) nonlinear distortion, iii) noise, and most important of all, an improvement in its stability against external influences, such as operating temperature fluctuations. We saw one example afforded by introducing an emitter resistor in the amplifier of Section 3.6.

One of the consequences of voltage shunt feedback, is a decrease of the input resistance, a subtle but noteworthy point and will become clear when we discuss the strength of electric field in the wires, leads and components of the voltage shunt feedback amplifier.

A detailed description of the effects of feedback in amplifiers is available in Ref. [17].

To accomplish negative feedback, designers first incorporate a means of sampling the output signal, which is then applied to a comparator section (not directly to the input !). The comparator receives another input from the signal generator. It compares the input signal against the sampled output signal, and generates an error (or difference) signal, which is applied to the input of the amplifier.

A single transistor amplifier using a bipolar transistor in the common-emitter configuration produces an amplified 180° phase-shifted version of the input. This makes the task of constructing a signal that can enable degenerative feedback, simple.

In this section, we make a qualitative study of the process of the feedback mechanism of a voltage shunt-feedback amplifier, built using a bipolar transistor in the common-emitter configuration, with the aim of introducing the reader to the nuances in the overall design. A detailed quantitative analysis of the voltage shunt feedback amplifier is available in Ref. [17].

Voltage shunt feedback is a basic principle, which finds application in a wide range of linear integrated circuits, especially operational amplifiers. The circuit processes described in this section will be useful in understanding the theory, and in the analysis of the operation of complex analog circuits that use operational amplifiers as building blocks.

A Voltage Shunt Feedback Amplifier

Consider the transistor amplifier of Section 3.7 Fig. 3.36 redrawn in Fig. 3.41.

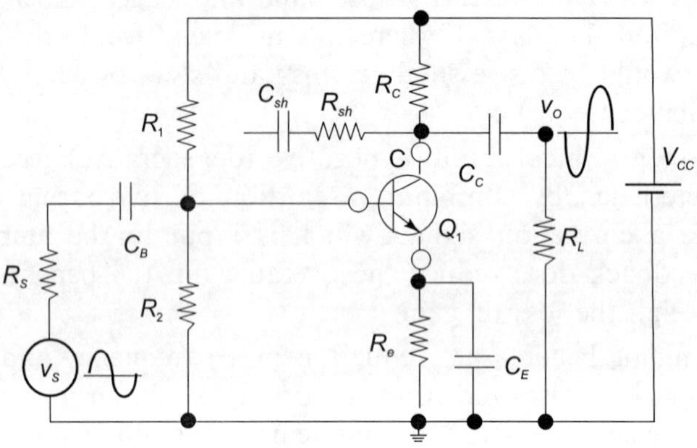

Fig 3.41 A transistor amplifier.

We shall not dwell upon the operation of the amplifier, because this is beyond the scope of this book. It should be sufficient for the purpose of this discussion to know that the amplifier biased in the "active region", strengthens the small input signal v_s and produces an output signal v_o which is larger in amplitude than the input signal, by a factor called the gain.

The amplifier in Fig. 3.41 is shown *without* shunt feedback. It does incorporate some amount of feedback provided by the emitter resistor R_e about which a description was given in Section 3.6.

The shunt feedback in the design, should be in a manner, such that the strength of the input signal to the amplifier be reduced, with reference to the strength of the output ac voltage.

For this purpose, it is necessary to *sample* and scale the output voltage, especially keeping in view, the large amplitude of the output voltage. We identify the collector terminal of the transistor to tap the output voltage from. This point has the advantage, of providing a ready-to-use negative feedback source as well, because the output voltage is a 180° *phase shifted* amplified version of the input signal voltage.

When the input signal is a positive going half of a sinusoid, the current in the collector-base-emitter bulk crystal increases in strength, and the output voltage at the collector reduces, and is negative going. For a description of the process that generates a negative going output voltage when current increases in the transistor crystal regions, refer to Fig. 3.24, Section 3.4.

To provide negative feedback to the amplifier, therefore, we introduce a series resistor (R_{sh}) and capacitor (C_{sh}) combination, at the collector terminal in the circuit diagram of Fig. 3.35. It is not shown connected, because we have to identify a possible comparator point (i.e. a node), where the sampled output ac component of the signal at the collector can combine with the input signal. In a shunt feedback, the feedback signal is connected to a node, which couples to the input of the transistor.

An exaggerated view of the input section of the transistor amplifier with the resistors R_e, R_c and R_{sh} is shown in Fig. 3.42.

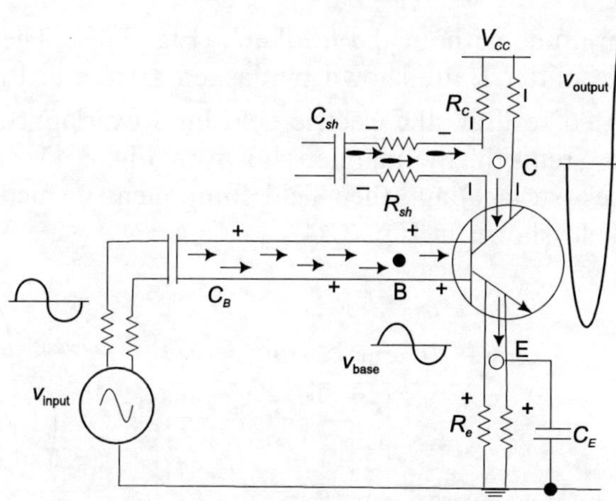

Fig. 3.42 Input section of a transistor amplifier incorporating voltage shunt feedback (feedback OFF) showing the electric field lines in the exaggerated views of resistors, wires and terminal leads.

The surface charges shown in Fig. 3.42 are indicative of the source charges that set up the *signal* component of the electric field lines, during the positive excursion of an assumed sinusoidal signal voltage. The negative charges shown on the collector and the lead to the resistor R_{sh}, indicate the reduction of the output voltage when the input voltage sinusoid is positive going.

The appropriate comparator point for shunt feedback will be, the Base terminal. By connecting the open end of the capacitor C_{sh} to the base terminal (B), we will be providing an electric field to the base terminal wire from a node, whose voltage is *reducing* in magnitude against the rising input voltage at the base (B) as shown in Fig. 3.43.

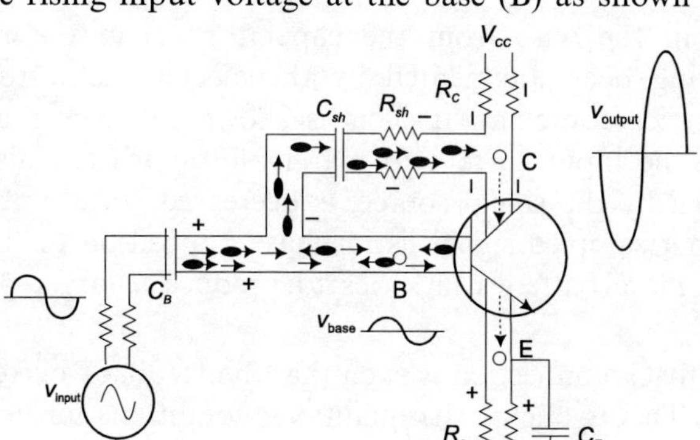

Fig. 3.43 Input section of a transistor amplifier incorporating voltage shunt feedback (feedback ON) showing the electric field lines in the exaggerated views of resistors, wires and terminal leads.

The electric field lines due to the feedback voltage from the collector terminal are shown in Fig. 3.43 by elliptically tailed arrows.

The normal arrows, depict the direction and field strength of the signal voltage source. Note the reduced amplitude of the output voltage with shunt feedback (Fig. 3.43), in comparison to the amplitude without the feedback (Fig. 3.42). The reduced field strength at the collector C and emitter E are shown by dashed arrows in Figs. 3.43 and 3.44.

We are not permitted to draw the electric field lines overlapped (Section 1.6) in some segments of the input circuit, in the manner shown in Fig. 3.43, which we did merely to show the source charges (surface) and their field components, which contribute to produce the resultant electric field shown in Fig. 3.44.

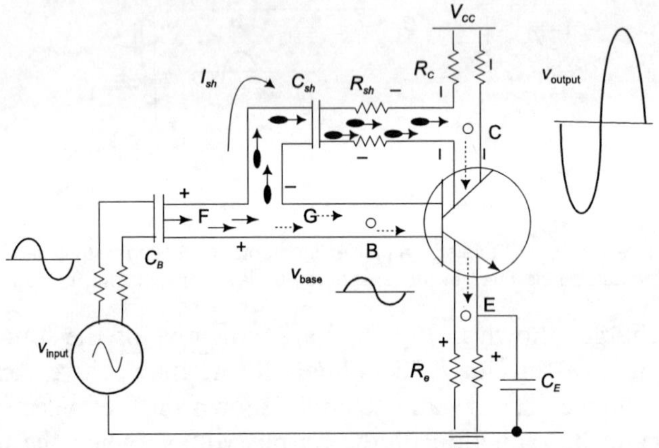

Fig. 3.44 Input section of a transistor amplifier incorporating voltage shunt feedback (feedback ON) showing the electric field lines in the exaggerated views of resistors, wires and terminal leads.

Decrease of Input Resistance

The electric field in the wire from the capacitor C_B in the region F (Fig. 3.44) is now stronger, having been augmented by the electric field from the shunt feedback resistor-capacitor series combination. This is shown by thicker arrows in the region F, and is a cause for the input current to marginally increase. This is the reason why the input resistance seen by the signal source, is decreased. Intuitively, this may be expected given that the output sampled signal, is in phase opposition to the input signal. Imagine connecting two or more batteries in series to produce a larger potential difference than that of one battery.

The circuit feedback connection is such that the feedback current I_{sh} is proportional to the output voltage. This is one of the qualifying conditions for an amplifier to be termed a "voltage-shunt feedback amplifier".

The Feedback Effect

The electric field in the wire from the node, where the shunt wire from the capacitor C_{sh} joins the wire from the capacitor C_B in the region G (Fig. 3.44), is now weaker having been reduced by the electric field from the shunt feedback resistor-capacitor series combination. The base voltage (v_{base}) reduces in magnitude (Figs. 3.43 and 3.44).

It is this weaker field that reduces the current into the base terminal (B) of the transistor, causing an emitter current, reduced from that without the shunt feedback. The output voltage is reduced in amplitude compared with the output voltage without feedback (Fig. 3.42), and this is shown by the smaller amplitude output voltage waveform in Fig. 3.44.

Computing the Values of the Shunt Feedback Resistor and Capacitor

The values of the shunt feedback resistor R_{sh} and the capacitor C_{sh} are chosen, such that the amount of reduction desired in the input voltage, produces the desired amount of output voltage.

In the following, a simplified analysis of the computation is provided. Expressions to obtain the values of shunt feedback resistor R_{sh} are available in Ref. [17].

A circuit diagram including values of the circuit components of a voltage shunt-feedback amplifier that was assembled, is shown in Fig. 3.45(a).

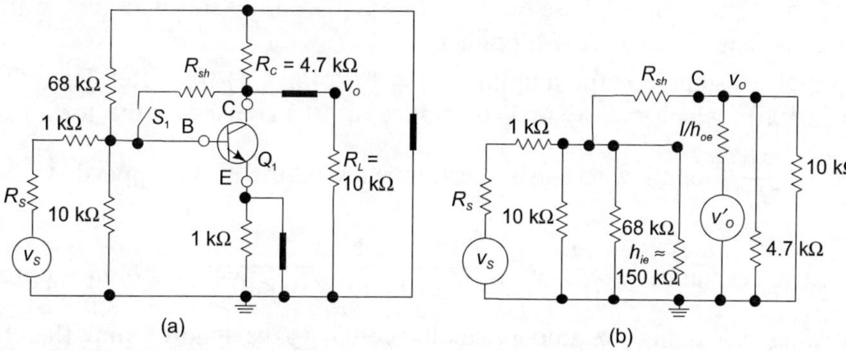

Fig. 3.45 Small-signal ac equivalent of a transistorised voltage-shunt feedback amplifier.

A transistor output circuit is usually modeled as a dependent current source, rather than, as a dependent voltage source, in the small-signal ac equivalent model, because the transistor supplies current which does not vary much when the collector-emitter voltage varies. It is determined to a large extent, by the *number of electrons injected* into the base by the emitter through both the junctions, to the collector terminal. The voltage that aids in the flow of drift current is the collector to emitter voltage, provided by the dc power supply V_{CC} (Fig. 3.41), *excepting* at the emitter junction. In the emitter junction the field is opposite in direction (as in forward biased diodes, see Figs. 9.13 to 9.16) to the field in the bulk regions and the collector junction. The emitter junction being forward-biased, the current in it is by the process of *diffusion*. The electric field due to the power supply V_{CC} is set up by surface charges in the transistor.

Both these processes, diffusion and drift, combine harmoniously in the steady-state to establish a current, and importantly, the current must obey the principle of conservation of charge. An electric field due to the power supply is set up by surface charges in the collector resistor, as well.

Using Norton's equivalent (Ref. [15]), the transistor is depicted as a dependent (on base current) current source ($h_{fe}I_1$ or $h_{fe}I_b$ Ref. [17]) with a parallel resistance (the output resistance of the transistor), as shown in Section 3.7 (Fig. 3.38). The current source model can be replaced by a Thevenin equivalent as a voltage source v'_0 with a series resistance (the output resistance of the transistor) (Ref. [15]), which we shall adopt for our analysis as shown in Fig. 3.45(b).

The equivalent circuit of Fig. 3.45(b) shows the output voltage v_o at the collector terminal, and the collector through which the output ac current flows superimposed over the dc quiescent current, and the load resistance through which the output ac current flows.

We are not interested in the value of the emf v'_o, which is rather difficult to determine. It is the output voltage v_o at the collector *C*, which we shall specify in a demonstration numerical example.

A circuit of a voltage shunt feedback amplifier was assembled around a transistor type BC107 with V_{CC} = 12 Volts, and the emitter resistor used was 1 kΩ. *All* the capacitors in the circuit were 10 μF and large enough to act as short-circuits at the frequency of 10 kHz of the input signal. These are therefore replaced, by short-circuits in the simplified small-signal ac equivalent of the amplifier.

The input resistance of the amplifier is $h_{ie} + R_e\ (h_{fe} + 1)$ from Ref. [17]. The capacitor C_{sh} used is 10μF, which has a small reactance at 10 kHz. Using the formula, capacitive reactance $= \dfrac{1}{2\pi fC}$ ohms where, *f* is the frequency of the signal applied, viz. 10 kHz, we obtain

$$\text{Capacitive reactance} = \frac{1}{2\pi fC} = \frac{1}{2\times\pi\times10\times10^3\times10\times10^{-6}} = 1.5 \text{ ohms approx.}$$

The value, 1.5 ohms, we find is small enough to be ignored and therefore justifies replacing C_{sh} by a short-circuit as shown in Fig. 3.45(a). Thus, the input resistance will be h_{ie}, the $R_e(h_{fe} + 1)$ term being effectively zero for ac signals.

To demonstrate the action of the amplifier without and with feedback, the circuit was energised and with input signal v_s applied, the values of the ac collector voltage and ac signal voltage at the base of the transistor were measured using an oscilloscope.

The frequency of the sinusoidal signal generator was set to 10 kHz and signal voltage to 17.7 mV(RMS).

The output voltage v_o measured 1.414 V_{rms} without feedback, set by switching OFF the shunt feedback switch S1. The voltage at the base terminal of the transistor measured 14 m V_{rms} (millivolts).

The voltage gain of the transistor amplifier is hence, $\dfrac{1414\,mV}{14\,mV} = 101$, i.e., the input at the base terminal is strengthened approximately 100 times.

At the base of the transistor, the input signal generator strength is attenuated slightly from 17.7 mV to 14 mV, due to the presence of the voltage divider circuit, comprising the series resistors of the source resistance R_s, the external 1 kΩ resistor tied to it and the parallel combination of resistors 10 kΩ || 68 kΩ || h_{ie} as shown in Fig. 3.46(a).

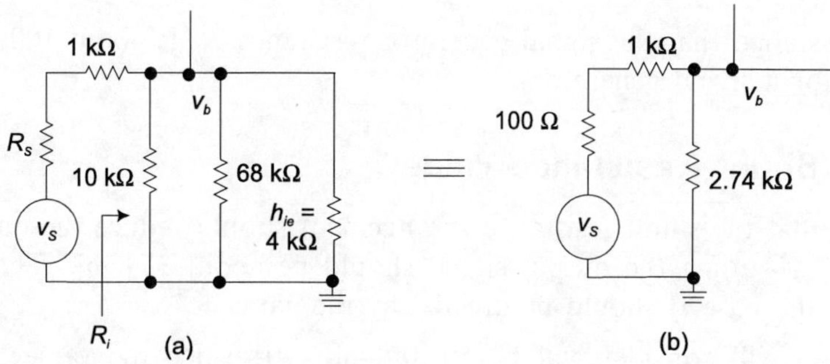

Fig. 3.46 Equivalent circuit of signal generator with a voltage divider.

The last resistor in the parallel combination viz. h_{ie}, is the ac input resistance of the transistor, and is 4 kΩ from published data of the BC107 transistor.

Note: For a transistor, connected in the common-emitter configuration, the h-parameter h_{ie} is the fraction given by a small change in the base-emitter voltage to the amount of *change* in base current Ref. [17]. The reader should be careful to avoid, being misled into assuming that h_{ie} should be a small value like the dynamic resistance of a diode, which is a few ohms or tens of ohms.

As a matter of fact, at a given forward biased voltage between the base and emitter of a transistor and the anode and cathode of a silicon diode, the diode current is several orders of magnitude larger than the base current of the transistor.

This is, because for an *n-p-n* transistor, the base current represents that component of the hole current from the base terminal, which recombines with electrons injected into it by the emitter, when the base emitter junction is forward-biased. The hole recombination current is not so large as to completely recombine with *all* the injected electrons, because most of the emitter injected electrons would have transited the narrow base region into the collector base junction. The tiny recombination current is a good indicator of the amount of collector current in the transistor.

Thus, the parameter h_{ie} is in the range of kilo-ohms. I urge the reader to compare the values of current in the *VI* characteristics of a silicon diode, with that of the input characteristics of a silicon transistor in the common-emitter configuration in Ref. [17].

The input ac resistance R_i as seen by the signal generator (Fig. 3.46(a)), after simplification using nodal analysis, works out to about 2.74 kΩ and is shown in Fig. 3.46(b).

Applying the voltage divider rule to the series combination of the resistors R_s, the external 1 KΩ resistor tied to it and the equivalent resistance 2.74 kΩ, we obtain,

$v_b = \dfrac{17.7 \text{ mV}}{(100\ \Omega + 1\ \text{k}\Omega + 2.74\ \text{k}\Omega} \times 2.74\ \text{k}\Omega \cong 13$ mV at the base terminal (B) which agrees reasonably with the measured 14 mV.

We have assumed that the signal generator resistance R_S is about 100 Ω, which is not unreasonable, for a signal generator.

Selecting a Shunt Resistance Value

If it is desired that the shunt resistance arrangement should reduce the output voltage by a factor of say 1/4, then the output signal should be about 354 mV. The corresponding input signal at the base B should be divided by the gain.

The gain we will consider will be 50 and not 101 since the value of h_{fe} the small signal current gain, is not constant with current in the small current ranges, of the BC107 transistor, and this impacts the voltage gain.

Therefore, $\dfrac{\text{output voltage}}{50} = \dfrac{354 \text{ mV}}{50} \approx 7.08$ millivolts. That is, asking for a reduction of the output voltage at the base by [354 millivolts (the output voltage which is the source of the feedback voltage) – 7.08 millivolts] = 347 millivolts.

To calculate the value of the shunt resistance R_{sh} we construct the equivalent circuit shown in Fig. 3.47(a) after plugging in, the calculated value of input ac resistance seen by the signal source.

Since the current through the transistor voltage source v'_o flows through the internal resistance $1/h_{oe}$ and also supplies current to the collector and load resistors, a computation of the internal resistance "looking into" terminal C (collector) is necessary, to account for voltage drops which these will produce, when the resistor R_{sh} is connected in the feedback circuit and current flows in it.

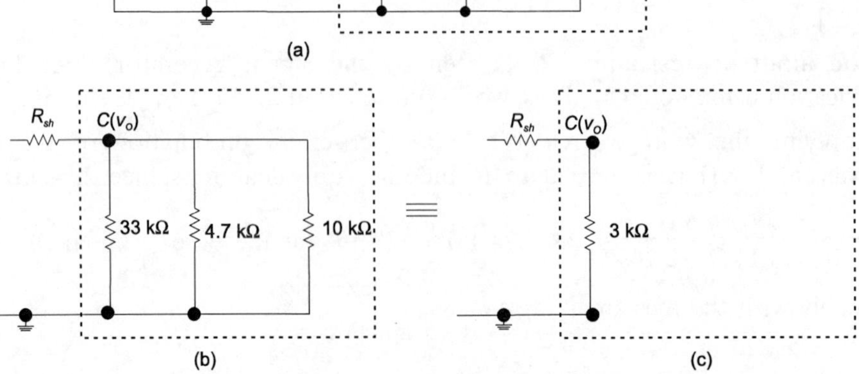

Fig. 3.47 Equivalent circuit to calculate the shunt resistance value.

The output resistance of the transistor is $1/h_{oe}$, where h_{oe} is a transistor h-parameter representing the output conductance in the common-emitter configuration. Its reciprocal $1/h_{oe}$ gives, the internal resistance of the transistor, between the collector and emitter. The transistor voltage source v'_o being 180° phase-shifted from the signal source, we use a minus sign to indicate its polarity (Fig. 3.47(a)).

We have chosen C_{sh} as 10 μF (reactance = 1.5 ohms), and replaced by a short-circuit in the ac equivalent circuit (Fig. 3.47(a)). The main purpose of the capacitor C_{sh} is to block away the dc component of the output signal to the shunt feedback comparator node, which it does, by charging up to the average value (dc component) of the output voltage.

Note: For a transistor, connected in the common-emitter configuration, the h-parameter h_{oe} is the fraction given by a small change in the collector current to the amount of *change* in collector voltage Ref. [17]. The unit of h_{oe} is of conductance, expressed in mhos or Siemens. For an *n-p-n* transistor, the collector current represents that component of the current due to electrons, which have transited into the collector base junction. Since the number of electrons which have reached the collector, is determined to a large extent by the number injected into the base by the emitter, there is not much change in the collector current with a change in the collector-emitter voltage, when the transistor is operational as an amplifier and has been biased accordingly. The collector current can be changed by injecting more or fewer electrons, by increasing or decreasing the forward base-emitter voltage. Thus, the parameter h_{oe} which is a conductance (in units of mhos or Siemens), is rather small, and its reciprocal which represents resistance is large, and is of the order of several kilo-ohms.

What is the physical significance of the output conductance h_{oe} of a transistor?

Figure 3.48 shows three views of the collector, base and emitter crystal regions of a silicon *n-p-n* transistor operating in the cut-off, active and saturation regions.

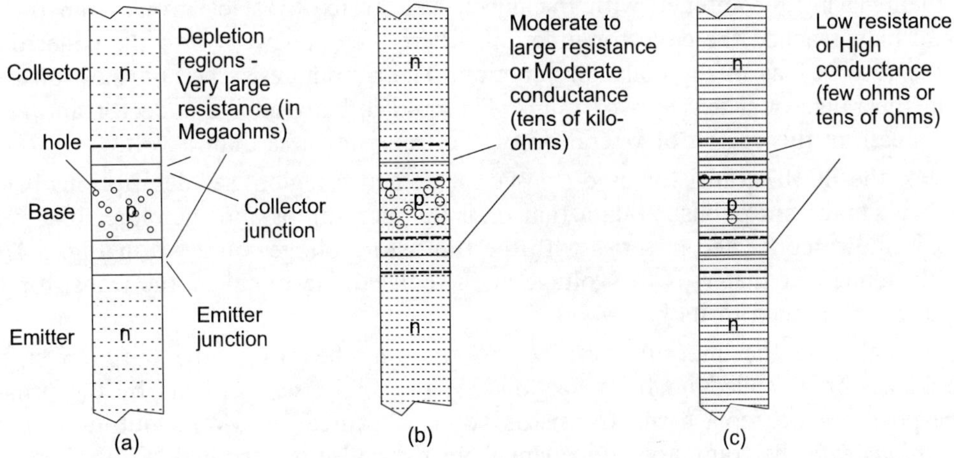

Fig. 3.48 (a) The electrons and holes in the emitter, base and collector regions of a transistor (open-circuit or *cut-off region*) (b) The electrons bridging the depletion regions of the collector and emitter junctions (*active region*) (c) Large numbers of electrons in the depletion regions of the collector and emitter junctions (*saturation region*).

In Fig. 3.48(a), the transistor is *open-circuit* or operating in the *cut-off region*, and the collector junction and emitter junction are depleted of carriers. Therefore, the transistor has a very large resistance between collector and emitter. Each of the depletion regions maybe not more than 0.0005 mm wide, compared to the overall length of say, 3 mm of a typical transistor. The width of the region is, 1/6000 times the length of the transistor, yet, that is small enough a gap, with none or a few mobile carriers, to cause the resistance to range in the order of a hundred kilo-ohms upto a few Mega-ohms. The emitter junction can be considered to be a *pn* junction diode whose resistance can be explained as we will in Chapter 9 Section 9.5.

Figure 3.48(b) shows the transistor operating in the "*active region*", in which the emitter junction is forward-biased and electrons from the emitter are injected into the base region. Note the large number of electrons which have bridged the depletion regions of both the junctions. A few holes in the base region combine with the injected electrons but, most of the electrons reach the collector junction, where they are swept into the collector region by a reverse bias at the collector junction. The output resistance h_{oe} of the transistor will be in the range of a few tens of kilo-ohms. Manufacturers usually provide values for h_{oe} of transistors operating in the "*active region*", the region in which the transistor is biased to achieve amplification. It should ideally be the value at the quiescent operating point, but manufacturers usually provide a minimum and a maximum or, a typical value at some specified operating point of collector current and collector-to-emitter voltage.

The collector to emitter voltage, of a transistor operating in the active region, will be a fraction say, one-fifth to one-half of the supply voltage.

Figure 3.48(c) shows the transistor operating in the "*saturation region*", in which the emitter junction is forward-biased *heavily* and a large number of electrons are injected into the base region from the emitter. Note the large number of electrons which have bridged the depletion regions of both the junctions. Very few holes in the base region remain, most of them having recombined with the injected electrons. Most electrons from the emitter would have reached the collector junction, where they are swept into the collector region by the reverse bias at the collector junction. The output resistance of the transistor will be in the range of a few tens of ohms. The voltage between the collector and emitter is very small, in this region of operation of the transistor, less than a volt.

For the BC107 transistor used (biased in the active region, see descriptions below Fig. 3.41), we find from published data that h_{oe} is 30 micro-mhos and therefore, $1/h_{oe}$ is 33 KΩ. This is shown connected in series with the transistor voltage source v'_o in Fig. 3.47(a). We should remember that v_o is the voltage at the collector terminal of the transistor with an input signal applied to its base.

To calculate the internal resistance of the amplifier, using the output resistance of the transistor. Recollecting from Section 3.5, the analysis of circuits by the principle of superposition, we replace the transistor voltage "source" v'_o by its output resistance $1/h_{oe}$ *looking into* the transistor equivalent from the collector terminal '*C*' as shown in Fig. 3.47(b). And, since the collector and the load resistances, 4.7 kΩ and 10 kΩ respectively, are effectively connected between the collector terminal C and circuit zero or circuit ground, it is easy to compute the equivalent resistance of the parallel combination of all the resistances between C and circuit zero viz. 33 kΩ || 4.7 kΩ || 10 kΩ. This works out to approx. 3 kΩ and is shown in Fig. 3.47(c). The equivalent resistance of 3 kΩ will be the internal resistance of the output voltage "source" equivalent v_o.

To calculate R_{sh}, we define $R'_{sh} = R_{sh} + 3\text{k}\Omega$ and use the principle of superposition of potential (Section 3.5) to compute the voltage drop across R'_{sh} and by equating this to the expected voltage drop across it, determine the required value of R_{sh}.

To compute the current in the shunt resistor R_{sh} we compute the current contributions made by each source v_o and then v_s and add up the currents.

To do this, we first replace the signal generator v_s by a short-circuit and replace the transistor voltage source v'_o by the output voltage source equivalent v_o (the targeted feedback incorporated reduced output voltage at the collector that we computed earlier to be –354 mV) and compute the current i_1, as shown in Fig. 3.49(a).

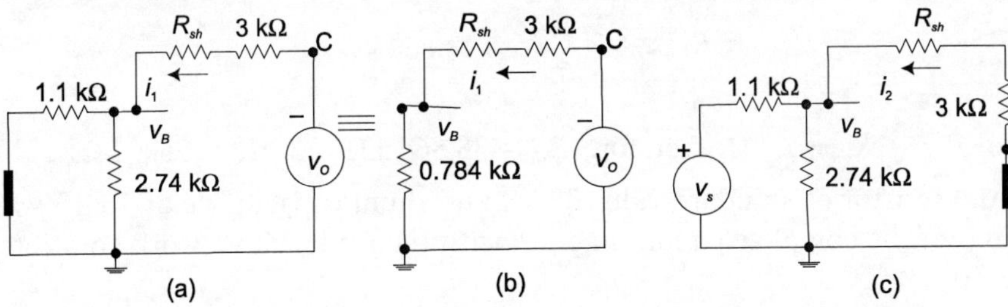

Fig. 3.49 Equivalent circuit to calculate the shunt resistance R_{sh}.

The internal resistance is now shown in series with the desired output voltage (source equivalent) v_o in Fig. 3.49. The current i_1 is given by

$$i_1 = \frac{v_o}{3\,\text{k}\Omega + R_{sh} + 1.1\,\text{k}\Omega \,||\, 2.74\,\text{k}\Omega}$$

$$i_1 = \frac{-354\text{ mV}}{R'_{sh} + \dfrac{1.1\,\text{k}\Omega \times 2.74\,\text{k}\Omega}{1.1\,\text{k}\Omega + 2.74\,\text{k}\Omega}} = \frac{-354 \times 10^{-3}}{R'_{sh} + 784\Omega} \tag{1}$$

where, $R'_{sh} = R_{sh} + 3\text{ k}\Omega$.

Since the parallel combination of resistances 1.1 kΩ and 2.74 kΩ give an equivalent resistance of 784 Ω, using this value we draw the simplified equivalent diagram Fig. 3.49(b).

Next, to compute the current i_2, we replace the output voltage source equivalent v_o by a short-circuit as shown in Fig. 3.49(c) and use the signal voltage strength of 17.7 mV(RMS).

$$i_2 = \frac{-17.68\text{ mV}}{1.1\,\text{k}\Omega + \dfrac{2.74\,\text{k}\Omega \times R'_{sh}}{2.74\,\text{k}\Omega + R'_{sh}}} = \frac{-(48.44 + 17.68 \times 10^{-3}\, R'_{sh})}{3.84\,\text{k}\Omega \times R'_{sh} + 3.014 \times 10^{6}} \tag{2}$$

The current is conventional and negative, because it will flow in the direction opposite to the assumed current direction of i_2.

We can ignore the current i_2, because it will make a very small contribution to the overall current in R'_{sh}.

Therefore, the voltage drop to be produced by the current i_1 is computed using eq. (1) by dropping the minus sign and then multiplying it with R'_{sh} and equating it to the required voltage drop of 347 millivolts. Thus,

$$\frac{354 \times 10^{-3} R'_{sh}}{R'_{sh} + 784\ \Omega} = 347\ \text{mV}$$

Solving for R'_{sh}, we obtain the following:

$354 \times R'_{sh} = 347 \times R'_{sh} + 347 \times 784$

or, $$R'_{sh} = \frac{272048}{7} = 38.864\ \text{k}\Omega$$

Since, $R'_{sh} = R_{sh} + 3\ \text{k}\Omega$, therefore $R_{sh} = 38.864\ \text{k}\Omega - 3\ \text{k}\Omega = 35.8\ \text{k}\Omega$.

A shunt resistor of standard value 33 kΩ and shunt capacitance of 10 μF was connected to the circuit. The complete circuit diagram with the feedback network connected, is shown in Fig. 3.50.

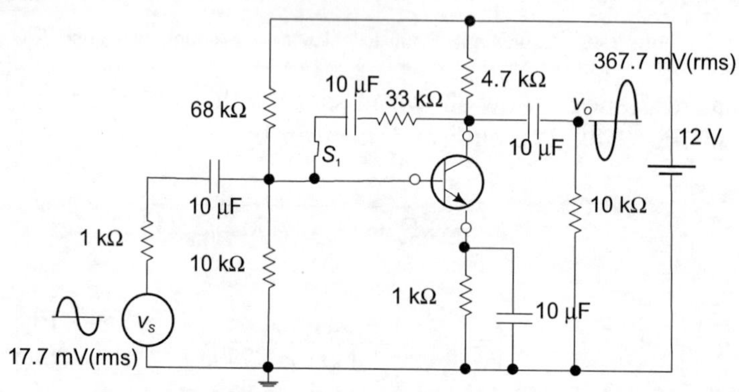

Fig 3.50 A transistor amplifier with voltage shunt feedback.

With the feedback switch S_1 closed (Fig. 3.50) and a sinusoidal signal of 17.7 millivolts(rms) at a frequency of 10 kHz applied, the output voltage obtained was about 367 mV(rms) against the expected 354 mV(rms), and input signal strength measured at the base was about 6.3 mV (rms) against the expected 7.08 mV(rms). The measured values are quite in agreement with the expected values, despite the several approximations which were made in the computations.

The 180° phase shifted collector voltage, is primarily responsible to reduce the field strength in the base terminal and wire, and it is the shunt connection which produced the necessary electric field configurations in the resistors, leads and wires (Fig. 3.44) that enable degenerative feedback of the amplifier.

The reader should also note that the shunt resistor (33 kΩ) is much larger in value than the parallel combination of resistors (1.1 kΩ || 2.74 kΩ) (Fig. 3.49(c)), which is 784 Ω (Fig. 3.49(c)), and it is that which enabled the output voltage to drop *heavily* to produce a small "sampled" voltage to be applied to the base terminal.

The reader is urged to think through the process of stabilization action from the field perspective assuming a change (increase or decrease) of the transistor's current gain parameter h_{fe}, noting the effect on the change in small signal collector current, it's effect on collector voltage, effect on shunt-arm current and the base current (Refer Section 3.6 also).

Operational amplifiers are integrated circuits that provide very large voltage gains (without feedback) ranging from 50,000 to beyond 400,000 and are mostly used with voltage shunt feedback to provide stable finite voltage amplification of weak signals along with a host of other benefits.[17] With such large "open-loop" gains, the feedback provided by external resistors in operational amplifiers is "heavy", to produce finite outputs.

Operational amplifiers have wideranging use in industrial, medical and space control applications.

SUMMARY

Commencing with the application of a sinusoidal voltage from a signal generator to an initially charged capacitor-resistor combination, we learned the processes which make the generator provide a closed path for the capacitor to discharge. To understand the voltage profiles in the discharge process, the principle of superposition of potential or voltage was applied. We learned that this principle is an extension of the principle of superposition of electric fields.

Next, the application of a small sinusoidal voltage signal to a charged capacitor-resistor combination was described, and the circuit processes that cause the signal to superimpose on the dc voltage across the capacitor were studied. We learned that the signal superimposes on the dc voltage, but because the capacitor has a closed path of the circuit to discharge through, the output is a signal superimposed on a slowly discharging (due to the resistor) voltage until, after a certain interval of time, the output across the resistor is the sinusoidal signal alone, the dc voltage of the capacitor having discharged to zero.

We also learned that if the signal voltage is much smaller in amplitude than the dc voltage across the capacitor, then the current through the voltage source does not reverse.

When a resistive divider network with a dc voltage source was used to provide a dc voltage across the capacitor, and a sinusoidal voltage signal applied to the CR combination, we observed that at the output across the resistor, the signal remains superimposed on the dc voltage and the dc voltage does not drop in magnitude, being sustained by the resistive divider and dc voltage source arrangement. This process was studied in quite some detail.

We learned that a capacitor is used in a similar manner, to couple time-varying signals generated by microphones to transistor amplifiers that need a small constant dc voltage across their base–emitter terminals, and the signal riding on that dc voltage is necessary to make the transistor amplify or strengthen the microphone signals faithfully.

Loadlines are used very often by network analysts to analyze networks graphically, and after reviewing the migration of surface charges on components such as resistors and diodes connected to one terminal of a voltage source, we studied their application involving simple networks with the Volt-Ampere characteristics of resistors, diodes and transistors. We learned how loadlines are used to determine the operating point of circuit devices.

The principle of superposition of potential is of tremendous importance in understanding the theory of circuits and in circuit analysis. We studied how the principle is used to determine unknown currents in networks of two loops. We then learned that steady negative currents in real batteries will cause them to heat up, and the observed current values in them will differ from computed values. Therefore, analysts prefer to use "ideal" voltage sources to determine the currents. We learned the definition and use of "ideal" voltage sources.

One commonly found amplifier is the common-emitter bipolar transistor, which is used to strengthen (amplify) weak signals of microphones and sensors used to measure and monitor temperature and pressure. To stabilize the amplification of such an amplifier, an emitter resistor is used, but if used alone, it reduces the strength of amplification of the input signals. Therefore, it is common practice to connect a capacitor across the emitter resistor, to restore the loss of gain. We studied how the principle of superposition of potential is applied to understand the process, which makes the capacitor bypass signals in the audio-frequency range and thus restore the amplification or gain.

The analysis of transistor amplifiers is performed using "small-signal ac equivalent circuits". We learned how the principle of superposition of potential enables analysts to neatly remove the portions of the amplifier network that set the operating point in them, and then using *h*-parameters, which characterize the behavior of transistors to ac inputs, draw up a "small-signal ac equivalent circuit", which retains only those aspects of the amplifier that produce amplification of input ac signals. This considerably simplifies computations to determine several important ac characteristics of an amplifier.

Stray or parasitic capacitances weaken signal inputs to amplifiers, especially at high frequencies, thus, rendering them incapable of generating amplified outputs at those frequencies. Applying a knowledge of the processes we learned in Chapter 2 and earlier sections in this chapter, on the principle of superposition of electric fields, we learned that the large currents that flow due to the stray capacitances between conductors and transistor terminals produce voltage drops and weaken signal voltages at the input terminals. The stray capacitance, which has a smaller reactance at high frequencies, causes the 180° phase shifted output voltage to directly couple with the input thereby, reducing input signal strength and causing the output voltage to drop. Further, the stray capacitance also bypasses to the emitter, the useful part of the signal required in the base emitter junction, thus causing a further reduction in the output voltage. In the same section, we learned how currents in components at high frequencies tend to flow radially rather than axially, thereby justifying the convention of treating the current density J as a vector.

We then learned generally how feedback samples the output of a system, which when fedback to a correction module, generates a correction signal. The correction signal when input to the system, can regulate it, such that the output remains within certain predefined limits or, in short can stabilize system operation. This was elaborated with the example of the motion of one's hand to grasp a glass of water. Similarly, the stability of amplifiers we learned, can be generally improved, by providing a kind of automatic negative feedback, which corrects for changes in the characteristics of transistors due to environmental (mainly temperature) variations during operation. Feedback reduces noise and distortion of the output signal. The technique of using negative voltage-shunt feedback for an amplifier was then described in detail, and using the principle of superposition of fields and of potential, the circuit processes that enable stabilization to be achieved were then explained using a numerical example. We also learned that the technique is applied to operational amplifiers, one of the major building blocks of analog integrated circuits, which find widespread use in numerous electronic circuitry designed for industrial, medical and space applications.

Questions

1. How does a charged capacitor discharge and then charge to the voltage of a small dc voltage when connected with a resistor as shown in the figure?

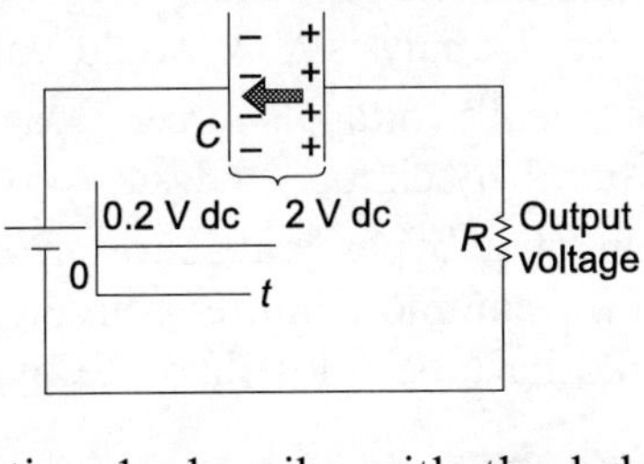

2. For the circuit of question 1, describe with the help of neat sketches the output voltage and the capacitor voltage with respect to time and what will be the voltage after a long time has elapsed.
3. A resistor *R* and charged capacitor C series combination is made to discharge by short circuiting their open ends. Describe with the help of neat sketches how the size of the capacitor and the resistor affect the time taken to discharge.
4. What is the effect of using a small value of capacitance for the capacitor *C* in the circuit of question 1 if the input is a small amplitude sinusoid? Explain with the help of neat sketches.
5. Describe with the help of neat sketches the output voltage with respect to time of the circuit of question 1 when subjected to a small amplitude sinusoidal voltage for a few cycles and after several cycles have elapsed.
6. Explain how the circuit of question 1 can be used with suitable modifications to couple ac signals to common-emitter transistor amplifiers where the requirement is to superimpose the ac signal on a dc voltage which biases the transistor in the active region.
7. Describe how the principle of superposition of electric fields may be used to validate the principle of superposition of potential. Is the principle valid to use in circuits? Explain.
8. Explain how the principle of superposition of electric fields simplifies the analysis of circuits.

9. A wire is connected to the positive terminal of a battery. Describe the process that makes the potential of the wire attain the same value as the positive terminal of the battery. If a resistor is connected to the open end of the wire what will be the potential of the resistor? Explain.
10. A wire is connected to the positive terminal of a battery. If a resistor and an insulator are connected to the open end of the wire what will be the potential of the resistor? What will be the potential of the insulator? Explain.
11. Describe the graphical technique used to determine the operating point of an external device connected to a simple resistive network. Explain the technique assuming the external device is (i) a resistor, (ii) a diode (in forward bias), and (iii) a transistor in the common-emitter configuration.
12. A simple network of batteries and resistors is connected and it is observed that there is a negative current in one of the batteries. What effects do you expect to observe with negative current in the battery?

    If you were asked to predict the amount of negative current through the battery in the network using network analysis how would you model the battery?
13. What is meant by an "ideal" voltage source? Draw the symbol of one and explain why such sources are used by circuit analysts.
14. Describe the operation of a bypass capacitor tied across the emitter resistor of a transistor amplifier in the common-emitter configuration. Explain how the capacitor restores gain that would have been lost had the bypass capacitor not been included in the circuit.
15. Explain with the help of diagrams how the principle of superposition enables the small-signal equivalent circuit to be drawn of transistor amplifiers especially with regard to replacement of dc voltage sources with short-circuits. What parameters are normally used in such small signal equivalent circuits?
16. What is attenuation? Discuss the factors that cause attenuation of voltage.
17. Draw the circuit of a transistor amplifier and clearly show the electric field lines between the terminals and in the connecting wires and leads. Describe how stray capacitances produce attenuation when the frequency of the input signal is high.
18. What is the source of the tiny pulsating currents that expand and contract the electron sea on the surfaces of the connecting wires to a transistor amplifier? What is the source of the base, emitter and collector currents in the transistor? Explain.
19. A transistorized amplifier is constructed using an *n-p-n* transistor in the common-emitter configuration. Describe how the output voltage at the collector terminal is 180° phase-shifted with reference to the input signal voltage at its base.
20. Draw the electric field lines in the main components of a common-emitter amplifier. Show how the strengths are altered when a negative voltage ac shunt-feedback is incorporated in the amplifier.

Additional practice questions are available in the CD with this book.

4

Electrical Energy

In this chapter, we will initially study the presence of and the mechanism of transport of electrical energy in simple circuits, qualitatively.

Beginning with a very simple and yet precise analogy of the energy in a rotating bicycle wheel to explain the concept of electric and magnetic energy, and how they combine to transport energy in circuits. The next section provides an overview of various theories of electrodynamics.

When studying the transport of energy in circuits, most among us think that the electrical energy is carried by individual electrons. But this is wrong. An energy audit of a simple electric circuit that reveals the energy storage, transport and dissipation components is carried out and is presented with the application of the Poynting Theorem. A brief introduction to various theories of electrodynamics: Weber's action-at-a-distance, Leigh Page's Emission, Maxwell and Lorentz's Classical and Dirac's Quantum theories that describe the interactions between charged particles, is provided in this section. In the following section, the energy dissipated component represented by the Poynting vector 'S', theories that is the cross product of the electric and magnetic fields and its equivalence to the wattage given by voltage times current, is presented. Electromagnetic wave propagation in space is by means of pulsating fields, and a brief description of the phenomenon based on Maxwell's theory that propounds the development of induced fields, is provided. In this, an important notion that electric fields and magnetic fields do not travel through space but merely pulsate, expanding into regions where there were no fields, is described.

4.1 WHERE IS THE ELECTRICAL ENERGY?[11]

Is Electric Energy carried by individual electrons? Wrong

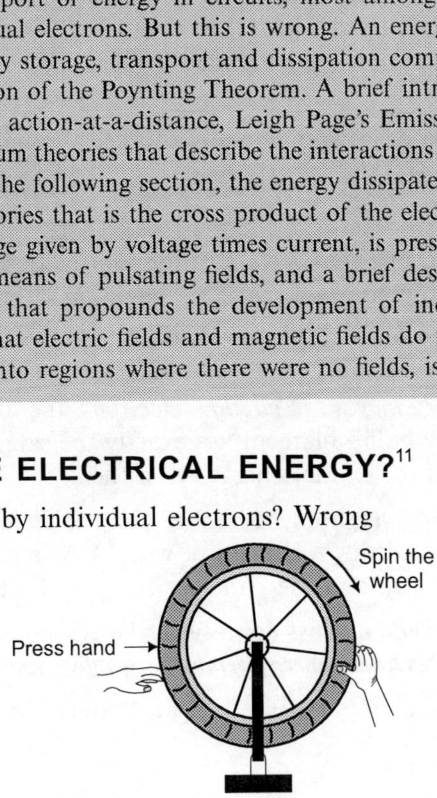

Consider a bicycle wheel mounted on a stand. Spin the wheel once using your ***right hand***. When you spin the wheel, your (right) hand ***injects energy***

into the entire wheel ***all at once***.

Remove your right hand.
Now put your ***left hand*** lightly against some part of the tire so the spinning wheel is *slowed* and *stopped by friction*.

Your *left hand gets hot*.
Your hand extracts energy from the *entire wheel*, all at once, and the whole wheel slows down.

Website: http://amasci.com/miscon/eleca.html#batt2

Would it be right to say that the "power" hand fills each rubber molecule with energy, that the molecules travel to the "friction" hand and dump their energy? Then the empty molecules return to the "power" hand and get refilled? No, of course not! If this were true then the energy would be forced to travel only as fast as the rubber. Your "friction" hand wouldn't experience any friction until those *magically energized* rubber molecules made their way around the rim.

Part of the wheel would be *spinning while part* would be de-energized and *unmoving* and this would be really a *strange sight to see*!!!!

Instead, one hand spins the wheel and fills the *whole thing* with kinetic energy, and the rubbing or pressed left hand *sucks* the energy back out at the *same* time. The wheel rotates, and *energy flows almost instantly* across the wheel, going from the "spin" hand to the "friction" hand.

Is electric energy carried by individual electrons? Wrong

A flashlight circuit is like the bicycle wheel. The *electrons* in the copper wire circuit are *like the rim of the wheel.*

Electrons are like an *invisible drive-belt* hidden inside the wires. *The battery* causes *ALL* the electrons in the loop of wire to begin moving *and* in this way, it *injects energy into the* WHOLE CIRCUIT all at once, just like a hand that spins a bicycle wheel.

As soon as the *battery moves the circuit's electrons*, the distant lightbulb lights up. The electrons moving into the bulb's filament are exactly the same as the ones moving out; the bulb doesn't change them or extract energy from them.

(Did your left hand do anything to *change* the rubber molecules as it rubbed on the bicycle wheel? It just slowed down the entire wheel.) A hand can extract energy from an entire bicycle wheel instantly, and the hand heats up by friction.

Same thing with the ***bulb*, it slows down ALL** the electrons throughout the entire circuit, and in this way ***the bulb extracts energy from the whole circuit as it lights up.***

This is depicted in the circuit comprising a battery, connecting wires and a bulb as shown in Fig 4.1.

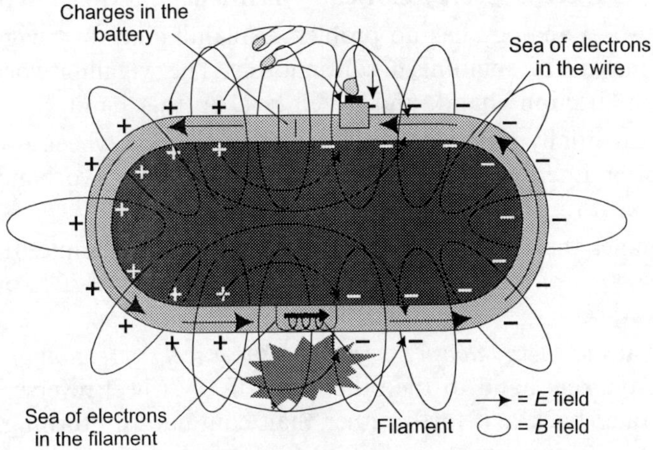

Fig. 4.1 The surface charges, electric and magnetic fields in the circuit.

When the circuit is closed there is an initial transient phase. During this phase, the surface charges are set up all around the circuit.

Positive charges from the positive terminal and negative charges from the negative terminal, enter the wires near the terminals.

The electric field set up by the charges gradually becomes uniform all around the circuit in the wires and more powerful in the filament of the bulb.

Shown in Fig. 4.1 are the surface charges, the electric field and the magnetic field.

After the initial transient is over, lasting a few picoseconds at most, all the electrons, the millions in the wires and the millions in the filament all begin moving with a start-stop motion.

The bulb lights up immediately as heat and light is generated by the collision of the electrons with the lattice ions. See also Section 1.18 and the descriptions related to Fig. 1.25 about how the electrons transfer their energy to the lattice ions.

Fig. 4.2 shows the electric field which exerts a force on the sea of electrons.

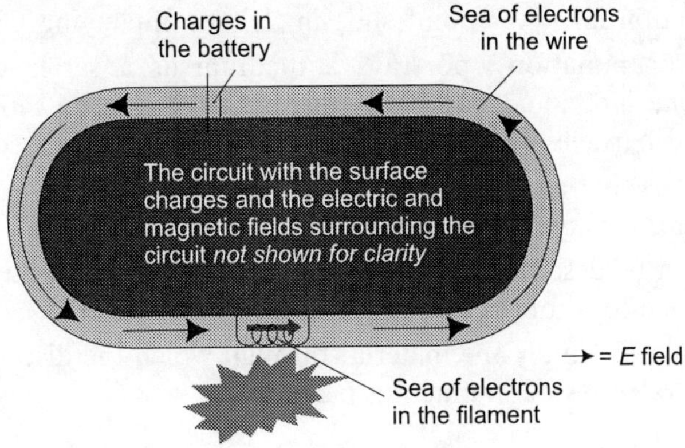

Fig. 4.2 All the electrons are pushed when the circuit is closed.

Is Electric Energy carried by individual electrons? Wrong

The bicycle wheel analogy has no problem explaining ac. Just wiggle the bicycle wheel back and forth instead of spinning it continuously. The wiggling wheel will rub upon the distant unmoving "friction" hand, and it will heat up that hand.

Energy can essentially travel instantly across the bicycle wheel, even though the wheel itself rotates slowly. Energy can travel instantly between the two hands, even if the wheel moves back and forth instead of spinning.

What determines the direction of this energy travel? It's simple. If one hand spins the wheel, *it throws energy out* into the wheel, and if another hand rubs on the tire, *it extracts energy* from the wheel.

Also, note that the *energy doesn't care about the wheel's direction of rotation.* The energy flows one way, from one hand to the other, even if the wheel reverses direction, and even if the wheel vibrates back and forth rather than continuously turning.

Another analogy using sound waves!

Would you say that your vocal chords place energy into air molecules, then each air molecule *zooms* out of your mouth at *720 miles per hour*, and eventually crashes into the ears of distant listeners?

That is silly. There is no supersonic wind coming out of our mouths.

Sound is carried by waves in the air, **not by air molecules** launched at immense velocities out of your mouth.

The air molecules remain at their place most of the time, but pass along the energy by wiggling and jiggling!

Electric energy is wave energy; electrical energy moves along columns of electrons like sound moves through the air, and when electrical energy flows across a circuit, the electrons **DON'T** flow along with it.

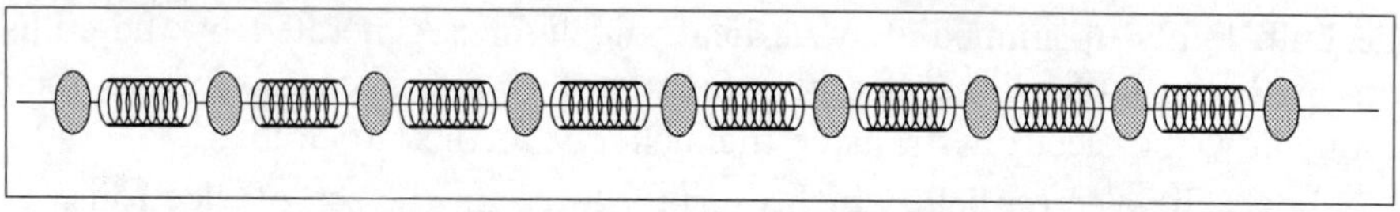

See animation in power point slide in the accompanying CD Chapter 4.

The above animation[14] portrays a medium as a series of particles connected by springs. As one individual particle is disturbed, it transmits the disturbance to the next interconnected particle. This disturbance continues to be passed on to the next particle.

The result is that energy is transported from one end of the medium to the other end of the medium, *without* the actual transport of matter.

A wave can be described as a disturbance that travels through a medium, transporting energy from one location to another location.

The medium is simply the material through which the disturbance is moving; it can be thought of as a series of interacting particles.

Is Electric Energy Carried by Individual Electrons? Wrong

Electrons do not push each other to carry the electric energy *unlike* in the case of air molecules that push each other and carry the sound energy.

In electric circuits, an electric field is set up by surface charges and this electric field after the very initial transient, **jolts all** the electrons into motion—**all around the circuit at the same time.**

4.2 THE NATURE OF ELECTRICAL ENERGY

Many things in nature wiggle and jiggle

We call a wiggle in *time* a *vibration* and...

a wiggle in *space* a *wave*...

A wave *cannot exist in one place* but must extend from one place to another.

Similarly, a vibration *cannot exist in one instant* but needs time to move to and fro.

Light and sound are *both vibrations* that propagate through space as *waves.*

But of two very different kinds...

Sound is propagation of vibrations through a material medium.

If there is no medium to vibrate, then no sound is possible.

On the other hand, light is a vibration of *nonmaterial* electric and magnetic fields, a...

...vibration of pure energy.

The *propagation* through space of these vibrating fields make up light waves

Although light waves *can* pass through many material media, *they need none*; light can propagate through a vacuum.

Sound and light are fundamentally different.

But the source of all waves, whether sound, light or whatever, is something that is *vibrating.*

For sound it can be the vibrating prongs of a tuning fork and for **light**, **the vibrations of electrons in an atom.**

Electrons are a wave medium,

and *electrical energy is wave energy.*

Remember, **electric current** is actually a flowing motion of charged particles.

Electric current is a very ***slow*** flow of charges.

On the other hand, **electric energy** is different. It is made of ***waves in electromagnetic fields*** and it moves VERY rapidly.

Electric energy moves at a ***different speed*** than electric current, so obviously they are two different things flowing in ***wires at the same time***.

In an electric circuit, the **path of** the **electric charges** is **circular**, while the **path of** the **energy is not**.

If electric current is **like wind**, then electrical energy is **like** some **sound waves**, and **electrons are like the molecules of the air.** For example, sound (energy) can

Is Electric Energy carried by individual electrons? Wrong

travel through a pipe if the pipe is full of air molecules, and electrical energy can flow along a wire, because the wire is full of movable charges.

Sound moves much **faster** than wind, correct? And electrical energy moves **much faster** than electric current for much the same reason.

Air in a pipe can flow fast or slow, while sound waves always move at the same very high speed.

Charges in a wire can flow fast or slow, while **electrical energy always flows along the wire at the same incredibly high speed.**

Whenever sound (energy) is flowing through a pipe, the air molecules in that pipe are vibrating back and forth.

When **waves of ac electrical energy** are flowing along a wire, the electrons in that wire are vibrating back and forth at 50 times per second (60 times per second in some countries)

The description of the sound waves and electrical energy traveling in pipes and wires are just an analogy, and sound waves ***are not exactly*** like electrical energy.

For example, *sound energy can flow inside an air-filled tube, while…*

…electrical energy ***is coupled*** with compression waves in the electrons of the wire

Electron-waves travel inside the wires coupled with the ***energy they carry in the invisible fields surrounding*** the wires.

We cannot see gravitational fields though they exist.

We can "see" electric fields and magnetic fields using grass seeds and magnetic compasses.

Electric currents in copper wires are a flow of electrons, but these electrons are not supplied by batteries. Generators do not 'generate' them. Instead, the electrons come from the wire. In a copper wire, copper atoms supply the flowing electrons.

The *millions of electrons* in a circuit were already there *before* the battery was connected.

They (*millions of electrons*) were even there *before* the copper was mined and made into wires!

Batteries and generators do not create electrons, they merely *pump* them and the electrons act like a pre-existing fluid which is always found within all wires.

Source: http://amasci.com/miscon/eleca.html#batt2

Most textbooks teach that the energy is trapped inside the electrons, and these electrons flow inside the wires. Doesn't this mean that electrical energy flows INSIDE the metal wires? No.

Because electrical energy is not trapped inside the electrons.

Instead, the energy is coupled with invisible *magnetic fields and electric fields* that *surround* the electrons, and these fields surround the wires.

Electrons don't flow fast like the energy does, instead they *ooze along slowly* (millions of them everywhere in the conductors... with a very high speed start-stop motion with a drift superimposed) to produce an electric current.

Let us consider an example of how much *kinetic energy*[27] is really there in the electrons in a copper wire of 1 mm^2 carrying a current of 1 amp powering a medium sized domestic appliance, say a refrigerator.

Each cubic centimetre of copper contains $8{\cdot}4830 \times 10^{22}$ free electrons per cc.

The mean velocity (drift) is 0.075 mm/sec or 0.00027 km/hr.

Taking the mass of the electron as 9×10^{-31}kg, the *power delivered* by the moving electrons as kinetic energy per second is....

a *teeny weeny* 1.5×10^{-20} watts (or joules/second).

Now, if this current (1 amp) is being delivered with a potential difference of 230 V then the power consumed is nearly a whopping 230 watts!

Or 10^{22} *times* the power of the moving electrons.*

We therefore have to *conclude* that *the real burden of the signal and energy transfer is being carried elsewhere, and **not** simply by the electrons in the wires.*

*http://www.st-andrews.ac.uk/~www_pa/Scots_Guide/audio/part6/page2.html

Since the observable *effects* of electricity in our daily lives especially the rotation of fans, are quite fast, the reader is advised to be careful in visualizing the *current* that powers it as being an *extremely slow* drift of millions electrons. The reversal of current at power line frequency (50 or 60 Hz) is fast, but the drift in either direction of the electrons is indeed slow. Doing this for a few days should enable the reader to get the correct visualization of currents, as a slow drift of millions of electrons in wires and circuits.

See the end of Section 1.19 on the aspect of the slow drift of electrons.

It is tempting to assume that the energy flows through the wires of the circuit, like water in a pipe. But, if we look carefully, we find that the energy density in the wires is *much too small* to account for the amount of energy in transit.

In one sense, the wires guide the energy, but do not carry it. For electromagnetic fields, the total energy density (in empty space) is u_{tot}

$$u_{tot} = u_{electric} + u_{magnetic} = \frac{1}{2}\left[\varepsilon_o E^2 + \frac{B^2}{\mu_o}\right] \tag{1}$$

and the speed of flow is *c*, the speed of light. We can also compute a momentum called the *electromagnetic momentum* that the fields themselves carry!

Check the rule v × B (Section 5.2) for a charge moving in a magnetic field. According to Härtel (Section 1.4.1, Hartel's "Teacher's Guide" in the accompanying CD), "In the case of the magnetic interaction, a single force acts on a moving charge carrier within a magnetic field which results in an acceleration of the charge carrier perpendicular to its velocity. There is no local reaction back onto the magnetic field. Such a single force or a single acceleration is a contradiction to Newton's Third Principle [and conservation of momentum]. For the magnetic interaction, this principle is only fulfilled when the reaction of the moving particle with the *complete system* is taken into account. The principle is, however, not valid on a local scale."

The calculation of the total energy is extremely complex, involving changes in the configuration of the fields and the electromagnetic field density.

The *rate* of flow of energy (power) in uniform wires in circuits involving the fields, is a more direct calculation, and is given by the Poynting vector to be described later.

We ignored the energy in the fields in the worked example in Section 1.7, because the velocities of the charged balls are quite small, the charge movement is sluggish because of the mass of the balls they are riding on. As a matter of fact, we can compute the energy by writing $E = E_1 + E_2$ (the total electric field is the sum of the field due to particle 1 and the field due to particle 2), and get for the energy $\frac{1}{2}\varepsilon_o(E_1 + E_2)^2$. Squaring out the binomial, we get three terms: one involving E_1 alone (the self-energy of charge 1), which we ignored because it is the same before and after, and hence is irrelevant to the calculation; one involving E_2 alone (to which the same remark applies); and a "cross-term" involving E_1E_2. This last, when integrated over all space, gives the interaction potential energy kqQ/r which we used in the example in Section 1.7. For details, see "Energy Stored in the Fields" in Appendix A and Section "Work and Energy in Electrostatics" in Ref. [8].

There is a tiny amount of electromagnetic energy radiated when the balls accelerate from rest; this is negligibly small in comparison with the kinetic energy of the balls.

However, when electrons are moved by an applied electric field in the *wires* (good conductors) in a circuit, they accelerate freely between collisions with the atomic cores, and so can attain large velocities that are yet non-relativistic (fraction of speed of light '*c*'), during those short intervals between collisions. Their total kinetic energy is quite negligible. But, *they* contribute to the *huge* amounts of electromagnetic energy transported. It may be said that the energy is entirely "potential" that we did in the example of Section 1.7, because their drift velocity, due to the start-stop motion, is tiny, for example, 1 meter/hour with a current of 115 amps, as we learned in Section 1.18.

But, wait a minute, does the expression $\frac{1}{2}\varepsilon_o(E_1+E_2)^2$ represent the energy being transported?

In the following, be careful to distinguish stored, transport and dissipated energies. Ref. [8] provides precise interpretations of Energy flow direction and their sense based on unit normal conventions of surfaces.

Suppose we have some charge and current configuration, which at time *t*, produces fields *E* (electric) and *B* (magnetic). In the next instant, *dt*, the charges move around a bit. This is the kind of movement that occurs in a wire carrying a current.

How much work, *dW*, is done by the electromagnetic forces acting on these charges in the interval *dt*?

According to the Lorentz force law (see Section 2.27), the work done on a charge *q* to move it a distance *dl* is

$$\vec{F}.dl = q(\vec{E}+\vec{v}\times\vec{B})\cdot\vec{v}dt = q\vec{E}\cdot\vec{v}dt \tag{2}$$

where v is the velocity of the charges. The second term in (2) is zero because the dot product of two vectors perpendicular to each other is zero.

Now,

$$q = \rho d\tau \tag{3}$$

and

$$\rho\vec{v} = \vec{J} \tag{4}$$

where ρ is the charge density in the wire, $d\tau$ is an infinitesimal volume element, and J the current density. Substituting ρ from (4) in (3), we get, $q = \frac{\vec{J}}{\vec{v}}d\tau$. Substituting this expression for q in (2), we get, the rate at which work is done on all the charges in a volume V is

$\frac{dW}{dt} = \int_{v}(\vec{E}\cdot\vec{J})d\tau$ where $\vec{J}$ is the current density established in a conductor.

Evidently, $\vec{E}\cdot\vec{J}$ is the work done per unit time, per unit volume, which is to say, the *power* delivered per unit volume and can represent either power dissipation or a power source, both denoted by p_d which can be negative if there is an external or internal source (e.g., a battery) supplying power to the volume. See https://ocw.mit.edu/courses/electrical-engineering-and-computer-science/6-013-electromagnetics-and-applications-spring-2009/readings/MIT6_013S09_notes.pdf.

The quantity $\vec{E}\cdot\vec{J}$ can be expressed in terms of the fields alone (for details see section Poynting's theorem in "Introduction to Electrodynamics" by David J. Griffiths) as

$$\vec{E}\cdot\vec{J} = -\frac{1}{2}\frac{\partial}{\partial t}\left(\varepsilon_o E^2 + \frac{1}{\mu_o}B^2\right) - \frac{1}{\mu_o}\nabla\cdot(\vec{E}\times\vec{B})$$

Putting this into the equation for $\frac{dW}{dt}$ above, and applying the divergence theorem to the second term, we have

$$\frac{dW}{dt} = -\frac{d}{dt}\int_{v}\frac{1}{2}\left(\varepsilon_o E^2 + \frac{1}{\mu_o}B^2\right)d\tau - \frac{1}{\mu_o}\oint_{S}(\vec{E}\times\vec{B})\cdot d\vec{a} \tag{5}$$

where S is the surface bounding the volume V and da an infinitesimal area element on this surface. This is Poynting's theorem where the first integral on the right is the total energy *stored* in the fields $U_{tot} = U_{electromagnetic}$ and the second term represents the rate at which energy is (dissipated in the circuit) *carried out* of the volume V (representing the volume of a wire say, in a circuit) across its boundary surface, by the electromagnetic fields.

The work done 'W' on the charges by the battery will *increase their* mechanical energy (kinetic, potential) as the atoms (millions of them !) stretch and the electrons (millions) are accelerated, and maybe collide with lattice ions to produce Joule heat. Also see Drude Model in Section 1.18. Poynting's theorem (5) says that the work done on the charges by the electromagnetic force is equal to the decrease in energy stored in the field, less the energy that flowed out through the surface as for example, in heating a resistive element.

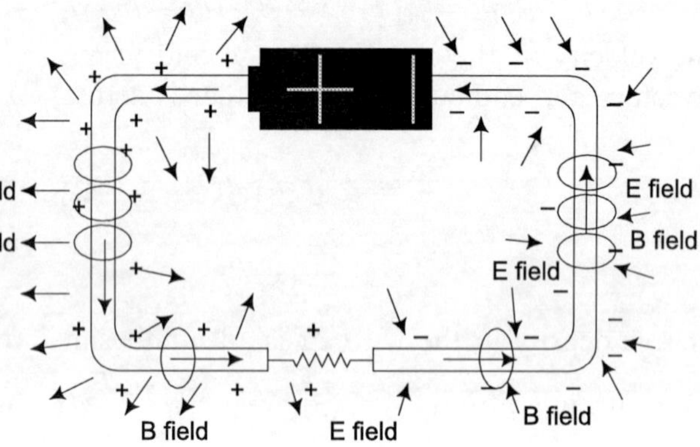

The energy *per unit time*, *per unit area*, transported by the fields (ignoring the stored fields) is called the Poynting vector *S*:

$$S = \frac{1}{\mu_o}(\vec{E} \times \vec{B}).$$

In Section 4.4, we will see how the Poynting vector '*S*' integrated over the surface of a wire $\frac{1}{\mu_o}\oint_S(\vec{E} \times \vec{B}) \cdot da$ gives us the familiar expression $V \times I$ for the power delivered to a load resistor. There is a nice article, *Energy transfer in electrical circuits: A qualitative account*, on the transport of energy in dc circuits in Am. J. Phys. 73(2), February 2005.

The transport of energy in circuits may be likened to a water delivery system with a storage tank, an inlet pipe near its bottom, connected to a source of water supply with a pump, and an outlet pipe also near the bottom which supplies water to a faucet, as shown in the figure. The work done by the pump to sustain the flow of water in the pipes represents the *dW*/*dt* term, the water storage represents the electromagnetic energy storage term $\frac{1}{2}\left[\varepsilon_o E^2 + \frac{B^2}{\mu_o}\right]$ (electric fields due to charges on the battery and surface charges, and magnetic fields due to moving charges), the flow of water represents the energy transport by the moving charges, and the frictional losses in the pipes represents the Poynting vector '*S*' term, the energy dissipated in the conductor and resistor.

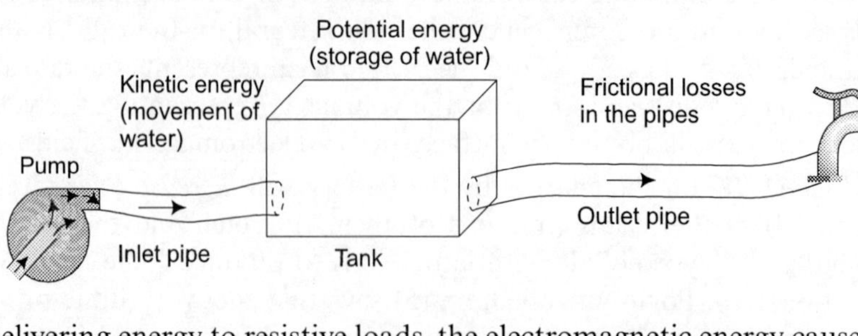

When delivering energy to resistive loads, the electromagnetic energy causes the electrons in the *resistors* (poor conductors) to gain large kinetic energies (and higher drift velocities in thin than in thick wires; see Fig 1.34) between collisions, which is *directly* transferred to the lattice during collisions, resulting in its dissipation as heat.

Whenever a pair of wires is connected to a battery or generator, the two wires *become oppositely charged*, and they become surrounded with an invisible *electrostatic field*.

This field **is** associated with the *"electrical energy"*.

During an electric current, the wires become *surrounded* with *magnetic field*.

This field **is** associated with the *"electrical energy"*.

Magnetic and electric fields exist in the empty space surrounding your *lamp cord*, and these *fields* are associated with the *'wattage'* that powers the light bulb.

Theories of Electrodynamics

The processes in the energy transfer mechanism that are based on Classical Electrodynamics described in this section, are accurate and precise enough to be able to analyze the power delivered in electric and electronic circuits. The Poynting theorem and Poynting vector accurately accounts the amount of power and energy inward on a surface over a volume as being equal to the sum of energy stored in the volume in electric and magnetic fields and dissipated (as in resistive loads) in the volume.

Section 4.4 discusses the application of the Poynting vector 'S' to a circuit.

We had called the "Electric" field as "an alteration of space", a vector field associated to each point in space. The reader should note that the field as an "alteration of space" is an aid to visualizing circuit processes, and is not the ultimate theory of fields.

Several "field" theories are propounded that describe the nature of charged particle interactions.

Section 6.6, Chapter 6 discusses the ideas of Classical Electrodynamics, Quantum Mechanics and Quantum Electrodynamics(QED). Quantum Electrodynamics was introduced by Paul Dirac in the late 1920's to explain certain aspects of the interaction between an electron and radiation which classical electrodynamics with the principles of Quantum Mechanics could not. A further refinement of QED led to Quantum Field Theory. According to Quantum Field Theory, the force between electrons is an exchange force arising from the exchange of virtual photons, and is represented by a series of Feynman diagrams. Each electron contributes to the field, and each one responds to the field. The electric field is quantized (in the form of photons). See "Introduction to Elementary particles" by David J. Griffiths, the link to which is available in the Links_to_animations_articles in the CD.

Charged particle interactions, according to Relational Mechanics Ref. [36], use the concept of "force" without a "field" of Weber's Electrodynamics, and is due among others to Leibniz and Mach. The theory explains some of the anomalies in the classic gravitational and electromagnetic theory.

Leigh Page in "The Emission Theory of Electromagnetism" (Ref. [39]), assumes a continuous emission of massless "moving elements", whose loci are Faraday's lines of force of the electric and magnetic fields. Moving elements are purely kinematical in nature, and they cannot be deflected or slowed down by matter. Also, they cannot of themselves impart energy or pressure to the bodies through which they pass. The moving elements do not exert forces of their own. They act merely as carriers of the lines of force, and as a vehicle for the transmission of energy in the form of radiation. The forces exerted by the electromagnetic waves that they carry are due to the lines of force. Page specified three vector functions of position in space and time; **E**, **H** and the velocity c of the moving elements, which carry the field. Vector c varies only in direction, as its magnitude is the constant speed of light. With moving elements, Page has proved the formalism of Maxwell's equations and Lorentz's force law using "the space-time transformation of the restricted relativity theory".

See "Fields and Momentum Conservation" in Appendix B and the topics "Coulomb Interaction". The reader may defer to view the video lectures recommended in the topic till the end of Chapter 8. See "Magnetic Interaction" in Section F.1, Appendix F also.

Härtel first prepared in 1984, "A Qualitative Approach to Electricity" and in the revised 2017 added "A Guide to Visualising Eletrodynamics using symbols and images". For visualizations of the interaction of accelerating charge carriers, Härtel adopted Page's theory. The guide with simple visualizations will help students make the transition from electrostatics to the study of moving charges and associated magnetic phenomenon consistent, without any loss of precision in cognitive assimilation of ideas. The guide (read Chapters 1 upto and including Section 3.4) is available in the folder "Teacher's Guide" in the accompanying CD.

A full description of all above theories is beyond the scope of this textbook. The topics in this book are based on classical electrodynamics with it's classical field theory approach. This is sufficiently accurate and precise to explain the theory of circuits and electromagnetic wave propagation. As the chapters progress, we will learn about situations where the classical approach was found unsatisfactory and theories described above were formulated.

4.3 IN A CIRCUIT WHY ARE TWO WIRES NEEDED?[11]

Let us answer this question with a question. When a circular belt is passing over two pulleys, why are two belts needed?

The answer: THERE ARE NOT TWO BELTS!

There is only one belt, and the belt is circular. It looks like there are two belts, with one of them flowing leftwards and the other one flowing right. But in truth, there is just one belt, and it is rotating.

So, why are two wires needed?

The answer: THERE ARE NOT TWO WIRES! Instead there is only one wire, but it is connected in a circle.

All metals are full of movable electrons, so when we connect a wire in a circle, we are forming a kind of "electric drive-belt", which can move inside the wire.

In Fig 4.3, is seen a simple circuit with a battery connected to a resistor, such as a light bulb. The battery converts its chemical fuel into waste products, and the resistor or the light bulb gets hot.

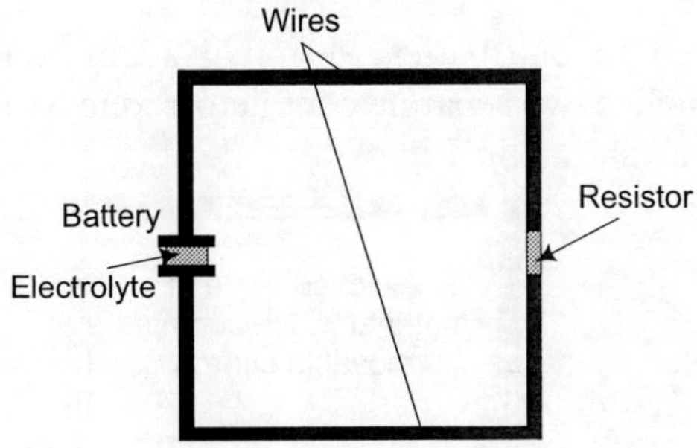

Fig. 4.3 A simple circuit.

In the circuit of Fig 4.3, the battery converts its chemical fuel into waste products, and the resistor gets hot, from the energy it receives in the form of electric and magnetic fields from the battery.

Refer : www.amasci.com

Electronics students commonly assume that electrical energy flows inside metal wires. Physics students know differently. Electrical energy only flows extremely slowly through metals. In fact, the energy sent out by batteries and generators is located in empty space: it takes the form of electromagnetic fields surrounding the wires.

Inductors store energy as a magnetic field outside the windings, capacitors store energy as an electric field in the empty space between the plates, and an electric circuit as a whole does both at once.

The energy which flows across a circuit is not moving through the interior of the metal wires.

Whenever a battery powers a light bulb, the battery spews electrical energy into space! The electrical energy is then guided by the wires, and eventually it dives into the light bulb, where it drives the charges against the resisting force of electrical "friction." An electric circuit is a transmission line for electromagnetic energy.

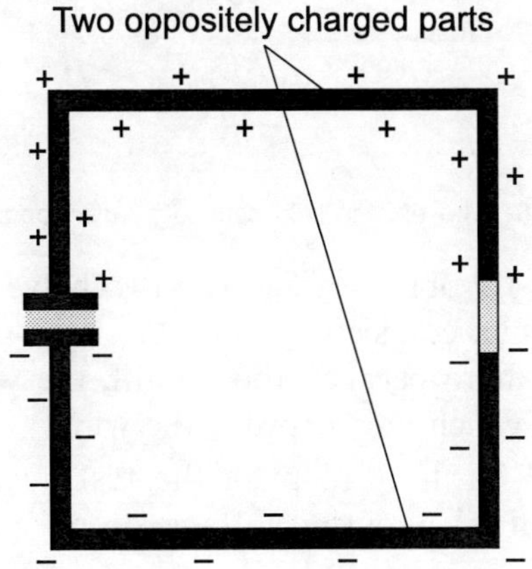

Fig. 4.4 Two charged conductors: *Voltage.*

Everything connected to one battery terminal acquires the same electrical potential (voltage). The circuit acts like two separate conductors, one with a positive charge imbalance and one with negative, as shown in Fig. 4.4.

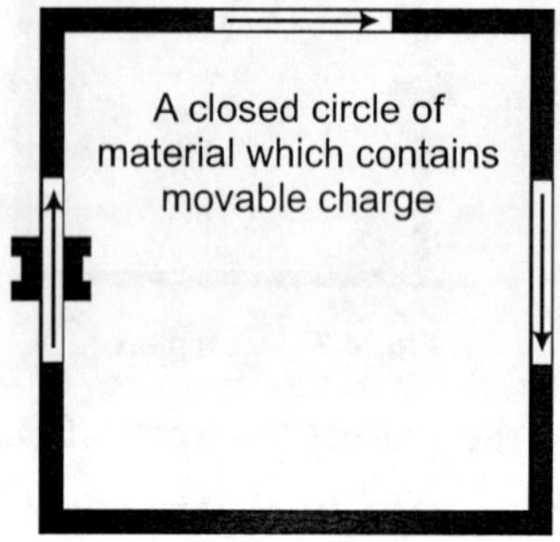

Fig. 4.5 The conductive path: *Current.*

All conductive materials contain movable charges. The resistor and the battery's electrolyte both are conductive. When we include them with the wires, we can see that an electric circuit is a complete circle which is full of "fluid" charge. It acts like a liquid flywheel; a flywheel hidden inside a closed ring of pipe, as shown in Fig. 4.5.

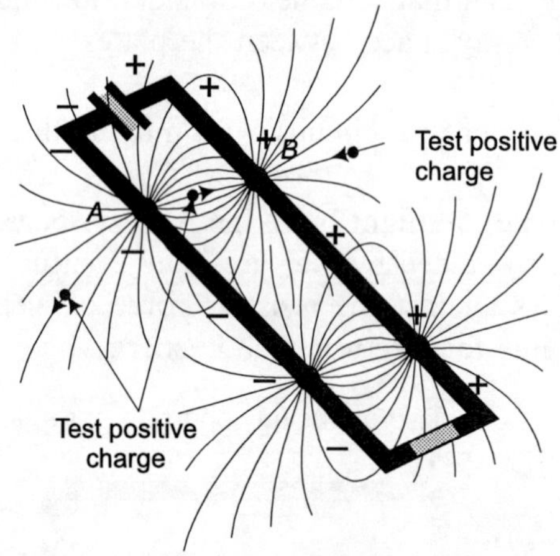

Fig. 4.6 The electric field caused by the opposite charges.

Again, in Fig. 4.6 is a 3D oblique view. The two halves of the circuit act as opposite-charged wires with e-field flux connecting them. We need to draw a third pattern between the two above, then draw more between those until, the whole wire is covered with bent sheets of electrostatic flux which arcs between the wires.

External work has to be done to push the test **positive** charge towards the wire connected to the + terminal and the voltage or potential difference from point A to B is

$$V = -\int_{A}^{B} E.dx$$

The magnetic (B) field caused by the current loop is shown in Fig. 4.7.

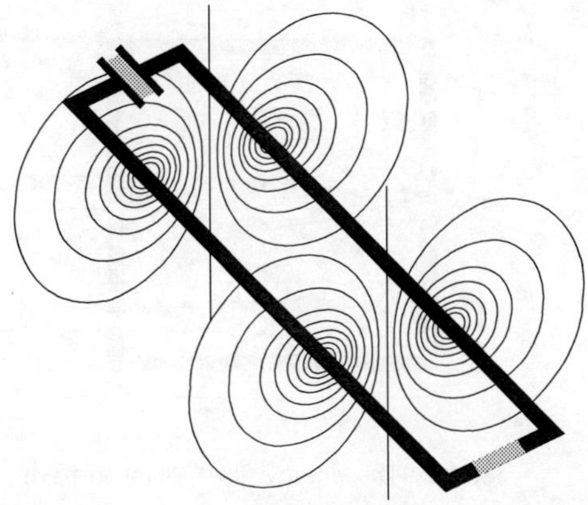

Fig. 4.7 The magnetic field caused by the current loop.

Shown in Fig. 4.7 is a three-dimensional oblique view of the magnetic field. To be more accurate, we need to draw more than just two patterns. Between the two patterns above, draw a third. Then between each of those draw more and more. The end result looks like "tubes" of magnetic flux surrounding the wires.

Figure 4.8 depicts a 3D oblique view of the two fields, e-field and b-field.

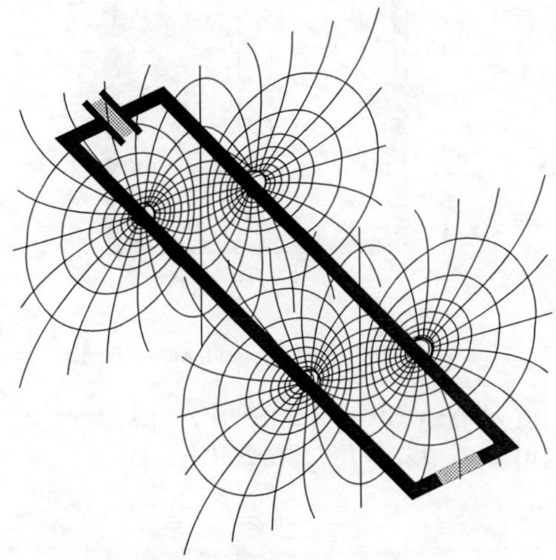

Fig. 4.8 A *e*-field and *b*-field together.

Add more and more patterns between two e-field and b-field patterns, until the empty space is packed full of "hair." Note that most of the flowing energy lies between the two wires... but quite a bit also surrounds the "cable pair" as a whole.

Electromagnetic energy flows out of the battery and into the empty space around the circuit. It flows parallel to the connecting wires, then it dives into the resistor, as shown in Fig. 4.9.

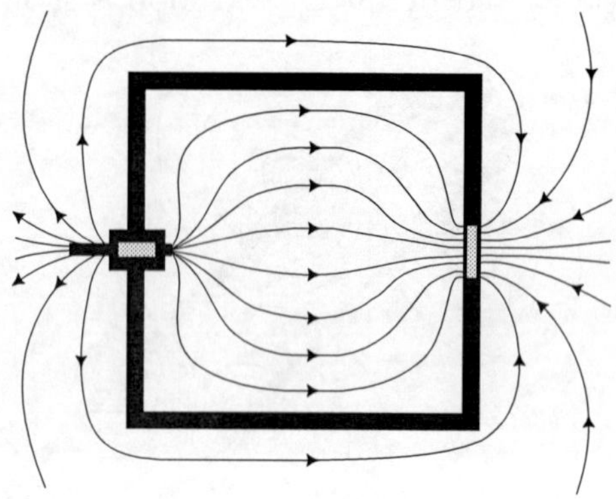

Fig. 4.9 The energy flow (Poynting field).

The field of energy flow is found by multiplying the e-field by the b-field ($E \times B$ vector cross-product) as seen in Fig. 4.9.

Figure 4.10 shows the Poynting energy flow field and the e-field together.

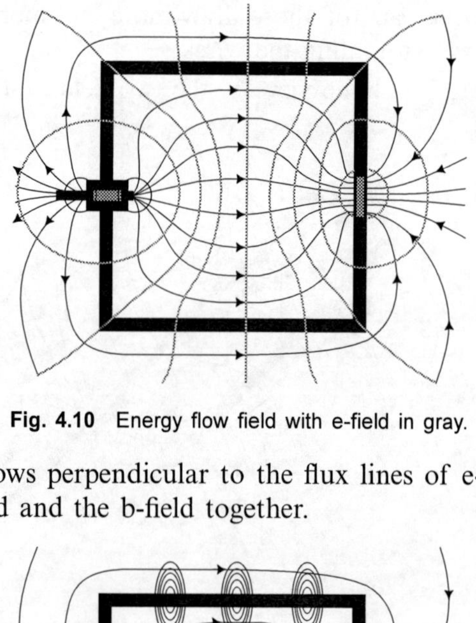

Fig. 4.10 Energy flow field with e-field in gray.

The energy always flows perpendicular to the flux lines of e-field. Fig 4.11 shows the Poynting energy flow field and the b-field together.

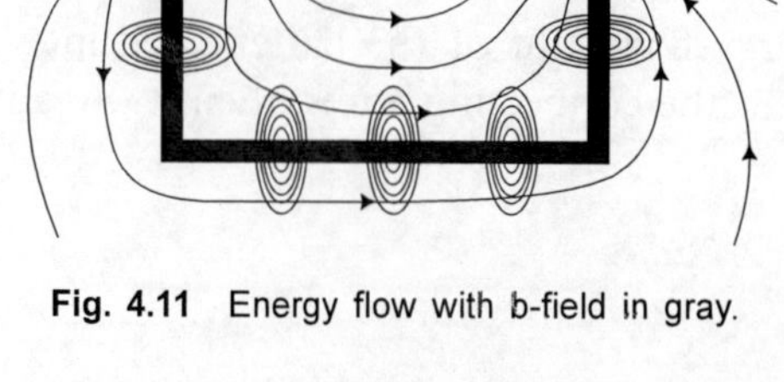

Fig. 4.11 Energy flow with b-field in gray.

The energy always flows perpendicular to the lines of b-field too, as shown in Fig. 4.11.

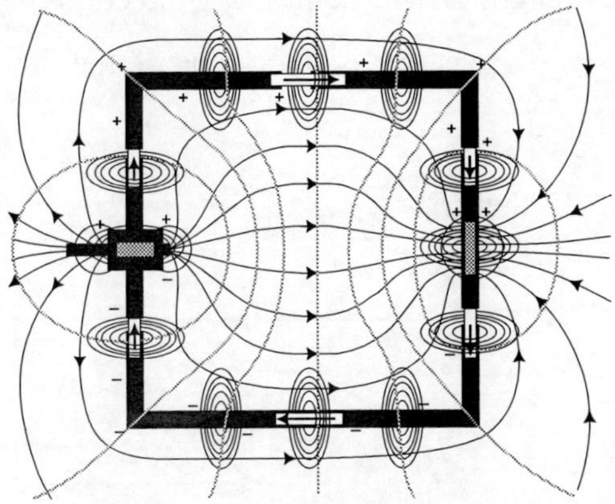

Fig. 4.12 A simple circuit?

When all the separate invisible phenomena are displayed together, you can see why "electricity" might be a bit hard to understand. And Fig. 4.12 only shows a two-dimensional slice; a sort of side view of the fields. The real fields are 3D and volume-filling, so an accurate drawing would look like a black glob of hairs.

4.4 THE POYNTING VECTOR

The *Poynting vector* describes the energy flux Joules $m^{-2} s^{-1}$ or Watts m^{-2} of an electromagnetic (em) field. It is named after its inventor John Henry Poynting.

It *points in the direction of energy flow* and its magnitude is the power per unit area crossing a surface which is normal to it.

It is given the symbol ***S*** (in bold because it is a vector), and in SI units, is given by:

$$\vec{S} = \vec{E} \times \vec{H} = \frac{1}{\mu}(\vec{E} \times \vec{B})$$

where, E is the electric field, H and B are the magnetic field strength and magnetic flux density respectively, and μ is the permeability of the surrounding medium.

For example, the Poynting vector near an ideally conducting wire is along the wire axis—so electric energy is flowing in space outside of the wire, as shown in Fig. 4.13(a).

The Poynting vector *tilts towards the wire for a resistive wire*, indicating that *energy flows from the em field into the wire,* producing resistive Joule heating in the wire, as shown in Fig. 4.13(b).

Application of the Poynting Vector for Energy Transported in a Transmission Line

Consider a long resistanceless transmission line of length 'L' delivering power from a battery to a resistive load. A steady current i exists as shown in Fig. 4.13(a).

The magnetic field strength H will be in the ϕ direction and at the surface of the wire it will have a value $H_\phi = \dfrac{i}{2\pi r}$ where r is the radius of the line. See Section 5.2 for details of this expression.

Let the voltage drop be V, then $E = \dfrac{V}{L}$.

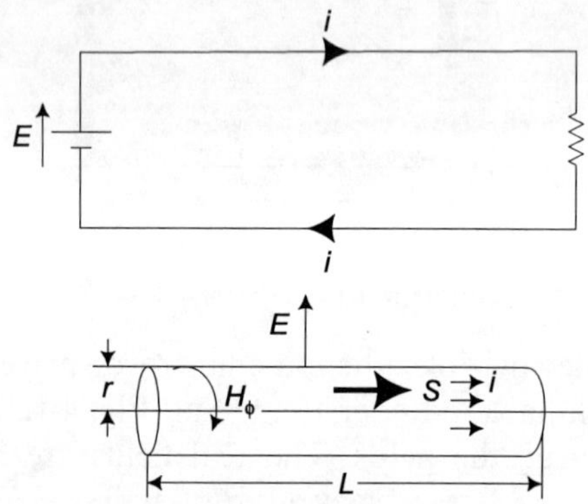

Fig. 4.13(a) Poynting vector to calculate the energy of a resistanceless transmission line.

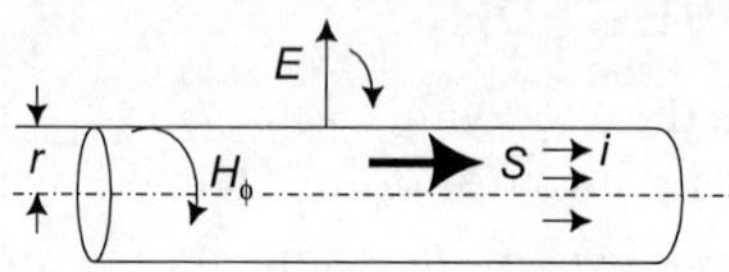

Fig. 4.13(b) Tilt of electric field on resistive wire towards the wire.

and $$H_\phi(r) = \frac{i}{2\pi r}.$$

The Poynting vector is integrated over the surface of the load. ***S*** is directed radially inward (Fig. 4.14), the caps at the ends do not make any contribution to

$$\oint S \cdot da$$

where, $da = 2\pi r dz$ is the surface area for an element length dz of the line.

Now, $$\oint S \cdot da = \int (E_Z \times H_\phi) . da$$

E_z and H_ϕ are mutually perpendicular to each other.

So,

$$\oint S \cdot da = \int_0^L E_Z H_\phi \, 2\pi r dz$$

Using the relations $E = \dfrac{V}{L}$, and $H_\phi(r) = \dfrac{i}{2\pi r}$ in the above expression, we obtain,

$$\oint S \cdot da = \frac{Vi}{L}\int_0^L dz = Vi$$

The quantity *Vi* represents *ohmic loss* in the resistor.

Therefore, we may conclude that the *loss* is supplied by energy *entering through* the surface of the resistive load.

Within the resistive load, the Poynting vector (not a physical entity) is directed radially inward (see Fig. 4.14) so that the ohmic loss equal to i^2R is supplied by the electric and magnetic fields within the resistor. However, in the free space surrounding the wires and in between them, the Poynting vector is directed axially towards the load (Fig. 4.13(a)).

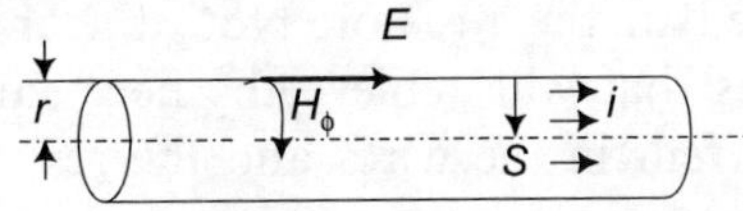

Fig. 4.14 In the load, a resistive conductor, the Poynting vector is directed radially inward.

This is a result of the electric field component E_l of E due to surface charges pointing mostly outwards (with a very tiny field component E_2 of E along the axial direction in the wire, components of the tilted field in Fig. 4.13(b)), so that the energy represented by $S = E \times H$ is mostly directed toward the load (resistor) and the energy flows through the dielectric (free space) *around the* wires coupled with the moving electrons inside and *not through the wires itself.* See Fig. 1.4 and also the article "Energy transfer in electrical circuits: A qualitative account" in the "Links_to_animations.pdf" in the CD alongwith the book.

According to Härtel "The kinetic energy of the free electrons is practically zero, the driving force - the EMF - is absolutely dominant".

Important Note

Only in a particular case involving the liberation of Joule's heat is the kinetic energy of electrons the intermediate form of energy through which the energy of the electromagnetic field is converted into heat. In other cases, the kinetic energy of electrons does not play any role.

See Figures 4.9 to 4.12 and verify the direction of energy flow between the wires, applying the right-hand rule for cross products described in Section 5.2.

A more detailed description of the manner in which the energy is transferred by a two-wire line called a two-wire transmission line used in common electrical and electronic circuits is given in Chapter 5, Section 5.22.

4.5 POWER FLOW WHEN A CAPACITOR DISCHARGES INTO A RESISTOR

We consider a simple RC circuit, where a charged capacitor (C) is made to discharge into a resistor (R) to which it is connected with a thick wire. A section of the wire offers a high resistance (R), formed by narrowing the wire. The electric field must be larger in the high-resistive region (and smaller in the low-resistance conductor portion) to have uniform current.

The plots of the field vectors and equipotential lines were made from computer-based simulation studies by Norris W. Preyer, Department of Physics and Astronomy, College of Charleston, Charleston, South Carolina, USA and reproduced with permission in the Figs. 4.15 and 4.16.

The Fig. 4.15(a) illustrates the orientation of the electric fields, in the space surrounding the circuit when quasi steady-state is reached, which will ensure that the energy placed in the capacitor is dissipated in the resistor. Note the intricate arrangement of positive and negative surface charges that will achieve the necessary electric field magnitudes and orientations to drive the current in the wires and the resistor.

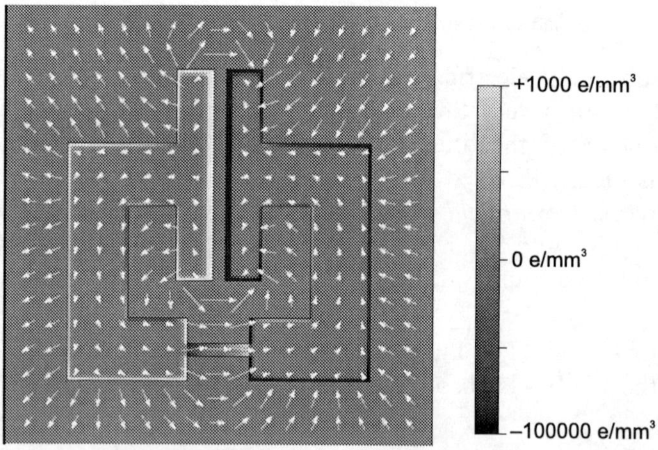

Fig. 4.15a The electric field vectors of a capacitor discharging in an RC.

The various gray shades represent the excess charge density, ranging between +/–5000e (charge of an electron)/mm^3. The charge density on the capacitor plates exceeds this range, which was chosen to prominently display the surface charges on the wires. The tiny arrows plot the square root of the electric field. The large field vectors between the plates have been omitted.

The electric field outside the conductors is important for ensuring that the Poynting vectors point in the correct directions, and that high-resistance regions have a greater energy flow than low-resistance regions. This model illustrates capacitor-like interface charges piling up on either side of the narrow wire resistor, which creates a powerful electric field inside the resistor, as shown in the magnified view of the resistor and the wire in the left picture of Fig. 4.15(b), which will produce uniform current in the entire circuit. The right picture in Fig. 4.15(b) shows only the surface charge with polarities, without the outline of the wires and resistors, to better illustrate the intricacy in the geometry of the surface charge and interface charge arrangement.

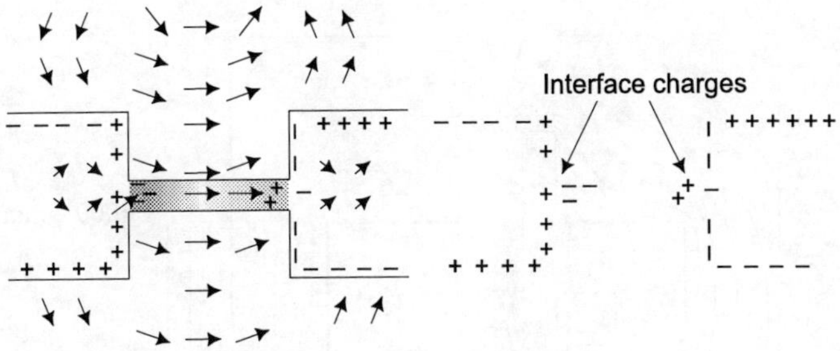

Fig. 4.15b A magnified view of the resistor and the wire showing the electric field vectors, surface and interface charges.

Notice how these interface charges at the ends of the resistor, increase the electric field and hence current density through the narrow region, and simultaneously decrease the field and current density in the wider wires. These and the surface charges on the wires will also vary, at every instant of the quasi steady-state to ensure that the conservation of current law is upheld.

The Poynting vector will be directed, into the resistor because a large amount of energy is absorbed here, and to a smaller extent into the conductor.

While the equipotential lines are interesting by themselves, these lines also show '*S*', the Poynting vector. Recollect from Section 2.26 that charged insulated conductors are equipotential on the surface, as well as in the interior. Similarly, in the electric circuit, there will be equipotential surfaces of varying magnitude along different cross-sections of the wire and resistor. An equipotential line (surface), is one for which no work is required for a test charge to move along (on) it. Since, E is the negative *gradient* of potential ($-\nabla V$), we interpret the *direction* of the gradient (or, in other words of E that set up the potential) as follows:

If one stands on the "surface" of potential, one would see the surface rising in some direction and sloping downward in the opposite direction. There *is* one direction, in which a short step will take you *lower* than a step of the same length in any other direction. *That* is the *direction* of E, or rather the *direction* of the gradient (represented mathematically by the symbol '∇') of potential.

A field flux line through any point and the equipotential surface through that point are perpendicular to one another, just as, on a contour map of hilly terrain, the slope is steepest at right angles to a contour of constant elevation.

Due to symmetry, B, the magnetic field, is perpendicular to the figure, so S is perpendicular to E and B. Hence, the equipotential lines are *parallel* to the Poynting vector (which is $E \times B$) and you can see most of the power going to the resistive regions.

The equipotential lines for the circuit with a discrete resistor are shown in Fig 4.16(a).

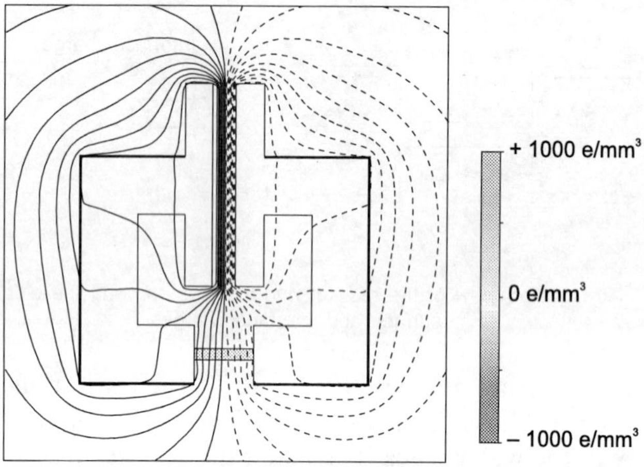

Fig. 4.16a The equipotential lines (parallel to the Poynting vector).

In Fig. 4.16a, the equipotential lines are thickly crowded in the resistive region, indicating a steep voltage change from one end to the other (see Section 2.6 for a description of variation of potential *across* these lines), and a large flow of power into the narrow-wire resistor. For the entire circuit, these lines are parallel to the Poynting vector, and are considered equivalent, in terms of orientation and density to the Poynting vector.

To better appreciate the representation using equipotential surfaces (lines) of the voltage, refer to Section 2.20 "Visualizing voltage", especially Fig. 2.53(b) for points in the interior of the conductor, wherein there is no work done penalty to be paid for a test charge to move towards the surface along this line.

As one moves *across* these lines, along a conductor or resistor, the potential changes, and this change is what is commonly called "voltage drop".

In the conductor regions, the lines are spread far apart indicating a very small voltage change from one end to the other, and a smaller flow of power into these regions.

For the sake of completeness, shown in Fig. 4.16(b) in the left picture, is a magnified view of the wire and resistor showing the equipotential lines, surface and interface charges, and in the right picture, the same view but without the wire, resistor and equipotential lines, which shows more clearly the intricate arrangement of surface charges which will produce the electric fields in the resistor. The full lines indicate a positive potential and the dashed lines indicate a negative potential.

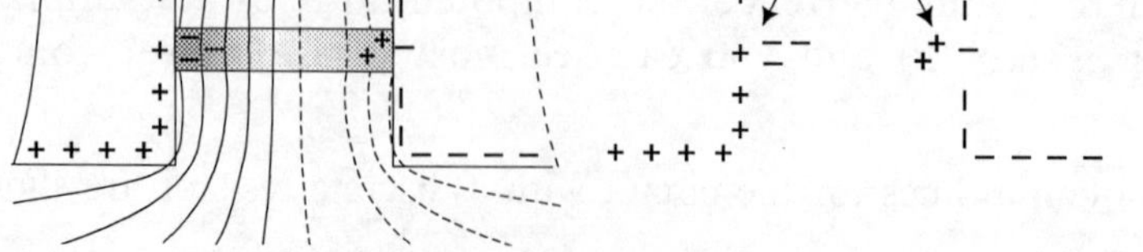

Fig. 4.16b A magnified view of the wire and the resistor showing the equipotential lines, surface and interface charges.

I urge the reader to visit the site http://web.archive.org/web/20120115080730/http://galaxy.cofc.edu/circuits.html for more examples and animations of these circuits.

4.6 AC ENERGY

When you plug a lamp into a wall socket, you *shouldn't imagine* that the ac energy is a mysterious invisible entity traveling back and forth inside the wires.

Instead you should think of ac energy as a mysterious invisible flow that comes out of the outlet, runs along the outside of BOTH wires, then dives into the filament of the light bulb.

Your *electric company* is sending out long "sausages" or tubes of electrical energy, the wires are guiding them, and your appliances (such as lamps, fans and TVs) are *absorbing* them.

4.7 HOW DOES THE LIGHT IN A DISTANT BULB TURN ON IMMEDIATELY WHEN YOU TURN ON THE SWITCH?

It is not necessary for charges to move physically to a distant place to cause a change at that place: a rearrangement of charges due to tiny movements produces a change, at the speed of light, in the electric field at a distance.

It is important to bear in mind, the effects do take some time to propagate (nanoseconds). The propagation is not instantaneous, but it is very fast.

When you turn ON the light switch in your room (making the final connection in a circuit), it could take hours on the average, for electrons in the switch to drift to the overhead light.

But the light comes on right away, because the rearrangement of surface charges in the circuit takes place at about the speed of light.

The final steady state of the circuit is established in a few nanoseconds, after which the electron sea....

...circulates slowly and majestically around and around everywhere in the circuit, including through the switch and the wires and the light bulb.

Most lighting actually uses "alternating current," in which case, the electron sea doesn't drift continuously but, merely sloshes back and forth very short distances, everywhere in the circuit.

Changing Electric Field Produces a Magnetic Field

We now provide an answer to an oft-asked question about the nature of light, which is electromagnetic (em) radiation (electric and magnetic fields at right angles to each other) produced by an accelerated charge. Electromagnetic wave propagation in space, requires that a time-varying magnetic field produce an electric field, and that a time-varying electric field produce a magnetic field. The former will be described in Section 5.8 and Section 5.12 on curly electric fields and Faraday's law.

A notion about the latter was introduced in Section 2.7 and Section 3.1, where we stated a *displacement current* was necessary for bookkeeping, to account for the "missing" current in the non-conducting space between the plates of a charging capacitor (See Section 2.7). The term was introduced by James Clerk Maxwell to Ampere's law (Section 2.28) to balance the expression of the curl of the magnetic field which is equal to the conduction current. The expression is a statement of the presence of a magnetic field when there is a current (Section 2.27). When the law is applied to the region between the plates of a charging capacitor, where there is no physical current, the term that Maxwell added takes effect which is that a changing electric field produces a magnetic field, and represents that the time-varying electric field between the plates is equivalent to a displacement current. See Links_to_animations_articles in the CD for links to Maxwell's original Treatise on Electricity & Magnetism. We will learn more about this in Section 5.22.

The *magnetic* field, in the non-conducting space between the plates is so tiny, in commonly used circuits at moderate frequencies that it cannot be easily detected. It is this magnetic field that is associated with a displacement current; there being, a magnetic field but no real current between the capacitor plates.

However, apart from curing the defect in Ampere's law (Section 2.28) when applied to charging currents in capacitors, Maxwell's term indicates that just as a changing magnetic field induces an electric field (Faraday's law), so a changing electric field induces a magnetic field. The magnetic field associated with the changing electric field becomes significant at high frequencies, and plays a crucial role in the propagation of electromagnetic (em) waves. We will learn more about these ideas in Section 5.8 to 5.12 and in Section 5.22.

Electric and Magnetic Fields do not Travel

If a charge is made to oscillate rapidly (which would involve accelerating the charge), a series of pulsating electric and magnetic fields, electromagnetic (em) waves grow outwardly transverse to the oscillating charge. The waves propagate or spread into a new region (where there were no fields), seemingly traveling outward from the charge. The fields themselves do not "travel" (it may be said that the wave travels or propagates), but rather, at one instant of time there are fields in one region of space, and at a different time there are fields in a different region (outward from the charge) of space. To visualize how the pulsations in strength of the electric field in one location and a different field in an adjacent location can make a wave seemingly travel, visit http://phet.colorado.edu/sims/wave-on-a-string/wave-on-a-string_en.html or http://www.animations.physics.unsw.edu.au/jw/waves_superposition_reflection.htm and view how waves seemingly travel on a rope while, the rope molecules "stay where they are", merely bobbing up and down. A detailed description will be provided on the above aspect of electromagnetic wave propagation in Chapter 10.

Then, view an animation of electromagnetic waves propagating around an oscillating dipole by following the instructions described in the "Links_to_animations_articles" pdf file in the accompanying CD to view the applet in www.falstad.com.

The electric field in an electromagnetic wave can also light a bulb fixed to a receiving antenna in an open circuit! Watch a live demonstration of this in a video lecture here https://www.youtube.com/watch?v=Hd29jEaGERk.

The reader is referred to Chapter 10 which describes the nature of light (electromagnetic waves), its production and the propagation of electromagnetic waves in space and in matter.

How does an opaque body (such as a cardboard) block light when we had stated that electric fields and magnetic fields penetrate through matter? (Chabay & Sherwood, 1995, used by permission).[2]

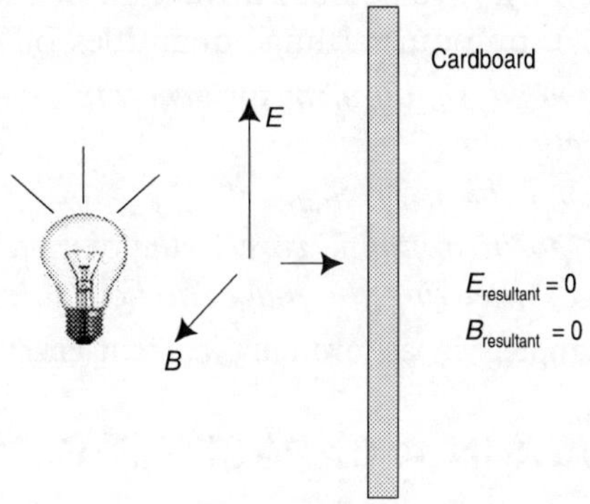

The electric and magnetic fields, radiated by the accelerated charges in the filament of the bulb, are certainly present to the right of the bulb and yet, there is no light on the other side of the cardboard, so the net field must be zero there.

There must be an additional source of electric and magnetic field somewhere, producing a field that adds to the radiative field, resulting in the zero net field.

The electrons in the atoms of the cardboard molecules are accelerated by the electric field of the light radiation in such a way that the re-radiated field (because this new field is from the cardboard) is equal and opposite to the incident time-varying electric and magnetic fields from the light bulb. The net field on the other side of the cardboard is the superposition of the two fields, and is zero.

Reradiation or Scattering is the phenomenon, by which molecules and atoms when accelerated by incident time-varying electric fields, generate electromagnetic waves in new directions; this phenomenon is described in more detail in Chapter 10.

Once the *net fields* of an electromagnetic wave traveling in space are reduced in strength or *attenuated* by any means say, by the cardboard in the figure, they will remain attenuated even as they propagate (remember fields do not travel) farther away on the other side of the cardboard. Chapter 10 Sections 10.4 and 10.5 provide descriptions of reradiated fields and their waves (such as light), the manner of their production and their propagation through media.

There are also links to animations and articles in the "Links_to_animations_articles" pdf file in the accompanying CD on the *Doppler shift* (see Section 5.22) wherein the frequency and wavelength of a wave-like signal - such as sound or light - depends on the movement of the sender and of the receiver.

Based on topics from "Introduction to electrodynamics" by David. J. Griffiths the following paragraphs provide a qualitative view of the magnitudes of electric and magnetic forces, a sort of comparison, though it must be understood that the forces act differently on charged particles.

Typically, electric forces are enormously larger than "magnetic ones". This has to do with the sizes of the fundamental constants ϵ_0 and μ_0.
How is it that we "notice" magnetic forces at all?

The answer, is that both in the production of a magnetic field (Biot-Savart law) *and* in its detection (Lorentz), it is the current (charge times velocity) that enters, and we can compensate for a smallish velocity (remember it took an hour for charges to move 1 meter in a wire carrying 115 amps), by pouring huge quantities of charge down the wire.

Ordinarily, this charge would simultaneously generate so large an electric force as to "swamp" or mask the magnetic one.

But, if we arrange to keep the wire "neutral", by embedding in it an equal amount of opposite charge at rest, "the positive atomic cores", the electric field cancels out (other than the one due to surface charges), leaving the magnetic field to stand alone.

This is precisely what happens in an ordinary current-carrying wire.

4.8 REACTIVE POWER IN A CAPACITOR CIRCUIT

In a circuit with a capacitor, engineers call the voltage multiplied by current, reactive power expressed in "reactive vars" and represents the power drawn from the voltage source to overcome the fringe field, and sink the current by the fringe field as the capacitor charges and discharges rhythmically in response to a sinusoidal voltage. Note that alternately during every quarter cycle, the power is drawn from or returned to the source by the capacitor as shown in Figs. 2.27 and 2.28. There is no net energy flow over one cycle. In this case, only reactive energy flows—there is no net transfer of energy to the load, the capacitor. This power does no useful work and it does not come for free, since the voltage source should be capable of overcoming the fringe field of the capacitor. Most among us ignore this aspect of the voltage source. It should have that capacity to establish the voltage by charging the capacitor against the forces exerted by the fringe field. Also, the capacitor delivers its stored energy every other quarter cycle to the voltage source when the fringe field overcomes the applied field. It is therefore, the combined energy exchanged between the voltage source and the capacitor that forms the Reactive power (measured in 'vars') which is present in a system containing reactive (inductive or capacitive) components and may be either produced or consumed by different generation or load circuit elements.

SUMMARY

In the opening section of this chapter, the electrical energy in a circuit we had learned, is not carried by the individual electrons but, by the fields. Using the analogy of a mounted bicycle tire that is spun by one hand, and the energy that is carried by all its rubber molecules generating heat in a hand which is pressed at a point diametrically opposite to the point where it was spun, we learned that the electrical energy in a circuit is present all around the circuit, since each and every one of the millions and millions of conduction band electrons execute a slow drifting motion. The energy is carried by the electric field due to the surface charges and the magnetic field of the electron current. Another classic analogy we studied is that of air molecules that carry sound energy. The vibrations of air molecules near the source, are conveyed to adjacent molecules, which in turn, convey their vibration energy to the next layer of air molecules and so on until, the entire slab of air molecules vibrate, and the last layer impinges on the eardrum. The energy travels, we learned, at the speed of sound, the air molecules not having moved far from their equilibrium positions.

The energy in circuits travels at nearly the speed of light, because changes in the electric and magnetic fields occur at the speed of light, and we learned that light is a vibration of pure electric and magnetic fields, a vibration of pure energy.

The electrons in wires existed since the time that the copper which makes up the wire was mined and converted into the wire. In short, when a circuit is energised, all these electrons that were existing all the time, receive a jolt from the electric field due to surface charges at the same time, and are set in motion everywhere in the wires simultaneously and thus, the electric field (due to surface charges), and the magnetic field (due to the association of a magnetic field with current), carry the energy from the source to the load. We showed how the kinetic energy of the electrons is too small to convey the huge amounts of energy from a voltage source to a refrigerator, and concluded that the energy indeed, must be in the fields themselves.

"The kinetic energy of the free electrons is practically zero, the driving force - the EMF - is absolutely dominant" we learned from Prof. Hermann Härtel. Only in the case of resistive heating is the kinetic energy of electrons the intermediate form of energy through which the energy of the electromagnetic field is converted into heat.

We then derived the Poynting Theorem which states, the rate of work done on all charges in a volume by the electromagnetic force is, equal to the decrease in energy stored in the field, less the energy that flowed out through the surface (Poynting Vector 'S'), as for example, in heating a resistive element. After being reminded that our description of the field as "an alteration of space" is one interpretation, we were then introduced to theories about the "stuff" of fields interpreted as Quantum Field and in another interpretation as a force in Relational Mechanics (without a field concept) and in Leigh Page's theory, as streams of moving elements. Härtel's guide to visualization of electrodynamics is based on this theory.

We then learned, in a stunning revelation that there is really only one wire in a circuit, and the colors we associate with the positive (red) and the negative (black) wires hide the simple fact that both the wires are made from the same material, copper !

Then, a plot of the electric and magnetic fields around a simple circuit showed how the energy transported by the fields is given by a cross product of the two fields viz. $\vec{E} \times \vec{B}$, which mathematically is a vector 'S' pointing in the direction from the source towards the load. Of course, there is no physical entity which we can call an energy vector; the vector 'S' merely represents the magnitude and direction of the flow of the energy of magnitude $|E| \times |B|$.

A derivation of the energy flow in a lossless transmission line, which conveys energy to a load, showed that the electric field is mostly normal to the line, but tilts inwards near a resistive load, and learned that the power delivered is given by $\oint S \cdot da = Vi$ where *da* is the surface area for an element length *dz* of the line.

We then reviewed the flow of power in a CR circuit, in which the capacitor C discharges into a resistor R, and showed how the direction of the electric field due to surface and interface charges points directly through the resistor, while noting that a path traced across equipotential lines drawn in the resistor, indicate the voltage "drop" that is present.

Next, we studied how a distant bulb lights up instantly when it is turned ON. We observed that surface charges set up the electric fields almost instantly everywhere in the filament and the connecting wires, to drive the electrons which are in the wires. These electrons, slosh forward and backward with a drifting motion, in response to the sinusoidal variations of the applied fields due to the voltage source.

We then made an important observation that the electric forces are much more powerful than the magnetic forces, but in an ordinary current carrying wire, the freely moving electrons being embedded in the lattice (leaving aside the surface charges), makes the interior mostly neutral and the magnetic field instead, stands out.

Opaque objects like cardboard, block out light, and this seemingly simple observation questions the notion we had that electric and magnetic fields penetrate matter. It was shown how the molecules of the cardboard, reradiate electromagnetic energy of incident light in many directions, and the combined effect of the waves is to produce a very tiny resultant electromagnetic field on the other side of the cardboard that makes us say, the cardboard blocks light.

When a sinusoidal voltage is applied to a capacitor, a sinusoidal current is established in the steady-state and power is drawn from the source represented by *vars*. It was observed that the power is delivered to the load during one quarter cycle and then returned the next quarter cycle of one full cycle, with the result that there is no net energy flow over one cycle. This, we learned, is reactive power.

Questions

1. Describe the various types of potential energy and how it is stored.
2. Derive the expression for the Poynting theorem $\frac{dW}{dt} = -\frac{d}{dt}\int_V \frac{1}{2}\left(\varepsilon_o E^2 + \frac{1}{\mu_o} B^2\right) d\tau - \frac{1}{\mu_o}\oint_S (E \times B)\cdot da$ with a description of all the energy processes applicable in a circuit making sure to identify the imparted energy, stored energy, and the dissipated energy denoted by the theorem.
3. Draw the electric field flux lines and the magnetic field lines for a simple series circuit comprising a battery, a resistor (dissipative element) and wires. Derive an expression for the energy transported by applying the Poynting vector
$$S = \vec{E} \times \vec{H} = \frac{1}{\mu}(\vec{E} \times \vec{B})$$
and show that $\oint S \cdot da = VI$, where V is the voltage of the battery and I the current

in the circuit. Describe the direction of the energy transport between the wires and in the load resistor.

4. Give brief descriptions of the following electrodynamic theories i) Action-at-distance ii) Classical iii) Emission, and iv) Quantum.
5. What are virtual photons and massless moving elements? Compare the nature of the electric field in Quantum Field Theory and Page's Emission Theory. Where is the origin of the force in a Coulomb interaction located. Hint: See "Coulomb Interaction" in the article in the folder "Teacher's Guide" in the CD.
6. Describe all the processes that occur when a distant electric bulb is turned ON by switching ON a voltage supply to it.
7. Typically, electric forces are enormously larger than "magnetic ones". Then, how is it that we notice and can measure magnetic forces in circuit experiments? Explain.
8. What are vars? Explain briefly with reference to a capacitor circuit subject to a sinusoidal voltage.

Additional practice questions are available in the CD with this book.

5

Faraday's Law and Transmission Line Fundamentals

After an introduction to the magnetic field, its interactions and detection, the initial sections of the chapter present the association of electric fields (non-conservative) to time-varying magnetic fields called Faraday's law. Self-induction is next treated and the subtleties in the processes involved that produce "induced emfs" presented.

The current in an inductor circuit lags an applied sinusoidal voltage and the processes that produce this steady state operation are then described. The behavior of inductive circuits when the frequency of a sinusoidal voltage they are subject to is varied, is explained.

The production of high voltages using inductors is important, because while this phenomenon is exploited in producing sparks for automobiles, they prove to be an irritant when switching inductive loads using electronic devices. This aspect is treated in following sections.

Transmission lines are used to transmit energy in electronic circuits, electric circuits, power transmission and also in communication systems. Primarily, the transmission is via electromagnetic fields and therefore, the section on transmission lines begins with a qualitative introduction to the Ampere-Maxwell law and is followed by a description of guided waves between parallel planes. The process of transmission of em waves by two wire lines is treated in some detail in the last portions of the section.

The lines exhibit peculiarities in behavior at high frequencies due to line inductances and capacitances and causes reflections of voltage and current. This is treated qualitatively and the phenomenon of *skin effect* is also described.

A description of reflected waves in short-circuited quarter-wave length long lines that behave as resonant circuits and find use as components in circuits operating at Ultra-High Frequencies is also given. The differential-mode and common-mode signals in transmission lines are important to understand and this aspect is treated in some detail in this section.

Notes on Attenuation, it's meaning and ways by which it reduces the strength of signals and electromagnetic waves are presented at the end of the chapter.

5.1 MAGNETIC INTERACTIONS

Magnetic interactions are *similar* to electric interactions:

1. There is *both repulsion* (as between like charges so between like poles) *and attraction* (as between unlike charges so between unlike poles of a magnet);
2. The *superposition principle is valid.* ($E_1 + E_2 + E_3$...), and *net field* of two or more magnets is the vector sum of the individual magnetic fields ($B_1 + B_2 + B_3$...), and

...Magnetic fields... *invisibly alter* the space where they exist, and they penetrate through matter.

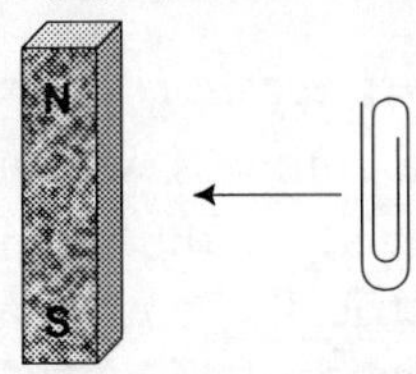

The space between the magnet and the metallic paper clip, is *altered invisibly* when the clip is attracted to the magnet. It is the same magnetic field that surrounds current carrying wires, as shown in the diagrams.

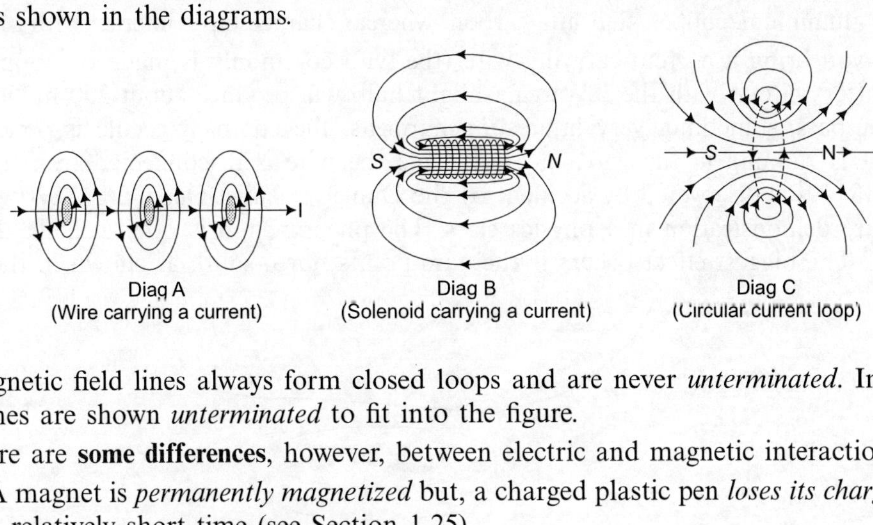

Diag A (Wire carrying a current) Diag B (Solenoid carrying a current) Diag C (Circular current loop)

Magnetic field lines always form closed loops and are never *unterminated.* In Diag C some lines are shown *unterminated* to fit into the figure.

There are **some differences**, however, between electric and magnetic interactions:

1. A magnet is *permanently magnetized* but, a charged plastic pen *loses its charge* within a relatively short time (see Section 1.25)...
2. A *magnetic field* around a current-carrying wire has a *curly* pattern that we don't see with electrostatic fields.

Magnetic effects are associated solely with ***moving charges*** whereas ***electric effects*** are evident with ***stationary and moving charges***.

A simple circuit, involving a battery, wire, and a flashlight bulb, is a convenient means of producing a supply of moving electrons. Despite the simplicity of the materials, the physics questions that can be investigated with such circuits are centrally important ones. If you have the appropriate equipment, you may wish to do these experiments yourself.

You can make a two-battery (D cell of 1.5 volt each fitted in a battery holder) circuit with a round flashlight bulb (1 watt, 2.5 volts), socket for the flashlight bulb, insulated "hookup" wire, some clip leads (wires with alligator clips at the end), and a liquid-filled compass (air-filled compasses typically don't work well because the needle tends to get stuck on its pivot).

We can use the magnetic compass as a detector of magnetic fields. Just as the deflection of a charged piece of invisible tape or the twisting of a permanent electric dipole indicates the presence of an electric field, the twisting of a compass needle indicates the presence of a magnetic field.

Detecting Magnetic Fields (Chabay and Sherwood, 2011 reproduced by permission)

How can you be sure that a compass needle is not simply responding to electric fields?

In observing the behavior of a compass, we find that:

1. The compass needle is affected by the proximity of objects made of iron or steel, even if these objects are electrically neutral (and therefore attract both positively and negatively charged tapes, Section 1.14). It is also affected by the presence of nickel or cobalt, though these are less readily available.
2. The compass needle is unaffected by objects made of most other elements, including aluminium, copper, zinc, and carbon, whereas charged tapes interact with these objects.

If you bring a current-carrying wire (the wire commonly is made of copper and the two-battery circuit with the 1 Watt flashlight bulb will produce about 300 milliamps) near the compass, something very interesting happens: the compass needle is deflected while current is running in the wire, but not while the wire is disconnected from the battery. This effect was discovered by accident by the Danish scientist Oersted in 1820 while doing a lecture demonstration in a physics class. The phenomenon is often called "the Oersted effect." The biggest effect occurs if the wire points north-south as shown in the figure.

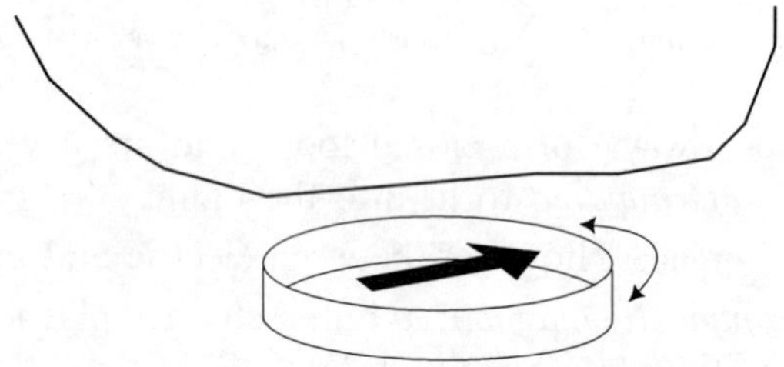

If a compass needle is originally pointing north, and a current-carrying wire aligned north-south is placed on the compass, the needle deflects. The deflection direction depends on whether the current runs northward or southward.

The compass needle doesn't deflect if the wire points east-west (see figure below).

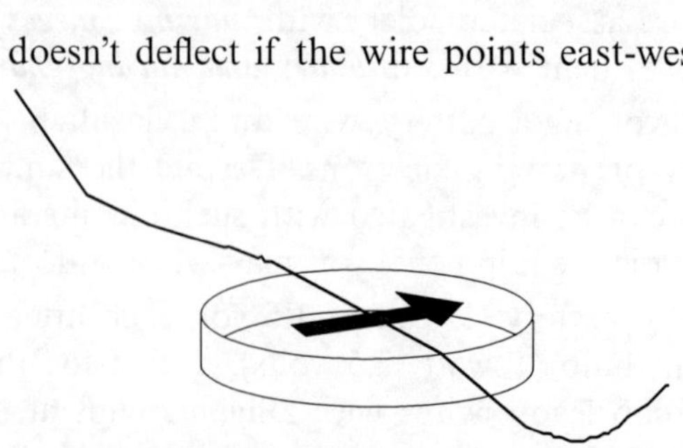

If a current-carrying wire aligned east-west is placed on a compass, the needle does not deflect.

From such experiments, one can draw the following conclusions:

1. The magnitude of the magnetic field produced by a current of moving electrons depends on the amount of current.
2. A wire with no current running in it produces no magnetic field.
3. The magnetic field due to the current appears to be perpendicular to the direction of the current.
4. The direction of the magnetic field due to the current under the wire is opposite to the direction of the magnetic field due to the current above the wire. This can be demonstrated by passing the wire under the compass.

What forces are responsible that enables the compass needle to align itself in the direction of the net magnetic field? We will learn in Chapter 8 that magnets (the compass needle is a tiny magnet) contain atoms, whose electrons, occupy orbital spaces, and their orbital motions and spins produce a magnetic field. The interaction of the magnetic field of the compass needle with the Earth's magnetic field or a field due to nearby wires, combined with the Earth's field, produces a net magnetic field around the compass needle. And, since the needle is balanced on a pivot, it swings and comes to a rest in the direction of the net field around it, where the potential energy of the magnetic "moment" is least.

A Model for the Observations

The moving electrons in a wire create a magnetic field at various locations in space, including at the location of the compass. The vector sum of the earth's magnetic field $\vec{B}_{\text{earth}}$ plus the magnetic field $\vec{B}_{\text{wire}}$ of a current-carrying wire makes a net magnetic field $\vec{B}_{\text{net}}$ (see figure).

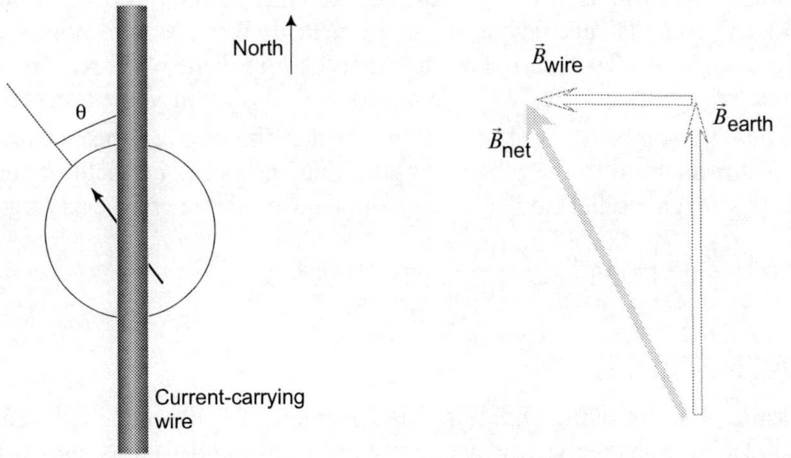

A compass needle points in the direction of the net magnetic field, which is the superposition of the magnetic field of the Earth and the magnetic field of the current-carrying wire that passes over the compass, aligned north-south.

Since a compass needle points in the direction of the net magnetic field at its location, the needle turns away from its original north-south direction to align with the net field.

Because the magnetic field made by the wire is perpendicular to the wire and is in opposite directions above and below the wire, the pattern of magnetic field made by the wire must look like that shown in the figure.

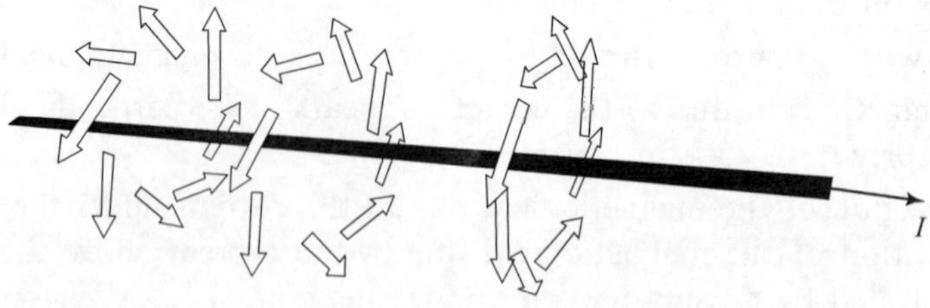

Magnetic field of a straight current-carrying wire.

It happens (conveniently) to be the case that the magnitudes of $\vec{B}_{\text{earth}}$ and $\vec{B}_{\text{wire}}$ are similar. If $\vec{B}_{\text{earth}}$ were much larger than $\vec{B}_{\text{wire}}$, then we would have almost no response from the compass.

What would we observe if $\vec{B}_{wire}$ were much larger than $\vec{B}_{earth}$?

If $\vec{B}_{\text{wire}}$ were much larger than $\vec{B}_{\text{earth}}$, then the compass deflection would be nearly 90°, since the net magnetic field would be primarily due to the magnetic field made by the current in the wire.

There is hardly any observable electric interaction of the current-carrying wire with nearby materials, because the wires have no net charge. (Actually, there are tiny amounts of excess charge on the surface, but too little to observe easily.)

Compass Deflection

A current-carrying wire is oriented north-south and laid on top of a compass. At the location of the compass needle the magnetic field due to the wire points west and has a magnitude of 3×10^{-6} T (Tesla). The horizontal component of Earth's magnetic field has a magnitude of 2×10^{-5} tesla. What compass deflection will you observe?

Solution: The compass needle will point in the direction of the net magnetic field, so from the vector diagram in the previous figure that shows the net field of the superposition of the Earth's magnetic field on the magnetic field of the current-carrying wire:

$$\theta = \tan^{-1}\left(\frac{\vec{B}_{\text{wire}}}{\vec{B}_{\text{earth}}}\right) = \tan^{-1}\left(\frac{3\times10^{-6}\,\text{T}}{2\times10^{-5}\,\text{T}}\right) = 8.5^\circ \text{ west}.$$

5.2 MAGNETIC FIELD

We now know that a moving charge produces an *invisible* magnetic field and to determine the direction of the magnetic field we apply the right-hand rule as shown in Fig. 5.1. We speak of a magnetic field due to a steady current and *not* of a single point charge as we did in Section 2.27. Steady currents (millions and millions of electrons everywhere in the wires moving with a start-stop motion with a drift superimposed, not piling up anywhere) produce magnetic fields that are constant in time just as stationary charges produce constant electric fields in space.

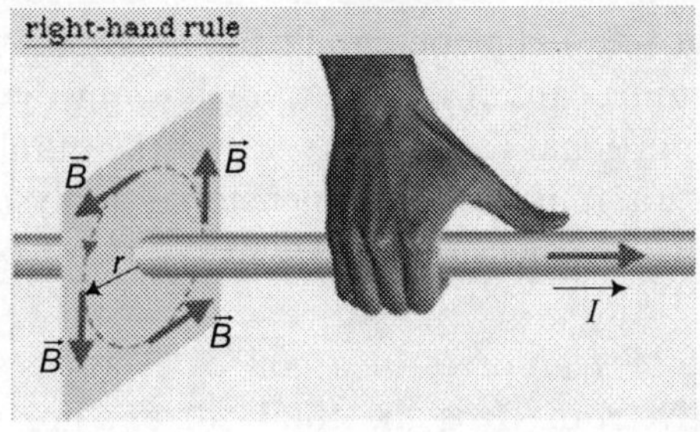

Fig. 5.1 The right-hand rule used to determine the magnetic field.

To apply the right-hand rule for determining the direction of the magnetic field produced by a steady current, imagine gripping the wire with your right hand so that your thumb points in the direction of current flow. Your fingers naturally curl in the direction in which the magnetic field circulates.

What forces produce the magnetic flux? Is it the current only? Or is its direction important?

Magnetomotive Forces (mmf) and Magnetic Field Strength, *H*

Experiments show that for a homogeneous medium, for a long wire carrying current I (Fig. 5.1), **B**, the magnetic flux per unit area is proportional to $\frac{\mu I}{r}$, where, r is the distance from the wire and μ is a constant which depends on the medium. When the experiment is performed in vacuum, it was found that the proportionality factor is $\frac{1}{2\pi}$ giving, $B = \frac{\mu I}{2\pi r}$. This may be written as $B = \mu H$ where $H = \frac{I}{2\pi r}$ Amps/meter.

The magnetic field strength H is thus defined by this relation in terms of the current which produces it and the geometry of the system. H, the magnitude of the magnetic field strength may be visualized as a "stress" or a force per unit length that produces a flux density through its medium. Many students confuse magnetic field strength H with the magnetic field or flux density B. In free space, the magnetic field strength H may be visualized as the *cause* and B, the flux density is its *effect*. In media, the expression is $B = \mu_o H + M$ about which we will learn more in Chapter 8, Section 8.28. The magnetomotive force (mmf) is a quantity which is defined as the line integral: $mmf = \int_a^b H \cdot ds$ along the path between two points a and b. If the path is circular of radius r about the current-carrying wire, with the wire at the center, H has the constant value $\frac{I}{2\pi r}$ and is directed along the path, so that $mmf = \oint H \cdot ds = I$ because the distance moved along the circle is $2\pi r$.

It is merely a coincidence that mmf is I in this particular case, and therefore cannot be generalized.

In an electric circuit, the current is due to the existence of an electromotive force. By analogy, we may say that in a magnetic circuit the magnetic flux is due to the existence of a *magnetomotive force* (mmf) caused by a current flowing through one or more turns. The value of the mmf is proportional to the current and the number of turns, and is expressed in *ampere turns*; but for the purpose of dimensional analysis, it is expressed in *amperes*, since the number of turns is dimensionless. Hence, the unit of magnetomotive force is the *ampere*.

If there is a current of I amperes in a coil of N turns, as shown in the figure of a toroid, the magnetomotive force F is the total current linked with the magnetic circuit, namely IN amperes.

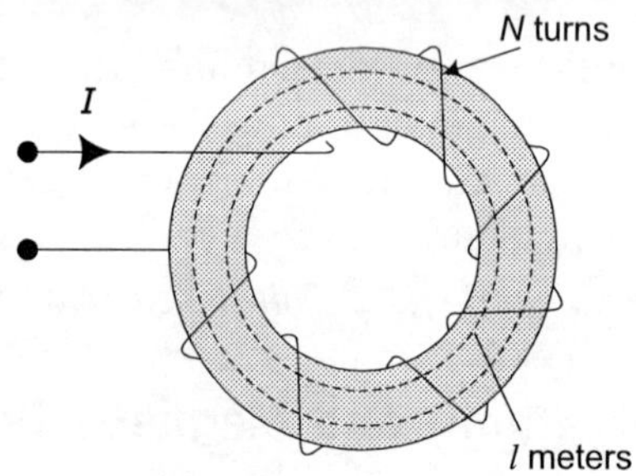

A toroid.

If the magnetic circuit is homogeneous and of uniform cross-sectional area, the magnetomotive force per meter length of the magnetic circuit is also the magnetic field strength H and if the mean length of the magnetic circuit is l meters,

$$H = \frac{N \times I}{l} \quad \text{amperes per metre, and}$$

$$H = \frac{F}{l} \quad \text{where the magnetomotive force } F = NI \text{ amperes.}$$

Biot-Savart Law

Careful experimentation shows that a *moving point charge* not only makes an electric field but also makes a magnetic field which is given by the Biot-Savart law. The law is named after scientists Biot and Savart, though both of them never worked with a "magnetic field". Faraday and Maxwell isolated the *expression* for a magnetic field from the force laws that Ampere, Biot and Savart had so rigorously investigated (Ref. [38, Sections 15.1 and 15.2]).

The Biot-Savart law for a single moving source charge gives the magnetic field B at the observation location as shown in Fig. 5.2 by

$$\vec{B} = \frac{\mu_o}{4\pi} \frac{q\vec{v} \times \hat{r}}{r^2}$$

in *tesla* where, v is the velocity of the source point charge q and $\hat{r}$ is a unit vector that points from the source charge to the observation location.

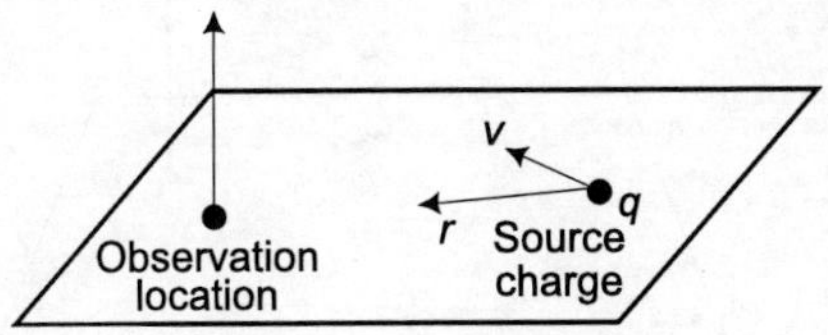

Fig. 5.2 Biot-Savart law to determine the magnetic field due to a moving charge.

Right-hand Rule for Vector Cross-product

(Chabay & Sherwood, 1995, used by permission)[2]

A simple method for determining the direction and magnitude of a vector cross-product is described below with reference to Fig. 5.3.

Keep your right hand open and flat (fingers and thumb in the same plane) with your thumb perpendicular to your fingers, and point your fingers in the direction of the first vector (v).

Now, bend your fingers toward the second vector (r), noting the angle of bend θ, which is always less than or equal to 180°.

You may have to rotate your wrist to get your hand into a position that lets you bend your fingers this way.

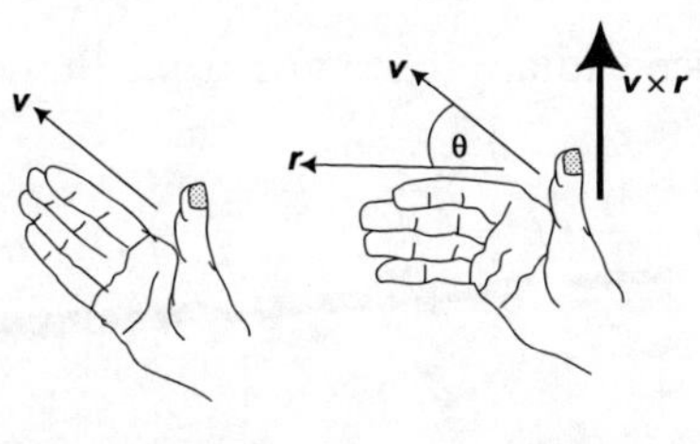

Fig. 5.3 The right-hand rule for the magnitude and direction of $\boldsymbol{v} \times \boldsymbol{r}$.

The thumb points in the *direction* of cross-product $\boldsymbol{v} \times \boldsymbol{r}$.

Magnetic Field of Current Distributions

We don't often observe the magnetic field of a single moving charge.

Usually we are interested in the magnetic field produced by a large number of charges moving through a wire in a circuit. The superposition principle is valid for magnetic fields, so we need to add up the contributions of the individual charges in the magnetic field.

The magnetic field due to a current in a small length of thin wire is given by $\Delta\vec{B} = \frac{\mu_o}{4\pi}\frac{I\Delta\vec{l} \times \hat{r}}{r^2}$ where $\Delta\vec{l}$ the length of the short wire, and r is the distance of the observation point from the wire as shown in the figure.

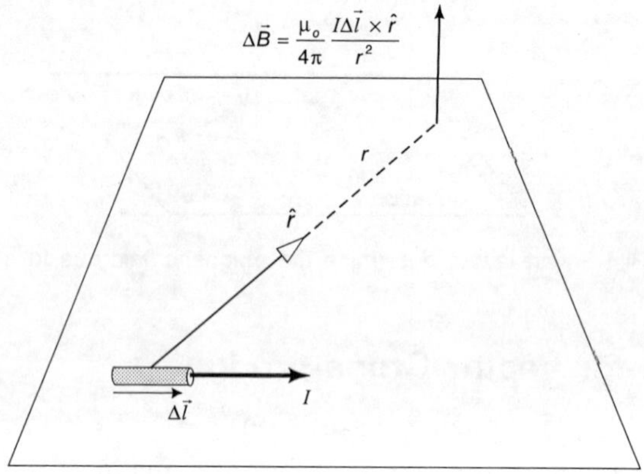

Magnetic field contributed by a short thin length Δl of current-carrying wire.

The direction of the field is found by applying the right-hand rule.

A derivation of this expression is given in "The Biot-Savart Law for a Short Thin Length of Wire" in Appendix A.

The magnetic field (in Tesla) of a long straight wire as shown in the figure is given by $B_{\text{wire}} = \frac{\mu_o}{4\pi}\frac{LI}{r\sqrt{r^2 + (L/2)^2}}$ where L is the length of the wire in meters, I is the current in amperes and r is the distance radially outward from the wire.

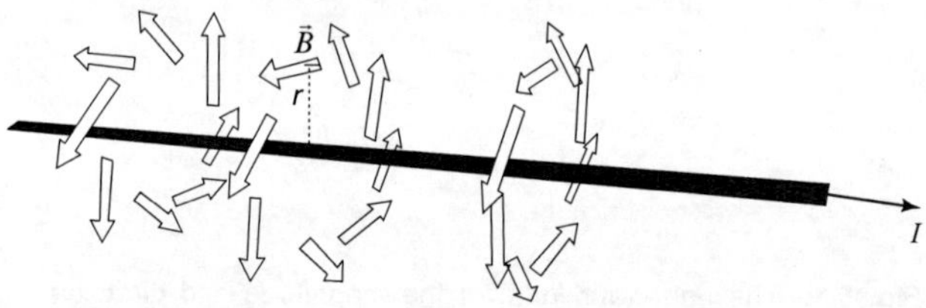

Magnetic field of a straight current-carrying wire, at selected locations a distance r radially outward from the wire.

A derivation of this expression is given in "The Magnetic Field of Current Distributions" in Appendix A.

Visit http://www.falstad.com/vector3dm/ and select Field selection: Current line and Display: Field vectors or Field lines to view the magnetic field of a current line.

The magnetic field calculated by applying the Biot-Savart law (Chabay & Sherwood, 1995, used by permission)[2] for a circular loop of radius a and conventional current I at a distance x from the center, *along the axis*, is as shown in Fig. 5.4:

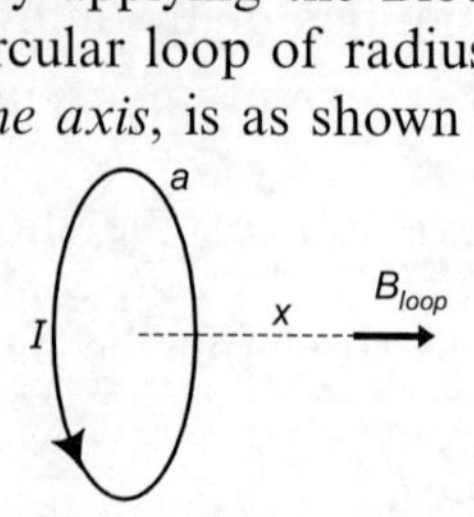

Fig. 5.4 The Biot-Savart law used to find the magnetic field B.

and is given by, $$\vec{B} = \frac{\mu_o}{4\pi}\frac{2\pi a^2 I}{(x^2 + a^2)^{3/2}}.$$

The magnetic field for a circular loop of radius a and conventional current I at a distance x from the center, *far away* along the axis, is as seen in Fig. 5.5, and

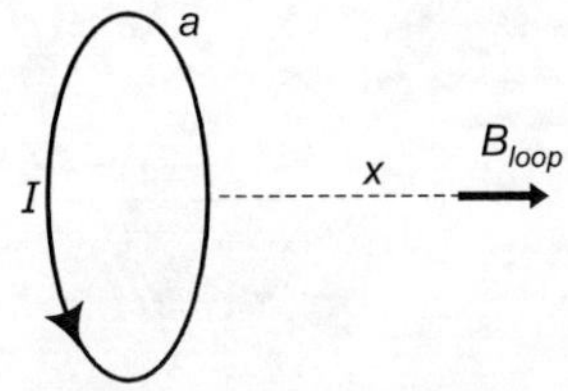

Fig. 5.5 The Biot-Savart law used to find B far away from the loop.

is given by $$\vec{B} = \frac{\mu_o}{4\pi}\frac{2\pi a^2 I}{x^3}.$$

In other words, the magnetic field falls away fairly rapidly by $1/(\text{distance})^3$ as we move away from a current carrying loop.

I urge the reader to study the derivation of the above expressions for the magnetic of a circular loop of wire given in "A Circular Loop of Wire" in Appendix A.

Visit http://www.falstad.com/vector3dm/ and select Field selection: Current loop and Display: Field vectors or Field lines to view the magnetic field of a current loop.

Magnetic Flux vs Magnetic Field

(Chabay & Sherwood, 1995, used by permission).[2]

The magnetic flux on a small area ΔA is $\vec{B}.\vec{n}\,\Delta A$ where, n is a dimensionless unit vector *perpendicular* to the area ΔA (n is called the normal in the surface). A diagram that illustrates the flux of a magnetic field B is shown.

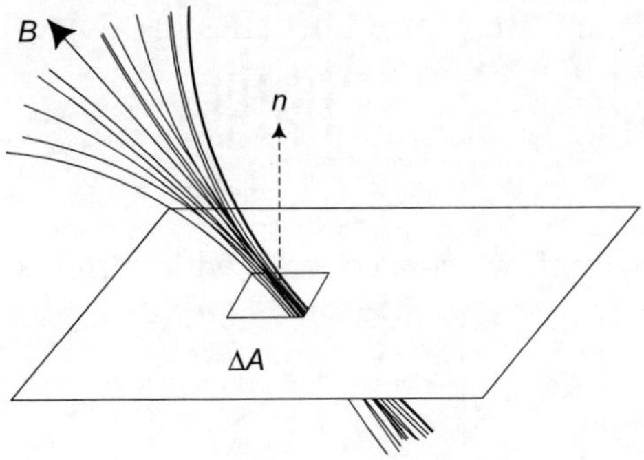

We add up all magnetic field contributions over an extended surface to get the magnetic flux $\phi_{magnetic} = \int \vec{B}\cdot\vec{n}dA$

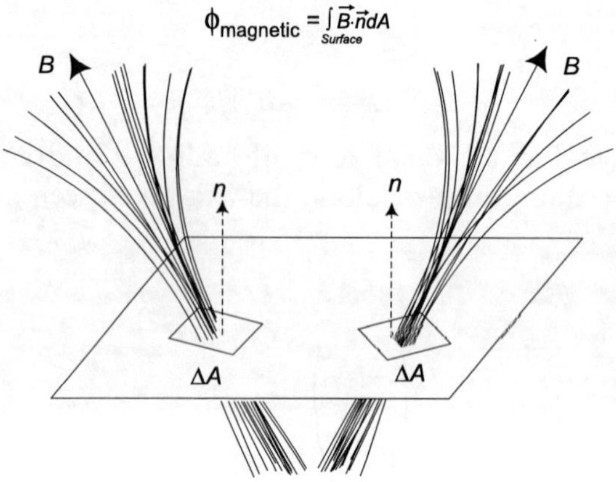

5.3 CHARGES AND ELECTRIC FIELDS AND MAGNETIC FIELDS

We have learned electric and magnetic interactions as separate phenomena. Charges ***make*** electric fields, and electric fields ***affect*** charges.

Moving charges ***make*** magnetic fields, and magnetic fields ***affect moving*** charges.

Yet, there is a deeper relationship between the two kinds of fields.

A ***stationary*** charge makes ***no*** magnetic field.

A ***time-varying*** magnetic field can ***produce*** an electric field.

The unit of magnetic field is the "tesla".

5.4 CURRENTS AND MMF

The effect of the current in a coil of wire is to provide an mmf (magnetomotive force) just as the chemical reactions within a car battery provide an emf.

The electric circuit and the magnetic circuit are analogous and just as low resistance copper wire connects the battery to the high resistance of the headlight bulb filament, so the ferrite core connects the coil with the air gap.

To produce a larger emf, we connect batteries or sources of emf...

... in series.

To produce a larger mmf, we connect coils with currents ...

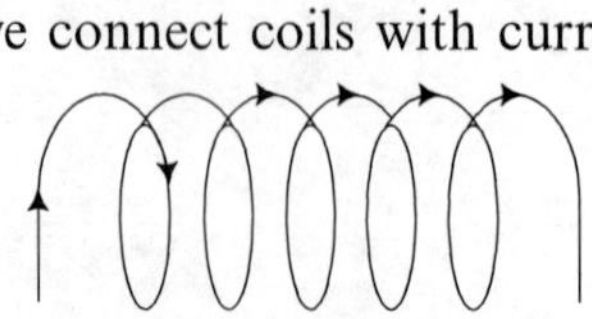

... in series.

mmf = No. of turns × current

The electric circuit with its *increasing emf* produces more current.

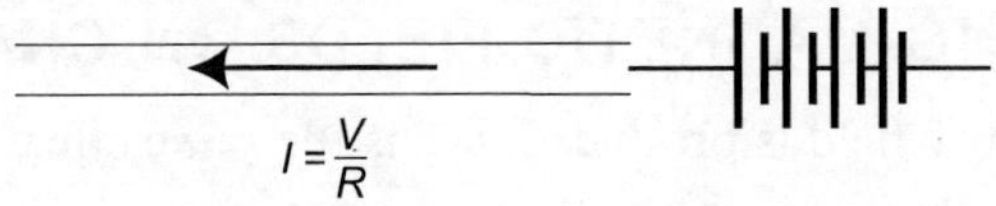

See "Superposition of Potential" Section 3.1.

The magnetic circuit with its increasing mmf produces more magnetic field lines and flux.

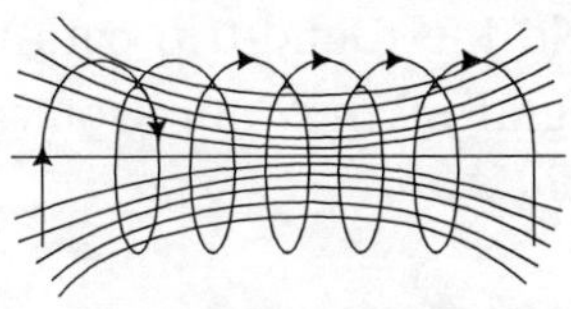

$$\textbf{Flux } \phi = \frac{mmf}{Reluctance}$$

Reluctance is a concept used in the analysis of magnetic circuits and is analogous to resistance in electric circuits. But, while a resistance dissipates energy, reluctance stores magnetic energy in the magnetic field. Reluctance is defined as the ratio of the magnetomotive force (mmf) that produces a magnetic flux ϕ in a magnetic circuit. It is expressed in ampere-turns per weber. Air and vacuum have high reluctance, while easily magnetizable materials such as soft iron have low reluctance. In later chapters, we shall see that the air gaps between the rotating structure and the stationary magnetic field sourcing coils in motors and generators are kept small to make the gap reluctance small.

5.5 PRODUCING ELECTRIC FIELDS

There are two different ways by which an electric field can be produced:

1. By charges according to ***Coulomb's law***.

When charges are separated, a Coulombic electric field is produced around them, for example, charges on the end plates of a battery and dipoles of atoms and molecules.

2. By ***time-varying magnetic fields***.

An electric field co-occurs with a time varying magnetic field, and the phenomenon is referred to as magnetic induction.

The historical term "magnetic induction" is often used to describe the phenomenon, and one says that the time-varying magnetic field "induces" the electric field (that we shall see has a different pattern in space from that produced by point charges). This is somewhat misleading, as the source of both the time-varying magnetic field and the curly electric field is varying (nonconstant) currents. Therefore, we will simply refer to the phenomenon as Faraday's law Ref. [2, 2010]. Also see Links_to_animations_articles in the CD for links to Faraday's original "Experimental Researches in Electricity".

5.6 EFFECT OF ELECTRIC FIELDS ASSOCIATED WITH TIME-VARYING MAGNETIC FIELDS ON CHARGES

No matter how an electric field is produced, it has the same effect on a charge q; Force = qE.

But the **non-Coulomb** electric field that occurs with a time-varying magnetic field has a *different pattern* in space than the **Coulomb** electric field $E = \dfrac{q}{4\pi\varepsilon_0 d^2}$.

In a "Coulomb" electric field the round-trip path integral is 'zero'.

But, a round-trip path integral of the "non-Coulomb" electric field associated with a time-varying magnetic field is *not zero*.

5.7 SOLENOIDS AND INDUCTORS

A long straight wire carrying a steady current I has a magnetic field surrounding it.

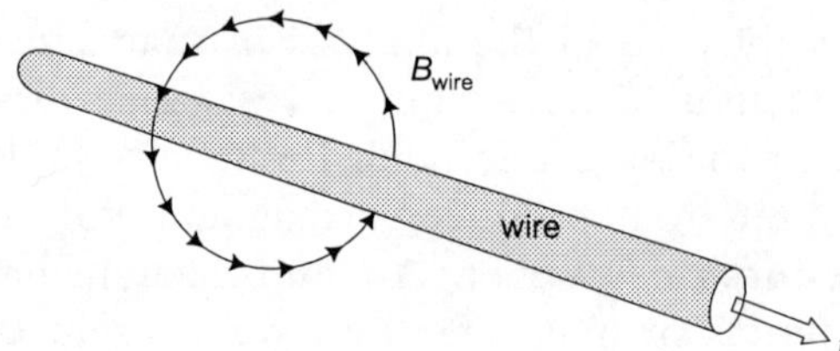

A long straight coil of wire can be used to generate a nearly uniform magnetic field similar to that of a bar magnet. Such coils, called solenoids (used as electromagnets and for actuation of mechanical valves) and inductors (used to smoothen out current variations in electronics and electrical circuits), have an enormous number of practical applications. The field can be greatly strengthened by the addition of an iron core. The figure shows the schematic of an inductor and inductors of different sizes, shapes and packaging.

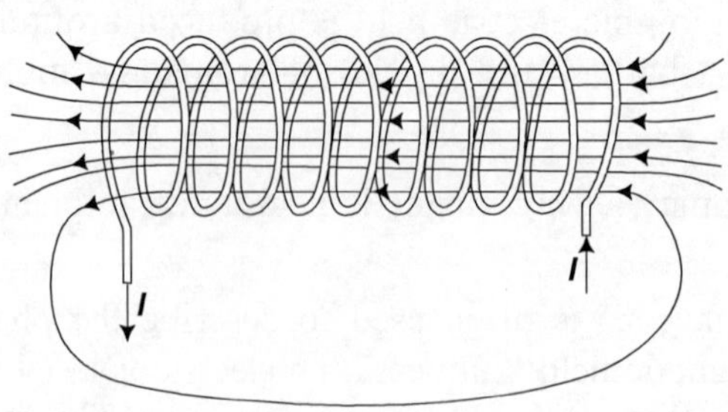

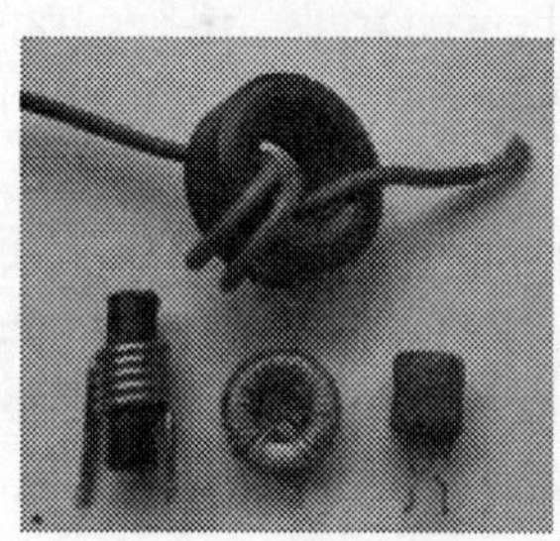

Consider a very long solenoid, a long hollow coil of current-carrying wire as shown in Fig. 5.6.

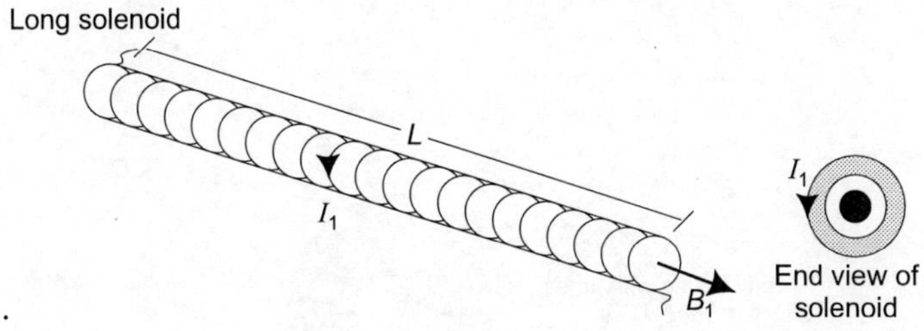

Fig. 5.6 Magnetic field inside a current carrying solenoid.

With a current I_1 in the solenoid of N turns and length L, the magnetic field B_1 inside the solenoid (not too near the ends) is

$$B_1 = \frac{\mu_0 N I_1}{L}$$

The magnetic field outside the solenoid is (not too near the ends) very small and for a very long solenoid can be taken to be "zero". The magnetic field of a coil of wire (one loop) is described in "Qualitative features of the Magnetic field of a loop of wire" in Appendix A and the field of a coil of many loops in "A Coil of Many Loops" and "Magnetic Field of A Solenoid" in Appendix A which I urge the reader to study.

Visit http://www.falstad.com/vector3dm/ and select Field selection: Solenoid and Display: Field vectors or Field lines to view the magnetic field of a Solenoid.

5.8 CURLY ELECTRIC FIELDS

A charge moving outside the central region of this solenoid will not experience an electric or magnetic force because the electric and magnetic fields outside the solenoid are essentially zero *if B_1 is constant in time* as shown in Fig. 5.7. (Chabay & Sherwood, 1995, used by permission)[2].

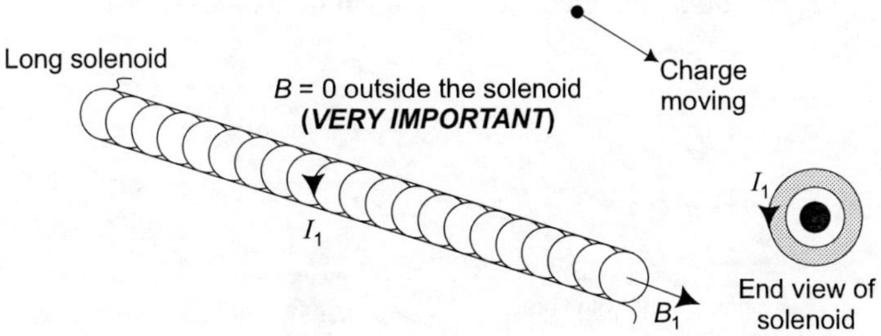

Fig. 5.7 Charge moving outside does not experience a force.

But something remarkable happens if the magnetic field B_1 inside the solenoid varies with time. There is still no magnetic field outside the solenoid, but, we observe an electric field both inside and outside the solenoid as shown in Fig. 5.8(a) and 5.8(b).

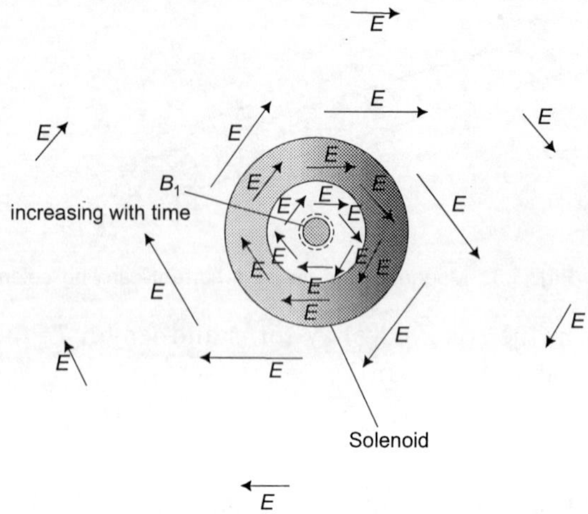

Fig. 5.8(a) Electric field co-occurring with a time-varying magnetic field.

Outside the solenoid, the electric field is proportional to $1/r$. Inside the solenoid, the electric field is proportional to r, the distance from the axis (smaller near the axis). These are indicated by the varying size of the electric field vectors inside and outside the solenoid.

This electric field *curls* around the axis of the solenoid.

This peculiar electric field is proportional to dB/dt, the *rate of change* of the magnetic field. This odd-looking electric field acts in the usual way:

a charge q experiences a force $F = qE$

A proton placed above the solenoid will be initially **pushed to the right** as shown in Fig. 5.8(b).

The curly electric field does not originate nor terminate on point charges and that is the reason why we call such an electric field a **"non-Coulomb"** field.

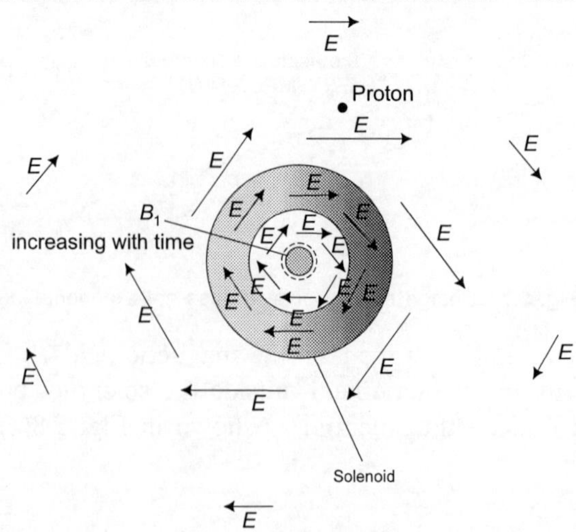

Fig. 5.8(b) Electric field co-occurring with a time-varying magnetic field.

Suppose we place a circular metal ring around the solenoid as shown in Fig. 5.9. (Chabay & Sherwood, 1995, used by permission).[2]

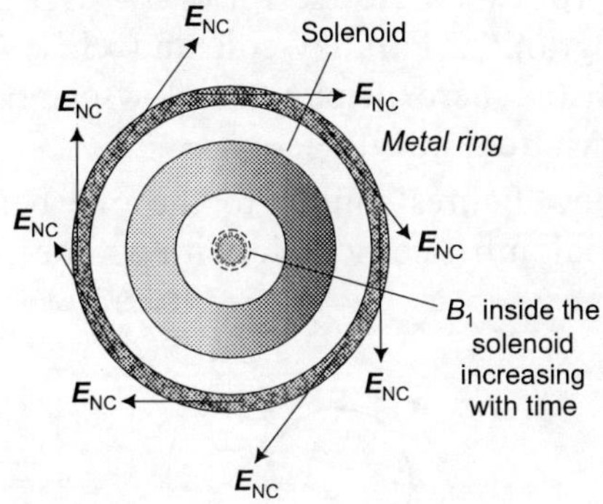

Fig. 5.9 Non-Coulomb electric field co-occurring with a time-varying magnetic field.

With the magnetic field pointing out *and increasing with time* the *non-Coulomb electric field* inside the metal will drive conventional current clockwise around the ring (Chabay & Sherwood, 1995, used by permission)[2].

The practical importance of the occurrence of a non-Coulomb electric field is that it can make current run in a wire just *as though a battery were present.*

Recollect what was said about the nature of the electric field in Sections 1.4 and 1.11: an electric field invisibly alters space and exerts a continuous and relentless force on charged particles causing them to accelerate, and is the cause of the current (counterclockwise start-stop movement of millions and millions of electrons everywhere in the wires with a drift superimposed) in the ring of Fig. 5.9. The electric field *fills* the ring.

In Chapter 2, we had asked a question: Is it possible to set up a configuration of charged objects in such a way that the electric field along a circular path looks like this? **The answer was "No".**

But...this curly non-Coulomb field E_{NC} field can be associated with a time-varying magnetic field.

Note: Some researchers of the phenomenon of the curly electric field appearing with a time-varying magnetic field, state that the rate of change of *B* is the effect, with the source of the effect being the curl of *E*! Because the distinction is not so important, Sherwood and Chabay in the third edition of their textbook "Matter and Interactions", John Wiley & Sons, 2010, state that the changing magnetic field and the curly electric field, co-occur.

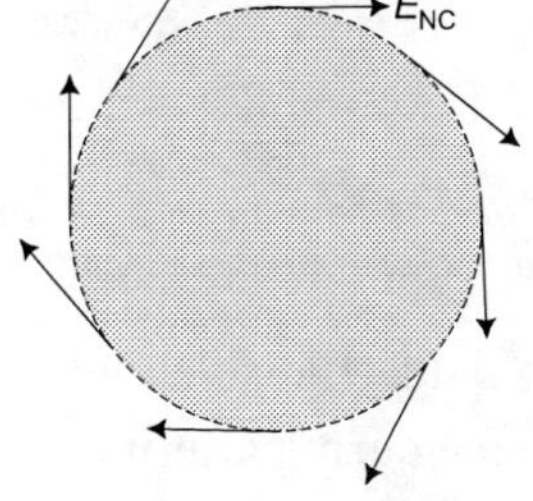

To view an animation of a curly electric field associated with the changing magnetic field of a coil visit http://www.glowscript.org/#/user/matterandinteractions/folder/matterandinteractions/program/22-Faraday-coil and click OK in the dialog. There are several nice animations listed here http://www.glowscript.org/#/user/GlowScriptDemos/folder/Examples/program/MatterAndInteractions.

Most Physics texts show figures similar to the one below to describe Faraday's law which associates a non-Coulomb electric field with a time-varying magnetic field.

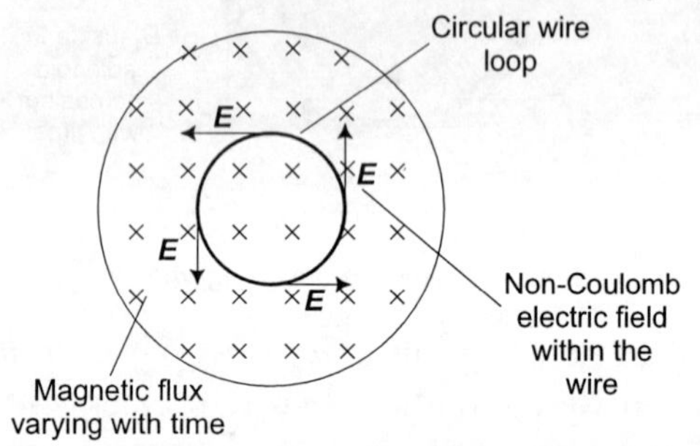

Notice how the circular wire loop is *totally embraced and is cut* by the changing magnetic flux.

But, see Fig. 5.10 in the next section and you will notice that what counts is the magnetic flux encircled by the circuit, and not the magnetic flux on a closed surface such as a box or sphere as we will see in section 5.12, Fig. 5.11b.

5.9 CURLY ELECTRIC FIELDS AND POTENTIAL DIFFERENCE

An electric field co-occurs in an area where there is no magnetic field with a time-varying magnetic field.

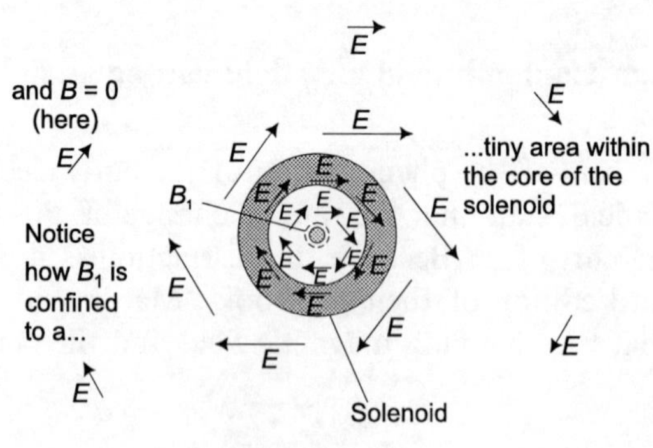

Fig. 5.10 Non-Coulomb electric field (in an area where there is no magnetic field) co-occurring with a time-varying magnetic field.

There is no magnetic field outside the solenoid.

Yet, an electric field which has a curly pattern co-occurs with the time-varying magnetic field (due to the varying current *in* the solenoid wire) outside the solenoid where, $B = 0$ or nearly zero. An electric field also occurs inside the solenoid.

Non-conservative fields

From: Massachusetts Institute of Technology lecture notes.[12]

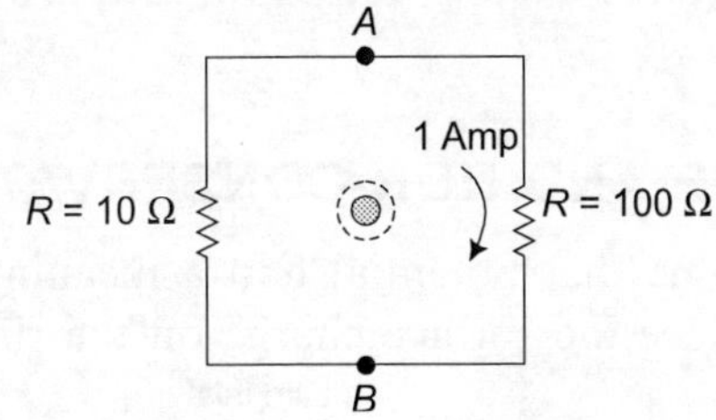

A ***magnetic field penetrates*** this circuit ***outwards***, and is increasing at a rate such that a current of 1 A is induced in the circuit (which direction?)

The ***potential difference*** $V_A - V_B$ is:

1. +10 V	2. –10 V	3. +100 V	4. –100 V
5. +110 V	6. –110 V	7. +90 V	8. –90 V

9. None of the above

The Answer is 9. None of the above.

The question is meaningless. There is no such thing as potential difference when a changing magnetic flux is present. This is because by Faraday's law, a non-conservative $\mathcal{E}$ co-occurs with the changing magnetic flux (that is, its integral around a closed loop is non-zero).

A live demonstration of nonconservative fields was given in a Physics lecture in Massachusetts Institute of Technology, Boston, USA available here https://www.youtube.com/watch?v=FUUMCT7FjaI for a circuit shown in the figure.

The two voltmeters connected across the same points *A* and *B* show different values for a short time when the magnetic field is increasing in the loop set up with the two resistors R_1 and R_2. The circuit was arranged to develop an induced emf of about 1 volt. The voltage across the resistor R_2 was recorded at 0.9 V and across R_1 at –0.1V, and are not equal, as they should have been according to Kirchchoff's law. The reason for this behavior is because the loop is in a non-conservative field and the results are not as intuitive as one would have expected.

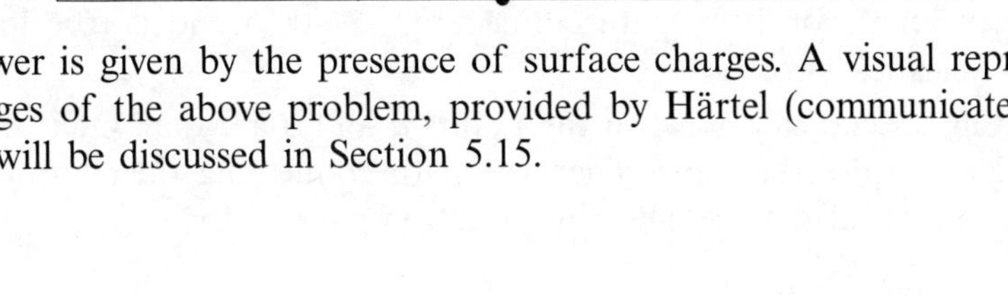

The answer is given by the presence of surface charges. A visual representation using surface charges of the above problem, provided by Härtel (communicated to the author, April 2017) will be discussed in Section 5.15.

A step-by-step derivation of the above astounding result is given here http://web.mit.edu/8.02/www/Spring02/lectures/lecsup3-15.pdf or http://web.mit.edu/8.02/www/Spring02/lectures.htm.

5.10 CONSERVATIVE AND NON-CONSERVATIVE ELECTRIC FIELDS

Let us for a moment assume that potential had a meaning in Faraday's law. If so, then imagine how you would place the surface charge on the ring.

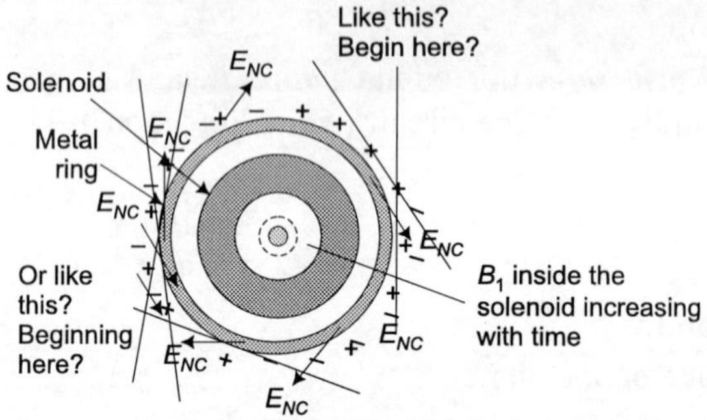

The symmetry of the ring makes it **impossible** to have a surface charge gradient along the ring.

Conservative and Non-Conservative Fields

The electric fields that occur in association with a changing magnetic flux are not associated with point charges. Although both kinds of electric fields exert forces on charges, there is a *difference between them*. The simplest manifestation of this difference is that flux lines of E associated with a *changing magnetic flux* can form *closed loops*.

Flux lines of E associated with *charges* always can be drawn to start on a positive charge and end on a negative charge.

5.11 CONSERVATIVE FIELDS (CHARGE SEPARATION—COULOMB FORCES)

We know from Section 2.21 that the potential difference between two points is

$$V_b - V_a = \frac{W_{ab}}{q_o} = -\int_a^b E \cdot dl$$

and we know that if potential is to have a real meaning then this integral (and W_{ab}) must have the same meaning for every path connecting a with b and is true for every case as explained in physics textbooks.

An interesting special case is when the points a and b are the same; the starting point and the ending point are the same. Then the path connecting them forms a closed loop and V_a must be identical with V_b and this equation reduces to

$$\oint_{a,b} E \cdot dl = 0$$

This happens when the fields are "*conservative*". See Section 2.19 "Conservative and non-conservative electric fields" and the topic "Allowable patterns of electric fields using charged objects".

Recollect from Section 2.17 "What is voltage?" that the round-trip path integral of the electric field due to a point charge (includes any static charge distribution and surface charges of circuits) is "zero", no matter which path is taken.

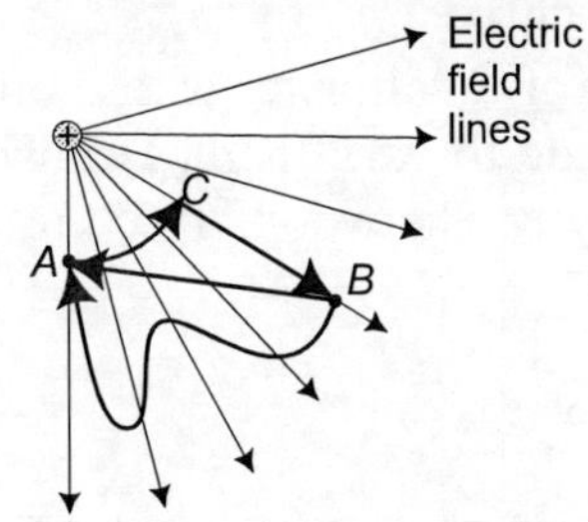

The work done to move a test charge from *A* to *C* to *B* and back to *A* is "zero".

Such distributions can have a *divergence* (the electric field flux lines diverge and curve around from the distribution originating and terminating on point charges) but never produce a curl that will close the field flux lines on themselves.

5.12 FARADAY'S LAW AND NON-CONSERVATIVE FIELDS

However, when changing magnetic flux is present,

$$\oint_{a,b} \cdots$$

is precisely *not zero* and is according to Faraday's law:

$$-d\phi_B/dt$$

where, ϕ_B is the total magnetic flux encircling the circuit as described in Section 5.2. Faraday's law is a major physical law and cannot be derived from other known physical principles. If we imagine a soap film stretched over the closed path around which we are calculating the path integral of the non-coulomb electric field E_{NC}, then ϕ_B is the flux on the area covered by

$$\oint E_{NC} \cdot d\vec{l} = -\frac{d}{dt}\int \vec{B} \cdot \hat{n} dA$$

the soap film. Then Faraday's formal version with ϕ_B as in Section 5.2, is

Electric fields associated with stationary charges are *conservative*, but those associated with changing magnetic fields are *non-conservative*.

Electric potential, which can be defined only for a conservative force, has *no meaning* for electric fields associated with *time-varying magnetic fields*.

Non-conservative fields can't have potentials associated with them.

As long as the curly non-conservative Faraday electric field occurs with a changing magnetic field *B*, the divergence of the field is zero. Such a field may have a non-zero curl and its field flux lines will close on itself, and if it is pure, that is, not influenced by a coulombic field due to point charges around, it will have a curl (due to the curly pattern), but its divergence will be "zero".

Compare the field as in Fig. 5.8 with the field due to point charges and static charge distributions of Fig. 2.63 and the figures in Section 1.6. Also refer Section 2.19 "Conservative and non-conservative electric fields".

What would be the reading of a voltmeter whose probes are connected together around a solenoid whose current is made to vary thereby causing the magnetic field B_1 to vary as shown in Fig. 5.11(a)? (Chabay & Sherwood, 1995, used by permission).[2]

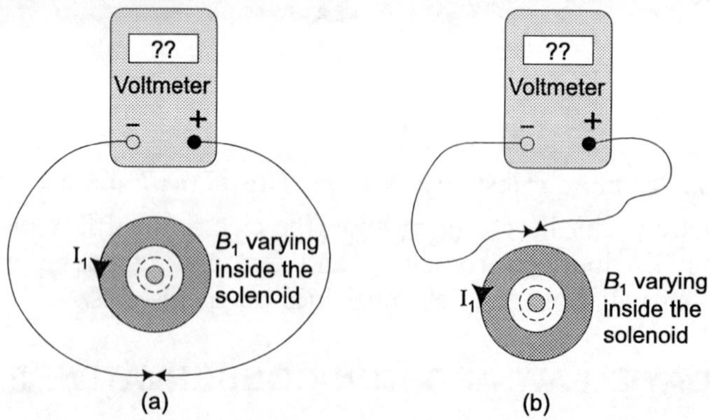

Fig. 5.11(a and b) emf due to a time-varying magnetic field. (contd...)

What would be the reading of a voltmeter whose probes are connected together around a solenoid as shown in Fig. 5.11(b)?

A voltmeter is an ammeter with a large resistance in series; a voltmeter may give a reading in the presence of a time-varying magnetic flux.

Since the electric circuit encircles the magnetic field inside the solenoid, the voltmeter will show a nonzero emf (*not* a p.d.) in Fig. 5.11(a) and is shown in Fig. 5.11(c).

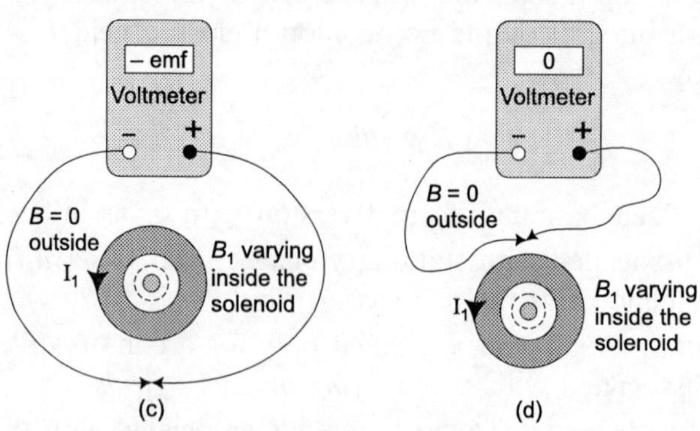

Fig. 5.11 (c and d) emf due to a time-varying magnetic field.

Due to the changing flux through the loop, a tiny current runs through the voltmeter from – to + and the reading is negative.

The voltmeter will show 0 volts when connected outside the solenoid and does not encircle it as in Fig. 5.11(b). Because there is no changing flux through the loop, there is no emf; therefore, the reading is zero as shown in Fig. 5.11(d).

Note that what counts is the magnetic flux *encircled* by the circuit, **not** the magnetic flux on a *closed surface* such as a box or sphere. A closed path like a circuit defines an open surface (imagine a soap film stretched over a closed path in space) whereas a closed surface defines a volume like the sides of a box that carves out a volume in space.

Being quantitative about the emf, experimentally it is found that the magnitude of the induced emf is numerically equal to this:

$|emf| = \left|\frac{d}{dt}(B_1\,\pi r_1^2)\right|$, where r_1 is the radius of the solenoid and B_1, the magnetic field in the solenoid.

The quantity $(B_1 \pi r_1^2)$ is called the "magnetic flux" Φ_{mag} (as we described in Section 5.2) on the area encircled by the circuit (the ammeter and its wires).

5.13 DIRECTION OF INDUCED EMF AND LENZ'S RULE

Consider an inductor with a current that is increasing such that the magnetic field due to it is pointing out and increasing, as shown in Fig. 5.12(a) (after Chabay & Sherwood, 1995).

We have seen the direction of the non-Coulomb electric field earlier. It is clockwise.

If the magnetic field points out of the page and is *decreasing*, the non-Coulomb electric field curls around in the opposite direction as shown in Fig. 5.12(b). It is not the direction of $\vec{B}_1$ that determines the direction of $\vec{E}_{NC}$, but rather *the direction of the rate of change* of $\vec{B}_1$.

In Figs. 5.12 a to d are shown all four cases, with the magnetic field pointing out or into the page, and increasing or decreasing with time (Chabay & Sherwood, 1995, used by permission).[2]

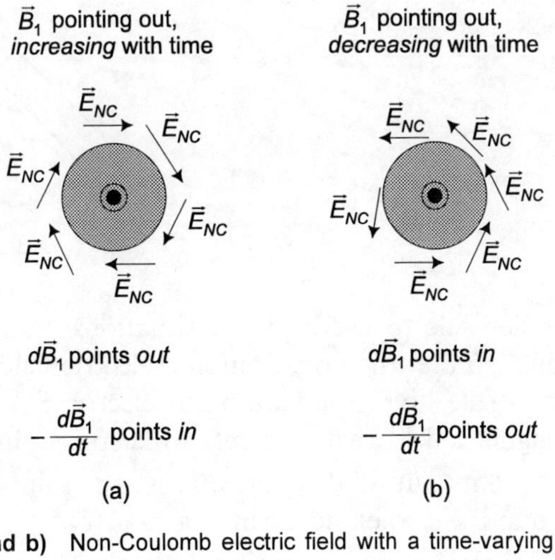

Fig. 5.12 (a and b) Non-Coulomb electric field with a time-varying magnetic field.

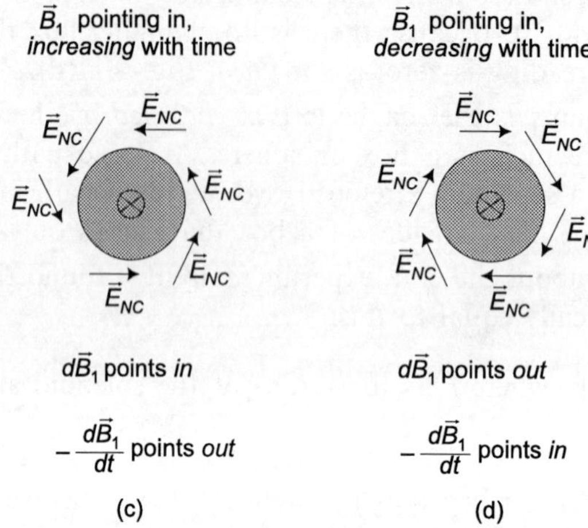

Fig. 5.12 (c and d) Non-Coulomb electric field with a time-varying magnetic field.

A right-hand rule covers all of these cases. With the thumb of your right hand pointing in the direction of $-\frac{d\vec{B}_1}{dt}$, your fingers curl around in the direction of $\vec{E}_{NC}$.

Note: A good way to find the direction of $-\frac{d\vec{B}_1}{dt}$ is to draw the magnetic field $\vec{B}_1(t)$ at a time t and the magnetic field $\vec{B}_1\,(t + \Delta t)$ at a slightly later time $(t + \Delta t)$, and observe the change $\Delta\vec{B}_1$. Then $-\Delta\vec{B}_1$ is the direction of $-\frac{d\vec{B}_1}{dt}$. See the figure below.

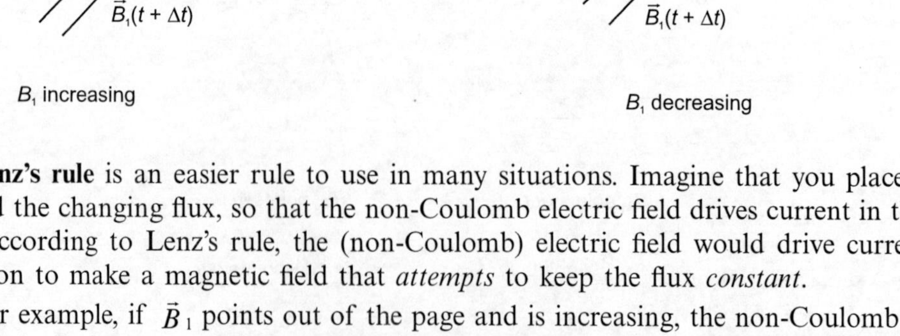

Lenz's rule is an easier rule to use in many situations. Imagine that you place a wire around the changing flux, so that the non-Coulomb electric field drives current in the wire, then according to Lenz's rule, the (non-Coulomb) electric field would drive current in a direction to make a magnetic field that *attempts* to keep the flux *constant*.

For example, if $\vec{B}_1$ points out of the page and is increasing, the non-Coulomb electric field curls *clockwise* around the solenoid as in Fig. 5.12(a).

If a wire encircles the solenoid, conventional current runs clockwise in the wire, and this induced current produces a magnetic field in the region. The direction of the magnetic field due to the clockwise current will be *into the page* at the center of the wire, decreasing the net flux.

This magnetic field tries to keep the magnetic flux constant, despite the increase in B_1. The induced current does not succeed at keeping the flux constant (unless the wire is made of superconducting material). In particular, if B_1 is increasing at a constant rate, the current I_2 in the ring is constant by Faraday's law, so the net flux does increase.

If on the other hand, B_1 points out of the page and is decreasing (the Fig. 5.12(b)), there is a counterclockwise non-Coulomb electric field which would drive conventional current counterclockwise in a wire, which would make a magnetic field out of the page, in the direction to try to keep the magnetic flux constant despite the decrease in B_1. (Again, the net flux will nevertheless decrease, unless the ring is superconducting).

The reader is urged to go through the examples Figs. 5.12(c) and Fig. 5.12(d) and verify that Lenz's rule does correctly summarize the experimental data for the direction of the non-Coulomb electric field.

5.14 DIFFERENCE BETWEEN "EMF" AND "POTENTIAL DIFFERENCE"

It is important to keep in mind that although the units of emf are volts, the emf is not a potential difference (Chabay & Sherwood, 1995, used by permission).[2]

Potential difference is a path integral of the electric field made by real charges. On the other hand, emf is the "non-Coulomb" work per unit charge, as is the case with chemical actions and say, a conveyor belt in the mechanical battery that cause a separation of charges by other than coulombic means. An example of separation of charges by coulombic means is, by the external charge that caused polarization in a metal resulting in a field due to surface charge as shown in Fig. 1.13.

Although the units of emf are volts, the emf **is not** a potential difference.

For example, the **emf** of a flashlight battery is 1.5 **volts**, and is associated with charge separation due to chemical reactions in which electrons are reactants or products.

In the mechanical battery described in Chapter 1, $\boldsymbol{F_{NC}}$**s** in the energy conservation equation represents the work done in moving one electron through the mechanical battery, or work per unit charge (joules per coulomb, **or volts**). **This is the emf.**

5.15 BOTH COULOMB AND NON-COULOMB ELECTRIC FIELDS IN A NON-UNIFORM CONDUCTOR

We have seen the co-occurrence of a *non-Coulomb electric field* with a time-varying magnetic field, and a current due to the resultant emf in a uniform circular metal ring.

If there is a thin section of ring, however, charge will *pile up* at the ends of the thin section, and there will be surface charge all along the ring (Chabay & Sherwood, 1995, used by permission).[2] This results in the production of a Coulomb electric field.

We had seen earlier that when a uniform metallic ring encircled a solenoid producing a time-varying magnetic field, there was a current driven around the ring and there were no surface charges along the ring.

Consider a non-uniform metallic ring encircling a solenoid as shown below in Fig. 5.13.

A $\boldsymbol{E_{NC}}$ (a non-Coulombic electric field) associates with the changing magnetic flux. Surface charges will pile up at the edges of the non-uniform section. These surface charges will produce a Coulomb electric field $\boldsymbol{E_C}$ which adds vectorially to the non-Coulomb electric field.

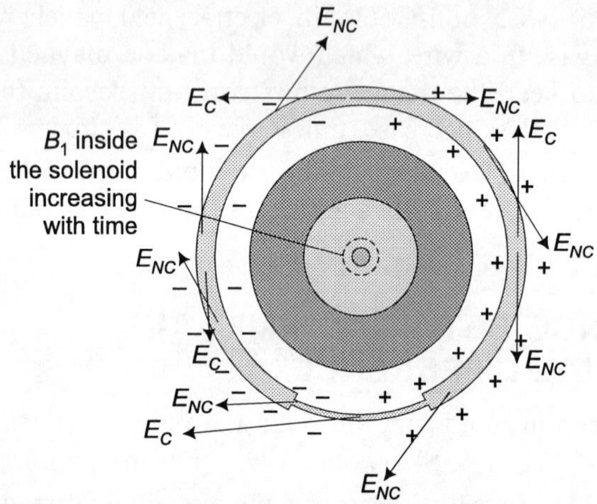

Fig. 5.13 Both Coulomb and non-Coulomb electric fields in a ring.

The *net* electric field (E_{NC} + E_C) is a combination of the two components fields E_{NC} and E_C, and what drives the electrons around the non-uniform ring.

In Section 5.10, Conservative and Non-Conservative fields, we had presented a problem of voltmeters across two same points of a conductor with resistors around a constantly changing magnetic field. Intuitively, we could imagine that surface charges have arranged themselves in the steady-state across both the resistors, producing unequal voltages across them.

But, if the change in **B** is linear for a certain Δt, and a constant current is flowing through the circuit containing R_1>R_2 (with internal strong electric fields) how is it possible, that there is during Δt only a rather weak electric field in the two conductors, connecting R_1 and R_2 ?

Hartel has made a series of figures (April 2017) which explain the puzzle in 3 stages which is reproduced below by permission. In the following figures Stages 1,2 and 3, it is assumed that the magnetic field is pointing into the page and is increasing at a constant rate.

In the first stage, the induced emf E_{NC} caused surface charges to accumulate across resistors R_1 and R_2 producing current in the circuit with unequal voltage drops across them and coulomb fields E_{C1} and E_{C2}.

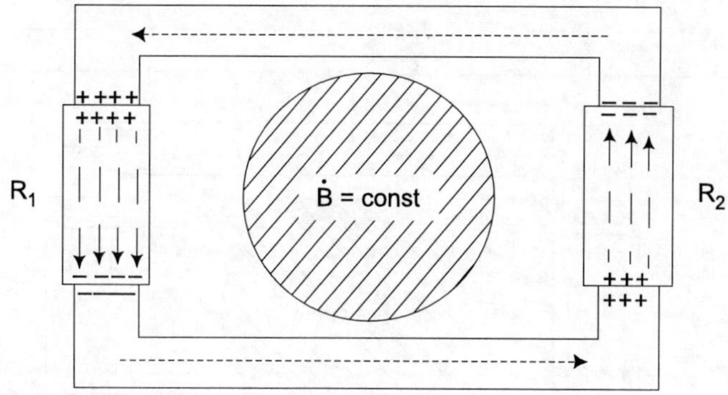

Stage–1 Induced emf produces surface charges with steep gradient across both resistors

The current around the circuit is driven by E_{NC} and in the resistors by the combination of E_{C1} and E_{C2} and E_{NC}.

During the second stage, surface charges develop a coulomb field E_C in the conductive connecting wires that is opposite to the non-Coulomb field E_{NC}, and the two combine vectorially to produce a small resultant field.

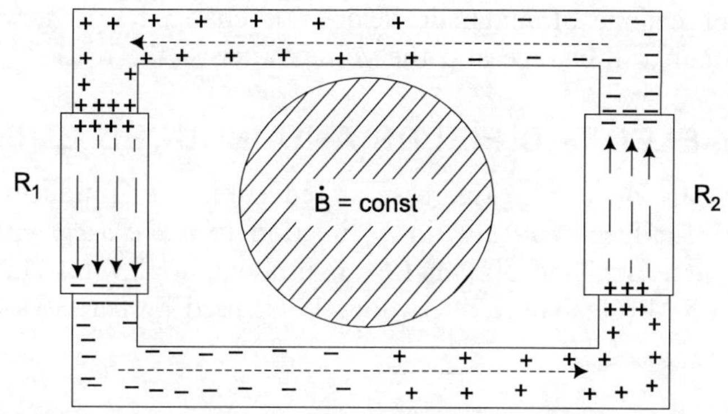

Stage–2 Surface charges with large gradient in the conductive arms produce a large Coulomb field that is opposite to the Induced emf non-Coulomb field.

In this manner, the charge distribution ensures that the current around the circuit is the same everywhere and satisfies the law of *current conservation.*

In the final third stage (shown with 2 voltmeters connected across same points of the circuit), the surface charge distribution is such that there is a translation into an arrangement in the connecting wires to the two voltmeters that record unequal voltages of opposite polarity across the two resistors.

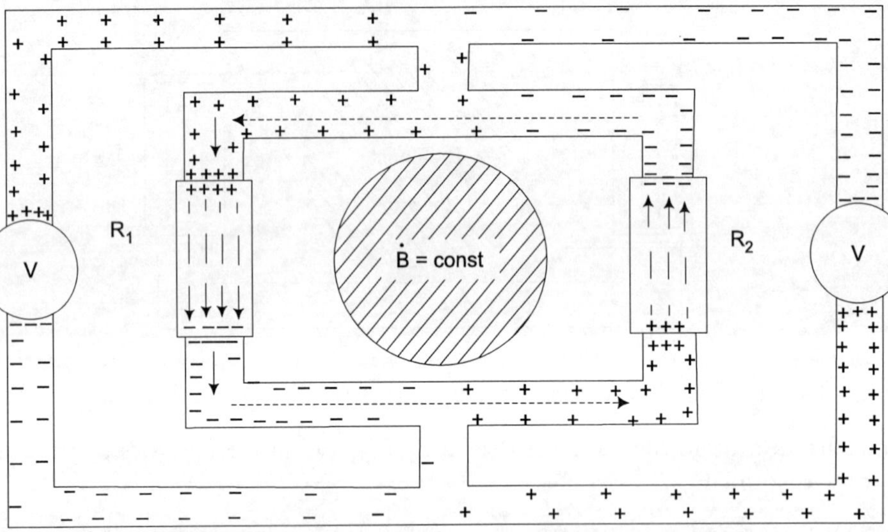

Stage–3 Surface charges adjust themselves so that Coulomb field of their distribution in the connecting wires combine with the oppositely directed non-Coulomb (induced emf) field to produce a weak field in them.

If the rate of change of magnetic field $\dot{B}$ is changed to a new constant, then the current in the circuit will change and the voltages across the resistors also will change.

5.16 CURLY ELECTRIC FIELDS AND INDUCED EMFs

An attempt to *change the current* (increase or reduce) in a coil induces an emf in *the same coil*. Since this self-induced *emf acts* in a direction to *make a current* that would *oppose* the flux *change*, there is a kind of *sluggishness* in any coil of wire. This is demonstrated in the circuit in Fig 5.14 (Chabay & Sherwood, 1995, used by permission).[2]

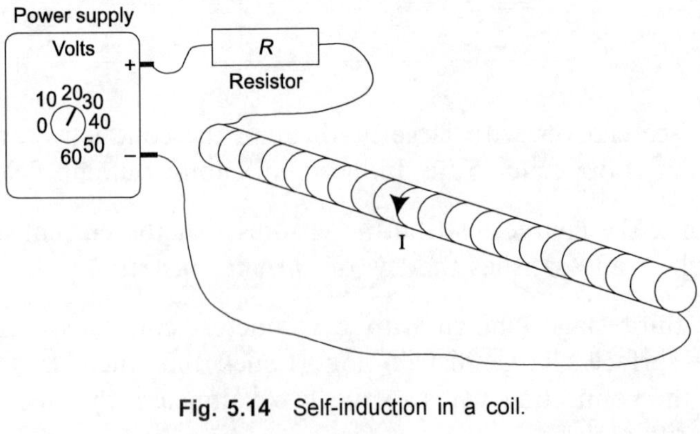

Fig. 5.14 Self-induction in a coil.

It is *hard* to change the current. The attempt to change the current in the coil induces an emf in the same coil. The self-induced emf acts in a direction to make a current that would oppose the flux change which is the attempt to increase or reduce the current, say, by increasing or decreasing the voltage in the circuit of Fig. 5.14. This makes the coil of wire act sluggishly. The current will change slowly in response. If there was no coil, the current would have changed *instantly* in response to a change in input voltage. Energy would simply be radiated away in the resistor, not to *return*. A resistor does not store energy. It takes it away from a circuit.

Now, let us go back to Fig. 5.14.

If the current is increasing, the **non-Coulomb electric field E_{NC} *curls clockwise*** as shown in Fig. 5.15.

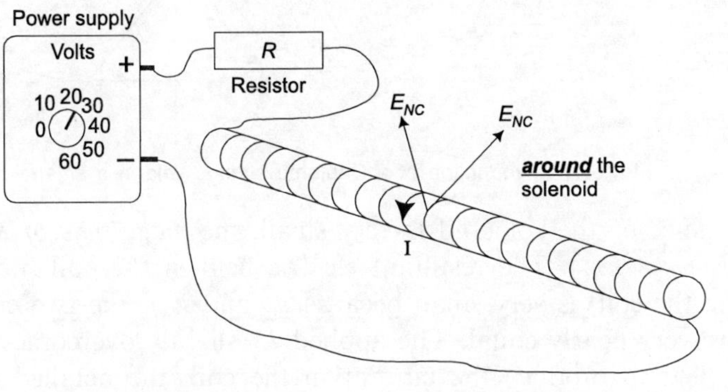

Fig. 5.15 Direction of the self-induced emf.

The electric field E_{NC} polarizes the solenoid and *surface charges* appear that produce a Coulomb electric field E_C as shown in Fig. 5.16.

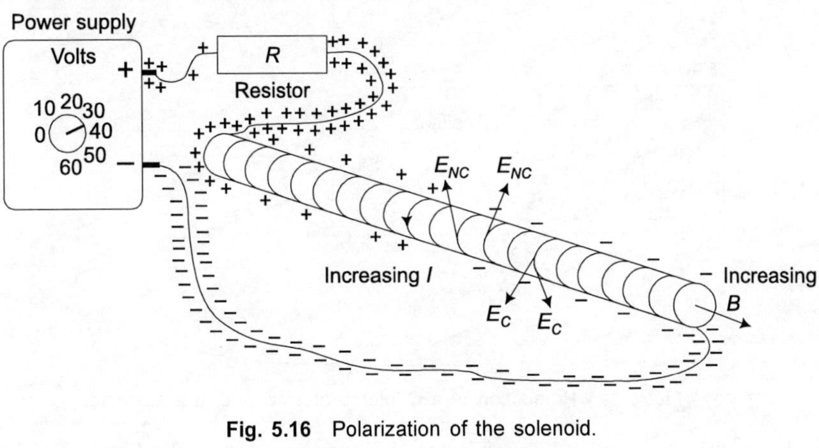

Fig. 5.16 Polarization of the solenoid.

If the current increase is more rapid, the non-Coulomb electric field E_{NC} grows and (remember that this is an electric field that co-occurs due to Faraday's law and not due to point charges) continues to polarize the solenoid and *surface charges* that appear *produce a large Coulomb electric field E_C*. In the final stage of development, a strong electric field is set up in the resistor and the field (not forgetting that an electric field is an invisible *alteration* of space) follows the wire and points opposite to the *non-Coulomb field* (Chabay & Sherwood, 1995, used by permission).[2] This is shown in Fig. 5.17.

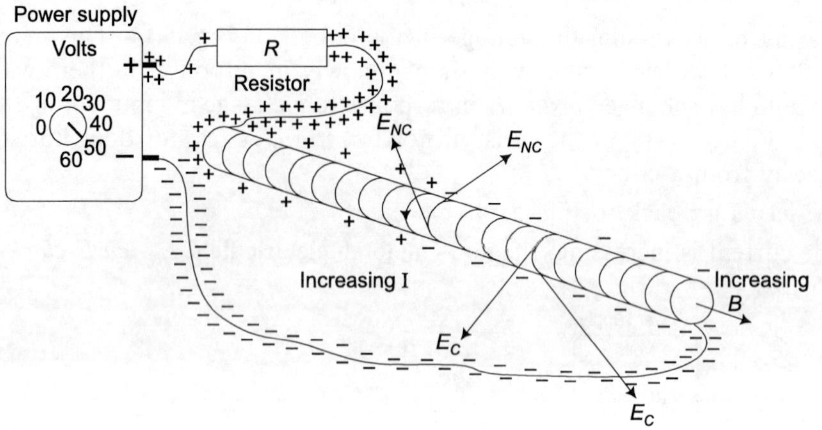

Fig. 5.17 Production of a Coulomb electric field in a solenoid.

If the resistance of the solenoid is very small, the *magnitude of the two electric fields is nearly equal* ($E_C \sim E_{NC}$). The resultant electric field in the coil (not forgetting that E_C and E_{NC} exist in the coil) is very small because E_C and E_{NC}, the two component fields that produced it, are very nearly equal. The applied E_C slightly overcomes the field due to the non-Coulomb field to produce the current in the coil, the detailed process of which is described further. Note the similarity to a battery: a *non-Coulomb* $emf_{induced}$ maintains a charge separation, and there is a potential drop ΔV *along the solenoid* that is numerically *equal to* $emf_{induced}$. If the wire resistance $r_{solenoid}$ is significant, we have $\Delta V = emf_{induced} - r_{solenoid} \times I$, just like a battery with internal resistance. This similarity to a battery is shown in Fig. 5.18.

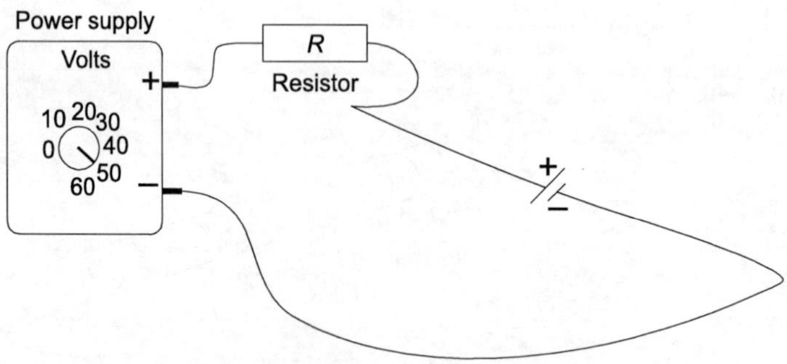

Fig. 5.18 Production of a Coulomb electric field in a solenoid.

It might not have occurred to you that a curly electric field cannot directly oppose the applied e-field due to the increasing emf which is attempting to change the current in the solenoid. The curly field polarizes the solenoid and shifts the electron sea resulting in surface charge production.

These charges in turn produce a **Coulomb** electric field inside the solenoid coil wire which follows the shape of the coil and opposes the **applied** emf. The electric field fills the coil wire. As the applied voltage grows (read the note below Fig. 1.23), it **overcomes** the self-induced (**opposing**) emf and **establishes** the current: that is the moment the energy is transferred. This is depicted in Fig. 5.19.

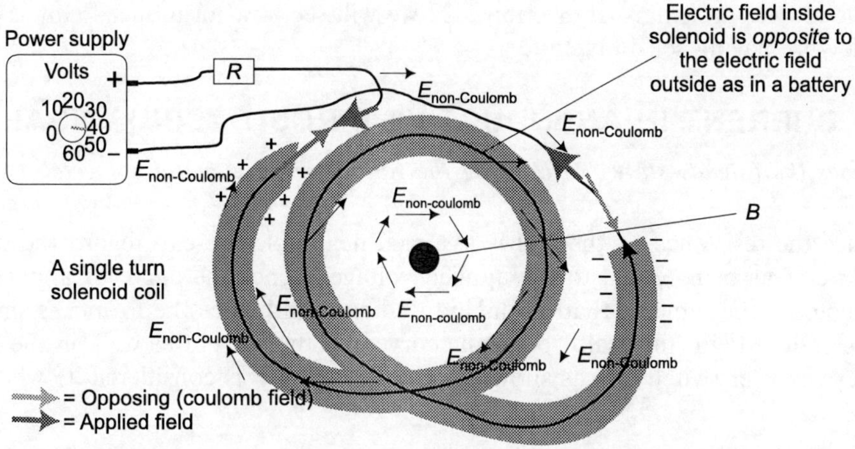

Fig. 5.19 Coulomb field of polarized coil opposes the applied field.

Note: While the description of the process which establishes a current in the solenoid makes references to an applied field and a Coulomb field and that these are present simultaneously, given the nature of the electric field (end of Sec. 1.6) and the superposition principle (See Sec. 2.9) that it obeys, it is neither possible to detect nor measure their individual strengths. The field in the wire is always a *resultant* of the applied and the opposing Coulomb field components, and a field exists in the resistor R as well. See Fig. 1.33.

When the drift velocity of the electrons is changed, accelerated or decelerated, to produce a change in current, it produces a change of magnetic field. A changing current indicates charged particles are either undergoing acceleration or deceleration.

To become familiar with the fields of accelerating charges and for a visualization of the self-induction process see Sections 3.4 to 3.6 in Härtel's article available in the Teacher's Guide in the CD. The reader may defer to read these sections in Härtel's article till the end of Chapter 7 where a visualization of the self-induction process will be discussed again.

The **emf induced** in a coil when its current is changed is given by Faraday's law:

$$v = L\,\frac{di}{dt}$$

where L = the inductance of the coil, and for a coil of N turns uniformly wound on a non-magnetic ring of uniform section, and A its cross-sectional area, as in the toroid shown in Section 5.2, is given by, $4\pi \times 10^{-7} \times \dfrac{AN^2}{l}$ henrys, where l is the length of the magnetic circuit in metres. A derivation of the expression of the inductance of a solenoid (a coil which is made up of a number of loops) is given in "Inductance of a Solenoid" in Appendix A.

Does a long straight wire have an inductance? Yes indeed! Remember, a magnetic field penetrates through matter. For one derivation of the inductance from the energy stored in the magnetic field and another assuming infinite multiply-connected loops which includes all the current in a straight wire, refer to a paper "Self-inductance of a long wire: infinite multiple loop approach".[29] In Section 5.22 we will see how an induced emf is produced in a single wire having self-inductance.

5.17 CURRENT IN AN INDUCTIVE CIRCUIT—SINUSOIDAL INPUT

Why does the (steady-state) current lag the voltage in an inductor subjected to a sinusoidal voltage?

Since the resistance of the coil is assumed neglibible, we can regard the whole of the induced emf as being equal to the applied voltage. Hence the curve of applied voltage to an inductor is the same as that of the induced emf; and since the former is sinusoidal, the wave of induced emf must also be a sine curve and the current as well, in the steady-state, a few cycles after switch ON as shown in the Fig 5.20 ('t' is considered '0' when $i(t)$ is '0').

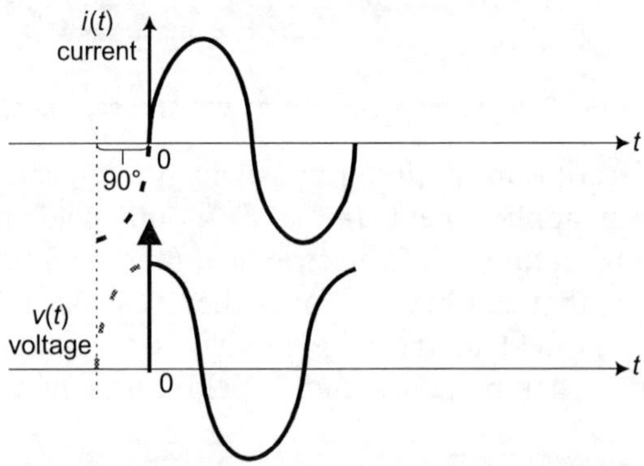

Fig. 5.20 Current lags the voltage in an inductive circuit.

At $t = 0$, the current rises from 0 at a quick rate; the induced emf is a maximum. A little later, the current continues to grow in magnitude, the induced emf is smaller because the rate of change of current is smaller than that at $t = 0$; the current is governed by $J = \sigma E_{net}$ where E_{net} is the resultant of the applied field and the Coulomb field due to polarization by the non-Coulomb electric field. As the applied voltage reduces in magnitude, the current is increasing because the induced emf is smaller.

When the applied voltage crosses zero and reversing, note that the current is a maximum and has ceased growing because the applied voltage is nearer zero but, has not yet begun to decrease; consequently, the rate of change of current is then zero. The induced emf will therefore have decreased from a maximum to zero. The energy stored in the magnetic field of the inductor will also be a maximum. As the current decreases, the energy in the magnetic field of the inductor reduces; the induced emf will be in the same direction as the current and the energy will begin to be *returned* to the source.

It is possible to trace the full voltage waveform following the above line of reasoning, and from the figure, it is observed that the induced emf, which is nearly equal to the applied voltage, attains its maximum positive value, a quarter of a cycle earlier than the current; in other words, the current in a purely inductive circuit lags the applied voltage by 90°.

In an inductor, the opposition to the applied voltage which is changing the current is *instantaneous* (the current change triggers the occurrence of the non-Coulomb field that by polarization then produces a Coulomb electric field which is in a direction opposite to the applied electric field) so, the current can only follow *after* the applied field has overcome the opposing emf. An inductor hates the change of high frequencies and thus has large voltages there, and for sinusoidal currents, the current lags the voltage by 90 degrees if the inductor is pure, and less if a resistance is in series with it; the inductor fights *before* current flows. If an inductor weren't to fight, you will get energy for free! A capacitor fights *after* the current deposits charge on its plates which develops the opposing fringe field; therefore the current leads the voltage in a capacitor circuit. It does not fight earlier, because it is *not* an inductor that stores energy in a magnetic field due to a current to do that.

To visualize the behavior of inductive circuits, it is helpful to consider a simple mechanical analogy (the most appropriate one, in my opinion) as described in Chapter 10 of "Electrical Technology" by Edward Hughes.[15]

Take a heavy metal cylinder **C**, as shown in the figure and roll it backwards and forwards using the flat palms of your hands, on a horizontal surface like a strong table between two extreme positions **A** and **B**.

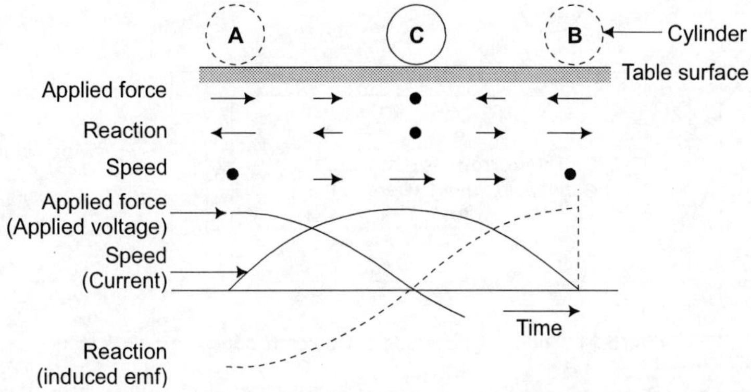

At first the speed is zero, but the force applied to the body is at its maximum, causing **C** to accelerate towards the right. You will need to reduce the applied force – as indicated by the length of the arrows in the Figure—until it is zero when **C** is midway between **A** and **B; C** ceases to accelerate and will therefore have attained its maximum speed from left to right. Immediately after **C** has passed the mid-point, the direction of the applied force is reversed and increased until the body is brought to rest at **B** and then begins its return movement. The reaction of **C**, on the other hand, is equal and opposite to the applied force and corresponds to the emf induced in the inductive circuit.

From an inspection of the arrows in the figure, it is seen that the speed in a given direction is a maximum a quarter of complete oscillation after the applied force has been a maximum in the same direction, but a quarter of an oscillation before the reaction reaches a maximum in that direction.

This is analogous to the current in a purely inductive circuit *lagging* the applied voltage by a quarter of a cycle. Also, it is evident that when the speed is a maximum the applied force is zero, and that when the applied force is a maximum the speed is zero, and that during the second half of the movement indicated in the figure, the direction of motion is opposite to that of the applied force. These relationships correspond exactly to those found for a purely inductive circuit in the steady-state.

When we start the cylinder rolling from rest from the center position **C**, the speed is zero though the applied force may be large; a transient period of a few cycles when the roller movement will not be perfectly sinusoidal is necessary before steady-state conditions result. Similarly, the current in an inductive circuit is zero at switch ON assuming that the sinusoidal voltage starts from zero and the current and opposing emf will make adjustments for a few cycles (a transient phase) before attaining steady-state.

An inductor has a non-Coulomb field on application of the input sinusoidal voltage. It then polarizes and develops the opposing coulombic electric field almost instantaneously to closely match the e-field of the applied input voltage because there is no resistance (which we assume), as shown in the Fig 5.21 …of course due to surface charges that are not shown in the figure. Read the **important note** below Fig. 1.23 before reading on.

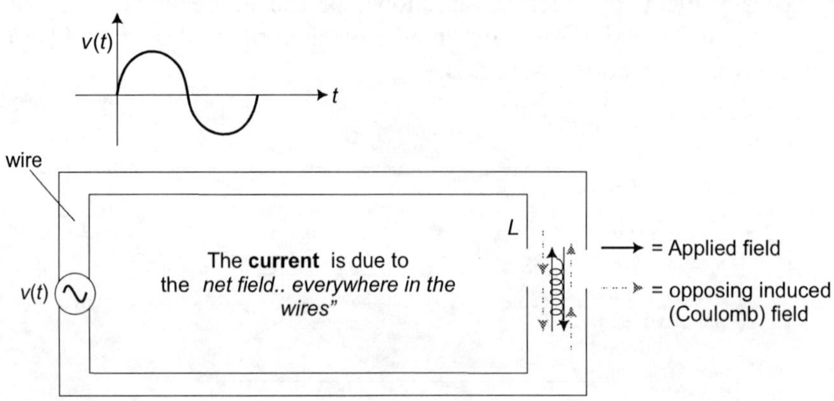

Fig. 5.21 Inductor develops a Coulomb opposing field instantly.

The existence of a *sinusoidal* current resulting from the application of a sinusoidal voltage to an inductor is a characteristic of the non-Coulomb electric field that is proportional to the rate of change in current causing a changing magnetic field. The relation between the "RMS" (square root of the mean of the squares) current and voltage of an inductor is given by:

$$\frac{V}{I} = 2\pi fL = \text{inductive reactance}$$

where, f = frequency of the sinusoidal voltage and L is the inductance of the inductor.

The inductive reactance is expressed in ohms, and as we studied in section 1.2 "Current", is useful to determine steady-state current values to sinusoidal voltage inputs. It is to be noted that though the value of the reactance is expressed in ohms, it was not obtained by computing a resistance, as of resistors. The current is a result of an opposing Coulomb electric field, which is a result of polarization by the non-Coulomb electric field associated with the changing magnetic field, and the current resulting thereof by applying the relation $J = \sigma E_{net}$ where E_{net} is the resultant field of the applied field and the Coulomb electric field.

What is meant by steady-state in an inductive circuit subject to a sinusoidal voltage input?

Inductive circuits, when subject to a sinusoidal voltage input, will exhibit a transient behavior for a few cycles initially before settling down into a steady-state. The remarks that we made regarding steady-state in capacitive circuits are also applicable to inductive circuits. Refer to Section 5.19 "Sinusoidal Steady-State".

If we increase the "*frequency*" of the **input voltage**, the "rate of change" of the input voltage and the applied field is "greater than" the rates obtained with applied voltages at lower *frequencies*.

At low frequencies, this causes a smaller induced opposing electric field and emf, therefore, large currents will flow within small intervals of time in the coils of the inductor. This is shown in Fig. 5.22. Further reduction in frequency, will cause the current to increase, and the induced electric field and the resulting polarization of the inductor, will be very tiny. There is barely an induced emf to oppose the input field.

In the limit, if the input is a dc voltage, the current will become so large that the inductor will burn out (unless the inductor coil was wound using a high resistance wire, as in dc coil operated relays and contactors used for switching ON and OFF motors).

At high frequencies, even as the current begins to grow and sets up a magnetic field, a powerful non-Coulomb electric field associated with the high growth rate of the current, *polarizes* the conductor (see Section 5.16) and produces an electric field *within* the inductor which opposes the applied field.

This process occurs at nearly the speed of light. The opposition to the applied field therefore, is also almost *instantaneous* and results in a net field E_{net} in the inductor which we use in the relation $J = \sigma E_{net}$ to compute the current.

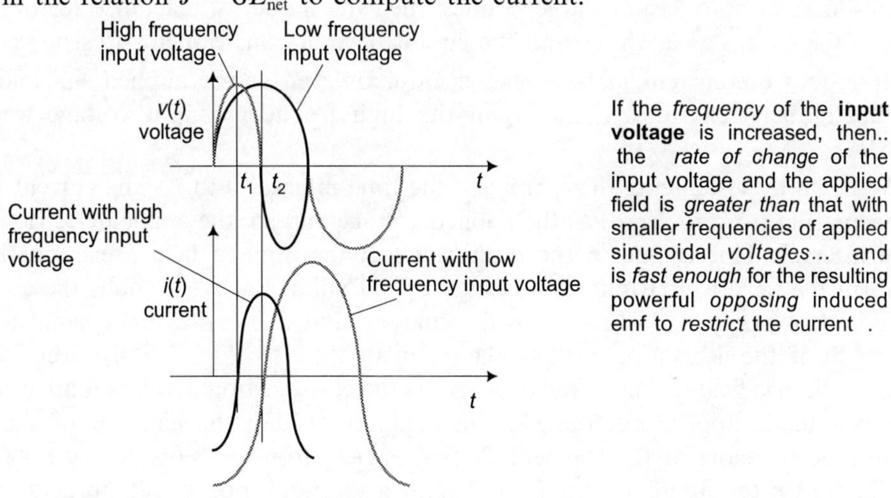

Fig. 5.22 Current when the sinusoidal input voltage frequency is increased.

The result: The current is "smaller than" that at lower frequencies of input voltage. Many students find it difficult to visualize the production of the non-Coulomb electric field as an instantaneous phenomenon and *seemingly wait* until the current is in the inductor, not realizing that it is the speed of the rise of the applied voltage that gets the current up affecting the rate of its growth (and consequently the *rate of growth* of the magnetic flux in the inductor). With high magnetic flux growth rates, the induced emf is stronger to oppose the applied field and is given by $|emf| = \left|\frac{d\phi_{mag}}{dt}\right|$ (as shown in Section 5.12). A few among them also make the mistake of matching the applied and the induced emf exactly. This is inconsistent with the observed presence of a current in the inductor. See Section 5.16. On the other hand, if the frequency is reduced, the non-Coulomb electric field and the resulting polarization of the inductor is smaller.

With smaller frequencies, the rate of change of the applied field is smaller and a smaller induced opposing emf is produced. The current is larger, determined by the relation $J = \sigma E_{\text{net}}$ where E_{net} is the resultant of the applied field and the *small* opposing induced emf due to the non-Coulomb field. Refer Section 5.16.

As we saw in the case of capacitors, it may appear that there are two relations (see "inductive reactance" in a previous paragraph) which the current seems to be required to satisfy: $J = \sigma E_{\text{net}}$ and $\frac{V}{I} = 2\pi fL$; here again, both relations are satisfied. Between two circuits whose resistances are insignificantly small (one cannot realize "zero resistance" in any practical circuit) yet different, the E_{NET} change in the wires of the two circuits is so *tiny* that the inductive reactance relation will appear not to have been violated, and becomes an *ideal approximation* (Refer "Ideal Approximation" in "Electronic Principles" by Paul Malvino[20]) while noting that E_{NET} regulates the total charge forward and reverse movement in the inductor wire, and is neither easily measured nor computed in a practical circuit. If the resistance were to assume larger values, then the inductive reactance relation does not apply to the circuit as a whole, and the circuit becomes an Inductor-Resistor circuit.

How does the current *increase* in the inductor even as the applied sinusoidal voltage is *reducing*? between times t_1 and t_2 in the high frequency input voltage waveform of Fig. 5.22?

We begin by visualizing the current in the time interval 0 to t_1. The current is negative and decreasing (towards zero) as the applied voltage and so, the applied electric field grows in the interval. This is due to the braking effect the applied field exerts on the charges that form the negative current. The rising applied voltage at $t = 0$, pulls the charges back ....like sort of applying a "brake" on the charges, and this causes the current to decrease. As a result of the decreasing current, the magnetic flux (see Fig. 5.12(d)) decreases and the Coulomb electric field (effect of polarization by the co-occurring non-Coulomb electric field), is opposite to the applied electric field, but weak, because of the slow rate of change of the current (and therefore of the magnetic flux) as is clear from the slope line at $t = 0$ shown in Fig. 5.23(a). In the figure, on the right is shown the weak non-Coulomb field by thin E_{NC} field vectors.

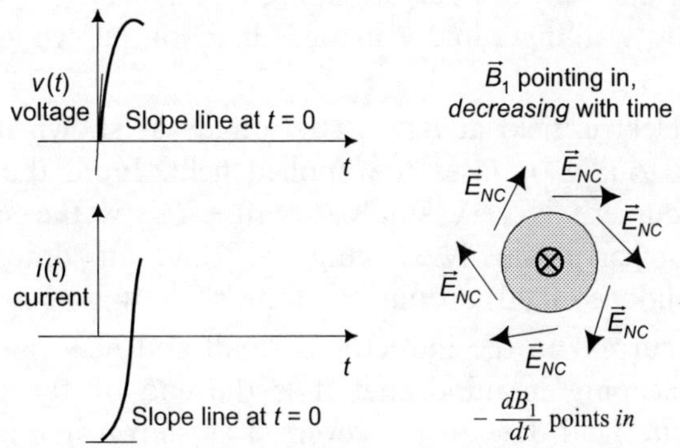

Fig. 5.23(a) The negative current decreases as the input voltage rises at $t = 0$.

In the following descriptions, reference to the co-occurring non-Coulomb electric field only will be made; the resulting Coulomb electric field due to polarization is implied.

As the applied voltage continues to rise, the braking effect is more powerful and this in turn causes the charges to slow down further; the rate of current decline is faster as can be seen if we draw a slope line at $t = t_A$, a time we choose midway between $t = 0$ and t_1 shown in Fig. 5.23(b). The associated electric field is stronger now and is shown by slightly thicker E_{NC} field vectors in the figure on the right in Fig. 5.23(b).

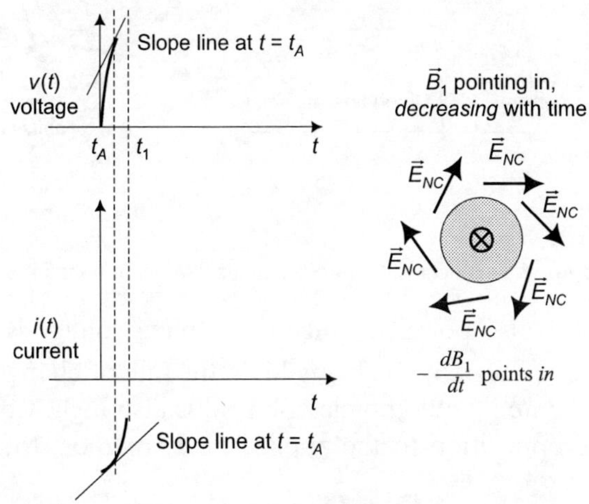

Fig. 5.23(b) Negative current when the input voltage is midway to its peak.

The associated electric field is stronger at $t = t_A$ but, because the applied field is stronger (note that the voltage has risen considerably at $t = t_A$), therefore, the current (negative) continues to decrease more rapidly towards zero. The rapid decrease of current is the reason for the strong associated field.

The reversal of the current, as it crosses zero, at $t = t_1$, is very eventful.

I have described this in Section 2.10 and urge the reader to once again visualize the rapid movement (start-stop) of the charges in one direction, their slowing down, halting and gradual movement in the opposite direction.

The acceleration at $t = t_1$ is huge assuming a reasonably high frequency and one can get a feel for this by walking rapidly in one direction, slowing down, halting and then walking backwards!

The associated electric field at $t = t_1$ is stronger, as shown in the figure on the right in Fig. 5.23(c), and totally opposes the applied field. From the relation $J = \sigma E_{NET}$, the magnitude of the current is very small (because E_{net} is the small result of two field components; one powerful slowly varying field and an almost equally powerful but oppositely directed induced field) around the time t_1, though its rate of change is very high!

At $t = t_1+$, the current in the inductor is small and now positive and still has a very steep rate of rise, keeping in mind that it is the end of the process of rapid current reversal. The magnetic field due to it is reversed as shown in Fig. 5.23(c). Checking with the corresponding Lenz rule Fig. 5.12(a), we find the non-Coulomb field does *not reverse*.

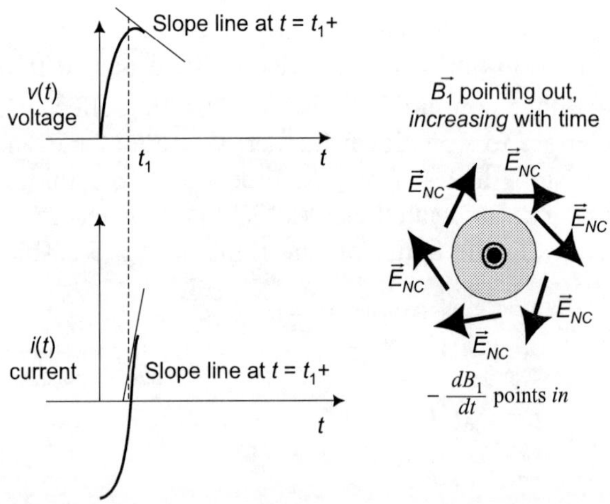

Fig. 5.23(c) A small current when the sinusoidal input voltage is near its peak.

The applied field is also powerful since the applied voltage is close to its peak value.

At $t = t_1+$, the rate of current rise being high, the rate of change of magnetic flux is high and the strength of the non-Coulomb electric field is also high. Consequent polarization of the inductor and the opposition to the applied field is almost total, resulting in the small current at $t = t_1+$.

Further, at $t = t_1+$, the applied field though large has a small rate of decline as is evident from the not-very-steep slope line at $t = t_1+$ in Fig. 5.23(c). The corresponding non-Coulomb electric field is large as evident by the steep slope line for the rate of growth of current (and therefore the high rate of growth of the magnetic flux) and this is shown by the thick E_{NC} arrows in Fig. 5.23(c).

The resultant electric field in the wires given by applied field—E_{NC} will be correspondingly small and so, the current given by $J = \sigma E_{\text{net}}$ will be small. See Note below Fig. 5.19.

The crucial point in the visualization is the slow rate of decline of the input voltage (and therefore, the applied electric field) at points near the time t_1. The decline of the applied field itself causes a decrease in the current and the rate of current growth reduces at the time $t = t_B$ which is say, midway between t_1 and t_2 as shown in the Figs 5.23(d and e).

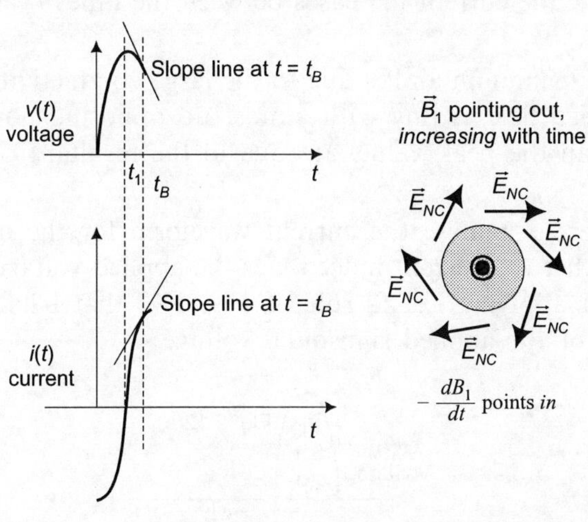

Fig. 5.23(d) The rate of growth of current is diminished at $t = t_B$.

At $t = t_B$, the current is still growing but its rate is diminished because of the reduced applied electric field. The co-occurring weaker non-Coulomb electric field E_{NC} is shown by thinner arrows in Fig. 5.23(d). The *resultant* electric field in the inductor wire is more than that at $t = t_1+$ in spite of the applied field having reduced in magnitude and the current given by the relation $J = \sigma E_{net}$ will therefore be more than that at $t = t_1+$.

As the input voltage reduces along the downward segment of the positive portion of the sinusoid, the applied field is near zero at $t = t_2-$. This is shown in Fig. 5.23(e). The current nearly "peaks" because $J = \sigma E_{net}$ and E_{net} is almost entirely the applied field close to zero.

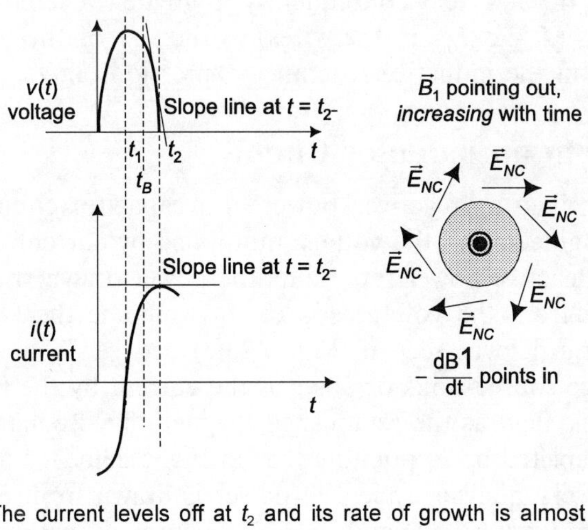

Fig. 5.23(e) The current levels off at t_2 and its rate of growth is almost zero at $t = t_2-$.

There is barely an electric field because the current is not changing (and therefore, the magnetic flux is not changing). The current levels off at t_2 and it remains constant and at this time the rate of change of voltage is also a maximum as seen from the slope line at $t = t_2$. See Fig. 5 23(e).

Thus, we see how the current increases between the times t_1 and t_2, even as the voltage is decreasing.

The current is a maximum and is due to the large conductivity σ of the inductor wire, even when the applied field is tiny with almost no opposing non-Coulomb electric field! Currents smaller than the peak value, are due to the resultant of more than one field in the wire.

I urge the reader to analyse the current waveform for the negative half-cycle of the sinusoid and verify that the current indeed lags the applied voltage by 90° as shown in Fig. 5.20 for an inductor applying Lenz's rule of Fig. 5.12 that I have indicated in the figure for all the portions of the applied sinusoidal voltage.

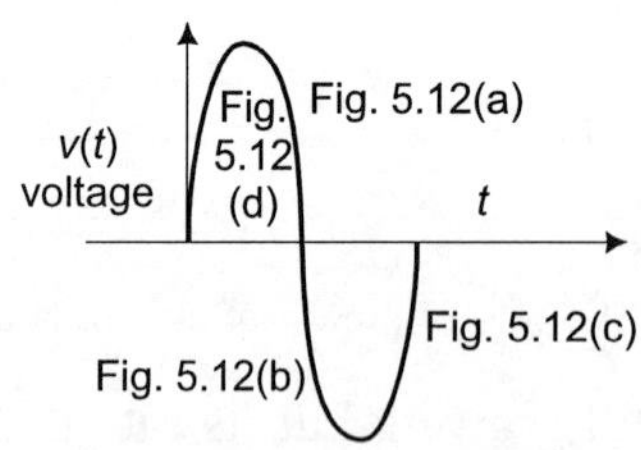

If the inductance of the inductor L is made smaller, then the non-Coulomb electric field and thence the Coulombic electric field by polarization, does not build as fast as it would have if L were to be larger. With a small L (imagine a fewer number of turns in the coil), the attempt to create a magnetic field is easier and causes the current to build faster with a smaller self-inducing (opposing) emf in a given interval of time as evident from the inductor current-voltage relation $i = \frac{1}{L}\int v \cdot dt$. The (opposing) Coulomb electric field is smaller and the net field consequently is greater. Therefore, at a fixed frequency, the current increases ($J = \sigma\ E_{net}$ in the wires) as the size of the inductor is decreased and this is encapsulated in the inductive reactance expression, $2\pi fL$.

Reactive Power in an Inductor Circuit

In Sec. 4.7, we described a "reactive" power in a capacitor circuit. Likewise, in a circuit with an inductor, engineers call the voltage multiplied by current, reactive power expressed in "reactive vars". Reactive power represents the power drawn rhythmically in response to a sinusoidal voltage from the voltage source to overcome the Coulomb field due to the non-Coulomb electric field as seen in Fig. 5.23(e) between t_1 and t_2 (B_1 pointing "out"), and when the voltage source sinks or absorbs the current by the Coulomb field due to the non-Coulomb electric field as the established magnetic field collapses and is established in the opposite sense depicted by B_1 pointing "in" in Fig. 5.23(a) for the time 0 to t_1. Note that alternately during every quarter cycle, the power is drawn from or returned to the source by the inductor as in Figs. 5.23(a) and 5.23(e). There is no net energy flow over one cycle. In this case, only reactive energy flows—there is no net transfer of energy to the load, the inductor. Most among us ignore this aspect of the voltage source. It *should* have that capacity to establish the current by developing an electric field against the forces exerted by the Coulomb field (due to the non-Coulomb field). This power does no *useful* work and it does not come for free, since the voltage source should be capable of *overcoming* the Coulomb field due to the non-Coulomb electric field of the inductor. Also, the inductor delivers its stored energy (in its magnetic field) every other quarter cycle to the voltage source when the Coulomb field (due to the non-Coulomb field) overcomes the applied field. It is therefore, the combined energy exchanged between the voltage source and the inductor that forms the Reactive power (measured in vars) present in a system containing reactive (inductive or capacitive) components and may be either produced or consumed by different load/generation elements.

5.18 APPLICATION OF A SINUSOIDAL VOLTAGE TO AN INDUCTOR AND RESISTOR COMBINATION—LR CIRCUIT

When an inductor is connected in series with a resistor R as shown in Fig. 5.24, and a sinusoidal voltage is applied to the combination, then the current is "*diminished*" in magnitude in comparison to the current that would result if the resistor were not there. A part of the input voltage will be used to push charges through the resistor.

With the inductor connected in series with a resistor R, the applied voltage $v(t)$ is split between the inductor and the resistor in a *slightly* more complex way.

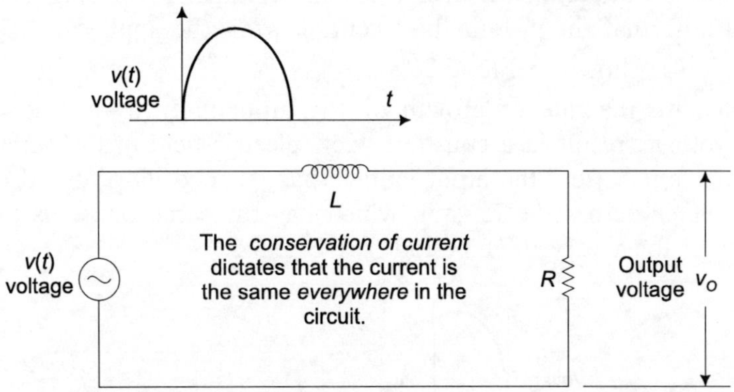

Fig. 5.24 Current when a sinusoidal voltage is applied to an inductor resistor combination.

Surface charges appear and bunch up at the junction of the inductor and resistance.

The initial current which is negative, is not as large as that in the inductor circuit (see Fig. 5.23(a)). It is restricted by the resistor in series.

All the processes that lead to the current lagging the voltage across the inductor... mind... despite there being a resistor in series... by 90 degrees... hold.

But, due to the resistor, the voltage *across* the inductor is not the same as the input voltage. It is smaller in magnitude.

This is shown in Fig. 5.25.

The ***initial*** negative current is not as large as that in the inductor circuit. It is restricted by the resistor in series.

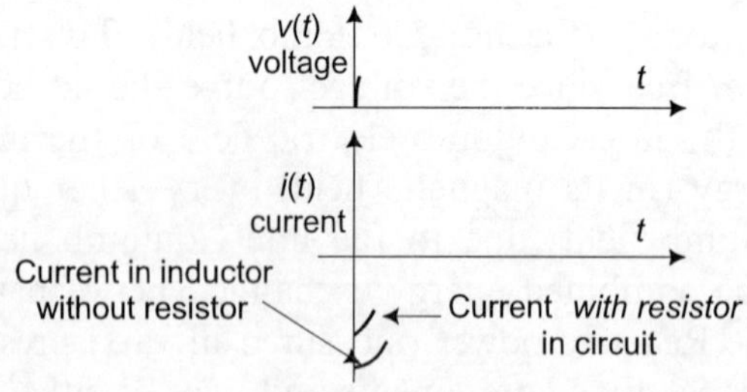

Fig. 5.25 Comparison of current in the inductor circuit (Fig. 5.21) and in an inductor-resistor combination circuit subjected to a sinusoidal voltage.

Consider the situation when steady-state is reached, and from a previous cycle the current through the inductor is negative and the input voltage is now rising from zero. See Fig. 5.25.

The initial negative current is reducing in magnitude and approaches zero, even as the voltage across the inductor is rising.

The current *levels off* and remains constant as the voltage across the inductor reaches zero and is just becoming negative.

Remember, the current is a *constant* when the *rate of change* of the voltage is a maximum across an inductor, which is when the voltage is reversing. See Fig. 5.23(e) and Fig. 5.29 and note that the instant this occurs is *after* the applied voltage peaks.

The situation is shown in Fig. 5.26.

At this instant, the rate of growth of the input voltage and the rate of growth of the inductor voltage, result in a constant "net" electric field in the wires which maintains the current constant "after" the input voltage has reached its peak. Also note that at this instant, the voltage across the resistor, which has the same phase as the current, is also constant, though for a very brief period (see Fig. 5.29).

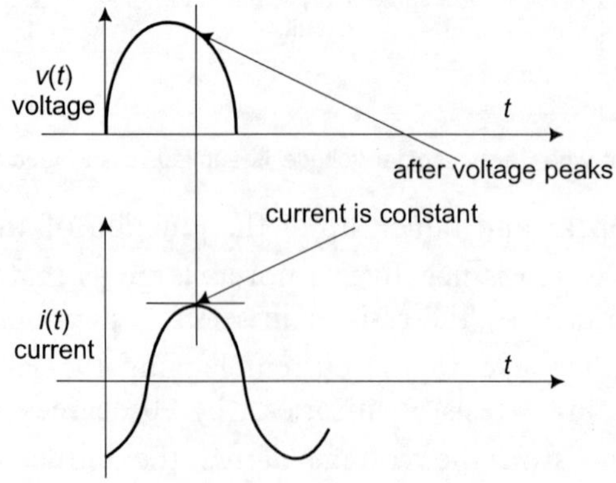

Fig. 5.26 Current peak in an inductor-resistor combination circuit subjected to a sinusoidal voltage.

Why does the current peak and remain constant briefly after the applied voltage has "peaked"? While this happens in the inductor circuit (see Fig. 5.22) when the applied voltage was crossing zero?

When there is a resistor, the current is reduced in the circuit, therefore, the rate of change of the magnetic flux will also be smaller in comparison with the circuit without a resistor.

Note from the current and voltage waveforms in Fig. 5.29, the negative current at $t = 0$ is smaller and decreases towards zero, earlier than it would have when there was no resistor in the circuit; with a resistor in circuit, it crosses zero, when the applied field (applied voltage) is smaller in magnitude than its peak value.

Also, note that the induced emf of the inductor when the current crosses zero is a maximum and is expected, since the rate of change of current, and therefore of the magnetic flux, is a maximum.

In the inductor-resistor circuit, as the current becomes positive and even as it continues to grow, the applied voltage peaks. The applied voltage decreases in magnitude after peaking but the *rate of change* of the current is also decreasing, and therefore, the non-Coulomb electric field E_{NC} decreases. Closer observation reveals that the rate of change of the input voltage and the rate of change of the voltage across the inductor is the same at this instant of time. See Fig. 5.29.

In other words, the rate of growth of the applied input field and the rate of growth of the associated non-Coulomb field and resulting emf due to polarization of the inductor wire (at different magnitudes though) is the "*same*" at this instant of time, as we shall soon see, resulting in a constant net field, therefore, the current remains *steady* since $J = \sigma E_{\text{net}}$ = a constant.

As the input voltage continues to reduce in magnitude, the current begins to reduce too, but since its rate of change is small, there is a weak non-Coulomb electric field.

However, in the inductor circuit (see Fig. 5.22), the instant when the current remains constant occurs when the input voltage and so the inductor voltage (being the same as the input voltage) is going through "zero" and their rates of change match at every instant.

In the inductor resistor circuit, as the applied field continues to reduce along the sinusoidal curve, it contributes towards establishing a resultant field in the inductor wire that firmly brings the current down.

Even as the applied voltage crosses zero and becomes negative, the current soon follows and also crosses zero, becoming negative, as seen in Fig. 5.27. Before the current becomes negative, it is the induced emf which sustains the current, as can be verified by checking the Figs. 5.23(d) and 5.23(e). With the applied voltage negative, and the current still positive, energy will be returned to the source. (The transport of energy and the processes involved are described in Chapter 4).

Further reduction in the input voltage makes the current cross "zero" and reverse under the influence of the dominant induced emf. The induced emf has assumed control over the current upto some point "after" the input voltage reverses, at 180 degrees. This is shown in Fig. 5.27.

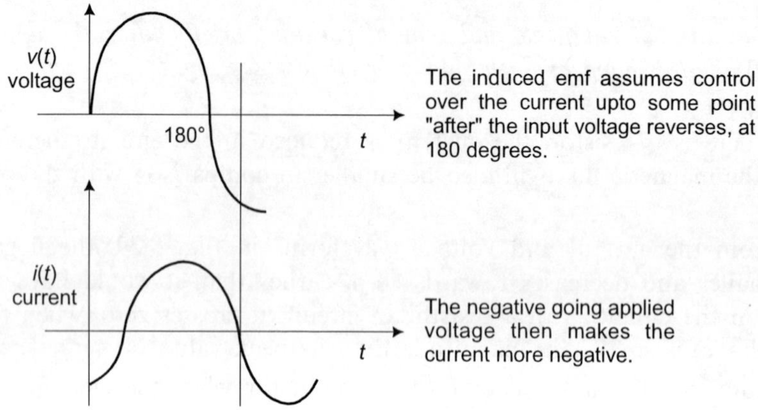

Fig. 5.27 The current in the inductor resistor combination subjected to a sinuoidal voltage.

After the current crosses zero and becomes negative, the negative going applied voltage then makes the current more negative.

Therefore, the current we say, "lags" the applied input voltage and by an amount less than 90 degrees.

The current in a circuit, comprising an inductor with a resistor, is smaller and lags the input voltage, by an angle less than 90 degrees.

The complete voltage and current cycle is shown in Fig. 5.28.

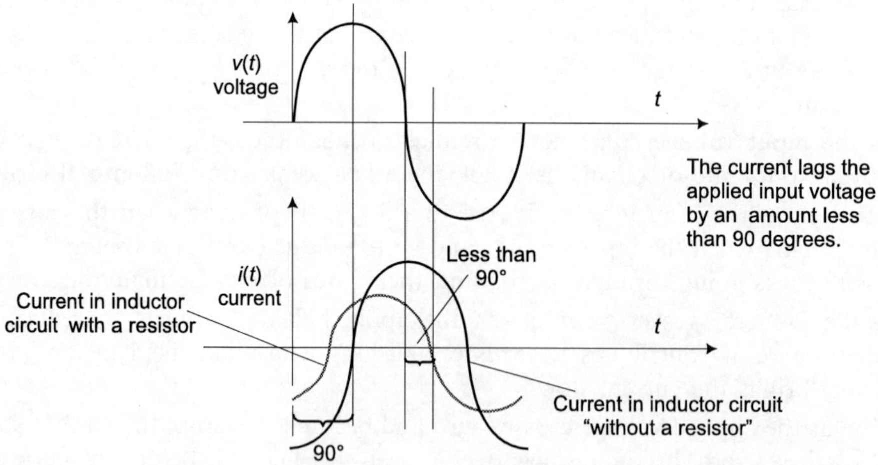

Fig. 5.28 Comparison of the current in the inductor resistor combination circuit with that in the inductor circuit (Fig. 5.20) subjected to a sinusoidal voltage.

If we measure the output voltage across the resistor of an LR circuit as in the figure, we will observe that the voltage across the resistor *lags* the input voltage by the same amount as the current, because they are in phase for a resistor. Therefore, the output voltage (across the resistor) *lags* the input voltage by the same amount as the current.

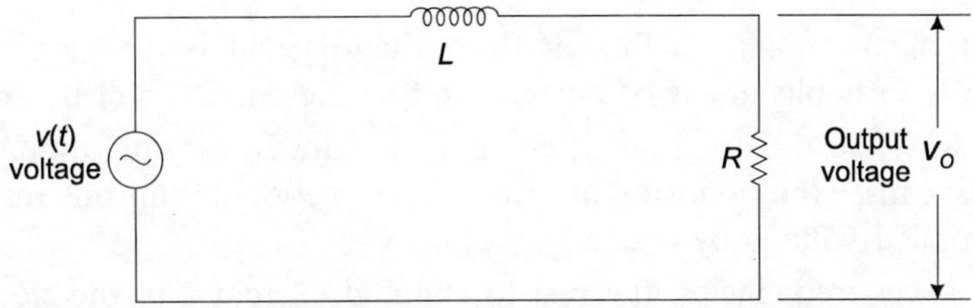

The current, which has to be the same all around the circuit, *lags* the voltage *across* the inductor by 90 degrees, a rule which is *never violated.*

Note the *slope* of the inductor voltage change *at the zero crossing point* of the input voltage is the *same* as the slope of the input voltage change after the input peaks, where the current is *constant*.

Also note that the output voltage is *smaller* in magnitude than the input voltage, as it should be.

The waveforms of input and output voltages and the voltage across the inductor are shown in Fig. 5.29.

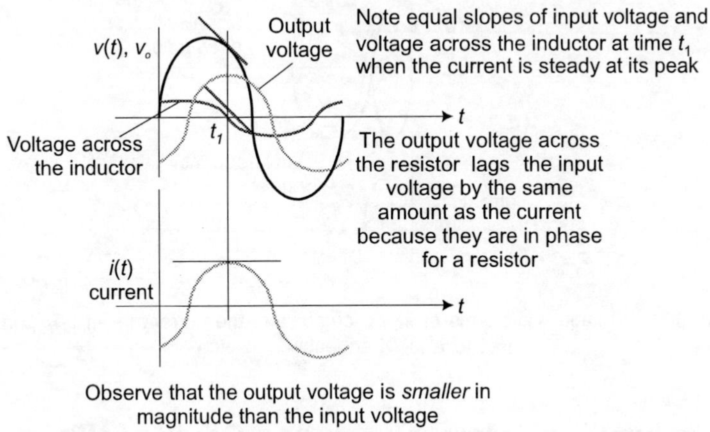

Fig. 5.29 Voltage waveforms of input, output and the current in a L-R combination circuit subjected to a sinusoidal voltage.

I urge the reader to review the comments made on visualizing currents in Section 2.5. One must not focus on the current in one element and then the other of a circuit. One has to think about what happens to every element in the circuit all at the same time.

5.19 APPLICATION OF SINUSOIDAL VOLTAGES AT VARIOUS FREQUENCIES TO AN L-R COMBINATION

What happens when the frequency of the input sinusoidal voltage is decreased?

If the frequency of the applied voltage is decreased, the associated non-Coulomb electric field strength will decrease, because the rates of current growth and decrease (and therefore that of the magnetic flux in the inductor) will be smaller. There will be less opposition to the establishment of current and so the current will be greater.

Together with the resistor limited current, the slowly varying applied field (due to its low frequency) causes the opposing induced emf to be weak, and this results in the current lagging the applied voltage by a small amount.

At much lower frequencies, the resistor limited current and the slow growth and fall of the applied voltage causes the rates of growth and fall of the current and the magnetic flux in the inductor to be weaker, and this results in the output voltage being almost in phase with the input voltage.

The waveforms for the circuit subject to a low frequency sinusoidal input voltage are shown in Fig 5.30.

The output voltage picked across the resistor is able to reproduce the input signal with less attenuation.

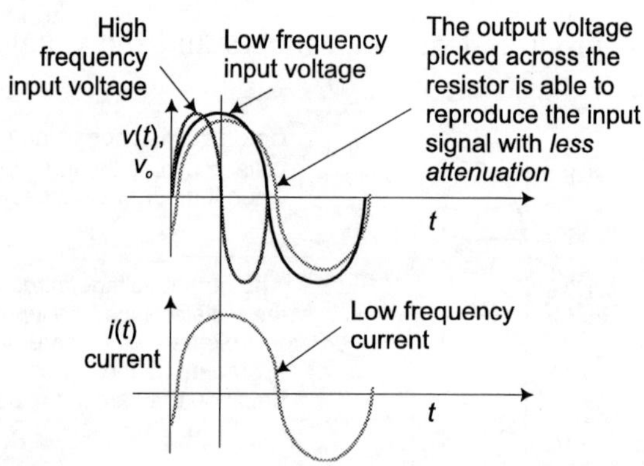

Fig. 5.30 Voltage waveforms of input, output and the current in an L-R combination subject to a low frequency sinusoidal voltage.

The L-R Combination—More on the non-Coulomb Field

The non-Coulomb field is *due* to the *rate* of growth (or decay) of current and the current is due to the applied field.

The current is limited by the resistor and the slowly varying applied input field. The current cannot grow to large magnitudes and its rate of rise is further limited by the opposing induced emf which builds up due to the *rate* of the changing magnetic flux which is again due to the current. At low frequencies, this *rate* being small, the induced emf is small.

And that is the reason why at *low* frequencies, the associated non-Coulomb electric field is unable to build enough strength to "oppose" the applied field.

Therefore, we say the *inductor* acts as a "short circuit" at low frequencies, as if the inductor was not there but a wire.

What happens when the frequency of the input sinusoidal voltage is increased?

If we *increase the frequency* of the input voltage, the strength of the associated non-Coulomb electric field will increase, because the rates of current growth and decrease (and therefore that of the magnetic flux in the inductor), are steeper. There is a stronger opposition to the establishment of current and so, the current is smaller, which is further restricted by the presence of the resistor, and the R-L circuit impedance will be large. The waveforms of output voltage and current, when a high frequency sinusoidal voltage is applied, are shown in Fig. 5.31.

While the resistor limited current causes the rates of growth and decay of the current to be gentler, the more rapid reversals due to the high frequency of the applied voltage (and therefore of the applied electric field) causes the opposing induced emf to be yet strong enough to cause the current to lag the applied voltage by almost 90 degrees as shown in Fig. 5.31.

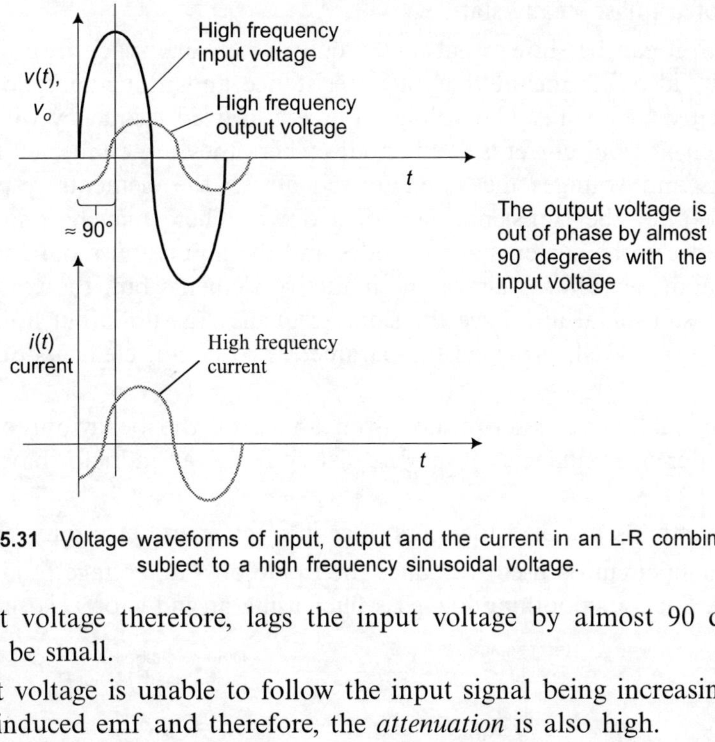

Fig. 5.31 Voltage waveforms of input, output and the current in an L-R combination subject to a high frequency sinusoidal voltage.

The output voltage therefore, lags the input voltage by almost 90 degrees and its magnitude will be small.

The output voltage is unable to follow the input signal being increasingly blocked by the inductor's induced emf and therefore, the *attenuation* is also high.

The potential difference across the inductor is frequency dependent and current flows *after* the induced emf/potential difference appears across the inductor coil; therefore, the current "lags" the applied voltage by a phase angle less than 90° but greater than 0°.

So, we say, the inductor acts an open circuit to high-frequency input voltages.

As the frequency of the applied voltage (e-field) is increased, the applied voltage is opposed by the induced emf of the non-Coulomb electric fields which are stronger (resulting in a powerful opposing induced emf due to polarization) thereby making it difficult to establish large currents, and the R-L circuit impedance (see Section 1.2 "Current") will be "high".

Sinusoidal Steady-State

Note that the current in the circuit of Fig. 5.24 and the voltage across the resistor and the inductor are sinusoidal. The current is phase-shifted with respect to the input voltage (Fig. 5.30). The quasi steady-state is applicable to the circuit as a whole since the input voltage is varying sinusoidally with time.

From switch-ON, the circuit will continue to develop these voltages and currents across and in the elements respectively, as long as the input voltage is sinusoidal. Assuming the input voltage is a pure sinusoid, there will be a transient period lasting a few cycles before a quasi steady-state when the voltages and current become purely sinusoidal, is established (Section 2.10).

Circuit analysts prefer to call this a "sinusoidal steady-state", which in a sense is a special case of a quasi steady-state.

In general, it can be shown that any circuit or network made from ideal resistors and inductors (by "ideal" is meant that their resistance and inductance do not change with time) and excited by a sinusoidal voltage source at a fixed frequency will produce voltages across the elements and currents at the nodes where they are connected together, and that these currents and voltages measured in and across the elements respectively will also be sinusoidal, after the transients have died down. There may be amplitude and phase differences between the currents at the nodes and the input sinusoidal voltage and between the currents and the voltages across the inductive elements but, the voltages and currents will be purely sinusoidal and have the same frequency (a subtle but noteworthy point) as the input voltage as well, provided the parameters of circuit elements do not change with time.

From the qualitative descriptions given above of the behaviour of an inductor to voltages of different frequencies, it may be observed that an inductor blocks high frequency inputs.

It therefore, finds use as a low pass filter in electric and electronic circuits.

A smoothing circuit is used to reduce the ripple of a dc voltage produced by rectifying an ac voltage. Such a smoothing low-pass filter using an inductor L is shown in Fig 5.32.

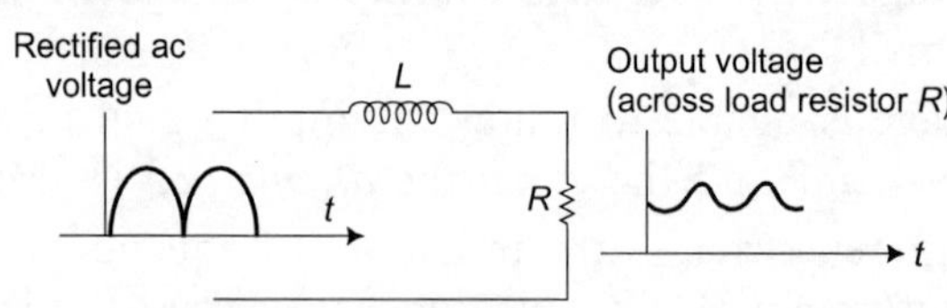

Fig. 5.32 Smoothing circuit that reduces the ripple of a rectified ac voltage.

Low pass filters are built using L-R circuits in simple applications while in applications that should meet stringent requirements, filters are built using electronic components such as transistors and operational amplifiers. Circuits built with these provide a sharp demarcation between the low frequencies that are passed and high frequencies, which are blocked. To view the output waveforms generated by computer simulation programs of low pass filters to signals which are a combination of several frequencies, the reader is referred to the "Links_to_animations_articles" file in the CD accompanying this book.

5.20 CURRENT WITHOUT A VOLTAGE!

Inductors – One Strange Phenomenon

Just as Isaac Newton's first Law of Motion ("an object in motion tends to stay in motion; an object at rest tends to stay at rest") describes the tendency of a mass to oppose changes in velocity, we can state an inductor's tendency to oppose changes in current: "Electrons moving through an inductor tend to stay in motion; electrons at rest in an inductor tend to stay at rest".

Hypothetically, an inductor left short-circuited will maintain a constant rate of current through it with no external assistance.

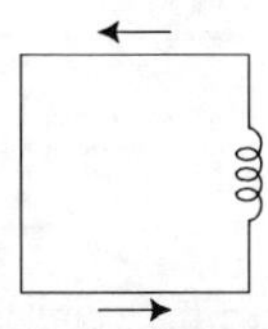

Current sustained with
the inductor short-circuited

The electric field will be "*zero*" everywhere around the inductor and also inside the inductor wire as, ***neither*** the current ***nor*** the magnetic field ***is changing***. There is no voltage across its ends, either.

We have a current without a voltage. Practically speaking, however, the ability for an inductor to self-sustain current is realized only ***with superconductive wire***, as the wire ***resistance*** in any normal inductor is enough to cause current to decay very quickly with no external source of power.

Just as a stationary electron has an electric field surrounding it, we see moving electrons have a magnetic field. The magnetic field sustains the current and the energy stored in the field is given by $\frac{1}{2}(LI^2)$ where L is the inductance and I the current in the inductor.

5.21 INDUCTORS—PRODUCTION OF A HIGH VOLTAGE

We are aware that a *current carrying inductor stores energy* in the magnetic field in the space surrounding the coil.

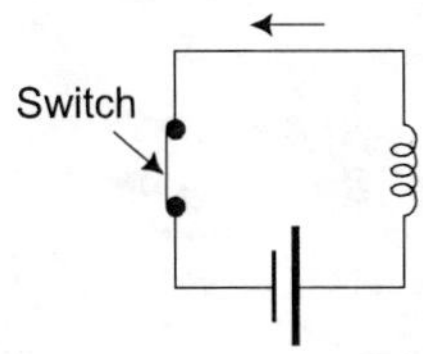

Fig. 5.33 Production of a high voltage using an inductor.

Consider the circuit of Fig. 5.33 which shows an inductor carrying a current with a switch in the circuit. If we *break* the circuit by opening the switch, we would be interrupting the current, which is a *manifestation* of the magnetic energy in the coil.

Breaking the circuit of an attempt to destroy magnetic energy ... but energy cannot be destroyed, so, the circuit reacts by inducing an emf.

Since the current is stopped abruptly, a large emf will be produced across the contacts of the switch and the air between the contacts of the switch will ionize.

An *arc* will appear across the contacts, and the energy of the inductor will be dissipated as electromagnetic radiation in the form of heat (infrared) and light (visible spectrum) as shown in Fig. 5.34.

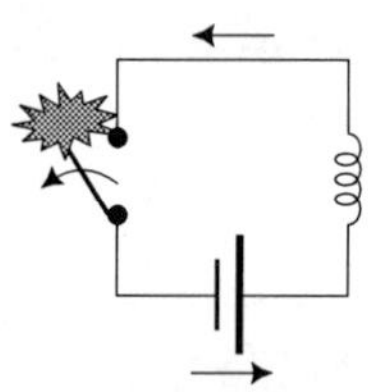

Fig. 5.34 Arcing across the contact resulting in a high voltage.

In Section 10.9, a description of how air ionizes to produce sparks, when the field strength is greater than 3×10^6 N/C, is provided.

This method of producing a high voltage is adopted to produce the sparks in spark plugs to ignite the fuel in motor cars, scooters and motorbikes using ignition coils.

The same method is used to initiate the discharge in fluorescent lamps or tubelights. Another way to interpret the production of the high voltage induced:

Since an attempt is made to stop the current abruptly, ***di/dt* is large** and so is the induced emf v.

The production of a spark is exploited in the spark plug circuit of petrol engine driven automobiles. However, this could cause catastrophic failures of sensitive electronic driver circuits like transistors when used to switch a relay coil.

See Section 9.13 for a description of how transistors used to drive relay coils are protected using a diode.

What is the sequence of events, at the microscopic level, if we **insert a resistor** in the inductor circuit made of **superconductive wire** with a current in it, as shown below in Fig. 5.35?

Fig. 5.35 Inserting a resistor in a current carrying inductor.

5.22 VOLTAGES, CURRENTS AND ELECTROMAGNETIC WAVES IN TRANSMISSION LINES

The right sequence of reading topics on the transmission of energy by conductors (electromagnetic waves) is i) Guided Waves ii) Transmission Lines, and then iii) Wave Guides.

The advantage of such an approach is because the understanding one would gain from topics on Guided Waves or waves between a pair of perfectly conducting planes, will be useful to understand the transfer of energy, by the commonly used two-wire transmission line and co-axial cable.

The wave nature of energy transfer is most evident in Wave Guides that are single-conductor guides that are usually rectangular or cylindrical in shape, and not infinite in dimensions. Wave guides will be treated in Chapter 10.

In Section 4.4 we had seen how energy flows, even though the fields were static, from the source to a load, and the power flow was given by the Poynting Vector $\vec{E} \times \vec{H}$.

When a high frequency signal voltage is applied to a transmission line, which consists of a pair of perfectly conducting parallel planes separated by a distance, energy transfers from the source to the load by pulsating fields between the planes; the fields don't travel.

We will learn later in this section that transmission lines, operated in the radio frequency range and at higher frequencies, are generally modeled and analyzed using distributed parameters (inductances and capacitances), because the quasi steady-state conditions (that had allowed us to write the capacitor reactance expression as $1/2\pi fC$ in the lumped regime of analysis in Section 2.10) breakdown in transmission lines of lengths which are greater than the free space wavelengths of the signal. Thus, transmission lines are commonly analyzed considering the propagation of energy by the electric and magnetic fields (electromagnetic waves), around the conductors of the line; fields don't travel, the wave travels.

The sustenance of an electromagnetic wave between the conductors is because of two time-varying field processes,

i. a changing magnetic field produces a curly electric field (Faraday's Law), and
ii. a changing electric field produces a curly magnetic field (Maxwell's added term to Ampere's Law Section 4.7).

A magnetic field is always curly. No magnetic field has been detected that is not curly.

Maxwell's Added Term to Ampere's Law or the Ampere-Maxwell Law

A steady magnetic field surrounds a current-carrying wire (Ampere's Law, Section 2.28) and a curly electric field is associated with a changing magnetic field (Faraday's Law, Section 5.8 to 5.12). James Clerk Maxwell was at a loss to balance Ampere's Law equation when this was applied to the current between the plates of a charging capacitor. He surmised that the electric field between the plates of the charging capacitor is changing and this somehow should produce an equivalent current, maybe virtual and not real. The fact that the current itself should be associated with a magnetic field made him realize that between the plates a magnetic field then should exist in the absence of a current in between; a magnetic field not associated with current and that would *seemingly* close the circuit !

He therefore modified Ampere's law (Chapter 4, Section 4.7), which is,

$$\frac{1}{\mu}(\nabla \times \vec{B}) = \vec{J} \text{ or } \nabla \times \vec{H} = \vec{J}$$

(in words, a curly patterned magnetic field is always associated with a current density $\vec{J}$ or, simply a curly patterned magnetic field is present around a current of density $\vec{J}$) to

$$\nabla \times \vec{H} = \vec{J} + \partial\vec{D}/\partial t \qquad \text{or,}$$

$$\nabla \times \vec{H} = \vec{J} + \varepsilon\partial\vec{E}/\partial t$$

where, ε is the permittivity of the medium.

The added term $\varepsilon\partial\vec{E}/\partial t$ associates a magnetic field with a changing electric field and this equation set is called, the Ampere-Maxwell Law.

Since Faraday's law associates a non-conservative electric field with a time-varying magnetic field, conversely, his added term $\varepsilon\partial\vec{E}/\partial t$ or, $\varepsilon\dot{\vec{E}}$ postulates the association of a magnetic field with a time-varying electric field, Maxwell theorized the possibility of the existence of electromagnetic waves in free space.

We will learn in Chapter 10 how electromagnetic waves propagate through free-space without the presence of charges.

The Fig. 5.36(a) shows the electric field and magnetic field components (shown by dots and pluses between the planes) of an electromagnetic wave propagating in the z direction between a pair of perfectly conducting parallel planes that are in the x-y plane separated by a distance 'a'.

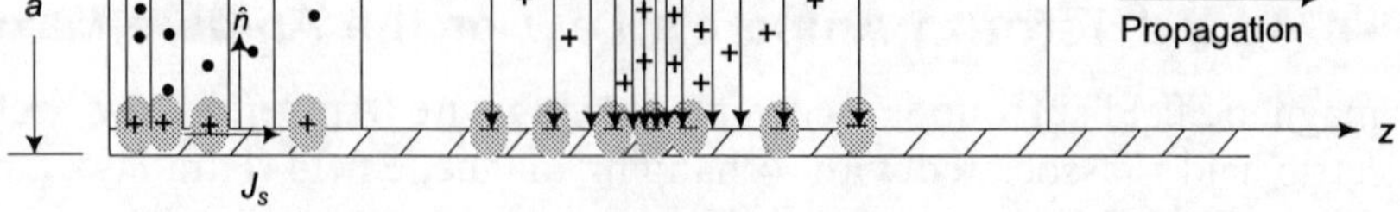

Fig. 5.36(a) A transverse electromagnetic (TEM) wave between parallel planes.

Fig. 5.36(a) shows the electric and magnetic fields at one instant of time. The electric field and the magnetic fields grow and reduce in strength at the same points (pulsate), creating a *wave*. The electric field points in the x-direction while the magnetic field points in the y-direction. There are no other components of the field in any other direction. Both the fields are transverse or perpendicular to the direction of wave travel, i.e., the z-direction.

The electric field E_x in Fig. 5.36(a) is non-conservative, but the field lines terminate on induced charges on the surface of the plane, i.e., the non-Coulombic field E_{NC} polarizes the plane. Interestingly, the surface charges appear by tiny current delivery from the interior of the conductor plane, and we will observe currents through the thickness of the plane! In Section 3.8, we had seen how currents can flow in directions other than along the axis of round conductors, because electric fields due to stray capacitance effects, are present in many directions. See "Electric field and current flow within the conductor" in Chapter "Guided Waves" Ref. [16].

There is also the surface current J_s, which manifests because of the magnetic field H_y and *not* by an electric field.

This requires some explanation. Intuitively, the presence of the surface current density J_s can be deduced from the Ampere-Maxwell relation in integral form $\oint \vec{H} \cdot d\vec{s} = \int \left(\dot{\vec{D}} + \vec{J}\right) \cdot d\vec{a}$.

Stated in words, the magnetomotive force (mmf) described in Section 5.2 around a closed path, where $d\vec{s}$ is an elemental length of the closed path is equal to the conduction current plus the time derivative of the electric displacement through any surface $d\vec{a}$, bounded by the path Ref. [16]. When the relation is applied to a closed path straddling the conductor and the space in between the planes in the *x-y* plane, it can be shown that there exists a finite surface current J_s (Fig. 5.36(a)), even when the conductor has infinite conductivity.

This aspect is important for two reasons. The infinite conductivity assumption prevents the existence of an electric field parallel to the plane conductor surface at the surface. We have enough evidence of the necessity for the presence of an e-field to drive currents in conductors with finite conductivity from Section 1.14 "Electric Fields in Matter"). So, effectively the infinite conductivity assumption makes the $\dot{\vec{D}} = \varepsilon \dot{\vec{E}}$ term contribute nothing, or zero to the mmf path integral. And, from Faraday's law if there is no electric field, then a magnetic field that produced a zero electric field cannot exist.

The Ampere-Maxwell relation thus leaves only a surface current J_s (current per unit width in the *z*-direction). This current adjusts itself to a finite value, and equal to a finite magnetic field just *outside* the surface. The magnetic field and surface current will be parallel to the surface, but perpendicular to each other. In vector notation, this is written

$\vec{J}_s = \hat{n} \times \vec{H}$, where $\hat{n}$ is the unit vector normal to the surface.

The unit vector is drawn in Fig. 5.36(a). The direction of the current may be determined by applying the cross-product rule of Section 5.2. The magnetic field indicated by dots near the unit vector, is pointing out of the page. The current J_s points to the right in Fig. 5.36(a) and is shown as J_{sz} in Fig. 5.36(b).

The characteristic of the surface current J_s is like the situation described in Section 5.20 "Current without a voltage".

A detailed mathematical treatment is available in the topic "Conditions at a boundary surface; Note on a Perfect Conductor" Chapter on Maxwell's Equations in Ref. [16].

Figure 5.36(b) shows an end-view of the fields between the planes.

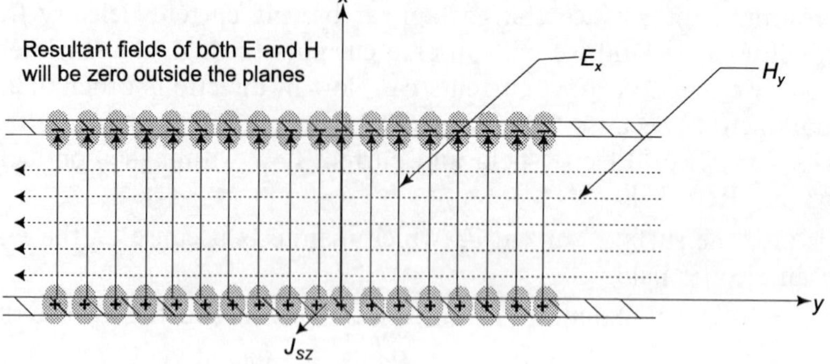

Fig. 5.36(b) A transverse electromagnetic wave (TEM) between parallel planes.

In one sense, the waves travel as uniform plane waves in free-space (that are described in Chapter 10), because there is no component of either the electric field or the magnetic field in the direction of propagation *z,* i.e., E_z and H_z do not exist. Further, the components E_x and H_y of the electric and magnetic fields, are transverse to the direction of propagation '*z*' and so, these waves are called Transverse Electromagnetic Waves or TEM waves.

If the planes are perfectly conducting and infinite in extent, and if the surface charges that are separated charges from the lattice structure appear and disappear producing a uniform field between the planes (maybe varying but uniform), the resultant electric and magnetic fields *outside* the planes will be zero, quite similar to the case of the tiny fringe field observed outside a parallel-plate capacitor. So, it is as if the travelling Transverse Electromagnetic Waves are trapped and *guided* between the planes. See the end of the Example in the topic "Capacitance" in Appendix A.

The TEM waves as shown in Fig. 5.36(b) we say, are therefore *guided* by the parallel planes in between them.

Figure 5.36(a) is a plot for a single instant of time. As time passes, the entire field configuration sweeps to the right (in the *z*-direction) with a velocity approximately equal to the speed of light in free space.

An animated view of TEM, TE (Transverse Electric) and TM (Transverse Magnetic) mode fields may be seen here https://en.wikipedia.org/wiki/Transmission_line and http://electronicsgurukulam.blogspot.in/search/label/Transmission%20Lines%20and%20 Waveguides.

How is it possible to represent a wave whose electric field component E_x is oscillating in strength at a fixed point in space and also seemingly travelling in the z-direction mathematically?

A wave is a function of both space and time. If the field component E_x of the wave is a sinusoid of amplitude E_o, the wave travel to the right is mathematically represented by a factor $e^{-j\beta z}$ which multiplies the field sinusoid where β, a constant, is called the wave number which has units of radians per meter. The propagation in the z-direction is a wave motion, the phase of the field E_x constantly changing, because of the phase factor $e^{-j\beta z}$.

The wave can be represented by the equation $E_x = E_o e^{j(\omega t - \beta z)}$ where, ω is the frequency of the signal voltage in radians per second. Taking the imaginary part of this equation, we have

$$E_x = E_o \sin(\omega t - \beta z).$$

The wave is time harmonic because i) we assumed time dependence $e^{j\omega t}$ to arrive at the equation, and ii) $(\omega t - \beta z)$ is the phase (in radians) of the wave; it depends on time t, and the space variable z.

One plot of the field component E_x, with the space variable z assumed constant versus time is that of a sinusoid of period $T = \dfrac{2\pi}{\omega}$. Another plot of E_x, with the time t assumed constant versus space variable z, is that of a sinusoid that takes a distance λ to repeat itself, where λ is the wavelength.

A sketch of the equation $E_x = E_o \sin(\omega t - \beta z)$ versus βz at times t = 0, T/4 and T/2 will reveal that the wave advances with time in the $+z$ direction.

A complete description of the mathematics for representing a wave with illustrations is available in the chapter Electromagnetic Wave Propagation, "Elements of Electromagnetics" by Mathew N.O. Sadiku, Oxford University Press, 2007.

The fields themselves do not travel, but the changes in the field produce the effect of a travelling wave, much like the travel one observes of the undulations of a rope fastened to a wall at one end, when the other loose end is moved up and down rhythmically.

Transmission line engineering recognises several important advantages in maintaining TEM wave propagation to transfer energy from a source to a load.

What are the other modes of transmission? Why is the TEM mode preferred over other modes of transmission?

The other modes of propagation are the Transverse Electric (TE) and the Transverse Magnetic (TM). In the TE mode, in addition to its transverse magnetic field component, one component of the magnetic field is parallel to the direction of propagation.

In the TM mode, in addition to its transverse electric field component, one component of the electric field is parallel to the direction of propagation. These modes can also be transmitted by a pair of perfectly conducting parallel planes.

However, these modes are generally not preferred, because they attenuate signals below a certain frequency, which is determined by the transmission line geometry. This could produce distortion of signals that comprise a mixture of signals of several frequencies.

However, these modes do find use in the transmission of energy in the microwave frequency range (300×10^6 Hz to 300×10^9Hz) in ovens, radar antennas, and striplines which transmit energy on printed circuit boards of electronic circuits.

On the other hand, the Transverse Electromagnetic (TEM) mode does not cutoff *any* signal frequency and transmits even dc, as we saw in the case of connections between the batteries and the lamp loads made by two wires (maybe not closely spaced), in Sections 2.20, Chapter 2 and Sections 4.3, 4.4 and 4.5 in Chapter 4.

There are a few disadvantages, however, such as the large attenuation at high frequencies (attenuation is described later in this section) due to dielectric and conductor losses. However, for low frequency transmission (below 300 MHz), the TEM mode of operation is considered suitable.

How is the production of a Transverse Wave started?

The motion we make which starts the wave determines what kind of wave it is. In other words, the direction of the field when launched initially determines the type of the wave. If the initial field (say, electric field) is oriented perpendicular to the direction of propagation of the wave, then the wave will be transverse.

The wave propagates by a sequence of pulsations of the electric and magnetic field strengths, the electric field generated as described by Faraday's law, and the magnetic field generated by Maxwell's added term $\varepsilon\dot{\vec{E}}$ to Ampere's law or simply, the Ampere-Maxwell law; the energy transmitted along a lossless conductive path is conveyed by the electric and magnetic fields in the region, the voltage and current being present merely as evidence of existence of the fields. A qualitative description of how the above two laws enable us to predict the generation of the electric and magnetic fields, whose strengths vary at fixed points in space producing a wavelike motion, is described in Chapter 10.

Transmission lines made from perfectly conducting parallel *planes*, are not usually seen in simple electric and electronic circuits, though they do make their appearance as parallel conducting *plates* (finite dimensions) in the newer antennas as feeding elements (striplines) that are sometimes fed by two-conductor co-axial lines.

It will be sufficient for the purpose of this brief treatment of transmission lines if, previous chapters on the nature and effects of electric field, surface charges and how they enable the process of conduction, and capacitance and inductance (even in straight wires) effects, are read.

Because changes in the electric field do not propagate instantly, but at the speed of light, and also due to the finite inductance and capacitance of the line and the presence of surface charges, these phenomena produce a behavior of voltages and currents in transmission lines of long lengths, which may be peculiar.

This is not a serious problem in simple circuits with small connecting wires that have small stray capacitances and inductances, because at signal frequencies below say, 1 MHz, the capacitances charge and discharge by "condensing" and "dispersing" the electrons on the surfaces quickly in response to the input signal voltage variations, setting the time varying potentials instantaneously at different points along wires and conductors.

To gain a correct perspective into transmission-line operation, we will study a peculiar set up of an infinitely long transmission line.

When a constant voltage is applied to an infinite transmission line without a load, a constant current can result. Suppose that there is a set of two wires (called a two-wire transmission line) of infinite length (visualize a line laid from the earth and extending into the depths of the galaxies) connected to a battery with a switch, as shown in Fig. 5.37.

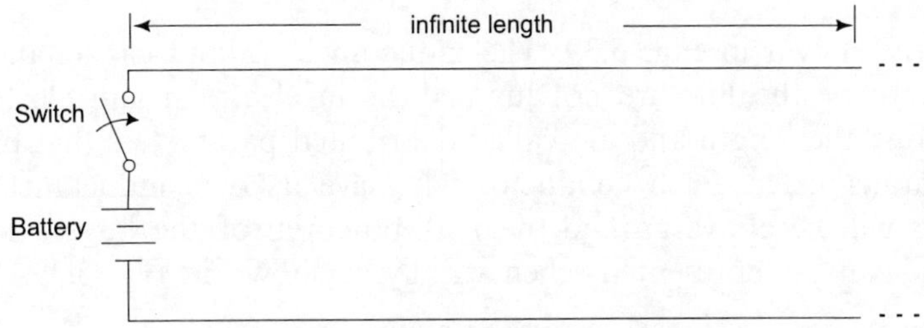

Fig. 5.37 Driving an infinite transmission line.

What happens when the switch is closed? With the line unterminated or open circuit, we would expect no current at all. However, recollect (Section 2.21) that capacitances are present between two conductors, between a single conductor and ground and between a conductor and a conductive shell at infinity. When a single wire is connected to one terminal of a battery, charges move from the positive terminal into the wire until the potential of the positive terminal and the wire are the same, and if there is another wire connected to the negative terminal, charges will move from the negative terminal into the wire until the potential of the negative terminal and the wire are the same.

For the transmission line, similarly, in the steady-state after closing the switch, the voltage across the line should attain the voltage of the battery, and until things settle down (the potentials are same everywhere in each wire) there will be charge movement or current in the wires until the lines attain the voltage of the battery.

Even if the wires were to be made of superconductive material (so that their resistance is "zero"), and the wire is considered to be lossless, there will be a capacitance along the wires' lengths. However, if the line is of infinite length and considered lossless, the current will be forever as described below!

Any pair of conductors separated by an insulating medium creates capacitance between those conductors, as can be seen in Fig. 5.38, which is a circuit representation of the parallel-plane transmission line.

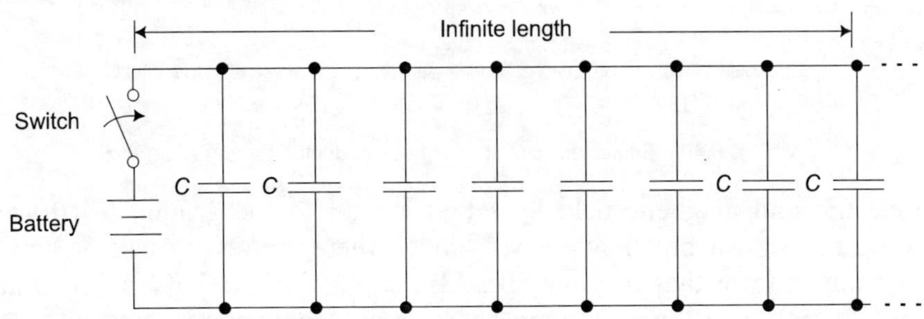

Fig. 5.38 Circuit representation of a parallel-plane transmission line showing line capacitance.

When the switch is closed, the capacitance between conductors will begin charging up, drawing current from the source, and with the line assumed lossless, the current should be infinitely large. However, the current drawn by a pair of parallel wires (not planes) will not be infinite, because there exists series impedance along the wires due to inductance (See

Section 5.16) as shown in Fig. 5.39. The inductance '*L*' and capacitance '*C*' that extend along the length of the line are not lumped, as in electrical and electronic circuits, but distributed along the length and are called distributed parameters that have units Henries/meter and Farads/meter. Each conductor will have its own inductance and capacitance, since the lines will be above ground (near '0' potential of the huge earth). Hence, a two conductor line can be represented schematically as shown in Fig. 5.39.

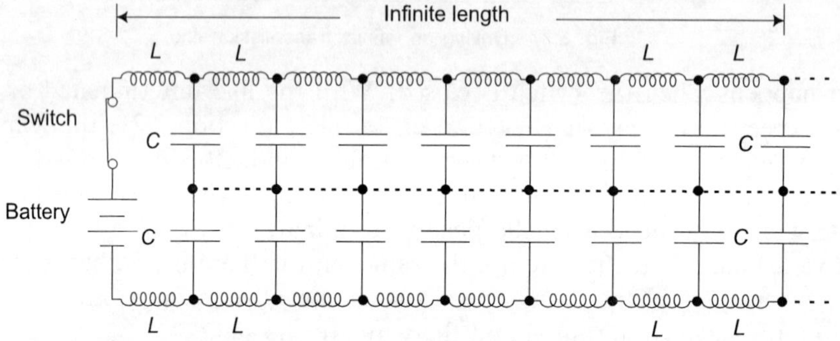

Fig. 5.39 Equivalent circuit of a two-conductor transmission line.

In several cases the ground effect is combined with the lines, hence, the two capacitances can be combined as a series combination and can be represented as shown in Fig. 5.40.

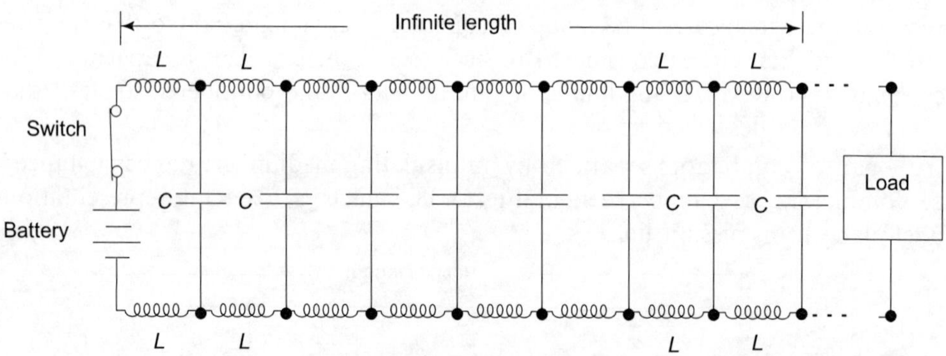

Fig. 5.40 Equivalent circuit of a two-conductor transmission line.

The electric and magnetic field lines that extend to the ground (earth) and between the lines are not shown but they are very much there, however weak or however strong, uniform or nonuniform that they may be. Also shown in Fig. 5.40, is a load at the end of the line which may be a lamp or a motor or an electronic circuit even.

A changing current carries charging current for the capacitance between the wires, developing an emf according to Faraday's law. Read the note below Fig. 1.23. Each wire develops a magnetic field as it carries charging current for the capacitance between the wires, and in so doing, drops voltage according to the inductance equation $e = L(di/dt)$. This limits the voltage rate-of-change across the distributed capacitance, preventing the current from ever reaching an infinite magnitude.

A "wave front" of voltage and current *change*, will propagate down the length of the wires, resulting in the distributed capacitance and inductance progressively charging to full voltage and current, respectively. We will learn at the end of Chapter 7 that charges when accelerated produce electric field lines with a circular pattern (circular waves) and that these propagate from the source to the load of a transmission line. The propagation using circular waves is vividly depicted in Hartel's article available in the Teacher's Guide in the CD which may be read after reading Chapter 8 of this text.

As a result of these processes, a constant current of limited magnitude flows through the battery source. Since the wires are infinitely long, their distributed capacitance will never fully charge to the source voltage, and their distributed inductance will never allow unlimited charging current.

Thus, this pair of wires without a load, will draw current from the source so long as the switch is closed, behaving as a constant load!

Visit http://www.allaboutcircuits.com/vol_2/chpt_14/3.html for a more detailed description.

To view the orientation of the electric field, bearing in mind that a lossless line precludes the existence of an electric field tangential to the line conductor which could cause a voltage drop, refer Section 4.4, "Poynting vector".

Transmission-line theory shows that for all types of TEM mode transmission lines (coaxial, parallel-wire), the basic equations of voltage and current variation over the line are

$$\frac{dV}{dz} = -(R + j\omega L)I, \text{ and } \frac{dI}{dz} = -(G + j\omega C)V.$$

A detailed derivation of these equations is available in Ref. [16].

Transmission-lines are usually analyzed for performance based on their voltages and currents, the field configurations and wave dynamics we studied in Fig. 5.36 giving us insights into the origins of these voltages and currents.

Examples of Two-conductor Transmission Lines

A common example of guided waves in transmission lines is a two conductor system; parallel two-wire or twin-lead cable and coaxial transmission lines. An example of guided waves in a single conductor system is, the waves in rectangular waveguides. The reader should note that waveguides are made from conducting plates of finite dimensions and come in rectangular box and cylindrical shapes, open on both sides or closed at one end. The propagation of waves in waveguides is treated in Chapter 10.

We had seen earlier (Fig. 5.36) how a pair of perfectly conducting parallel planes can support a Transverse Electromagnetic (TEM) wave, and guide the energy to the source with no loss. It is impossible to produce such a transmission line economically, and besides such a line if produced would become unwieldy and cumbersome to install and operate. Two-conductor transmission lines (Figs. 5.39 and 5.40) made from copper wires, offer a feasible alternative, which by careful selection and design can deliver performance with fewer overheads.

The following paragraphs describe the operation of such transmission lines and the manner in which they support the TEM mode. In particular, we will study two of the common types of two-conductor transmission lines; the twin-lead cable and the coaxial cable. While the TE and TM modes require a certain minimum separation (measured in wavelengths) between the conductors for propagation to be possible, only for the TEM wave can the separation be small compared with a wavelength. This statement holds for practical transmission lines such as the twin-lead cable, the coaxial cable or parallel wire lines, and it is for this reason that only the TEM mode need be considered at low frequencies, which is at power (50 Hz or 60 Hz), audio (20 Hz to 22 KHz) and radio frequencies below 200 or 300 MHz. All other modes would require impractically large cross-sectional dimensions of the guiding systems.

Figure 5.41(a) shows a short piece, of an antenna to TV 300 ohm twin-lead cable, also known as biaxial cable (or just 'biax' for short).

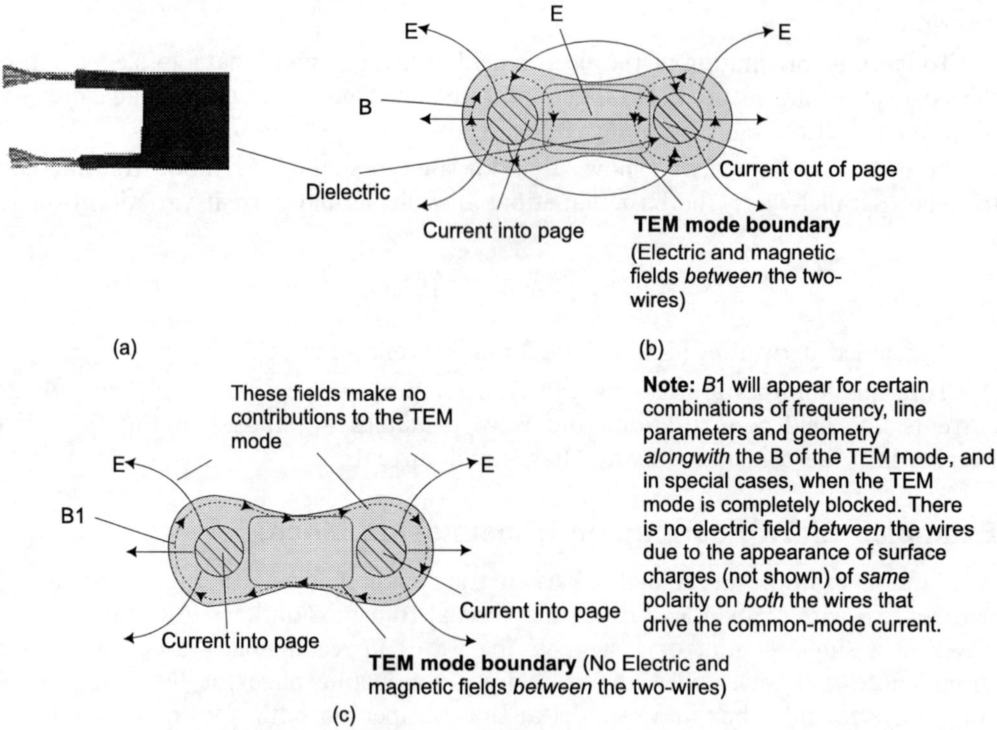

Fig. 5.41 (a) a twin-lead cable (b) fields of a twin-lead cable in Transverse Electric and Magnetic field (TEM) mode (end-view) (c) fields of a twin-lead cable with the Transverse Electric and Magnetic fields prevented from appearing producing a common-mode current (See note also).

The twin-lead cable is a two-conductor flat cable used as a transmission line to carry radio frequency (RF) signals. It is constructed from two multistranded copper or copperclad steel wires, held a precise distance apart by a plastic (usually polyethylene) ribbon. The close uniform spacing of the wires is the key to the cable's function as a parallel transmission line; any abrupt changes in spacing would reflect radio frequency power back toward the source.

Figure 5.41(b) shows the electric and magnetic fields at one instant of time when the twin-lead cable is excited by a radio frequency signal generator. The orientation of fields between the two wires shown in the dashed Transverse Electromagnetic (TEM) boundary in Fig. 5.41(b) resemble the orientation of the transverse electric and magnetic fields in the perfectly conducting parallel-plane transmission line of Fig. 5.36(b).

The fields which are present outside the dashed Transverse Electromagnetic (TEM) mode boundary around the twin conductors of Fig. 5.41(b), do not support TEM mode of propagation.

The surface current J_s of the parallel plane line (Fig. 5.36(a)) and the surface line current of a twin-lead line (Fig. 5.41(b)) that are associated with magnetic fields, are differential-mode currents. That the conductors are lossless does not preclude the existence of currents that do not dissipate energy! However, lossless conductors are never seen in practical transmission lines unless the wires are superconductive, in which our working definition of current as a start-stop motion of millions and millions of electrons moving everywhere within the wires with a drift superimposed, will no longer be applicable.

The current shown in Fig. 5.41(b) from the source is into the page in the left conductor towards the load and returns in the right conductor out of the page towards the source. Since the forward and return currents are opposite to each other in any cross-section of the line, they are called differential-mode currents.

The close spacing of the lines and the differential mode of current flow also ensures that the superposition of the radiated fields from the two conductors, results in a cancellation or large reduction of radiated field strengths, in the space around the transmission line. This tends to reduce the generation of signals that can interfere with signals in other nearby transmission lines. We will learn about this in Section 10.10.

How does a Transverse Electric and Magnetic field (TEM) wave start in a transmission line?

Whenever a source is connected to a circuit and the final connection made, a few electrons leave the wire and enter the positive terminal of the source, and a few electrons simultaneously leave the negative terminal and enter the wire. This initial movement triggers the production of the transverse electric field and as the charges move along the wire towards the load, a magnetic field encircles the wires. This is how the TEM mode acquires the '$-\beta z$' term into the wave equation: $E_x = E_o \sin(\omega t - \beta z)$. See descriptions of the mathematical representation of a wave given in the questions below Fig. 5.36(b).

A snapshot view of the fields of the twin-lead line in Fig. 5.41(b) shows the wave propagating in a direction, which is transverse or perpendicular to both the fields which is into the page of the Fig. 5.41(b).

In short, the direction of the field initially determines the type of the wave or its mode, which in this case, is the Transverse Electromagnetic (TEM) mode.

At the end of Section 5.16 we had indicated that a single wire has an inductance. It is instructive to apply Lenz's rule at the beginning of the line to visualize how this inductance produces an induced emf.

Figure 5.42(a) shows a truncated section of a long straight twin-lead cable with the self inductance of the cable shown as imaginary tiny inductors, for ease in understanding the development of an induced emf when the cable is energized.

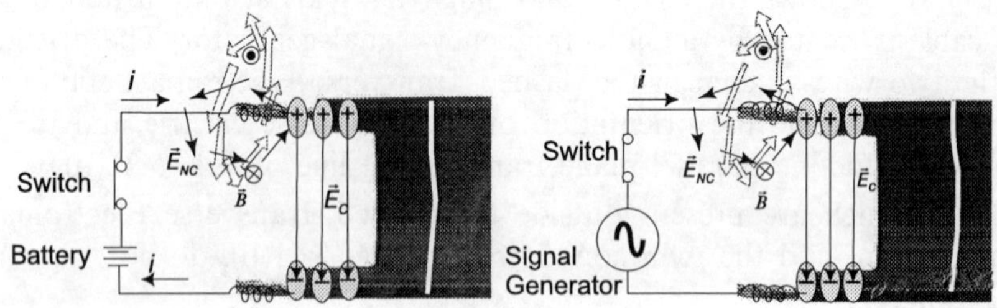

Fig. 5.42 (a) Generation of Induced emf in a twin-lead cable when a dc voltage is applied at the transmission end. (b) Application of a sinusoidal signal and the launch of the transverse electromagnetic wave.

When the switch is closed, positive charges enter the upper lead and negative charges enter the lower lead. As the applied voltage rises, a current i is initiated associated with a magnetic field B, which encircles the wire, shown using block arrows in Fig 5.42. The rising current i is associated with a non-Coulomb field $\vec{E}_{NC}$ whose direction is determined by the application of Lenz's rule of Section 5.13 (symbols indicating the magnetic field pointing *into* and out of page are shown in Fig. 5.42). The non-Coulomb field polarizes the lead and a positive charge develops on the upper lead. In the meanwhile, a similar process on the lower lead connected to the negative terminal of the battery, causes the wire to polarize and a negative charge appears on the lower lead. A nascent coulomb field $\vec{E}_C$ develops as shown by the field line $\vec{E}_C$ which originates on positive charge on the upper lead and seemingly terminates on the negative charge on the lower lead. More and more field lines (not shown) may be drawn between the two wires to show the growing field strength which develops within nanoseconds. Since the circuit is energized by a dc voltage that is constant, both the fields, i.e., the electric and magnetic, will be of constant strength in the steady-state and correspond to the situation described in Sections 4.3 and 4.4. The magnetic field is not shown around the negative lead, in Fig. 5.42(a), for the sake of clarity.

The reader is urged to verify that the induced emf on the negative lead of the twin-lead cable in Fig. 5.42(a) does oppose the applied electric field on that lead.

Figure 5.42(b) shows the process of generation of an induced emf, when a sinusoidal signal voltage is applied to the twin-lead cable. The process is similar to that when a dc voltage is applied to a battery. With a sinusoidal signal applied to the twin-lead line however, within a few instants of the formation of the Coulomb electric field, the time-varying nature of the sinusoidal voltage forms non-Coulomb fields $\vec{E}_{NC}$ which pulsate between the leads as indicated in Fig. 5.42(b) and terminate on charges on the leads. The $\vec{E}_{NC}$ field reminds us that it's origin is due to a time-varying magnetic field (Faraday's law). The associated pulsating magnetic fields (Ampere-Maxwell law) are not shown for clarity in Fig. 5.42(b). The fields are not constant in strength and will be pulsating. They attain the shape shown in Figs. 5.36(a) and 5.36(b) soon and the wave glides towards the load. The propagation is termed Transverse ElectroMagnetic. The question naturally arises as to what *is* the voltage drop, given the non-Coulomb nature of the field for a twin-lead cable when it carries a sinusoidal signal voltage. This will be described later in "What is the "voltage" of a lossless transmission line?"

Differential and Common-mode Currents in Twin Lead Cables

In special applications, by arranging the length of the twin-lead line to a quarter (shorted at the load end) or half (opened at the load end) free space wavelength at signal frequency, it is possible to create a very high impedance line. This suppresses the differential line current almost entirely. However, since the two leads of the line are closely-spaced, a *common-mode* of current flow can exist, i.e., the currents of both conductors flow in the same direction.

Quarter-wave line length and half-wave line length transmission lines are used for the purpose of impedance matching, and these are generally not used to deliver power to a load.

The special case of full common-mode operation is illustrated in Fig. 5.41(c). These common-mode currents reduce to zero at the load and source ends. No currents flow in the source and load because there is no potential difference (common-mode voltage) across them. Surface charges of same polarity on both wires drive the common-mode current. Thus, there is no electric field between the two wires, but diverges around them. The common-mode signal energy will either dissipate in the line resistance or be radiated away (as we shall learn in Chapter 10). We will now see how common-mode currents are produced.

In general, both differential mode and common mode currents can exist in an isolated two-conductor line such as the twin-lead cable and is due partly to the fields (E and B1) that are present outside the wires as shown in Fig. 5.41(c). This can be expected, since the conductors are not infinite planes, but wires of finite dimensions and the fields are not effectively confined between the wires. Note the current directions of both the wires pointing into the page in Fig. 5.41(c). It would be instructive to apply the Lenz's rule to the right-hand wire (Sections 5.13 and 5.14) assuming that the current in the left-hand wire (Fig. 5.41(c)) is increasing. You will find that the associated Coulomb field in the right-hand wire also points into the page, making the current into the page as well, and the current is therefore common-mode.

The currents observed in the circuits using a battery and wires and resistors of Figs. 1.24, 1.25, 1.29, 1.30 and 1.33 are differential-mode, because conventional current flows from the battery positive to the load resistor, and returns to the negative terminal of the battery. Similarly, the currents observed in the RC circuits Figs. 2.15 (voltage pulse), Fig. 2.21 and Fig. 3.1 (observe the positive cycle of the sinusoidal voltage) are differential-mode because conventional current flows from the positive terminal of the source (pulse or signal generator) to the load resistor, and returns to the negative terminal of the source. The current in the RL circuit of Fig. 5.14 is also a differential-mode current. In the analysis of these abovementioned circuits since our focus was on the basic circuit processes and currents, we ignored a discussion of the presence and effects of common-mode currents.

Under what conditions can common-mode currents be present?

The common-mode of current maybe present in twin-cable lines depending upon the frequency, conductor spacing and other geometrical factors. External electromagnetic fields such as electromagnetic interferences can *induce* currents on transmission lines. The components of external fields can couple or excite the transmission line and distort the desired signal waveforms being transmitted by the source to the load. The incident electric field will give rise to surface currents that may be both, common-mode and differential-mode.

A transmission line, which is carrying weak analog signals (signals which vary with time) say, from a sensor to a signal processing electronic circuit, is normally connected to an amplifier, which is used to strengthen the signal. Such signals normally produce a differential-mode current in the line and the amplifier is usually configured to amplify the differential signal.

While designers and users of electronic equipment may take care by shielding and grounding the equipment to minimize the coupling of external fields to the sensitive portions of the circuit, there is always the case when some external field does get coupled in, producing *unwanted* common-mode signals. In industrial applications, where large loads are switched ON and OFF, arcing across switchgear and sudden current interruptions can produce electromagnetic interference (unwanted common-mode noise) in sensitive electronic circuitry.

When measuring the emf of a thermocouple (industry) or electrocardiogram potential from a patient's body surface (medical) in noisy environments, the sensor lead voltages and currents are differential but electromagnetic interfering noise appears as an offset on *both* input leads, making it a common-mode voltage signal. The induced common-mode voltage will raise or lower the voltage of both the conductors of a twin-lead cable. Such induced common-mode voltages will also be amplified by the amplifier and can result in noisy signals, signal distortion or even clipping of signals since these signals will drive the amplifier into non-linear regions of operation.

To suppress such externally induced undesired common-mode signals, differential amplifiers are commonly used at the receiving end. These amplifiers feature a high common-mode rejection ratio which amplifies only the differential-mode signal voltage at the amplifier input and suppresses the undesired common-mode signal.

For readers who have been introduced to transistorized differential (also called difference amplifier) amplifier[17], it may seem that since the common-mode current is zero in the source and load impedances, as we had mentioned in a previous paragraph, there will be no effect produced in the amplifier. The common-mode voltage which persists at the load, which is the differential amplifier input (differential *pair* input which are *both* changed in potential due to the line induced common-mode voltage) conveyed by the *two* lines of the source, is usually absorbed by the high resistance of the emitter in the differential pair[17] or, by the reverse-biased collector junction of the constant current source (usually a transistor in the common-base configuration) in the emitter of the differential amplifier pair[17], thereby providing a high common-mode rejection ratio.

A power point presentation "Basic Action of a Differential Amplifier-Heart of the Opamp" including animations which describes the principle and action of a differential amplifier is available in the CD with this book.

A comprehensive description of differential- and common-mode currents is available in Albert A. Smith, *Radio Frequency Principles and Applications, The Generation, Propagation, and Reception of Signals and Noise*, Wiley-IEEE Press, 1998.

One classic and ingenious application of a transmission line operated with a large common-mode current, is a folded-dipole antenna.[16]

Other common example of two-conductor transmission lines, are coaxial cables that are used to connect home video equipment, in ham radio setups, and in measurement electronics and the consumer cable modem market for broadband Internet access. Micro coaxial cables are used in a range of consumer devices, military equipment, and also in ultra-sound scanning equipment. In electronic engineering laboratories, to view voltage waveforms at circuit nodes, a coaxial cable is often used to connect the node to an oscilloscope input terminal. Fig. 5.43(a) shows a photo of a section of a coaxial cable.

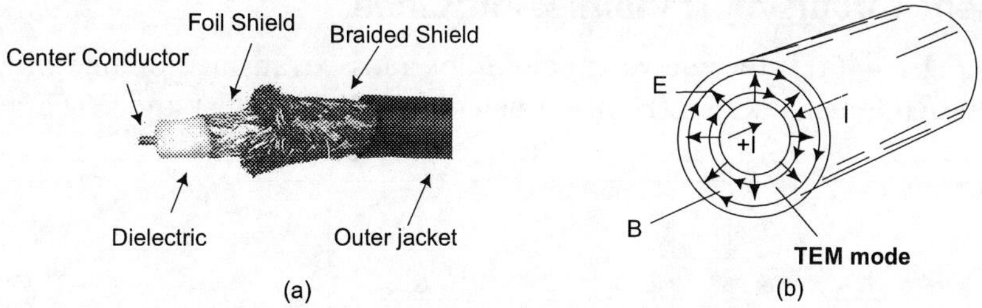

Fig. 5.43 (a) A coaxial cable and its principal parts (b) fields of a coaxial cable Transverse Electromagnetic (TEM) mode (end-view).

The energy, conveyed by the electric and magnetic fields, is guided along the line by confining the electromagnetic wave to regions between the inner and outer conductors of coaxial cables, as shown in Fig. 5.43(b). There is a penetration of the fields in the conductors as well. The mode of propagation is essentially Transverse Electromagnetic (TEM). A current of $+I$ amperes will be carried on the inner conductor and a current of $-I$ amperes on the outer conductor, as shown in Fig. 5.43(b). A flux path surrounding the outer conductor encloses zero net current, and therefore there is no flux (magnetic) external to the outer conductor. It may seem that the coaxial line offers simplicity in supporting perfect TEM mode propagation. However, leakage of the input signal through the shield, remembering that a magnetic field penetrates through matter, can also produce common-mode currents on the outer surface of the shield. The direction of common-mode current on the shield may be verified, by drawing the magnetic field lines B outside the cable and applying the right-hand rule for current in Fig. 5.43(b). The current will be in the same direction as the center conductor, i.e., $+I$. These processes are referred to as differential-mode to common-mode conversion.

For the two-conductor line in Fig. 5.41(c), the common-mode currents on the conductors are equal in magnitude and flow in the same direction. The term common mode is also referred to as antenna (folded dipole) mode and codirectional mode.

Incident and Reflected Waves on Open- and Short-circuited Two-wire Lines

When the field waves strike the end of an open-circuited two-wire conductor line, the magnetic field must become zero, because the current is zero at an open circuit. The energy conveyed by the magnetic field cannot be dissipated, so it will appear as additional energy in the electric field (Faraday's law), causing an increased voltage to appear. The increased voltage, then sets up a returning current wave down the line.

If the line were short-circuited at the receiving end, it is the electric field which is forced to zero by the condition of zero voltage. The transferred energy causes an increase in the magnetic field (see the term Maxwell law in the beginning of this section), which in turn induces a returning voltage wave down the line. Reflections are undesirable, because a portion of the received energy is rejected by the load and further, returning waves result in additional losses (I^2R) in lossy lines. The electromagnetic waves are transmitted along the surface of the conductors, and the portions that penetrate the wires, cause energy losses.

Equivalent Circuit of Transmission Lines

Shown in Fig. 5.44(a) is the equivalent circuit of a lossy transmission line with its distributed parameters R(Ohms/meter), L(Henries/meter), G(Mhos/meter) and C(Farads/meter).

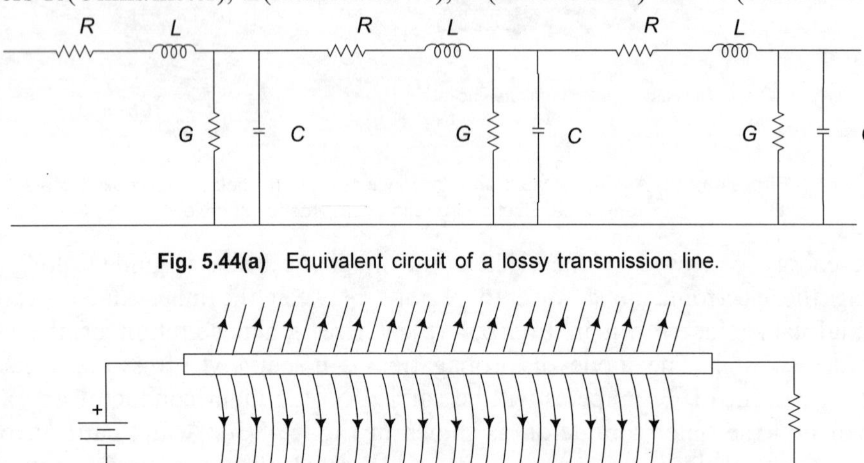

Fig. 5.44(a) Equivalent circuit of a lossy transmission line.

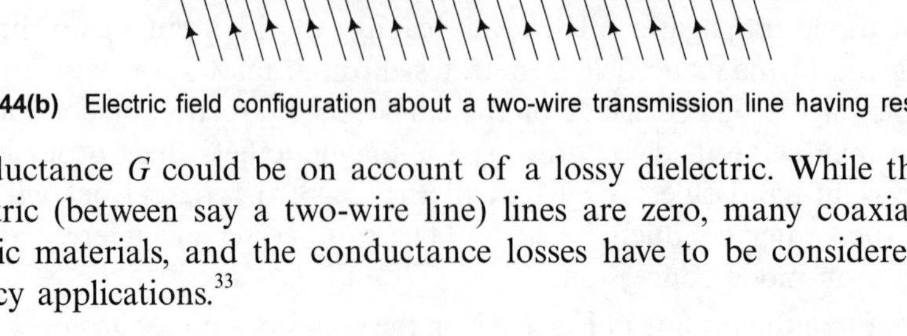

Fig. 5.44(b) Electric field configuration about a two-wire transmission line having resistance.

The conductance G could be on account of a lossy dielectric. While the shunt losses of air dielectric (between say a two-wire line) lines are zero, many coaxial lines employ solid dielectric materials, and the conductance losses have to be considered especially in high frequency applications.[33]

The configuration of the electric field about a two-wire line will appear somewhat as illustrated in Fig. 5.44(b) when there is a resistance drop in the conductors. The curvature, near the surface of the wire, is due to the voltage drop along the wire.

A transmission line is said to be *lossless* if the conductors of the line are perfect and the dielectric medium separating them is lossless, i.e., in the equivalent circuit of Fig. 5.44(a), $R = G = 0$.

If the end of the line is terminated in a resistance, then surface charges of opposite polarity will appear and disappear at either ends of the load resistance corresponding to the time varying voltage of the source. Pulsating electric field lines that were seemingly gliding along the line will, at the end of the line, tilt towards the load resistance, and a pulsating field will appear inside the resistance delivering power to the load resistance. See the "The Poynting Vector", Section 4.4.

What is the "Voltage" of a *Lossless* Transmission Line?

The equivalent circuit representation of a transmission line may be misleading.[16] In the equivalent circuit there exists a voltage drop $L(di/dt)$ along each unit length of line. In the actual line, since E tangential to the surface of a perfect conductor is always zero (the charges arrange themselves in the manner described in Section 1.16), the voltage drop *along* the surface of the line is necessarily zero.

Fig. 5.45 shows the electric and magnetic fields of a perfectly conducting parallel-plane transmission line carrying a high frequency signal.

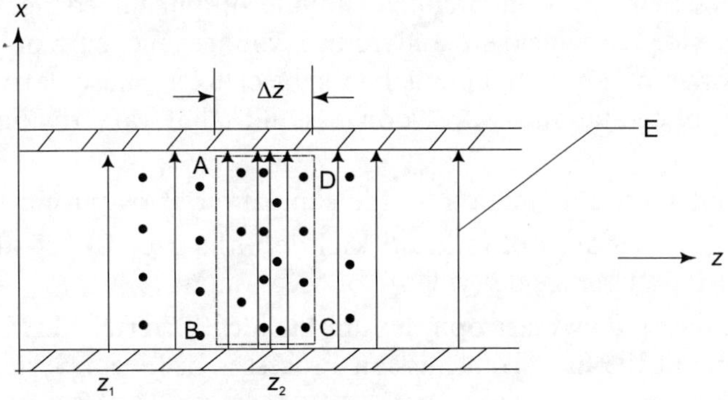

Fig. 5.45 Parallel plane transmission line.

Even if the conductors are imperfect so that an E parallel to the surface of the conductors is possible, the only voltage drop along the line would be that due to the current flow through the surface impedance (comprising inductance per unit length) which is very small, and the drop is ordinarily very small. What the $L(di/dt)$ drop represents in the actual line is, the change per unit length of the *transverse* voltage between conductors. This is equal to the induction voltage $- d\Phi/dt$ around the closed path ABCDA in Fig. 5.45 for the parallel-plane line.

The current flowing into a section of conducting plane at say, z_1 does not generally equal the current flowing out at z_2, seemingly violating Kirchoff's current law (the sum of currents flowing into a node is zero). This inequality exists because any section of parallel planes exhibits capacitance, which conveys a displacement current $\partial E/\partial t$ between the two planes.

Recollect from the descriptions with reference to Fig. 5.36 that the electric field is non-Coulombic (that terminates on the conducting plane) and therefore, we can add the voltage contributions around the loop as $E{\cdot}dl$ and this should be equal to the voltage induced by a changing magnetic field, i.e., H. It so happens that the field terminates on surface charges on the planes, but that does not prevent us from computing the voltage around a closed path $ABCDA$ and equating this to the changing magnetic field (Section 5.12), i.e., $\oint E \cdot dl = -d\phi / dt$ where ϕ is the magnetic flux encircling the closed path. We can write, therefore, $V_{AB} + V_{BC} + V_{CD} + V_{DA} = -j\omega B_y\, a\Delta z$ where, in the sinusoidal steady-state (Section 2.11 and Section 5.17), ω is the frequency in radians of the electromagnetic wave or signal frequency, B_y is the magnetic field strength (in Tesla), 'a' is the separation between the planes and Δz is the distance along the z-direction.

For perfectly conducting planes, the tangential component of E is zero (remember that the surface current J_s which was described with reference to Fig. 5.36(a) is equal to the magnetic field $\vec{J}_S = \hat{n} \times \vec{H}$), therefore, $V_{BC} = V_{DA} = 0$ which leaves $V_{CD} - V_{BA} = -j\omega B_y\, a\Delta z$.

Note that there can be two different voltages between the same pair of planes at different positions z_1 and z_2.

Standing Waves in Transmission Line

Transmission lines that are open-circuited (infinite output impedance) and short-circuited at the receiving end, have incident and return voltage and current waves. The incident and returning waves of an open-circuit line subject to a sinusoidal voltage will result in a standing-wave. Note standing-waves don't "stand", but vary rhythmically in amplitude with time.

To view an interactive animation of the simulation of a transmission line,

i. visit http://www.astrophysik.uni-kiel.de/~hhaertel/TL/TL-doc/index.htm and download the "User guide (video)" or view online

ii. Next, visit http://www.astrophysik.uni-kiel.de/~hhaertel/TL/TL-tutorial/TL-uk.pdf and download the file "Transmission Processes in Linear Systems.pdf"

iii. Then visit http://www.astrophysik.uni-kiel.de/~hhaertel/TL/TL.zip and download the WinRAR zip file "TL"(reproduced here with permission from Hermann Härtel, University Kiel, Germany). Then right click select Extract files choosing the option for the extracted file location. Run the extracted Java file program 'TL' executable Jar file. The documentation file "Transmission Processes in Linear Systems", User-guide video "TL-uk", and the "TL" jar file are available in the "Transmission Line Simulation Program" folder in the CD.

   Java can be downloaded and installed from www.java.com.

iv. Read the descriptions of the progress of the pulses (Section 2.1.3 to 2.1.7) and follow the instructions to interactively set the Java applet (iii) to get a feel for the surface charge movements on a transmission line.

Note that the current is not uniform throughout the line as is observed in simple electric and electronic circuits of small dimensions.

A very nice depiction of Standing waves may be seen in the java applet TL executable Jar file.

In the left bottom panel, click the ▶ button and observe the sinewave voltage and current moving down the line.

The default settings are

Source: The source settings may be viewed in the interactive dialog by clicking the source resistor symbol once. Sinusoidal voltage, freq. = 300 MHz and Source resistance of 50 ohms.

Line: The line settings may be viewed in the interactive dialog by clicking any one of the lines once. Uniform line with R = 0 ohms/m; R(p) (or G in Fig. 5.44 = 0); L = 166.7 nanohenries/m(nH/m); C = 6.7picofarads/m; Z (or characteristic impedance) = 50 ohms.

Load: The load settings may be viewed in the interactive dialog by clicking the load impedance at the end of the lines once. Load type: Resistor; 50 ohms.

Make a note of the amplitude of the voltage and current amplitude of the sinewave on the screen. Now, click one of the lines and in the line parameters, set L = 166000.7 nH/m and observe the amplitude of the voltage wave and the current wave. Note how *large* the amplitude of the voltage wave is in comparison with the amplitude when the parameter L was 166.7 nH/m.

Next, click the "Defaults" button on the Line parameters dialog to restore the line to default values.

Then, select the Load type to open circuit (click the radio button of the open-circuit symbol). Observe the *standing* wave set up along the line. Note that the current is zero at the beginning and the end of the line. The voltage is a maximum at points where the current is a minimum or zero, indicative of a purely reactive nature of the line when it is terminated by an open-circuit. Observe the difference in phase between the current and voltage waveforms. Answer the question "How much is the phase difference between V and I?"

Standing waves shown in textbooks are usually static and hide the dynamics which is brought out richly in these animations.

There are several more combinations of line parameters and loading types that can be studied in the applet of the webpage. For instance, click the source, line and load parameters to default and observe the voltage and current at each of the loads 100Ω, 200Ω, 300Ω, 400Ω, 500Ω and 600Ω. What did you observe happening to the waveforms?

We will discuss one of the other important terminations, the short-circuited (at the receiving end) line. Select the Load type to "short-circuit" (click the radio button of the short-circuit symbol). When the line is terminated by a short-circuit, a standing wave is set up along the line with the voltage zero at the beginning and the end of the line. The current is a maximum at points where the voltage is a minimum or zero, indicative of the purely reactive nature of the line when it is terminated by a short-circuit. Observe the difference in phase between the current and voltage waveforms. Answer the question "How much is the phase difference between V and I?" Observe the dynamics of the standing wave.

Transmission of Radio Frequency Signals Over a Line

While it seemed (Section 2.6) that the electric field propagated incredibly fast in the lumped element RC circuit of Fig. 2.6, the peculiar behaviour of transmission lines is more pronounced with high frequency input voltage signals (remembering speed = frequency × wavelength), because the wavelengths are shorter which we shall see in "Speed of the voltage wave". If the line is assumed to be lossless (infinite conductivity, physically impossible to produce), there will be no attenuation of the input voltage at the output, but due to the line inductance and the capacitance between the two lines, a phase shift between the input and output sinusoids will appear.

Such line effects are very important to be aware of in their design, and if the length of the transmission line significantly alters the input impedance of the line, then the current into the load from the source will be very small and the power transferred to a load at the end of the line will be small, despite the transmission line assumed to be lossless.

At high frequencies, energy from the source is transferred at quicker rates to the circuit as if being hurled into it like a javelin by a javelin thrower!

Will the surface charges appear and disappear fast enough keeping track of the wave dynamics?

The time initial free charges dissipate in, is a characteristic time, and is called the relaxation time. The relaxation time reflects the familiar fact that if you put some free charge on a conductor, it will flow out to the edges.

The relaxation time of charges is now known to be of the order of about 10^{-14} seconds (Ref. [8]), which is roughly the average time between collisions of an electron in the lattice. Assume that some surface charge has appeared on the surface plane in response to the peak (T/4 second) field demand of a sinusoidal electric field. As the field strength reduces after T/4 seconds, the charge should reduce in magnitude. If the frequency of the electric field wave is say, 200 MHz or 200×10^6 cycles per seconds, then its time period would be $\frac{1}{200 \times 10^6} = 5 \times 10^{-9}$ seconds and a quarter of that would be 1.25×10^{-9} seconds. The demand made by the field variations on the surface charge configurations is far slower, than the speed with which the charges can respond, which is denoted by the relaxation time which is a thousand times faster.

We may consider still shorter periods like say, one-tenth the time period and the surface charges would appear and disappear fast enough. So, transmission lines of reasonable length with proper design can handle frequencies upto a few Megahertz but not more than about 400 Megahertz.

It is apparent that at higher frequencies (more than a few hundred Megahertz) the effects of relaxation time ($\tau_{relaxation}$) will begin to assume significance, and the quasi steady-state (quasistatic) conditions which are described by the reactance expressions $\frac{1}{2\pi fC}$ and $2\pi f$L, associated with lumped capacitors and inductors (Sections 2.10 and 5.17) begin to breakdown. Sharp changes in signal magnitudes will definitely not be transmitted satisfactorily.

As a matter of fact, scientists and engineers devised the "wave" models of transmission lines, as a back to basic physical entities of electric and magnetic fields approach, to analyze transmission line behavior at radio frequencies and higher, in which the distributed parameters and the length of the line make their appearance.

Several of the macroscopic models breakdown in the radio frequency range and higher range of frequencies.

In one thought experiment, Edward Purcell (Ref. [22]), imbeds two sheets of charge, a negative sheet and a positive sheet of area A, opposite to one another a distance s apart, in a conductor, specifically in an n-type semiconductor of cross-sectional area A (we will learn about semiconductors in Chapter 9, but for the present it is sufficient to understand that it behaves like a conductor with negative charge carriers, electrons), creating a capacitor of capacitance $\dfrac{A}{4\pi s}$.

What will make these charges disappear? Do negative charges move from the sheet neutralizing the positive charges when they arrive at the positive sheet?

Surely not – if that were the process, the time required would be proportional to the distance between the sheets. Remembering that charges move only infinitesimal distances (Section 1.16), what happens is that the entire population of negative carriers that fills the space between the sheets, is caused to move by the electric field due to the excess charges of the sheets.

Only a very slight displacement of this cloud of charge suffices to remove excess negative charge on the negative sheet, while providing on the positive sheet the extra negative charge needed to neutralize the positive sheet.

Within a conductor, in other words, neutrality is restored by a small readjustment of the entire charge distribution, and not by a few charge carriers moving a long distance. Read the **important note** below Fig. 1.23.

If the semiconductor has a resistivity ρ, then the resistance between the sheets will be $\dfrac{\rho s}{A}$ and the time constant RC (Section 2.7) $= \left(\dfrac{\rho s}{A}\right) \times \dfrac{A}{4\pi s} = \dfrac{\rho}{4\pi}$.

For a semiconductor with resistivity 10^{-5} ohm-cm, $\dfrac{\rho}{4\pi}$ (in CGS units, the unit of resistivity, is seconds and, 100 ohm-cm = 1.11×10^{-10} seconds Ref. [22]) works out to about 10^{-19} seconds, which is several orders of magnitude shorter than the mean free time of a conduction electron, which we found (Section 1.18, "Neutrality in current carrying conductors") to be, $1/10^{14}$ seconds or 10^{-14} seconds. Incidentally, this is a special case of a capacitor (a positive sheet and a negative sheet), in which the (conductive) discharge path is in the short distance between the sheets ! We will come across another special case of a capacitor when we discuss diffusion capacitance in diodes in Section 9.14. For the CGS system of units refer to "Links_to_animations_articles" pdf file in the accompanying CD.

The relaxation process is governed by the conservation of charge, Gauss's law and Ohm's law and the relaxation time derived using these (See http://web.hep.uiuc.edu/home/serrede/P436/Lecture_Notes/P436_Lect_07.pdf and http://ocw.mit.edu/courses/electrical-engineering-and-computer-science/6-013-electromagnetics-and-applications-spring-2009/readings/MIT6_013S09_notes.pdf), is given by the formula $\frac{\varepsilon}{\sigma_C}$ (SI units) where, ε is the permittivity of the medium (conductor, if the medium is a conductor), and σ_c is the conductivity of the medium (conductor, if the medium is a conductor). It is interesting to note that the expression of the relaxation time $\frac{\varepsilon}{\sigma_C}$, when converted to CGS units (Ref. [8]) by substituting $\frac{1}{4\pi}$ for ε and ρ for $\frac{1}{\sigma}$ gives, $\frac{\rho}{4\pi}$ which is the expression in Purcell's derivation that was developed using the RC time constant expression.

The relaxation time ($\tau_{relaxation}$) for copper calculated using the formula is, 4.5×10^{-19} seconds. As a relaxation time, 10^{-19} seconds makes no sense. Purcell adds that present day theories do not shed much light on events on a time scale as short as that. That is why, the relaxation time can be independent of the size of the system.

The experimentally measured relaxation time in pure copper is 4.0×10^{-14} seconds which is ~5 orders of magnitude larger than the calculated charge relaxation time, 4.5×10^{-19} seconds. The problem here is that (the macroscopic) Ohm's law is simply out of its range of validity on such short time scales. Therefore, it makes more sense to consider the longer of the two times, the relaxation time and the mean time between collisions, i.e., 10^{-14} seconds as the relaxation time, which we did in Section 1.16.

Two additional facts here are that both, ε and σ_c are frequency dependent quantities, that become increasingly important at the higher frequencies $f = \frac{2\pi}{\omega} \sim \frac{1}{\tau_{relaxation}}$ associated with short-time scale, transient-type phenomena. So, in reality, if we are willing to wait a short time $\Delta t \sim 1$ *picosecond* $= 10^{-12}$ seconds, then any free charge density accumulated inside a good conductor will have dissipated away/damped out, and from that time onwards, zero free charge density can be safely assumed. Visit http://web.hep.uiuc.edu/home/serrede/P436/Lecture_Notes/P436_Lect_07.pdf for more details.

In any case, engineers prefer to operate two-wire transmission lines at frequencies not exceeding about 400 MHz, because then the charges which appear due to the rising portion of say, a sinusoidal signal and which should disappear as a result of the signal reducing in magnitude will faithfully transmit the signal variations. This becomes especially important with signals consisting of a range of frequencies since the phase delays may be different for different frequency components and will result in signal distortion as we shall see in "Operating frequencies and propagation delay in circuits" later in this section.

What about the amount the charges have to shift, to produce the surface charge?

In Section 1.25 and in Appendix A "Amount of the Electron Cloud Shift...", we had seen that the distance the electrons have to shift to produce a significant amount of surface charge, is of a fraction of 10^{-15} meters.

The extension from the lumped (inductors, capacitors and resistors) regime to the distributed regime (transmission line inductance, capacitance and resistance) is treated in a very intuitive manner in Chapter 2 "Planar Microwave Engineering" by Thomas Lee, Cambridge University Press, 2004 in which we learn that one of the remarkable properties of distributed systems (characterized by a transmission line), is that the tiny inductance and tiny capacitance do not cause a bandwidth (range of frequencies a line can transmit) reduction, but result in a *propagation delay*.

In a circuit comprising lumped elements *R*, *L* and *C*, the surface charges are set up so quickly, by the near speed of light propagation of changes (See Section 2.6) in the electric field at signal frequencies that the voltage measured at nodes say, at *any* point on connecting wires is uniform. The current is also mostly uniform at any cross-section along the connecting wire.

However, this may not be the case with transmission lines operating at high frequencies, when the voltage (remembering voltage is due to the accumulation of surface charges) and current are found to vary along the length of a lossless transmission line. In the link to Java applets http://www.astrophysik.uni-kiel.de/~hhaertel/TL/TL.zip and documentation http://www.astrophysik.uni-kiel.de/~hhaertel/TL/TL-tutorial/TL-uk.pdf see Section 2.1.4 "Displacement of a voltage pulse on a double line" and Section 2.1.5 "Displacement on a zone with drifting electrons on a double line".

Operating Frequencies of Lines and Propagation Delay in Circuits

Transmission lines are used primarily to transfer energy from one point to another. At frequencies above 150 MHz upto about 3000 MHz, quarter-wave long lengths of lines are used as circuit elements because at these frequencies ordinary lumped-circuit elements are difficult to construct.[16]

The validity of the lumped element model holds, when a circuit fulfills design expectations, ignoring the finite time it takes signals, to propagate around a circuit, as we shall see later in "Speed of the voltage wave". In good circuit layouts, usually, the characteristic length of a circuit is much smaller than a tenth of the circuit's operating wavelength.

The operating (free-space) wavelength λ is given by the formula $c = f\lambda$, where c is the speed of light in vacuum $\approx 3 \times 10^8$ meters/second. At 1 MHz, the free-space wavelength is 300 meters. At 150 MHz, the free-space wavelength is 2 meters and most electronic circuits found in electronic equipment have an operating length not exceeding 2 meters definitely.

Whenever the propagation time of a signal around the circuit is not significant to the application, lumped elements can be used to construct the circuit. This is the case when the propagation time is much less than the period of the signal involved. For example, a signal of 100 MHz would have a time period of T = $1/100\times10^6$ seconds = 0.01 microseconds or 10 nanoseconds. If a signal were to propagate between the two farthest points, counting twists and turns, in the fabricated circuit in less than that time, the lumped element model is valid. So, for the circuit of operating length of 2 meters, the 100 MHz signal would take approximately (experimental measurements indicate it takes about 1 nanosecond for a signal to travel down a wire of 1 meter), 2 nanoseconds which is $1/5^{th}$ the time period, the lumped model would be appropriate.

However, with increasing propagation time there will be an increasing error between the assumed and actual phase of the signal which in turn results in an error in the assumed amplitude of the signal.

For example, a circuit operating with a 3000 MHz signal, whose free-space wavelength is 0.1 meters and time period $1/3000 \times10^6$ seconds = 0.00033 microseconds or 0.33 nanoseconds, the delay of 2 nanoseconds of a 2 meter operating length circuit will certainly not be acceptable.

There is a nice video about wires in circuits here, https://www.youtube.com/watch?v=O_up46ic3Do&index=8&list=PLYvv55hF1HIk3VZptrkXNhl6a3nkcwC-7.

Lumped elements at such high frequencies show up parasitic effects that can hardly be avoided, and these effects begin to impact circuit performance adversely.

However, circuit designers recognized that with distributed parameters applied to small element circuit analysis, the voltage and current effects of small sizes of transmission lines can be exploited to advantage finding use as circuit elements.

A detailed treatment of the voltage and current behavior of small lines operated at high frequencies is beyond the scope of this textbook.

Lenz's Law Applied to Open- and Short-Circuited Lines—The Quarter Wave Line

We will now demonstrate the application of Lenz's law to determine the effects of small, finite length open- and short-circuited lines, on voltages and currents. Specifically, we will treat the effect a quarter wavelength line has, on voltages and currents in it. The quarter-wave length long line shorted at one end, has special significance since it approximates a parallel resonant circuit (see the topic "UHF lines as circuit elements" in the chapter "Guided Waves" in Ref. [16]).

Figure 5.46(a) shows a two-wire transmission line with one end open- and the other end short-circuited.

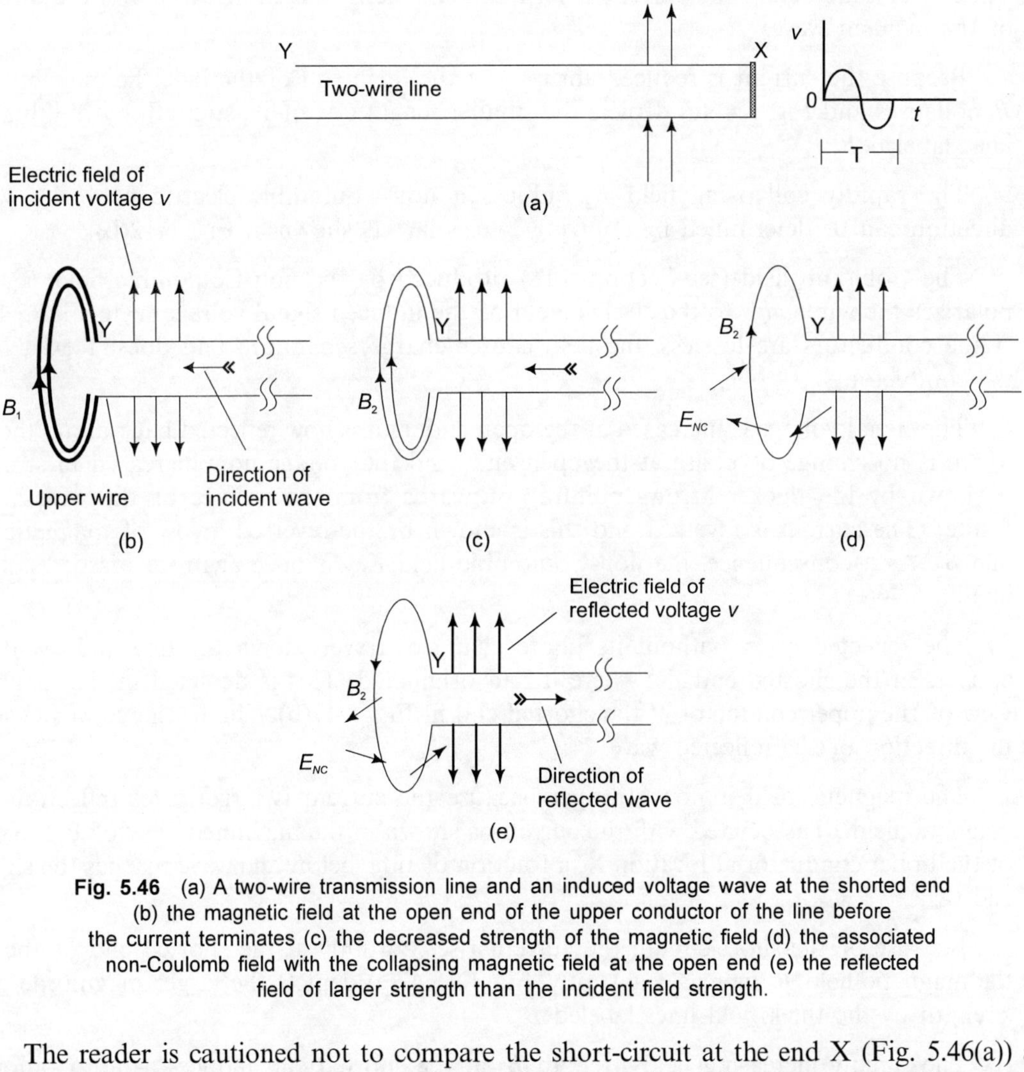

Fig. 5.46 (a) A two-wire transmission line and an induced voltage wave at the shorted end (b) the magnetic field at the open end of the upper conductor of the line before the current terminates (c) the decreased strength of the magnetic field (d) the associated non-Coulomb field with the collapsing magnetic field at the open end (e) the reflected field of larger strength than the incident field strength.

The reader is cautioned not to compare the short-circuit at the end X (Fig. 5.46(a)) as if it were a short piece of wire on a simple electric or electronic circuit, where the voltage should be 'zero' across the short, and field lines cannot be as shown! Figure 5.46(a) is of a transmission line of length comparable to the wavelength λ of an arbitrary signal being transmitted.

If a voltage is induced near the short-circuited end marked 'X' in Fig.5.46(a), there will be a voltage wave sent down the line shown by the field arrows of the incident voltage v. At the open end 'Y', the surface charges on the line cannot appear, and their flow or current is therefore terminated abruptly. The magnetic field B_1 associated with the current is shown in the magnified view of Fig. 5.46(b) of the upper conductor at location Y, a fraction of time before the current is terminated. The feathered arrow shows the direction of the incident wave.

Because the current is reduced abruptly at the open end of the line, the magnetic field B_1 collapses and Fig. 5.46(c) depicts the smaller magnitude of its strength by the thin field lines labeled B_2.

The rapidly collapsing field B_2 induces a non-Coulombic electric field E_{NC} whose direction can be determined by applying Lenz's law as shown in Fig. 5.12(b).

The Coulomb field (see Section 5.16) produced by the non-Coulombic field E_{NC} that polarizes the wire, *adds* to the electric field of the induced signal voltage '*v*' in Fig. 5.46(d). If the conductors are lossless, the field causes charge separation and doesn't exist inside the conductor.

The signal voltage is increased at the open end, and is now reflected back along the line. There is no change of phase at the open end, and the voltage now increased in strength, is shown by the thicker arrows pointing outwards from the surface of the wire in Fig. 5.46(e). The current is reversed and this is shown by the reversed arrow of magnetic field line B_2. As a consequence, the non-Coulombic field E_{NC} is once again reversed, as shown in Fig. 5.46(e).

The reflected wave, without its phase changed, travels down the line and eventually appears at the shorted end 'X' where it had originated. This is depicted in the magnified view of the upper conductor at the shorted end in Fig. 5.47(b). The feathered arrow shows the direction of the reflected wave.

The magnetic field is now stronger because the current is larger after reflection. The magnetic field B_3 associated with the current is shown in the magnified view of Fig. 5.47(b) of the upper conductor at location X, a fraction of time before the wave reaches the shorted end 'X'.

Because of the short circuit, the current rises abruptly at the shorted end of the line, the magnetic field B_3 increases abruptly. Figure 5.47(c) depicts the larger magnitude of its strength by the thick field lines labeled B_4.

The rapidly increasing field strength B_4 at the shorted end induces a non-Coulombic electric field E_{NC}, whose direction can be determined by applying Lenz's law as shown in Fig. 5.12(c).

The Coulomb field (Section 5.16) produced by the non-Coulombic field E_{NC} that polarizes the wire, opposes the electric field of the reflected "enhanced at the open end" voltage '*v*' in Fig. 5.47(d). If the conductors are lossless, the field causes charge separation and doesn't exist inside the conductor.

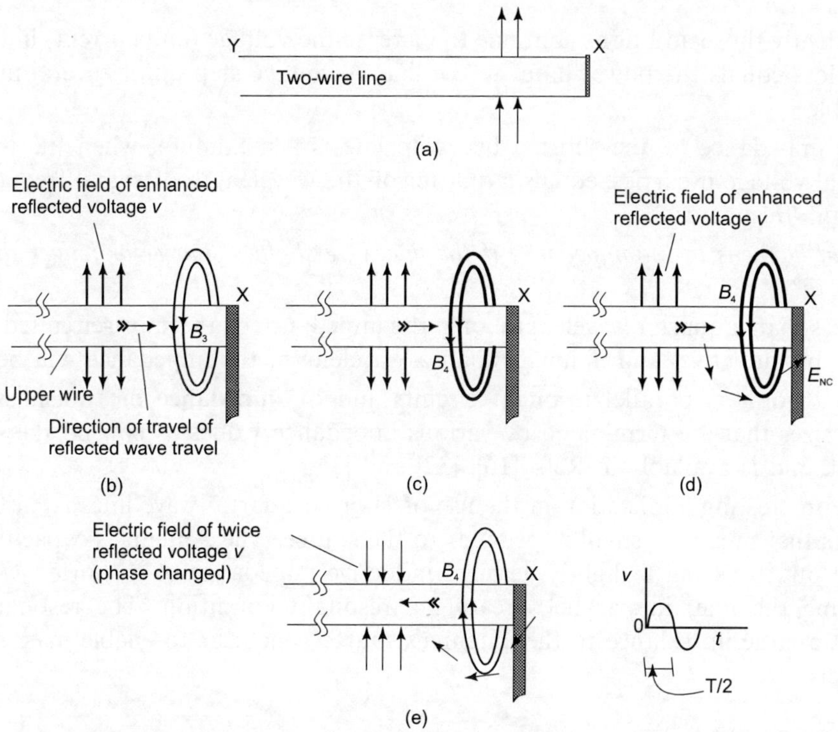

Fig. 5.47 (a) A two-wire transmission line and the reflected (enhanced) voltage wave at the shorted end (b) the magnetic field at the shorted end of the upper conductor of the line before the wave reaches the shorted end (c) the increasing magnetic field strength at the shorted end (d) the increasing magnetic field strength and its associated non-Coulomb field (e) the change in phase by the Coulomb field produced by polarization of the wire by the non-Coulomb field.

At the shorted end, the reflected wave is once again reflected, but there is a change of phase, a reversal as a matter of fact, because the powerful E_{NC} and the accompanying Coulomb field (not shown), will cause an accumulation of negative charge at the shorted end. The magnetic field B_4 is also reversed as seen in Fig. 5.47(e) causing the non-Coulombic field E_{NC} to change sense, polarize the wire and produce a coulomb field which adds to the (original) negative induced source voltage. The electric field of the wire is shown by thin arrows pointing inwards on the surface of the wire in Fig. 5.47(e).

Suppose that the line is $\lambda/4$ meters long, where λ is the wavelength of the induced sinusoidal voltage whose frequency is f, λ being given by the formula $c = f\lambda$, where c is the free space speed of light.[16]

In such a case, the wave would have traveled exactly the distance $\lambda/4$ meters and returned to the shorted end in the time $T/2$ seconds (see graph in Fig. 5.46), where $T = 1/f$ is the time period of the sinewave.

Note, that the twice reflected wave will be in phase with the original induced voltage and so, adds directly to it.

Evidently those additions continue to increase the voltage (and current) in the line until, the I^2R loss equals the power input to the line. A voltage step up of several hundred times is possible.

The impedance of the shorted two wire line is a maximum, when the wavelength of the signal voltage it carries, equals a quarter of the wavelength of signal frequency because the circuit "resonates".

What happens to the impedance if the line were slightly shorter or longer than a quarter wavelength?

If less than a quarter wavelength long, the impedance of the short-circuited $\lambda/4$ two wire line will be inductive, and if longer than a wavelength, the impedance will be capacitive.

The theory of parallel resonant circuits and of impedance of transmission lines of various sizes that are terminated by various impedance values, is beyond the scope of this textbook and is available in Refs. [16], [32] and [33].

One interesting application in the use of shorted quarter-wave lines is, in the design of antennas that present a small impedance to the source. The Antenna's capacitive reactance is tuned out, by using a slightly detuned inductive impedance of a shorted $\lambda/4$ "stub" (by shortening the line) in parallel, creating a resonant condition. The resonant condition presents a large impedance to the antenna's source generator to enable maximum transfer of power.

Skin Effect

A transmission line operating at high (radio) frequency, carries less current in the interior of the conductor than the outer regions, and this phenomenon is known as *skin effect.* The reader is cautioned *not* to generalize this and conclude that conduction in simple circuits is confined to the *skin* of conductors. The *skin effect* is only a high frequency phenomenon and we have seen in Section 1.12 that the electric field *fills* the wire in simple circuits operating at dc to moderate frequencies, and so, the current *fills* the wire. For a qualitative treatment of the *skin effect*, which is due to unequal impedances due to a larger inductance at the center of the conductor in comparison to the inductance away from the center, see "Networks, Lines and Fields".[33] Also see "Skin Effect" in Appendix A.

Transmission Line Model of a Lumped Capacitor

In Section 2.7 on the charging of capacitors, we had depicted a growing electric field across every elemental area of the plates simultaneously implying that the current entering at one point and instantly distributing itself across the whole capacitor plate. This may not be necessarily happening in the manner shown and instead an alternate model in which the current spreads out over tiny pie-shaped wedges producing a staircase type charging waveform has been proposed by engineers. Details are provided in the article "Displacement Current" link to which is available in the Links_to_animations_articles file and in Härtel's guide to visualizations in the "Teacher's Guide" in the CD.

Speed of the Voltage Wave

The *speed* of a voltage wave in a transmission line is less than the speed of light in free space, and it can be sometimes as slow as 0.5 times the free space speed of light. The speed of light in vacuum "is the same for all inertial observers, regardless of the motion of the source" [Ref. 8], therefore, the question of the speed of the wave exceeding that of light can never arise. The slowness of the wave in propagation in transmission lines, is due to the distributed inductances and capacitances of transmission lines which prevent the current and the voltage to rise and fall at every point along its length in tune with say, an applied sinusoidal voltage of arbitrary frequency.[32] It may be noted that the speed is *not reduced* by reradiation or scattering (Section 4.7 and Section 10.4), but reduces primarily due to the distributed inductance and capacitance which cause a phase change of signals *along* the line impacting their *wavelength* and the *wave speed*. The frequency is never affected; it is the "beat" of the driving signal.

Recollect from Section 4.2 that a wiggle in *space* is a wave; a wave is a function of both time *and* space, and the transmission line and its surrounding space is the space or medium through which the fields of the signal voltage across the terminals of the signal generator propagate as a *wave*. Therefore, there is a variation of an assumed sinusoidal voltage with both time *and* distance (or space variable) along the line. We were not concerned about *distance* in circuits built using lumped elements.

With long lines the story is different. The phase *difference* as the wave travels along the line, is given by a constant (determined by the line parameters) times the distance of a point along the line, where the wave speed is being measured. This constant determines the amount by which the wave speed is reduced. Look at it this way, the source signal's electric field (sinusoidal time-varying) and the line inductance's Faraday's law induced (opposing) electric field superimpose so, a smaller time-varying field results, and the capacitance needs time to charge (with smaller doses of surface charge) and be registered as a voltage, all this at every elemental length of the line. Extend this process along the length of the line along zillions of elemental lengths, and it is easy to see how a voltage wave with *smaller* than speed of light wave speed is produced even as, changes in electric field, travel at the speed of light.

The wavelength will be shorter along the line than that of an electromagnetic wave in free space with the same frequency, because for each cycle of the voltage, the associated field *wave* traverses a smaller distance, the process repeating for subsequent cycles with a constant smaller wavelength along the line. The speed of the wave equals frequency (of the signal voltage, which is say, fixed) times the wavelength and will be smaller than that of a corresponding free-space wave of the same frequency. The frequency of the line is not affected being the driving signal frequency. See Ref. [32] for graphs of wave propagation.

A two-wire line of nonmagnetic material with air spacing is shown in Fig. 5.48.

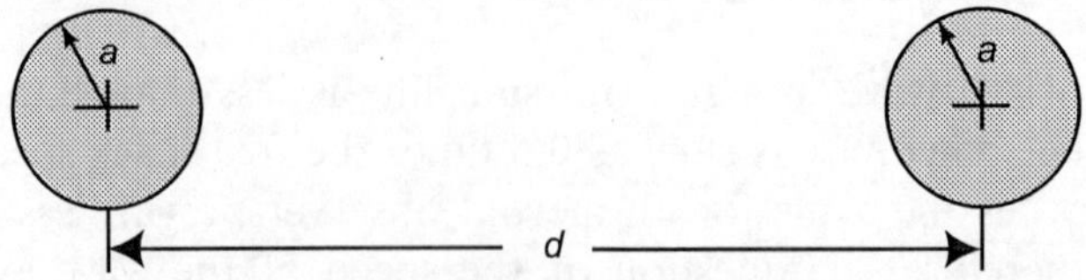

Fig. 5.48 Two-wire transmission line.

The distance between the two wires each of radius '*a*' is '*d*', the wave speed (Ref. [33]) '*v*' is given by

$$v = \frac{1}{\sqrt{LC}}, \text{ where } L \text{ and } C \text{ are line constants (distr. Parameters)} \tag{1}$$

and

$$LC = \mu_o \varepsilon_o \left[\frac{1}{4 \ln \frac{d}{a}} + 1 \right] \tag{2}$$

where, μ_o is the permeability of free space and ε_o is the permittivity of free space. Substituting for LC in (1), we obtain the speed of the wave, v

$$\text{v} = \frac{1}{\sqrt{\mu_o \varepsilon_o \left[\frac{1}{4 \ln \frac{d}{a}} + 1 \right]}} \text{ m/sec} \tag{3}$$

The first term on the right in Eq. (2) and the first term in the bracket under the radical of Eq. (3) are present because of the *internal* inductance of the wires[33] (where *internal* inductance is due to the flux inside the conductor from the center to the radius a, and external inductance is due to the flux in the space outside the conductor). The expression for the speed of light or of any electromagnetic wave[32] in free space is

$$v_{free\ space} = \frac{1}{\sqrt{\mu_o \varepsilon_o}} \tag{4}$$

and if the internal inductance is reduced by skin effect, the wave speed v can be increased as evident from Eq. (3), to reach the limiting condition of $\frac{1}{\sqrt{\mu_o \varepsilon_o}}$, which is the speed of light in free space. See Section 10.2 on "Units of the constant $\frac{1}{\sqrt{\mu_o \varepsilon_o}}$. If the limiting condition is achieved,[33] the electric and the magnetic fields will seemingly glide on the surface of the conductors without penetration and therefore, will travel at the speed of light.

In practical transmission lines, where the wave speed does reduce due to the inductance and capacitance, the visualization of the marching band slowing down in the muddy portion of the field described in Section 10.5 is appropriate but, note that in the transmission line this is due to the distributed inductance and capacitance and not due to reradiation.

Transmission Line Effects on p-c Boards

In the design of p-c boards with interconnections of digital integrated circuits, high frequency effects show up, which can be analyzed using transmission line principles. For an informative article on this visit http://www.edn.com/design/analog/4340135/As-edge-speeds-increase-wires-become-transmission-lines or http://ecee.colorado.edu/~mcclurel/EDN_Transmission_Lines.pdf.

See also "Links_to_animations_articles" in the CD accompanying this book for a link to an article on the use of low *k* dielectrics in integrated circuits.

Visit http://en.wikipedia.org/wiki/Transmission_line#Short to view animations of currents and voltages of receiving end open and short-circuited transmission lines.

A description of high frequency considerations is available at http://www.antenna-theory.com/tutorial/txline/transmissionline.php#txline.

Special Applications of the Transmission Line

Before proceeding ahead, it will be instructive to refer to Section 2.13 on the R-C combination, Section 5.19 (Fig. 5.32) on the L-R combination and their use as low-pass filters. The response of a low pass RC circuit to sinusoidal inputs is described in Ref. [34] which is also useful to read.

How Lossless Lines also Produce Attenuation or Amplification

Lumped pure reactance (L and C) T-section and m-derived section filters attenuate signals[33] *not by* dissipative losses in resistances (assuming zero resistance) and may be attributed to the last terminating impedance, which will have a large or small impedance depending on frequency. This will produce either an attenuated or a raised magnitude of the output voltage for an applied input sinusoidal voltage.

Lossless transmission lines, assuming zero distributed resistance and zero distributed conductance like the transmission line of Fig. 5.40 and Fig. 5.44, have a series inductance (of the order of nanohenries per meter) and a parallel capacitance (of the order of picofarads per meter) for each incremental length. It is the impedance of the entire line, usually employed in $\frac{\lambda}{4}, \frac{\lambda}{2}, \lambda, 2\lambda, \ldots$ lengths (λ = wavelength of transmission) and terminated by appropriate impedance, which determines whether output signals will be attenuated, or raised in magnitude (amplification).

Transmission lines of appropriate length may be used at high frequencies (tens to hundreds of megahertz), as efficient impedance matching sections and as reactance elements in wave filters that attenuate signals in specified frequency ranges. They may be employed as *resonant circuits* and may be analyzed and proportioned to achieve desired frequency selective properties.[33]

There is a good description of a *lumped* element "L-Section Filter" comprising a series inductor and capacitor in the chapter *Applications of Diode* in Ref. [17], a reading of which will give a nice perspective view on the effects of distributed inductances and capacitances in transmission lines.

One other classic application of a section of transmission line that exploits the phenomenon of the reduction of wave speeds in lines is, as a *delay line* that delays an input signal. Such delay lines are used in oscilloscopes which are used to observe fast waveforms, when the input signal that triggers the sweep is delayed slightly, before being applied to the vertical-deflection circuit. A transmission line can also be used to generate a voltage *pulse* from a step voltage![34] See "Links_to_animations_articles" pdf file in the accompanying CD for more animations.

Attenuation

The term "attenuation" is used in electrical and electronic engineering, in the most general sense to denote a decrease of the amplitude, or magnitude of coherent or incoherent electromagnetic waves, or electrical impulses without specifying, (1) what quantity shall be used to measure the decrease (2) whether the decrease is space-dependent, time-dependent or both, (3) the cause of the decrease or, (4) the conditions under which the decrease is to be measured. See http://www.nist.gov/calibrations/upload/mn137.pdf.

Consider the series resistor combination which forms an attenuator shown in Fig. 5.49.

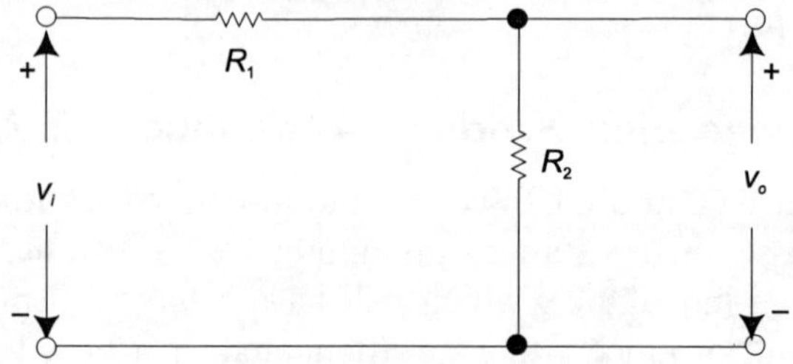

Fig. 5.49 A simple resistive attenuator.

The resistive attenuator multiplies the input signal by the ratio $a = v_o/v_i = R_2/(R_1 + R_2)$ which will be always less than 1, where v_o is the output signal. The attenuator *reduces* the magnitude of voltage of signals of *any* frequency.

When the signal voltage v_i is applied to the circuit, surface charges are set up which are not shown in the circuit. A steady-state is reached, and the output voltage v_o is observed to be attenuated.

For the low-pass filter (Section 2.13), the amplitude of the output is smaller than the input at high frequencies and it may be said that the voltage of the signal is "attenuated" at high frequencies.

The circuit of Fig. 5.50(a) passes low frequencies readily, but attenuates high frequencies because the reactance of the capacitor C decreases with increasing frequency.[34]

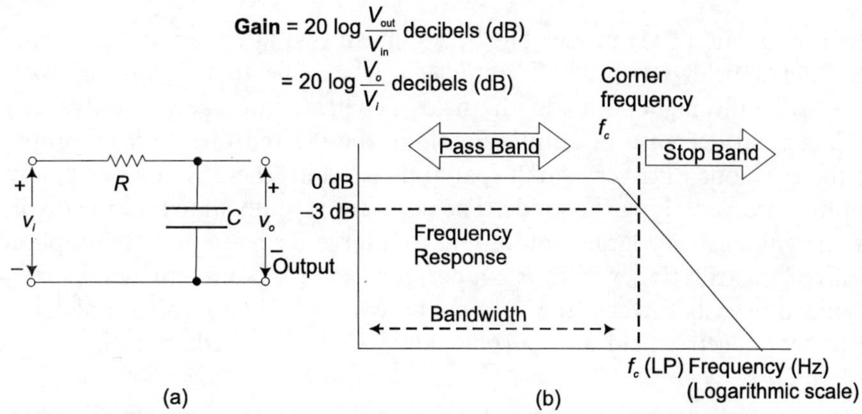

Fig. 5.50 (a) A low pass circuit (b) Plot of Gain vs. frequency of the RC low pass circuit.

See Section 2.10 for how the reactance of the capacitor decreases as the frequency is increased. The attenuation was effected by the small fringe field which develops across the capacitor at high frequencies because at higher frequencies, charges cannot "stick" to the plates and develop the output signal voltage v_o. The quantity attenuated, is the signal voltage v_i.

The attenuation is indicated as Gain below 0 decibels (dB) in Fig. 5.50(b) which are negative values of decibels (dB), where the 0 dB level refers to a sinusoidal signal in the low frequency range passed without attenuation.[34] See also Section 2.11 on the reactance of a capacitor.

Attenuation of a Wave

It is possible to decrease the energy of a wave by placing dissipative, reflective, or scattering objects in its path. We will learn about the phenomenon of scattering in Section 10.4.

In the case of an electromagnetic wave radiated by an antenna (which will be discussed in Chapter 10), and traveling over the earth (ground wave), the wave amplitude (electric field or magnetic field) decreases as one goes farther from the antenna due to spreading out of the radiation ----inverse distance effect. Here the quantities under consideration are the strength of the electric and magnetic fields, and the decrease is space-dependent. We will learn more about this in Section 10.10.

A space dependent decrease might also be caused by a lossy transmission path observed by detectors located at different points along the path.

A rectangular waveguide shown in Chapter 10, Figs. 10.38 and 10.39 is used to *guide* electromagnetic *waves* and these are confined inside the waveguide due to reflection from the waveguide walls.

Waveguides propagate electromagnetic waves by guiding them along rectangular or circular cross-sectional conducting plates and cylinders with silver, gold or copper plating on the inner walls to reduce dissipative losses.

Some authors[16] prefer to use the term "attenuation" to the decreasing strengths of the field components in waveguides that operate in either Transverse Electric (TE) or

Transverse Magnetic (TM) mode. They distinguish two attenuation constants. The first of these, is defined below the cut-off frequency, when due to the *standing*-waves produced by the electric and magnetic fields, the fields penetrate into the guide (walls) producing a sharp exponential decrease in amplitude along the desired direction of propagation. But, because there is none or a very small spatial phase-shift, there is none or barely perceptible propagation observed along the guide. The constant is quite large and the dissipation in the walls of the guide is very large, amounting to a large decrease in wave amplitude. A graph of the *current* distribution within a copper conductor at different depths, as a 100 MHz wave is guided over its surface (the currents have a vertical and horizontal component), can be seen under "Electric field and current flow within the conductor" in Chapter "Guided Waves" Ref. [16].

The second attenuation constant *above* the cut-off frequency is small. Above the cutoff frequency, a wave *with* a spatial phase-shift propagates along the guide with an exponential decay of the amplitude but now, the nodal points of the electric fields occur on the walls. Thus, the stronger (antinodes) field components are in the waveguide. Additionally, the attenuation constant is small because the depth of penetration of the field is inversely proportional to the square root of conductivity as well as the square root of the frequency so, at higher frequencies the strengths of the reflected fields is more while keeping the associated surface currents small.

See also Section 7.14 on "Microphones" on how engineers produce high frequency, high power signals for satellite communication.

5.23 NEUTRAL ATOMS

What is *fascinating* about an atom is how it remains "neutral" despite it being a *collection of sources* of powerful electric and magnetic fields. The electrons at very high *speeds* (the velocity of the electron of a hydrogen atom is about 2.19×10^6 meters/second and is yet only about 1/137th the speed of light), orbit around the central core (in the classical view), so perfectly balanced that the "*net*" electric field is very close to *zero* outside the atom. Because of this astounding assembly of sub-atomic particles....for all time.....the e-field **never becomes** significantly non-zero even for the tiniest fraction of time.....the cancellation of the fields outside is ever so precise.

If it were not so, we would be living in a "*shocking*" world ! and.....

...you would be shocked by anything you touch !

It is this "*dance*" (that I would like to imagine) that Neils Bohr and Schrodinger and Fermi have tried to encapsulate in their theories, equations and solutions using *quantum wells* and *potential wells*.

Little wonder then that *seeing atoms* has been a fascination for all....laymen, engineers and scientists !

But, alas *we can probably never see this dance* and probably never ever can, as described in the article in the link http://web.archive.org/web/20170107155922/http://www.americanscientist.org/issues/pub/capturing-quantum-corrals

Scientists doubt that we'd ever be able to observe an atom, or anything else smaller than a few hundred nanometers, because of the limitations of the optical microscope. This microscope uses light to make an image, so its performance is constrained by the nature of light.

Light behaves as a wave, and for an optical microscope to form a clear image of a sample, *the sample has to be larger than a wavelength of light* (the distance between successive wave crests or other identical parts of a wave).

The wavelengths of visible light range from about 700 nanometers to about 400 nanometers. But atoms are a mere 0.1 *nanometers to 0.5 nanometers in diameter*.

Seeing atoms

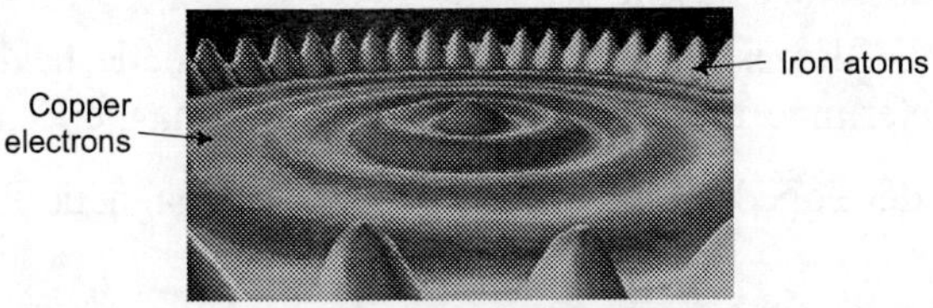

This intriguing nanoscale image, called the Quantum Corral, was created by Donald Eigler, Crommie, and Christopher Lutz in 1993 at IBM's Almaden Research Center. The team first used a scanning tunneling microscope (STM) to move iron atoms (which you see as turquoise mounds) one by one to form a corral on a copper surface. Inside the corral, trapped copper electrons, which ordinarily would move freely, form standing waves—waves that appear to be standing still.

After the iron atoms were in place, the team used the STM to scan back and forth over the sample to collect information about the contours of the surface, and the computer displayed the information as a black-and-white topographic map.

An STM image of Platinum is available in Section *Direct Imaging of Atomic Structure* in Chapter "Crystal Structure" and of the quantum corral in Section *Scanning Tunneling Microscopy* in Chapter "Surface and Interface Physics" in Introduction to Solid State Physics by Charles Kittel, Publ. Wiley, 7th ed. Also see Links_to_animations_articles in the CD for more links to interesting articles on the quantum corral.

SUMMARY

In the beginning of this chapter, we learned the fundamental properties of the magnetic field, which we stated, is an alteration of space surrounding a moving charge, and its interactions. The interactions are the repulsion of like poles and the attraction of unlike poles. Like the electric field, a magnetic field obeys the superposition principle. One other important property of the magnetic field we observed, is that it penetrates through matter. The magnetic field is usually depicted using curly patterned magnetic field lines, noting that magnetic field lines which are divergent, have never been observed.

One of the important aspects of a magnetic field is that it is associated solely with moving charges (currents in circuits, for example), an observation which has far reaching implications. After studying the effects current-carrying wires produce on compasses, we learned that its direction is given by the right-hand rule, and its strength by the Biot-Savart law, using which we made the observation that the field at a distance *d*, from a current-carrying wire drops away at the rate of $1/d^2$. We learned about the magnetic field strength H and the magneto-motive force (MMF) and derived expressions for them. Then, we derived expressions for the magnetic field due to other current distributions; a circular loop of wire, and a coil of many loops. The magnetic field configuration due to current distributions is best depicted using vectors (arrows) of appropriate lengths, which represent the magnitude and point in the direction of the magnetic field at points in space. We then learned the use of the right-hand rule to obtain the direction of the magnetic field B at a point away from a current carrying wire.

Since the magnetic flux is an aggregation of the magnetic field over a surface, and is a useful quantity to determine the electric field, when the magnetic field varies with time in a region of space, we developed a simple formula for the magnetic flux, $\phi_{magnetic} = \int_{surface} \vec{B}.\vec{n}dA$ to compute it.

While it may be said that an electric field is produced by charges, in reality, it simply exists with the charge. We studied another way to produce an electric field; by time-varying magnetic fields or Faraday's law. While the field of a charge is divergent, the field associated (we discarded the term "produced") with a time-varying magnetic field is, curly patterned. This was demonstrated by changing the current in a solenoid (a tightly wound coil of several turns of conducting wire) and noting that a curly patterned field appears even as the current is being changed. We called the curly electric field, a co-occurring field with a time-varying magnetic field.

We learned that the curly electric field shared the same properties as the electric field of a charge such as, exerting a force on charged particles and penetrating through matter. We then made the important observation that the co-occurring curly electric field is non-conservative, and that it cannot be associated with a potential since a round trip path of its integral is not "zero".

With examples of voltmeter probes forming loops surrounding and outside a solenoid as illustrations, we made the observation that it is the magnetic flux encircled by the circuit and not the flux on a closed surface which counts for the induced emf due to Faraday's law. Next, we learned the application of Lenz's rule to determine the direction of the associated non-Coulomb field, an important exercise since this aspect will help in understanding the behavior of inductive circuits.

We then learned that while the units of emf and potential difference (p.d.) are the same, viz. volts, there is a difference between the two; p.d is the path integral of the electric field made by real charges while, emf is the non-Coulomb work per unit charge as in the case of chemical batteries and the hypothetical mechanical battery, we devised in Chapter 1.

There is a subtlety involved in the production of an induced emf by a curly electric field associated with a time-varying magnetic field. The non-Coulombic curly electric field associated with a current change in an inductor or a solenoid producing a time-varying magnetic field, polarizes the conductor. Surface charges appear along and these produce a *Coulomb* field in the inductor wire, which alters the original surface charge configuration due to the source. It is the path integral of this Coulomb field which bears a resemblance to a battery which opposes the emf of a source, which is designated the *induced emf.*

The study of an inductor subject to a sinusoidal voltage involved determining the direction and strength of the induced emf, and it was observed that the current lags the voltage applied, which maybe intuitively understood as a sort of frictional force of the induced emf (acting instantaneously) of the inductor which prevents an applied sinusoidally varying voltage from setting up a current in it. However, as the applied voltage changes, the net field in the inductor, is just sufficient to cause the current to develop with a lag between the current and the voltage. We studied the process in detail with sketches of the Non-coulombic field and piecewise sinusoidal waveforms of voltage and current. We learned that in a circuit with a pure inductor, the current lags the voltage by 90 degrees in the sinusoidal steady-state.

We also learned that if the frequency of the sinusoidal voltage is increased, the current is smaller, because the induced emf is more powerful due to the rapid voltage change of the applied field. The attempt being to establish a large current change in the coil, the time-varying magnetic field is more rapid and a more powerful associated curly electric field produces a stronger polarization of charges in the coil, therefore, a larger induced emf. The resultant of the applied field and the induced emf being smaller, the current is smaller. An inductor offers a reactance given by $2\pi fL$ where L is the inductance of the solenoid.

The inductance L we understood, depends on the geometry of a current distribution, and for a long coil of many loops is $4\pi \times 10^{-7} \times \dfrac{AN^2}{l}$ Henries.

We learned to clearly distinguish the nature of the opposition made by resistors and inductors in circuits. While a resistor opposes the flow of mobile charges either when there are more lattice imperfections or across boundaries where there are fewer charges available, an inductor opposes the applied voltage (field) directly by the induced emf, which is due to the Coulomb field produced by polarization of the inductor wire by the non-Coulomb curly field associated with a time-varying current. In the relation $J = \sigma E_{net}$, a resistor affects $\sigma = qn\mu$, while an inductor *affects* E_{net}. In the sinusoidal steady-state, it is the regulating effect of E_{net} that led to the expression for reactance $2\pi fL$.

Incidentally, an inductor is not very often energized by a dc voltage unless it is made from high-resistance wire.

With a sinusoidally varying voltage applied to an LR combination, we learned after a detailed study of the process that the current lags the voltage by an amount which is less than 90 degrees. We then made the observation that it is this process which makes the LR circuit, a filter.

Since slowly varying or low frequency sinusoidal voltages are slow enough not to be able a generate a significant induced emf, the low frequency voltage produces smaller voltage "drops" across the inductor, and most of the applied voltage drops across the resistor. The current in the circuit is large and the voltage across the resistor is large and we say, the signal across the resistor is attenuated, although the attenuation itself is not very large.

With rapidly varying or high frequency sinusoidal voltages, the time-varying current effects are more rapid, and associated with a more rapid time-varying magnetic field there is a larger associated non-Coulomb field magnitude which causes a stronger polarization in the inductor and the induced emf is large. This results in large voltage drops across the inductor, the current becomes smaller. Therefore, the high frequency signals produce very small voltage "drops" across the resistor and we say, the signal across the resistor is *attenuated* and the attenuation is quite large.

If signal voltages comprising a combination of several frequencies are applied to the LR circuit, we can "tap" the low-frequency content of the signal across the resistor even as the rapidly varying portion of the sinusoidal voltage is busy trying to overcome the induced emf of the inductor. And, the high-frequency content of the signal can be "tapped" across the inductor.

We learned that LR circuits are used as low-pass filters to reduce the ripple content at the output of rectifying circuits, which convert ac voltages to a steady voltage (dc).

An ideal inductor made from a superconductive wire can sustain a current forever without a voltage, and this amazing phenomenon is attributed to the association of a magnetic field with a current (Ampere's law).

We studied how an ignition coil (a form of inductor) is used to produce high voltage arcs across spark plug tips in automobiles, by interrupting a current in a coil. The current break induces a very powerful emf across the breaking contact, because the current is associated with the magnetic energy in the coil, and interrupting the current amounts to destroying the energy stored in the magnetic field. The air between the contact tips ionizes when the electric field strength exceeds 3×10^6 N/C, and a spark is initiated. The exact process of arc production could be attributed to one among several causes, details of which we will learn in Chapter 10.

Transmission lines are found everywhere wherever electromagnetic energy is transported; in homes, factories, power grids and space and military applications and wherever electrical and electronic circuits are in operation.

The two laws, Faraday's law and the Ampere-Maxwell law, in their original form as we had learned in Section 4.7, seemed to indicate that they are applicable to the propagation of waves (comprising time-varying electric and magnetic fields) in free space, and not surprisingly, the two main laws are applicable to transmission lines as well.

A study of electromagnetic wave propagation between a pair of perfectly conducting parallel planes was then made, and we observed that the energy propagates by time-varying electric and magnetic fields. We learned that there are three modes of propagation; TE, TM and TEM and we brought our attention to the TEM mode, since in transmission lines this is the most popular mode with a small distance between conductors. For the TE and TM mode to be sustained, the geometry of the conductors and their spacing is economically not feasible. The TEM mode does not have a lower nor an upper cut-off frequency, but high frequency effects, such as, phase distortion and propagation delay prevent its use at high frequencies.

Next, we studied with appropriate illustrations, how the wave (not the fields themselves), seemingly glides between the planes as it propagates down a pair of parallel planes and plates that are appropriately dimensioned. We learned, that ordinary twin-lead cables used to connect an antenna and a television set and coaxial cables, propagate the em waves that carry energy between their conductors, by the TEM mode, and therefore, we had given importance to its study.

We dwelt for some length in the manner by which the wave, which consists of pulsating transverse electric and magnetic fields, is initiated by the source. This put into proper perspective, what we had learned in Chapters 1 and 2 about the processes which occur, when a voltage source is connected to one wire with one terminal or two wires to the two terminals of the source.

We then studied the manner in which an induced emf is generated at the ends of a twin-lead cable connected to the source, with the line characterized by distributed inductances and capacitances.

We touched upon the differential and common mode currents in transmission lines. We learned that in conservative circuit analysis, we always consider the differential mode currents, since this is the dominant one. The common-mode of current maybe present however, in ac transmission lines in circuits, which depends upon the frequency, conductor spacing and other geometrical factors apart from external influences such as noise and electromagnetic interferences. These can give rise to common-mode signals that distort the signals of interest in instrumentation applications and therefore, an introduction to this mode was made.

We mentioned one technique which prevents common-mode signals from superimposing on the outputs from amplifiers, which are used to strengthen weak differential signals from sensors. This is, to use operational amplifiers with large common-mode rejection ratios.

We then studied one more commonly found transmission line, the coaxial cable. While the coax seems to offer the perfect solution to contain signals in the TEM mode, they are also prone to conduct common-mode signals due to signal leakage through the shield.

Open-circuited transmission lines reflect voltage and current waves from the open end, and we learned that these may sometimes combine with the forward traveling wave to produce standing waves or waves of voltage and current that vary rhythmically at fixed points on the line, without progressing. The current is different at different points on the line, seemingly violating the current conservation rule, but this is expected, given the length of the line and the finite time (near speed of light and not instant) of propagation of the em wave.

We then learned that transmission lines are represented by distributed parameters; inductance and capacitance, which enable calculations of the impedances, attenuation and velocity of wave propagation along the line to be made.

One important aspect of a lossless transmission line in which the conductors are perfect, is that there is no voltage drop associated with resistance. However, a voltage does exist which can be computed by integrating the electric field due to the time-varying magnetic field of the TEM wave around a closed path touching the space between the conductors.

Transmission lines behave peculiarly when high frequency signals are being transmitted and slow the speed of propagation with increasing frequency. Signals comprising several frequencies will arrive at different phases and this causes phase distortion, which is one of the most serious effects which destroys signal integrity. Stray effects in transmission lines produce strong currents in lines, increasing voltage drops and reducing signal strengths. We had learned of these effects in Section 3.8. At high frequencies, we also learned, that the current in the wires crowds near the surface of conductors, moving away from the center because of the larger inductance around the center of the wire in comparison to that near the surface.

The wavelengths of electromagnetic waves are smaller in transmission lines in comparison with the free space wavelength, but the frequency remaining the same, the speed of the wave is smaller than that of light. Next, we studied in some detail the effect the "relaxation time" has to limit the upper-frequency limit of transmission of signals by lines and noted that two-wire transmission lines are seldom used at frequencies above 400 MHz since above this frequency phase delays will produce unacceptable signal distortions and signal losses in the line will become significant.

Lossless lines can produce attenuation or amplification of the input signal at the load, and this we learned, depends on the length of the line expressed in wavelengths and the terminating impedance.

We then studied in some detail the propagation delay in circuits operating at high frequencies and how these can produce significant error between the assumed and actual phase of the signal which in turn results in an error in the assumed amplitude of the signal.

We then learned that to meet desired performance objectives of high-frequency circuits, circuit designers often use quarter-wave long lengths of lines as circuit elements because at these frequencies ordinary lumped-circuit elements are difficult to construct and their parasitic effects are difficult to suppress.

Next, we studied in some detail using Lenz's Law the process which makes a quarter-wave length long two-wire transmission line to resonate because of reflected waves and can produce a voltage step-up upto several hundred times.

The skin-effect, we learned is a high-frequency phenomenon when signal currents are stronger on the outer regions of a conductor rather than in the interior. We were cautioned not to visualize circuits operating at any frequency including dc conduct current only in the outer regions. The skin-effect is only a high-frequency phenomenon.

We then studied in some detail the speed of voltage waves in transmission lines and how the distributed line inductance and capacitance affect the speed, making signals travel at speeds less than the speed c. We were cautioned to note that the frequency of the signal is not changed being the signal source frequency but the wavelength and thereby the speed is altered (reduced).

Attenuation causes the fields of transmitted signals to lose strength reducing the energy they carry. We learned that while attenuation by a resistive attenuator can be explained as the voltage drop set up by surface charges producing weaker fields and voltage at the loads, attenuation in capacitive and inductive circuits is subtler, and is related to the fringe field and induced emf effects, respectively. These aspects were discussed under the special topic of Attenuation towards the end of the chapter.

Finally, we learned that electromagnetic waves also undergo attenuation in which their energy is decreased by the phenomenon of "Scattering" by obstacles in the path of the wave and sometimes by lossy transmission paths.

Questions

1. What is a magnetic field? Describe its nature.
2. State the Biot-Savart law and write the expression for the magnetic field of a short thin length of wire at a distance r from the observation point from the thin length of wire. What are the units of the magnetic field?
3. Using the Biot-Savart law derive an expression for the magnetic field at a point perpendicular to the wire at a distance x from it. What are the four steps you followed to derive the expression? Explain.
4. When you bring a current-carrying wire down onto the top of a compass, aligned with the original direction of the needle and 5 mm above the needle, the needle deflects by 10 degrees. (a) Show on a diagram the original direction of conventional current in the wire and the direction of the additional magnetic field made by the wire underneath the wire, where the compass needle is located. Explain briefly. (b) Calculate the amount of current flowing in the wire. The measurement was made at a location where the horizontal component of the Earth's magnetic field is $B_{earth} \approx 2 \times 10^{-5}$ tesla.
5. Derive an expression for the magnetic field at any point along the axis of a single loop carrying a current I.
6. Sketch the magnetic field at locations along the axis and above and below a coil of many turns. At what locations is the magnetic field very strong. At what locations is the magnetic field very weak. Give reasons why this is so.
7. How is a curly electric field produced? Describe with the help of neat sketches. Why is this field called non-conservative? Explain.
8. Consider an inductor with a current that is increasing such that the magnetic field due to it is pointing out and increasing as shown in the figure.

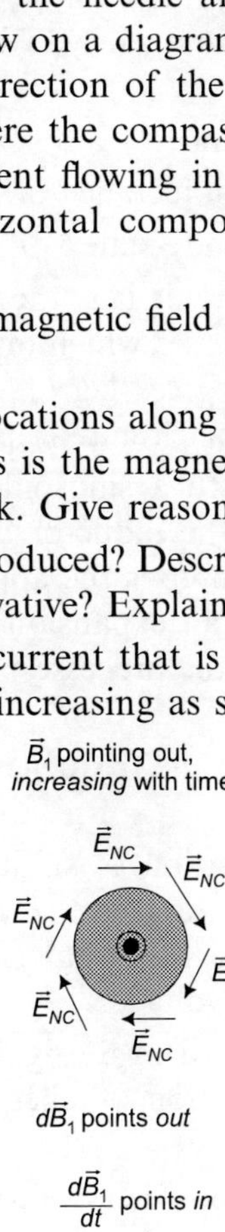

The non-Coulomb electric field that will result is indicated in the figure. Sketch the non-Coulomb electric field for the 3 cases with B_1 pointing out (increasing), pointing in (increasing) and pointing in (decreasing).

9. An otherwise uniform metallic ring with a thin section surrounds an inductor whose current is varied. Sketch the effect this will have on the potential of the ring. Describe the nature of the electric field inside the ring.
10. In the problem of Section 5.10 of two resistors connected to each other at both ends surrounding a constantly changing magnetic field, sketch the surface charge configuration as it develops in the circuit with electric field vectors of the coulomb and non-Coulomb electric field vectors drwan separately in the wires and resistors. Describe how these fields combine such that current conservation is maintained in the circuit.
11. A varying dc voltage is applied to an inductor. Describe the processes which lead to the production of a Coulomb electric field in the inductor.
12. Derive an expression for the magnetic field of a solenoid (or inductor) having N turns and of length L. The radius of a loop of the solenoid is R. If the radius R is smaller than the length L, what can you say about the strength of the field inside the solenoid at its center?
13. What are the units of inductance of a solenoid?

    Derive an expression for the inductance of a solenoid.
14. Explain briefly "sinusoidal steady-state" of a circuit with an inductor.
15. The current in an inductor circuit lags the input by 90 degrees. Describe with the help of neat sketches the processes which maintain this lag angle in the steady-state. Is it possible to achieve 90 degrees lag angle between the current and the voltage accurately in a typical inductor circuit? Explain giving reasons.
16. When an inductor is subject to a sinusoidal voltage, it develops a non-Coulomb electric field which polarizes the inductor and produces surface charges such that their Coulomb electric field opposes the applied electric field of the voltage source. Since the surface charges are an expansion and contraction of the electron sea of the inductor, explain in detail the processes which maintain neutrality in the interior of the inductor.
17. Describe the effects of varying the frequency of a sinusoidal voltage to an inductor with respect to the current.
18. An inductor resistor combination is subject to a sinusoidal voltage. With neat sketches explain how the current lags the voltage by an angle that is less than 90 degrees. If the frequency is varied, what effect does this have on the current and the phase with respect to the voltage?
19. How are high voltages produced in the ignition circuit of an automobile? Explain with reference to the conservation of energy principle and the expression for the induced emf.

    Explain why we are not shocked when we touch objects considering the fact that all matter is made of atoms comprising electrons and protons and the potential very close to them can be of the order of thousands of kilovolts.

20. The unit of resistance and inductive reactance is ohms. Describe how the two are different qualitatively.
21. What is meant by reactive power in a circuit containing an inductor? What is the direction of its flow during an entire sinusoidal cycle of an applied sinusoidal voltage to the inductor?
22. What was the modification Maxwell made to Ampere's law to account for the missing current between the plates of a capacitor?
23. Describe how an electromagnetic wave propagates between a pair of planes. Is there a component of either of the fields in the direction of propagation? What is the mode of transmission?
24. Describe surface current in a pair of parallel-plane transmission line and how it appears on the surface of the planes. How is it denoted in vector notation?
25. Do the fields of the electromagnetic wave travel? Describe how the wave starts and then progresses along a perfectly conducting parallel plane transmission line. Write the equation of the electric field of a sinusoidal wave which propagates in a transmission line.
26. How many modes of propagation of em waves are possible in a perfectly conducting pair of parallel planes?
27. What makes the TEM mode of propagation preferable over the TE and the TM modes? State the devices which support these modes of propagation.
28. Describe how an em wave propagates along a two-conductor transmission line. Draw the equivalent circuit of such a line.
29. Give examples of two-conductor transmission lines. How do these differ in the field configurations of propagating em waves from the field configurations of em waves between a pair of perfectly conducting parallel planes?
30. What is meant by "self-inductance" of a straight wire?

    A twin-lead cable carries a steady current '*i*'. Describe the process of generation of an induced emf when the constant voltage is applied to it. What is the nature of the electric field between the wires? Explain.
31. A twin-lead cable carries a sinusoidally varying voltage. Describe the process of generation of an induced emf when a signal generator is turned ON. What is the nature of the electric field between the wires? Explain.
32. What are differential and common-mode currents? Operationally, which current mode is present in two-wire transmission lines? With a neat sketch of a twin-cable line describe this mode.
33. Explain how coaxial cables can have undesirable common-mode currents. What is the factor in the cable that causes this?
34. What is meant by common-mode rejection ratio with reference to amplifiers? Which special amplifier is used to suppress the common-mode voltage from being amplified?
35. State the conditions under which transmission lines will not radiate em waves. Sketch the electric field configuration about a transmission line having resistance.

36. Under what conditions does the common-mode current appear? Explain.
37. Describe how reflected voltage waves occur in transmission lines.
38. What is the "voltage" of a lossless transmission line? Explain with a sketch the origins of this voltage. Is it different qualitatively from the voltage one measures in a simple electric circuit built with resistances?
39. Explain how surface charges appear and disappear on the surfaces of transmission lines and state the amount of shift of the electrons that produces surface charges. What is the source of the charges that maintain neutrality in the conductors as expansion and contraction of the electron sea of the line conductors occurs? Generally, upto which frequency of the em wave is support possible by a transmission line?
40. Describe how the induced voltage of a $\lambda/4$ two-wire line can be stepped up. What is the impedance of such a line? What is the impedance if the line were shorter or longer than a quarter wavelength?
41. What is the speed of the voltage wave of a transmission line? Is it slower or faster than the speed of light? Explain.
42. What are distributed inductances and capacitances of a two-wire transmission line? How do these affect the speed of transmission of voltage and current waves?
43. Does the terminating impedance have an effect on the output voltage of a transmission line?
44. How should lossless lines be configured and terminated to produce attenuation and amplification of a signal voltage? Explain.
45. What is attenuation? How does a series resistor combination attenuate an input voltage?
    Is the attenuation frequency dependent?
46. Sketch the circuit of a low-pass RC circuit and describe how it attenuates signals. What circuit process is responsible for producing attenuation. What frequency of signals are attenuated by such a filter?
47. What is meant by "attenuation" of a wave? Describe qualitatively how electromagnetic waves are attenuated.

Additional practice questions are available in the CD with this book.

6

Magnetic Forces

The first section of this chapter deals with the forces between current elements according to Ampere and Grassmann which led to the formulation of the Lorentz force law. Following sections describe the processes which enable a current carrying wire to move when placed in a magnetic field. The fact that magnetic forces do no work, is explained by a dramatic experiment in which, a mass attached to a current carrying loop, when placed in a magnetic field, hangs in mid air. As proof, calculations of the magnetic force are also provided. The magnetic field, as a relativistic effect, is one among several interpretations of charged particle interactions.

The final section briefly traces the development of the theory of Classical Electrodynamics including a brief discussion on the Action-at-a-distance and Aether theories. This is followed by a discussion on the Lorentz Force Law from the viewpoint of Special Relativity with pointers to Appendix F where a discussion of the Lorentz transformation equations is provided. The equations are used to demonstrate how the fields of moving charges transform leading to a definition of another field, the magnetic field by a velocity-dependent part of the force which acts on a charge in motion.

Pointers to sections of Härtel's article "A Qualitative Approach to Electricity: A Guide to Visualization of Electrodynamics using symbols and images" are made at appropriate locations in the text that will enable students to make the transition from electrostatics to electrodynamics (Faraday's Law and Lorentz Force Law from the Theory Special Relativity viewpoint) smoothly.

6.1 MAGNETIC FORCES

Magnetic fields *affect* moving charges. The magnetic force in a charge q, moving with velocity v in a magnetic field B, is

$$F_{mag} = q(\vec{v} \times \vec{B})$$

This is known as the *Lorentz force* law.

After detailed experimental investigations, Ampere (who also studied Biot and Savart's work minutely), in 1824 arrived at a highly accurate expression (without a "field" concept) for the force between current elements (See Ref. [38] and Appendix F Section F.1) which satisfied Newton's third law when applied to any system of their interaction. Grassmann, also separately formulated force expressions between current elements, which when applied for the case of a force exerted by a current in a closed circuit on a current element of another circuit, satisfied Newton's Third Law, but failed in most other situations. Maxwell and

Lorentz in 1861 and 1895, respectively expressed Grassmann's force in terms of a magnetic field (Ref. [38]). This is what came to be called the Lorentz force law. A detailed treatment is provided in Section F.1 in Appendix F.

The experiments described below provide evidence of the magnetic force according to the Lorentz force law. Reproduced by permission Pearson Publications Ref. [8].

Two wires are hanging from the top, a few centimeters, as seen in Fig. 6.1(a). When the current is turned ON they *jump* apart evidently *repelling* one another, as shown in Fig. 6.1(b).

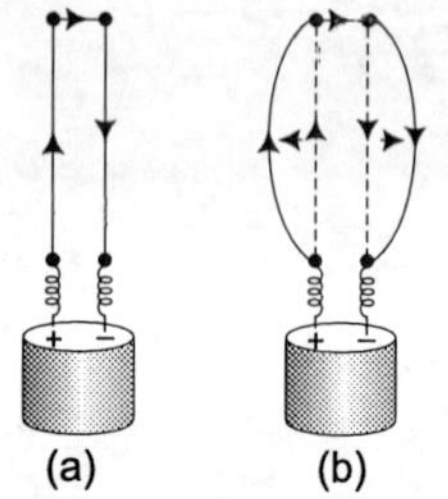

Fig. 6.1 Force in a current carrying wire.

That the wires will repel one other can also be shown based on the Lorentz transformation (See Appendix F). In Fig. 6.2 is shown a system of coils that can carry a current to produce a magnetic field.

An **electron** is made to move by an electric field (not shown) towards a **coil system** with a velocity v.

A current is switched on in the coil(s), and a horizontal magnetic field is produced.

The moving electron is *deflected* vertically by the horizontal *magnetic field.*

$$F_{mag} = q\vec{v} \times \vec{B}$$

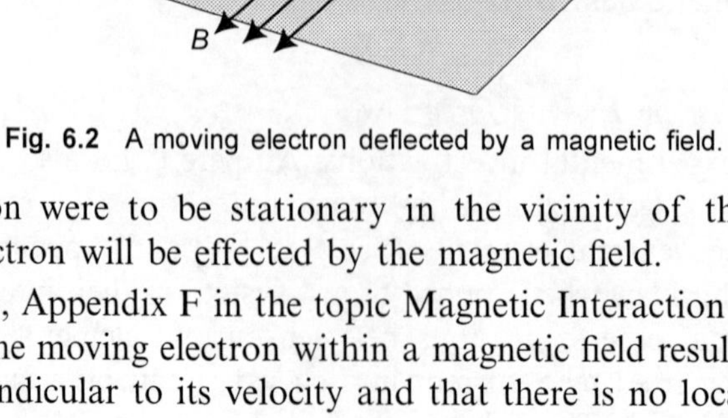

Fig. 6.2 A moving electron deflected by a magnetic field.

Note: If the electron were to be stationary in the vicinity of the magnetic field, no movement of the electron will be effected by the magnetic field.

Note: In Section F.1, Appendix F in the topic Magnetic Interaction we will learn that the force which acts on the moving electron within a magnetic field resulted in an acceleration of the electron perpendicular to its velocity and that there is no local reaction back onto the magnetic field. A single force is a contradiction to Newton's Third Principle which is only fulfilled when the reaction of the moving particle with the complete system (interaction between the moving charges as origin of the magnetic field and the single moving charge) is taken into account.

In Fig. 6.3 are shown 2 wires in close proximity to one another, with a current I in each of them. The charges move with a velocity v.

Current in wire 1, produces a magnetic field B around wire 2, which can be determined using the Biot-Savart law.

The force on the charges having velocity v in wire 2 is given by $F = q\vec{v} \times \vec{B}$.

The direction of the force is determined by applying the right hand rule for vector cross products (see Section 5.2).

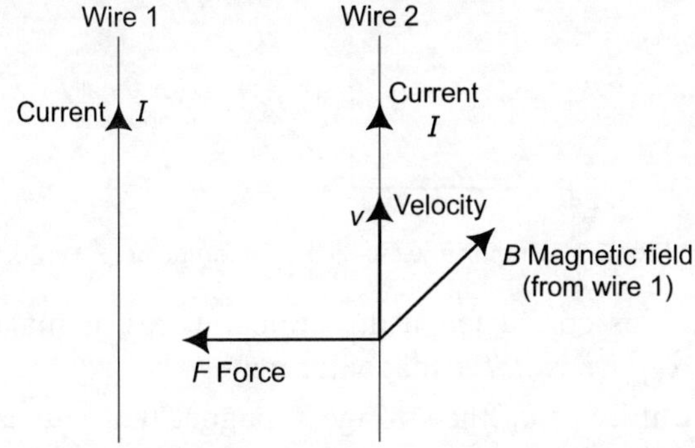

Fig. 6.3 Forces of two current carrying wires in close proximity to one another.

A single wire with a current in it cannot exert a force on itself! There is no net longitudinal force either. An internal tension can exist. See Section 6.1, Links_to_animations_articles in the accompanying CD.

It is the *external* magnetic field, which applies a force on a current carrying wire as shown in Fig. 6.4.

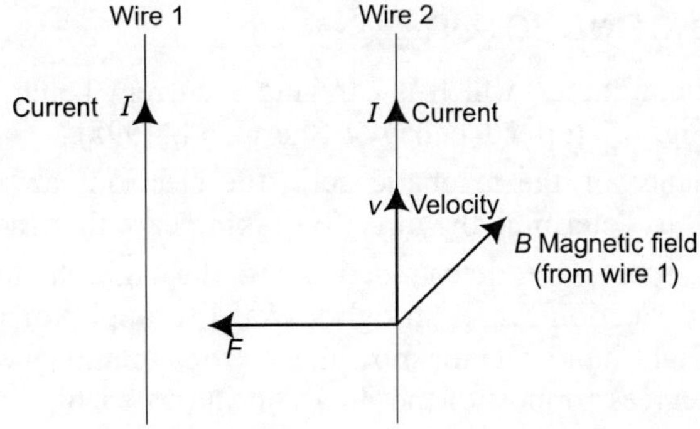

Fig. 6.4 Force is due to external magnetic field.

A similar force will result due to a magnetic field set up by wire 2 on wire 1, and the result is *attraction*. This is shown in Fig. 6.5.

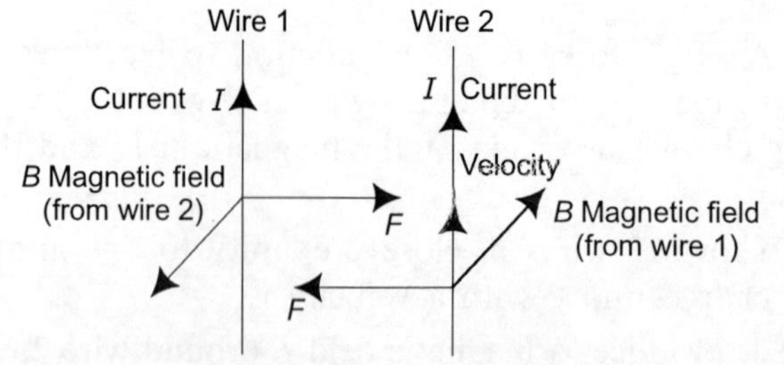

Fig. 6.5 Force of attraction of two current carrying wires in close proximity.

Expressed vectorially, as shown in Fig. 6.6, the magnetic force F is given by

$$F = q\vec{v} \times \vec{B}$$

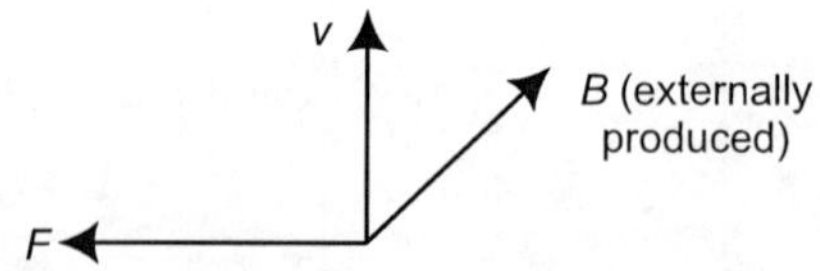

Fig. 6.6 Graphical view of the force vector $\vec{F} = q\vec{v} \times B$.

If the experiment was conducted using copper wires, it makes the results even more astounding! Because, *copper* is *not* a magnetic material!

You can *test* this out by using the strongest magnet that you can find to pick up copper materials such as copper coins or copper wires. The magnet will *not be able to attract these copper items*, thus showing that copper is a *non-magnetic* metal.

Then, why does the wire move?

To answer the question we begin (see next section) by describing the action of a magnetic field on a current carrying conductor.

6.2 A MAGNETIC FIELD ACTING ON A CURRENT CARRYING CONDUCTOR

Consider a metallic conductor, which is carrying a current being subjected to a magnetic field, as shown in Fig. 6.7 (after Chabay & Sherwood, 1995).

Under the influence of the magnetic field, the electrons are deflected downward, as shown (apply $qv \times B$ to determine the direction taking care that the particles are electrons).

Note: The direction of force determined using the right hand rule for vector cross-products described in Section 5.2, is straightforward to apply for moving positive charges. When applying the right hand rule for moving electrons in a magnetic field, the direction of $v \times B$ will be 180 degrees (opposite) the direction indicated by the thumb.

As time goes on, more and more electrons moved due to the *action of the magnetic field* will pile up on the bottom of the wire...with a corresponding deficiency of electrons on the top.

However, this piling-up won't continue forever...

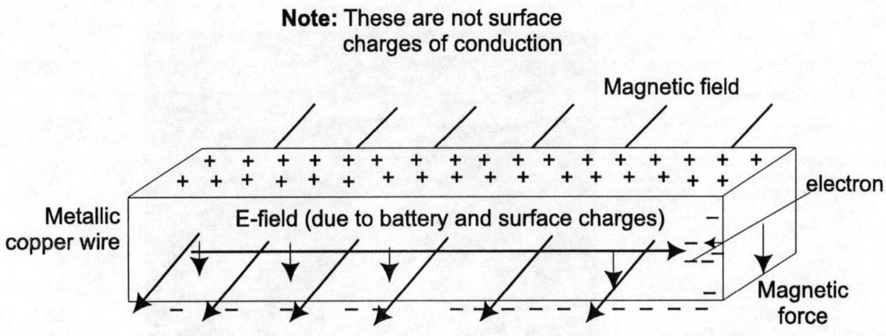

Fig. 6.7 Charges pile on the top and bottom of the wire.

The extra surface charges produce an *electric field* between the upper and lower surface of the metallic wire.

The extra surface charges exert an electric force qE (shown in grey in Fig. 6.8 and on electrons this will be *upward*) on the *moving* electrons, which in the steady state, just balances the magnetic force that caused their movement to the bottom, qvB and so, $qE = qvB$.

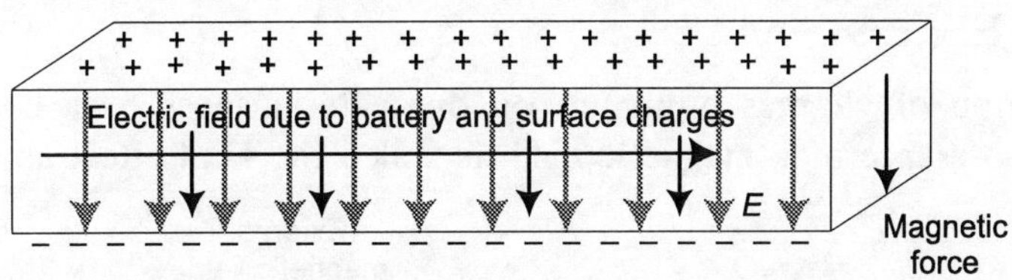

Fig. 6.8 Extra surface charges exert a force qE on the moving electrons.

Since qE becomes equal to qvB, the moving electrons now continue through the wire without being deflected as shown in Fig. 6.9.

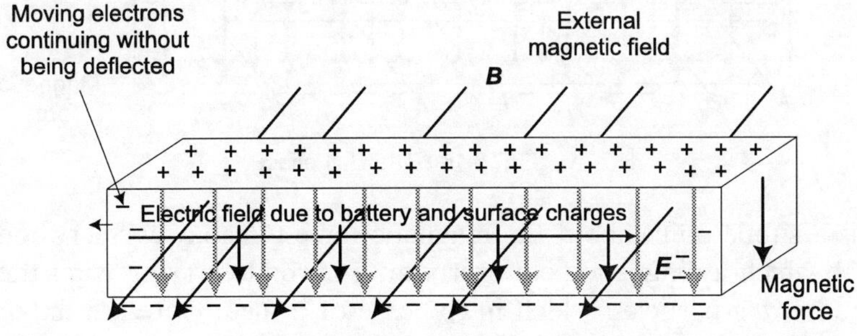

Fig. 6.9 $qE = qvB$ and the moving electrons continue without being deflected.

6.3 BALANCED FORCES WITH MOVEMENT

Another example of balanced forces with movement: the carrom player in the image below pushes the striker...the downward force of gravity is balanced by the *upward reaction* of the carrom board surface!

The striker moves horizontally between these vertically *balanced opposing forces* on the board.

And, that is exactly how the electrons move through the wire (not forgetting that it is a start-stop motion of millions and millions of electrons everywhere in the wire with a drift superimposed) between the electric force and its balanced oppositely directed magnetic force, in the current-carrying wire when immersed in a magnetic field.

6.4 HALL EFFECT

The phenomenon of charges piling up on opposite surfaces of a conductor carrying a current in the presence of a **magnetic field**, is called the **Hall effect** as shown in Fig. 6.10.

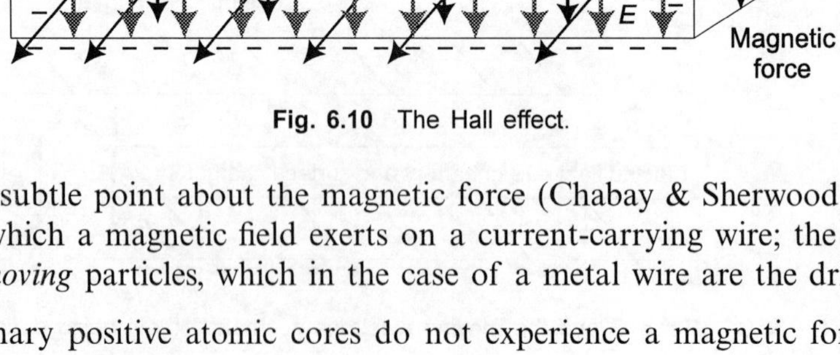

Fig. 6.10 The Hall effect.

There is a subtle point about the magnetic force (Chabay & Sherwood, 1995, used by permission)[2], which a magnetic field exerts on a current-carrying wire; the magnetic force only acts on *moving* particles, which in the case of a metal wire are the drifting electrons.

The stationary positive atomic cores do not experience a magnetic force. So why do we see the entire wire move?

We have seen how the extra surface charges exert an electric force qE on the moving electrons, which in the steady state just **balances the magnetic force** qvB.

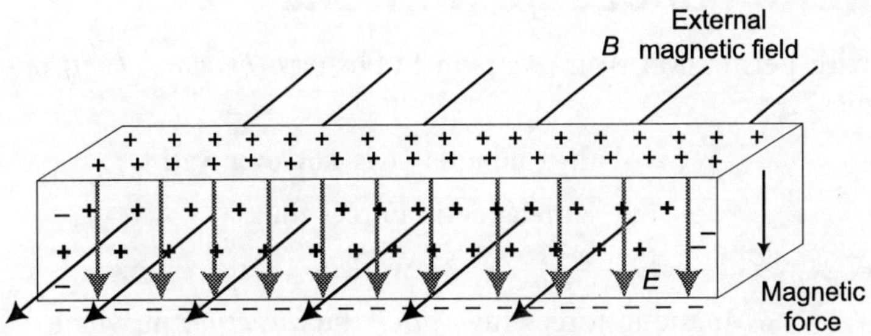

The extra surface charges also exert an electric force qE on the "*stationary*" positive atomic cores.

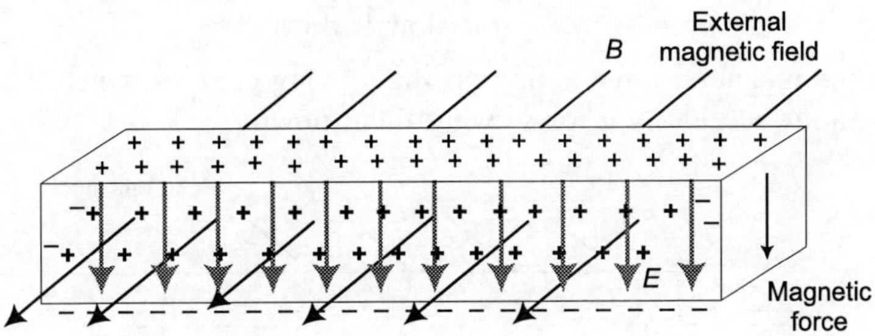

Fig. 6.11 The electric force acts on the positive atomic cores.

And, this electric force is an unbalanced force (as shown in Fig. 6.11). There is no counter to this force, because the atomic cores initially are not moving, and hence are not subject to a magnetic force.

And, that is the reason *the wire moves*; an *electric force* acts on the positive atomic cores, as shown in Fig. 6.12.

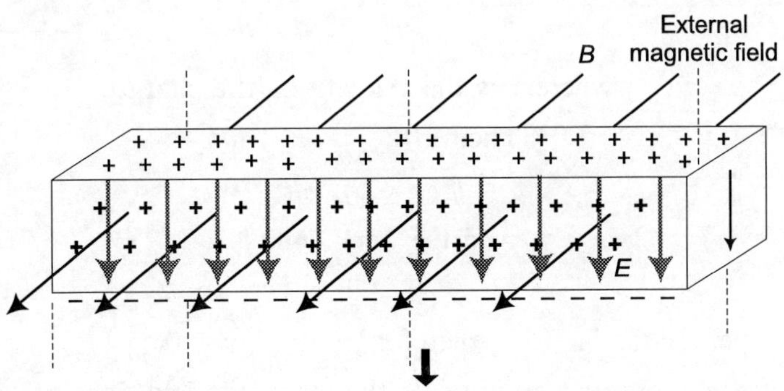

Fig. 6.12 The electric force acting on the positive atomic cores causing it to move.

The electric force due to the Hall-effect surface charges, forces the wire as a whole to move in the direction of $I(d\vec{L} \times \vec{B})$ (where dL is an elemental length of the conductor of length L carrying a current I), and the magnitude of the force is equal to this force.

6.5 MAGNETIC FORCES DO NO WORK

Reproduced with permission from Pearson Publishers, *David J. Griffith's*, "Introduction to Electrodynamics".

Wait a minute! It is not over yet!

Magnetic Forces do...

***no* work**

Magnetic forces may *alter* the *direction* in which
a charged particle moves ... but...
...they cannot... *repeat*... cannot
speed it up
... nor slow it down.

Consider the magnetic force acting on the moving charges which caused these charges to pile up, without accelerating, or slowing them down.

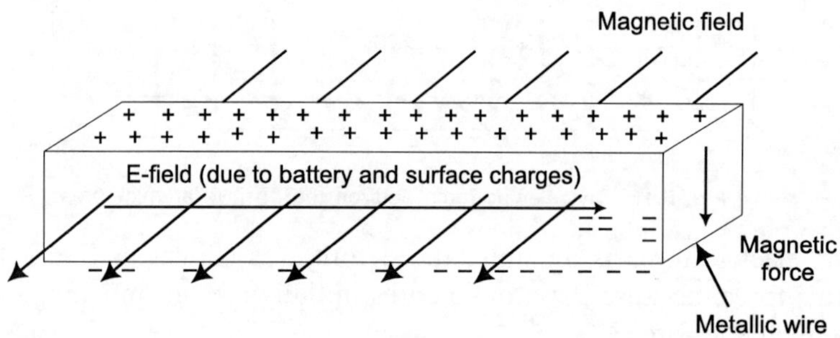

Compute the work done by the magnetic field as follows:

If the charge 'q' moves an amount dL in time dt, then the charge q moves by

$$dL = v.dt$$

where v is the velocity of the charge.

$$\text{Then, } dW_{mag} = F_{mag} \cdot dL$$

$$= q\,(v \times B).\,vdt = 0$$

And the work done is...

0!!

Note: $v \times B$ is perpendicular to v, therefore, $(v \times B)\cdot v = 0$.

If the charge 'q' moves an amount dl, the magnetic field does not do any work! = 0!

Not Convinced? Observe Carefully

The *force* is always acting in a *direction...that is perpendicular* to the motion of the charge or electron, as shown in Fig 6.13. Note how the magnetic force acts *perpendicular* to the motion of the electrons, pushing them down.

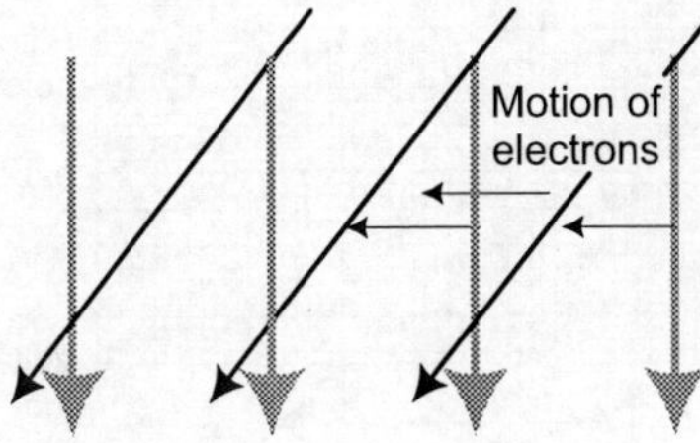

Fig. 6.13 The magnetic force acting perpendicular to the motion of charge.

For work to be done, the movement *along the direction of the force* only counts.

$$W_{mag} = F_{mag} \cdot dL,$$

or the dot product

$$= F \cos\theta\, dL$$
$$= FdL \cos 90°...$$
$$= 0.$$

Then, when an electromagnet is used to lift a heavy weight, does the magnetic field *not* do the work?

We seek an answer by studying an experiment.

We begin by defining a current. The current in a wire is the *charge per unit time* passing a given point.

A line charge λ (mobile charges per meter) traveling down a wire at speed v constitutes a current, as shown in Fig 6.14.

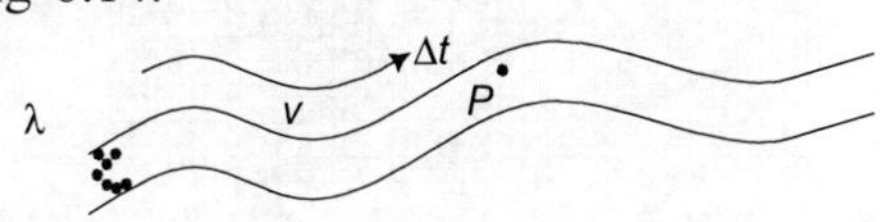

Fig. 6.14 Definition of a current.

The current $I = \lambda v,$

because a segment of length $v\Delta t$, carrying a charge $\lambda v \Delta t$, passes point P in a time interval Δt.

The magnetic force on a segment of current-carrying wire is evidently,

$$F_{mag} = \int (\vec{v} \times \vec{B}) dq = \int (\vec{v} \times \vec{B}) \lambda dl = \int (I \times \vec{B})\, dl$$

Since I and dl both point in the same direction, we can write this as

$$F_{mag} = I \int (d\vec{l} \times \vec{B})$$

Typically, the current is constant along the wire, and therefore, I comes outside the integral,

$$F_{mag} = I \int (d\vec{l} \times \vec{B})$$

Consider an experiment, in which a rectangular loop of wire supporting a mass m hangs vertically, with one end of the loop in a uniform magnetic field B, as shown in Fig. 6.15.

What should be the current I in the loop that will balance the gravitational force downwards?

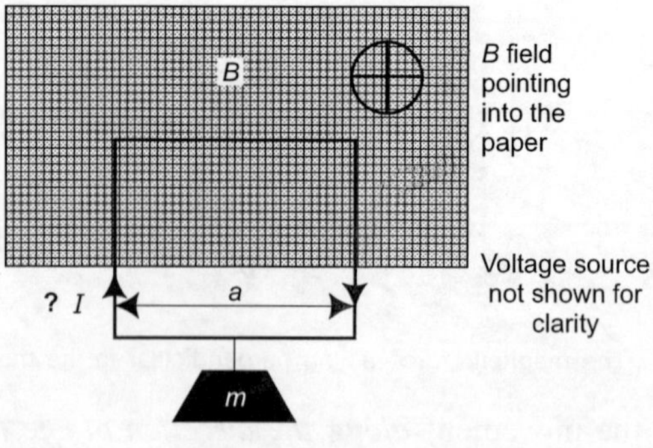

Fig. 6.15 A current carrying loop in a magnetic field suspends a mass.

First, the current must be *clockwise* so that the magnetic force on the horizontal segment is upwards.

Since,
$$F_{mag} = I \int (dl \times B)$$

Then,
$$\boldsymbol{F_{mag} = IBa}$$

where, a = width of the loop because the integral on all the small current segments Δl is 'a' as shown in Fig. 6.16.

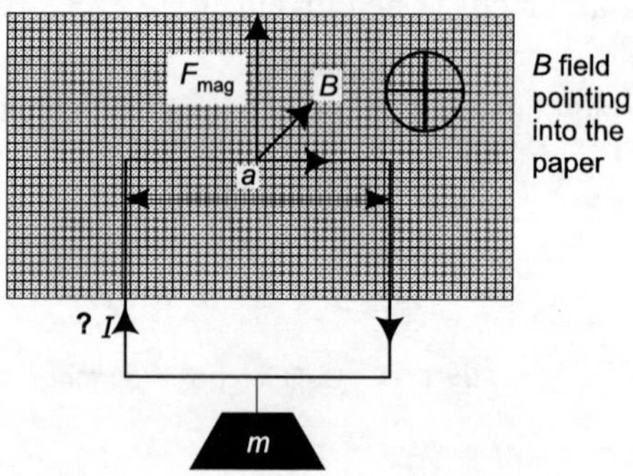

Fig. 6.16 Current must be clockwise so as to apply a force F_{mag} upwards.

The magnetic forces on the two vertical segments cancel as shown in the figure below.

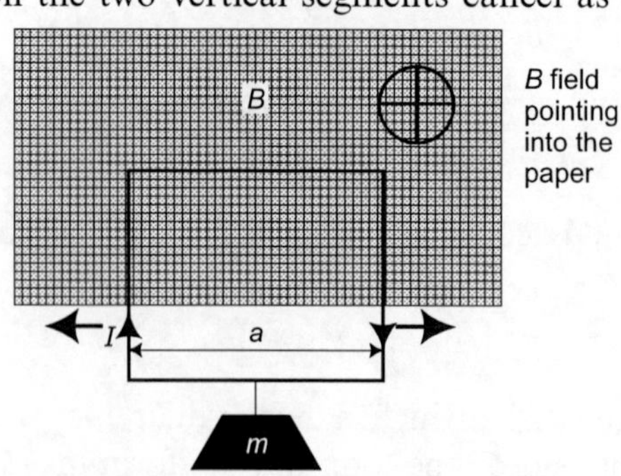

For F_{mag} to balance out the weight mg:

$$F_{mag} = IBa = mg$$

So, we have, $I = mg/Ba$

For the current I, then both the forces are in *perfect balance* and the weight is hanging in mid air! This is shown in Fig. 6.17 (see side view).

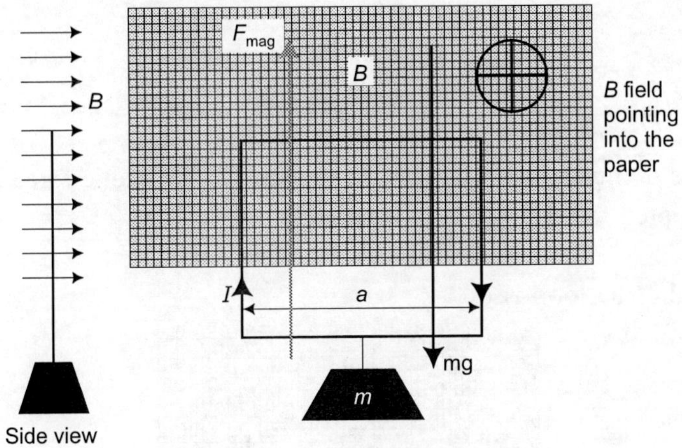

Fig. 6.17 The mass is hanging in mid air!

At a microscopic level, the force holding the weight up, is due to the *electric field* force on the positive atomic cores in the wire, and this force dvelops by the Hall effect phenomenon of separation of charges.

What happens when we *increase* the current?

The upward magnetic force increases, and there is a corresponding increase in Hall effect surface charges, and the electric field strength increases its force on the positive atomic cores.

This force now exceeds the downward force of gravity, and the loop rises.

And that is the reason why the loop rises. An electric force acts on the positive atomic cores as shown in Fig. 6.18.

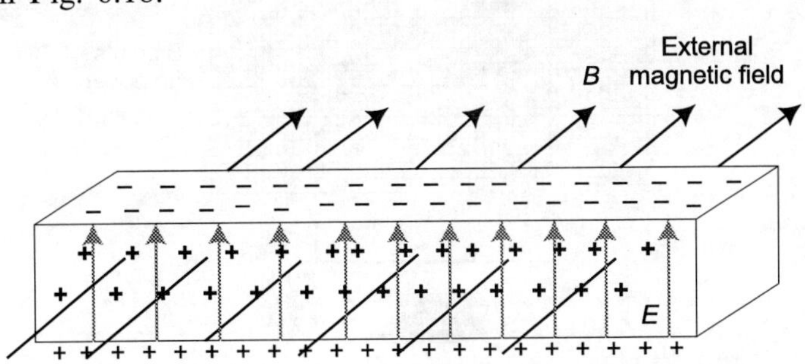

Fig. 6.18 An electric force acts on the positive cores raising the loop.

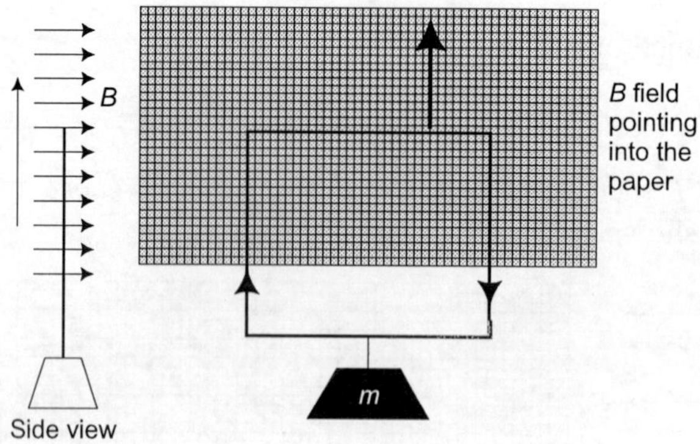

Somebody's doing the work and it looks as if the magnetic force is responsible. We are tempted to write

$$W_{mag} = F_{mag}h = IBah$$

where h = height, the loop rises.

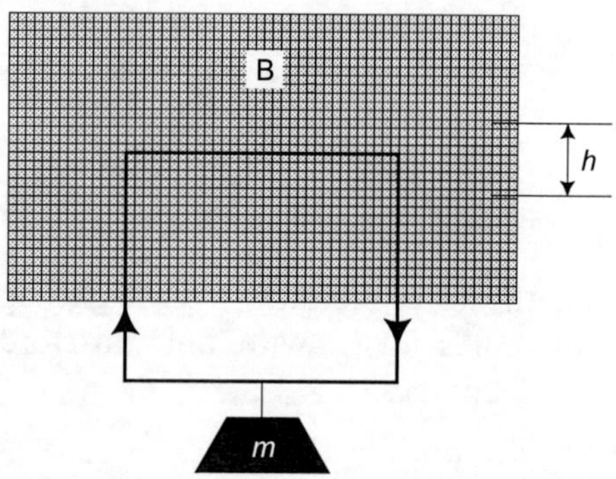

But we know that magnetic forces never do work.

So what is going on here?

When the loop starts to rise, the charges in the wire **are no longer** moving horizontally (with speed w) only, as shown in the figure below.

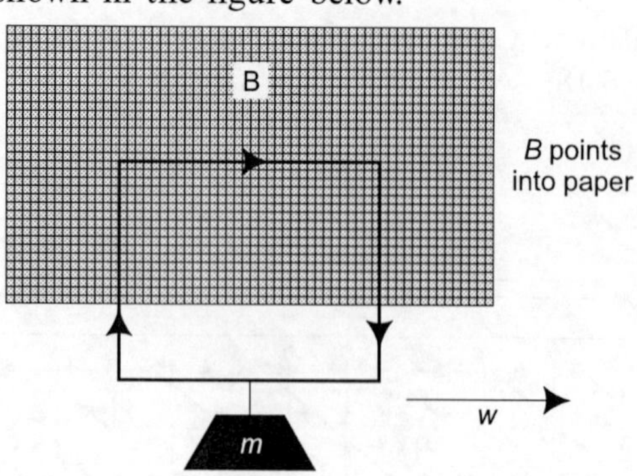

Their (charges) velocity now acquires an upward component *u*, which is the upward speed of the loop in addition to the horizontal component associated with the current *I* as shown in Fig. 6.19.

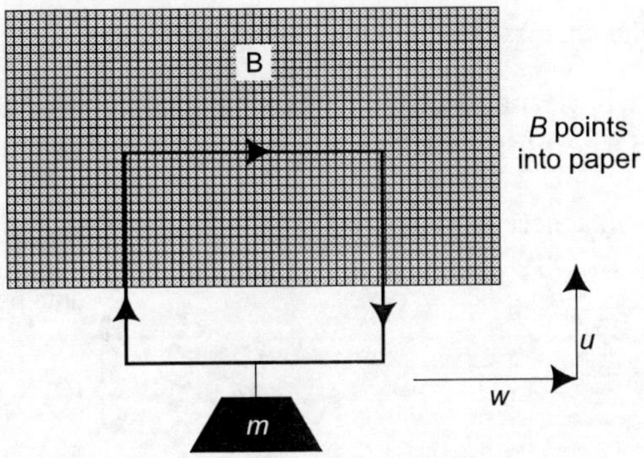

Fig. 6.19 Upward component '*u*' of velocity of charges.

But the magnetic force which is always perpendicular to the velocity (resultant), check this, no longer points straight up, but *tilts back*.

Figure 6.20 shows the velocity components and the resultant velocity '*v*' of the moving charges and Fig. 6.21 shows all the components of the velocity plotted inside the wire.

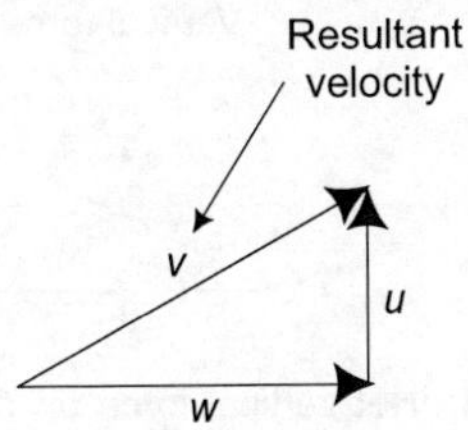

Fig. 6.20 Resultant velocity of the charges.

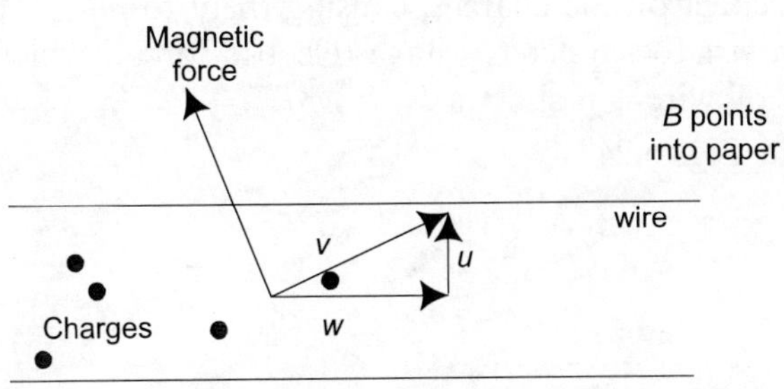

Fig. 6.21 The velocity components of the charges plotted inside the wire for clarity.

To know how it feels to be an electron—subjected to such a force:

Imagine that you are walking on a long narrow carpet and the carpet ahead is lifted all of a sudden! See Fig. 6.22.

Fig. 6.22 How the charges "feel" the magnetic deflecting force and their resultant velocity.

The magnetic force is perpendicular to the net displacement of the charge (which is in the direction of v) and therefore, it does no work on q.

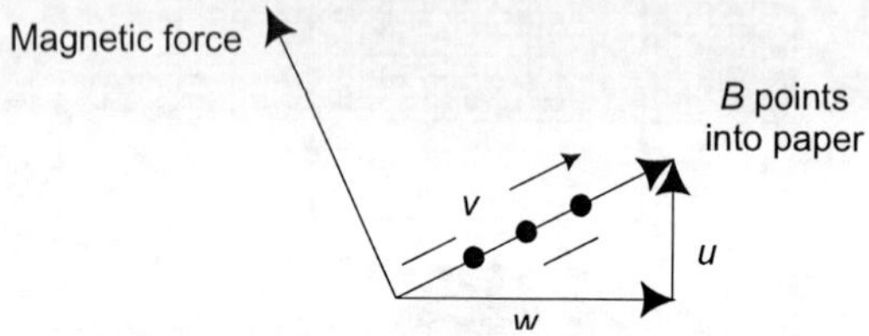

It *does* have a vertical component (qwB), indeed. The net vertical force on all the charges (λa) in the upper segment of the loop is (see Fig. 6.23) given by,

$$F_{vert} = \lambda a w B = IBa$$

where, λ is the charge per unit length, I = current, B = magnetic field, a = width of the loop and w = horizontal speed of the charges.

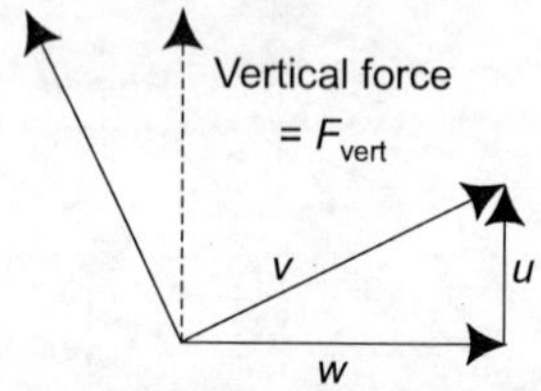

Fig. 6.23 Net vertical force on all the charges.

But now it also has a horizontal component quB, which *opposes* the flow of current, imposing a *braking* effect on the charges, causing them to *pile up* at one end, and produce an electric force *against* the applied voltage (electric force), which had set the charges in motion in the loop of wire—in short a *back emf.*

This is shown in Fig. 6.24.

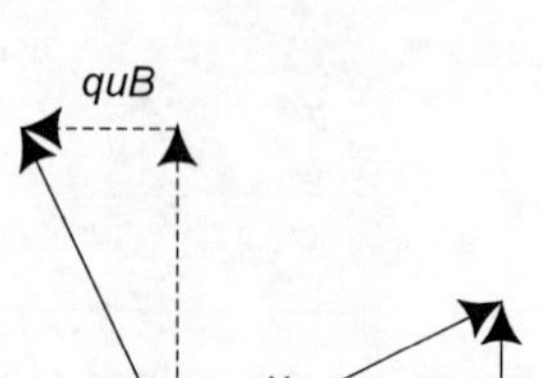

Fig. 6.24 Horizontal component of net force on all the charges.

Whoever is in charge of *maintaining* that current, therefore, must now push those piled up charges along—against the backward (braking) component of the magnetic force—against the back emf as shown in Fig. 6.25.

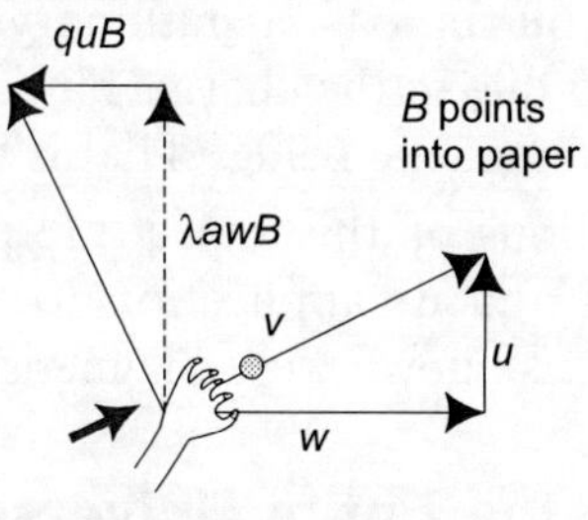

Fig. 6.25 Pushing the charges forward.

The total horizontal force on the top segment of the loop is

$$F_{horiz} = \lambda auB$$

where, λ is the charge per unit length.

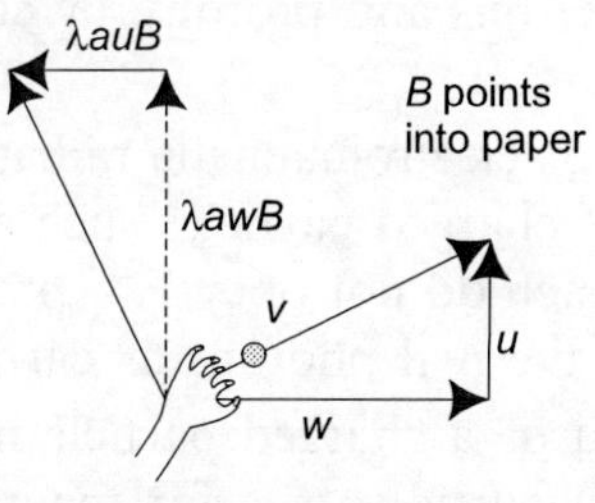

In a time *dt*, the charges move a (horizontal) distance *wdt*. So, the work done by this agency, such as, a battery or a generator is:

$$W_{battery} = \int F_{horiz}\, wdt = \lambda aB \int uwdt = IBah$$

where *h* is the height the loop rises, and because $I = \lambda w$ and integral $udt = h$.

This is what we attributed (wrongly!) to the magnetic force and wrote $W_{mag} = IBah$.

Was work done in the process?
Absolutely!
Who did it?
Not me, said the "magnetic field"...

Then, who did it?
It was the Battery!

6.6 THE MAGNETIC FIELD AND RELATIVISTIC ELECTRODYNAMICS

In many situations one can consider magnetic fields to be a relativistic consequence of electric fields. For example, if you run past a stationary charge you will observe a magnetic field along with the electric field due to the charge, but if you stand still you will observe the electric field and not observe a magnetic field.[31] That is the reason, magnetism is *relativistic*.

In Appendix F demonstrations of this aspect are provided. Beginning with equations of the Lorentz transformation and its application to a charge moving near a current to derive the Lorentz force law, it is shown how magnetism is relativistic.

The Magnetic Field Fills the Entire Universe

Magnetic fields are a relativistic effect of electric fields. There are two different representations for light–as waves in the electric and magnetic fields, and as motion of particles–photons.

We had briefly touched upon Quantum Electrodynamics and Quantum Field Theory in "The Theories of Electrodynamics" in Section 4.2, which postulate the existence of quantum "fields", in which electrons and protons, by continuously emitting and absorbing (virtual) photons, form a field.

The photon, the quantum of electromagnetic radiation is the force carrier (or particle) for the electromagnetic force of charged particles such as electrons and protons, even when static, via "virtual photons" which do not obey E = p*c (E is energy, p = momentum and, c is the speed of light), which the real photons as carriers of light, do.

And, since the electric field of a charged particle fills the entire universe, so whatever is true for electric fields, they will hold true for magnetic fields as well (as a relativistic effect). So, if photons can reach all over the universe, so does the electric field and so also the magnetic fields.

The story of how the connection between thermodynamics and electromagnetic radiation in the form of photons was made is described briefly in Section 10.9 "The birth of Quantum Physics" and Links_to_animations_articles in the CD.

Whatever be the fields of charge carriers, classical or quantum, and whether lines of force or photonic, the electric and magnetic fields, fill the entire universe.

There are other theories on charged particle interactions (Section 4.2). In 1924, Leigh Page proposed an Emission Theory of Electromagnetism which assumes a continuous emission of "moving elements", whose loci are Faraday's lines of force (Section 4.2) of the electric field and an "apparent" (which is wiped away by an observer's state of motion) magnetic field, and an "intrinsic" (produced by acceleration of a charge) magnetic field. Härtel's guide to visualize electrodynamics based on Page's theory is available in his article in the folder Teacher's Guide in the CD and a reading from the beginning of the article upto Section 3.4.3 would be useful for students to make the transition from electrostatics to electrodynamics.

Weber's force between charged particles depends on the relative velocities and accelerations of charged particles and is a type of action-at-a- distance Theory. All theories indicate magnetic interactions over the entire universe. There are more details in "Fields and

Momentum Conservation" in Appendix B. The reader may defer to view the video lectures recommended in the topic till the end of Chapter 8. Section F.5, Appendix F discusses the several theories of interaction between charge carriers.

Classical Electrodynamics and Electromagnetism–How they came about

When Ampere gave the final touch to his force expressions, he did not think of a "magnetic field". The concept of a magnetic field was mentioned by Michael Faraday. Maxwell later introduced this in the formulation of the set of equations now known as "Maxwell's equations" and Hendrik Lorentz in the Lorentz Force law. It was believed till the year 1905 that light waves propagated in a medium called the aether, until the Michelson-Morley experiment (1887) disproved the existence of the aether (See Section F.1 in Appendix F). This made Einstein abandon the concept of the aether in 1905, which Maxwell earlier had believed carried the undulations [or fluctuations or perturbations or signals] of the electromagnetic fields, and signaled the beginning of (early 20th century) of Classical Electrodynamics and Electromagnetism. Classical field concepts were begun to be used extensively to explain all known phenomena of electricity and magnetism.

The Lorentz Force Law from the Viewpoint of Relativity

After Einstein declared that the aether was superfluous, the Lorentz (co-ordinate) transformation equations (which were used by Einstein in the Special Theory of Relativity with it's postulate of the absoluteness of the speed of light), were applied to describe the magnetic field as a relativistic *effect* of the electric field.

I urge the reader to view a set of power point slides with animations that this author prepared which traces the development of Classical Electrodynamics beginning from Action-at-a-distance and the Aether theories. The Power Point Presentation is available beginning slide #59 of "Chapter 6" which is included in the CD along with this book.

The Lorentz contraction, which is a consequence of Lorentz's co-ordinate transformation (which formulates the equivalence of inertial frames of reference, a postulate of the Special Theory of Relativity) is used to explain the force between a moving charged particle and a current carrying wire.

It will be instructive for students to read Sections F.1 to F.3 in Appendix F to gain familiarity with the Lorentz set of equations and it's use to determine the fields of charges at rest, in motion and the force between current carrying wires on a moving charge. Particular emphasis should be given on the way the fields transform (Eqs. 24 to 26, Section F.2, Appendix F) to an observer in the frame F' moving with uniform velocity v using the example of two sheets of charge described in Section F.2, Appendix F, which led to the definition of the magnetic field as a vector given by Eq. 34. Eq. 33 (and Eq. 34) in Section F.3, Appendix F to describe the Lorentz Force Law Section F.4, Appendix F, from the viewpoint of the Theory of Special Relativity. It is especially important to become familiar with the configuration of the field of a charge in motion which may be sketched as depicted in Fig. 10.40 of this text and also in Figs.3.6 and 3.7 in Härtel's article "A quantitative approach to electricity" available in the Teacher's Guide in the CD.

In the same article, Prof. Härtel has explained the repulsion between two steady-current carrying wires, which we had described using the Lorentz Force law in Section 6.1 using the Theory of Special Relativity. Beginning by showing the enhancement of the field perpendicular to the velocity (of a charge in motion) and a reduction in the forward and backward direction in Sections 3.2.3 - 3.4.3, Härtel explains the interactive force when pairs of charges interact and then shows how wires with currents interact and develop a repulsive force.

Note: The reader should understand the differences in the meanings and application of different theories in electrodynamics; Classical Electrodynamics, Quantum Electrodynamics, and Quantum Field Theory.

The Theory of Special Relativity is applicable to the electric and magnetic fields of **Classical Electrodynamics** where the fields are assumed to be represented by lines of force which are continuous and vectors denoting their strength at discrete points in space. In Classical Electrodynamics, photon-like *quantization* of fields is *not* implied. However, *quantizing* using Quantum Mechanics (or Quantum Physics) is allowed in Classical Electrodynamics to describe the energy levels of electrons of atoms by the introduction of Planck's quantization of energy in the solutions to Schrodinger's wave equation and the Density of States function of conduction band electrons of metals and semiconductors (Refs. [17,30], Appendix B and E).

In the realm of **Classical Electrodynamics** of interacting charges, we measure the force of repulsion, say of like charges, by suspending two charged pith balls as Coulomb did to arrive at the force law between charged particles. The electrical repulsion is attributed to the electric field surrounding them. Even when the charges were smaller, like electrons (say), the force of repulsion is attributed to the field surrounding them. However, with such small objects, as in the case of electrons and protons, such experiments cannot be performed.

Quantum Electrodynamics attempts to explain their interaction, by means of photons or packets of energy. For example, when two electrons approach one another, a photon passes between them; one emits the photon and the other absorbs it, and thus the two electrons repel one another.

When it became necessary to explain certain aspects of the interaction of an electron with radiation, especially in defining the structure of the electron at high energies, the photon-like *quantization* of the electric and magnetic fields themselves was made which is described by **Quantum Field Theory**. In Section F.5, Appendix F, it is seen by calculation, that there are such large numbers of photons in a signal received at a point a kilometer away emanating from a cellphone that the classical notion of the electric and magnetic fields as lines of force or as vectors in space in Classical Electrodynamics is a limit of large quantum numbers of a quantum-mechanical description in terms of real or virtual photons, and that in the domain of macroscopic phenomena and even some atomic phenomena, the discrete photon aspect can usually be ignored.

The Theory of Special Relativity is also applicable to Quantum Electrodynamics (QED) and Quantum Field Theory (QFT).

This textbook does not treat QED and QFT in any detail and is restricted in scope to the application of The Theory of Special Relativity to Classical Electrodynamics to show how the magnetic field is relativisitic by a transformation of fields (Appendix F).

We will learn more about these aspects in subsequent chapters.

SUMMARY

Beginning with the Lorentz Force Law statement of the magnetic force on a moving charge q in a magnetic field, we learned how the charge is deflected by a magnetic field. It was Ampere who proposed a force expression between two current elements applicable to all systems of interacting current elements. We learned that Grassmann also separately gave a force expression between current elements which failed to satisfy Newton's Third Law in some situations, unlike Ampere's expression which satisfied Newton's Third Law in all situations. Maxwell and Lorentz separately formulated what came to be called the Lorentz force law with the magnetic field whose direction was defined using Ampere's term for the Directrix in his force expression between current elements.

A description of the forces between two current-carrying wires, made of copper which is not a magnetic material, made us ponder about the Lorentz force law and how the magnetic force, which cannot do any work since it acts perpendicular to the direction of motion of the charge, can make a wire move.

The answer, we found, lies in Hall Effect charges which accumulate on one surface deflected by the magnetic field while exposing lattice ions on the other side. The resulting electric field exerts a force on the moving electrons, balancing out the magnetic force and the current of electrons continues along the wire, not being deflected anymore. The electric field also exerts a force on the lattice ions, which are stationary, and it is this electric force, which is unbalanced (because the lattice ions are stationary) that made the wire move.

The rest of the chapter was devoted entirely to show how a a mass suspended from a current carrying wire when placed in a magnetic field, rises. A detailed study of the forces that the current loop experiences revealed that it was the battery that did the work of raising the suspended mass, and not the magnetic field.

We then learned that magnetic fields are a relativistic effect of electric fields and that in Quantum Field theory, the photon is a force carrier mediating the two fields. And, since the electric field of a charged particle fills the entire universe, then the magnetic field also fills the entire universe. We also learned that all the other theories including those which may be derived from Action-at-a-distance theories such as Weber's Electrodynamics, and L. Page's Emission Theory, also indicate that the magnetic field fills the entire universe.

Ampere's force expression (Section F.1, Appendix F) represents an action-at-a-distance concept without a magnetic field. It was Faraday who mentioned the idea of a "field" which was later formalized by Maxwell and Lorentz. We then learned that the Lorentz Force law can be derived based on the postulates of Special Relativity using Lorentz's transformation equations.

At the end of the chapter, we learned the scope of the subject areas of Classical Electrodynamics, Quantum Mechanics (or Physics), Quantum Electrodynamics and Quantum Field Theory. We learned that unlike in Classical Electrodynamics where the fields are considered as lines of force or as vectors in space, Quantum Electrodynamics quantizes the fields themselves which became necessary to explain the interaction of an electron with radiation.

Questions

1. After writing Ampere's force expression between two current elements in the electrodynamic system of units use appropriate conversions to convert to the International system of units Ref. [38]. Do Ampere's expressions obey Newton's Third Law? Write Grassmann's force expression and discuss whether it conforms to Newton's third law. Explain the situations where the expressions of Ampere and Grassmann are applicable in conformance with Newton's Third Law. Refer Section F.1, Appendix F.
2. Write Ampere's force expression for the force exerted by a closed circuit S on a current element of another circuit. Then, write Grassmann's force expression for the same situation. Compare the two and prove their equivalence. Refer Section F.1, Appendix F.
3. From which basic force expression did Maxwell and Lorentz formulate the Force law incorporating the magnetic field? Explain. Explain how magnetic fields affect moving charges and state and explain "Lorentz law". Refer Section F.1, Appendix F.
4. Describe the processes that cause two current-carrying wires in close proximity to each other to attract.
5. What are Hall Effect charges? Explain. With the help of neat sketches explain how a current-carrying wire moves when placed in a magnetic field.
6. Explain why magnetic forces do no work using "Lorentz law".
7. A current-carrying rectangular loop (of width 'a') of wire carrying a current I is placed in a perpendicular magnetic field B. A mass m is suspended from the loop. Describe all the processes that cause the loop to rise when the current is increased and to prove that magnetic forces do no work show that the work done to raise the weight is due to the voltage source.
8. What is the extent of a magnetic field? Explain how the magnetic field is *relativistic*.
9. Name all the famous scientists who believed that the Aether carried electromagnetic perturbations and when was the idea of aether discarded. Explain.
10. What is an inertial frame? Can a frame moving with constant velocity be inertial? Explain.
11. What are the postulates on inertial frames and speed of light in The Theory of Special Relativity?

12. Write the Lorentz transformation and the inverse Lorentz transformation (see Section F.2 in Appendix F). Explain Lorentz contraction and state how this is different from the common observation of lengths of moving trains from a railway platform. What is time dilation? Explain.
13. With the example of two stationary sheets of charge show how the electric field transforms to an observer in the moving frame F' moving with uniform velocity v. Refer Section F.2, Appendix F.
14. Sketch and show how the electric field of a charge moving with uniform velocity is relativistically compressed. Refer Section F.3, Appendix F.
15. Symbolically sketch the field of a single charge carrier when a) stationary, and b) moving with uniform velocity. Then draw the correct diagram using relativistic effects of the interaction between two carriers moving at right angles. Hint: Refer Sections 3.3 and 3.4 in Härtel's "Teacher's Guide" in the accompanying CD.
16. Symbolically sketch the interactions between a single electron and a current carrying wire when the electron a) is moving parallel to the current, b) is moving anti-parallel, c) moving at right angles towards and away from a current, and d) is near a conducting loop using relativistic considerations. Hint: Refer Section 3.4 in Härtel's "Teacher's Guide" in the accompanying CD.
17. After tracing the origin of Quantum Electrodynamics and Quantum Field Theory explain why these topics were introduced in the study of Electrodynamics. What are the force carriers of electrons and protons? Briefly describe the force carriers of stationary and moving electrons and protons.
18. Explain why John David Jackson considered the discrete photon aspect of electric and magnetic fields **E** and **B** "can be ignored or at least glossed over".

    Calculate the number of photons per second per meter^2 in the radiation from a cellphone antenna operating at 1 GHz with a radiating power of 300 mW at a distance 1 Kilometer from the phone. Does the discrete nature of the photons become evident from their number? Hint: See Section F.5, Appendix F.

Additional practice questions are available in the CD with this book.

7

Motional emf and Generators

The application of Faraday's law, which is an important law of Physics is the theme of Chapter 7. The chapter begins with an introduction to the phenomenon of the co-occurrence of a non-coulomb electric field with a changing magnetic field and its use in determining the induced emf when a coil in which the current is varying is placed near another. After describing another experiment, in which an induced emf is detected when one coil with a fixed current is moved towards another coil whose ends are short-circuited, the important concept of motional emf is explained and gradually developed to describe how generators and microphones operate. Often, it may seem that Faraday's Law and Lorentz's force law can be used to predict induced emfs in circuits with inductors and magnets. Subtle differences in the two approaches are highlighted and explained, and a new approach devised by Hermann Härtel based on the theory of special relativity is indicated for readers. Similarities in physical aspects of sound waves in air, and electrons moving in a vacuum, led engineers to develop devices that produce high frequency and high power electromagnetic waves for transmission to satellites and this aspect is treated in the section on microphones.

In the last section, the student is introduced to the idea of mutual-induction and self-induction by means of accelerating charge carriers due to Hermann Härtel and pointers to relevant sections of his article "A Qualitative Approach to Electricity: A Guide to Visualization of Electrodynamics using symbols and images" are provided.

7.1 THE MAGNETIC FIELD—A REVIEW

Magnetic interactions have some similarity to electric interactions:

1. There is *both* attraction and repulsion.
2. The *superposition principle* is valid.
3. The interactions pass *through matter*.

The *differences* between magnetic and electric interactions are:

1. A magnet is *permanently* magnetised while a charged plastic pen loses its charge within a relatively short time (see Section 1.25).
2. The magnetic field around a current carrying wire has a *curly* pattern, which is *not* seen with electric fields of point charges.

7.2 THE PHENOMENON OF ELECTRIC FIELD ASSOCIATED WITH A TIME-VARYING MAGNETIC FIELD

Consider two coils of wire *near each other*, positioned on a common axis, and one coil is connected to a *power supply* with an ammeter in series as shown in Fig 7.1 (after Chabay & Sherwood, 1995).

The second coil is *short-circuited at both its ends* with an ammeter in series. There is *no voltage* source in the second coil.

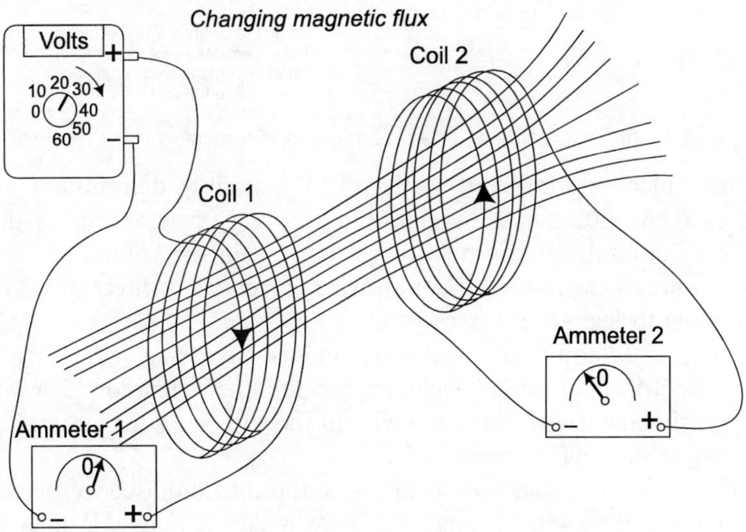

Fig. 7.1 The phenomenon of magnetic induction.

When we increase the current in the coil 1 at a steady rate, there is a steady negative current measured by ammeter 2, as shown in the Fig 7.1.

If we keep the current constant in coil 1, there is no current in the second coil (not shown in Fig 7.1).

And, if we decrease the current 1 at a constant rate, there is a large steady positive current in ammeter 2.

7.3 LENZ'S RULE FOR THE DIRECTION OF THE INDUCED EMF

Recollect our observations in Chapter 5 that a changing magnetic field of a solenoid or coil is associated with a non-Coulomb electric field as shown in Fig 7.2.

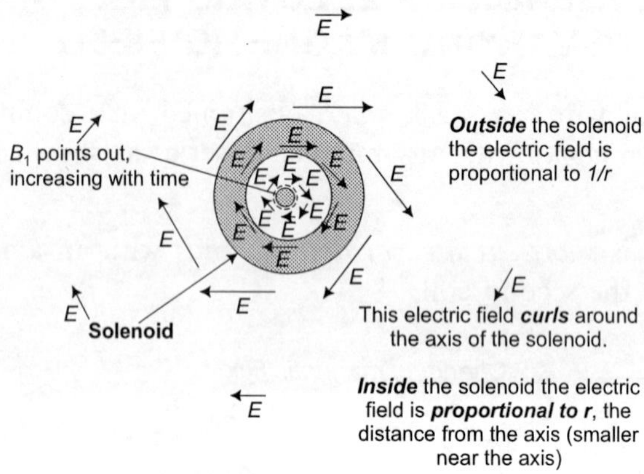

Fig. 7.2 The co-occurrence of an electric field with a time-varying magnetic field.

Lenz's rule, which we studied in Section 5.13, is used to determine the direction of the induced emf in many situations. Imagine that you place a wire around the changing flux, so that the non-Coulomb electric field drives current in the wire.

The non-Coulomb electric field would drive current in a direction to make a magnetic field that attempts to keep the flux constant.

For example, if B_1 points out the page and is increasing, as shown in Fig. 7.2, the non-Coulomb electric field curls clockwise around the solenoid. If a wire encircles the solenoid, conventional current runs clockwise in the wire, and this induced current produces a magnetic field in the region. See Fig 7.3.

In what direction is the magnetic field inside the solenoid, due to the clockwise current?

Current flowing clockwise through the wire makes a magnetic field into the page at *center* of wire, decreasing the net flux.

7.4 CORRECT APPLICATION OF LENZ'S RULE

If a metallic ring encircles the solenoid, then the "induced" current driven by the non-Coulomb electric field produces a magnetic field into the page at center of the wire, *decreasing* the net flux.

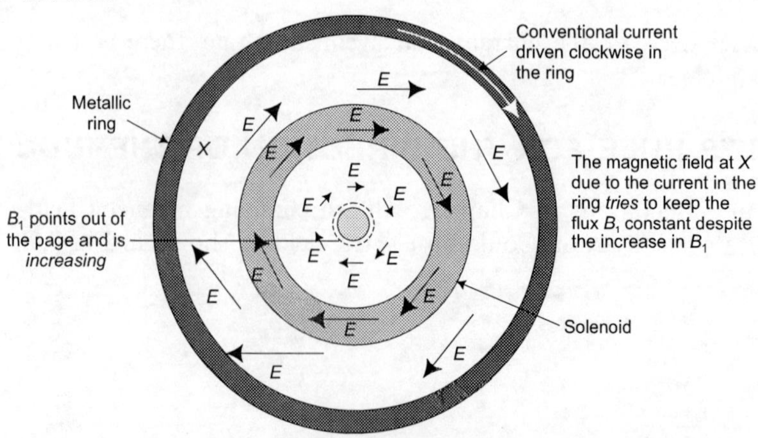

Fig. 7.3 Correct application of Lenz's rule.

Applying this rule, it should be that if the current is increased at a steady rate in the coil 1 in the circuit of Fig 7.1, there will be a steady negative current measured by ammeter 2, and this current makes a steady magnetic field B_2 to the left inside the coil and so the emf and current should be constant (Chabay & Sherwood, 1995, used by permission).[2] This is demonstrated in the experiment of Fig 7.4.

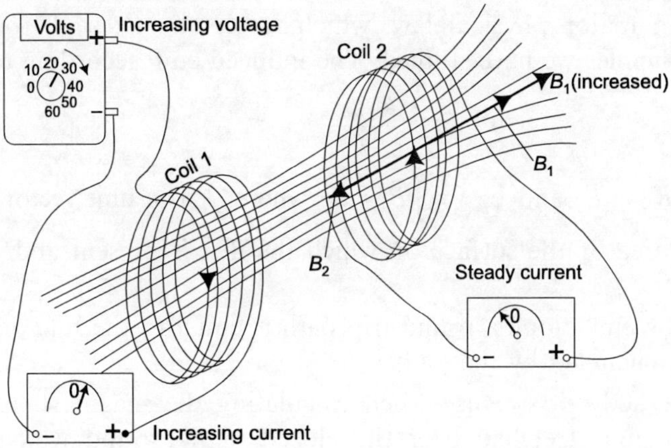

Fig. 7.4 Steady current in coil 2 when the current in coil 1 is increasing at a constant rate.

7.5 INDUCED CURRENT WHEN THE SECOND COIL IS TILTED

Is the *induced emf* and therefore, the current in the second coil affected when the second coil is *tilted at an angle* θ? (after Chabay & Sherwood, 1995).

When the second coil is tilted by an angle θ, the **flux** *involves* (that induces the emf) the *perpendicular* (to the coil face) component of $\boldsymbol{B}_1$, and therefore, the induced current in the second coil is reduced by the factor **cos** θ (Chabay & Sherwood, 1995, used by permission).[2] This is shown in Fig 7.5.

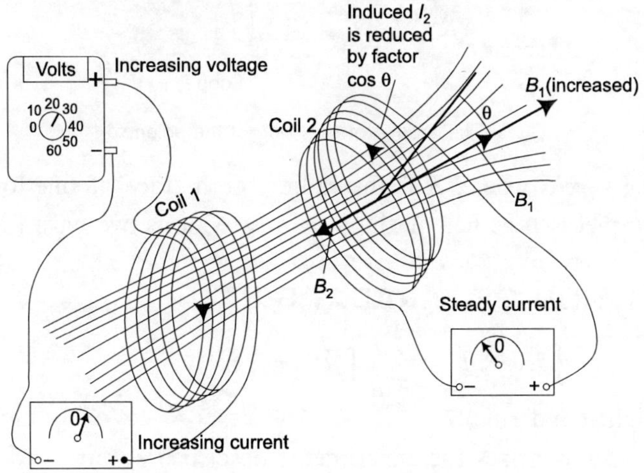

Fig. 7.5 Coil 2 tilted at an angle θ, causes induced current to reduce by factor cos θ.

7.6 FARADAY'S LAW

The principle of inducing an emf in the second coil, due to a changing current in the first coil on the common axis, is used in transformers to *transform* voltage.

The phenomenon discovered by Michael Faraday, and known as Faraday's law of *induced emf* is a major physical law, which *cannot be derived* from any of the other fundamental principles we have studied. The induced emf according to the law, is given by

$$\text{emf} = -\frac{d\phi_{mag}}{dt}$$

where emf $= \oint \vec{E}_{NC} \cdot d\vec{l}$ and $\phi_{mag} \equiv \int \vec{B} \cdot \hat{n} dA$ where, $\hat{n}$ is a unit vector perpendicular to the area ΔA, a tiny area of the surface on which the flux is present and which in the integral becomes dA.

In words: The emf along a round-trip path is equal to the rate of change of magnetic flux on the area encircled by the path.

What area exactly do we use when calculating the magnetic flux in Faraday's law? Imagine a soap film stretched over the closed path around which we are calculating $\oint \vec{E}_{NC} \cdot d\vec{l}$. We calculate ϕ_{mag} on the area covered by that soap film.

A closed path, like a circuit (or a line) defines an open surface (like the soap film stretched) whereas, a closed surface, defines a volume like boxes, spheres and tetrahedrons! (Section 5.12).

To determine the open surface for a solenoid, we imagine a closed path like a thin disk stepping across one loop of the solenoid whose area is that of the face as shown in the figure.

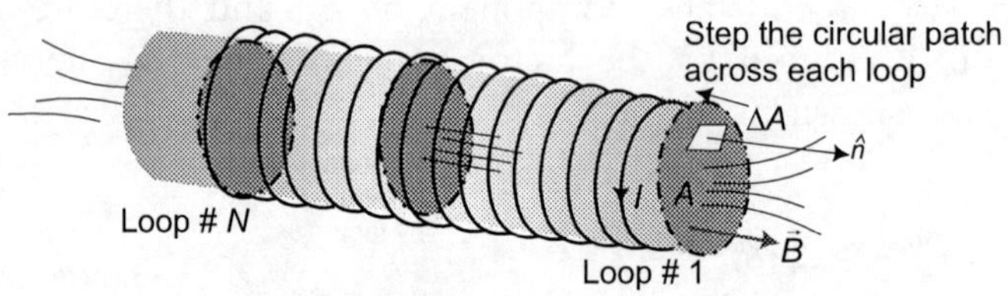

The area A of one loop of the solenoid

The area is $A = \pi R^2$ where, R is the radius of the face of one loop of the solenoid.

Putting these pieces together of the previous sections, we have the following form:

FORMAL VERSION OF FARADAY'S LAW

$$\oint \vec{E}_{NC} \cdot d\vec{l} = -\frac{d}{dt} \int \vec{B} \cdot \hat{n} dA$$

(sign given by right-hand rule)

Faraday's law summarizes the experiments described so far:

1. A bigger emf is associated with a faster rate of change of magnetic flux (if the magnetic flux is constant there is no emf).

2. If there is a bigger area with the same perpendicular magnetic field there is a bigger emf.

Faraday's Law Applied to Moving Coils

We have seen a non-Coulomb electric field associated with a *time-varying magnetic flux*.

But wait, this is *not the only way* to produce a varying flux.

With a *steady* current in the first coil, you can *move* the first coil closer to the second coil. This is shown in Fig 7.6.

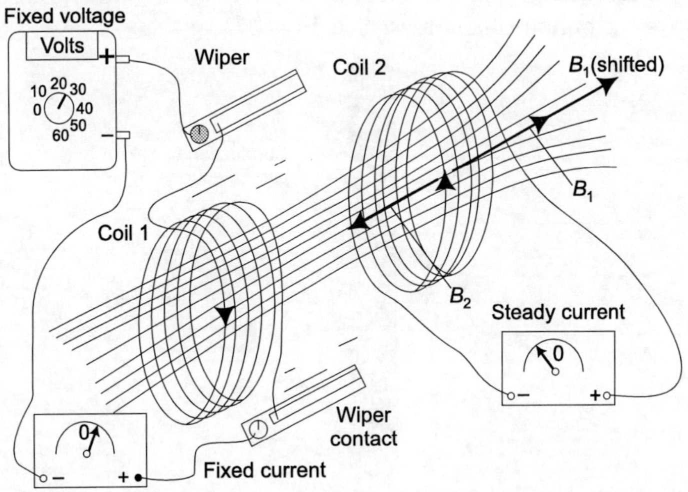

Fig. 7.6 Induced emf by moving the 1st coil with a steady current towards the 2nd coil.

The magnetic flux inside the second coil increases, and while the flux is increasing there is an emf (and a current in the second coil). It does not matter which coil moves: in both cases there is a change of magnetic flux through coil 2 and an induced current.

Bar magnets when moved near a coil will also produce an emf in the coil. See "Faraday's law and Moving Coils or Magnets" in Appendix A for a description of this phenomenon.

We next study the very important phenomenon of motional emf which is fundamental to the way generators and microphones operate.

7.7 MOTIONAL EMF

We have seen that a *changing* magnetic field is associated with a curly (non-Coulomb) electric field.

However, there is one more way that magnetic flux can *change....or does it really?* We shall study this.

And, this is the way that many types of electric generators, galvanometers, loudspeakers, microphones, and motors operate.

While, in this method Faraday's law still correctly predicts the induced emf, there is a *subtlety involved in the process.*

We shall demonstrate one more phenomenon of charge by observing what happens when charges are moved—simply moved.

Can you *move* a bunch of charges by holding them in your fist?

It is really very *simple.*

Simply grab *any* piece of wire, even *without a current in it,* you have a bunch of jumpy energetic conduction band electrons. Move the wire, and lo and behold, you have successfully moved a bunch of charges! See Fig 7.7.

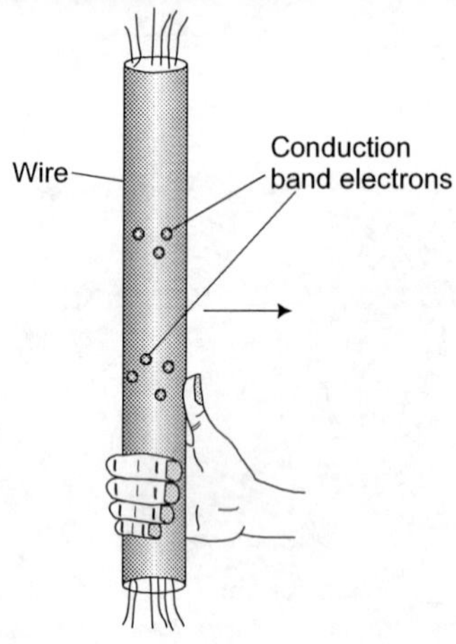

Fig. 7.7 Moving a fistful of charges.

Now, stretch your imagination and don't you think *whatever you move contains some charges* or not? Simple, was it not?

Consider the case of pulling a *metal bar* at a constant speed v, along frictionless *metal rails* through a uniform constant magnetic field B pointing into the paper, as shown in Fig 7.8. A resistor of, *resistance* R is connected *between* the *rails.* As the bar is pulled along the rails, the resistor will heat up (Chabay & Sherwood, 1995 used by permission).

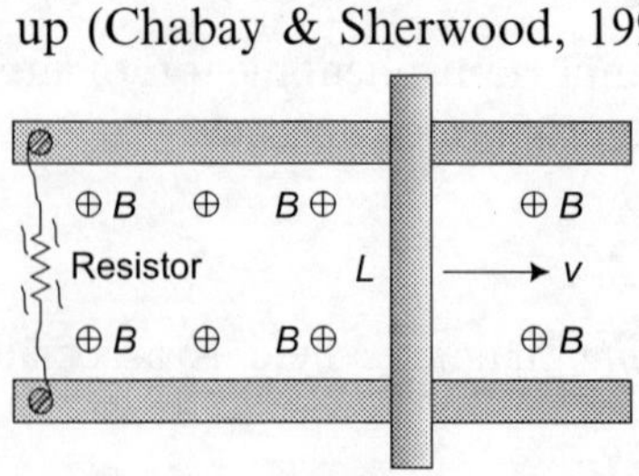

Fig. 7.8 Moving a bar on rails immersed in a constant magnetic field.

The *additional flux* enclosed by the circuit loop consisting of the moving bar, rails, and resistor is

$$\text{additional flux} = \Delta\varphi_{\text{magnetic}} = B\Delta A = BL\Delta x = BL(v\Delta t)$$

where, L = length of the bar and Δx is a small incremental change in the distance the bar moves when it is pulled with a velocity v.

The area, the loop comprising the resistor R, the rails and the bar encloses, is A, and ΔA is a small incremental change in the area when the bar moves a distance Δx.

And, the rate of change of flux

$$= \frac{d\varphi_{magnetic}}{dt} = \frac{(BLv)\Delta t}{\Delta t} = BLv$$

This is also equal to the emf generated in the circuit.

The amount of current through the resistor is

$$I = \frac{emf}{R} = \frac{BLv}{R}$$

Faraday's law is very general and any *change* in magnetic flux, whether *due to* a change in magnetic field or, a *change in area*, produces an emf equal to the rate of change of flux.

In fact, both may be present simultaneously, as when you drag the metal bar along the rails, *and* the magnitude of the uniform magnetic field *is increasing* as shown in Fig. 7.9.

The flux is ***BA***, and the rate of change of the flux is due *in part to the change* in the magnetic field, and *in part to the change* in *the area* enclosed by the changing path.

The *resistor* will heat up!

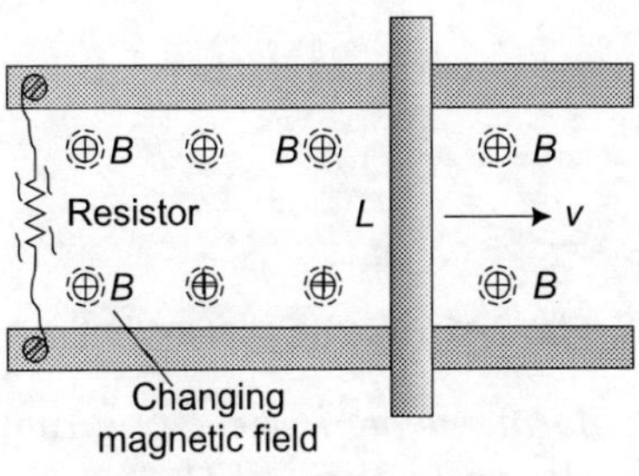

Fig. 7.9 Moving a bar on rails immersed in a magnetic field that is changing.

The equation of the rate of change of flux....

$$\frac{d\varphi}{dt} = \frac{d}{dt}(BA) = \frac{dB}{dt}A + B\frac{dA}{dt} = \frac{dB}{dt}A + B\frac{Ldx}{dt}$$

Change of B, fixed path — Change of path

$$= \frac{dB}{dt}A + BLv$$

The calculation is a bit mysterious. Physically, *where* did the emf represented by BLv come from if it is *not* associated with a *changing* magnetic field?

We will solve this *puzzle* now.

Let us take another look at how emf was produced when the *magnetic field* was constant in our experiment, and the *area* of the loop enclosed by the circuit *was changing*. See Fig. 7.10.

We have a conducting metal bar and we are moving it.

Note carefully that we are moving the jumpy *energetic conduction band electrons* in it as well. See Fig. 7.11.

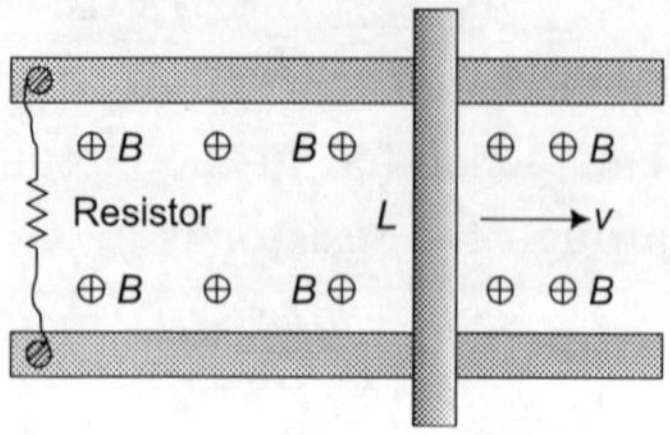

Fig. 7.10 Moving a bar on rails immersed in a magnetic field.

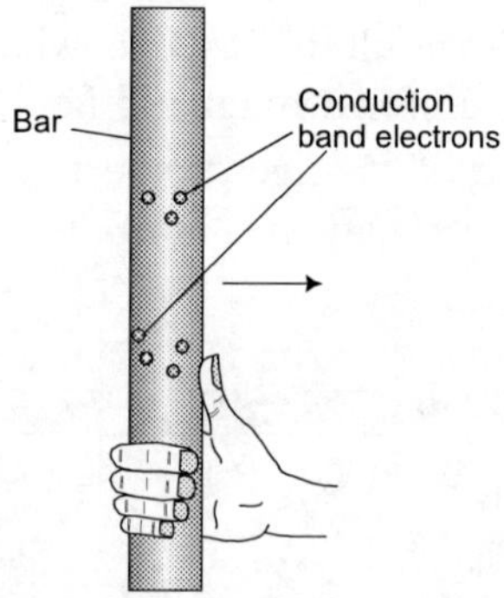

Fig. 7.11 Moving charges in the bar.

Now let us stick in a *uniform* ***magnetic field*** pointing ***into*** the area where the bar is moving and move the bar as shown in Fig. 7.12.

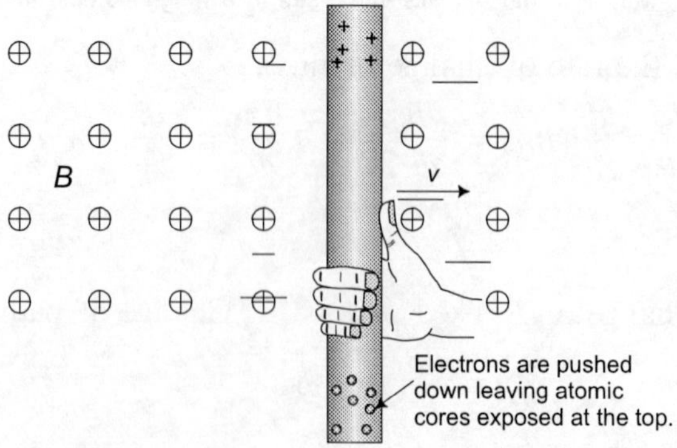

Fig. 7.12 Polarizing the bar by moving it on rails immersed in a uniform magnetic field.

As a result of the magnetic force (due to the presence of the uniform magnetic field), the free electron sea shifted, and the bar polarized and a Coulomb electric field develops as in Fig. 7.13.

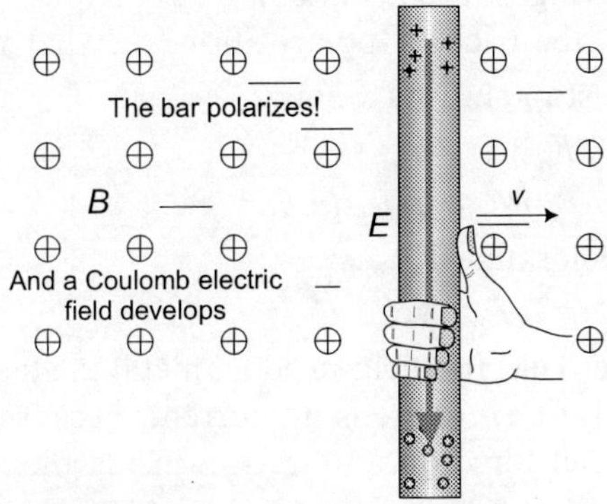

Fig. 7.13 Development of a Coulomb electric field inside the bar.

Move the bar the ***other way*** and it is oppositely polarized as shown in Fig. 7.14.

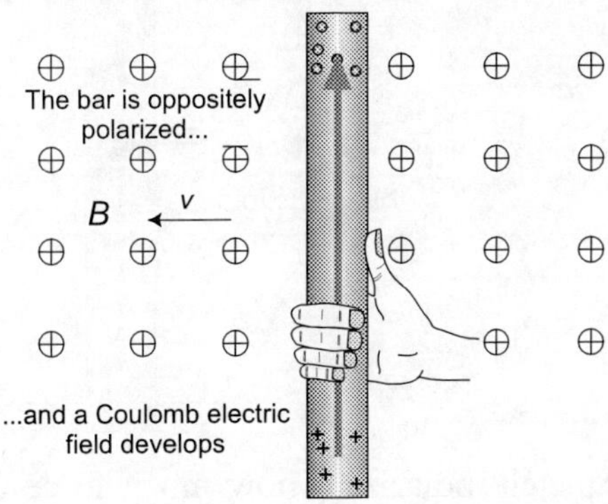

Fig. 7.14 Move the bar the other way and the bar is oppositely polarized.

Therefore, the *induced emf* is due to *separation of charges* in the bar as it moved. This we call motional emf, the *BdA/dt* term in the emf equation....which can result even when the magnetic field is *not changing* or is constant.

When does the *electron sea* **stop shifting?** Precisely, when the $qv \times B$ magnetic force (dark arrow pointing downwards) is balanced by the Coulomb electric force as shown in Fig. 7.15.

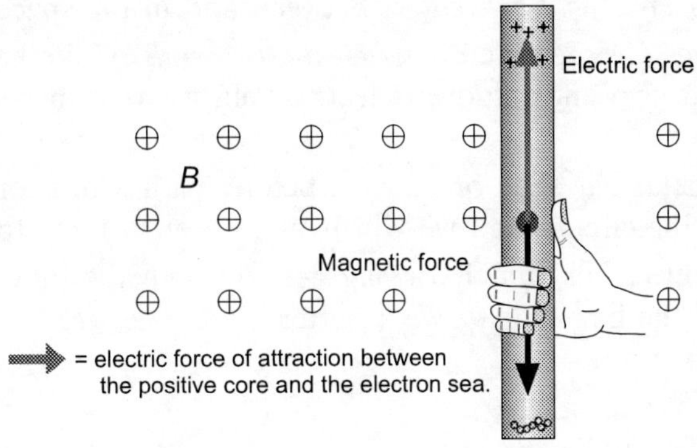

Fig 7.15 The sea stops shifting when the magnetic force equals the Coulomb electric force.

If the **bar** is kept moving at a constant rate ***v***, it is possible to calculate the Coulomb *electric* field produced by the trick of polarization, or what we call separation of charges.

The electron sea will stop shifting when

$$F_{electric} = F_{magnetic}$$

or,

$$qE = qvB \sin 90°$$

using the rule $v \times B$ to determine $F_{magnetic}$

Therefore,

$$E = vB$$

Note that there is a nonzero electric field inside a metal in static equilibrium (bar moving at constant speed, no acceleration). There is no current, because the electric force is balanced by a magnetic force: the net force on an electron inside the metal is zero.

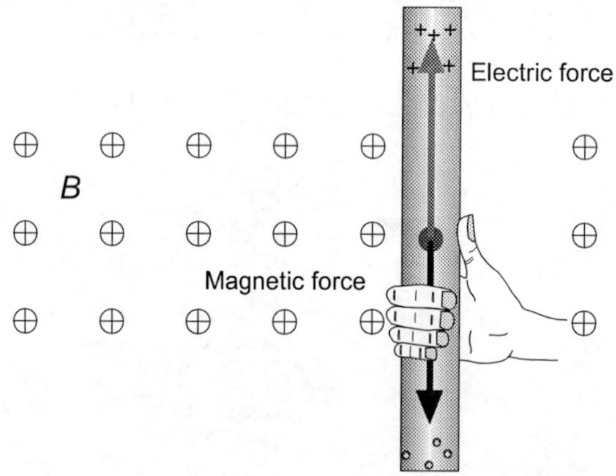

Once the bar is completely polarized, how much force must we apply to the bar to keep it moving at constant speed?

None! Since the bar is moving at constant speed, the *net* force on it must be zero. No force is needed to keep it moving, since no other forces act on the bar.

And the *polarized* bar acts like a battery.

A power point presentation with animations showing the above actions is available in the slides of Chapter 7 in the CD alongwith this book.

In a battery, charges are *separated* by chemical reactions, and an *emf* develops between its plates, and an *electric field as well*, in between and in the space surrounding.

Similarly, charges are separated *between the two ends* of the bar, and an *emf* develops therefore *between* its two ends, and an electric field as well, in between and in the space surrounding.

If we let this metal bar slide on rails as before, with a resistor connecting the rails, a current will run in the circuit, and we will have to exert a force to keep the bar moving.

Also note that electrons run the *wrong way* through the moving bar, moving toward the *negative* end of the bar as if *inside* a battery.

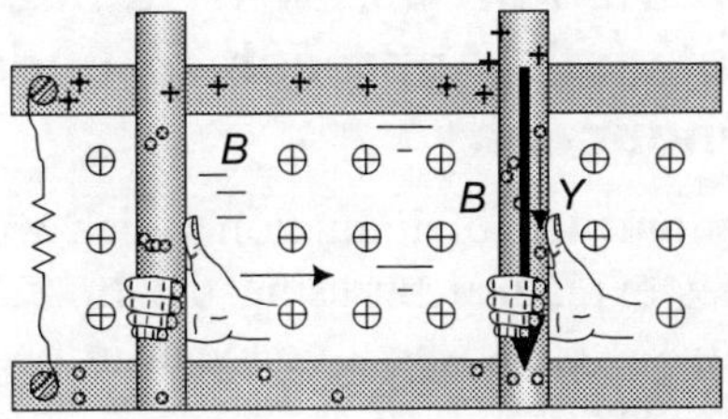

Fig. 7.16 The electrons move the wrong way inside the bar.

Why do the electrons move the wrong way inside the bar (shown by the arrow marked Y in Fig. 7.16)?

The current through the *resistor* is *continually depleting* the charges on the ends of the moving bar, with the result that the electric field inside the bar, is always *slightly less* than is needed to balance the magnetic force.

And since the upward electric force qE is *slightly less* than the downward magnetic force qvB, so the electrons move towards the negative end of the bar.

A *reduction* of internal field due to *depletion* of terminal charges and replenishment by the *separation of charges inside the bar*.... electrons stripped from immobile *positive atomic cores*...motional emf process in generators as shown in Fig. 7.17.

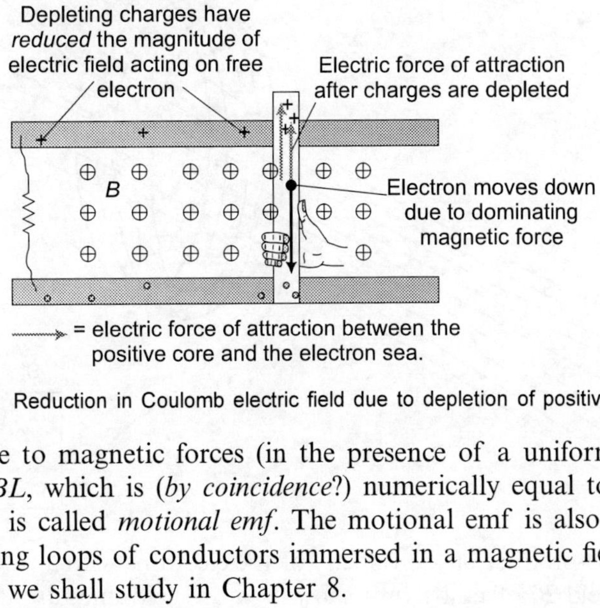

Fig. 7.17 Reduction in Coulomb electric field due to depletion of positive charges.

So, the emf due to magnetic forces (in the presence of a uniform *constant* magnetic field) is equal to vBL, which is (*by coincidence*?) numerically equal to $d\varphi_{magnetic}/dt$ for this situation. This emf is called *motional emf*. The motional emf is also termed a back emf when current carrying loops of conductors immersed in a magnetic field develop a torque as in motors which we shall study in Chapter 8.

In a moving component of a current-carrying circuit, a motional emf develops due to magnetic forces, and it can be shown that this motional emf is always numerically equal to the $d\varphi_{magnetic}/dt$ associated with the change in path surrounding the flux.

Faraday's Law or Lorentz Force Law? Traditional Explanation of Induction and Relativistic Approach to understanding the phenomenon of Induction

In some textbooks the phenomenon of induction which was explained in Chapter 5 using Faraday's Law, is sometimes explained by using the Lorentz force law. The reader would have observed that there does seem to be a connection between the two and he would be correct in making such a connection.

According to Härtel, the phenomenon of induced emf in a coil moving relative to a second coil with a steady current (Fig. 7.6) or a stationary permanent magnet in Härtel's experiment (see the right Fig. 3.3, Section 3.2.1 in Härtel's article "A Qualitative Approach to Electricity" in the Teacher's Guide in the CD), can be explained by the *Lorentz force law*.

For example, in the experiment shown in Fig. 7.6, imagine the coil 2 is moved towards coil 1 with a velocity v. Electrons in the coil 2 will be set in motion towards the left with velocity v and shown in the figure is one such electron in an exaggerated hashed segment of coil 2 of the right end-face.

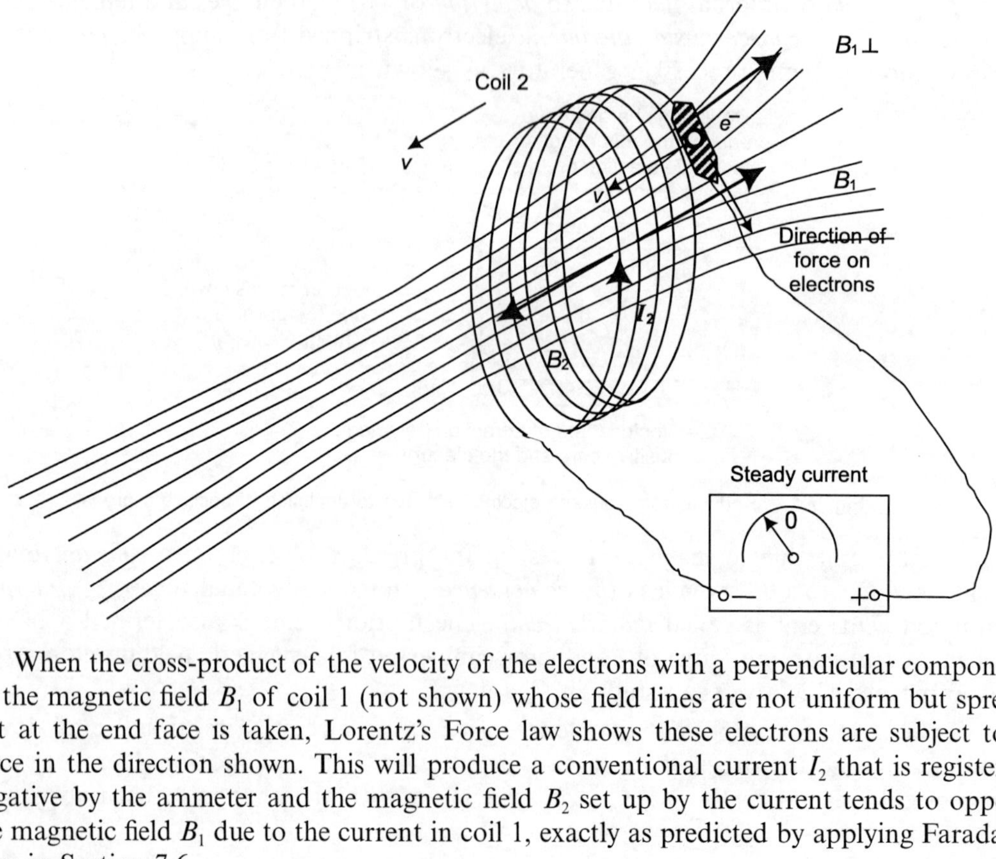

When the cross-product of the velocity of the electrons with a perpendicular component of the magnetic field B_1 of coil 1 (not shown) whose field lines are not uniform but spread out at the end face is taken, Lorentz's Force law shows these electrons are subject to a force in the direction shown. This will produce a conventional current I_2 that is registered negative by the ammeter and the magnetic field B_2 set up by the current tends to oppose the magnetic field B_1 due to the current in coil 1, exactly as predicted by applying Faraday's Law in Section 7.6.

However, the phenomenon of induced emf when both coils are stationary and the current in one coil is changed (Fig. 7.1) requires the application of *Faraday's Law only*. The same is true for the experiment in the left Fig. 3.3 in Section 3.2.1 of Härtel's article. (The magnetic field in Härtel's descriptions is of a magnet, not a current carrying coil, but that doesn't affect the observed phenomenon).

In the case of Fig. 7.6, with coil 1 moving towards the stationary coil 2, the flux was increasing in coil 2 and we made use of Faraday's Law to determine the induced emf and current in coil 2. Otherwise, one would have to introduce the velocity of a moving magnetic field line to find a v (relative velocity of the field lines with respect to the stationary coil 2) for the expression $v \times$ B, whereas there is no way that the velocity of a magnetic field line can be defined or measured.

Also, while it may seem that Faraday's law can explain all situations of induction, there is one situation presented by Feynman, Leyton and Sand, *Lectures on Physics*, Chapter 17, Vol. 2, Addison-Wesley Publishing Co. of a copper disc rotating in the field of a permanent magnet, which cannot be explained using Faraday's law. This special case is also described in Section 3.2, Fig. 3.4 in Härtel's article available in the Teacher's Guide in the CD.

While Härtel and others including Richard Feynman and Edward Purcell, do not question the correctness of the application of Faraday's Law, the question of treating Faraday's Law (Section 7.6) as a *causal* relation $\text{emf} = -\frac{d\,\phi_{\text{mag}}}{dt}$ as presented in textbooks, makes it difficult for students who, in many instances, see this as counter-intuitive and sometimes question "How can an electric field suddenly make it's appearance?" when attempting to gain a cognitive insight of the phenomenon.

Härtel's article presents a new approach considering relativistic effects and he states: "the Theory of Relativity states as a basic principle that length, time, and mass depend on the velocity of the system for which these values are defined. All this is a consequence of the postulate that the Speed of Light can only be reached approximately by any material object. No object can ever move faster than the Speed of Light. As a consequence of this postulate, it can be shown by calculation that the field around a charged particle changes in a specific way". These ideas are also presented in Section F.2, Appendix F "The fields of moving charges according to the Theory of Special Relativity".

The reader who may have understood these effects after reading the topic "The Lorentz Force Law from the viewpoint of Relativity" in Section 6.6 will not find difficulty in understanding the new more encompassing approach discussed by Härtel in Sections 3.3 to 3.7 of his article in the Teacher's Guide in the CD, which is comprehensive.

7.8 MOTIONAL EMF AND SURFACE CHARGES

In Fig. 7.18 is shown the motional emf with surface charges in the external circuit. It is the surface charges that cause a current to flow in the circuit, the rails and the bar.

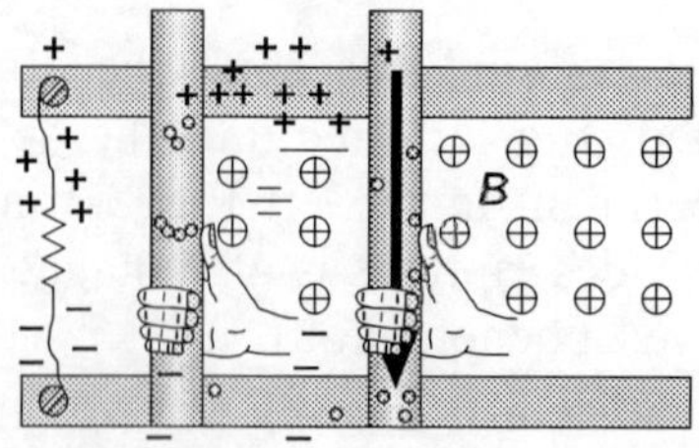

Fig. 7.18 Motional emf with surface charges in the external circuit.

The above principle of producing an emf is used in the design of *generators* and *microphones*.

To produce an emf, all you have to do is simply wave an object containing electrons and immobile positive cores in a magnetic field.

The electrons of the electron sea move and an electric field is established between regions where they bunch up and regions where the lattice ions are exposed. Therefore, an *emf* is created between the two regions.

But, is it *physically possible* to drag a metal bar on metal rails to generate an *emf*? How long should the rails extend? How long a *magnet* will you use to generate the magnetic field *B?*

This seemingly difficult problem is easily overcome by *folding* over the rails to form a continuous ring or a split-ring commutator, and locating the sources of magnetic field around a cylinder.

That is how electric *generators* are constructed.

We will now observe how an emf is produced in such a machine by *rotating* a rectangular wire loop fixed on an axle in a magnetic field.

7.9 GENERATING AN EMF IN ROTATING MACHINES

Suppose we turn a crank to rotate a loop of dimensions $w \times h$ at a constant angular velocity $\omega = d\theta/dt$ inside a uniform magnetic field B. Refer to Fig. 7.19.

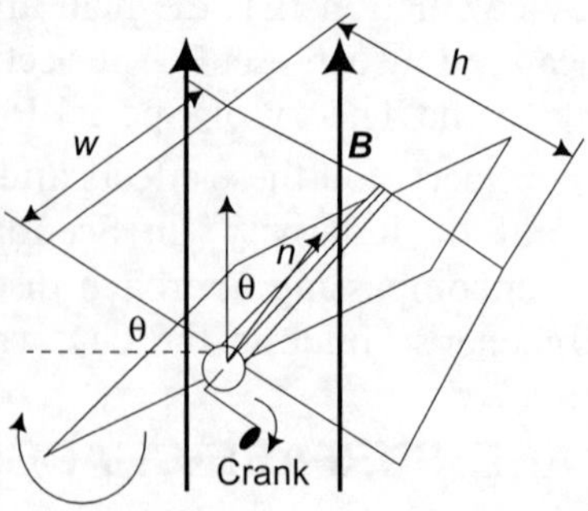

Fig. 7.19 Rotating a rectangular current carrying loop in a magnetic field.

When the loop is at an angle θ from the horizontal, the normal n to the loop is at an angle θ to the magnetic field B.

The area of the loop is $A = wh$.

From the expression derived in Section 5.2, the total flux Φ = sum of $B{\cdot}n\Delta A = B\cos\theta\,(wh) = B\,wh\cos\theta$, where $B\cos\theta$ is the *projected* area on the horizontal.

Since, $$\text{emf} = \frac{d\Phi}{dt} = \frac{d}{dt}(Bwh\cos\theta)$$

$$= Bwh\,\frac{d}{dt}\cos\theta = Bwh\,(-\sin\theta)\frac{d\theta}{dt}$$

or $$\frac{d\Phi}{dt} = Bwh\,(-\sin\theta)\,\frac{d\theta}{dt}$$

The rate of change of the angle is the angular velocity in radians per second,

$\omega = \dfrac{d\theta}{dt}$ And, since the emf $= \left|\dfrac{d\Phi}{dt}\right|$,

therefore, the emf $= \left|Bwh\,(-\sin\theta)\dfrac{d\theta}{dt}\right| = \omega Bwh\,(\sin\theta)$

7.10 POLARITY OF EMF IN ROTATING MACHINES

When a loop is driven with a force of rotation F_{rot}, it rotates and reaches the position X (Fig. 7.20(a)), and the emf developed is determined using the relation emf = $\omega Bwh(\sin\theta)$ which was derived in Section 7.9.

The force F_{rot} can be resolved into two components, F_1 and F_2 perpendicular to each other, to determine the *direction* of velocity of the charges. This is shown in Fig 7.20a.

The component F_1 does *not* contribute to the development of the emf, since it is *in line* with the magnetic field.

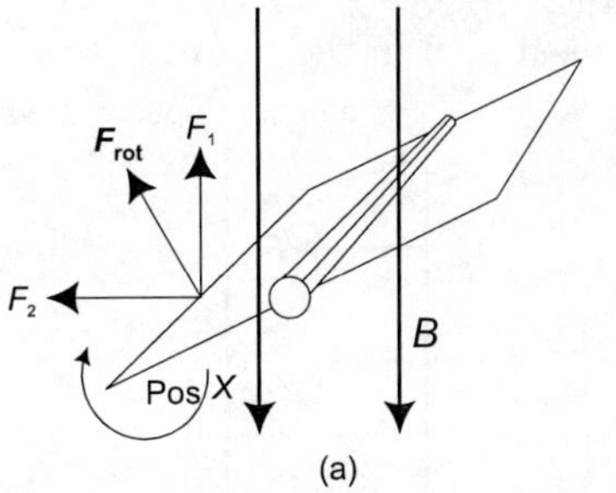

Fig. 7.20a Resolving the force causing rotation into components.

The emf developed by the component F_2 will be as shown in Fig. 7.20(b) using $v \times B$ applied to the movable charges (which are electrons) inside the wire with a *velocity component* v in the direction of F_2 and thus, we determine the *direction* of the magnetic force and the resulting separation of charges.

Note: When applying the hand rule for vector cross products described in Section 5.2 to moving electrons in a magnetic field, the direction of $v \times B$ will be 180 degrees (opposite) the direction indicated by the thumb.

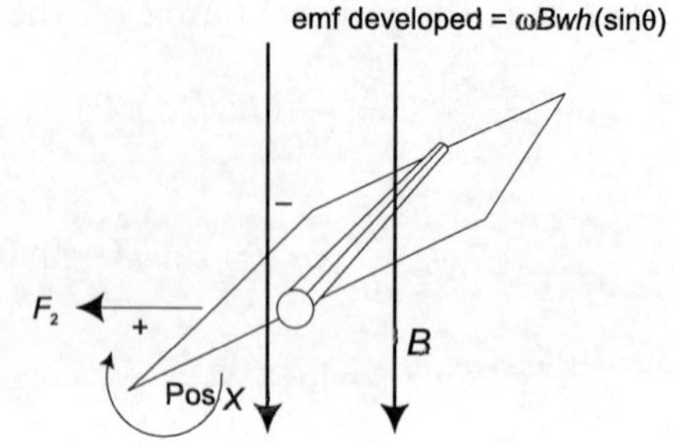

(b)

Fig 7.20b Determining the direction of movement of electrons and resulting emf.

A similar analysis when carried out on the loop, after it has rotated to position Y as shown in Fig. 7.21, shows that the *emf* developed is due entirely to the *velocity of charges* (*electrons*) in the direction of F_3.

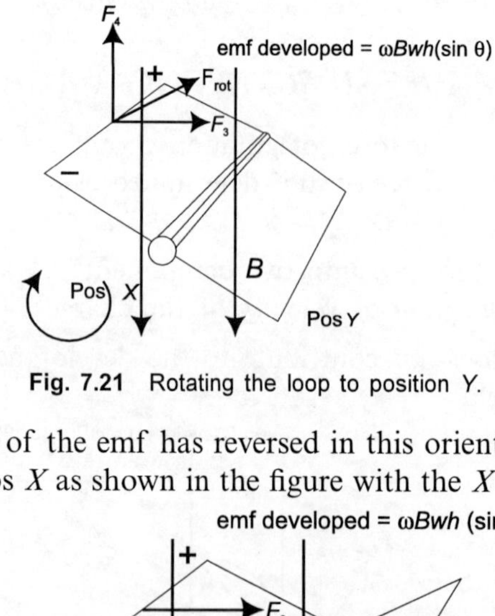

Fig. 7.21 Rotating the loop to position Y.

Note that the *polarity* of the emf has reversed in this orientation in comparison to the previous orientation of Pos X as shown in the figure with the X and Y positions combined.

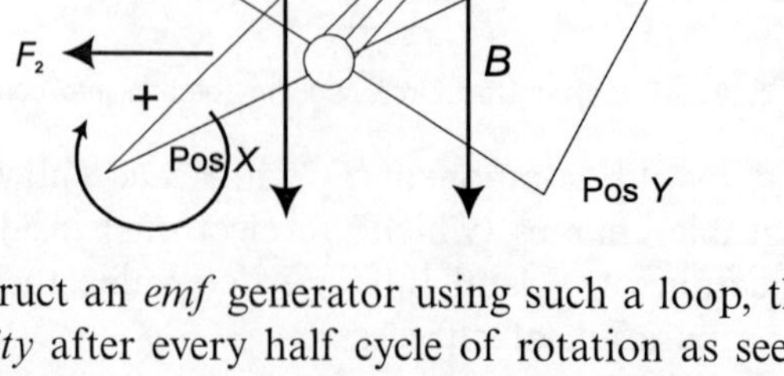

If we were to construct an *emf* generator using such a loop, then it would generate an emf that *changes polarity* after every half cycle of rotation as seen from the Fig. 7.22.

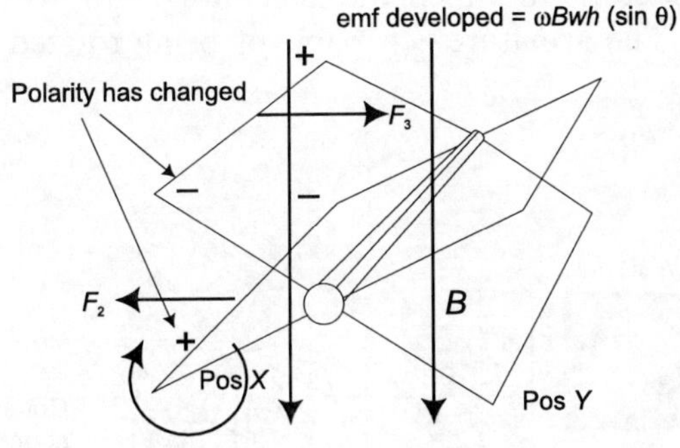

Fig. 7.22 The polarity change at the ends of the conductor when moved from position X to position Y.

7.11 UNIDIRECTIONAL EMF

The key to develop an *emf* with an *unchanging or unidirectional* polarity, is to make electrical connections to the coil in such a way that just as it crosses the position where its polarity is changing, we *reverse* the connection to the coil.

This is achieved by using a *split-ring commutator*, a device which is coupled to the rotating axle (or shaft) of the generator. The two segments (or splits) of the commutator, collect the *emf* automatically while *reversing* the coil connections where the emf polarity changes. This will be described in Section 7.12.

Armature and Commutator

Figure 7.23 is a photograph of an armature and commutator of a dc machine disassembled from the machine (reproduced with permission Chuck Morrison www.morrisonprairie.com) Generator operation will be discussed in the next section.

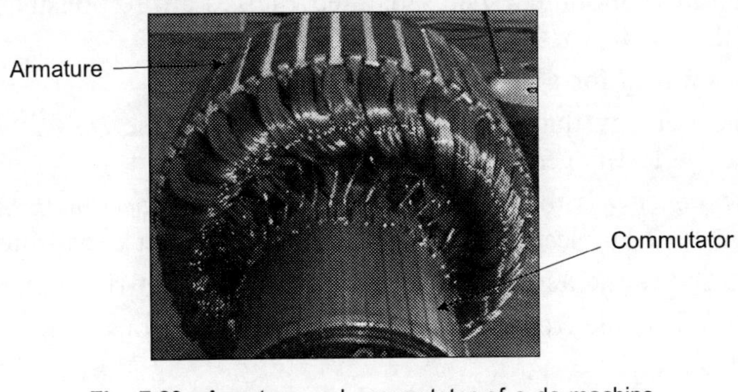

Fig. 7.23 Armature and commutator of a dc machine.

7.12 GENERATOR OPERATION

Figure 7.24 shows the production of emf in a generator....using the essential components.

The generator consists of a yoke with field poles, in which coils produce a magnetic field directed towards a cylindrical structure called an armature, which carries the main conductors. The conductors, in which the generator's emf will be induced, are placed in slots of the armature. The armature is capable of being rotated on an axle by a drive such as a *turbine*.

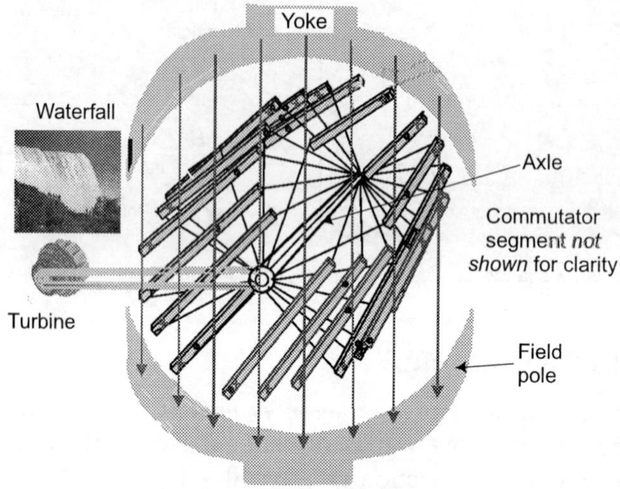

Fig. 7.24 Generator operation.

Figure 7.24 is an attempt to capture the essence of a dc generator operation. The armature and the commutator rings are not shown for the sake of clarity.

Shown are the yoke and field poles with the magnetic field and the armature conductors.

The generator is shown being driven by a turbine via a belt and pulleys arrangement. The turbine is driven by water falling in a waterfall.

In the conductors, the charges are shown being separated (electrons and exposed positive atomic cores) as the conductors rotate in the magnetic field driven by the axle on which the armature is mounted. The separated charges are responsible for the production of the emf in the conductor.

What constitutes a *load* for a generator?

It is the electrical energy that is produced as a result of the *conversion* from mechanical energy of the drive to the generator.

The electrical energy is the *combination* of the *emf produced with the current* in a *load*, such as, a *resistor*, or an electrical network, or a motor that constitutes the *load*.

Figure 7.25 shows the action of a generator with a split-ring commutator.

The armature is made from copper conductors which consist of positive atomic cores and free electrons.

Note how the emfs produced on opposite sides of the coil (due to separation of charges) add up like *batteries connected in series*, and the action of the commutator reverses the coil connections, so that the *emf remains unidirectional* at the terminals of the generator; the *load is a lamp*. The commutator is made from a conductive ring split in two halves, and is fixed on an insulated cylinder coupled to the axle. The commutator rotates along with the axle of the generator. The generated emf is picked off the commutator using two leaf springs pressing against it. In large generators this is achieved by pressing carbon brushes against the commutator *segments*.

In the series of figures from Figs. 7.25 to 7.28 is shown the rotation of the armature and the process of commutation. The separation of charges is shown in the conductors in the Figs. 7.25 to 7.28.

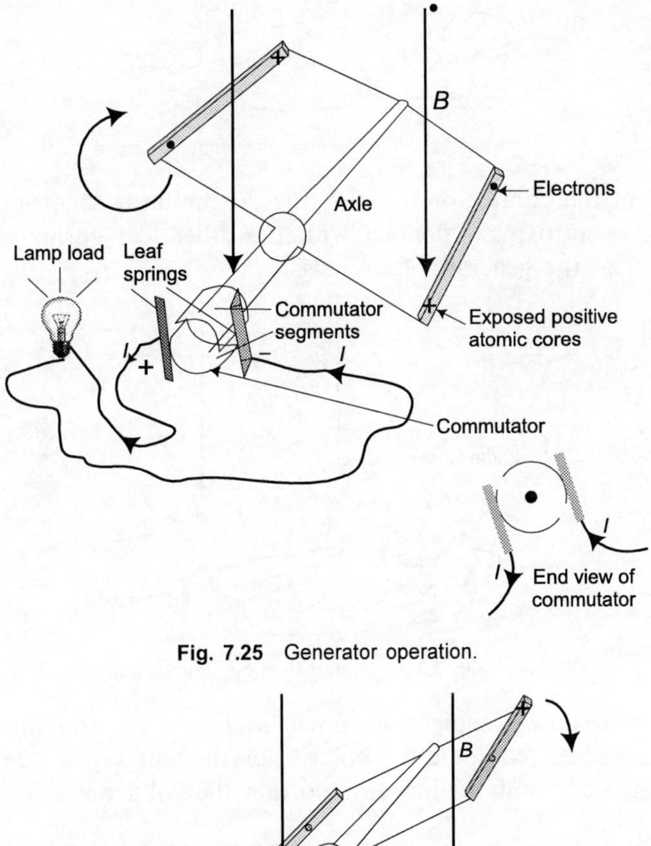

Fig. 7.25 Generator operation.

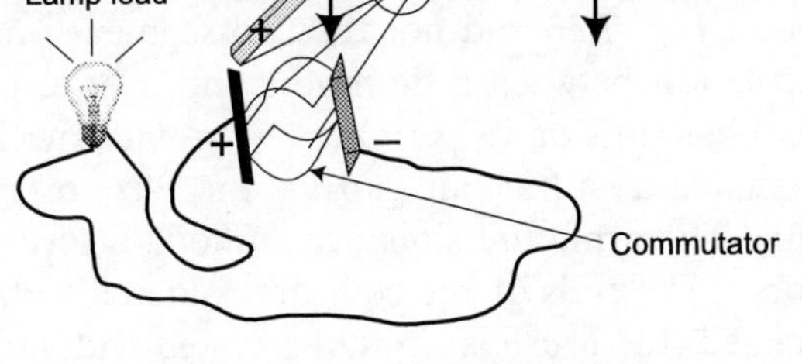

Fig. 7.26 About half rotation later.

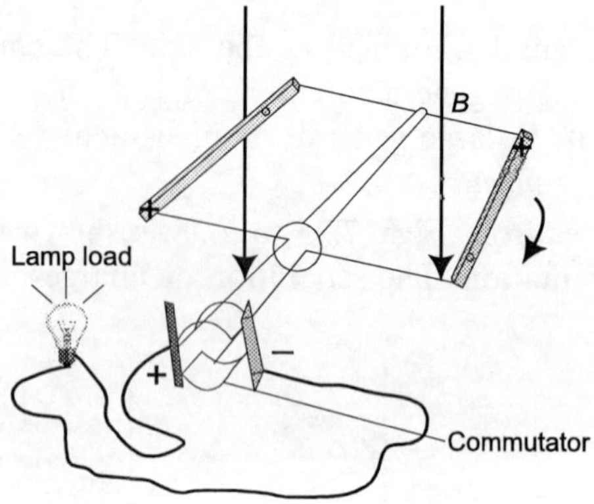

Fig. 7.27 Just before commutation occurs.

As the emf in the conductor reverses, the commutator segment in contact with the leaf spring slips over to make contact with the other leaf spring and this maintains a *unidirectional* emf at the generator terminals.

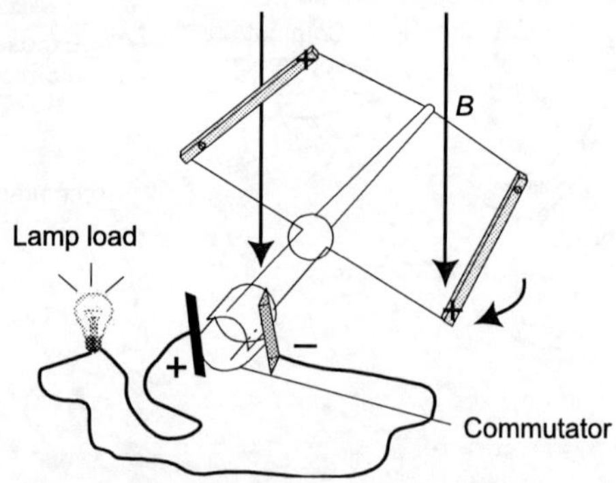

Fig. 7.28 Commutation has completed.

Commutation is now complete as shown in Fig. 7.28, and though the emf in the conductors has reversed (compare the polarity of the emf in the conductors in Fig. 7.28 with Fig. 7.27), the commutator has ensured that the voltage applied to the lamp load is unidirectional.

Most commutators in dc machines for industrial applications are multi-segment in construction as shown in Fig. 7.23 and not the two-segment one as in Fig. 7.25. There is no difference in construction between a dc motor and a dc generator. The armature (see Fig. 8.46 of a dc motor) has slots on its periphery to provide mechanical security for several coil windings and to enable a shorter air gap for the flux to cross between the pole face and the armature teeth. The armature slots accommodate several coils, made from a few hundreds of conductors.[15] The ends of the coils are soldered to the segments against which carbon brushes are pressed (the brush is about a segment wide) to collect the generated emf from all the conductors, which may be connected in series to increase the terminal voltage, or in parallel, to increase the current capacity though with a smaller terminal voltage.

7.13 MAGNETIC FIELD LINES

In Fig. 7.24, the magnetic flux is shown *unterminated* for lack of space. Remember magnetic field lines are never unterminated and must always form *closed loops.* The air gap is necessary to separate the rotating and stationary structures, and kept small to provide a low reluctance (Section 5.4) so that the magnetic field generated in the poles find a path through to the armature conductors.

In a real generator how do the magnetic field lines form loops?

Magnetic field lines follow the path of *least reluctance*, similar to the way in which current takes the path of least resistance.

In dc generators and motors, the path of *least reluctance* lies in the frame or yoke of the motor and the field lines form *closed loops* taking the *following path:*...the field poles, the air gap, inside the armature structure and through the air gap again and into the frame or yoke and return to the field poles as seen in Fig. 7.29.

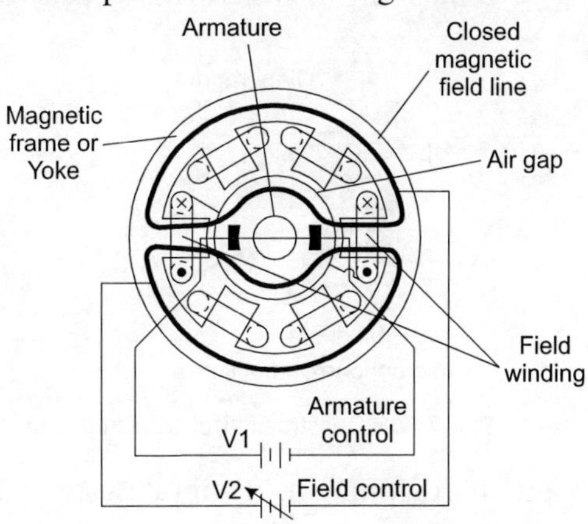

Fig. 7.29 Magnetic flux path in dc machines.

7.14 MICROPHONES

A microphone is an electromechanical device that uses vibration to create an electrical signal or emf proportional to the vibration, which is usually an air pressure wave such as that produced by voice or sounds.

Sound is produced by a rapid variation in the average density or pressure of air molecules above and below the current atmospheric pressure. These minute changes in atmospheric pressure are referred to as **sound pressure**, and the fluctuations in pressure, as **sound waves**. A vibrating sound source, such as a loudspeaker cone, causes the surrounding air molecules to bounce off each other with a force proportional to the disturbance. This creates ripples of more dense (higher pressure, called compression) to less dense (lower pressure, called rarefaction) air molecules. The back and forth oscillation of pressure produces the sound wave, which impinges on our eardrums and which we perceive as sound.

To view an animated presentation of the motion of air molecules in open and closed pipes visit http://www.miyazaki-catv.ne.jp/~yuasa1436/PlF4.htm.

The softest perceptible sound, corresponds to a pressure variation of 0.00002 psi (pounds per square inch), which is less than a billionth variation around normal atmospheric pressure of 14.7 psi, and the loudest sound that can cause ear damage is 0.002 psi variation. For the sake of comparison, a bicycle tire tube is inflated to a pressure usually in the range of 25-100 psi.

Microphones operate based on the same principle as generators.

In *generators*, note that all the conductor loops are arranged such that the direction of motion of the conductors is a *twist*, so the direction of the uniform and constant magnetic field on opposite sides of a *loop* was the same.

There is one subtle difference between a generator and the *microphone*.

In a generator, the loops of coils can twist, so that when these are moving in a *constant magnetic field*, every loop produces an *emf* that when added over all the loops amounts to a large emf. This is similar to connecting batteries in series to produce a large voltage.

Microphones consist of a *diaphragm*, which is attached to a *coil of wire* as shown in Fig 7.30.

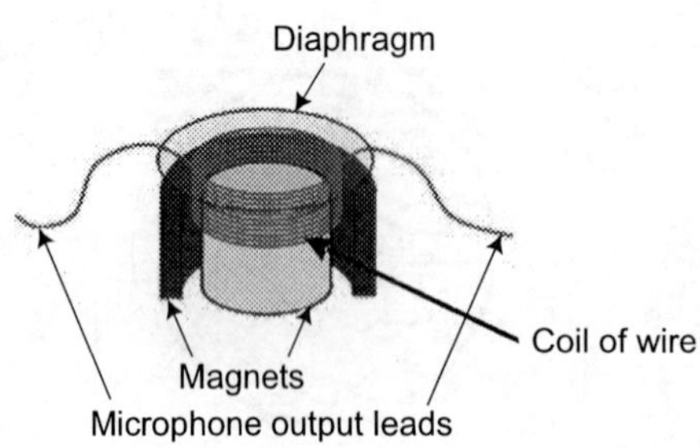

Fig. 7.30 Construction of a microphone.

When sound waves hit the diaphragm, it vibrates and vibrates the entire coil along with it.

As there is no *twisting* action of loops of wire, microphones require that the magnetic field be oriented uniformly all around using *circular magnets* or other such devices such that the emfs produced by opposite sides of *all the loops* are additive, thus *maximizing* the emf generated.

In Fig. 7.31 it is seen that the magnetic field is directed radially inward around the microphone through the coils of wire. Therefore, the emfs generated in every segment of the coil when it moves, is additive.

Microphone Operation

Figure 7.31 shows a view of the generation process of emfs in a microphone.

Shown in the figure are the coils (grossly exaggerated using rectangular cross-section wires) and the diaphragm that vibrates when sound waves strike it. The diaphragm is attached to the coil and so, causes the coil to vibrate alongwith it, leading to the generation of an induced emf at the ends of the coil, the audio frequency electric signal.

The circular magnet is of special construction such that its magnetic field is directed radially inwards.

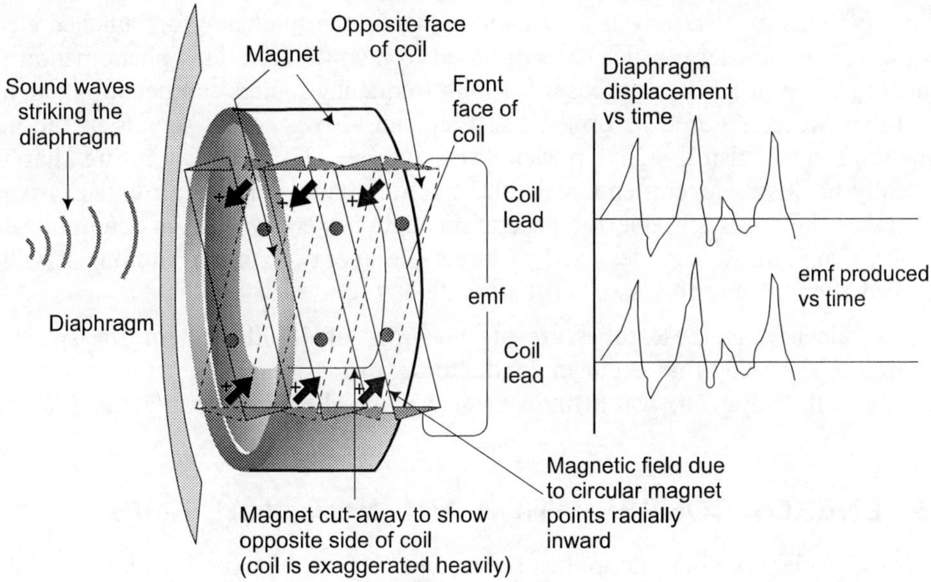

Fig. 7.31 Generation of emf by the microphone.

Summarizing the Effects of Motional emf

In the presence of a uniform *constant* magnetic field, the motional emf is due to magnetic forces on the moving electrons, causing them to accumulate at one end of a conductor while exposing lattice ions at the other end.

Of course, the *electrons* that are *shifted* create a negative potential while, the *immobile positive cores* are simply "uncovered" to create a positive potential.

And calculations from Faraday's law predict this *induced emf* correctly...even though it is <u>not</u> associated with a changing magnetic field.

Note on the production of high-frequency and high power electromagnetic waves for transmission to satellites

Incidentally, one of the other reasons why conductors do not transmit high frequency signals efficiently (treating the signals as compressions and rarefactions of electrons on conductor surfaces and remembering that a voltage is possible by the accumulation of charges), is because of the loss of signal strengths (attenuation) in lossy transmission lines and waveguides as the charges move around the surfaces in response to high frequency signal voltages (Section 5.22).

Generation of high frequency high power signals and, coupling these to antennas and, between communication equipment is a challenging task. Frustrated in their attempts to amplify (strengthen weak signals) and generate high frequency (GigaHertz, 10^9 Hertz) electric signals using conventional vacuum tubes, even after minimizing stray capacitances and inductances, engineers found the answer by observing the compressions and rarefactions of air molecules in sound waves. They designed Klystron tubes that create compressions and rarefactions of *electrons* in a *vacuum,* in which the tendency for bunched electrons to disperse and waste energy is least, as opposed to a conductor. This phenomenon is termed "bunching", and in Klystron tubes weak high frequency signals can be amplified, and when these tubes are configured as oscillators, they can also generate signals in the microwave frequency range, raising signal power levels to several kilowatts (noting that there are practically no losses to contend with, the electrons being moved around in a vacuum !). This makes the tubes suitable for generating high power signals for communication with satellites. Coupling waveguides used to direct the energy to other equipment will need to be cooled though, due to losses on their walls !

Electron velocities in these tubes are of the order of 1/10th that of the speed of light, compared to the 1/200 that occur in conductors. For an interesting article on the history of invention of the Klystron, visit http://www.slac.stanford.edu/cgi-wrap/getdoc/slac-pub-7731.pdf.

7.15 ENERGY CONSERVATION AND MOTIONAL EMF

We next consider whether energy has been conserved in the experiments, in which a metal bar was moved along rails immersed in a magnetic field.

If the metal bar in Fig. 7.32 slides *effortlessly* along at a constant speed v, and the *resistor* heats up the surroundings, we would be getting *something for nothing*!

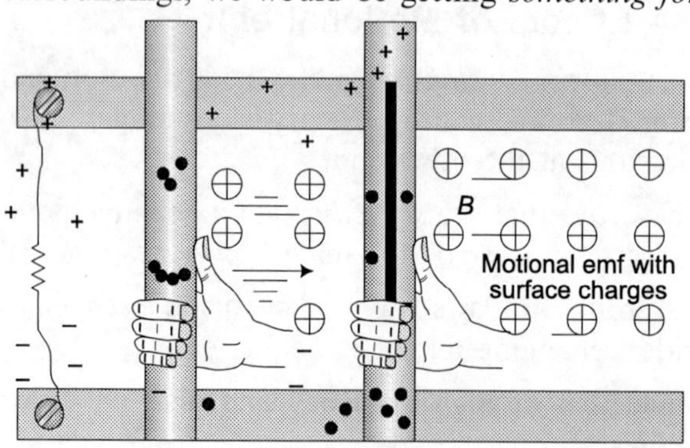

Fig. 7.32 Metal bar sliding on rails immersed in a magnetic field.

There must be an energy input into this system. It turns out that we must exert a force on the bar to keep it moving at constant speed, because there is a magnetic force on the moving bar, due to the current I in the bar.

The bar now has a current flowing through it, and feels a magnetic force to the left, in the direction of $dl \times B$, since conventional current runs upward through the bar. This is shown in Fig. 7.33.

Of course, the force is a result of the Hall effect charges producing an electric field (not shown in the figure), which exerts a force on the positive atomic cores in the bar.

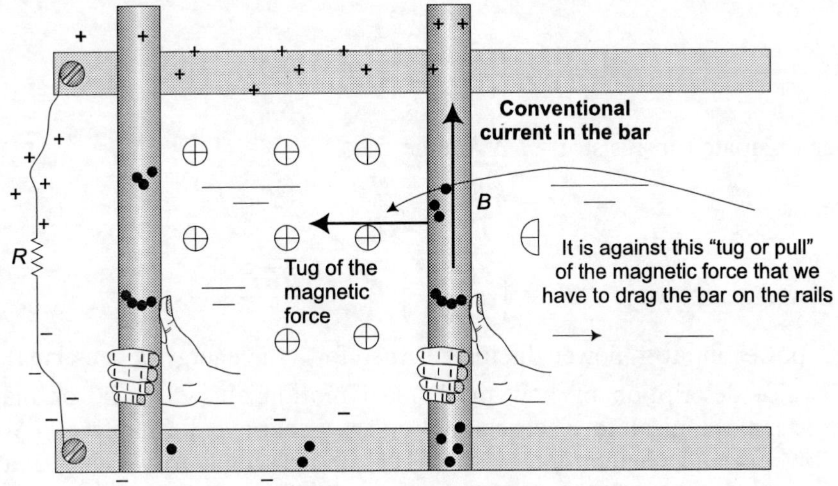

Fig. 7.33 Tug of the magnetic force.

It is *against* this *tug or pull* of the magnetic force that we have to *drag* the bar on the rails!

Now you know why electricity is not *cheap!*

It comes at a price!

The price we have to pay to "overcome" the magnetic tug.

Now, we can understand the role of the *turbine* in our generator.

It *overcomes* the tug of the generator conductors when current is in them, delivering power to the electrical load.

A little maths now! To calculate the amount of magnetic *tug* on the bar on rails (see Section 6.5)

$$F_{magnetic} = \left| \int_{bar} Id\vec{l} \times \vec{B} \right| = \int_{bar} IBdl \sin 90°$$

where, I is the direction of conventional current upwards in the bar, dl is an elemental length of the bar of length L.

$$= IBL = \left(\frac{vBL}{R}\right) BL = \frac{v(BL)^2}{R}$$

where, R is the load resistance, and $I = vBL/R$ is from the expression derived in Section 7.7.

So, we must apply a force of magnitude $v(BL)^2/R$ to the right to keep the bar moving.

In a short time Δt, we move the bar a distance Δx. We do an amount of work $F\Delta x$ in a time Δt, so we have to supply energy at the rate of $F\Delta x/\Delta t = Fv$.

So, power input Fv is numerically equal to the power output of the resistor.

$$Power\_input = F\,\frac{\Delta x}{\Delta t}$$

$$= Fv = \left(\frac{v(BL)^2}{R}\right)v = \frac{(vBL)^2}{R}$$

Power dissipated in resistor= $I\,\Delta\text{Voltage} = I^2R$

Since, $$I = \frac{emf}{R} = \frac{BLv}{R}$$

we get, power dissipated $= I^2R = \left(\frac{v(BL)}{R}\right)^2 R = \frac{(vBL)^2}{R}$

Since, power input = power dissipated, therefore, the energy **is conserved!**

For a nice description of how mechanical motion alters electric or magnetic fields and converts mechanical to electrical power see Section 6.3 Chapter 6, Actuators and Sensors, Motors and Generators, of the MIT article, a link to which is available in the links_animations_articles file in the CD.

Visualizing Self-Induction

Referring to the discussion on the phenomenon of induction using Faraday's law and Lorentz force law at the end of Section 7.7, and as a continuation of the study of the relativistic effects of electricity and magnetism, it would be useful to review the configuration of fields of accelerating charge carriers. While a more formal treatment will be given in Chapter 10, Sections 3.5 and 3.6 of Härtel's article "A qualitative approach to electricity" in the Teacher's Guide in the CD provides a qualitative approach that would be useful to read. I urge the reader to sketch symbolically the electric field pattern of accelerated charges and the circular waves (loci of massless moving elements which trace the classical electric and magnetic lines of force Ref. [39]) they produce depicted in Figs. 3.18, 3.19, Fig. 3.21 and Fig.3.22 in Härtel's article. The visualization of self-induction is treated elegantly in Section 3.6 of Härtel's article in the Guide and vividly portrayed in Figs. 3.23, 3.24 and 3.25 in the Guide which it would be instructive to sketch.

SUMMARY

We had discussed Faraday's law quite extensively in Chapter 5, in relation to time-varying currents (and their associated time-varying magnetic fields) in stationary coils.

In this chapter we learned, after reviewing the nature and properties of magnetic fields and studying from an observation of a current in a short-circuited coil 2 close to and along the same axial direction of another coil 1 in which the current was made to vary, the association of an electric field with a time-varying magnetic field.

We also learned that the current in coil 2 is steady, if the current varies at a constant rate in coil 1, as a confirmation of Faraday's law: (v = $L\, di/dt$).

We also learned how to place a coil around a solenoid that gives a correct application of Lenz's Rule, to determine the direction of the curly patterned electric field associated with a magnetic field, and how the current driven in this coil associates a magnetic flux which is in a direction to reduce the original magnetic flux. Of course, it does not succeed to reduce the original flux to zero; to achieve that the coil would have to be made using superconductive wire !

By slightly modifying the experiment with two coils, such that coil 2 is tilted from its axis, the current in it is reduced, and we learned that it is the component of flux produced by coil 1 that is perpendicular to the face of coil 2 which determines the induced emf.

We then learned that while Faraday's Law may be applied correctly to predict induced emf sometimes Lorentz Force Law can also be used, for example of a coil moving with respect to a stationary coil. However, there are some situations that cannot be explained even by Faraday's law. With a new approach based on the Theory of Relativity, the phenomenon of induction can be explained for all situations that arise in circuits.

Next, we studied the important application of Faraday's law to moving coils. All electric generators operate on the principles we learned in this section.

When a length of conductor which consists of millions and millions of conduction band electrons in a giant structure of lattice ions is moved, the electrons also move. And if the conductor were immersed in a magnetic field, we studied in detail, the process by which the electrons polarize the conductor, with an accumulation of electrons at one end and an exposure of a like number of lattice ions at the other end. This is what we termed "motional emf".

When a conductor mounted on a conducting pair of rails with a resistor across the rails at one end, and with the entire structure immersed in a magnetic field, was moved, a current was registered in the resistor, and the circuit as a whole. We learned that this was due to surface charges that developed on the ends of the conductor, along the rails and across the resistor, as the conductor was moved. This is how an emf could be produced or, in other words, we could generate electricity.

Generators are used to generate electricity, and the long rails which we had used to demonstrate the effect was not an efficient means to continuously generate an emf. Instead, the rails were bent into a tight circle (commutator) with the conductors mounted on a rotating structure (an armature) and a provision to generate the working magnetic field using field coils, and the emf picked off the commutator. The production of a unidirectional emf was resolved, by terminating the conductors on a split-segment commutator and collecting the emf at a fixed point on it. Modern generators are constructed in this manner.

We also studied the path of the magnetic flux in a modern generator and noted that indeed it had a closed path, as expected of a magnetic field which always has a curl and is never divergent.

We then studied in some detail the process by which a microphone is used to generate an electrical signal (emf) that is time-varying when air pressure waves strike a diaphragm that vibrates. We learned the process by which a coil, attached to the diaphragm, and immersed in a magnetic field, generates the microphone signal output whenever the diaphragm vibrates when struck by the air pressure waves of a sound.

In the last section of the chapter, we learned that when current flows in the load connected to the commutator, a force which overcomes the magnetic force of the conductors, is necessary for generator operation.

From the expressions that we derived for the forces the conductor exerts, we learned that energy is conserved when the generator operates.

We also learned to visualize the phenomenon of self-induction by plotting the circular waves produced by accelerating charge carriers.

Questions

1. Describe the phenomenon of magnetic induction with reference to two separate coils placed along the same axis a little distance separating them and when the current in one of the coils is changed with the other coil short-circuited.
2. Which one among Faraday's Law and Lorentz force law explains induction in all situations? Explain.
3. Explain why Faraday's law is counter-intuitive, and describe an approach and the theory behind the approach, which is comprehensive. Hint: See Fig. 3.4 and then see Sections 3.5 and 3.6 of Härtel's article "A qualitative approach to electricity" in the Teacher's Guide in the CD.
4. What is motional emf? With the help of neat sketches describe the processes which produce this.
5. Derive an expression for the emf produced in rotating machines using the example of a loop of dimensions $w \times h$ rotating at a constant angular velocity $\omega = d\theta/dt$ inside a uniform magnetic field B.
6. With the help of neat sketches showing a rectangular loop rotating in a magnetic field that causes an emf whose polarity changes, explain how a unidirectional emf can be produced.
7. Using appropriate expressions, prove that the energy is conserved when a bar of length L is pulled at a uniform speed v over two metal rails, two ends of which are connected together using a resistor of resistance R and the rails are immersed in a magnetic field B.
8. Sketch symbolically i) the electric field of an accelerated charge carrier, and ii) circular waves of positive and negative charges with different directions of acceleration. What are the circular waves made up of? Hint: Refer Ref. [39] and Section 3.5 of Hartel's Teacher's Guide in the CD.
9. Sketch symbolically i) the interaction between parallel accelerated charge carriers using circular wave symbols, and ii) the interaction between opposite parts of the circuit using circular wave symbols. With the help of these sketches explain the effect on the magnitude of self-inductance when two current-carrying wires are near and far away from each other. Hint: Refer Section 3.6 of Hartel's Teacher's Guide in the CD.

Additional practice questions are available in the CD with this book.

8

Magnetic Moments, Motors and Loads

The first section of this chapter describes the Lorentz force law and the magnetic field of a circular loop. The ideas of magnetic moments and their behavior when left to rotate freely in magnetic fields, are then developed. This is followed by a description of the magnetic force that develops when a magnetic moment, a current carrying loop of wire, is placed in a magnetic field. Then, is shown how it is this phenomenon, which leads to the production of torque in motors. One section is devoted to a description of the concept of load torque, the effect of its imposition on motor torque, and how this causes the motor to draw power from the power supply. The phenomenon of how it is the 'back emf' in motors, which regulates the motor torque to match load torque is described in some detail.

In following sections, calculations for the potential energy of magnetic moments, descriptions of the structure of magnets and magnetic interactions with matter are provided.

In the final section of the chapter, the Lagrangian from which the Lorentz force law may be derived, and the Hamiltonian, which describes the conservation of energy of moving and interacting charges are both mentioned with references to video lectures and texts. Then, pointers to sections in Appendix F on how fields of moving charges transform from the viewpoint of Special Relativity to show that magnetism is relativistic, are provided. A pointer to a section in Appendix F which describes the repulsion between two current-carrying wires from the viewpoint of Special Relativity is also provided.

The ideas of self-induction in Hartel's article are once again emphasized at the end of the chapter so that students may learn to depict the interactions of accelerated particles which will aid in the visualization of the phenomena of both mutual-induction and self-induction.

8.1 MAGNETIC FORCE AND MAGNETIC MOMENT

When an electric charge is placed in a magnetic field it experiences a force as we learned in Section 6.1.

The magnetic force in a charge q, moving with velocity v in a magnetic field B, is

$$F_{mag} = q\ (v \times B)$$

This is known as the *Lorentz force* law, which was formulated without the magnetic field by Ampere and Grassmann separately. Maxwell and Lorentz in 1861 and 1895, respectively (Sections 15.2 and 15.3, Ref. [38]) defined the magnetic field from one of the terms in Grassmann's force expression between current elements. See Appendix F Section F.1 on Ampere and Grassmann's force expressions and their applicability to systems from the standpoint of Newton's Third Law.

In the presence of both electric and magnetic fields, the net force on q would be

$$F_{mag} = q[E + (v \times B)]$$

The magnetic field for a circular loop of *radius a* and conventional *current I,* at a distance x from the center, **far away** *along the axis*, is (Chabay & Sherwood, 1995, used by permission)[2]

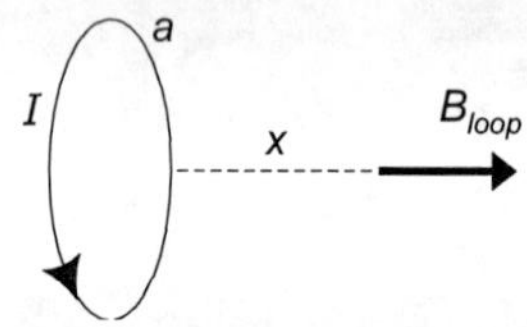

Fig. 8.1 The magnetic field for a circular loop.

$$\vec{B} = \frac{\mu_o}{4\pi}\frac{2\pi a^2 I}{x^3}$$

In this formula, the quantity $\pi a^2 I$ in the formula is called the **magnetic moment.**

It is usually referred to by the letter μ. Some authors call this the magnetic dipole moment and refer to it with the letter ***m***.

More generally, $\mu = IA$

where, A is the area of the loop and is equal to πa^2 for circular loops.

With this definition of the magnetic moment, we write B_{axis} as

$$B_{axis} = \frac{\mu_o}{4\pi}\frac{2\mu}{x^3}$$

8.2 MAGNETIC MOMENT AS A VECTOR

The ***magnetic moment*** of a current carrying loop is considered to be a *vector* μ whose **direction** is the direction of the **magnetic field** at the center of the coil. This is shown in Fig. 8.2.

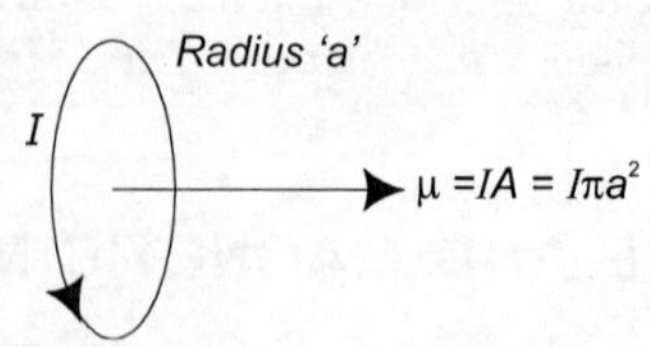

Fig. 8.2 The magnetic moment of a current carrying loop.

8.3 MAGNETIC FORCES ON A MAGNETIC MOMENT

A magnetic field can cause a current-carrying coil of wire to *twist.* (Chabay & Sherwood, 1995, used by permmission).[2]

Consider a rectangular current-carrying loop of wire in a uniform magnetic field (that is, the magnetic field has the same direction and magnitude throughout this region).

The loop is pivoted on a horizontal axle. If the conventional current I is as shown in Fig. 8.3, then the magnetic forces on each of the four sides of the loop will be as shown.

The forces are determined by $F = qv \times B$ applied to a wire.

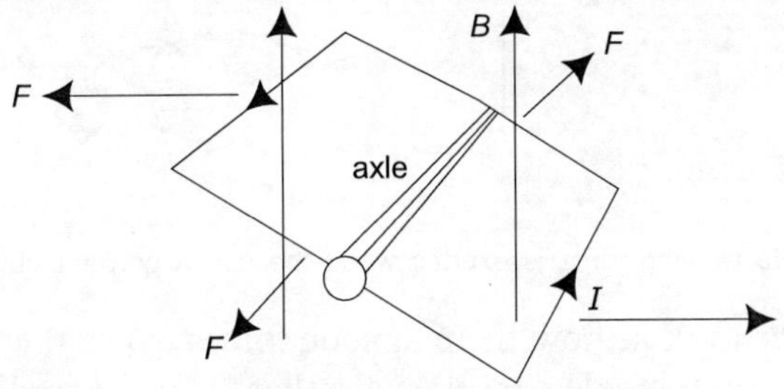

Fig. 8.3 The magnetic forces on a magnetic moment.

A simple method for determining the direction and magnitude of a *vector cross product* $v \times B$.

Keep your right hand open and flat (fingers and thumb in the same plane), with your thumb perpendicular to your fingers, point your fingers in the direction of the first vector (v) as shown in Fig. 8.4.

Now, *bend* your fingers toward the second vector (B), noting the angle of bend θ, which is always less than 180°. You may have to "rotate" your wrist to get your hand into a position that lets you bend your fingers this way.

The thumb points in the direction of cross product $v \times B$. For a 3D view of the cross-product visit http://www.glowscript.org/#/user/matterandinteractions/folder/matterandinteractions/program/17-cross-product.

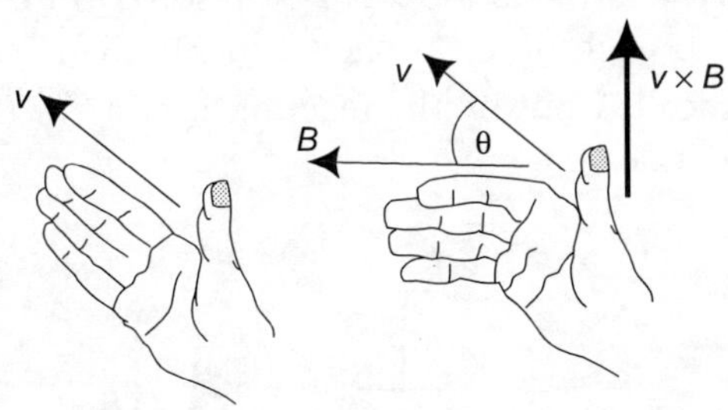

Fig. 8.4 Determining the magnitude and direction of $v \times B$.

This rule may be applied to verify the magnitude and direction of the forces given by $qv \times B$ in the previous Fig. 8.3.

When the plane of the loop is perpendicular to the magnetic field, the magnetic forces ***don't*** exert any twist as shown in Fig. 8.5. Why? Using the vector product rule, check how the forces only tend to pull the loop apart.

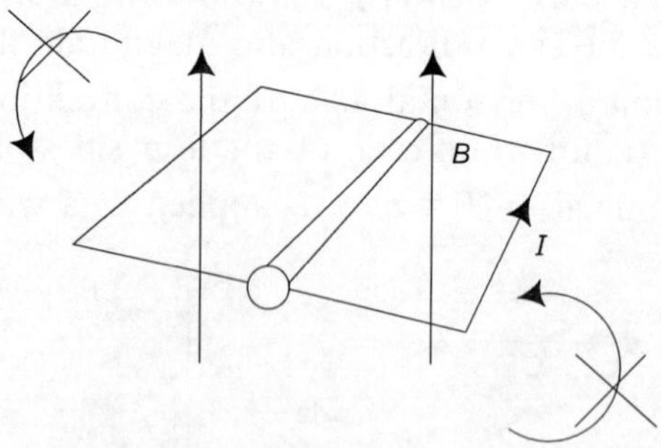

Fig. 8.5 No twisting force is exerted when the coil is perpendicular to the field.

Let us take a closer look at how the magnetic forces cause the loop to twist. The forces at the top and bottom of the axle cancel each other. The left and right side forces will act as shown in Fig. 8.6.

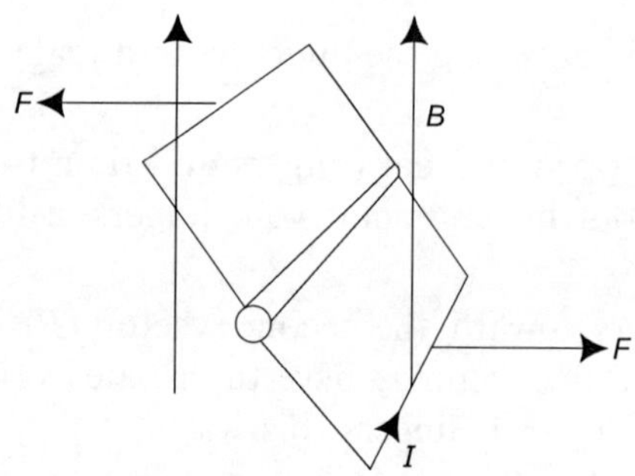

Fig. 8.6 The forces acting on a current carrying rectangular loop in a magnetic field.

The force acting on the left side loop can be resolved into 2 component force vectors, F_1 and F_2 *perpendicular* to each other as shown in Fig. 8.7.

Component F_1 is cancelled out by the reaction forces $2 \times (F_1)/2$ along the arms of the loop connected with the axle.

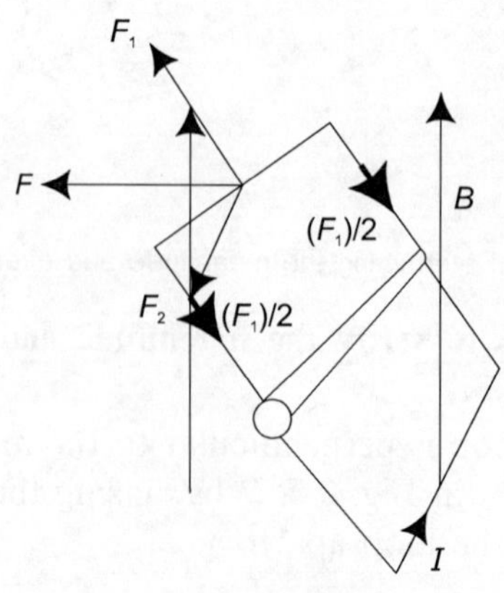

Fig. 8.7 Resolving the (left) force *F* into its components.

This leaves F_2 capable of imparting a movement to the left edge of the loop as shown in Fig. 8.8.

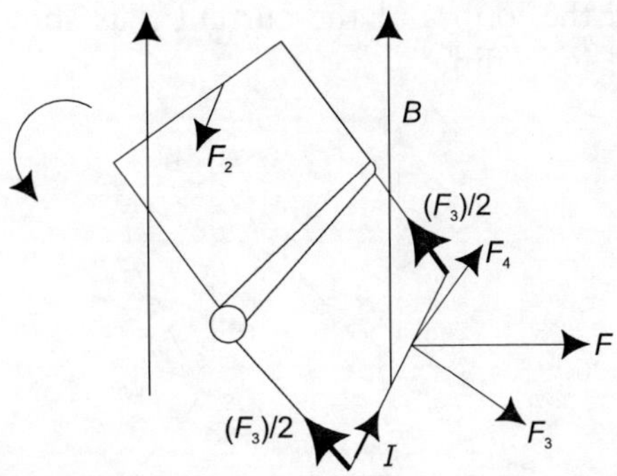

Fig. 8.8 Resolving the (right) force F into its components.

A similar analysis of forces carried out on the right side of the loop, shows component F_4 capable of imparting a movement to the right edge of the loop as shown in Figs. 8.8 and 8.9.

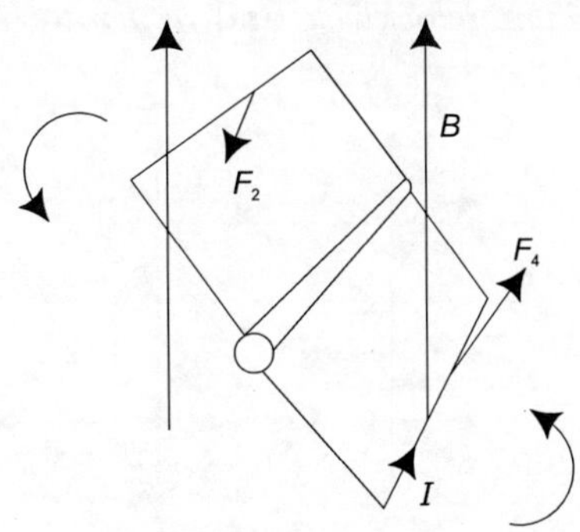

Fig. 8.9 The forces F_2 and F_4 capable of rotating the loop.

We see in Fig. 8.10 that the two components F_2 and F_4 rotate the loop *counterclockwise.*

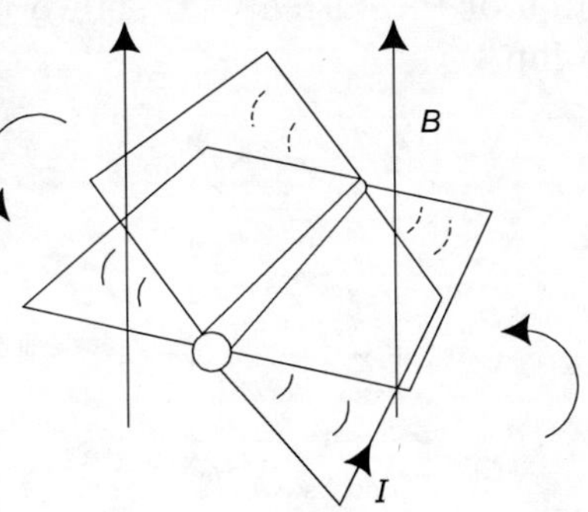

Fig. 8.10 The forces F_2 and F_4 rotate the loop counterclockwise.

8.4 STABILITY OF THE MAGNETIC MOMENT

Given the orientation of the loop and the current I as shown in Fig. 8.11, if you *nudge* the loop slightly, will the loop flip?

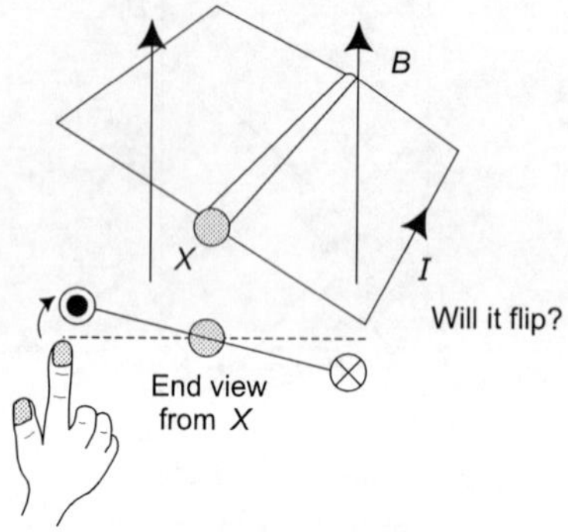

Fig. 8.11 Will the loop flip?

The loop is stable because if it is *rotated slightly* from the horizontal, the magnetic forces will cause it to turn *counterclockwise* and *return* to its original position as seen from Fig. 8.12.

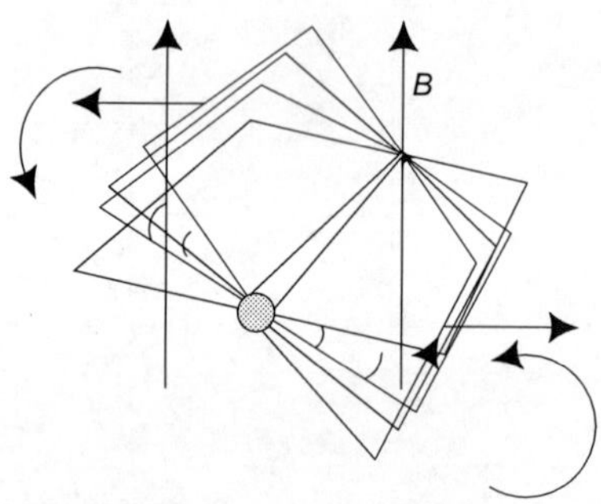

Fig. 8.12 The loop is stable.

If we *reverse* the direction of the current I as shown in Fig. 8.13 and if you *nudge* the loop slightly, will the loop flip?

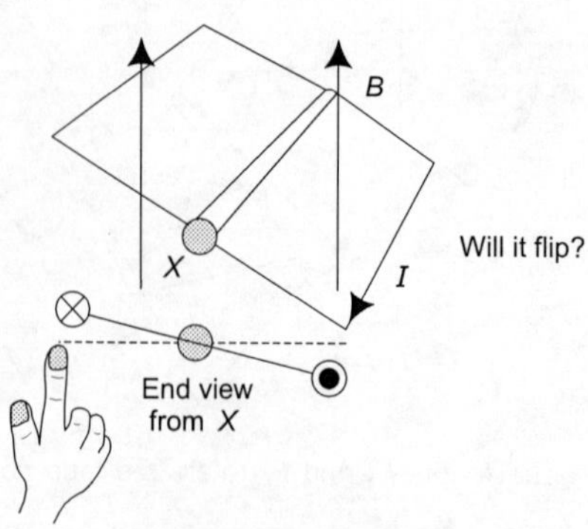

Fig. 8.13 Will the loop flip with the current reversed as shown?

This loop is unstable because if it is rotated slightly the magnetic forces as shown in Fig. 8.14 will cause it to *continue rotating* in the same direction and ***not*** return it to its original position.

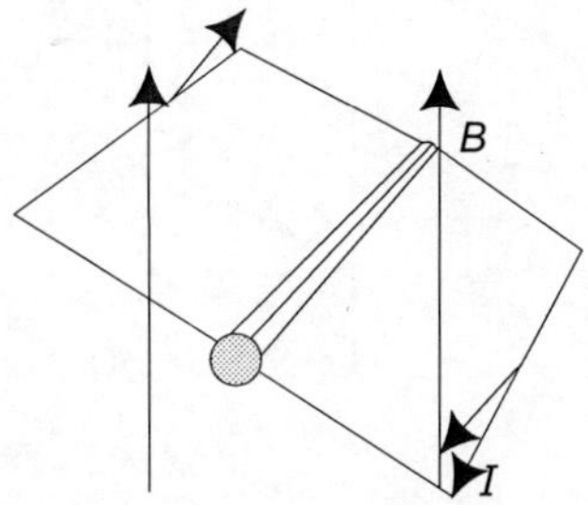

Fig. 8.14 The magnetic forces that act on the edges of the loop.

The loop will continue to rotate as shown in Fig. 8.15.

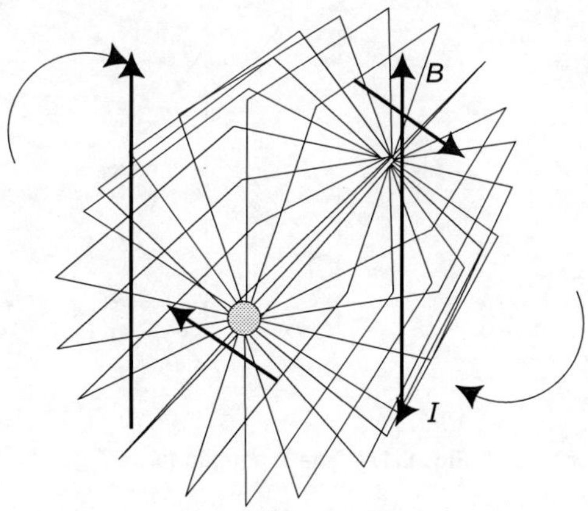

Fig. 8.15 The loop will continue to rotate.

The **magnetic moment (vector)** μ (see Sections 8.1 and 8.2) is ***not*** a piece of fiction that merely makes the formula for B look cleaner,

$$B_{axis} = \frac{\mu_o}{4\pi}\frac{2\mu}{x^3}$$

It is a vector that tells us what exactly a current carrying rectangular loop does when placed in a magnetic field and it is nudged.

One rectangular loop has an area $A = wh$, so its magnetic moment is $\mu = IA = Ihw$ as shown in Fig. 8.16. When we say a loop has a magnetic moment it means that the loop is a current-carrying loop.

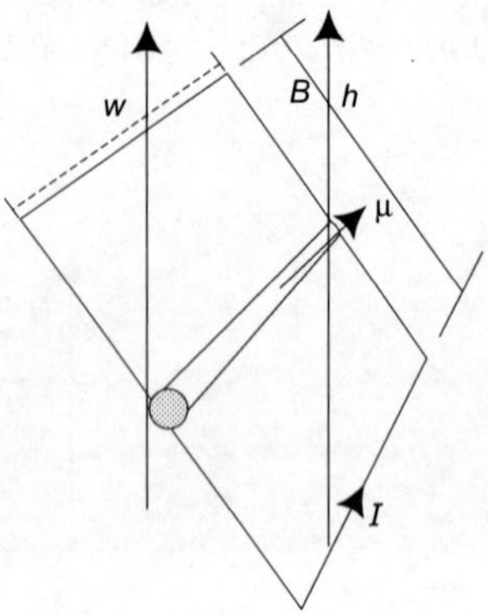

Fig. 8.16 The magnetic moment of a current carrying loop.

For the given current direction, the magnetic forces F_{mag} developed are shown in Fig. 8.17.

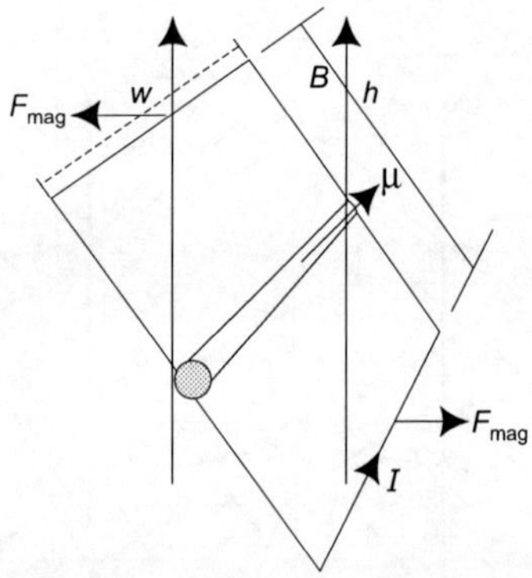

Fig. 8.17 The magnetic force F_{mag}.

And the coil *tends* to twist (counterclockwise here) to make μ line up with *B* as shown in Fig. 8.18.

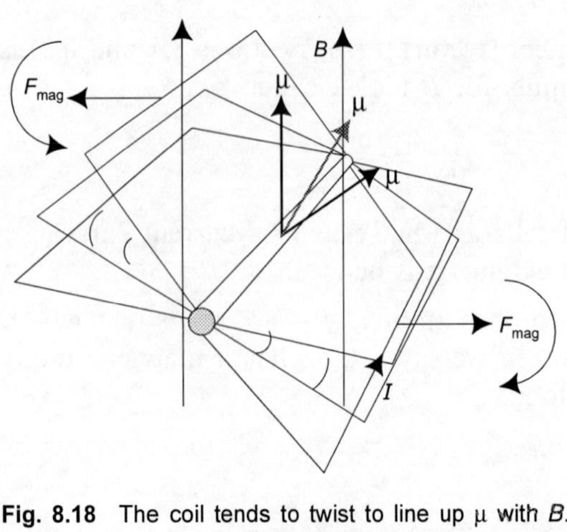

Fig. 8.18 The coil tends to twist to line up μ with *B*.

8.5 MAGNETIC MOMENTS OF ATOMS IN MAGNETS

Every atom consists of an electron occupying an orbital space around a positive atomic core: the neutron and the proton, as shown in Fig. 8.19.

The electron is a "moving charge" and some among these have a *non-spherically* symmetric probability distribution viz. the *'p, d, and f' orbital.*

The electron of an atom is a "moving charge", and it produces a magnetic field. There is an associated non-zero magnetic moment with the orbitals of the electron.

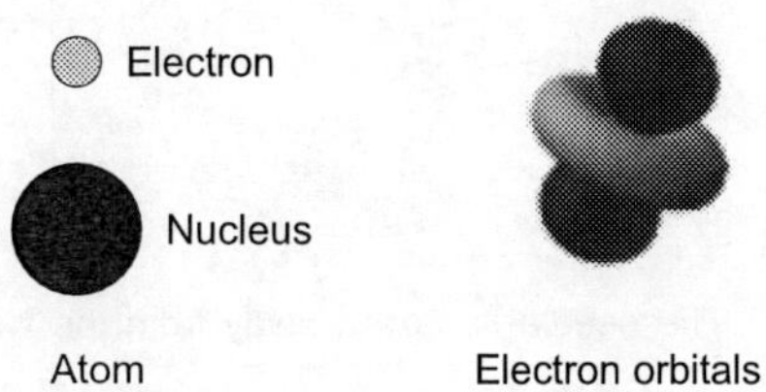

Fig. 8.19 An atom and the electron orbitals.

The Fig. 8.19 shows electron atomic orbitals also. The term *orbital* has become known as either the *mathematical function* or the *region* generated with the function. Specifically, atomic orbitals are the possible quantum states of an individual electron in the electron cloud around a single atom, as described by the mathematical function. Refer "Stable Atoms" in Appendix B.

In addition, some electrons also spin and these spinning balls produce a magnetic field.

In *many* materials, an atom has *no net orbital or spin* magnetization. But, in a few materials, notably *iron, nickel, cobalt and some alloys of these elements*, the orbital and spin motions in neighboring atoms *line up with each other* and therefore produce a sizable magnetic field, though not significant enough for practical applications and these are called *ferromagnetic* materials.

8.6 A MAGNETIC COMPASS IN A MAGNETIC FIELD

Consider a magnetized light-weight compass needle (balanced on a nearly frictionless pivot point) as shown in Fig. 8.20(a). What is the *direction* of its atomic magnetic moments? The direction is shown by the arrow marked 'μ'. It points in the same direction as the needle.

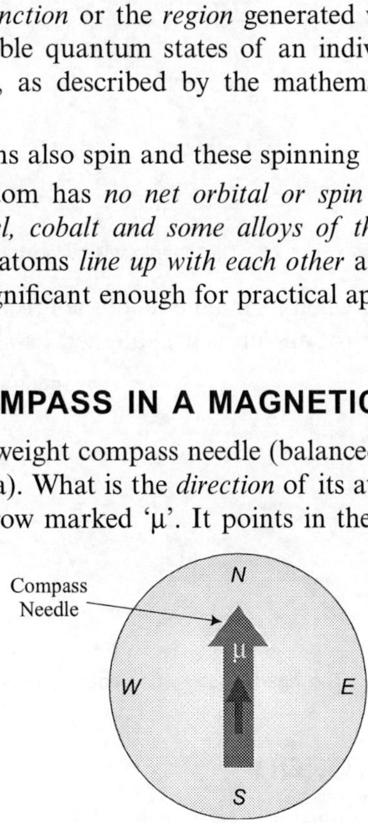

Fig. 8.20(a) A magnetized compass needle

If the needle is moved away from the North as shown in Fig. 8.20(b) and *then released* what will happen?

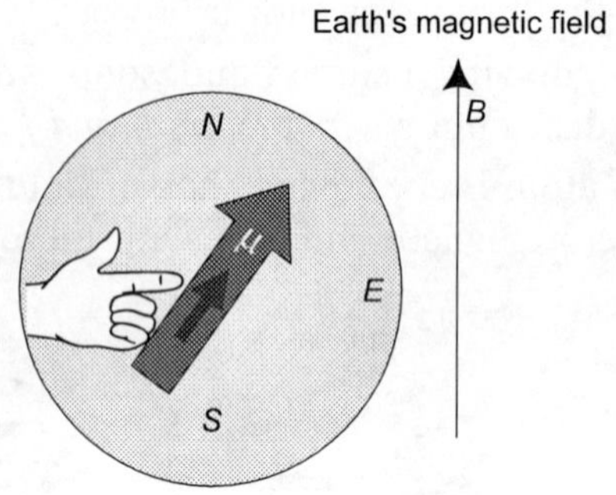

Fig. 8.20(b) The needle is moved away from the North and released.

The needle *swings back* to point North again as shown in Fig. 8.21.

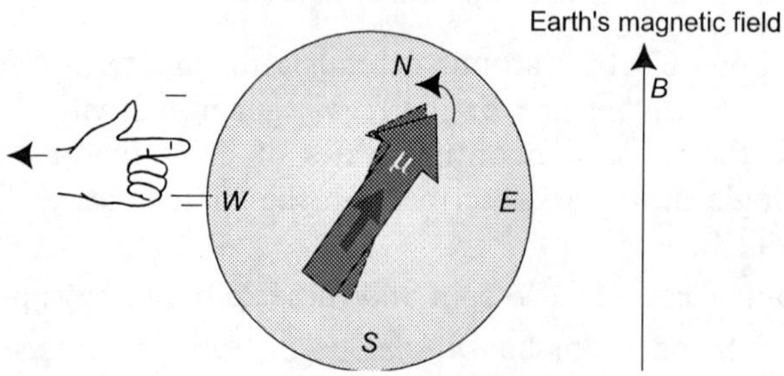

Fig. 8.21 The needle will swing back.

The needle is most *stable* when its magnetic moment is aligned with the earth's magnetic field, so the compass needle rotates until it is aligned as shown in Fig. 8.22.

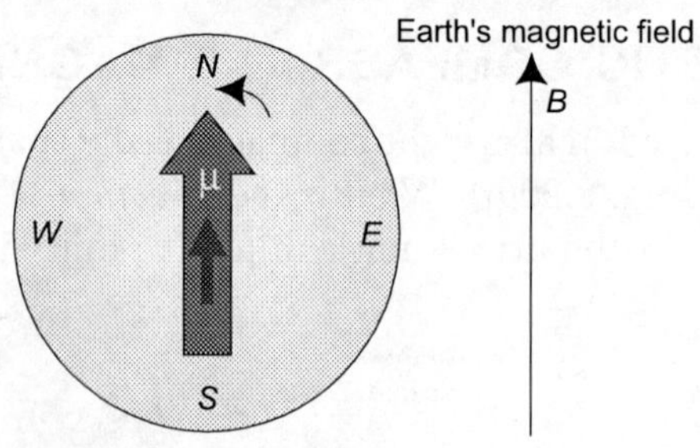

Fig. 8.22 The needle has swung back to point North again.

8.7 QUANTITATIVE TORQUE

The fact that a current-carrying coil *twists* in a magnetic field is very important. (Chabay & Sherwood, 1995, used by permission).[2]

The quantitative amount of *twist* is formally called the *torque* which is provided by each of the magnetic forces around the axle and is equal to the distance from the axle (the *lever arm*) times the component of the force *perpendicular* to the lever arm.

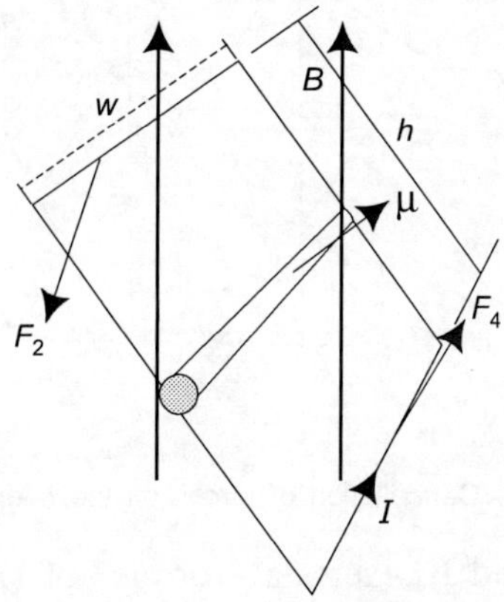

Fig. 8.23 The torque in a current carrying loop in a magnetic field.

Figure 8.23 shows a rectangular loop of length '*h*' and width '*w*' carrying a current *I* placed in a constant magnetic field *B* (Chabay & Sherwood, 1995 used by permission).

Using the relation for F_{mag} (Section 6.5, Fig. 6.15, $F = qv \times B$), the *perpendicular* component of each force is $F_2 = F_4 = F_{\perp} = IwB\sin\theta$, the lever arm is $h/2$ and 'θ' is the angle by which the loop is twisted from the horizontal as shown in the inset (side view) in Fig. 8.24.

The twist applied to this rectangular loop is due to the magnetic forces on the *w*-lengths of the loop.

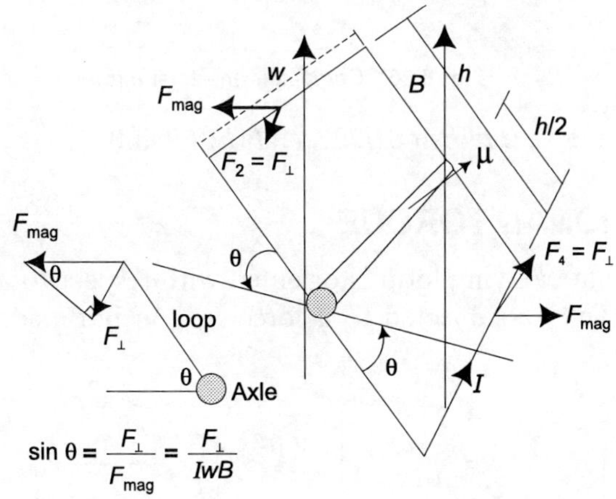

Fig. 8.24 Calculating the torque of a current carrying loop in a magnetic field.

The other **component** of *F*, and the forces on the *h*-length of the loop, just tends to *stretch* the loop without causing a twist as shown in Fig. 8.25.

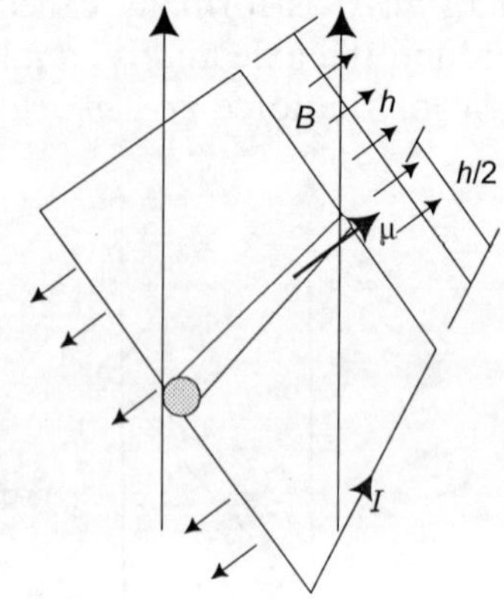

Fig. 8.25 Cancellation of forces on the *h*-length of the loop.

Each of the forces F_2 and F_4 exerts a "torque" of $(F_{\perp})(h/2)$ as shown in Fig. 8.26.

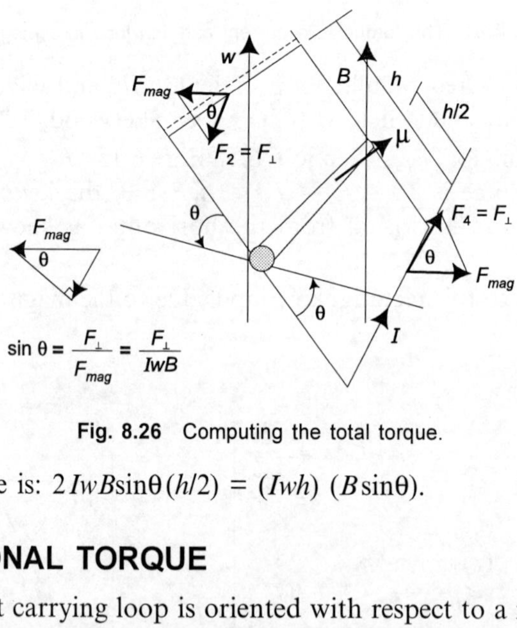

Fig. 8.26 Computing the total torque.

So, the total torque is: $2IwB\sin\theta\,(h/2) = (Iwh)\ (B\sin\theta)$.

8.8 UNIDIRECTIONAL TORQUE

If a rectangular current carrying loop is oriented with respect to a magnetic field as shown in Fig. 8.27, then it will be subjected to a force F_{mag} on both sides of the loop.

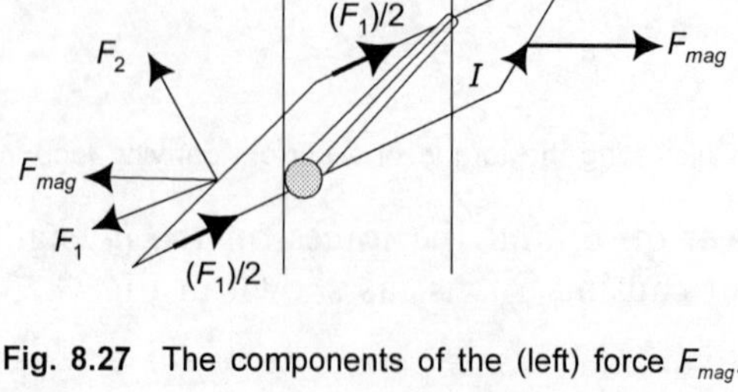

Fig. 8.27 The components of the (left) force F_{mag}.

The force F_{mag} acting on the left side loop can be resolved into 2 component force vectors F_1 and F_2 *perpendicular* to each other. Component F_1 is canceled out by the reaction forces $2 \times (F_1)/2$ along the arms of the loop connected with the axle as shown in Fig. 8.27.

This leaves F_2 capable of imparting a movement to the left edge of the loop. A similar analysis of forces carried out on the right side of the loop shows component F_4 capable of imparting a movement to the right edge of the loop as shown in Fig. 8.28.

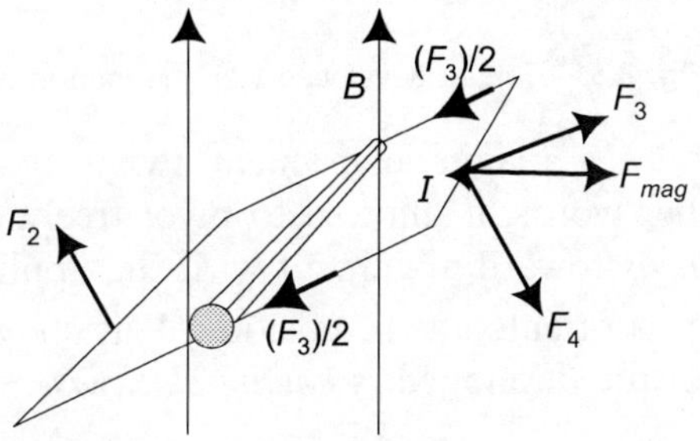

Fig. 8.28 The components of the (right) force F_{mag}.

Thus, the loop twists clockwise and attains the position as shown in Fig. 8.29 and will continue to rotate clockwise as shown in Fig. 8.30.

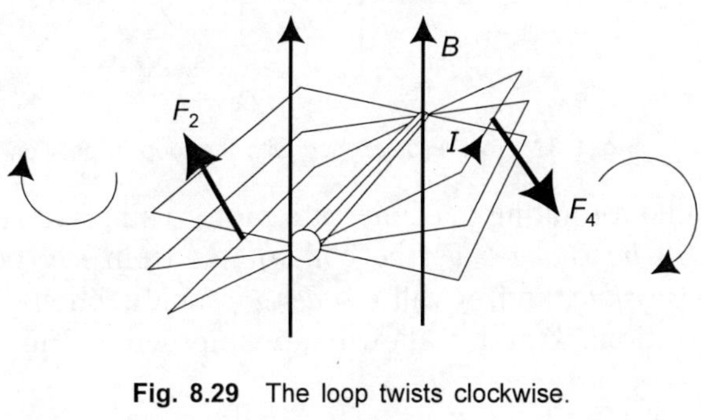

Fig. 8.29 The loop twists clockwise.

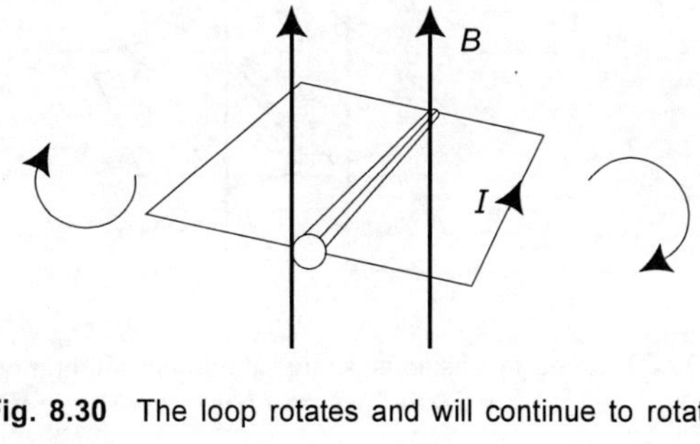

Fig. 8.30 The loop rotates and will continue to rotate.

Note that when the loop is oriented at position *X*, with the current direction as in Fig. 8.31, the torque developed is *clockwise*.

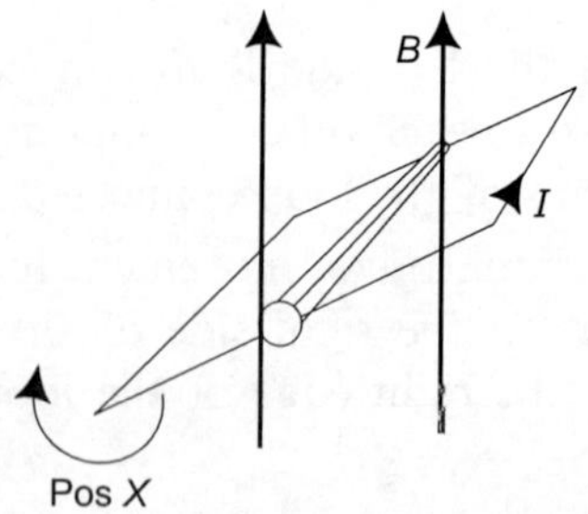

Fig. 8.31 Torque developed with the loop in position *X*.

This happens, because a magnetic moment twists to its most *stable position* in a magnetic field, or in other words, if allowed to pivot freely, a magnetic moment such as a current carrying loop *turns* toward alignment with the applied magnetic field.

Whereas, if the loop is oriented as in position *Y* as shown in Fig. 8.32, with the same current direction, the torque developed is *counterclockwise*.

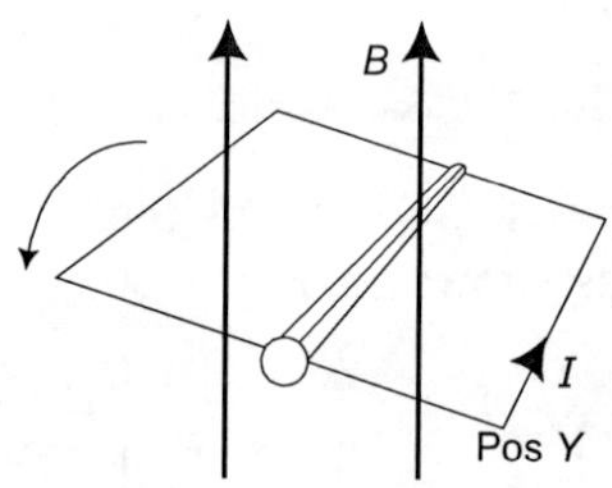

Fig. 8.32 Torque developed with the loop in position *Y*.

If we wish to build a rotating machine, or a motor using such a loop, and if the current direction *were not to be changed* as the coil rotates from say position *X* to *Y*, *uniform rotation will not be possible*, and, it will *not develop* unidirectional torque with the torques developed in the positions *X* and *Y* shown superimposed in Fig. 8.33.

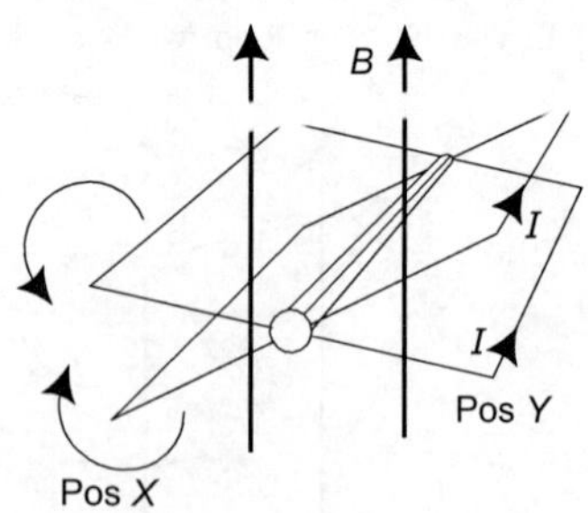

Fig. 8.33 Torques in positions *X* and *Y* (shown superimposed) will not develop unidirectional torque.

The key to develop unidirectional torque is to make electrical connections to the coil in such a way that *just as it is coming* to its stable position, we *reverse* the direction of the current.

Electric dc motors employ a *commutator* to change the direction of current as the shaft rotates, similar to the commutator that was used to maintain a unidirectional emf in generators (Section 7.12). A description of commutator operation in a motor is given in Section 8.10.

Torque as a Vector

For a rectangular current (I) carrying loop ($w \times h$) in a magnetic field B, the torque is (Iwh) (B sinθ), and $\mu \times B = (Iwh) \times B = (Iwh)(B \sin\theta)$ ...(1)
where θ is the angle the loop makes with the horizontal, as we learned in Section 8.7.

The *torque* can also be expressed as a vector $\mu \times B$ (See Eq. 1) and depicted in Fig. 8.34 (after Chabay & Sherwood, 1995).

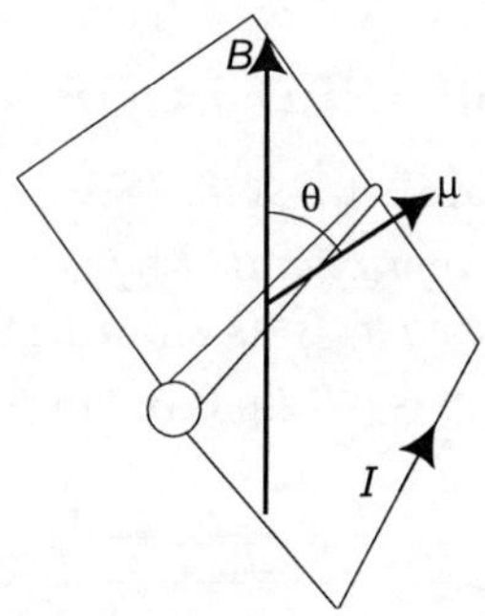

Fig. 8.34 Torque can be expressed as "μ × *B*".

The *torque* vector $\mu \times B$ has a magnitude $|\mu|\ |B| \sin\theta = \mu B \sin\theta$ and a direction along the axle around which the loop rotates as shown in Fig. 8.35.

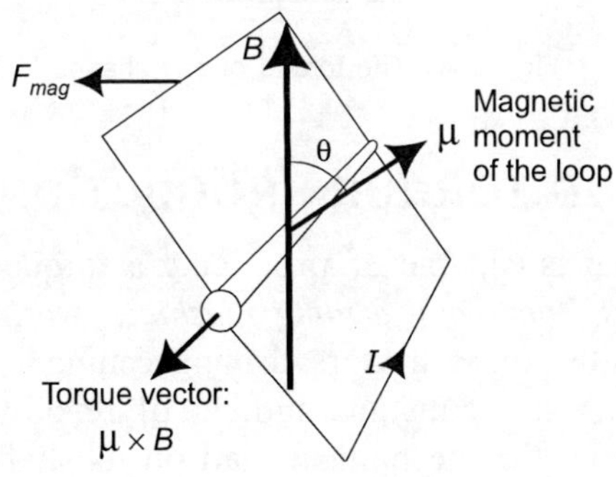

Fig. 8.35 The torque vector.

The force F_{mag} (see Fig. 8.35) is a maximum when the loop is in line with the magnetic field B. This may be surprising when one compares with the movement of a compass needle which always aligns its length with the magnetic field (see Fig. 8.22). But, remember that it is the *magnetic moment* that aligns, and while this is longitudinal for a compass needle, it is perpendicular to a current-carrying loop

(Section 8.2). That F_{mag} is a maximum can also be seen by substituting $\theta = 90°$ in the expression for the forces F_2 and F_4 (Refer Fig. 8.29): $IwB \sin\theta = IwB$.

The forces F_2 and F_4, as a matter of fact, reduce as the loop rotates and become zero when θ is 0° or perpendicular to the magnetic field.

If a real machine were to be built, the loop will be carried past this point as it rotates, by the momentum it had gained as it approached this point.

The non-uniformity of the force developed as it rotates creates torque pulsations and causes vibrations in the machine structure.

To remedy this, motor designers arrange to provide a radially directed magnetic field to the loop conductors so that the force that is developed is a maximum at every point of their travel. The radial magnetic field is normally provided using suitably fashioned pole faces. See the pole head shapes of a real motor in Fig. 8.46.

8.9 TORQUE OF ANY SHAPED LOOP

The torque for a rectangular loop is $\mu B \sin\theta$. (See below Fig. 8.34).

A loop of *any shape* can be *approximated* by a set of current-carrying rectangles along whose adjacent sides the current *cancels* as shown in Fig. 8.36.

The *torque* of *any* kind of loop is $\mu B \sin\theta$ because *any kind* of loop can be *approximated* by a set of rectangles.

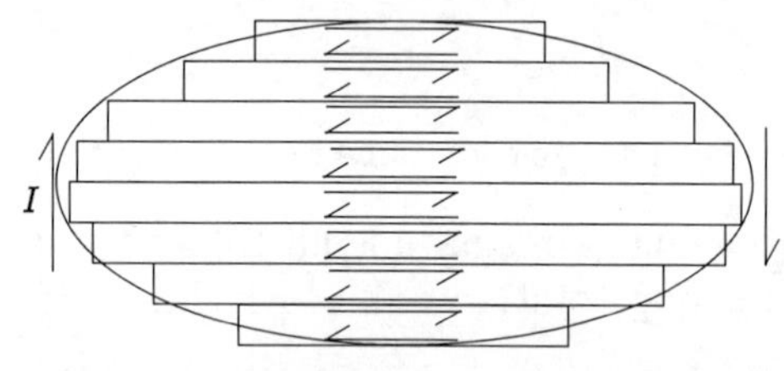

Fig. 8.36 The torque of any shaped loop.

8.10 ROTATION AND TORQUE PRODUCTION IN MOTORS

The *motor* is a device that is capable of producing a **torque** and is used to power *pumps, fans, blowers, compressors, conveyors, drilling machines, washing machines and many other mechanical loads.* Generally, most motors driving ceiling fans, pumps and compressors, operate for long durations at a constant number of revolutions per minute (rpm) when, the load torque imposed by the mechanical load on its shaft is fairly constant.

However, the rpm may vary (maybe slightly or by large amounts), even during normal operation, as in the case when the load torque varies, for example when the pressure of gas pumped into a tank is raised by a motorised compressor or, when more weights are placed on a conveyor belt driven by a motor. See Chapter "Angular Momentum" in Ref. [41].

In this section, we will see how a *motor* rotates and produces a torque.

The dc motor (see Fig. 8.46) consists of coils of conductors (windings) inserted into slots of an armature which is cylindrical in shape and made from metallic stampings. In the center of the armature is the motor shaft to which the stampings are fixed. The armature shaft is mounted on the frame using bearings. The yoke that is fixed to the frame carries the magnetic poles, which have windings that when energized produce a magnetic field in the air gap between the pole faces and the armature. Motors are designed for base or flange mounting on machines, and the motor must be securely installed to a rigid foundation or mounting surface to minimize vibration and alignment maintained between the motor and shaft load. If not rigidly mounted, the motor *reaction* torque will cause the motor *frame to turn* in a direction *opposite* to the direction of rotation of the motor, and can apply tremendous forces that could be dangerous; motors that are not mounted should never be energized.

The constructional details of a typical dc motor are not shown in Fig. 8.37 which focuses on depicting commutation to generate unidirectional torque.

In Fig 8.37 is shown a single loop of a small dc motor powered by a battery with the shaft of the armature fitted with a split-ring commutator. The commutator is made from a conductive ring split in two halves or segments and fixed on an insulated cylinder coupled to the shaft, which rotates along with the shaft of the motor. The voltage from the battery is applied through the commutator leaves to the two ends of the loop.

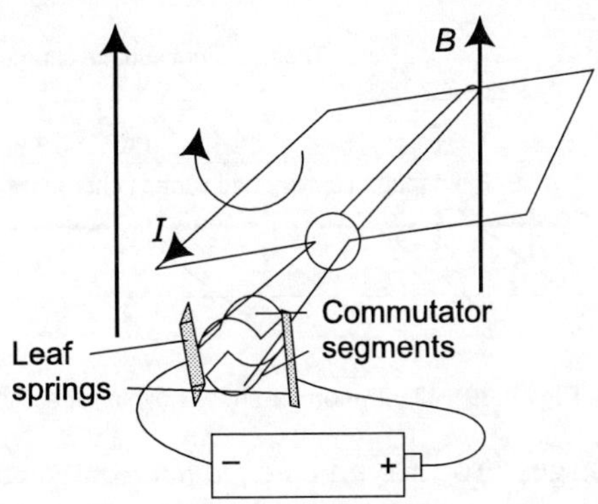

Fig. 8.37 The loop and commutator position before commutation.

The armature conductor carries a current as a result of application of the battery voltage and is shown in the position just before commutation occurs.

Figure 8.38 shows the loop and commutator position after commutation and also shown is the direction of the current which has reversed.

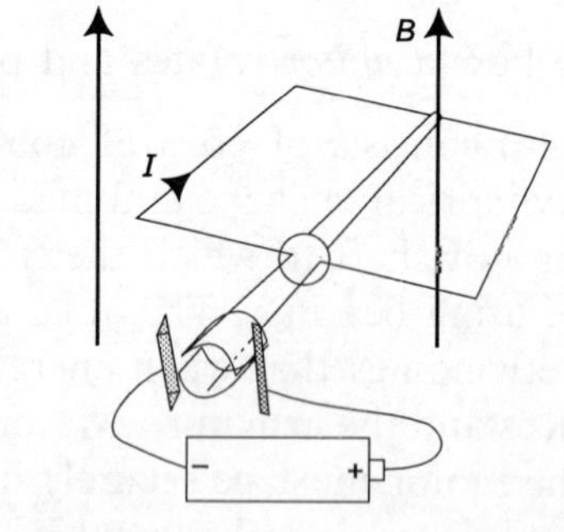

Fig. 8.38 The loop and commutator position after commutation.

The direction of the current having reversed in the conductors, the torque produced is unidirectional.

Remember, copper is *not* a magnetic material. Then, how does the conductor move?

Recollect from Section 6.2, how a metallic current-carrying conductor when subject to a magnetic field will cause the electrons to be deflected downward as shown in Fig. 8.39 (apply $qv \times B$ to determine the direction taking care that the particles are electrons).

As time goes on, more and more electrons moved due to the *action of the magnetic field*, will pile up on the bottom of the wire....with a corresponding deficiency of electrons on the top.

However, this piling-up won't continue forever.

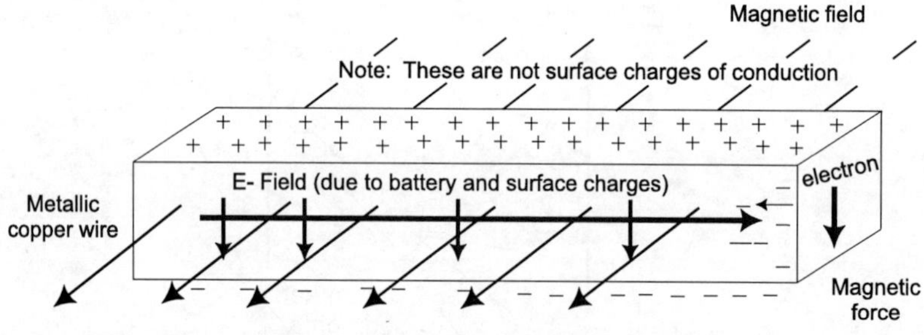

Fig. 8.39 Illustration to show how a wire moves.

The extra surface charges produce an *electric field* between the upper and lower surface of the metallic wire.

The extra surface charges exert an electric force qE (shown in gray in Fig. 8.40, and on electrons this will be "upward") on the *moving* electrons, which in the steady state just balances the magnetic force qvB that caused their movement to the bottom, and so, $qE = qvB$.

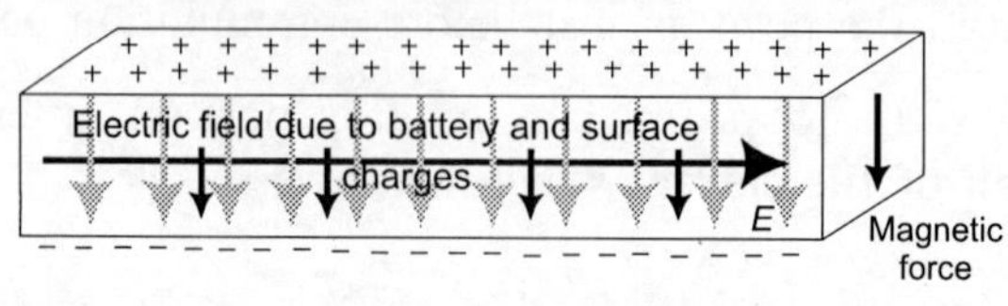

Fig. 8.40 Forces on the electrons are balanced.

Because, qE becomes equal to qvB, the moving electrons now continue through the wire without being deflected as shown in Fig. 8.41.

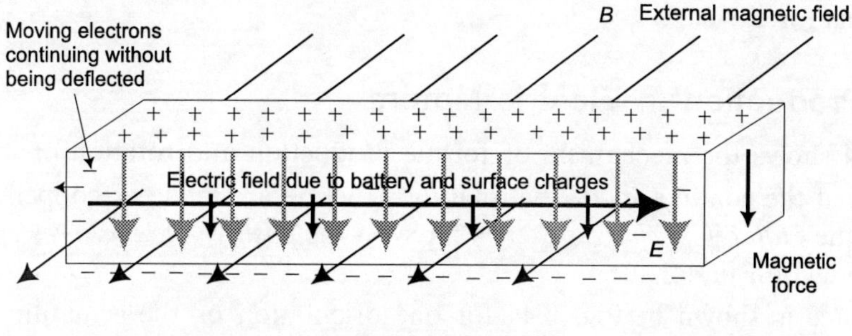

Fig. 8.41 Electrons continue to move without being deflected.

We have seen how the extra surface charges exert an electric force qE on the moving electrons, which in the steady state just **...balances** the magnetic force qvB.

The extra surface charges also exert an electric force qE on the *stationary* positive atomic cores as shown in Fig 8.42.

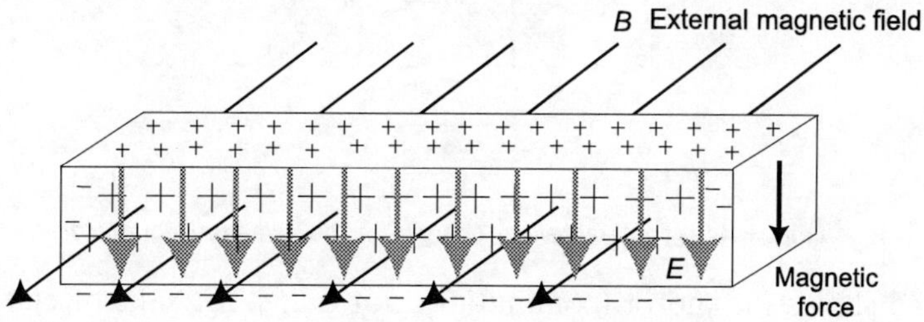

Fig. 8.42 The electric force acts on the positive atomic cores.

This electric force is an *unbalanced* force. There is no counter to this force, because the atomic cores initially are not moving and hence are not subject to a magnetic force.

And that is the reason why *the wire moves*; an *electric force* acts on the positive atomic cores as shown in Fig. 8.43.

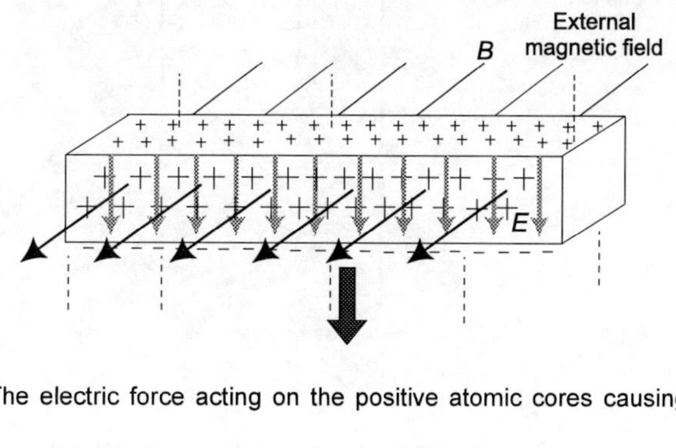

Fig. 8.43 The electric force acting on the positive atomic cores causing it to move.

The electric force due to the Hall effect surface charges forces the wire as a whole to move in the direction of $I(d\vec{L} \times \vec{B})$ where, dL is an elemental length of the conductor of length L carrying a current I, (also see Section 6.5) and the magnitude of the force is equal to this force.

Torque Production in Electric Motors

Figure 8.44 shows the mechanism of torque production and rotation of motors.

Note that the conductors of the rotating structure are made of copper and highlighted is the way the *Hall effect charges* exert a **force** on the *positive atomic cores* and is *responsible* for the production of torque.

This force is shown in Fig. 8.44 for one orientation of the armature and one set of conductors.

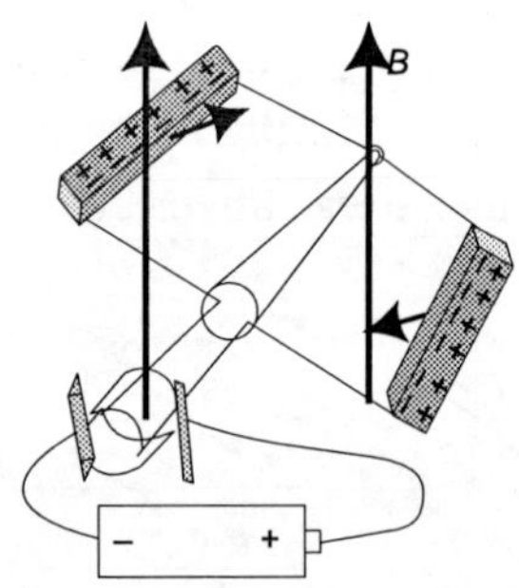

Fig. 8.44 The Hall effect charges in the armature conductors.

Figure 8.45 shows the motor (commutator not drawn for sake of clarity) driving an air blower load via a belt on a pulley fitted to the motor axle, with the forces in the conductors shown by arrows.

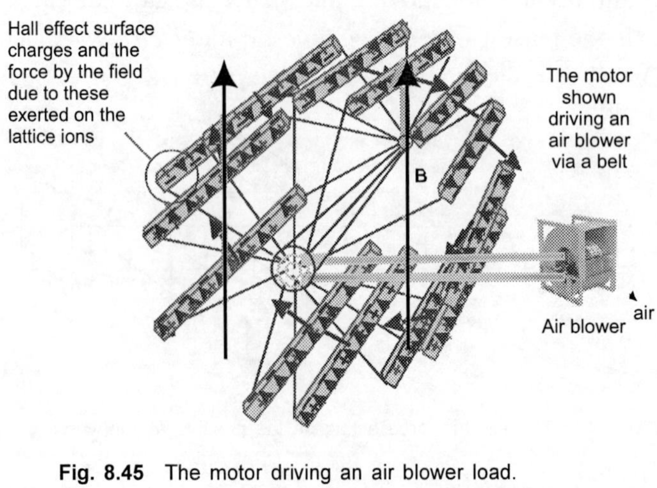

Fig. 8.45 The motor driving an air blower load.

A cut-away view of the principal components of a dc motor is shown in Fig. 8.46.

The magnetic flux path of *least reluctance* lies in the frame or yoke of the motor and the field lines form *closed loops* taking the *following path:* the field poles, the air gap (necessary to separate the rotating and stationary structures), inside the armature structure and through the air gap again and into the frame or yoke and return to the field poles.

The air gap is kept small to provide a low *reluctance* (Section 5.4) path for the magnetic field generated in the poles through to the armature conductors. The motor conductors are placed in a structure commonly called the *armature* in a dc motor and the torque produced rotates the structure and the shaft.

The more the number of conductors and these may be connected in series or parallel then, for a given current in them, the larger the torque the motor is capable of developing.

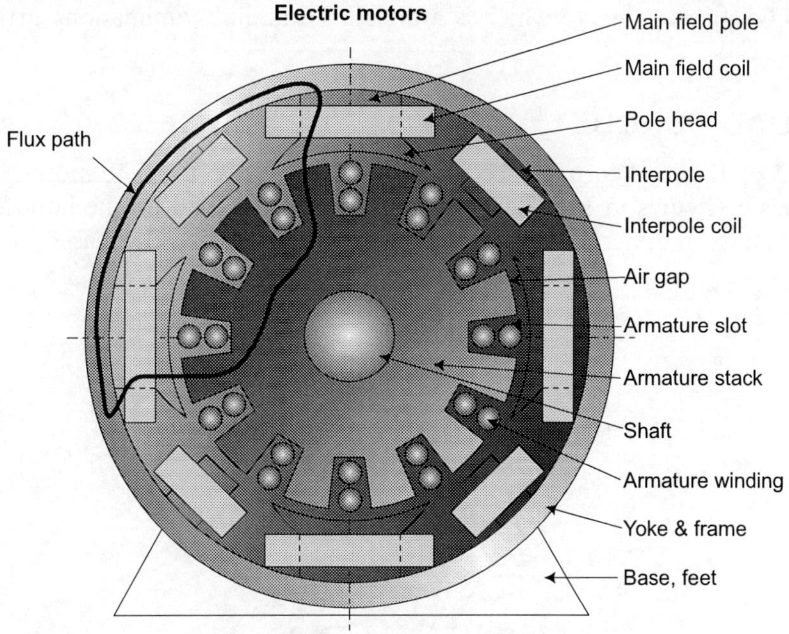

Fig. 8.46 Magnetic flux path in a dc motor.

The commutators of dc motors (not shown in Fig. 8.46) for industrial applications are multi-segment in construction as shown in Fig. 7.23. The two-segment commutator shown in Fig. 8.37 is found only in motors used to power small toys. As described in Section 7.12, hundreds of coil ends placed in the slots of the armature are soldered to the segments of the multi-segment commutator. The coils themselves are connected in series or in parallel, in a manner such that the motor is capable of delivering large power to mechanical loads.

A mechanical load, such as a pump or fan or blower, compressor or conveyor belt or other mechanical device, has a pulley mounted on its shaft and is driven by a belt running on a pulley mounted on the motor shaft. The load maybe directly coupled also to the motor as shown in the photograph of a motor directly coupled to a blower in Fig. 8.47.

Fig. 8.47 Motor directly coupled to a blower.

For a nice description of how electrical power into motors could be transformed into mechanical power see Section 6.3 Chapter 6, *Actuators and Sensors, Motors and Generators*, of the MIT article, a link to which is available in the links_animations_articles file in the CD.

8.11 LOAD TORQUE

The action of the load torque is to impose a *braking effect* on the motor shaft as shown in the series of figures in Fig. 8.48 which show how the load torque imposed is increased as the tightness of the grip (of a hand) on the motor shaft is increased.

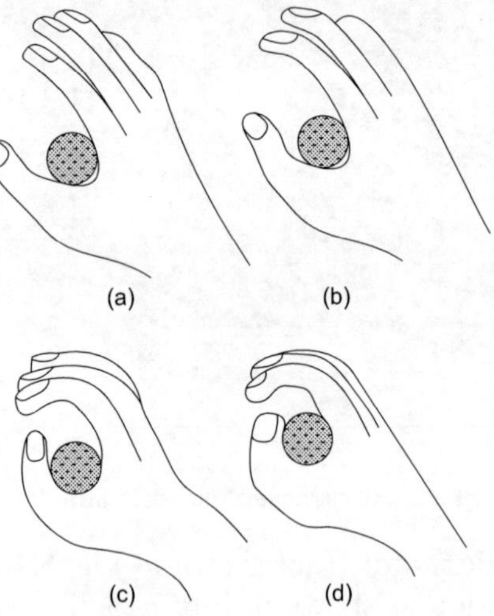

Fig. 8.48 Imposing a load torque on the motor shaft (increasing from *a* to *d*)

Warning! Do not try this anytime! You will lose your hand!

This provides a symbolic interpretation only of the effect of the load torque on the motor shaft.

A motor is *normally used* by *coupling its shaft* to some mechanical device (the *load*) such as a pump, a compressor, a fan or a blower.

The devices extract energy from the motor and the motor *converts the electrical energy* (electromagnetic energy, voltage and current supplied by the battery or voltage source) from the power supply *into kinetic energy of rotation* of the motor shaft and the shaft of the coupled device.

8.12 MOTOR TORQUE AND LOAD TORQUE

A motor develops a torque as it rotates, which is small when there is no load, that is, when no pump or compressor or any other device is coupled to it. The load torque is small when, the motor shaft is not coupled to a pump or fan or compressor, and it consists of only the frictional and other small losses of the motor bearings and heating of windings (also called a no-load condition). This is shown in Fig 8.49a by different sized arrows representing the strength and direction of load and motor torques.

The motor always develops a torque, which *meets* the load imposed torque, resulting in a *steady* rpm. If it were to be *more* than the load torque, then the motor would *accelerate* and steady state cannot be achieved.

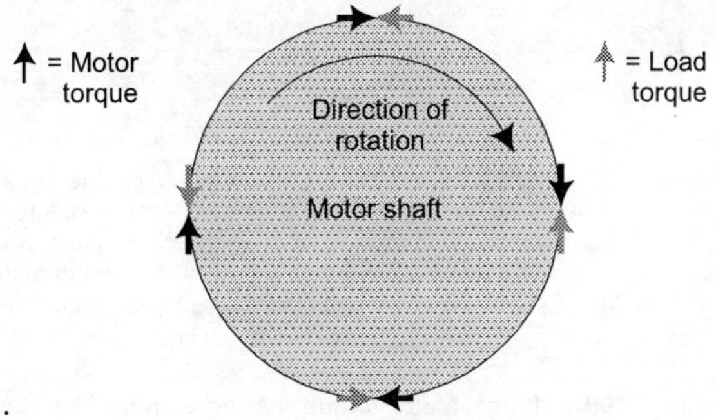

Fig. 8.49(a) A small load torque applied to the motor shaft.

The no-load rpm is large and greater than the *rated* load rpm where, the rated load is the torque the motor is capable of delivering when operating normally at its full capacity.

When a load (pump, fan or compressor) is coupled to the motor, then the load torque is large at the time of starting due to load standstill inertia, and the motor develops more than the running torque at the time of starting. This excess torque serves to accelerate the standstill motor as shown in Fig. 8.49b.

Conventional direct starting of motors always cause a large inrush current for a few seconds that provides the excess torque to accelerate the motor. Switchgear such as contactors, overload (overcurrent) and short-circuit protectors which carry the motor current, should be so selected as to allow for this short duration inrush current without tripping. If the devices prevent inrush current, the motor will not accelerate and will surely stall or not start even. More on starting current and the back emf that determines its strength will be discussed in Sections 8.17 to 8.22. The reader is urged to view the power point slides (Chapter 8) with animations on motor and load torque included in the accompanying CD.

When using resistive current limiters or electronic starters, a careful match of the consequent lower motor starting torque with the load torque is necessary to ensure that the motor starting current is yet strong enough to develop the torque required to accelerate the motor from standstill.

The standstill load torque will vary with the type of the load and also the construction of the rotating structure of motors. Long slender rotors (smaller diameter) will impose a smaller load torque as compared to rotors that are short and of large diameter. The load torque of machine tool type loads will be large, and will be small for fan and pump loads.

Usually, a few seconds after starting, and thereafter, the motor meets the load torque matching it exactly and the rpm remains steady. This is shown in Fig. 8.49c. When properly selected the load torque at steady rpm will be equal to the full load torque of the motor.

Should the motor not match the torque, it will accelerate or decelerate, and when the torques match, the rpm will remain steady.

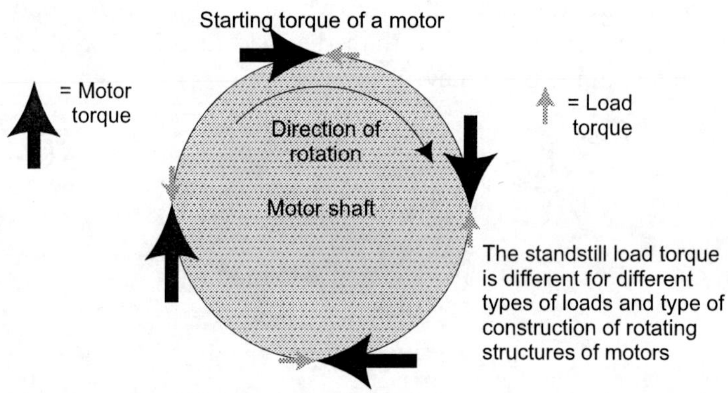

Fig. 8.49(b) Large motor torque that accelerates the load from standstill.

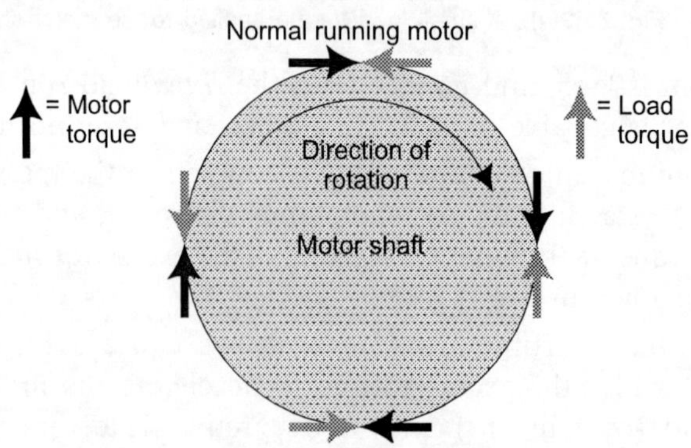

Fig. 8.49(c) Normal operation with full load torque.

If the load torque were to *decrease,* the motor will speed up or accelerate as shown in Fig. 8.50a. Such a situation will occur, for example, when parts are removed from a motor-driven moving conveyor belt or a load of parts that is rising on a hoist driven by a motor. The motor will then develop a reduced *torque* to meet the reduced load and will operate at a steady higher rpm. This is shown in Fig. 8.50b.

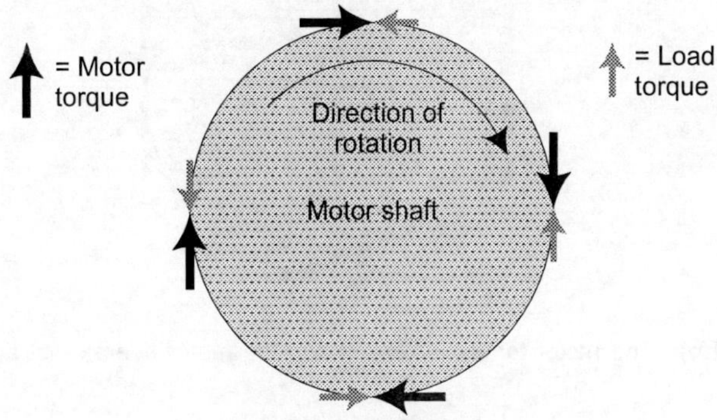

Fig. 8.50(a) In a normal operating motor, the load torque is decreased.

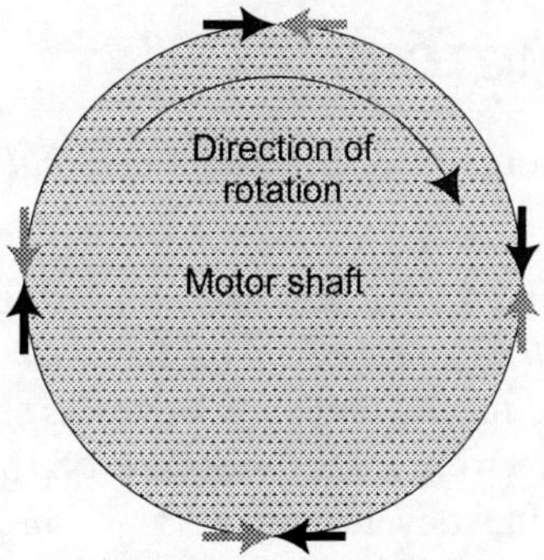

Fig. 8.50(b) The motor accelerates to a larger rpm and develops a smaller torque.

If the load torque were to *increase,* as shown in Fig. 8.51a, as will occur when more number of parts are placed on a conveyor belt driven by the motor, it will slow down or decelerate and in the process will develop an *increased torque* to meet the increase in the load torque and a steady *lower* rpm will be reached. This is shown in Fig. 8.51b.

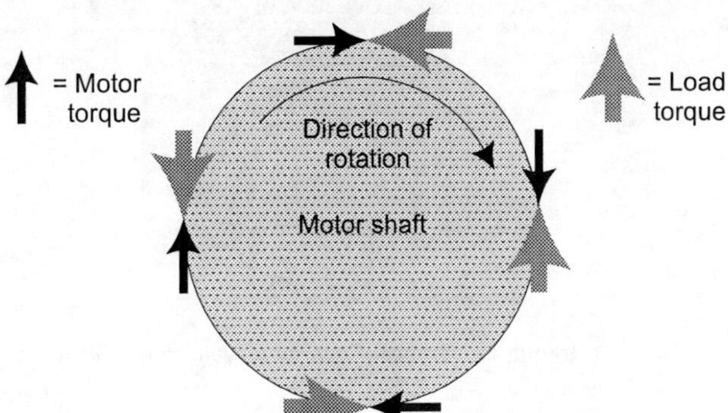

Fig. 8.51(a) The motor decelerates because of the increased load torque.

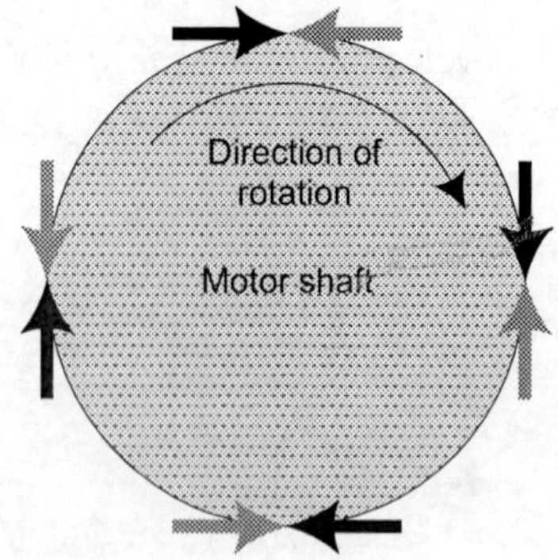

Fig. 8.51(b) The motor torque increases and the motor operates at a reduced rpm.

We shall study in Section 8.17 how the motor regulates its torque production to meet reducing or increasing load troques.

8.13 OVERLOADED MOTORS

What will happen if due to some reason a *large demand* is made on the torque developed by the motor?

Such situations will occur when loads are *not* properly maintained, especially due to *insufficient grease in large bearings or loss of oil in sintered bush bearings* in fan mountings and compressor housings. The motor will slow down as shown in Fig. 8.52a, and eventually may even *stall* if the load were to *jam*, and because of the small resistance of the winding, the motor will try to continue developing a *large torque*. The large current drawn by the motor, to meet the increase in load torque, will flow in the windings which may be *insufficiently* thermally insulated, and the motor will heat up and eventually burn out, as shown in Fig. 8.52b.

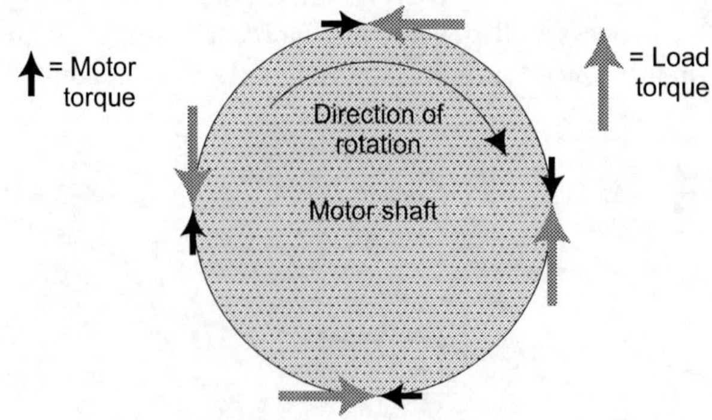

Fig. 8.52(a) The load torque is increased to a large value and the motor rpm is reduced.

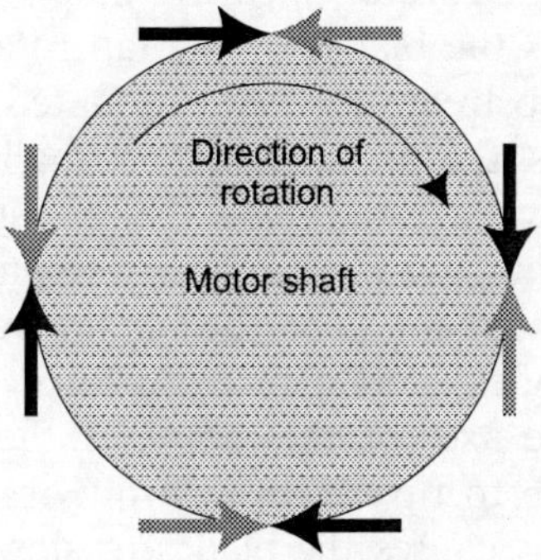

Fig. 8.52(b) The motor torque increases and the motor heats up.

8.14 MOTOR AND LOADS

The load torque that is *imposed* by the mechanical device such as a pump, a compressor, a fan or a blower has the effect of slowing down the motor shaft, which would be otherwise rotating at a *greater no-load speed.*

The devices extract energy from the motor, and the motor *converts the electrical energy* (electromagnetic energy, voltage and current supplied by the battery or voltage source) from the power supply *into kinetic energy of rotation* of the motor shaft and the shaft of the coupled device.

In short, the motor feels the *tug* of the mechanical device.

When we receive the *electricity bill* for operating motors, the charge it contains, is the price we pay for running the motor with a mechanical device. It is the price of the *tug* felt by the motor and the torque it develops to meet the *load.*

The load torque of motors generally is that of compressors, machine tools, fans, blowers and conveyor belts. Such loads are coupled to the motor using pulleys and belts, direct in-line flange couplings or using gear boxes that provide a reduction in rpm at the load shaft.

Electric motors *drive* mechanical loads. We see numerous examples in our daily lives of motors rotating at *a constant* speed such as the fan motors at home.

It is instructive to study how a motor drives a load, such as the ceiling fan (Fig. 8.53) or *axial* fan, and what the forces are that act on the motor shaft by the load which it is rotating.

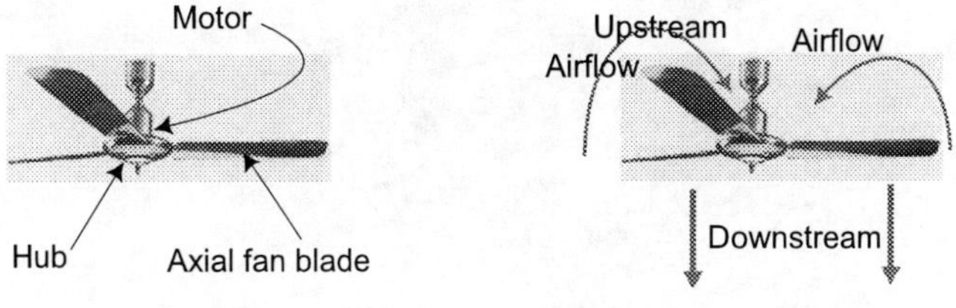

Fig. 8.53 The movement of air produced by a ceiling fan.

It is force of the load (or torque since the motor shaft executes a rotating motion) that we will describe now. As the blades of the fan rotate, forces develop on them due to the flow of air molecules. Upstream (above a mounted ceiling fan as shown in Fig. 8.53 on the right), near the blades, the pressure will be small or negative since the blade would have pushed air molecules downstream (below the ceiling fan). The negative pressure causes the air above and surrounding the fan to rush into the upstream zone of the fan. The same does not happen downstream since the blades push the air down. The accelerated air molecules leaving the blades exert a reaction force. Since the blades are constrained to rotate by the hub and the axes of the motor of the fan, the reaction force along the length of the blade from hub to tip causes a torque in a direction that is opposite to the direction the fan rotates. This causes the blades to slow down. See the symbolic effect of the braking action of the load torque on the motor shaft in Fig. 8.48. The motor in turn, draws power from the supply mains to meet this load torque, and keeps the blades rotating by applying a torque, thereby producing continuous rotation at a fixed rpm during normal operation (Fig. 8.49(c)). More details of this we shall see in Sections 8.21 and 8.22.

8.15 TESTING MOTORS AT DIFFERENT LOAD TORQUES

Electrical equipment manufacturers commonly use a *brake dynamometer* to test and rate the various sizes of varying capacities of motors.

The Brake dynamometer shown in Fig 8.54, applies a torque (load) on the shaft of the motor that is under test, using a belt which can be tightened or loosened, by a nut and screw mechanism. Spring balances suspended from a frame to the ends of the belt, record the amount of tension in the belt and to calculate the amount of load torque applied. For a given setting of the nut, the tightness of the belt is fixed, and the brake applies a constant torque. The motor is energized, and it's current and speed recorded. In this manner, the current and speed values are recorded at various load torque settings, and the motor characteristics obtained.

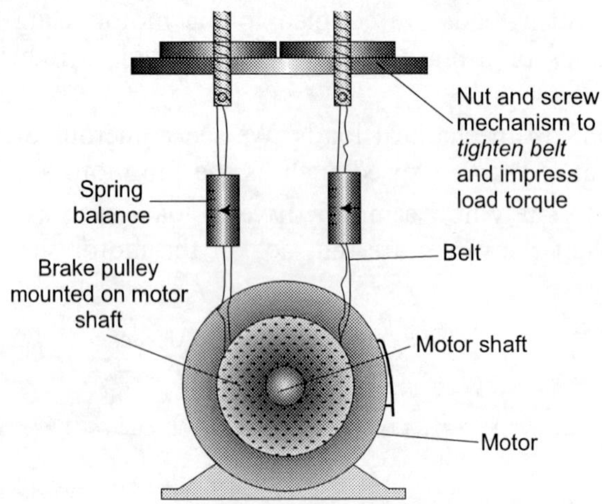

Fig. 8.54 Brake dynamometer.

8.16 DIFFERENT MOTOR SIZES AND RATINGS

Shown below are photographs of motors of various ratings and sizes, the more the power rating, the larger the motor.

Tiny 0.03 HP motor

0.75 HP motor

2 HP Motor

15 HP motor

Giant 200 HP moter

8.17 BACK EMF

A motor always attempts to develop the amount of torque that will meet the load torque.

How does it do it? Nature uses one trick which we shall see enables it to generate the required amount of torque.

To understand this, look at the torque equation: torque = $(Iwh)(B\sin\theta)$. Because, B, w and h are constant, the quantity that may be "varying" is the motor current I. How can the motor current I be varied? if, V the applied voltage is constant, as in say, dc operated motors.

In large motors, the motor current is *less* than volts applied/resistance of the winding conductor, and is exactly (volts applied *minus* the *back emf*)/resistance of the winding.

The *back emf* is the trick nature uses, which has a *regulating effect* in determining the amount of current the motor draws, to develop a torque which meets the load torque, and this "back emf" is a natural consequence of the induced emf process when a conductor *moves* in a magnetic field.

It must be borne in mind that in large motors, the current given by volts applied/resistance of the conductor is quite large, because the resistance of the conductors is small, and this large current is drawn when the rotor does not rotate and is at standstill when there is no back emf. As the motor *picks up speed*, it is the *back emf* that *reduces* the effective voltage which a conductor is subject to and therefore, the current drawn is consequently *much smaller* than that it would draw when at standstill. The governing relation is $J = \sigma E_{net}$.

The *back emf* is in reality the *motional emf* produced when a conductor moves in a magnetic field, which we learned in Sections 7.7 and 7.8. This is shown in Fig. 8.55.

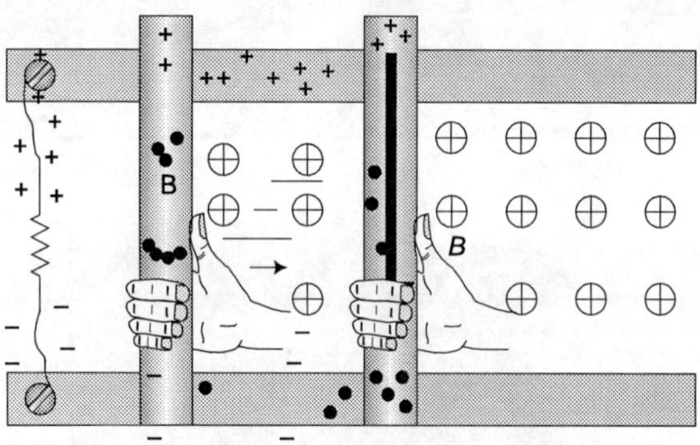

Fig. 8.55 Back emf is a type of motional emf.

As a result of the magnetic force (due to the presence of the uniform magnetic field), the free electron sea shifts and the bar is polarized, producing the emf as shown in Fig. 8.55.

In a motor, the armature conductors carry the current due to the applied voltage, and because of the constant magnetic field (produced by the field poles) in which these are immersed, produce an electric force on the stationary atomic cores in the conductors due to the Hall effect phenomenon (see Section 6.4). This action causes rotation of the armature, and the process is called *motoring*. Now, as the conductors rotate in the magnetic field (like pulling the bar in Fig. 8.55), the electron sea in the conductors shifts laterally, and because of the presence of the magnetic field, the electron sea experiences a magnetic force which polarizes the conductors to produce the motional emf. This emf is in a direction that *opposes* the applied voltage and is hence called, *back emf*.

The motor current is governed by the relation $J = \sigma E_{net}$, where σ is the conductivity of the conductors (that does not change), and E_{net} is determined by two electric field components; one, due to the applied voltage and the second, due to the motional emf. Because the back emf (or, motional emf) increases with increasing rpm of a motor, the motor current reduces in magnitude, as it picks up speed.

If the load torque were to increase slowing the motor down, the back emf will be smaller, and the motor draws more current causing an increase in torque to meet the load. Refer Section 8.12 and Section 6.5.

8.18 MOTOR CURRENT—START TO NORMAL OPERATING SPEED

When the motor starts from standstill, it draws a large current (because there is no back emf, a form of motional emf), to develop the torque necessary to overcome the inertia of the motor. This is shown in Fig. 8.56. As we studied in Section 7.7, motional emf is given by the expression vBL, where v = velocity of the conductor, B is the external magnetic field and L the length of the conductor. From this expression we deduce that there is no back emf developed in the conductors when the motor is at standstill and about to start because $v = 0$. This is also to be expected, since the magnetic field deflects only moving charges.

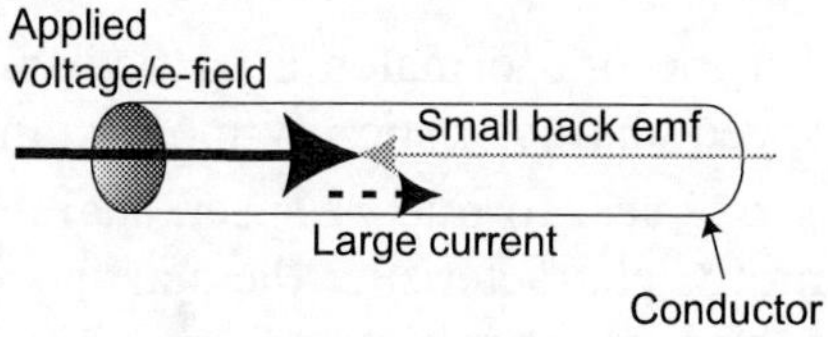

Fig. 8.56 Starting current.

After the motor accelerates from standstill, the initially drawn large current reduces because the back emf grows. This is because the conductors would now have attained some velocity and the electron sea in them is deflected by the magnetic field. The current is governed by $J = \sigma E_{NET}$. After the motor develops a large steady rpm, it draws a small current, because the back emf would have grown in magnitude and nearly equals the applied voltage. This is shown in Fig. 8.57.

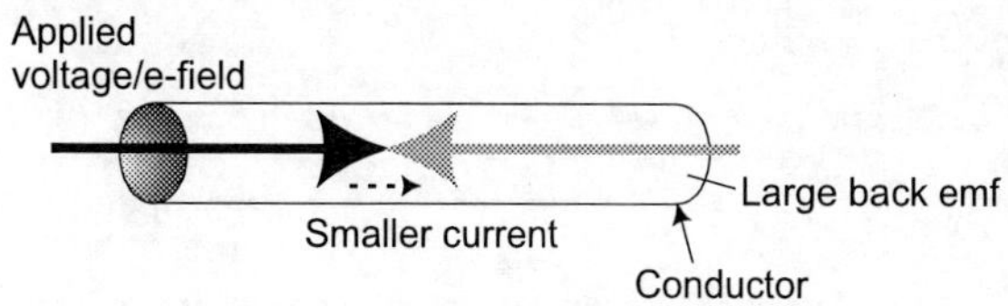

Fig. 8.57 Normal operating current.

The primary process when a motor is switched ON, is the development of the motor driving torque, which bears a relationship with the flux (total magnetic field in the air gap) and the armature current.

The torque is *independent* of the speed, and this aspect is not apparent when the formulas are scanned for torque and speed.

The formulas mask the physical processes involved; of how the driving force applied to the load is the torque developed by the motor, and the speed of the motor depends on this torque.

8.19 MOTOR SELECTION

In most applications, motors are selected largely based on motor rating or *capacity in hp and the rpm.*

Sometimes, when the load inertia is more, a *larger* motor is *preferred* which accelerates quickly to the *operating speed* at which speed, the motor matches the load torque and does *not spend too much time* in drawing a large "starting current" and consequent heating.

Care should be exercised when making such large motor selections, because too rapid an acceleration means tremendous forces are impacted (short of hitting the load with a sledge hammer!) on the load and may cause damage.

8.20 MOTOR TORQUE VS SPEED CHARACTERISTIC

Equipment designers select motors to match a particular type of load. This they achieve by using the torque vs. speed characteristics of motors and loads.

Shown in Fig. 8.58 is a typical torque vs. speed characteristic of an ac motor. Observe the two regions of the motor characteristic: the starting region where, the motor spends little time after starting, and an operating region where, the motor normally operates, continuously.

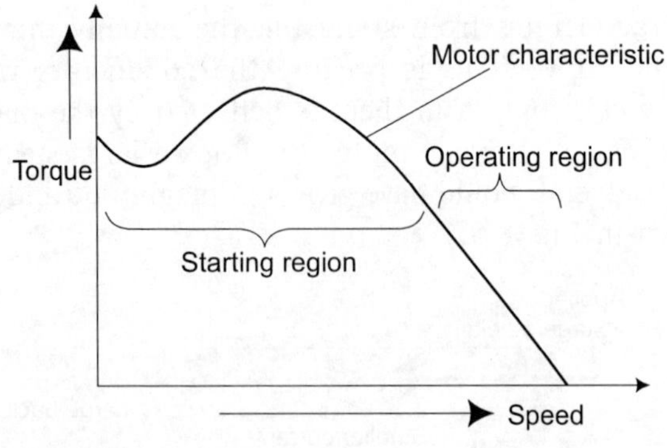

Fig. 8.58 Motor torque vs speed characteristic.

In Section 8.22 is given a description of the significance of the two regions and how motors are selected by matching the load torque with the motor torque.

A motor is useless unless it drives a load which could be a blower, fan, spindle of a lathe or a conveyor belt. It is necessary to understand the type and characteristics of such loads and the following section describes some of the more common loads found in industrial applications.

8.21 LOAD TORQUE VS SPEED CHARACTERISTIC

We have studied the characteristics of motors, and since these are used to drive loads, it is important to study loads and their characteristics in order to be able to match motors to suit loads.

Loads such as fans, pumps and compressors apply varying levels of *squeeze* or torque at different speeds at which they are made to rotate by the motors. They do not always apply a constant torque at different speeds when in operation.

The load torque vs. speed characteristic helps equipment designers to select motors appropriately.

For most loads, the motor is made to operate *continuously* after starting, at the point where, its developed torque *matches* with the load torque therefore, it is useful to study the load torque vs speed characteristic.

Some common types of loads will be shown to illustrate.

Machine tools load motors during cutting, drilling and milling operations as shown in Fig. 8.59. Unlike fans, compressors and pumps, the torque applied by machine tool cutters, drills and grinders to the motor shaft *does not* vary with motor rpm or speed. The torque vs. speed characteristic is a straight line in the graph of Fig. 8.59.

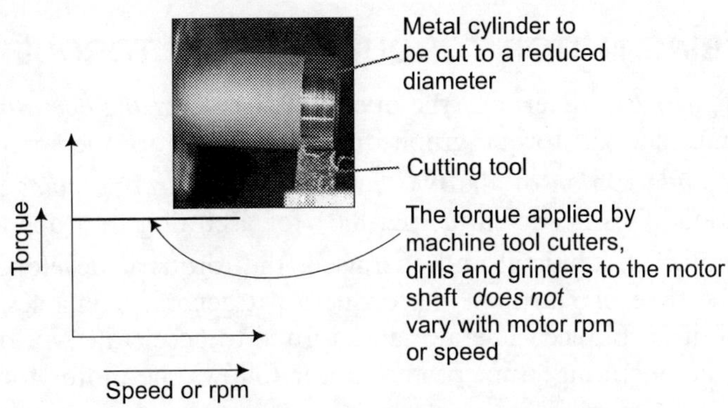

Fig. 8.59 Machine tool (load) vs. speed characteristic.

On the other hand, the torque vs. speed characteristic of axial fans, centrifugal pumps and blowers have a varying characteristic and follow a square law, as shown in Fig. 8.60. The torque is proportional to the square of the speed for such loads.

The torque applied by the fan and blower to the motor varies in *squared proportion* with motor rpm or speed.

The power consumed by such loads varies in cubic proportion with speed because the power consumed is proportional to torque × angular velocity.

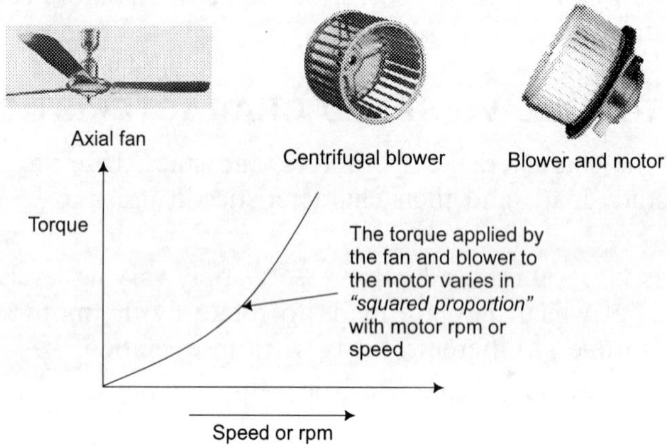

Fig. 8.60 Axial fans, centrifugal fans and blowers (load) vs speed characteristic.

These characteristics are obtained by coupling the devices such as fans, blowers to a *motor whose rpm can be varied* and from the corresponding power drawn at different speeds, the torque of these devices is computed and plotted.

In the case of machine tools, the *motor whose rpm can be varied* is coupled with a large cylinder and cutting forces increased by pressing the cutting tool at various forces, recording the power drawn at different speeds and the torque computed.

8.22 MATCHING MOTOR TORQUE TO LOAD TORQUE

Equipment designers can determine the operating speed, or *the operating point* very simply by superimposing the load torque graph on to the motor torque speed characteristic.

The *point of intersection* of the two curves is the *operating point* for the given motor –load combination. This is shown in Fig. 8.61 for a cutting tool operation.

With the motor at standstill and energised, the torque it develops is T_A as shown in Fig. 8.61. At the time of starting, the torque T_A it generates, enables it to accelerate the load with which it is coupled. The available torque to accelerate will be $T_A - T_{load}$. As the rpm increases, the operating point of the motor follows the motor torque curve.

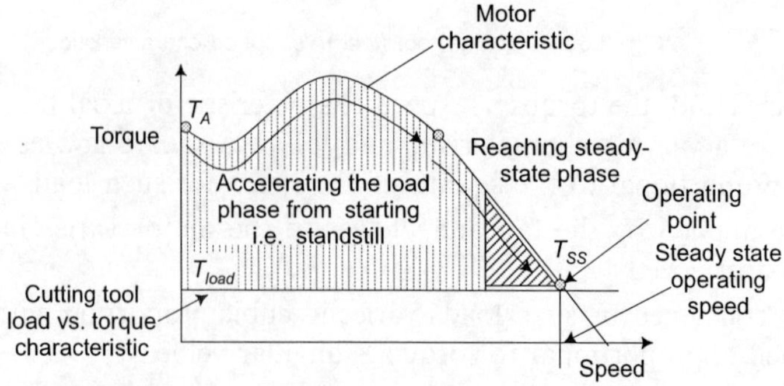

Fig. 8.61 Motor torque and speed variation from starting to steady-state operation.

Subsequently, after a few seconds as the back emf comes up, the motor torque gradually reduces to T_{ss}, the steady-state value, and the rpm becomes steady at the point of intersection of the load torque curve with the motor curve as shown in Fig. 8.61.

Assume that the designer selects a motor having *not enough* capacity for the given cutting tool torque; the cutting tool characteristic intersects the motor characteristic in the *starting region* (shown hashed in Fig.8.62) and the *operating point* is in this region.

Since, the motor is not designed to operate continuously in this region, it will heat up and eventually burn.

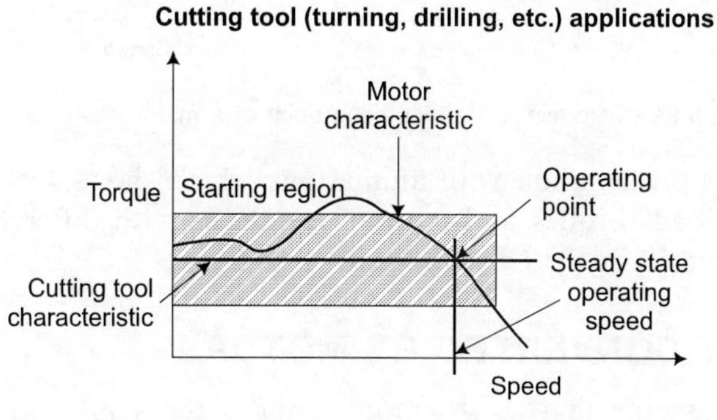

Fig. 8.62 Determining the operating point of a motor (wrong selection).

A more appropriate selection of the motor (larger capacity) is shown now in Fig. 8.63 where, the *point of intersection* of the two curves, the operating point lies in the operating region (shown hashed) of the motor.

The motor can now run *continuously* without heating up.

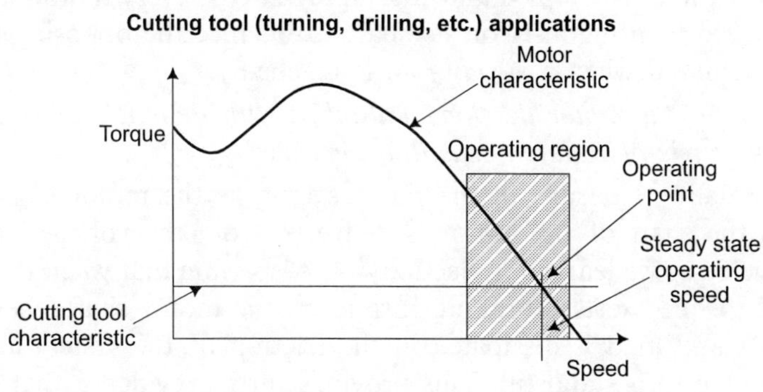

Fig. 8.63 Determining the operating point of a motor (correct selection).

Fig. 8.64 shows how the steady-state speed of the motor and load combination for fans, blowers and centrifugal pumps, is determined from the motor and load torque-speed curves. With such loads whose torque vs. speed characteristic follow a square-law, the starting regime is not demanding (because the load torque is zero at the time of starting) but, care must be exercised in not selecting too large a motor as too high an acceleration can damage the load.

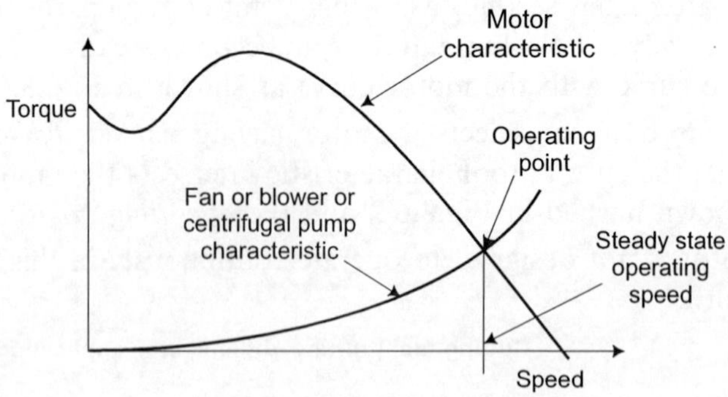

Fig. 8.64 Determining the operating point of a motor (fan or blower load).

A power point presentation with animations which shows the action of a motor, it's matching against load torques and current variations with different loads is available in the slides of Chapter 8 in the CD alongwith this book.

8.23 ENERGY CONVERTED BY MOTORS

It is not right to assume that as the motor rotates the shaft and the coupling with the load, it does *not* deliver energy to the load because the *rpm is steady.* This is a common erroneous assumption made by students.

The reasoning that because there is *no acceleration* of the shaft and since, energy = torque × angle, and torque = force × lever arm = mass × acceleration × lever arm, therefore with zero acceleration, *no energy is delivered* by the motor, is erroneous !

As long as the load is coupled and the supply is ON, every motor strains and develops the necessary forces continuously on its shaft to balance the opposing forces of the load; the *balance* is achieved when a *steady rpm* is reached.

And, the *torque the motor develops multiplied with the angle of rotation or distance it travels along the circle of rotation constitutes energy.*

The energy delivery occurs all the time as long as the motor is ON and is rotating. The energy to the shaft of the motor is delivered from the voltage source powering the motor by the fields as we learned in Section 4.4. Also, note that when the motor is switched OFF, the load torque exerts a braking torque on the motor shaft, the kinetic energy of the motor shaft and load is dissipated in the friction of the system and the motor with its load both come to a standstill. This provides further evidence that the motor delivers energy to the load when switched ON and is running. Read the important note at the end of Section 4.4.

For a nice video description of the operation of DC motors visit https://www.youtube.com/watch?v=LAtPHANEfQo or http://www.learnengineering.org/2014/09/DC-motor-Working.html

For a video description of the operation of AC Induction motors visit https://www.youtube.com/watch?v=LtJoJBUSe28 or http://www.learnengineering.org/2013/08/three-phase-induction-motor-working-squirrel-cage.html

8.24 LOUDSPEAKERS

Loudspeakers operate on the principle of magnetic forces developed when a current carrying conductor is placed in a magnetic field.

Figure 8.65 shows the construction of a loudspeaker. It consists of a single moving coil, whose ends are connected to the amplified electrical speech or music signal source, using flexible wire. The coil is attached to a light weight paper cone, which is supported at its inner and outer edges by circular, pleated paper *springs*.

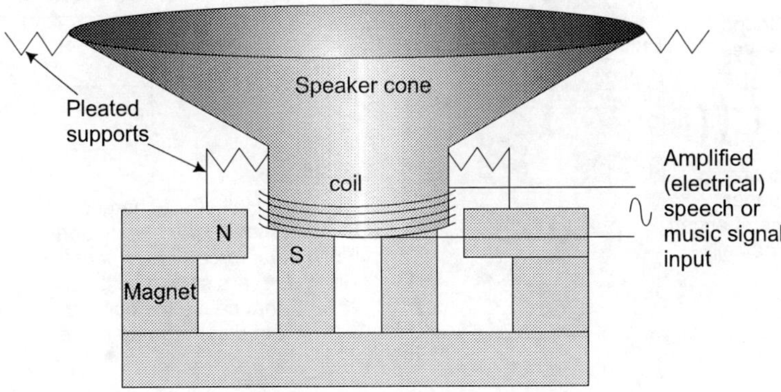

Fig. 8.65 Construction of a loudspeaker.

The coil moves in the field of a permanent magnet, which is usually shaped to produce maximum force on the coil. This is shown in the cutaway view in Fig. 8.66.

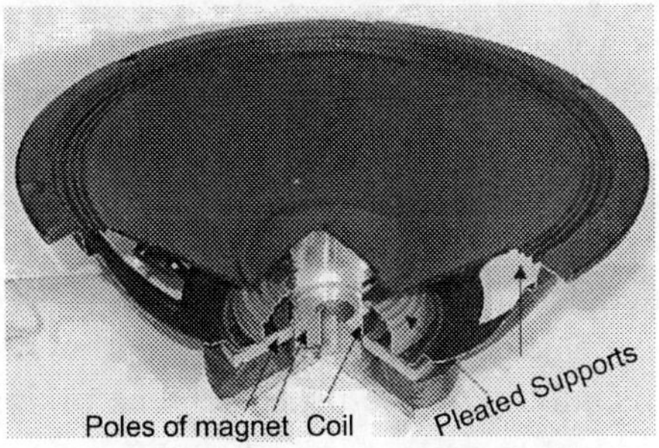

Fig. 8.66 Cutaway view of a loudspeaker.

Figure 8.67 depicts how forces that make the cone vibrate, are produced by a loudspeaker.

Shown in the figure, are the coils (grossly exaggerated using rectangular cross-section wires) and the diaphragm, which is made to vibrate by the coil to which it is attached. The coils are immersed in a magnetic field produced by a permanent magnet.

The permanent magnet is specially made, and is shaped like a cylinder. It's shape and magnetization is such that its' magnetic field is directed radially inward, penetrating the coil.

When the coil vibrates, due to the Hall effect forces that develop by the applied audio signal emf and the radially directed magnetic field, it causes the diaphragm to vibrate along with it.

This causes sound waves to be propagated.

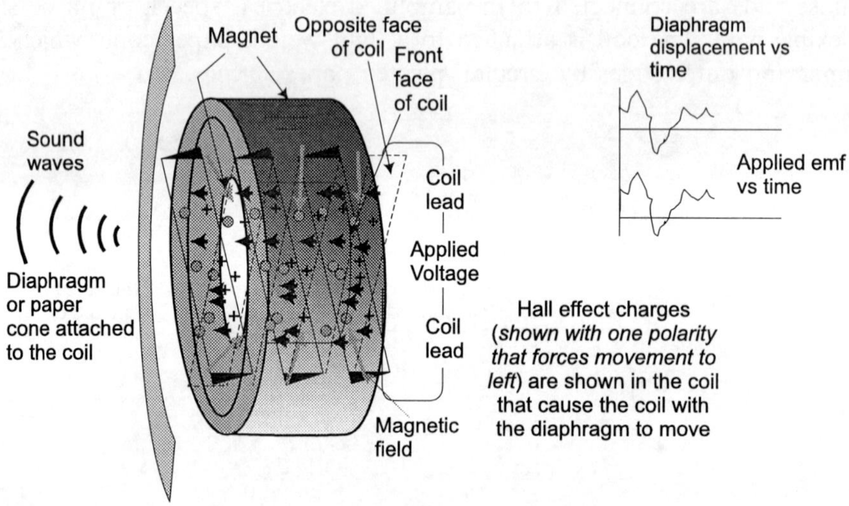

Fig. 8.67 Cutaway view of a loudspeaker.

To view an animation of how a loudspeaker produces sounds visit http://www.mediacollege.com/audio/01/sound-waves.html.

An amplifier, such as a class A transistor amplifier, which produces powerful electric signals with little distortion from weak microphone electrical signals, may be used to deliver power to a loudspeaker connected as the load resistance to the amplifier. A loudspeaker with a small load resistance, if connected directly in the output circuit of a transistorized power amplifier, causes the quiescent DC current of the transistor to pass through this resistance and results in a considerable wastage of power. There will also result an impedance mismatch between, a loudspeaker whose resistance is a few ohms, and the output resistance of the transistor, which usually is of the order of tens of kilo-ohms.

A transformer coupling, with the primary of the transformer in the collector circuit and the loudspeaker connected across its' secondary, will provide a "reflected impedance" whose value can be adjusted by the turns ratio of the transformer so that the impedance seen by the collector may be increased to a few hundred or even kilo-ohms. Such a transformer coupling will mitigate the effects of an impedance mismatch, and will also enable a large amount of power to be delivered to the loudspeaker with a controllable amount of power dissipation in the transistor.

The reader is referred to a power point presentation "Transformer-coupled Audio Power Amplifiers" with animations in the accompanying CD that describes the design, construction and operation of a transformer coupled audio power amplifier.

8.25 POTENTIAL ENERGY OF A MAGNETIC MOMENT

In Chapter 1, we had studied different types of potential energy and to this we now add one more type; a magnetic moment.

If allowed to pivot freely, a *magnetic moment* turns towards alignment with the applied magnetic field. See Section 8.3.

This means that an aligned magnetic moment has *lower potential energy*, and the system tries to *go to the lower potential energy configuration.*

Recollect from Chapter 1 that by separating the charges of a neutral atom (electrons and positive atomic cores), we could store the *work done* energy in the *electric field* of this *system* of charges—comprising the two charges.

....like stretching a spring....

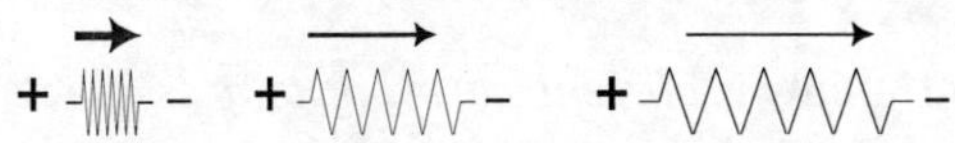

....and when you *released* the *negative charge*.....

.......the *electrostatic energy stored in the system of 2 charges was released* and converted to *kinetic energy* in the *negative charge*......

***.......like releasing a stretched* spring.....**

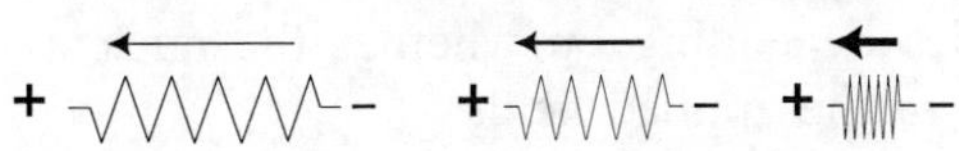

Analogously, observe carefully how when a current carrying loop in a magnetic field is displaced away from its equilibrium position, it "*attempts*" to return to the equilibrium position.

And, you have to hold your finger against the loop applying a *"force"* to return to a horizontal equilibrium position.

This is shown in Fig. 8.68a.

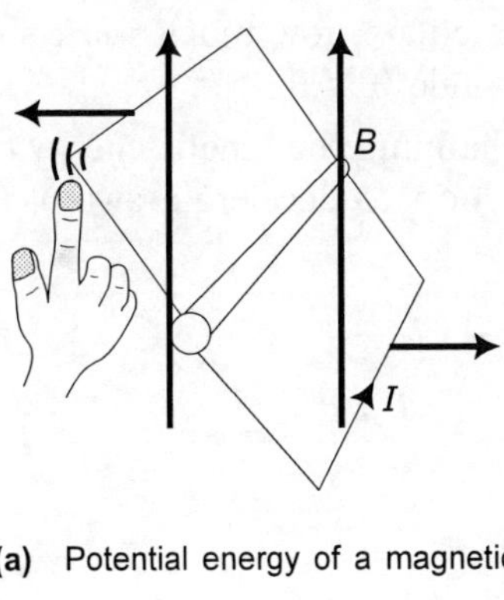

Fig. 8.68(a) Potential energy of a magnetic moment.

If you remove your finger, the magnetic forces will cause it to turn *counterclockwise* and *return* to a more stable horizontal position as seen in Fig. 8.68(b).

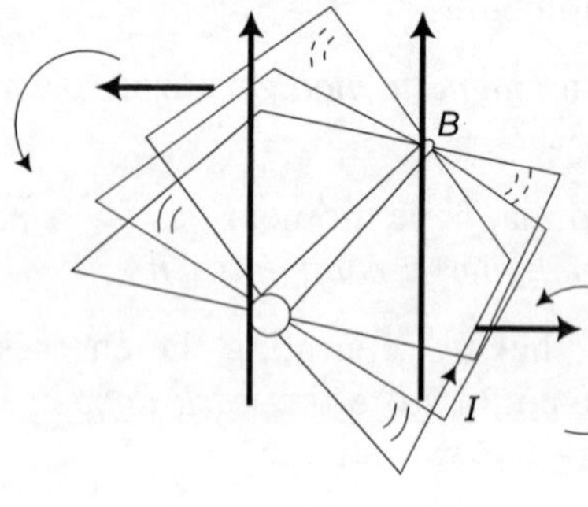

Fig. 8.68(b) Potential energy of a magnetic moment.

When a magnetic moment, a *current carrying loop* is placed at an angle in a magnetic field, it stores energy—potential energy—just like *separated opposite charges* and like a *stretched spring.*

This picture of the *potential energy* of a magnetic moment in a magnetic field is very important in atomic and nuclear physics.

In magnetic imaging equipment used in hospitals, the patient lies inside a large solenoid. In the presence of the magnetic field of the solenoid, nuclei in the patient's body have slightly different energies depending on whether the nuclear magnetic moment is *in the direction of* or *opposite to* the applied field.

Sensitive equipment detects transitions between these two slightly different levels.

Calculating the Potential Energy (PE) of a Magnetic Moment

The potential energy stored by a mass m raised to a height h above the ground is given by mgh. Similarly, we can calculate the potential energy of a current-carrying loop in a magnetic field (Chabay & Sherwood, 1995 used by permission).[2]

Consider a rectangular loop (length h and width w) as shown in Fig. 8.69a with current I as in Fig. 8.23, and let us calculate how much **work** should be done to move it from an initial angle $\theta = \theta_i$ to a final angle $\theta = \theta_f$.

We move slowly, ***without*** changing the **kinetic energy** of the loop (that is by **not** applying a **force** that would cause the loop to accelerate away). (Chabay & Sherwood, 1995, used by permission).[2]

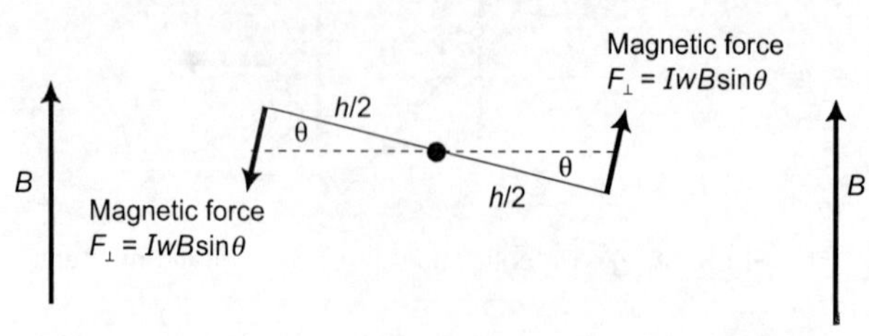

Fig. 8.69(a) A current carrying loop in a magnetic field.

We would have to exert a varying force $\boldsymbol{F_{us}} = IwBsin\theta$ (Section 8.7, Fig. 8.24) at both ends of the loop, *perpendicular* to the loop, to *balance* the perpendicular component $F_\perp$ of the magnetic force on the loop as shown in Fig. 8.69b.

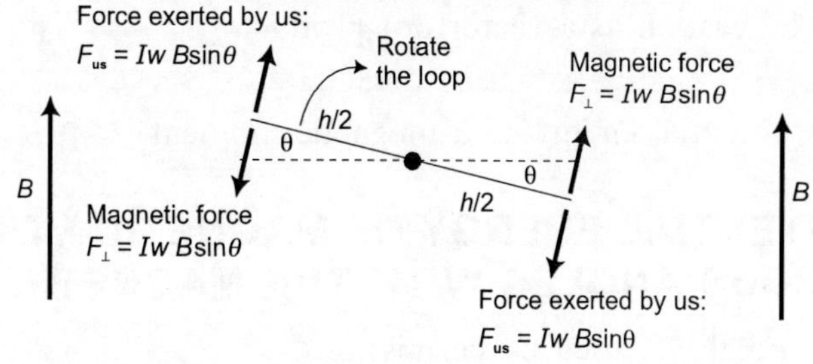

Fig. 8.69(b) Force exerted by us to balance the loop.

As we rotate the loop through a small angle $d\theta$, both of the forces that we exert act through a distance $\frac{h}{2}d\theta$ because arc length is radius ($h/2$) times angle ($d\theta$ measured in radians) as shown in Fig. 8.69(c).

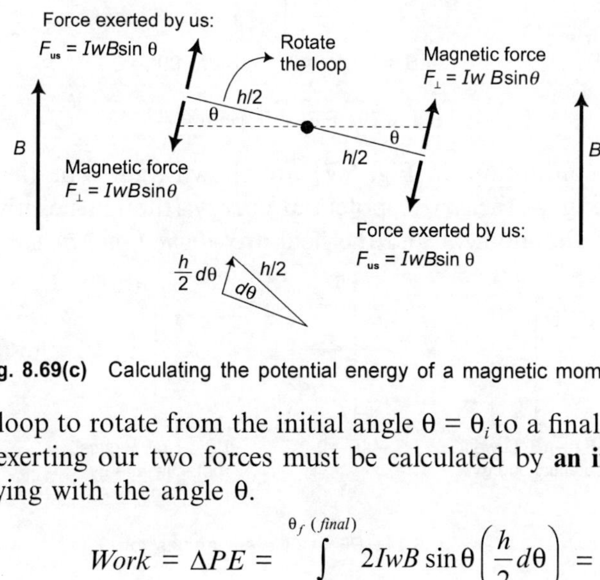

Fig. 8.69(c) Calculating the potential energy of a magnetic moment.

If we force the loop to rotate from the initial angle $\theta = \theta_i$ to a final angle θ_f, the **amount of work** we do by exerting our two forces must be calculated by **an integral**, because $\boldsymbol{F}$ is not a constant varying with the angle θ.

$$Work = \Delta PE = \int_{\theta_i\,(initial)}^{\theta_f\,(final)} 2IwB\sin\theta\left(\frac{h}{2}d\theta\right) = IwhB\int_{\theta_i}^{\theta_f}\sin\theta d\theta$$

$$Work = \mu B[-\cos\theta]_{\theta_i}^{\theta_f} = -\mu B\,[\cos\theta_f - \cos\theta_i]$$
$$= \Delta(-\mu B\cos\theta)$$

where, $\mu = I \times$ area of the loop $= Iwh$ (see Section 8.1).

Or,

$$Work = \Delta\,(-\mu B\cos\theta)$$

It is customary to define the *potential energy* (PE) of a **magnetic moment** in a magnetic field **by** choosing the "zero" of potential energy in such a way that one writes:

$$\text{P.E.} = -\mu B\cos\theta$$

This can also be written as a vector dot product:

$$\text{P.E.} = -\mu\cdot B$$

Therefore, the potential energy of a magnetic moment is $-\mu\cdot B$.

8.26 THE POTENTIAL ENERGY OF MAGNETIC MOMENTS ALIGNED AT VARIOUS ANGLES WITH THE MAGNETIC FIELD

(Chabay & Sherwood, 1995, used by permission).[2]

Convention requires calling the potential energy ***zero*** when $\theta = 90$ degrees, since cos $(90°) = 0$ as shown in Fig. 8.70.

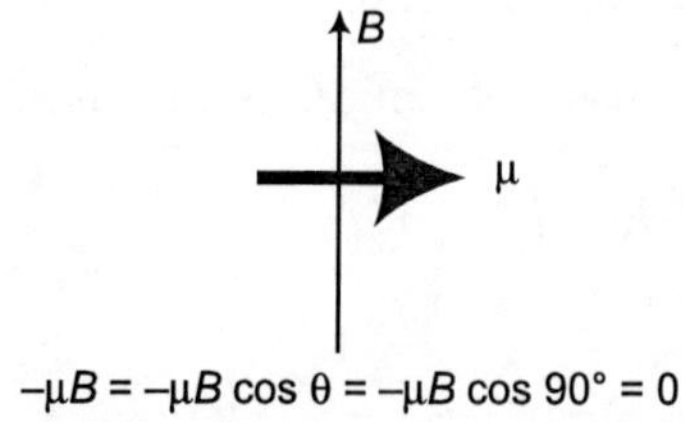

Fig. 8.70 P.E. = 0 when θ = 90°.

Applying this convention, in Fig. 8.71 are shown figures of the alignment between μ and *B* corresponding to the *lowest* potential energy (though negative) to the *highest* and this is *lowest* when the moment and the field are *aligned and in the same direction.*

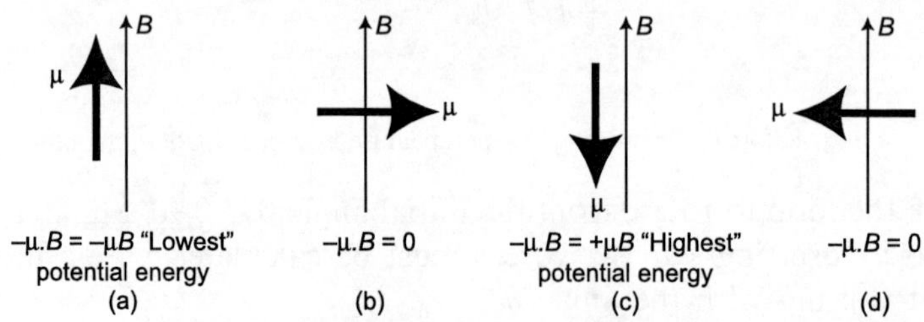

Fig. 8.71 PE at different angles for 'θ'.

The potential energy is the highest when the moment and the field are aligned in opposite directions as in Fig. 8.71(c).

8.27 MAGNETS AND MAGNETISM

We have seen the *twist* of a magnetic moment in a *uniform* magnetic field.

How can one explain the force on a magnetic moment in a magnetic field?

For example, how does a magnet *attract* a paper clip made of iron or steel?

It is not hard to show that the answer lies in placing a magnetic moment in a *non-uniform* magnetic field. Let us first see how magnetic fields are produced in magnets at the atomic level.[28]

Cause of Magnetism

Magnetic fields are created by moving charges. Any time a charge moves (current) a magnetic field is created. Magnetic force is created by two magnetic fields occupying the same space.

A magnetic field is defined as any region where another magnet would "feel" a force.

What are magnets? (atomic view)

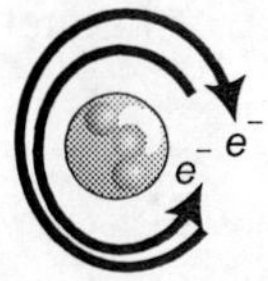

Most materials have paired up electrons spinning in the opposite directions. The field created by one moving charge is canceled out by the other, and so, no magnetic field is created.

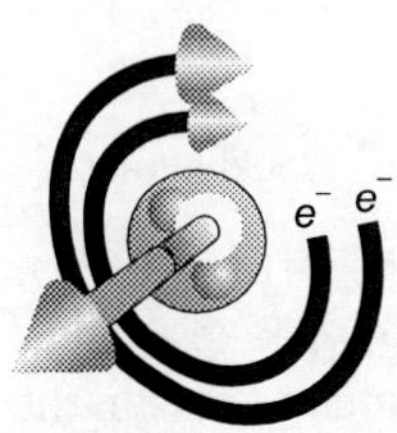

Some materials like iron, nickel, cobalt, or neodymium have a single electron or paired up electrons spinning in the same directions. The magnetic field created by one does not cancel the other. An atomic sized magnet is created.

Atomic to life size?

The atomic magnets line up to form domains within a sample. But, the lining up may not be uniform. Therefore, the metallic sample is not magnetized.

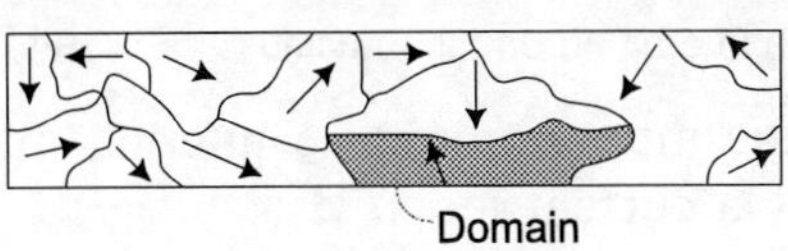

Magnetic domains can be encouraged to line up to form a magnet say, by the application of a strong magnetic field.

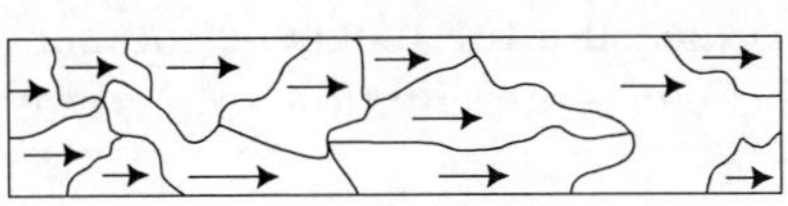

Conventions and Definitions

The direction of a magnetic field is defined as the direction that the north pole of a compass (magnet on a swivel) would point when placed in the field. The field lines drawn are imaginary lines of force-flux lines. The closer they are, the stronger the field.

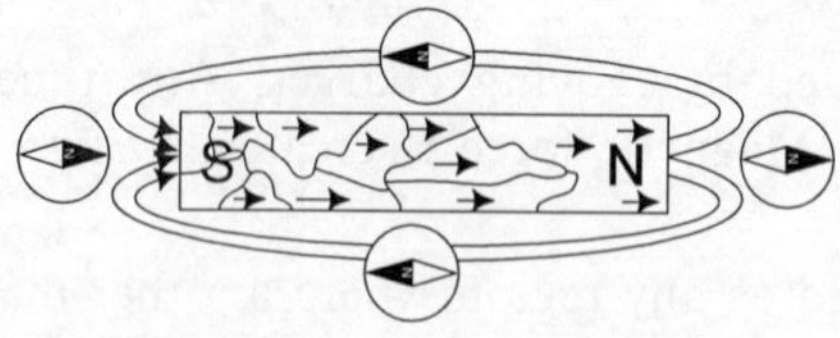

Permanent magnets

Magnets always have two poles-called "magnetic dipoles". There is no such thing as a magnetic monopole!

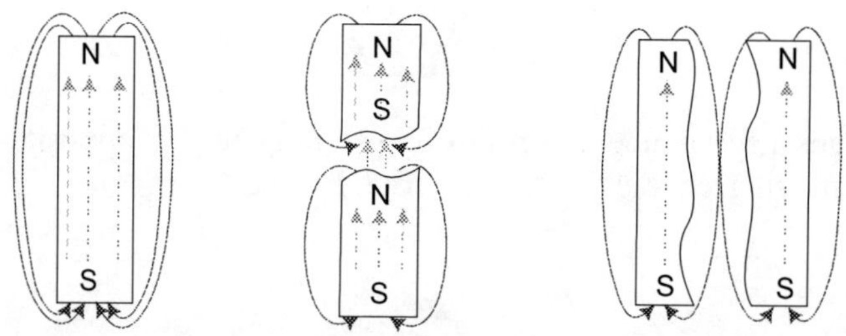

Each atom consists of electrons orbiting about a central positive *nucleus* along with the electrons spinning about their own axes. Thus, an internal magnetic field is produced by electrons orbiting around the nucleus as in the left figure below or electrons spinning as in the right figure below.

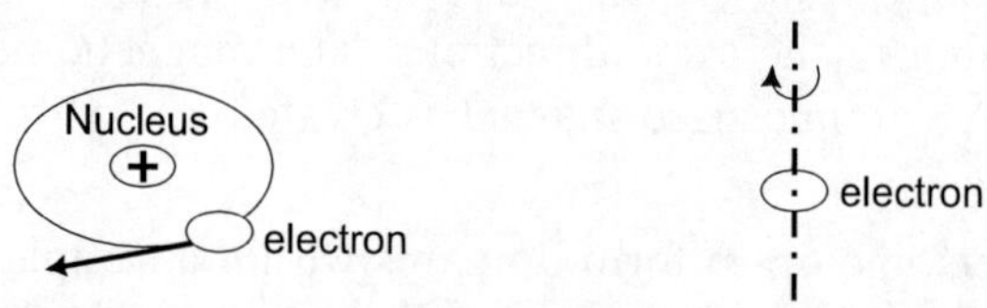

Electron in orbit around the nucleus

Electron spin

Both of these electronic motions produce internal magnetic fields B, similar to the magnetic field produced by a current loop.

The equivalent current loop has a magnetic (dipole) moment of $\vec{m} = I_b S\vec{a}_n$, where S is the area of the loop, I_b is the bound current and $\vec{a}_n$ is a unit normal vector.

Since it is impossible to measure the magnetic moment of individual atoms, physicists use the *bound current* equivalent representation to analyze the effect of magnetic fields in matter.

8.28 MAGNETIC FIELDS IN MATTER

Like electric fields, magnetic fields penetrate through matter.

Unlike electric polarization, which is almost always in the same direction as E the applied field, certain materials acquire a magnetization parallel to the external applied magnetic field B (paramagnets) and a few, opposite to the external applied magnetic field B (diamagnets).

In the presence of the external magnetic field, matter becomes magnetized; that is, upon microscopic examination, it will be found to contain many tiny dipoles, with a net alignment along some direction[8]. There are two mechanisms that account for this polarization: 1. Paramagnetism, in which the dipoles associated with the spins of unpaired electrons experience a torque tending to line them up parallel to the field, and 2. Diamagnetism, in which the orbital speed of the electrons is altered in such a way as to change the orbital dipole moment in a direction opposite to the field.

A few substances called ferromagnets (the most common example is iron), retain their magnetization even after the external field has been removed, and for these, the magnetization is not determined by the present field, but by the whole magnetic "history" of the object.

Permanent magnets made of iron (see Section 8.27) are the most familiar examples of magnetism, though from a theoretical point of view they are the most complicated.

Diamagnetism and paramagnetism are extremely weak, and it would take delicate experiments and a powerful magnet to detect them at all. The weakness of the effect, is one of the reasons we are able to calculate the magnetic fields in copper wires, easily.[8]

A magnetic dipole experiences a torque in a magnetic field and for a rectangular current loop of length a and width b carrying a current I, in a uniform field $\vec{B}$, the torque is given by

$$\vec{N} = \vec{m} \times \vec{B}$$

where, $m = Iab$ is the magnetic dipole moment of the loop.

The figure on the left below shows a piece of material of volume Δv before a magnetic field is applied and the right figure shows the magnetization M (the magnetic dipole moment per unit volume summing the magnetic dipole moments of all the atoms in the infinitesimal volume $\Delta v/\Delta$v)after a magnetic field B is applied.

$$\vec{B} = 0; \quad \vec{M} = 0$$

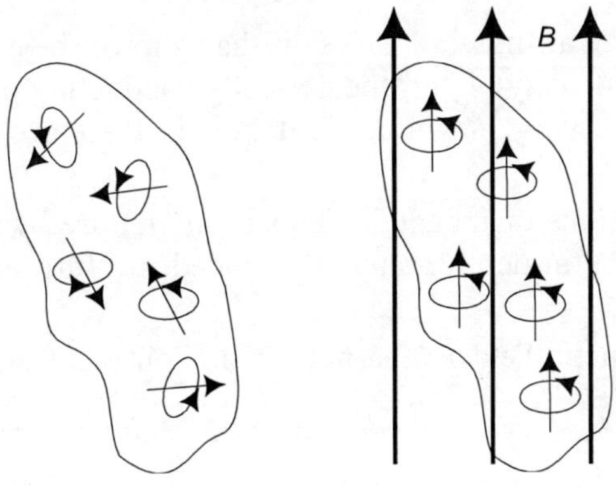

As we learned in Section 8.4 the torque is in such a direction as to line the dipole up parallel to the field. It is this torque that accounts for paramagnetism. While one could rightly expect paramagnetism to be a universal phenomenon, the laws of quantum mechanics dictate that the electrons within a given atom lock together in pairs with opposing spins, and this effectively neutralizes the torque on the combination. As a result, paramagnetism normally occurs in atoms or molecules with an odd number of electrons, where the "extra" unpaired member is subject to the magnetic torque. Even here the alignment is far from complete, since random thermal collisions tend to destroy the order.

Surface Current Density

According to the picture of magnetic polarization outlined above, the application of an external magnetic field to a material causes the formation of magnetic moment within the material.[16] It can be shown that[16] the volume distribution of magnetic dipoles is equivalent to a surface current density $\boldsymbol{J_s}$ and a volume current density $\boldsymbol{J}$ given by

$$\boldsymbol{J_s} = \mathbf{m} \times \mathbf{n} \equiv \boldsymbol{J_b}$$

and

$$\boldsymbol{J} = \nabla \times \mathbf{m}$$

With the characteristic rigor of David J. Griffiths[8], $\boldsymbol{J_s}$ is rightly $\boldsymbol{J_b}$, indicating the bound nature of the surface current over which there is no control. This result is to be expected, since the magnetic properties may be derived by regarding the material as being made up of many small current loops. In the interior of the material, incomplete cancellation of the currents in adjacent loops, results in a net volume current density J. Similarly, at the surface of the material, the currents are not canceled, giving a net surface current density.[8,16]

Note: In Section 5.2, it was shown that $B = \mu_o H$ in free space. In magnetic media, the total current is $J_{tot} = J_f + J$, where J is the bound volume current density, and J_f is the free current density due to an applied voltage source. The free current might flow through wires imbedded in the magnetic substance or, if the latter is a conductor, through the material itself. Then, from Ampere's law (Section 2.28), we can write

$J_{tot} = \dfrac{1}{\mu_o}(\nabla \times B) = \left[J_f + (\nabla \times m)\right]$, where m is the magnetic polarization. This may be written $\nabla \times \left(\dfrac{1}{\mu_o} B - m\right) = J_f$. Designating $H \equiv \dfrac{1}{\mu_o} B - m$, we obtain $B = \mu_o H + m$.

In Section 5.22 on Transmission Lines, we had shown the existence of a surface current (really surface current density) J_s, on the perfectly conducting planes which is an associated current of the time-varying magnetic field and is therefore, different from the bound current J_b.

The above descriptions of magnetic fields in matter are by no means complete and the reader is referred to the in-depth analysis provided in "Introduction to Electrodynamics" by David J Griffiths.[8]

In a uniform magnetic field B, the net force **F** on a current loop is zero:

$$\vec{F} = \vec{I} \oint (d\vec{l} \times \vec{B}) = I \left(\oint d\vec{l}\right) \times \vec{B} = 0$$

the constant B comes outside the integral, and the net displacement $\oint d\vec{l}$ around a closed loop vanishes. In a non-uniform field this is no longer the case.

For example, suppose a circular wire is placed away from a bar magnet as in Fig. 8.72(a), the magnetic field fringes and B has a radial component, and there is a net force on the loop.

We will now show how when a circular current loop (assume that the loop is a coil and is a rigid body and R is the radius of the circular loop) is placed near the north pole of a bar magnet along the axis of the magnet, it experiences a force. Such a situation is shown in Fig. 8.72a (Chabay & Sherwood, 1995, used by permission).[2]

The magnetic field of the bar magnet diverges with an angle θ as shown and so is *non-uniform*.

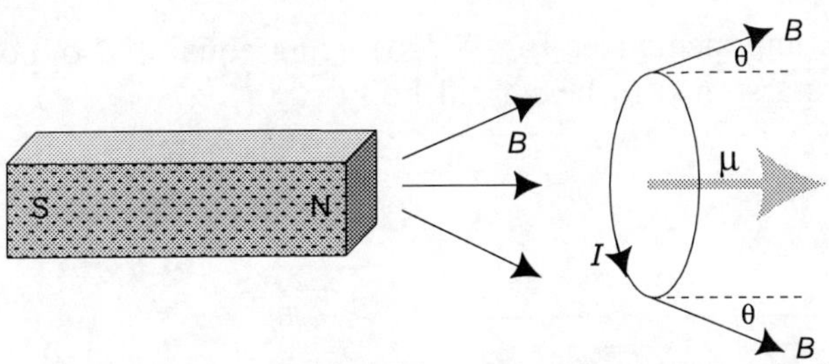

Fig. 8.72(a) Field of the magnet is non-uniform near the magnetic moment.

Each segment dl of the ring experiences a magnetic force $d\vec{F} = I\ d\vec{l} \times \vec{B}$ as we learned in Section 6.5 and is shown in Fig. 8.72b.

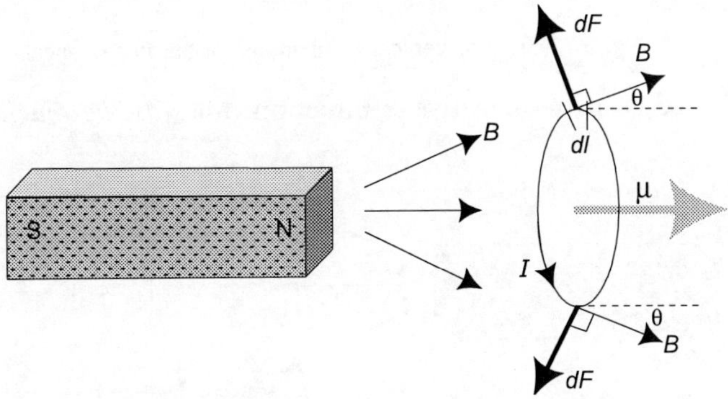

Fig. 8.72(b) Magnetic force on an elemental segment dl.

The horizontal component is $dF\cos(90° - \theta) = dF\sin\theta$ as shown in Fig. 8.72c.

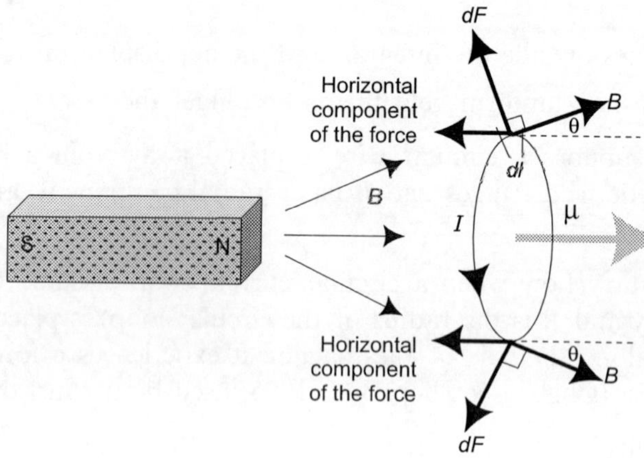

Fig. 8.72(c) Horizontal component of the force on a magnetic moment.

The vertical components (see Fig. 8.72d) being equal and opposite on the ring *cancel* because the coil is assumed to be a rigid body.

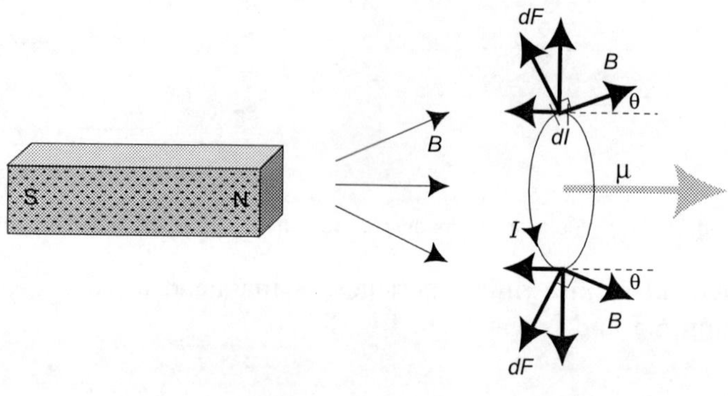

Fig. 8.72(d) The vertical components of the force cancel.

So, the net force is in the negative x direction, *towards the magnet.* This is shown in Fig. 8.72e.

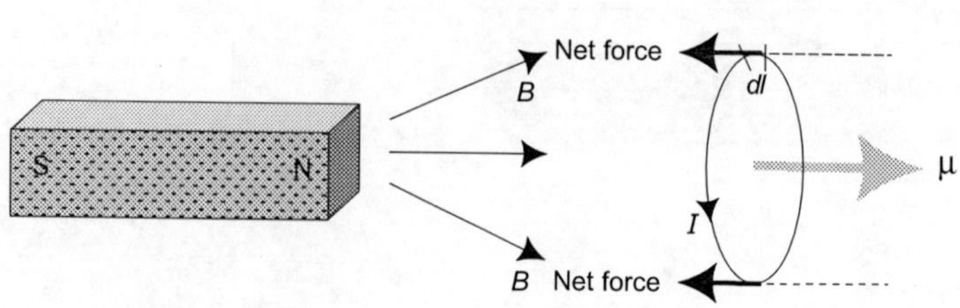

Fig. 8.72(e) Net force causes the ring to move towards the magnet.

The resultant force on the ring causes it to move towards the magnet. The net force on the magnetic moment, which is the current carrying loop of radius R, can be calculated as follows:

$$d\vec{F} = I d\vec{l} \times \vec{B}$$

$$F_{net} = F_x = \int_{loop} (IdlB \sin 90°) \sin\theta$$

$$= IB\sin\theta \int_{loop} dl = IB(2\pi R)\sin\theta$$

Since, $\mu = IA = I(\pi R^2)$ (Refer Section 8.2)

$$F = \left(\frac{R}{R}\right) IB\,(2\pi R)\,\sin\theta = \frac{(I\pi R^2)2\,B\sin\theta}{R}$$

The net force F is, therefore

$$\frac{\mu(2B\sin\theta)}{R}$$

Force on a Magnetic Moment with the Magnet Reversed

(Chabay & Sherwood, 1995, used by permission).[2]

If either the magnet is reversed or the *direction of the current* in the coil is reversed (but not both), as shown in Fig. 8.73a then, by resolving the elemental force dF into components, we note that the components vertical to the axis of the coil cancel out because the coil is assumed to be a rigid body.

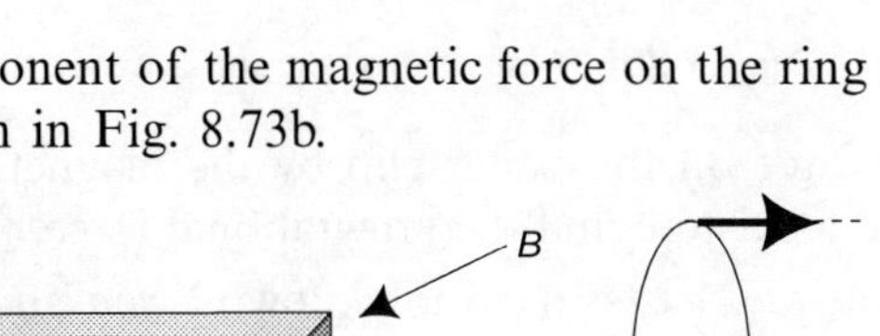

Fig. 8.73(a) The magnet is reversed.

The horizontal component of the magnetic force on the ring *is reversed* (it is now *away* from the magnet) as seen in Fig. 8.73b.

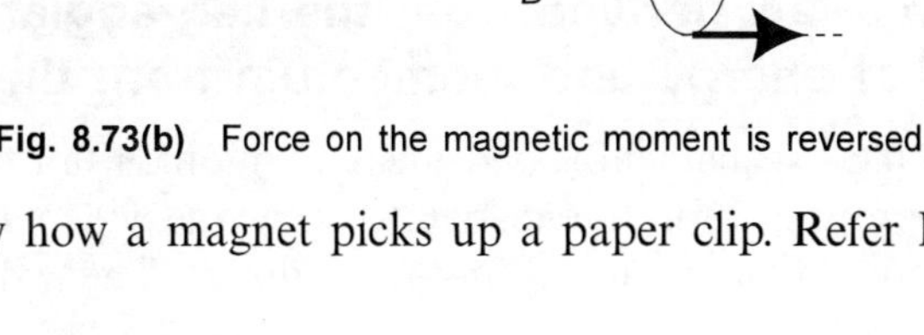

Fig. 8.73(b) Force on the magnetic moment is reversed.

We will now study how a magnet picks up a paper clip. Refer Fig. 8.74a.

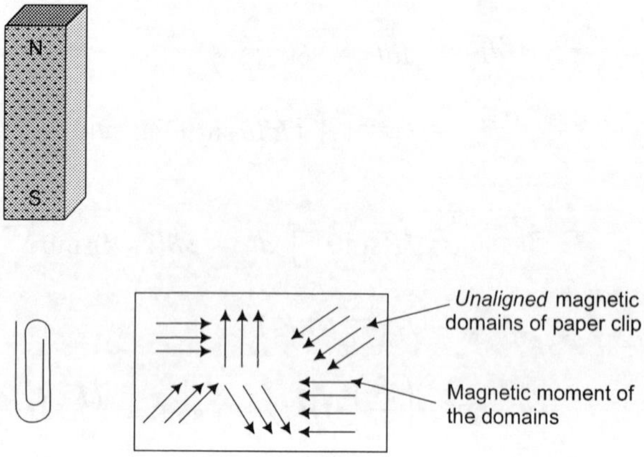

Fig. 8.74(a) The unaligned magnetic domains of a paper clip.

As shown in Fig. 8.74a, initially, the paper clip is *not* magnetized and all the magnetic domains inside it are pointing in random directions. The net force on the paper clip by the magnet is "*zero*".

However, the field of the magnet causes a torque on the magnetic moments of the iron atoms, and the domains tend to *align* with the magnet's field and are shown in Fig. 8.74b.

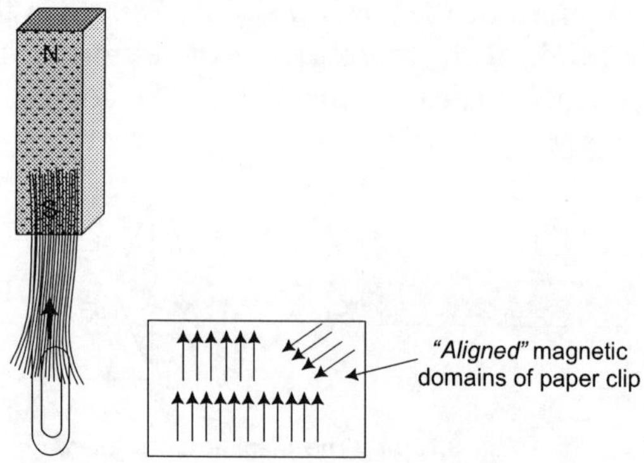

Fig. 8.74(b) A magnet picking up a paper clip.

Now there is a net force on the paper clip by the magnet. When the magnet is close enough for the magnetic force to equal the gravitational force, it can *pick up* the paper clip.

According to leading physicists, it is not known "*who*" *does the work* when magnets pick up metallic paper clips.

The Lorentz Force law derived from the Lagrangian and the principle of conservation of energy and momentum from the Hamiltonian

In Sections 4.2 and 6.6, we had indicated ideas of momentum conservation in the fields and it would be appropriate for the reader to view the video lectures recommended in "Fields and Momentum Conservation" in Appendix B. The video lectures by Professor

Leonard Susskind, Stanford University, USA provide a comprehensive introduction to the derivation of the Lorentz force law from the Lagrangian function applied to a charged particle in motion in a magnetic field, and of conservation of momentum and energy based on the Hamiltonian.

Application of the concept of the Magnetic field as a Relativistic effect

Readers who have read through Sections 3.3 to 3.6 of "A Qualitative Approach to Electricity: A Guide to visualizing Electrodynamics using symbols and Images" in the Teacher's Guide in the CD should review the application of the Lorentz transformation that shows how the fields transform in Sections F.2 and F.3 (Appendix F) and which finally shows that the velocity-dependent part of the force that acts on a charge in motion is the magnetic field, or that magnetism is a relativistic effect. Then, a reading of Section F.4 is recommended that explains the repulsion of two current-carrying wires (experiment in Section 6.1, Chapter 6 of this book) from the Special Relativity viewpoint. In Section 5.22 and Section 7.15 "Visualizing self-induction", we had discussed the production of circular waves by accelerated charges. While details of their production will be discussed in Chapter 10, it would be instructive to review Section 3.5.2 of Härtel's article in the "Teacher's Guide" in the CD. Following this, for vivid and precise representation of transient phenomenon in transmission lines, it is recommended to read Sections 3.8 on Transmission Line wave propagation progressing to Section 3.9 on Electric Circuits and Transmission Line, in Härtel's article.

In all the previous and in forthcoming chapters, the electric field and magnetic field form a central theme in explaining circuit processes. Because most present-day engineering curricula, especially in Universities in India, do not take such an approach in electricity and magnetism and circuit theory courses, the author with Prof. V. Kamaraju, Retd. Principal, JNTU, Kakinada, India, and Vedula Kavya, Ineda Systems, Hyderabad, India, prepared a set of guidelines that can assist in modifying traditional courses in Electricity and Magnetism in engineering curricula. This is available in the paper "Unifying Electric Theory and Circuits in Engineering Curriculum", Intl Journal of Engineering Technology Science and Research, Volume 4, Issue 9, Sept. 2017, ISSN 2394 – 3386, online http://www.ijetsr.com/images/short_pdf/1504360766_45-56-ieteh840_ijetsr.pdf.

SUMMARY

The chapter began with a description of magnetic forces and the Lorentz Force Law followed by the magnetic field due to a circular loop at a point far away from the loop. In the formula for the field of the circular loop, the Magnetic Moment 'μ' (= IA) was defined.

The magnetic moment, we learned, is a vector whose direction is that of the magnetic field at the center of the coil, which is normal to the face of the loop.

The magnetic moment associated with a rectangular coil, mounted on an axle and free to rotate, when subject to an external magnetic field, simplified the study of motors that was described later in the chapter.

By resolving the forces on the different segments of a rectangular loop in a magnetic field B, we learned that the coil (loop) will have a tendency to rotate such that, the direction of its magnetic moment 'μ' is aligned with the direction of the magnetic field B. When the alignment occurs, there will be no further rotation, and the loop will be in equilibrium. In equilibrium, the magnetic field of the loop being perpendicular to the face of the loop, the face of the loop would be perpendicular to the external magnetic field B, or simply stated, the field of the coil (loop) and the external magnetic field B would both be aligned.

We then studied how to test the stability after alignment, and reached the conclusion that the loop is most stable, where its magnetic moment 'μ' is aligned with B.

All atoms have a magnetic moment because of their moving electrons which occupy orbital spaces and also spin. Both these movements produce magnetic fields and in most substances, the moments are so randomized that there is no resultant magnetic moment. However, in some substances called ferromagnetic, the atomic magnetic moments line up so that there is a strong moment.

Because magnetic moments tend to align with external magnetic fields, a compass needle which is free to rotate, rotates until it aligns in the direction of the earth's magnetic field.

We then derived an expression for the torque or the twisting force of a magnetic moment. The torque will exert a force until the magnetic moment (which is the current-carrying loop itself) is aligned with the direction of the external magnetic field. To make such a magnetic moment rotate past its equilibrium point, we need to reverse the current in the coil just before it reaches equilibrium.

The torque, or the twisting force, is a vector $\mu \times B$ whose direction is that of the axis of the axle, on which the magnetic moment say, a coil is mounted.

We then studied how to utilize the phenomenon of the tendency of a magnetic moment to align with an external magnetic field, to build electric motors, which can supply energy and power to the load.

The problem, of having to reverse the current through the coil, to generate a unidirectional torque was resolved, by fitting a commutator split into two segments on the axle, which supplies the current from an external source by brushes pressed against its segments.

We also learned that the conductors of motors usually made of copper are non-magnetic and the forces that they develop are due to the Hall effect charges in them when a current is established.

A construction of a motor with several conductors has a tendency to develop more torque with fewer torque pulsations and this was discussed in some detail.

A detailed study of the effects of several types of loads, constant with speed, and varying with speed was made, and the effect that the loads have on the motor and load (since a load such as a fan or the impeller of a pump is coupled to the motor shaft and rotates with it) speed. It was observed that the motor always meets the load torque requirement by generating the necessary motor torque. This, we learned, the motor does automatically, by the process of back emf generation which regulates the current, producing the right amount of motor torque.

We learned how a load torque is produced with the example of a ceiling fan pushing air molecules and the reaction forces of the molecules of air when pushed by the fan blade tips. The reaction force is transmitted along the blades to the rotating motor shaft, which meets this reaction (load) torque of the blades by drawing an appropriate amount of current from the supply and maintains the combined fan load and motor rpm.

The back emf generated by a motor is a form of motional emf. We studied in some qualitative detail how the motor back emf regulates the amount of current drawn from the supply as it makes the transition from start to rated rpm.

Motor selection is an elaborate design task. Loads are tested separately on special rigs with variable speed motors and, their torque vs. speed characteristics in the form of curves, prepared. Motors, similarly, are tested by applying known loads and their characteristics in the form of curves, prepared. A superimposition of the two simplifies motor selection.

Loudspeakers operate on the same principle as motors with the difference that its coil is constrained to move *along* the axis of the coil. Coil current, supplied by an audio amplifier, makes the coil immersed in a magnetic field to move rapidly back and forth a paper cone attached to it, following the signal strength variations. The air pressure waves, produced by the paper cone are the sounds we hear.

A study of the energy of a magnetic moment at different orientations to the direction of a magnetic field revealed that it was a maximum, not surprisingly, when the moment was pointing directly opposite to the direction of the field.

Permanent magnets contain magnetic domains that line up to form a magnet. Magnetic fields affect the atomic level magnetic dipoles of matter to produce an effect called polarization. We learned about these effects in diamagnetic, paramagnetic and ferromagnetic materials.

The magnetic polarization effect by external magnetic fields causes the formation of a magnetic moment within the material. The volume distribution of magnetic dipoles is equivalent to a surface current density J_s and a volume current density J, and we learned that the surface current density J_s, is termed bound current density J_b. The bound current density should not be confused with the surface current density J_s, associated with a time-varying magnetic field, which is present on the surface of a transmission line that we had discussed in Section 5.22.

It was pointed out that the distinction be clearly understood.

A study of the net force of a current loop in a uniform magnetic field, we found, is zero. But, in a non-uniform magnetic field, this is no longer the case. With an example of a bar magnet and a current-carrying ring some distance away, we showed, by resolving the forces of the ring that there will be a resultant force on the ring acting towards the magnet. Had the field been uniform (not divergent), the resultant forces on all segments of the rings would have canceled, leaving a zero resultant force.

We then described how when a bar magnet picks up a paper clip the magnetic dipoles in the clip align with the bar magnet's field, producing a net force on the paper clip. When the magnet is close enough, the attractive force overcomes the gravitational force, and the magnet picks up the clip. We also learned that it is still unknown as to who does the work when a magnet picks up a paper clip.

At the end of the chapter, we learned that the Lorentz force law can be derived from the Lagrangian equation and the conservation of energy from the Hamiltonian. We also learned how to use the Lorentz contraction and time dilation aspects in the Theory of Special Relativity to describe the repulsion between two current-carrying wires.

Questions

1. With reference to a loop of radius 'a' carrying a current I, define the magnetic moment of the loop and give its magnitude and direction as a vector.
2. Given the situation of a current carrying loop as shown in the figure, state whether the loop will flip or not.

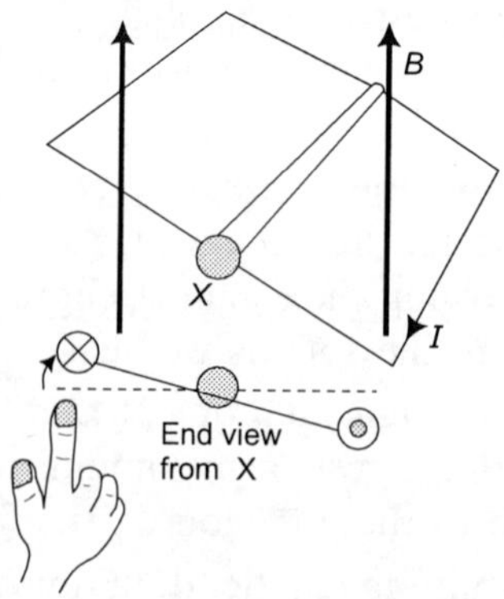

3. What is the magnetic moment of a magnet due to? Explain the action of magnetic needles placed in magnetic fields.
4. Derive an expression for the torque developed by a current-carrying loop placed in a magnetic field.
5. What are Hall effect charges? Explain how they make a current-carrying wire placed in a magnetic field to move.
6. With the help of neat sketches showing a rectangular current-carrying loop in a magnetic field which develops a torque whose direction changes, explain how a unidirectional torque can be produced.
7. Describe the operation of a dc motor.
8. Explain what is meant by "load torque" and how this influences the back emf and the current consumed by a motor.
9. Describe how a motor drives a ceiling fan motor. Is there a reaction torque for which the motor consumes power from the supply to drive the ceiling fan blades? Describe the production of reaction torque by the blades as they rotate.
10. What are bound currents in a material and how are they produced? Are these produced by application of a magnetic field or an electric field to a material?
11. Describe some common industrial load torques.
12. Derive an expression for the potential energy of a magnetic moment for a rectangular loop (length h and width w) placed in a magnetic field of strength B. Determine the

position the moment, if left free to rotate, will assume where the potential energy is a minimum.

13. How do magnets exert a force on a current carrying loop? Explain with the help of sketches. Who does the work when a magnet is used to pick up paper clips?
14. Is the electric field real? Do the concepts of E and B fields precisely explain the operation of circuits and circuit components? Explain.
15. What was Newton's idea about space and time ? Describe the meaning of Relativity of objects in motion.
16. Describe in simple terms Leigh Page's Emission Field Theory. How many types of magnetic fields did Page describe existed? Does Page's Theory conform to the Principle of Relativity?
17. Describe the repulsion and attraction between two current carrying wires using the Lorentz contraction. Hint: See Section F.4 in Appendix F.
18. Two square sheets separated by a small distance are charged uniformly with opposite charges of density σ in an inertial frame *F*. If *F*' is a different frame moving in the $-\hat{x}$ direction, show that $E_z' = \frac{E_z}{\sqrt{1-\beta^2}} = \gamma E_z$ and $E_x' = 4\pi\sigma' = 4\pi\sigma = E_x$.

    Qualitatively interpret the equations.
19. Derive the general Lorentz force law $\mathbf{F} = q\mathbf{E} + \frac{q}{c}v \times \mathbf{B}$ (CGS units) assuming a test charge moving with velocity v parallel to a current carrying wire.
20. Using the example of a test charge moving with uniform velocity parallel to a current carrying wire, show that magnetism is relativistic.
21. What are the various interpretations for the velocity in the Lorentz Force Law from the viewpoint of Relativity? Which interpretation would correctly describe the velocity?
22. In Classical Mechanics, explain what is "Action" and the "Principle of Least Action".
23. Beginning with the Lagrangian equation of motion of a charged particle in a magnetic field, derive the Lorentz force law. Then express the momentum in terms of the Lagrangian, and using this derive the Hamiltonian. Hint: See Appendix G in Kittel's "Introduction to Solid-State Physics".
24. Derive the expression for the momentum in the electromagnetic field that accompanies a particle moving in a magnetic field and show that the total momentum $\mathbf{p}$ = Mv + QA/c. Hint: See Appendix G in Kittel's "Introduction to Solid-State Physics".
25. What is the "Quantum field" theory for particle interactions? To explain which structure was the theory introduced?

Additional practice questions are available in the CD with this book.

9

PN Junctions, Diodes and Ohmic Contacts

The initial sections of the chapter provide descriptions of conduction processes in semiconductors followed by a description of energies of charges in atoms. This is then followed by energy transfer processes at junctions, the [illegible] of the potential barrier and the behavior of a junction to applied forward and reverse bias voltages. Two sections are devoted to explain how the potential barrier V_0 of a junction cannot be made zero by a forward bias with a worked example.

Junction diodes are used to rectify ac voltages to dc and this application is treated in some detail. A junction diode fabricated from semiconductor material should have metallic leads for making connections to other components of electronic circuits. Since the leads form a junction at the interface with the semiconductor, there is a possibility of a barrier being formed there and of unwanted rectification. The last section deals with this aspect and how ohmic contacts can be formed to prevent unwanted rectification and also the processes that prevent the potential barrier from being measured using a voltmeter.

9.1 SURFACE CHARGES IN A SIMPLE CIRCUIT WITH A DIODE

We review the development of surface charges in a simple circuit consisting of a battery, wires and a resistor. This is shown in Fig. 9.1.

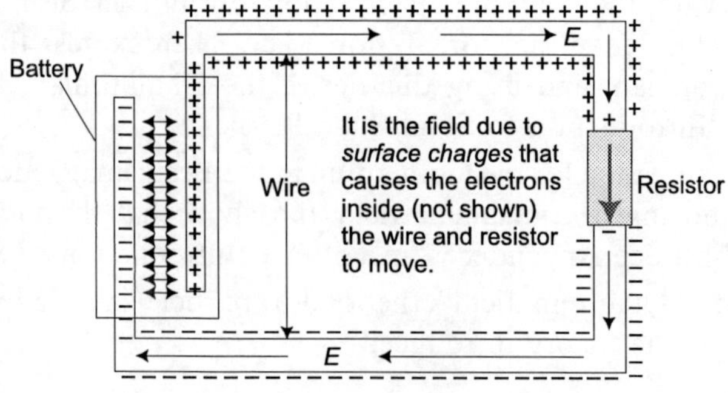

Fig. 9.1 Development of surface charges.

We now add a diode to the circuit and retain the resistor to restrict the overall current. The initial establishment of surface charges takes a few nanoseconds. This is shown in Fig. 9.2.

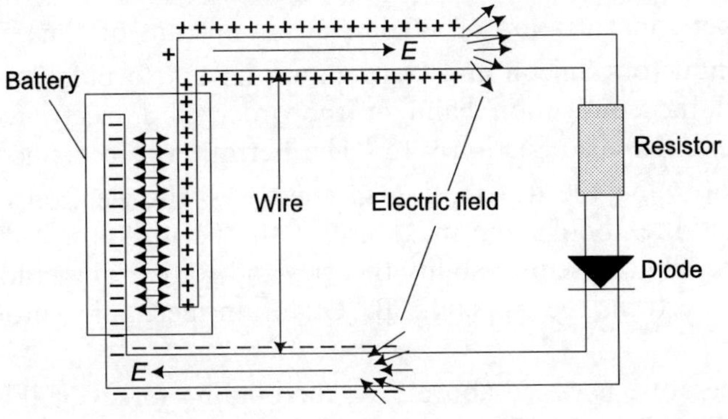

Fig. 9.2 Development of surface charges—intermediate stage.

In the final stage of development a strong electric field is set up in the resistor and the field across the junction in the diode is reduced (which we shall describe in a later section) as shown in Fig. 9.3 (not forgetting that an electric field is an invisible *alteration* of space).

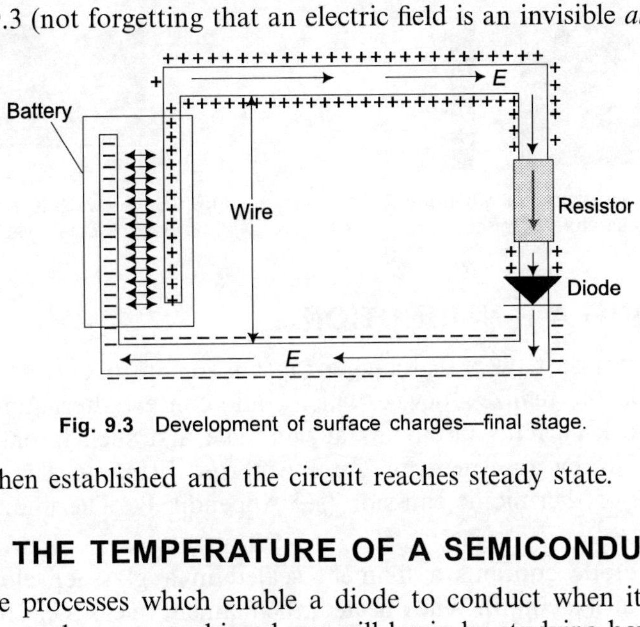

Fig. 9.3 Development of surface charges—final stage.

A current is then established and the circuit reaches steady state.

9.2 RAISING THE TEMPERATURE OF A SEMICONDUCTOR BAR

To understand the processes which enable a diode to conduct when it is forward biased and to block current when reverse-biased, we will begin by studying how a semiconductor conducts. A silicon bar at zero kelvin is an insulator as shown in the visualization in Fig. 9.4.

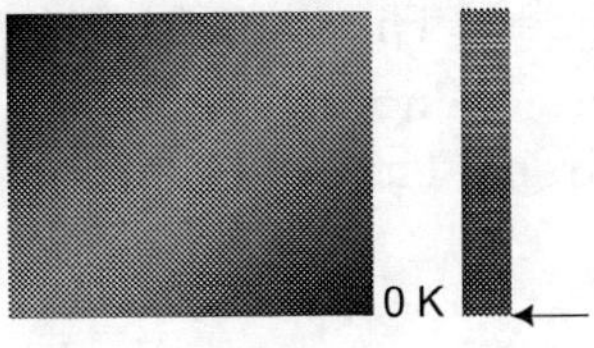

Fig. 9.4 A silicon bar at 0 K.

However, when its temperature is raised above zero kelvin, it turns into a **beehive of activity** as shown in the visualization in Fig. 9.5. To gain an intuitive understanding of temperature, read "What is temperature" in Appendix B.

There is a **net** concentration of carriers at any instant of time above zero K.

The semiconductors Silicon and Germanium have a forbidden energy region between the valence and the conduction band of the order of 1 electronvolt, where no mobile carriers are found. See also Section 9.15 "The Fermi level in Semiconductors".

The laws governing the distribution in energy of the electrons in the semiconductor are given in several textbooks and make use of the Fermi-Dirac probability function. For notes on the necessity to use probability theories and statistical methods that predict energy distributions of electrons see Appendix B "Quantum mechanics, probabilities and orbitals of electrons".

As the temperature is raised above 0 K, the binding forces of the atoms....the covalent bonds break and make, and Electron Hole Pairs (EHPs) are continuously generated and annihilated—and there is a net concentration of carriers.

There is always a net concentration at any instant of time....maybe not the same set of EHPs, but there is *always* a net concentration as seen in the visualization in Fig. 9.6.

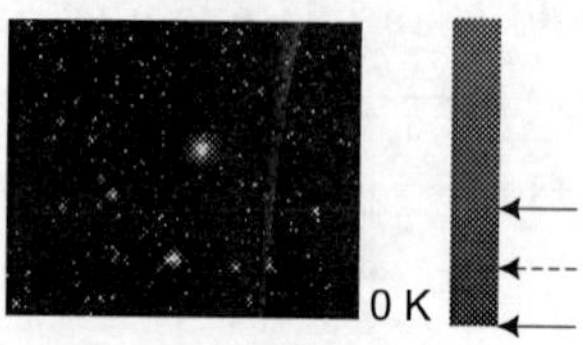

Fig. 9.5 A silicon bar when the temperature is raised.

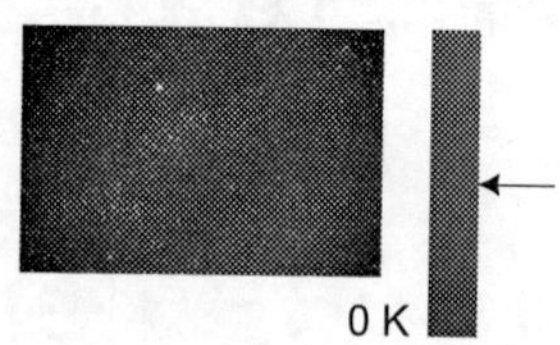

Fig. 9.6 Above 0 K a net concentration of electron-hole pairs always exists.

9.3 CREATING A P-N JUNCTION

Since the 1880s, physicists were searching for means to control current direction in electrical circuits with a view to realize a device, which could convert alternating voltages into direct voltages, and block currents in circuits at will. The first such invention was the vacuum tube diode in 1904 by an American, John Ambrose Fleming. The vacuum diode relied on the priniciple of thermionic emission (see Appendix B "Thermionic Emission") for its operation.

A vacuum diode contains a filament sealed in a glass envelope, which has been evacuated of all air (vacuum). When heated, the filament releases electrons into the vacuum creating a negatively charged cloud or space charge. These electrons will be drawn to a metal plate inside the envelope, if the plate (also called the anode), is positively charged relative to the filament (or cathode). The result is a flow of electrons from filament to plate.

The vacuum diode cannot conduct in the reverse direction because the plate is not heated and does not emit electrons. The device operates as a diode: a device that conducts current only in one direction.

The tubes were critical to the development of electronics and do still provide service in radio broadcasting stations but were very bulky and fragile and proved cumbersome in radio receivers which need to be portable. That was soon to be changed after the second world war with the discovery of semiconductors.

In their attempts to realize diodes using crystalline materials, physicists realized the impossibility of making the positive lattice ions of materials to move that could have properties of controlling current direction!

Note: It is not possible for the lattice ions to move in metals but (positive) lattice ions of metallic salts dissolved in solution can move as was seen in Section 1.8. Also, visit http://www.science.uwaterloo.ca/~cchieh/cact/c120/electrolyte.html "Electrolyte and Battery" for a description of the movement of positive ions or anions as they are called. In "dry" cells (http://everything2.com/title/dry+cell), the electrolyte is a *moist* paste which allows ion movement. However, note that the resistivity of Copper is ten million times smaller in magnitude (see Appendix E "The third process of the influence of the electric field on the energies of electrons") than that of a NaCl (2 g/100 milliliter) aqueous solution.

The Drift Speed of Electrons in Silicon

The drift speed of electrons in silicon is about two and a half times that of copper which is 0.00027 meters/second with an applied field strength approximately 0.005 Volt/meter (Section 1.18). In pure silicon, the electron has a drift speed of 0.00068 meters/second with an applied field strength of 0.005 Volt/meter.

The *p*-material and *n*-material

To produce a material of only positive mobile charges, one possibility was to produce a *p*-type material by cleverly doping a semiconductor bar with trivalent impurities. Such a material could not be totally devoid of conduction electrons but that was a small price to pay for realising vacant valence band states, having a positive quality, which could be filled by an electron supplied from a conductor or an *n*-material touching it.

This electron that filled such a valence band state could, on application of a properly directed electric field "hop" along to a neighboring say, silicon atom at the valence energy level, and not at the conduction energy level and in a similar fashion to another silicon atom and so on. The hole attracts an electron from a neighboring sillicon atom's valence state leaving a hole exposed in the silicon atom, which in turn, attracts another electron from its neighbouring silicon atom thereby generating an apparent movement of positive charge in the opposite direction, or a movement of the hole. This action of an electron filling the vacant state is called *recombination*. Recombination is the merging of a free electron and a hole. Presumably, the term recombination was used for EHP annihilation in intrinsic material where the process occurs all the time.

The term *recombination* is used even when a neutral trivalent impurity atom creates a positivity (or empy valence band energy state) in the intrinsic semiconductor lattice that attracts an electron, which it never really lost in the beginning, for combination. *Recombination*, in a *p-n* junction diode is the important *link* between the free electron movement at the conduction band energy level and the *seemingly* opposite movement of holes (because positive charges or ions cannot move) at the valence level that we shall see in Section 9.6. This maintains current conservation in a *p-n* junction diode and enables its rectifying properties. Remember only electrons are mobile and not lattice ions.

An *n*-type semiconductor, is made by doping a semiconductor bar with pentavalent impurities that results in a profusion of conduction band electrons.

The movement of conduction band electrons that are attached to the lattice with a very weak coulombic electric force of an *n*-type material, cannot be construed as a positive charge movement in the opposite direction, because the conduction band electrons are everywhere, being *free*, and do not allow the creation of positive pockets. However, in the *p*-type zones wherein the electron attached to the lattice with a stronger coulombic electric force, hops at the valence level from one positive pocket to the next (nearer to the lattice ion), we can surely say there is a positive charge or hole movement in the direction opposite to the valence electron movement.

The making of *p*-material and *n*-material created a milestone in the realization of a device, the *p-n* junction diode, which could block current in one direction and allow current in the opposite direction and could operate at room temperatures and in compact dimensions. With its invention, equipment for telecommunication, radio, television, computers, medical electronics and industrial control were no longer the same. These became compact, smarter and more affordable.

To create a *p-n* junction, **neutral** atoms of dopants are added to a silicon or germanium bar as shown in Fig. 9.7: first, with a mask on the *n*-side when the *p*-dopants are added to the bar, and then with a mask on the *p*-side when the *n*-dopants are added to the bar. The fabrication of the *p-n* junction described is oversimplified and the methods adopted for manufacturing junctions are described in several textbooks.[30]

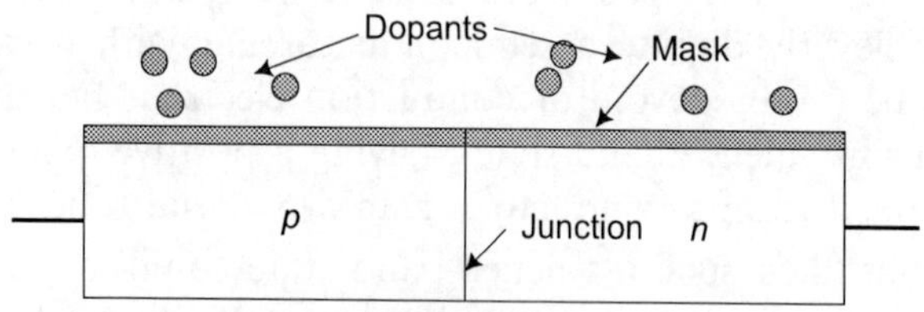

Fig. 9.7 Creating a *p-n* junction.

Immediately the mobile charges on either side of the *p-n* junction diffuse to the opposite side as shown in Fig. 9.8. "Diffusion" describes the spread of particles through random motion from regions of higher concentration (e.g., holes in *p*-region) to regions of lower concentration (*n*-region where very few holes are present).

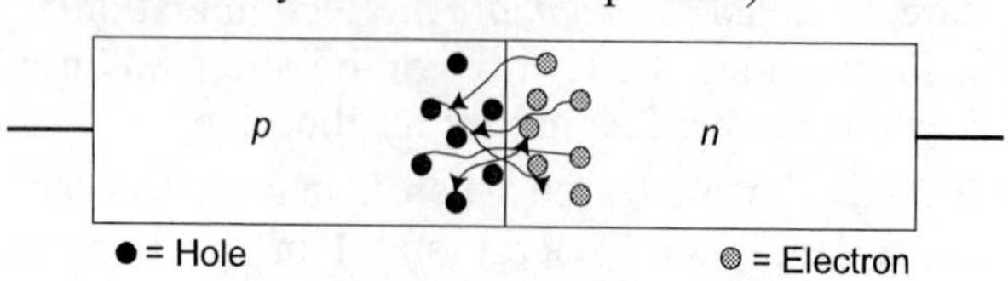

Fig. 9.8 Electrons move to the *p*-side.

In reality, it is the electrons that move to the p-side (the n-side electrons shift and breakaway to recombine with p-side holes) exposing (immobile) lattice positive charges on the n-side. The electrons in excess there now, create a negative potential (at the p-side).

Why do the electrons move to the *p*-side and holes to the *n*-side?

Whenever a semiconductor is non-uniformly doped, or junctions are formed between different semiconductors which are crucial to the various types of electronic and optoelectronic devices made using semiconductors, an important concept should be established here regarding the demands of equilibrium, when there can be no net current in any direction.

That concept is "*no discontinuity or gradient can arise in the equilibrium Fermi level* E_F".

Imagine having two steel cylindrical glasses of water at different temperatures, and because water is a liquid, the temperature in each glass can be the same everywhere within each glass at a molecular level; quickly, either by stirring the water, or slowly, by the process of natural convection.

When you heat up a pan of water, the heat from the stove causes the molecules in the pan to vibrate faster causing the pan to heat up. The heat from the pan causes water molecules to move faster and heat up. So, when you heat something up, you are just making its molecules move faster. Heat is the total amount of energy possessed by the molecules in a piece of matter. This energy is, both kinetic energy and potential energy.

If a hot high-energy atom or molecule comes into contact with a cool low-energy atom or molecule, the excited atom or molecule will lose some of its energy to the cool atom or molecule. The two atoms or molecules will settle into an energy level that's between where they each started out. That level is called Thermal Equilibrium.

Now, when the glasses are brought into contact, their energies exchange, and then the temperatures will equalize—the energy exchange stops, when there is no discontinuity in their energy level. The temperature is the same in both the glasses and thermal equilibrium is attained.

Within a semiconductor, which is really a crystal, there can be no evening out of the energies of individual electrons (potential energies, really); their energies are spread over the acceptable levels with a distribution, or in short, they possess distributed energies.

We cannot stir up the semiconductor material and make the energies of the electrons same everywhere as we could do with water molecules in the steel glasses. See Appendix B on how electrons occupy specific energy levels, with fixed potential and kinetic energies.

Distributed energies are characterized by a *characteristic* energy or a *representative* energy and this we call the Fermi-level E_F. Consider two materials in intimate contact such that electrons can move between the two as shown in the energy diagram in the figure, when the p-type and n-type material join to form a p-n junction.

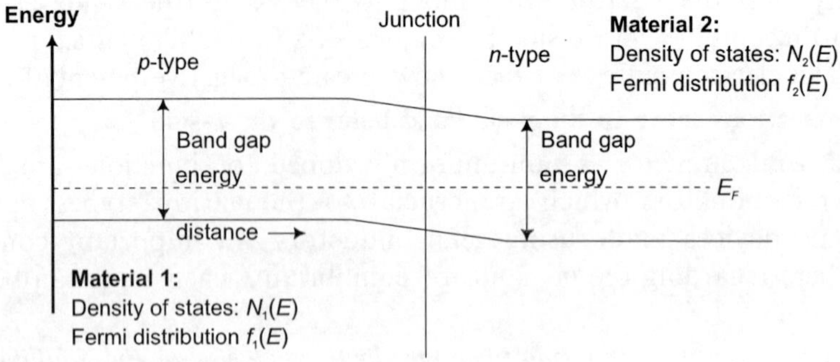

A *p-n* junction energy band diagram a short time after formation

Consider a time very shortly after the depletion layer begins to form ("before" the bands bend in the energy diagram above). It takes 1.1 eV (the band gap energy) of thermal excitation to create a mobile electron in the *p*-region of the silicon. Once mobile, this electron can "fall" through the potential energy difference (downhill) into the *n*-region. In contrast, at this time an already mobile electron in the *n*-region needs very little kinetic energy K from thermal excitation to be able to cross into the *p*-region, where it can "drop" 1.1 eV to bind with a silicon atom that is missing an electron (a hole). The smaller energy requirement means that initially it is much more likely for electrons to move from the *n*-region to the *p*-region than from the *p*-region to the *n*-region.[31]

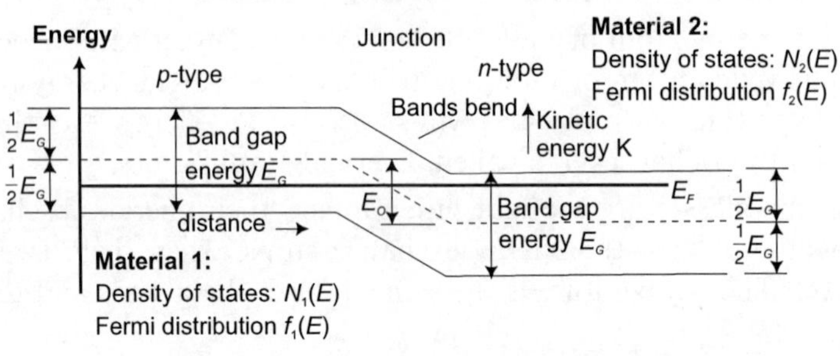

A *p-n* junction energy band diagram in dynamic equilibrium

The net flow of electrons from the *n*-region to the *p*-region increases the potential energy difference as is shown with the energy bands bent in the diagram above in the dynamic equilibrium situation. Now, it still takes 1.1 eV of thermal excitation to create a mobile electron in the *p*-region. Once free, this electron can "fall" through the potential energy difference into the n-region. Similarly, an already-mobile electron in the *n*-region must obtain kinetic energy K of 1.1 eV of thermal excitation to cross into the *p*-region, where it can drop 1.1 eV to bind with a silicon atom that is missing an electron (a hole). There is no advantage to moving in one direction or the other. In dynamic equilibrium the two processes have equal rates.[31] It makes little sense to consider that the energy acquired by the electrons are potential and not kinetic, as the diagram for the junction in dynamic equilibrium suggests that an electron will require kinetic form of energy to move to unoccupied available energy levels up the potential energy ladder.

For the two materials in intimate contact forming a *junction*, then at equilibrium, since the net motion of electrons should be zero, the equilibrium Fermi level must be constant throughout; else, it will result in a motion of the electrons, until the Fermi level becomes constant again.

Since nature changes gradually and continuously in space and time, there is a local change in the energy of electrons at a semiconductor junction, due to space charge effects. This exchange of charge *redistributes* the energies of those charge carriers which have been exchanged, giving a curvature to the energy vs. distance diagram near the junction and the bands appear to bend when the Fermi levels equilibrate as seen in the figure above. The general shape of the charge distribution results in the formation of a transition region at the junction which consists of immobile ions, negative in the *p*-side because of electrons *filling* the neutral acceptor or *p*-type atoms, and positive in the *n*-side because of electrons *leaving* the neutral donor or *n*-type atoms.

9.4 TRANSITION REGION

Electrons *away* from the junction on the *n*-side, *shift less* exposing lattice positive charges, and at the junction they fill the nearby holes in the *p*-side exposing a negativity there. These actions result in the production of an electric field that has maximum strength at the junction which decreases progressively away from the junction. The region where the immobile charges (these are positive ions on the *n*-side because neutral pentavalent atoms lost electrons and negative ions on the *p*-side because neutral trivalent atoms gained electrons) lie exposed is called the *transition region* as shown in Fig. 9.9, wherein a "dipole" is formed and there is an electric field pointing to the *p*-side from the *n*-side.

This field may be contrasted with the electric field between the plates of a charged capacitor that is *uniform* in the space between its plates caused by tight and closely spaced charges on its plates. In a *p-n* junction, the immobile positive and negative ions on either side of the junction are typically a hundred atoms in width and cause the electric field due to these to *vary* with distance as shown in Fig. 9.9.[17] Visit http://jas.eng.buffalo.edu/education/pn/biasedPN2/BiasedPN2.html or http://web.archive.org/web/20170327235000/http://jas.eng.buffalo.edu/education/pn/biasedPN2/BiasedPN2.html for an animated interactive depiction of the electric field in the *p-n* junction.

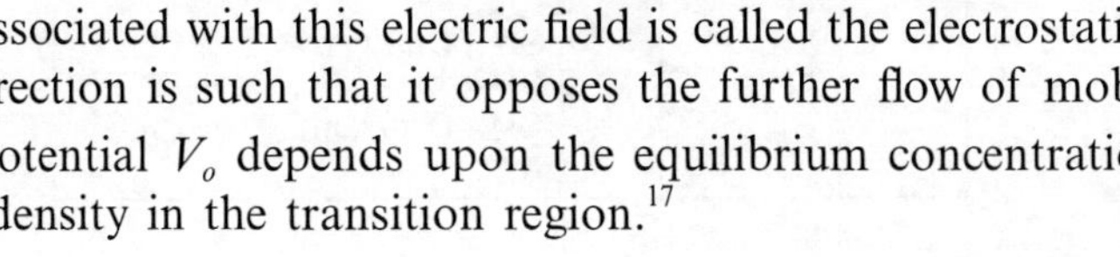

Fig. 9.9 The transition region of a *p-n* junction.

The voltage associated with this electric field is called the electrostatic potential barrier V_o, because its direction is such that it opposes the further flow of mobile charges.

The barrier potential V_o depends upon the equilibrium concentrations and not at all upon the charge density in the transition region.[17]

Note: The *p*- and *n*- impurity concentrations are so adjusted when manufacturing *p-n* junction rectifier diodes described in this chapter that the barrier potential V_o corresponding to the barrier potential energy E_o in electronvolts never exceeds the forbidden gap energy E_G (see figure "A *p-n* junction energy band diagram in dynamic equilibrium" in Section 9.3).

In contrast to rectifier diodes, in *tunnel* diodes, the impurity concentrations may be very high, and so the barrier potential energy E_O *can* exceed the forbidden gap energy E_G. A description of the operation of tunnel diodes is beyond the scope of this textbook and the reader is referred to Ref. [17]. The barrier potential V_o of a junction diode is shown in Fig 9.10 (for simplicity the field variation in the transition region is not shown henceforth).

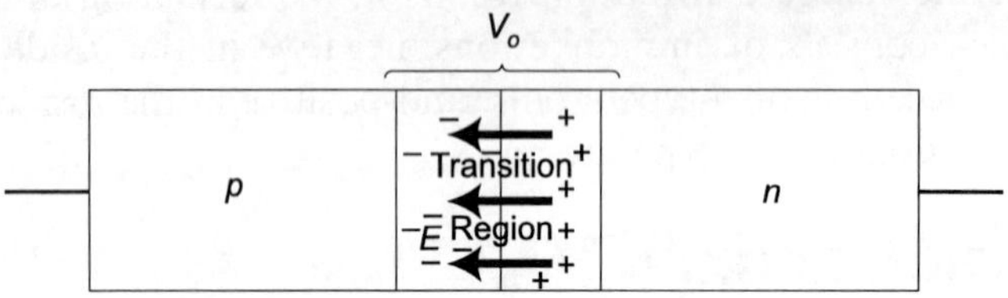

Fig. 9.10 The electrostatic potential barrier V_o.

The **electric field *E*** in turn causes a drift current of holes from the *n*-side (from thermally generated EHP pairs in the *n*-side) and electrons from the *p*-side (from thermally generated EHP pairs) to the opposite side as shown in Fig. 9.11.

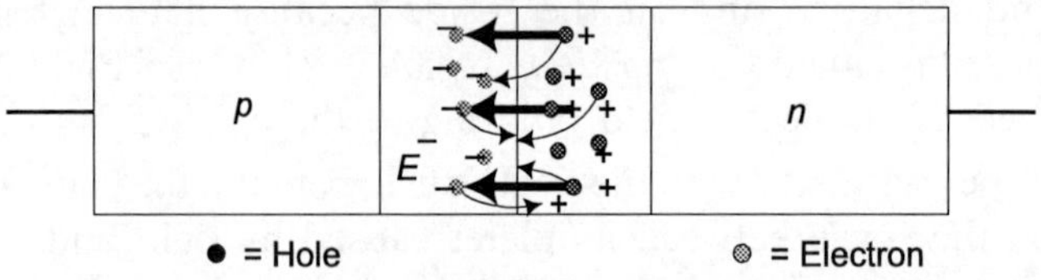

Fig. 9.11 The drift current in the transition region.

The drift current is opposite to that of the diffusion current and is of an equal magnitude so that the net current of the open-circuited diode is zero; otherwise, charge will continue to pile up and the diode will become a live wire!

9.5 DYNAMIC EQUILIBRIUM IN AN OPEN-CIRCUITED *P-N* JUNCTION

If we assume that the temperature is *not* anywhere near absolute zero and is above it, the diffusion current will always remain and so will the production of the electric field and the resultant drift current...all in balance...also called "dynamic equilibrium" as shown in Fig. 9.12.

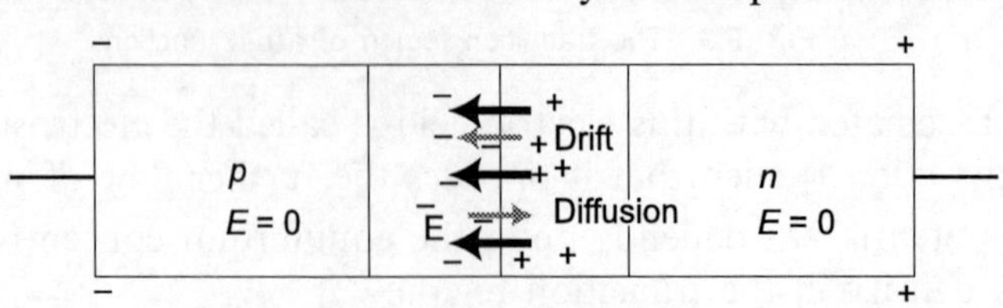

Fig. 9.12 Dynamic equilibrium in the open-circuited *p-n* junction.

The diffusion current which is a *forward* current (from *p* to *n*) is due to the *mobile* holes and electrons in the doped *p*- and doped *n*-regions respectively that would have gained sufficient kinetic energy from random thermal collisions (the only source of energy being the warm surroundings for the atoms in an open-circuited diode) to move through the transition region *against* the electric force. The drift current which is a *reverse* current (from *n* to *p*) as a result of an electric field (see "Origins of built-in electric field" Appendix E) is due to *thermally generated* holes and electrons in the *n*- and *p*-regions respectively that have survived long enough to wander near the edges of the transition region and to be swept across it by the electric field.

Note that we have not applied any voltage nor made any connections to the diode. It is open circuited yet the diffusion current will always remain and so does the production of the electric field and the resultant drift current..."dynamic equilibrium". There are two reversible processes at work here; diffusion, and drift (Section 2.24).

There are no mobile charges in the transition region. This is why the diode has a high resistance end to end when left open-circuited. The *controlling portion* of the diode is the *p-n* junction transition region, which is a mere 10^{-4}cm (or 10^{-6}m) in the overall length of say 13 to 18 mm (or 13×10^{-3}m) of the diode, which is roughly a fraction 10^{-6}m/13×10^{-3}m = 0.077 thousandth of a meter!

In equilibrium, the net electric field in the material outside the depletion layer or transition region of an open-circuited *p-n* junction is zero. This is so, because as long as there is a nonzero electric field, the mobile charges (holes in the *p*-region and electrons in the *n*-region) will move in a direction to polarize the material in such a manner that will lead to a zero net electric field. This is another example of the fact that inside a conductor in equilibrium, the net electric field must become zero through polarization shown by the tiny charges on the outer surfaces of the junction in Fig. 9.12. The tiny amount of charge on the outer surfaces plays no significant role in the behavior of the *p-n* junction, so we will omit showing these charges in later diagrams.

We can create an imbalance in the two types of current, the forward and the reverse currents, to tilt in favor of the forward current by altering the potential difference across the diode with a battery. When a diode is forward biased, the thermal generation drift current (or *reverse leakage current*) is very tiny and still in the *reverse direction* to the forward diffusion current which will become large in a forward biased diode as we shall see in Section 9.6.

9.6 A DIODE IN FORWARD BIAS AND RECOMBINATION

When a voltage is applied to forward bias a diode, surface charges are arranged in a gradient and these produce a resultant electric field inside the diode developing a voltage across the diode. This is shown in Fig. 9.13.

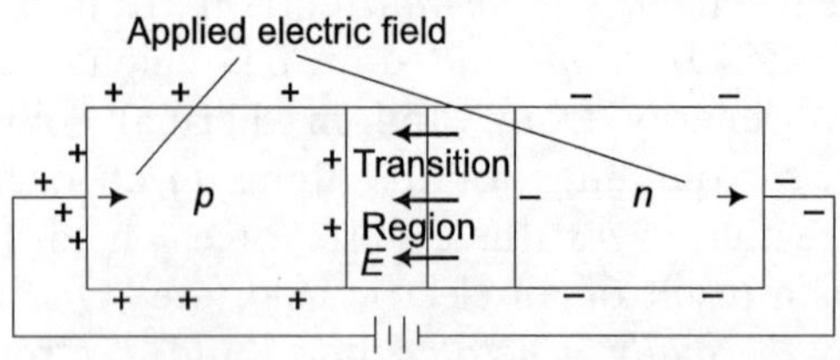

Fig. 9.13 A diode in forward bias.

The electric field in the transition region is the net (resultant) field due to the applied field and the electrostatic potential barrier V_o, (which is lowered, because some of the immobile charges are neutralized, or the stretch of the dipoles is reduced due to the voltage source). This is shown by the smaller length arrows in the transition region compared with their lengths in Fig. 9.12.

When a forward bias voltage V which is *less than* V_o (the potential barrier voltage), is applied to a diode, the *transition region width* is reduced and the holes in the *p*-side gain potential energy (see Section 9.7 how the empty states in available energy levels are filled with respect to Figs. 9.23 and 9.24 and also "The electric field and its influence on the energies of electrons" in Appendix E) and a tremendous amount of *kinetic energy*. The electric field in the neutral regions is very small (shown by the small arrows *near the diode contacts* in Fig. 9.13) and when deriving the diode equation assumed ideal, the electric field in the neutral regions is *considered to be zero*, and the applied voltage is assumed to drop entirely across the junction. If a field and associated voltage drop were to be taken into account in the neutral regions, the diode equation becomes complex, and introduces a complexity that is really needless, when the relation derived without this field, is fairly accurate.

With the potential energy they gain from the applied voltage (see Section 9.7 how the empty states in available energy levels are filled with respect to Figs. 9.23 and 9.24) and large kinetic energy, the *holes overcome* the electrostatic potential barrier and swiftly begin crossing the junction, as shown in Fig. 9.14.

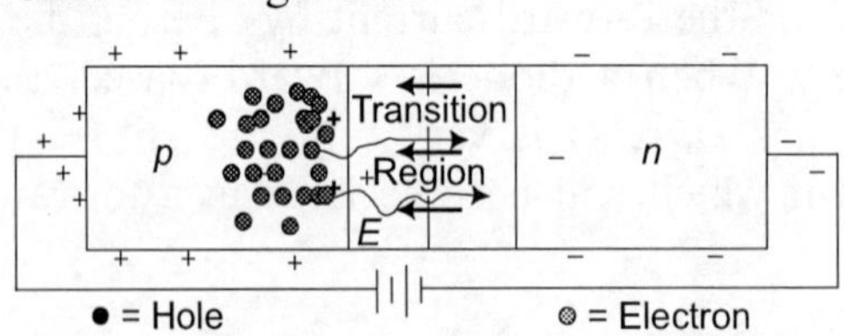

Fig. 9.14 Holes from *p*-side cross the junction.

Note that the holes seemingly are flowing against the electric field in the transition region. Their *large kinetic energy* enables them to *overcome* the potential barrier that was lowered by the applied forward bias voltage within the transition region. This is shown in Fig. 9.15. The process is one of drift within the *p*-region and within the transition region, and diffusion at the edge of the transition region near the *n*-region and a little into the *n*-region.

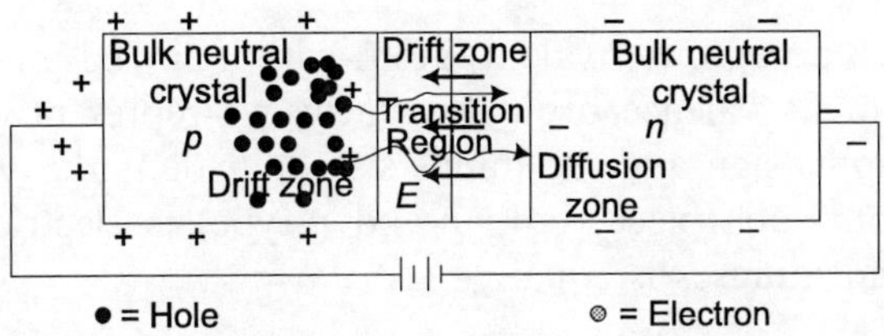

Fig. 9.15 Holes cross the junction because of their large kinetic energy.

The **electrons** also cross the junction because of their large kinetic energy due to the forward bias as shown in Fig. 9.16. It may seem here too that the electrons are flowing against the potential barrier that was lowered by the applied forward bias voltage in the transition region. They have sufficient potential (see Section 9.7 how the empty states in available energy levels are filled with respect to Figs. 9.26 and 9.27) and kinetic energy to *overcome* the potential barrier within the transition region. The process is one of drift within the *n*-region and within the transition region, and diffusion at the edge of the transition region near the *p*-region and a little into the *p*-region.

The high resistance of the transition region which was earlier devoid of mobile charges is seemingly brought down by the presence of the carriers crossing over from either side due to the forward bias.

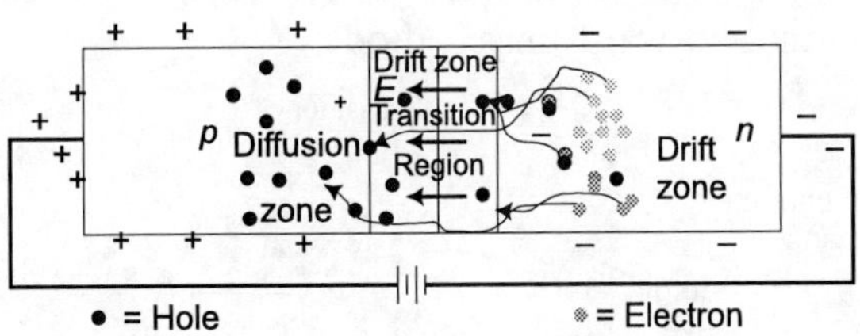

Fig. 9.16 Electrons also cross the junction because of their large kinetic energy.

The holes contribute to conduction when they are moving in the bulk *p*-regions and across the transition region and a little into the *n*-region, when they soon begin recombining with valence electrons.

For an animation of how the holes move in the lattice see http://www.ece.utep.edu/courses/ee3329/ee3329/Studyguide/Shockwave/Fundamentals/Demos/acceptor.html.

After a free electron leaves the negative terminal of the battery, another enters the right end of the diode. Another electron from the *n*-region reaches the junction. When the battery voltage is greater than a few tenths of a volt, the free (conduction band) electron has enough energy to crossover the depletion layer.

Recombination is the mechanism by which the conduction band electrons (*n*-side) can fill the valence band holes (*p*-side). This makes available valence electrons (transition region, *p*-side and a few in the *n*-side) that enable the continued movement of holes to the right in the *p*-side and a little into the *n*-side as shown in Fig. 9.16 and 9.17a.

In Fig. 9.17a, the *moving* free electron for the sake of clarity is shown with a '•'.

We represent each acceptor (trivalent) atom with a circled minus sign and the plus sign outside is the hole in its valence orbit. Similarly, we represent each donor (pentavalent) atom with a circled plus sign and the minus sign outside is the *free electron* it contributes to the semiconductor. Each piece (both *p*-and *n*-type) is electrically neutral because the number of pluses and minuses is equal.

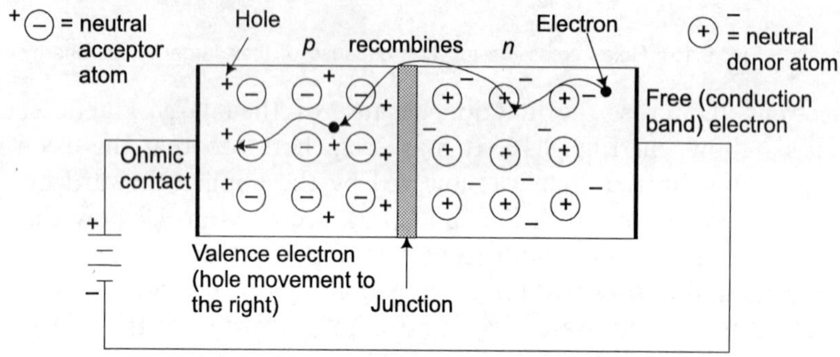

Fig. 9.17(a) Recombination in a forward biased diode.

Forward bias lowers the potential barrier at the junction and gives free electrons more energy equivalent to higher energy levels. Fig. 9.17(b) gives a comprehensive view of the movement of carriers in a forward biased diode.

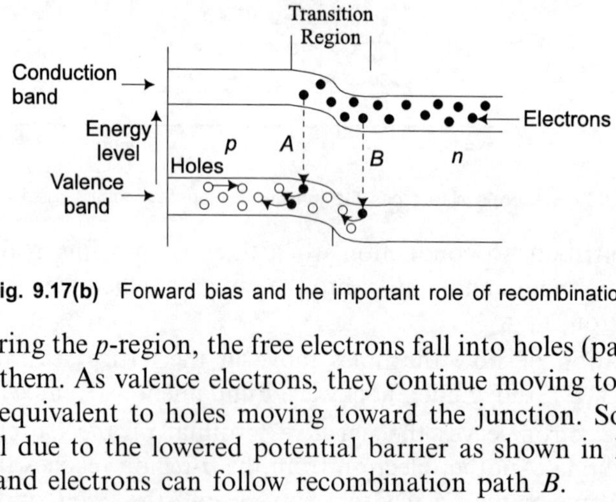

Fig. 9.17(b) Forward bias and the important role of recombination.

Soon after entering the *p*-region, the free electrons fall into holes (path *A*) due to mutual attraction between them. As valence electrons, they continue moving toward the left end of the crystal; this is equivalent to holes moving toward the junction. Some holes penetrate the *n*-region as well due to the lowered potential barrier as shown in Fig. 9.17(b). In this case, conduction-band electrons can follow recombination path *B*.

Regardless of where the recombination takes place, the result is the same. Free electrons move in a steady stream toward the junction and fall into holes to which they are attracted, near the junction. The captured electrons (now valence electrons) move left in a steady stream through the holes in the *p*-region. The movement of electrons in the *n*-region is similar to the movement of electrons described during conduction in metallic conductors in Section 1.18; millions and millions of electrons and holes in the *n*- and *p*-regions participate in the process of conduction. The electron that leaves the *p*-terminal (having left a hole in the *p*-lattice) to drift toward the battery positive terminal is different from an electron that fills a hole at one point in the *p*-region and which is different from another electron that fills a hole at another point in the *p*-region and so on at millions of locations in the *p*-region. The overall snapshot at every instant during forward bias in the steady state, is of millions of electrons filling holes throughout the *p*-region of the diode at various cross sections of the *p*-region, resulting in an apparent movement toward the right in Fig. 9.17 of holes toward the junction and into the *n*-region. Hole movement in the *p*-region thus caused by the action of millions and millions of electrons filling holes and creating holes in the atoms they left is observed only within the *p*-region and extends slightly into the *n*-region at the junction.

In this way, a seemingly *continuous* flow of electrons (to the left in Fig. 9.17(a)) through the diode is observed.

In a forward biased diode, there is ever present a thermally generated *reverse leakage* current that was described in Section 9.5 which is very small compared to the forward current due to the mobile holes and electrons that forms the forward current in a forward bias. The *reverse leakage* current will be manifest when the diode is reverse biased as will be described in Section 9.11.

Note that there are interface charges at the boundary of the transition region and the *p* and *n* layers...to produce the necessary driving electric field that lowers the barrier as shown in Fig. 9.18.

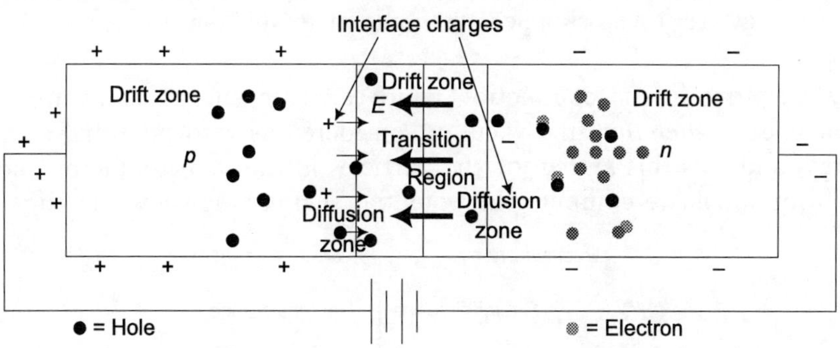

Fig. 9.18 Interface charges at the boundary of the transition region.

It may seem here too that the electrons are flowing *against* the potential barrier in the transition region; they cross-over due to their large kinetic energy. The holes then move by a *diffusion* process (that is statistical in nature) from a high concentration region (*p*-region and the transition region they find themselves in) to a low concentration region (*n*-region). See Fig. 9.15. The same holds true for the electrons in the *n*-region. They too move into the *p*-region by a *diffusion* process. See Fig. 9.16.

9.7 POTENTIAL AND KINETIC ENERGIES OF THE CHARGE CARRIERS OF A JUNCTION UNDER FORWARD BIAS

The distance between the conduction band edge E_c, and the energy of a free electron outside the crystal (called the vacuum level labeled E_{vacuum}) is quantified by the electron affinity, χ multiplied with the electronic charge q as shown in Fig. 9.19.

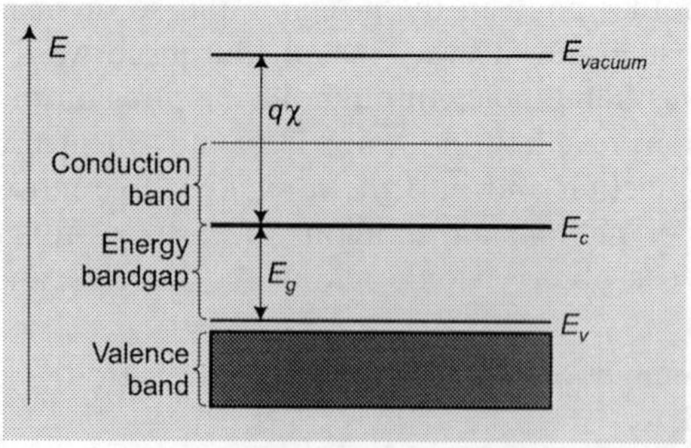

Fig. 9.19 Potential energy of an electron with reference to E_{vacuum}.

The electron affinity $q\chi$ is important for selecting the right metals when making metal to semiconductor ohmic (or nonrectifying) contacts which we shall discuss in Section 9.15.

It is usually the practice to set the vacuum as the zero potential energy level. If the potential energy is set to zero at infinity, the usual convention, then bound (conduction and valence) electron states have negative potential energy (see Fig. 9.20) which means some amount of energy has to be supplied to the atom to shift an outermost electron from its orbital space.

A few electrons in the conduction band can gain energies that are positive. They possess energies *greater than* the vacuum considered as zero potential energy (reference zero energy) and it is this *system* of the particles in the semiconductor and no particles in vacuum for which we compute the potential energies, as shown in Fig. 9.20.

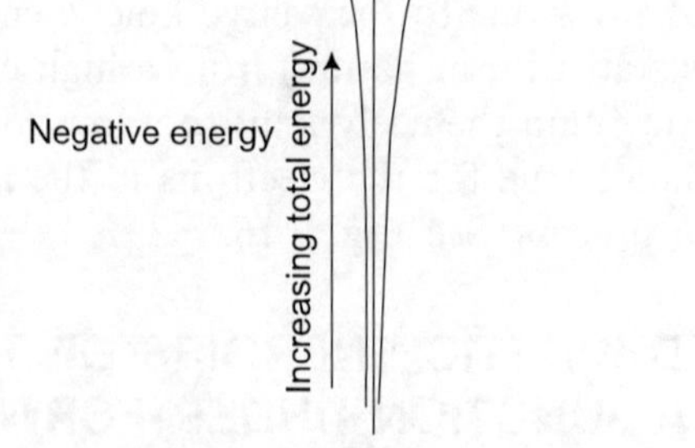

Fig. 9.20 Potential energy variation with distance of an electron from the nucleus.[19]

All potential energies have an *arbitrary* zero point: the point we designate a zero value for the potential energy. For semiconductors, physicists normally specify the conduction band edge (bottom) as the reference potential energy (zero) for electrons.

Potential energy is relative to that datum or zero point. If work is done by external agents (heat, light and voltage) on the charge to move it against the forces of attraction from fields, its potential energy is increased above that datum. See Fig. 9.44 also.

If, on the other hand potential energy is extracted or converted into kinetic energy from the charge by letting it fall with the electric force (of attraction say between electrons and lattice ions), then the potential energy becomes smaller and smaller and possibly even below the datum (or reference) value, and hence negative.

Negative energy is a matter of convenience to represent confinement, like when the electron is confined within the atom due to its attraction by the nucleus (consisting of protons and neutrons and not including the electrons), else it should be at infinity or in vacuum, with zero potential energy! The potential energy profile for the simple case of a hydrogen atom as shown in Fig. 9.20 may be computed from e^2/r where e^2 is the product of charge of the nucleus and electron and r is the radial distance of the electron from the nucleus. The electric field which gives rise to the electrostatic force of attraction between the nucleus and electron is *not shown* in Fig. 9.20.

The potential energies (negative) of electrons are typically spread over a few tens of eV (electron volts) in conductors and semiconductors which are commonly used in the manufacture of electric and electronic components. The upper conduction band and the lower valence band in semiconductors are separated about an eV or slightly more and insulators have gaps of about 10 eV. In energy band diagrams, the energies of *inner electrons* close to the nucleus (See "Inner electrons and the binding energies of electrons" in Appendix B) should be shown below the lowest valence band energy level but these are usually not shown, because in electricity and electronics it is the valence and conduction bands that are important. As was mentioned in Section 1.20, physicists and engineers prefer the small energy unit eV (electron volt) to express electron energies over the joule. In electric and electronic circuits, millions and millions and millions of electrons in tiny energetic amounts (expressed in electron volts), work harmoniously as we learned in Chapter 4, transporting and converting energy that is measured in joules, which is a large unit of measure of energy.

Positive energy of electrons can result even when bound to the crystal and is *mostly kinetic* because Total E = potential E (small negative) + kinetic E (large positive) as shown in Fig. 9.21 where we see the electron gain kinetic energy and thus possesses a net positive energy.

A description of the relationship between the energy levels and orbits of electrons is given in "Energy levels of electrons" in Appendix B. For a comparison of the *binding* energies indicative of the strength of force of attraction to the nucleus of innermost and outermost electrons of Silicon see Appendix B "Inner electrons and the binding energies of electrons".

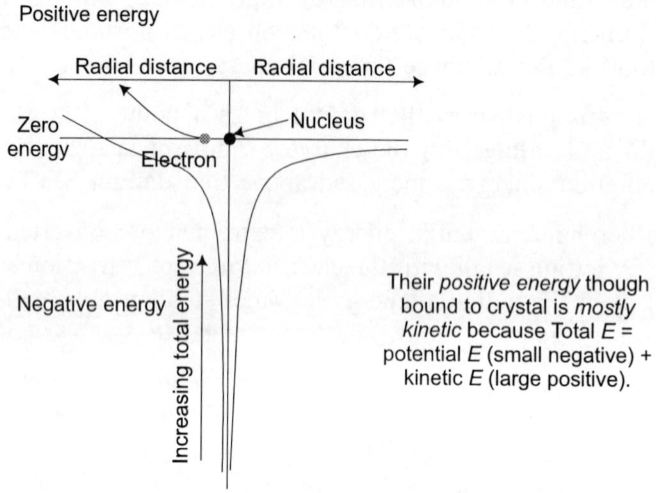

Fig. 9.21 Positive total energy of an electron.

That is how the electrons in the conduction band possess a total positive energy. This is depicted in the energy band diagram in Fig. 9.22.

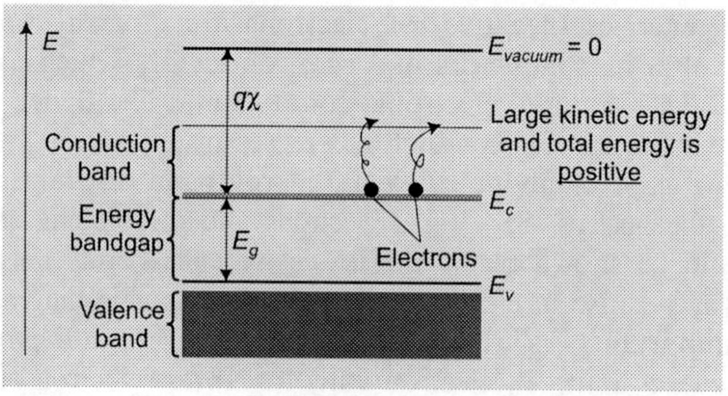

Fig. 9.22 Positive energy of an electron.

Figure 9.22 also shows the electron affinity $q\chi$ of a semiconductor which is the energy required to remove an electron from the bottom of the conduction band E_C to the vacuum level. This parameter is important from the viewpoint of selection of metals when making metal to semiconductor ohmic or rectifying contacts which we shall discuss in Section 9.15.

How Holes Surmount the Potential Barrier

We will now study how holes in the *p*-region of a *p-n* junction surmount the potential barrier under the influence of a forward bias voltage.

When a *p-n* junction is left open-circuited, the **holes** in the *tail* band energies will be as shown in Fig. 9.23. There are several possible unfilled states.

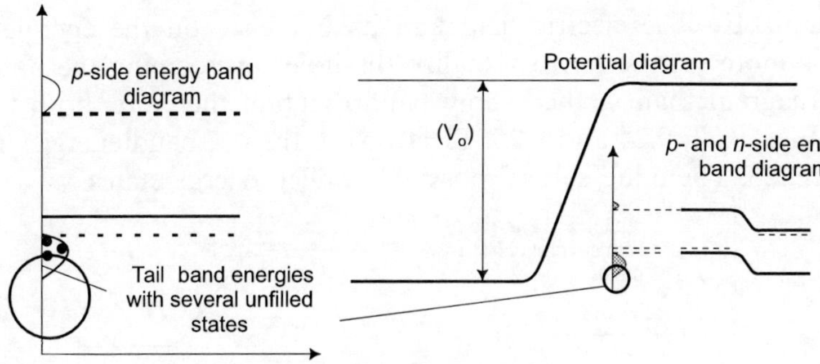

Fig. 9.23 Holes in the tail band.

When a forward bias voltage V_f with its accompanying e-field is applied then several of the unfilled states are filled and the filled energy band states are shown in Fig. 9.24. How do holes *fill* these states? When a valence (bound) electron gains energy, it moves to a higher level within the valence energy band. This is equivalent to a hole transiting *down* the vacant energy states to gain energy, thus *filling* a tail band energy, and so giving the atom a stronger positive quality.

Note: $V_o - V_f$ refers to the potential *at* the transition region—the *true* junction and not in the bulk of the diode.

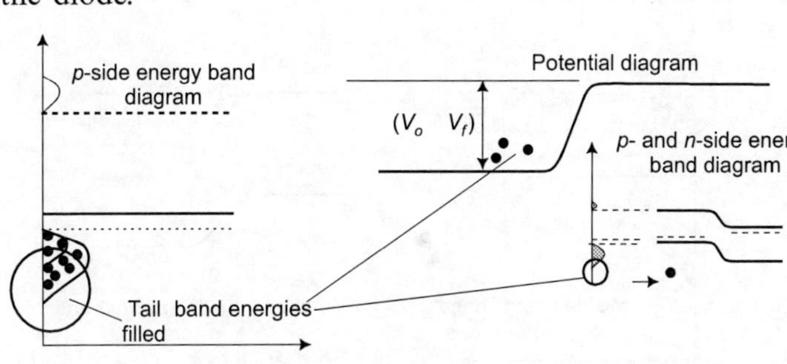

Fig. 9.24 Filled energy bands in the tail band after the forward bias is applied.

Due to their large kinetic energy the holes in the tail band energies can surmount the potential energy barrier as shown in Fig. 9.25.

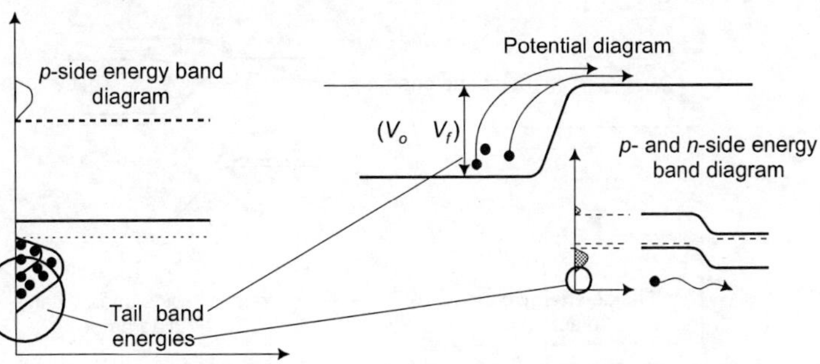

Fig. 9.25 Holes surmounting the potential energy barrier.

See Appendix E "The electric field and its influence on the energies of electrons". Intuitively, it is more instructive to visualize the holes overcoming the potential barrier in the potential diagram than in the energy band diagram therefore, both the diagrams are shown in the Figs. 9.23, 9.24 and 9.25. Electrons in the *tail* band energies for electrons are shown in Fig. 9.26. There are several possible unfilled energy states.

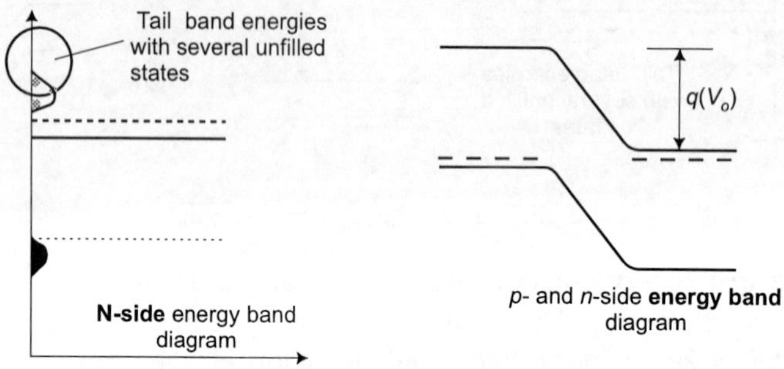

Fig. 9.26 Tail energy bands for electrons.

Similarly, electrons in the tail band energies for electrons which were filled by the effect of the applied forward bias voltage V_f and accompanying e-field, can surmount the potential/potential energy barrier due to their large *kinetic energy*. The filled energy bands are shown in Fig. 9.27.

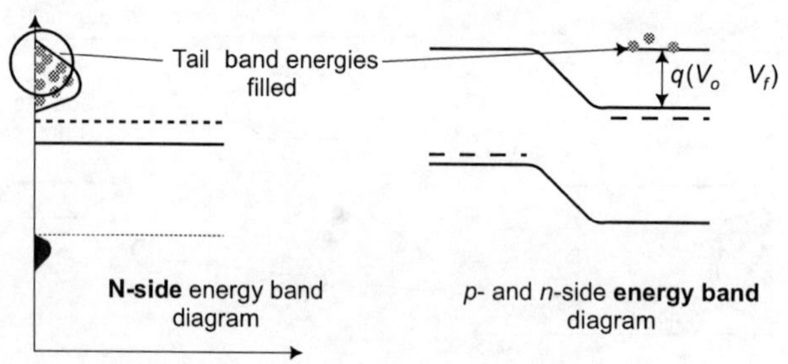

Fig. 9.27 Filled energy bands in the tail band after the forward bias is applied.

The electrons surmounting the lowered potential energy barrier with their large kinetic energies is shown in Fig. 9.28.

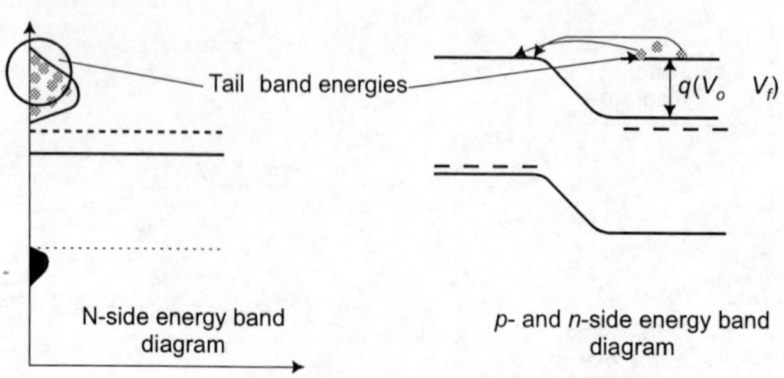

Fig. 9.28 Electrons surmount the potential energy barrier.

See Appendix E "The electric field and its influence on the energies of electrons".

9.8 VOLTAGE DROPS IN A FORWARD BIASED DIODE

The potential barrier of a *p-n* junction can never be brought down to zero ....nor ever *reversed*, however strong the forward biasing voltage. The question then arises: *where does the applied voltage drop?*

Consider a silicon *p-n* junction with doping concentrations $N_A = 2.5 \times 10^{15}$ cm^{-3} and $N_D = 5 \times 10^{16}$ cm^{-3} at 300 K with the intrinsic concentration $n_i = 1.5 \times 10^{10}$. (Problem in "*Electronic Devices and Circuits*" by Milman, Halkias and Satyabrata Jit).[17]

Using $V_o = V_T \, ln \dfrac{N_D N_A}{n_i^2}$, we obtain V_o = **0.703** Volts at 300 K and $I_o = 9.17 \times 10^{-15}$ Amps. Using the formula for current $I = I_o(e^{V/V_T} - 1)$, we obtain for a forward bias V of 0.6 V, V_T = 0.026 V at 300 K or 27°C, current I = **0.1045** milliamps. Note that a forward bias voltage of **0.6 V** which is *less than* V_o = 0.703 V has caused a current of **0.10 mA** in the diode. For a forward bias voltage **0.69 V <** V_o **= 0.703 V**, we obtain I = **3.40** milliamps!

An *increase* in forward bias voltage of 0.09 V has caused the forward current to *increase* nearly **33 times to 3.40 *milliamps*** and the bias voltage is yet less than the barrier potential. With V = **0.7029 V < 0.703 V (**V_o**),** I **= 5.41 milliamps (54 times the value when** V **was 0.6V)!** A plot of the current-voltage characteristic is shown in Fig. 9.41 (Section 9.13). If a small diode, like say, the 1N625 it will surely burn out with this current!

We assumed that the applied voltage appears across the junction entirely ignoring the voltage drops across the bulk of the *p* and *n* regions. It is reasonable to assume that as the applied voltage approaches V_o = 0.703 V, and the current increases, some voltage will begin to drop across the bulk *p* and *n* regions (the electric field in these regions increases) and across the resistance of the ohmic contacts to uphold the current conservation law while, the part of the applied voltage that lowers the barrier potential V_o, makes the voltage *across the junction close to zero*, but not equal to zero, and a *vanishingly thin depletion region* remains, which demarcates the *p* and the *n* regions with holes (from the *p*-side) and electrons (from the *n*-side) recombining all around.

Thus, the diode forward current increases rapidly near $V = V_o$, and the barrier (contact) potential V_o is the limiting forward voltage for the junction.[30] Note the use of the words "diode" and "junction", where while the former refers to the diode end to end, the latter refers to the junction of the diode only.

If the forward voltage applied to the diode exceeds V_o, the ohmic effects are significant, but the **junction voltage itself will always be less than** V_o.

For the small current diode considered in our example, the total bulk resistance of the neutral region can be represented by R_S and we can write the junction voltage V as $V = V_{applied} - I(R_S)$ where $V_{applied}$ is the voltage applied to the diode. When the current has increased (with $V_{applied}$) and reaches 5.41 milliamps, there is an increasing drop in R_S, and the junction voltage decreases. The reduction in junction voltage lowers the level of injection so that the current increases more slowly with increased (forward) bias.

Since the effects indicated by R_S in $V = V_{applied} - I(R_S)$ are more pronounced at high injection levels, the injected carriers can reduce R_S significantly. Therefore, up to a certain current level of the diode, the ideal *exponential* diode characteristics are applicable after which the characteristic becomes nearly linear mainly dominated by the bulk region as shown in Fig. 9.41. Consequently, at these current levels there will be an increase in the ohmic losses in the bulk regions[30], and above 5.41 milliamps or so, the governing *I–V* relation may be written $I = I_O\,(e^{\frac{V - IR_S}{V_T}} - 1)$.

Normal Operation Region of a Diode

Ohmic losses are purposely avoided in properly designed devices by appropriate choices of doping and geometry. Therefore, deviations of the current in the ohmic effect regions as shown in Fig. 9.41 generally appear only for very high currents, *outside* the normal operating region of the device.[30] The barrier potential V_o it may be noted is never larger than 1.1 volts, the forbidden energy E_G expressed in volts (see Note below Fig. 9.9 Section 9.4) in the case of silicon diodes, no matter whether the diode is rated for small or large currents.

For an intuitive understanding of V_T see Appendix B on the "Einstein relation".

A description of the circuit processes when a *p-n* junction is forward biased and a simplified derivation of the current expression of a *p-n* junction diode is provided in Appendix E.

9.9 A DIODE AND ITS OPERATION

One intuitive way to understand the operation of a diode is to visualize the operation of a **check valve** for water, which allows water to flow in one direction only. Such valves are used to prevent the downward flow of water back into a ground-level sump or reservoir when the pump that raised the water to an overhead tank is shut off. Such a pumping system is commonly found in multi-storeyed buildings. With very little or no fluid flow the valve is held closed by the ball pushed against the seat by the spring and water cannot flow to the other chamber. The spring displacement maybe likened to V_o, the potential barrier in the transition region of a diode as shown in Fig. 9.29a.

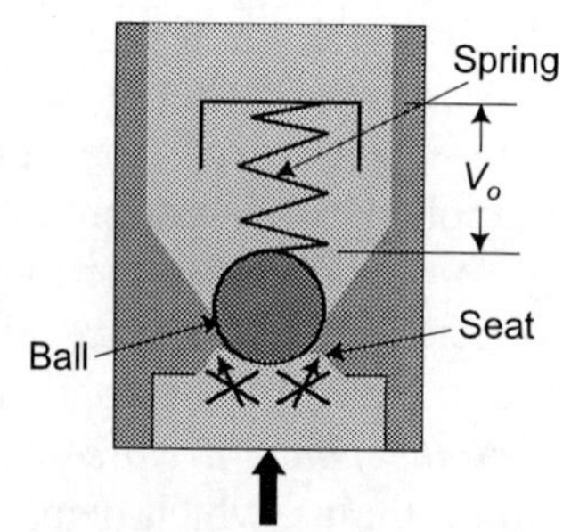

Fig. 9.29(a) Water check valve with little or no fluid flow.

With a large fluid flow, the **ball** is raised by the water pressure and this enables the water to flow into the other chamber.

The spring is compressed and the *reduced* spring displacement may be likened to V_o–V_f the *lowered* potential barrier in the transition region of a diode as shown in Fig. 9.29b.

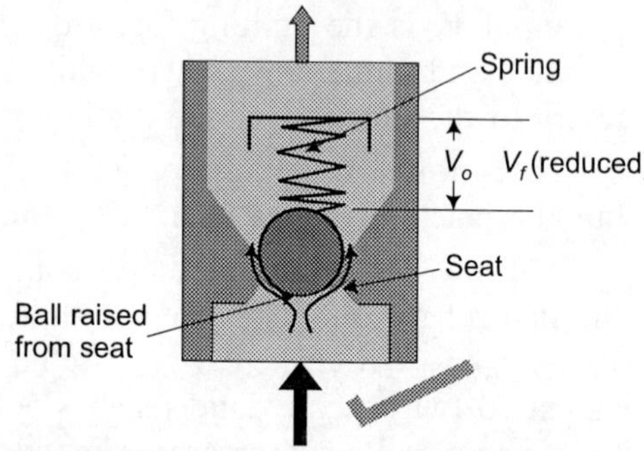

Fig. 9.29(b) Water check valve with large fluid flow.

With large flows V_o–V_f is small. With small flows V_o–V_f is large.

For operation as a check valve, a **spring** is necessary. For operation as a diode, a ***p-n* junction** is necessary. As long as the valve is operating, the spring will **always** have a displacement, however small.

Similarly, as long as the diode is operating the junction will always have a barrier potential, V_o, however small.

The length of the seat, ball and orifice assembly of the valve maybe a mere 2 cm, yet it exercises on/off flow control of water over a piping system several meters long. It should not escape the reader's attention to the similarity with the diode transition region that is a mere 10^{-6}m in length (Section 9.5) yet, it exercises on/off flow of current in a circuit that is maybe several centimeters long!

One of the *vexing* questions that students seek an answer to is, how much is V_o–V_f? For large applied forward voltages? Greater than V_o? say, greater than 1 V?

The answer: V_f *adjusts* itself to a value such that the conservation of current law is upheld. And this will not make $V_o = 0$. In other words, $| V_f | < | V_o |$. The bulk of the applied voltage in case of large forward biasing voltages, drops in the bulk of the diode material ....***away*** from the transition region. See the worked example in Section 9.8.

V_f *adjusts* itself to the value such that the sum of the large diffusion current and the small drift current *in the transition region* equals the drift current *in the bulk region where most of the applied voltage is dropped.*

Any deviation will cause a pile-up of charges and this will automatically cause a correction of V_f to the right value so that the current is uniform around the circuit.

That is the reason why the *VI* characteristics of diodes plotted in data sheets and electronic manuals are upto about 0.6 V to 1 V for silicon diodes and seldom beyond (See Fig. 9.41). And if applying voltages greater than, say, 1 V, a current limiting resistor is added in series with the diode.

When the forward bias voltage approaches V_o (typically about 0.6 to 0.7 volts for a silicon diode), the current will become large and then begin to increase linearly mainly dominated by the bulk region (see *VI* characteristic in Fig. 9.41).

As a practical matter we can never reduce the barrier to zero, because as the current increases without limit, the bulk resistance of the crystal, as well as the resistance of the ohmic contacts, will limit the current and it is no longer possible to assume that all the voltage V appears as a change across the *p-n* junction, as described after the numerical example in Section 9.8.

One intuitive way of understanding this, is when the forward bias voltage is *much less* than V_o, the current is determined or rather restricted in a manner of speaking almost entirely by the diffusion components. The majority carrier concentrations are usually large in the bulk neutral regions, so that only a small electric field is needed to drive the drift currents in these regions.

However, with forward bias voltages that approach and exceed V_o, the drift component has to *catch up with and supply* the flood of diffused carriers to uphold the conservation of current law and this consumes larger portions of the applied voltage. A significant applied electric field (shown by long arrows in the figure) therefore appears in the bulk neutral regions reflecting the increased voltage drops to sustain the large currents in these regions.

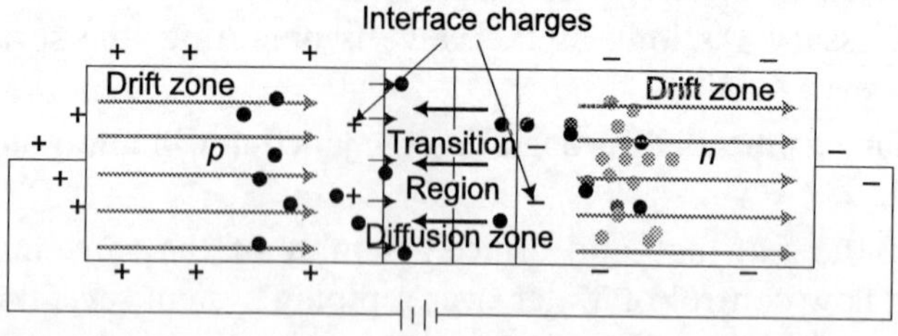

We conclude then, as the forward voltage V becomes comparable with V_o, the current through a real *p-n* diode will be governed by the ohmic-contact resistances and the crystal bulk resistance.

To better understand *p-n* junction operation, look at the check valve again. Take away the spring and ball, and you will be left with a hollow piece of pipe.

Analogously, in the *p-n* junction, if there were no V_o, you would have a bar of semiconductor material of opposite polarity charge carriers across a dividing line.

A word of caution: It is possible to adjust the acceptor N_A and donor N_D levels of a *p-n* junction such that the potential barrier is **zero** or V_o = zero. Since, $V_o = V_T \ln \frac{N_D N_A}{n_i^2}$ using $N_A = 1.5 \times 10^{10}$, $N_D = 1.5 \times 10^{10}$, $n_i = 1.5 \times 10^{10}$, we note that $V_o = 0$.

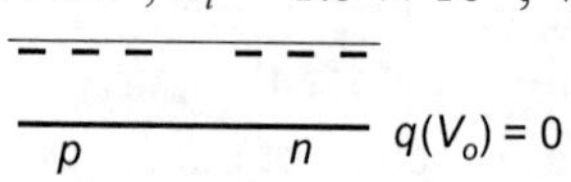

However, an **ever so slight forward bias** will trigger conduction as seen in the figure.

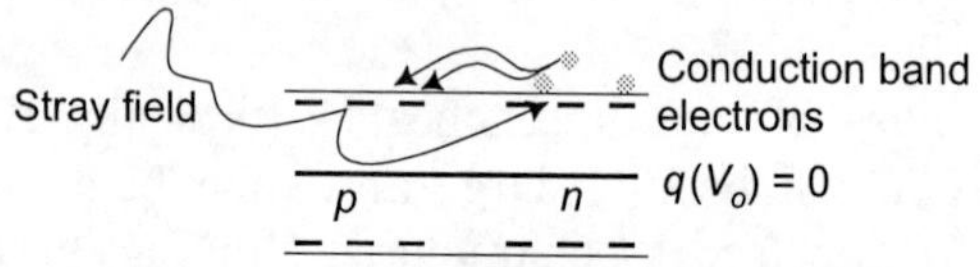

Visit http://jas.eng.buffalo.edu/education/pn/pnformation3/index.html orhttp://web.archive.org/web/20170328000334/http://jas.eng.buffalo.edu/education/pn/pnformation3/index.html and enter the n_i, N_A and N_D values to check that V_o is indeed zero. This can achieve an ideal diode characteristic; when forward biased, the diode will begin conducting with a "zero" voltage drop and block reverse current with the application of a non-zero reverse bias voltage.

There is no change in the processes of the diode with ideal characteristics, both in forward and reverse bias, except that there is no barrier voltage for the holes to contend with and a mere voltage difference > 0 between the *p*- and *n*-sides is sufficient to trigger conduction in the forward bias and a voltage difference < 0 to block current in the reverse direction.

But there is little to be gained with such an ideal diode...even with a likelihood of its manufacture being successful—like the water valve with a ball *but without a spring*.

In noisy environments, circuits using diodes with such ideal characteristics will be so susceptible to interfering inputs (for example, conducting at unwanted times) that they would soon become a nuisance.

For most commercially available diodes, the diode (the bulk region and the junction) in forward bias behaves like a battery of voltage $V\gamma$ (also called the *cutin* voltage, which encapsulates the barrier potential and the geometric properties of the junction) connected in *series opposition* with a resistance $R_{forward}$ as shown in Fig. 9.30. In some textbooks, $R_{forward}$ is also called the average ac resistance, r_{av}.

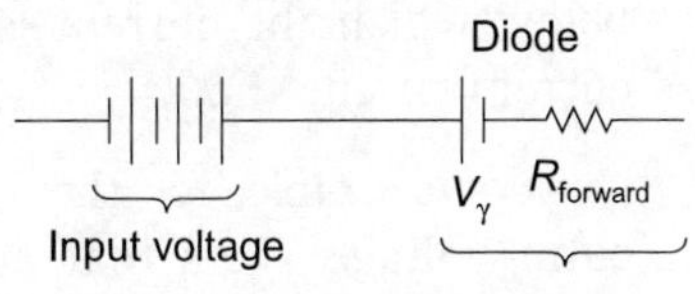

Fig. 9.30 Diode equivalent circuit.

VI Characteristics of Conductors (Resistors) and Diodes

A resistor (conductor) may be modeled when it is in thermal equilibrium with its surroundings (meaning a few minutes after switch ON), by a Volt vs. Current characteristic. Since the number of conduction band electrons is usually a fixed number (given the material of its construction) which does not change with applied voltage, the characteristic will be a straight line through the origin. Consequently, the resistor (conductor) has a constant resistance when in thermal equilibrium with its surroundings. See "What determines Bulb Brightness" in Sect 2.20 and the characteristics of resistors in Fig. 3.27.

A diode, on the other hand, cannot be modeled as a straight line or a constant resistance because the number of conduction band electrons and holes injected into the depletion region (which determines the conductance of the diode) depends on the amount of forward bias voltage and this would cause the conductance (or its reciprocal, resistance) to be different at different bias voltages. Thus, the diode Volt vs. Current characteristic is curved as shown in Fig. 9.41. So, diodes are modeled differently from ordinary resistances and manufacturers specify an ac average resistance for diodes, which is the amount of incremental change in current (more carriers, electrons and holes injected) over the amount of incremental change in potential difference across the diode, when a fixed amount of current is flowing in the diode. The diode resistance is hence, the gradient of the *VI* curve at some fixed point on the curve. See Fig. 3.28 also.

Note that the voltage $V\gamma$ opposes the forward biasing applied voltage, whatever be the applied voltage strength. Typically, $V\gamma$ is 0.2 volts for germanium and is 0.6 volts for silicon diodes. For applied voltages < $V\gamma$, the current is very small, and is less than 1 percent of the maximum rated value. Beyond $V\gamma$, the current rises very rapidly as may be seen in the forward *VI* characteristics of these diodes.[17] But remember, the barrier potential V_o of the depletion region is never reduced to zero when the diode is operated within rating. The *cutin voltage* is simply the potential where the current is drastically changed from a very low value (~0) to a high value (like a switch) and is *only* realized from the piecewise linear characteristics of a commercial diode.

When visualizing the current in forward bias, note that the movement of the holes from the *p*-side is a current, and the movement of electrons from the *n*-side is a current, as well.

The *movement* of charges continues—moving and recombining *on and on and on*—and a *steady state* is established with the current *limited* by the *resistor* in the circuit.

$V\gamma$, the cutin or offset voltage is defined for a commercial diode as that *applied* voltage below which the current is very small, say, less than 1% of the maximum rated value of current of the diode.

For example, for the 1N4007 diode, the maximum rated current is 1000 mA and the cutin voltage $V\gamma$ is that applied voltage that will result in the current just beginning to exceed around 1% of 1000 mA = 10 mA.

$V\gamma$ is *measured* and found to be typically around 0.2 volts for Germanium diodes and around 0.6 volts for Silicon diodes.

V_o is the potential barrier that comes into existence due to a built-in electric field when a *p-n* junction is formed. The barrier ensures that the net hole current due to diffusion and drift is zero under dynamic equilibrium. V_o *cannot be measured* but is estimated from doping concentrations and its typical values are 0.3 volts to 0.7 volts for normal diodes. V_o is a potential barrier expressed in volts for a *p-n* junction in a diode.

Further, V_o the potential barrier cannot disappear.

If a forward bias voltage V_f, greater than $V\gamma$ (of course) but less than V_o were applied, the diode current will be large but the potential barrier may be lowered to a small value.

If V_f is greater than V_o, then the current will be quite large, V_o does not disappear but is reduced to an extremely tiny value and the excess voltage of V_f over V_o is dropped across the bulk *p* and *n* regions as shown in the figure below.

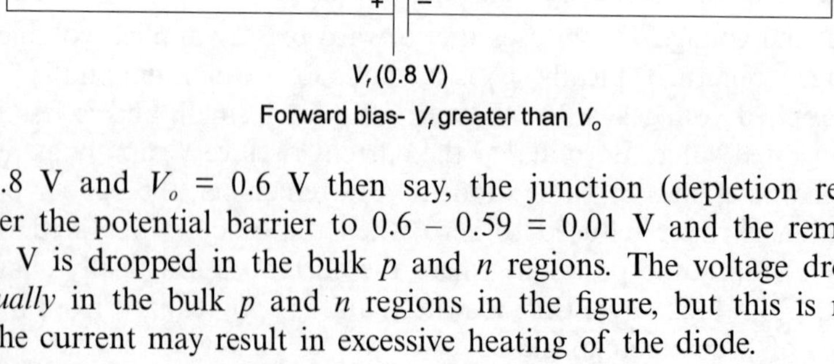

Forward bias- V_f greater than V_o

If V_f = 0.8 V and V_o = 0.6 V then say, the junction (depletion region) absorbs 0.59 V to lower the potential barrier to 0.6 – 0.59 = 0.01 V and the remaining 0.8 V– 0.59 V = 0.21 V is dropped in the bulk *p* and *n* regions. The voltage drops are shown distributed *equally* in the bulk *p* and *n* regions in the figure, but this is not necessarily always true. The current may result in excessive heating of the diode.

In most electronic circuits incorporating diodes which operate with supply voltages in excess of V_o say, 3 V, 5 V and 10 V and so on, a current limiting resistor is introduced as shown in the figure.

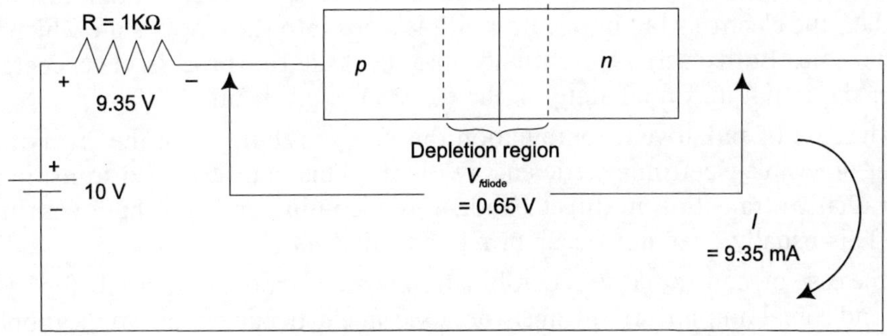

A diode circuit with a current limiting resistor

For example, if the battery applies 10 volts (dc) and if the open-circuit barrier voltage of the diode were to be V_o = (say) 0.72 V, then the diode absorbs 0.65 V to lower the barrier to 0.72 V – 0.65 V = 0.07 V. The current in the circuit in the steady-state will be 10 V – 0.65 V divided by the resistance 1 kΩ = 9.35 milliamps.

The drop across the resistor will be 9.35 mA × 1 kΩ = 9.35 volts. Thus, 10 V (battery) = 9.35 V (resistive drop) + 0.65 V (across the diode). Note that what we measure across the diode is 0.65 V and not 0.07 V (the height of the lowered barrier).

All of the above processes occur during the very initial transient period of the order of picoseconds before steady-state is established.

Note that the barrier height *adjusts itself* to a convenient value by a feedback mechanism and V_{fdiode} (0.65 V) self-adjusts to a value, such that V_{fdiode} (0.65 V) < V_o (0.72 V) so that the current is conserved around the circuit.

In the event that R is small and I is quite large, there will result some voltage drop in the bulk region, V_o will become tiny but can never disappear.

$V\gamma$ and V_o of large current diodes

In large current diodes, the cut-in voltage $V\gamma$ may be more than the barrier potential voltage V_o due to voltage drops across the quasi neutral regions and the forward voltage drop across the diode terminals and at rated current can be as large as 1.9 V with 12 Amps flowing through the diode even though the barrier potential of the diode is $V_o < 1.1$ V for silicon diodes (See end of Section 9.8 and note above Fig. 9.10, Section 9.4).

Forward Diode Current is Due to Recombination

Under forward bias, the diode current is due to recombination.[18] This recombination can occur within the quasi-neutral regions, within the depletion region or at the metal-semiconductor ohmic contacts. The thermal equilibrium EHP generation and recombination continues in the background but that is *swamped* by the dominant injected current components. The generation current becomes manifest when the diode is reverse biased and has swept away all excess mobile charges.

For recombination to occur, electrons must lose energy to fill the holes. How do they lose energy?

Recombination of electrons and holes is a process by which both carriers annihilate each other: the electrons fall in one or multiple steps into the empty state which is associated with the hole. Both carriers eventually disappear in the process. The energy difference between the initial and final state of the electron is given off.

In the case of radiative recombination this energy is emitted in the form of a photon or a packet of visible electromagnetic energy (light). This band-to-band transition is typically also a radiative transition in direct bandgap semiconductors as in light emitting diodes or LEDs. This usually does not occur in rectifier diodes.

In the case of non-radiative recombination, which is more commonly found in rectifying diodes and signal amplification transistors used in electronic circuit applications, the energy difference is passed on to one or more phonons (another name for the vibrating lattice ions) in the form of heat (more intense lattice vibrations) due to a collision with the momentum conserved.

In rectifying *p-n* junction diodes, the level does not change by a direct transition from a conduction energy level to a valence energy level but via a *trap* assisted process much like descending to a lower floor using a few steps. This is accomplished by allowable energy levels or traps created in the band gap by impurities within the semiconductor material.

In a *p-n* junction diode, there is **no generation** of **additional** electron-hole pairs due to the applied forward bias voltage under low-level injection conditions (see the computation of injected hole concentrations $P_n(0)$ for various forward bias voltages after the expression (20) in Appendix E); the thermal equilibrium concentrations were established and that was over when doping was completed and the junction had formed.

For a fixed forward bias, the current components therefore, are due to the **movement** of the open circuit thermal equilibrium number of the holes (acceptor) and the electrons (donor)...and are decided by the number of injected minority carriers.

The *movement* of holes and electrons continues...moving and **recombining** (*n*-side conduction band electrons recombining into *p*-side valence band energy levels) ***on and on and on*** and a ***steady state*** is established. The valence electrons in the bulk *p*-region hop along deep into the *p*-side and move on toward the *n*-side ohmic contact and deplete the positive charges on the voltage source positive terminal. The voltage source sensing the drop in voltage commences charge separation internally, and deposits electrons on its negative terminal.

Potential barriers will form at the two metal to *p*- and *n*-type semiconductor contacts because of the two dissimilar materials involved in the contacts as shown in the figure.

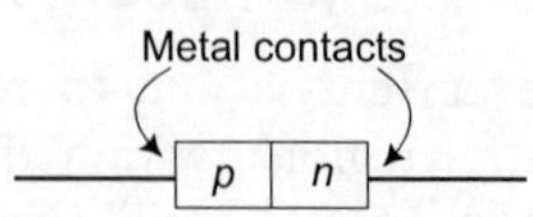

These will result in rectifying junctions at the metal contact to semiconductor contact surfaces. The processes which lead to the formation of these barriers and methods to make the contacts ohmic are described in Section 9.15.

It should be borne in mind that a *high level* of hole current is present only in the *p*-region and slightly into the *n*-region.

There is *no* hole current at the diode contacts or within the wires connecting it to the source of emf. This way, the circuit action is continued and a steady state is observed.

Given a forward bias, the currents will change only if the thermal equilibrium conditions were to change. For example, if the diode was allowed to heat up without provision of a heat removal mechanism, the elevated temperatures could result in additional electron-hole pair generation and the forward current for a given forward bias voltage will increase.

The currents in forward bias under constant thermal equilibrium conditions are therefore due to the *movement* of the holes and the electrons moving and **recombining *on and on and on*.**

9.10 A DIODE AS A RECTIFIER

A diode when used as a half wave rectifier enables the production of a positive voltage when a *sine wave* is applied to it. The process of rectification is shown in Fig. 9.31.

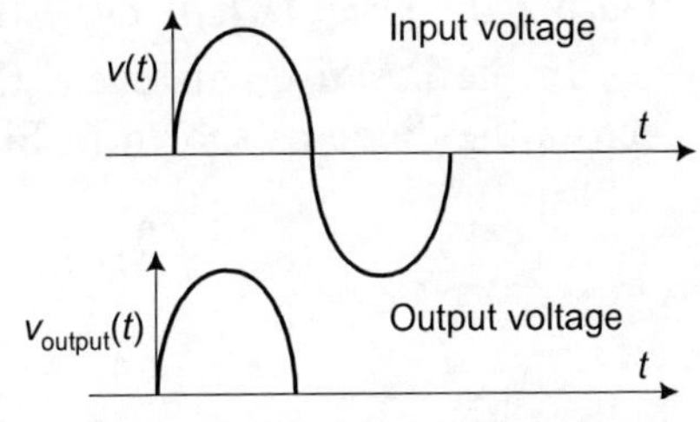

Fig. 9.31 Rectification process.

The diode blocks out the **negative swinging** voltage of the input as seen in Fig. 9.31. A half wave rectifying circuit can be used to achieve rectification.

Such a circuit is shown in Fig. 9.32.

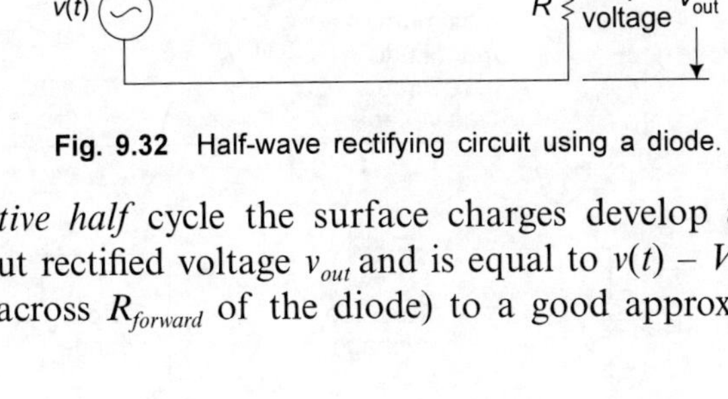

Fig. 9.32 Half-wave rectifying circuit using a diode.

During the *positive half* cycle the surface charges develop as shown in Fig. 9.33, resulting in the output rectified voltage v_{out} and is equal to $v(t) - V_\gamma$ (the cutin voltage and neglecting the drop across $R_{forward}$ of the diode) to a good approximation for all currents within rating.

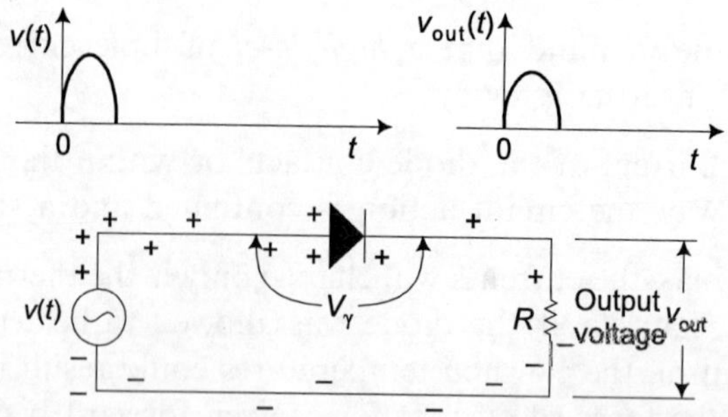

Fig. 9.33 The positive cycle applied to the rectifying circuit.

But what happens during the negative cycle?

9.11 REVERSE BIASING A DIODE

During the *negative* half cycle there is a rearrangement of surface charges.

In the negative half cycle, the surface charges cause the appearance of a reverse bias across the diode as shown in Fig. 9.34.

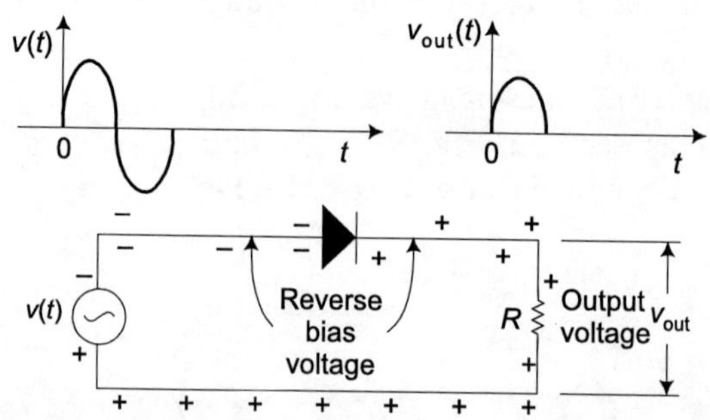

Fig. 9.34 Reverse biasing voltage during the negative cycle.

It should be borne in mind that the *maximum magnitude* of the reverse voltage across the diode is the peak negative cycle voltage of the input. This is shown in Fig. 9.35.

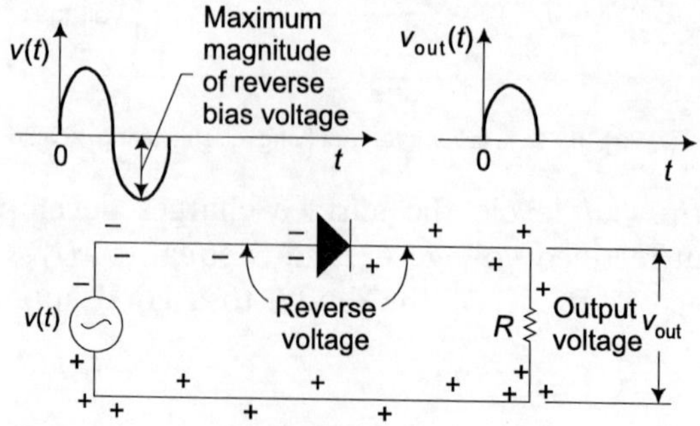

Fig. 9.35 Maximum magnitude of reverse bias voltage during the negative cycle.

Under reverse bias, the current is due to generation of electron-hole (EHP) pairs. Several mobile charges of the extrinsic *p*- and *n*-materials would have been absorbed by the reverse bias as they move away from the junction and nominally a zero current results.

Actually, a small current does flow because a small number of electron-hole pairs are generated throughout the crystal as a result of thermal energy (even room temperature is sufficient as a source of thermal energy) contributing to what we call *generation* or *leakage* current as was described in Sections 9.5 and 9.6.

We have observed the output voltage of the half-wave rectifier with the load resistor connected. See Appendix C "Dielectric in a uniform electric field", for a discussion of why it is difficult to characterize the output voltage with the load resistor *not* connected.

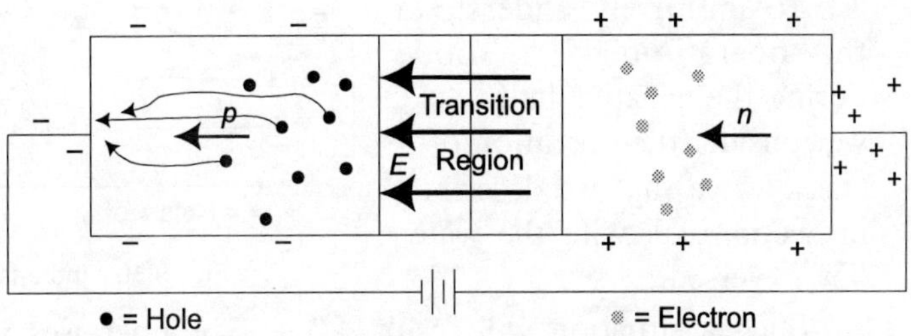

Fig. 9.36 Removal of a few mobile charges by the reverse bias.

When a diode is reverse biased, the electric field acts to move the *mobile* charges away from the junction and some remain in the diode; the voltage source *"absorbs"* the ones that are moved as shown in Fig. 9.36. More immobile charges are exposed at the junction in comparison with the number when the diode was open circuited.

The battery pulls some electrons out of the *n*-region, leaving a thicker positive portion of the depletion layer. Similarly, the battery pulls some holes out of the *p*-region leaving a thicker negative region of the depletion layer. The battery does not pull out *all* the mobile holes and electrons from the diode; merely a fraction of these are pulled out to cause the depletion layer to "drop" the applied reverse bias voltage across the diode. There are still plenty of mobile carriers remaining in the bulk regions. Even with large reverse biasing voltages, the transition layer does not extend from the junction to the metal contacts at the ends of a diode !

The *slightly* thicker depletion (transition) layer is sufficient to produce a larger average electric field in the layer to "drop" the large reverse bias.

The voltage *across* the diode is the *full* reverse voltage V_m applied plus V_o (the electrostatic potential barrier voltage) as shown in Fig. 9.37.

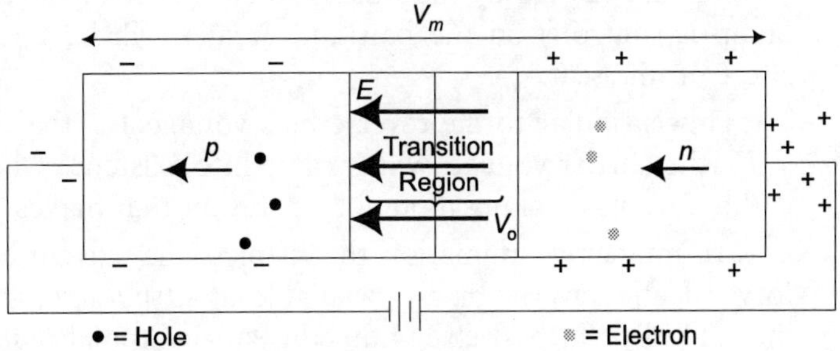

Fig. 9.37 The voltage across the diode is $V_m + V_o$.

The reverse biased diode causes an increase in the width of the transition region as shown in Fig. 9.38.

The electric field in the transition region due to the exposed immobile charges, is not uniform and is a maximum at the junction progressively decreasing toward the bulk of the neutral *p*- and *n*-regions and becomes very small within the bulk of the *p*- and *n*-regions (with large numbers '*n*' of mobile carriers) as seen in Fig. 9.38 and is just sufficient to ensure that the current given by $neA\mu E$ (expression of electron current in Section 1.20 is multiplied by the electron charge '*e*' to obtain amount of charge passing a point per second or current in amperes) to be same throughout the diode.

To intuitively understand the operation of a diode during the negative half cycle, we consider the operation of a check valve that we described in Section 9.9, with the water flow reversed.

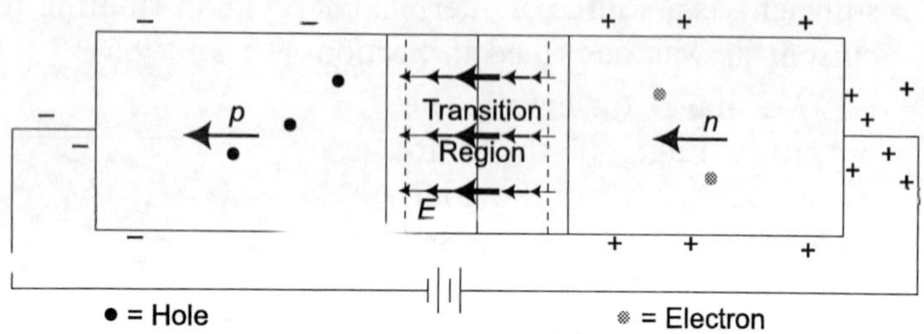

Fig. 9.38 Increase in the width of transition region.

Such a situation will result, after a pump raises water to an overhead tank from a ground level reservoir and the pump is shut off. The operation of the valve is shown in the figure below.

The water in the pipes from the overhead tank will exert a downward pressure on the check valve after the pump is shut off. This causes the ball to exert a pressure and the spring to stretch, aided by the head due to the column of water in the pipes running upto the overhead tank located a few tens of metres from the ground, thereby closing the orifice in the seat completely.

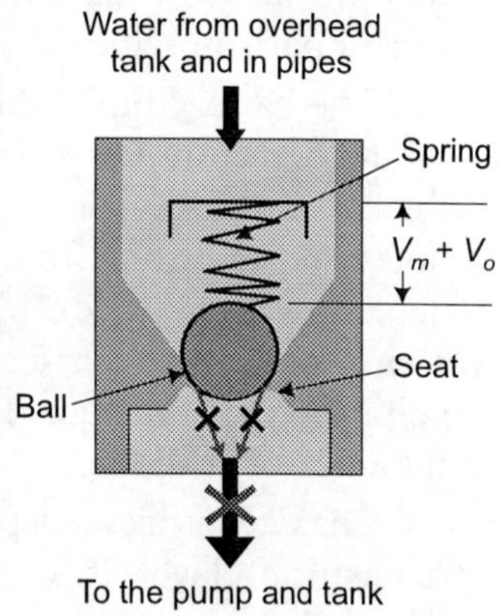

This prevents the water, all that was pumped into the overhead tank (!), from draining back into the ground level reservoir. It is in this mode that the check valve is most useful—preventing water flow in the reverse direction.

If the water was pumped to a great height, greater will be the pressure exerted by the water present in the pipes (after the pump is shut off) on the ball and greater will be the sealing effect on the seat.

This is similar to the reverse bias voltage V_m (that is applied to reverse bias the diode) + V_o (the barrier voltage, which came into existence when the junction was formed). And it is this current blocking mode of operation that makes the diode a useful device in circuits.

An interactive animation that depicts the current components of a *p-n* junction under forward and reverse bias is available at http://jas.eng.buffalo.edu/education/pn/biasedPN/index.html or http://web.eng.fiu.edu/~npala/Applets/education/pn/biasedPN/index.html.

I urge the reader to visit http://jas.eng.buffalo.edu/education/pn/biasedPN2/BiasedPN2.html or http://web.eng.fiu.edu/~npala/Applets/education/pn/biasedPN2/BiasedPN2.html and view the variation of the electric field with distance of a *p-n* junction under reverse bias.

After the reverse bias is applied, the diode is in thermal equilibrium after a small duration.

Remember that there is continuous production and annihilation of electron hole pairs when a semiconductor is in thermal equilibrium (dynamic) and net p_{no} and n_{po}, thermal equilibrium concentrations of holes in the n-side and electrons in the p-side regions respectively, exist.

The electron hole pairs generated in thermal equilibrium at the *edges of the transition region* move toward the diode terminals urged by the electric fields to recombine there…with charges from the voltage source as shown in Fig. 9.39.

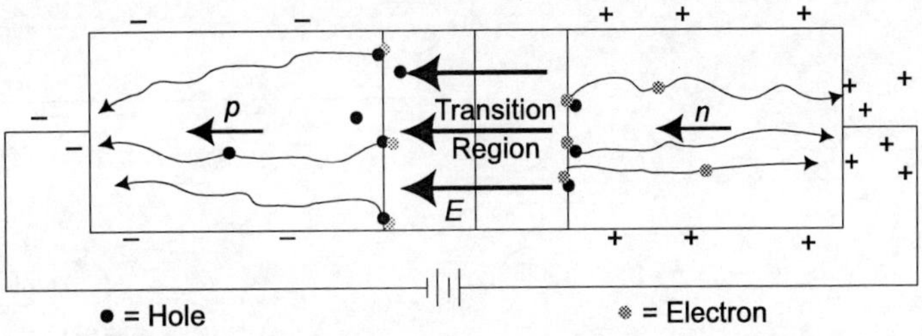

Fig. 9.39 EHPs migrate toward the diode terminals

This is what we call the reverse saturation current I_o.

The reverse saturation drift current I_o is relatively insensitive to the height of the potential barrier as seen in the negative voltage portion of the Diode characteristic (compressed scale) of Fig. 9.41. We normally think in terms of material with ample carriers, and therefore we expect the reverse drift current to be proportional to the electric field. The reason for this anomaly is the fact that the drift current is limited not by *how fast* carriers are swept down the barrier, but rather *how often*. See the last paragraph in Section 1.23 on current is the *number* flow rate of electrons.

9.12 DIODE – PEAK INVERSE VOLTAGE (PIV)

The diode chosen for half wave rectification duty should be capable of withstanding the *peak negative voltage* of the sinewave voltage input as shown in Fig. 9.40.

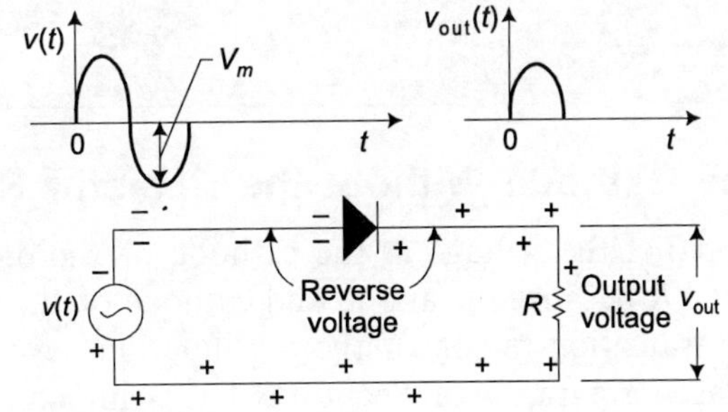

Fig. 9.40 In the reverse cycle, the maximum voltage across the diode is V_m.

9.13 PEAK INVERSE VOLTAGE (PIV) RATINGS OF DIODES

Diodes are chosen such that their peak inverse voltage (PIV) is ***greater*** than the peak negative voltage V_m of the applied sinewave, else the diode will breakdown and may short-circuit. The diode characteristic is shown indicating the PIV voltage in Fig. 9.41.

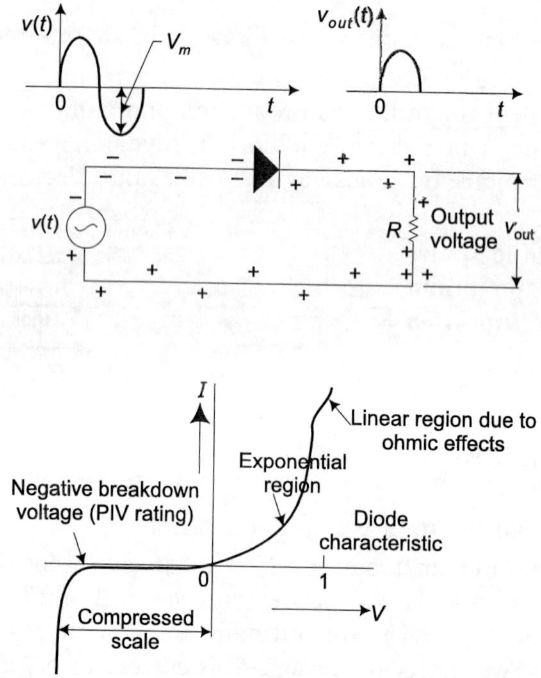

Fig. 9.41 The diode PIV rating should be greater than V_m.

The table shows the part numbers, PIV ratings, reverse saturation currents and applications of a few commercially available diodes.

Diode part no.	*Peak inverse voltage (PIV)*	*I_o (Reverse saturation current)*	*Application*
1N4002	400 V (can be used with 230 VAC)	30 μA	dc power supply
1N4007	1000 V (can be used with 230 VAC)	30 μA	dc power supply
1N3605	40 V	50 nA	Small signal rectifying
1N914	110 V	20 nA	Small signal rectifying

The Voltage at the Cathode Without the Resistor R

It is not possible to predict the voltage at the cathode of the diode in the circuit of Fig. 9.41 without the resistor R because the anode and cathode of the diode are separated by an insulator, which is the transition region of the *p-n* junction devoid of mobile charges. For reasons given in the last paragraphs of Section 3.4 it is impossible to predict the voltage at the cathode. Also see Fig. 3.23(d).

If net charge were to be present on the circuit and the voltage source, then there is still no manner by which the voltage of the cathode can be predicted because the transition region which is a region with no mobile charges (an insulator) will not provide charges to even out the potential everywhere. Also see why it is difficult to characterize the no-load output voltage (without a filter capacitor) of a half-wave rectifier circuit in "Dielectric in a uniform electric field", Appendix C.

Electronic engineers and technicians call the static electric charges on components and electronic equipment simply "static". The potential developed by these excess charges can range in the kilovolts but do not have the energy to be fatal to personnel handling electronic equipment. Technicians take special care when soldering electronic components on printed circuit boards and when assembling them into equipment to ensure that all static (excess) charge is removed by "grounding" (Section 2.25) the tools they handle including their hands. This brings the potential due to excess charges to more tolerable levels of tens of volts or less.

The Diode as a Protection Device for a Transistor that Switches a Relay

The relay shown in the sketch consists of a coil which forms a part of an electromagnet that attracts a metal piece called the armature, which is pivoted. The armature carries one contact, the moving contact called the common or COM. The other contacts are stationary on the relay structure, and these are the normally open (NO) contact and the normally closed (NC) contact.

When the relay coil is energized, the moving contact closes with the stationary NO contact, and may be used in circuits to switch ON large loads like fan motors and industrial valve actuators. Line voltage from a power supply is connected to the COM contact and the load connected to the NO contact. The second terminal at the load is returned to the neutral wire.

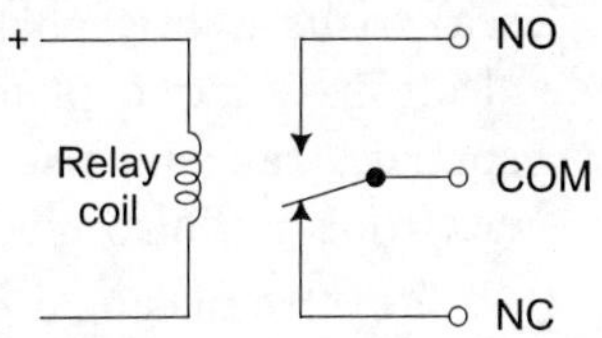

The contacts open when the coil is de-energized. In electronic control circuits, the relay is usually switched ON and OFF using an electronic device such as a transistor. See the figure.

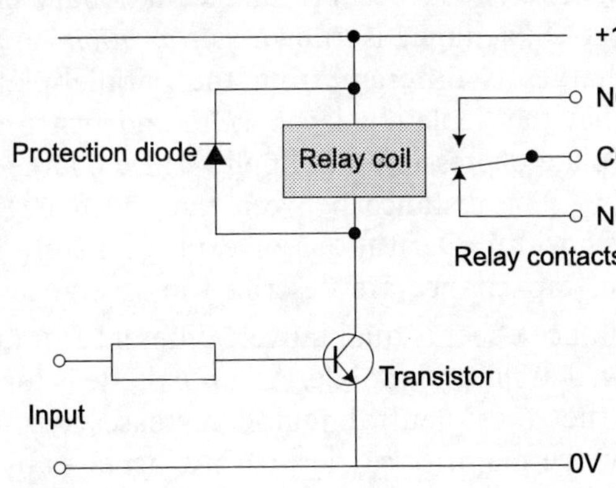

The transistor may be turned ON by applying a suitable positive voltage at its input terminals and may be turned OFF by applying zero volt at the input.

At the moment when the relay coil is deenergized by applying a zero volt input to the transistor, the attempt being made to stop current flow through the coil which is an inductor, $d\phi/dt$ or the rate of change of flux in the relay coil is high (see Section 5.12).

This results in the production of a large induced emf across it that could destroy the transistor with which it is connected. Transistors are usually not capable of handling voltages beyond a few hundred volts while the induced emf could exceed these values.

To protect sensitive transistors operating in this manner, a protection diode is connected as shown in the figure. The diode does not conduct when the transistor is ON because it is reverse biased, but conducts (forward-biased) whenever the relay coil is de-energized when the transistor is switched OFF. It *free-wheels* the current through the coil sustained by the induced emf allowing the magnetic field to die away quickly rather than instantaneously. So, the magnetic energy in the coil is dissipated in its resistance and the diode whose resistance and voltage when conducting is small. This prevents a sudden rise in the voltage that could damage the transistor.

9.14 TRANSITION AND DIFFUSION CAPACITANCES IN DIODES

It will not be proper not to include in this chapter a reference to the capacitive effects in diodes. Capacitance effects are not solely associated with the classic lumped capacitor elements such as the disk type parallel-plate capacitor as was described in Chapter 2. These capacitors when charged cause separated point charges (shifting electrons in the electron sea) to be tightly packed and remain stationary on the surfaces of its conductive plates, electrons on one plate surface and lattice ions on the other. In *p-n* junctions, different spatial arrangements of unlike charges such as positive and negative lattice ions, holes and electrons can also give rise to a capacitive effect.

Reverse biasing a *p-n* junction causes majority carriers to move away from the junction, thereby uncovering more immobile charges or lattice ions. This uncovering of unlike charges on either side as the thickness of the space-charge (or transition) region *increases* with reverse voltage is a capacitive effect, because unlike stationary charges accumulate over a distance with voltage. The capacitance is known as *transition* capacitance. The distribution pattern of separated charges is different from the parallel-plate capacitor (see Section 9.4). The difference is that the depletion layer *width* and hence the capacitance is *voltage dependent*. The parallel plate expression still applies since charge is only *added* at the *edge* of the depletion regions.[18] The distance between the added negative and positive charge equals the depletion layer width. Derivations of expressions of the non-uniform net field, its pattern and transition capacitance are described in several textbooks.[17]

The diffusion capacitance effect is qualitatively different from the transition capacitance in that the charges involved are not immobile. As a *p-n* diode is forward biased, the minority carrier distribution in the quasi-neutral region increases dramatically. In addition, to preserve quasi-neutrality, the majority carrier density increases by the same amount due to carriers supplied from the contact terminals. This effect leads to an additional capacitance called the diffusion capacitance. Unlike a parallel plate capacitor, the positive and negative charge is not separated, as if on plates. Instead, the electrons and holes are separated by the energy *bandgap everywhere in* the crystal upto a certain distance on either side of the junction. Nevertheless, these voltage dependent charges yield a capacitance just as the one associated with a parallel plate capacitor.

It should suffice here to state that the electrons and holes that wait to recombine give rise to a resultant electric field around them giving rise to the diffusion capacitance effect. There is ever present a voltage dependent store of charges waiting to recombine, which are all the elements giving rise to a capacitance; holes and electrons separated by a distance. The recombination current is due to a real *flow* of charge between separated charges that comprise the *diffusion* capacitance. This is one more difference between the diffusion capacitance and the capacitance of a parallel-plate capacitor; there is no *physical* current between the plates of the latter.

When a junction is forward-biased, the injected minority carriers (holes and electrons) on either side of the junction must recombine with their opposite types to maintain current conservation. Recombination is a very orderly event governed by energy and momentum conservation principles. It requires several conditions to be fulfilled by both the carrier types before the energy give and take, takes place, when it occurs, and there is no margin for error. A description of the conditions is beyond the scope of this textbook. For details the reader is referred to "Solid State Electronic Devices" by Ben G. Streetman and Sanjay Kumar Banerjee.[30]

The derivation of expressions for diffusion capacitance is available in several textbooks.[17] The derivation of an expression for the electric field, however, is difficult, because the unlike charges involved are neither on the surface as in parallel-plate capacitors nor arranged spatially like the immobile charges in the space-charge region, but are *within* the lattice, the bandgap separating them.

When a diode is driven from the reversed condition to a forward state or in the opposite direction, the transition and diffusion capacitances cause the diode response to be accompanied by transients and an interval of time elapses before the diode recovers its steady state.[17]

The transition and diffusion capacitances of a *p-n* junction diode are the important parameters that limit its application in high-frequency switching circuits. With the increasing availability of computers, networks, communication and wireless communication networks fueled by the growth of the internet, speed is of the essence in carrying data, messages and information to the growing multitude of users. The need for faster speeds in communication networks is what makes device designers and manufacturers to look at newer materials for device fabrication in smaller dimensions to minimize capacitances.

It would be instructive to read Section 3.8 on how stray capacitances limit the operation of transistors at high frequencies. The processes are similar to those that make transition and diffusion capacitances also limit the operation of transistors at high frequencies. This can then be followed by reading a more detailed treatment on transition and diffusion capacitances in Refs. [17 and 34].

9.15 OHMIC CONTACTS AND WHY V_0 CANNOT BE MEASURED BY A VOLTMETER

If a depletion region must be avoided at the metal wire to *p* and *n* material interfaces, ohmic contacts should be formed. A review of how the Fermi Levels equilibriate at the junction of *n*-type and *p*-type semiconductors in Section 9.3 resulting in the formation of a transition region in Section 9.4 indicates that when two dissimilar materials are joined, metal and semiconductor, there is a formation of a depletion region possible which would rectify the applied signal voltages to a diode. Therefore, there is a need to make ohmic or nonrectifying contacts between the metallic leads and the semiconductor materials which would have a linear *IV* characteristic with minimal resistance and no tendency to rectify signals.

When metal contacts are made to the *p* and *n* materials of a *p-n* junction diode, the Fermi levels of the metal and a semiconductor must align. Before we discuss the manner of their alignment, we need to understand Fermi levels in metals.

We begin by studying a graph of electrostatic potential-energy in a metal.

Consider a single isolated nucleus of a metal atom consisting of protons and neutrons. Assume that V is the potential at a distance r from the nucleus. Since it is only differences of potential that have any physical significance, it is convenient to choose zero potential at infinity, then the potential energy at any point, of the system comprising the nucleus and the electron, is negative as shown in the 2 dimensional plot for a single atomic nucleus in Fig. 9.42(a).

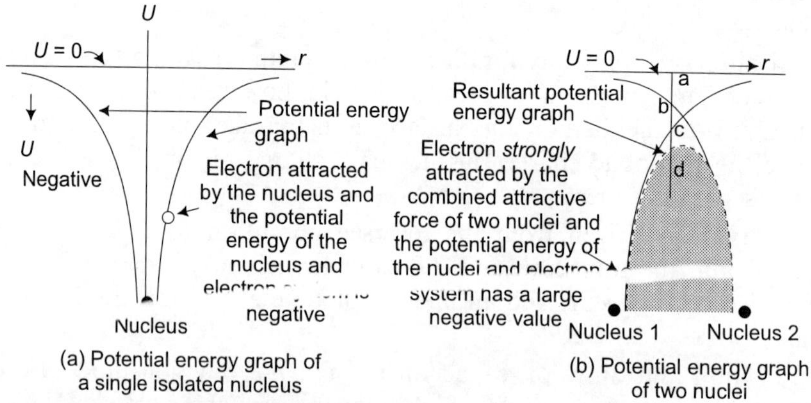

Fig. 9.42 Potential energy graphs of nuclei

Since the potential V equals the potential energy U per unit charge, then $U = -qV$ where q is the charge of an electron. The minus sign is introduced since q represents the magnitude of the (negative) electronic charge. In the figure, the vertical scale represents U, and the horizontal scale gives the distance r from the nucleus. The electron is attracted to the nucleus as shown.

With the zero potential chosen at infinity, then the potential energy at any point in space is negative and represents confinement (see also Section 9.7) of an electron, due to electric attraction by the nucleus. Because the fields due to the protons and electrons always exist, there is a resultant electric field in the lattice but is not shown due to its complexity, yet it definitely exists. As a matter of fact, the potential energy graph is due to the electric field which permeates the space around the protons and the electrons. If there were no electric fields, the potential energy graph cannot be drawn.

When two atoms combine, different types of bonding can occur: covalent, ionic, and metallic. Silver, iron, platinum, gold, and copper all form metallic bonds. Unlike covalent bonding, metallic bonding is non-directional. The strong bond consists of positively charged metal atoms in fixed positions, surrounded by delocalized or free electrons. These free electrons are often referred to as "a sea of electrons". In Fig. 9.42(b) is shown the potential energy graph (dashed curve) at every point in space (two-dimensional view) of the resultant U due to two nuclei as if these were the only nuclei present in the metal, constructed by summing the potential energy of individual nuclei (solid curve).

There is a stronger negative U at points around the midway point between the nuclei (since $ad = ab + ac$) in comparison with that of the single nucleus at the same distance (Fig. 9.42(a)).

If an entire row of nuclei is singled out from the metallic lattice, then following the same type of construction as above, but considering the small influence of other nearby nuclei, an energy distribution somewhat as illustrated in Fig. 9.43 is obtained.

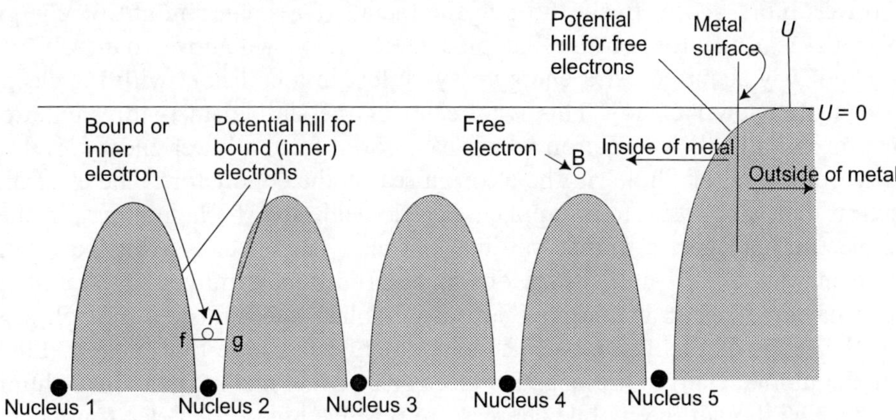

Fig. 9.43 Potential energy graph of a row of nuclei from a metallic lattice

At the surface of the metal (the exact position of the "surface" cannot be defined) located at a small distance from the last nucleus, Nucleus 5 in the row, since there are no nuclei to the right of Nucleus 5, there can be no lowering and flattening of the potential energy curve such as prevails in the region between the nuclei. Note that a potential hill or barrier exists at the surface of the metal.

Note the hill at the surface that requires an electron to gain sufficient energy from some external source to overcome. An *inner* electron possessing a total energy corresponding to the level A, collides and rebounds from the potential hills at points f and g that are present in the interior of an atom as well. The "inner" electron, being attracted strongly by powerful electric fields to the nuclei, lacks the *huge* amount of positive energy required to penetrate the hill around the level A, and move away from the nucleus being of the order of hundreds and thousands of electron volts (See "Inner electrons and the binding electrons of electrons" in Appendix B), and not the 1/100 of an electron volt which is the order of the average thermal energy required for valence electrons. This inner electron, at level A, is therefore a "bound" electron. These electrons in the solid are bound to the atomic cores, and electrons in these core states are not mobile. In electrical and electronic engineering, the electron at level A is not of interest because it is least affected by external applied fields of moderate intensity.

Important: The Outermost Electrons of Large Numbers of Metallic Atoms

It is the electron which possesses potential energy at level B that is most important for engineers, because these are the free electrons that are sufficiently far away from the nuclei and roam around freely in the lattice of the metal, if the energy band is not full. Note that the millions of electrons in level B occupy finely split energy levels, called a band. And, these bands maybe partially filled. At no point is the free electon's energy entirely potential, since it lies in the conduction band and its velocity is never zero.

It travels more or less freely through the metal. If a small amount of energy is added to electrons near the top of the occupied levels, they will move to nearby unoccupied levels. The newly acquired extra energy may be lost in a collision, with the electron falling back down to a lower energy. This is an echo of the simple start-stop model of electron conduction in a metal. An electron which has moved from a lower energy level to a higher energy leaves a kind of "hole" in the electron sea of the conductor ! The electron moves in a direction that is opposite to an applied electric field, and the hole moves in the direction of the electric field, like a mobile positive particle. The effect is a bit like the motion of a bubble in water: the water moves down, but the bubble moves up. A full description of the behavior of electrons requires advanced solid state and quantum physics which is beyond the scope of this textbook. The experimental facts are that most metals behave as though the mobile charges are negative electrons. A few metals, including aluminium and zinc, behave as though the mobile charges are positive holes![31] An electron when it reaches the surface of the metal, collides with the potential energy barrier there, since it lacks the energy to penetrate the barrier and move farther away from the metallic ions (comprising the nucleus and the inner or "bound" electrons).

If it were to move out, surely, it will not go unnoticed and you would be shocked if you were to touch the metal body even it were isolated ! When it encounters the barrier, its kinetic energy is reduced to zero and it is turned back into the body of the metal. So, the electron at level B is "free" to move around as long as it is within the body of the metal.

The resultant electric field due to millions of nuclei and millions of bound and free electrons is fairly complex and is not shown in the figure. At distances close to the nuclei and the bound (inner) electrons, the potential varies appreciably, but this is a very small portion of the volume of the metal. See "The Average Field Inside Matter", Section 2.26 and "Inner electrons and the binding energies of electrons" in Appendix B.

Figure 9.43 is redrawn in Fig. 9.44(a) showing the plateau (a plateau is an area of fairly level high ground) of the slowly varying potential far away from the nuclei around the potential hills; the area near and above the plateau is a fairly large percentage of the overall volume of the metal and is an equipotential region.

It is sufficient for the purpose of this discussion, to be aware that the field distribution within the fairly large volume of the metal near and above the plateau is equipotential (resultant electric field is nearly zero) which gives the metal its neutrality despite the millions of nuclei and electrons that make it up. The free electrons contribute to maintain the region equipotential (Section 2.26, "The average field inside matter") by moving around into regions everywhere above the plateau, where the potential varies due to fluctuations maybe in temperature or whatever.

When given additional energy from an external (applied) electric field, electrons at level *B* and above can drift in a preferred direction and produce currents (Section 1.17).

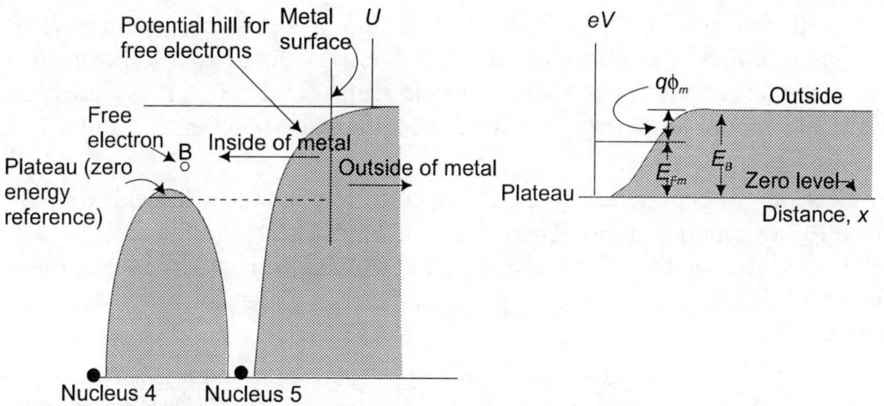

(a) Potential energy curve of the last nucleus near the surface (b) Potential energy barrier at the surface

Fig. 9.44 For a free electron there is a potential energy barrier at the surface

The inner electrons of an atom, there are 28 inner electrons in copper (Section 1.13), are so tightly bound (their energies are of the order of hundreds of electronvolts) that they are of little use in electrical conduction. This also means that the free electrons have a very small probability for occupying any of the inner orbital spaces, the resultant attractive forces on the free electrons having been reduced due to the resultant weak electric field, which is the combination of the fields due to the protons(29) and inner electron system(28). The vertical scale of Fig. 9.44(a) is shown compressed for the inner bound electron energies that range in keV. Therefore, turning our attention to the free electrons, and assuming that they are mostly in the equipotential region above the plateau, Fig. 9.44(b) is drawn without having to carry a large potential energy variation in the keV of the inner electrons. Since, now we need account merely the tinier energy variations of the free electrons only, which will be at most a few tenths of eV, we fix the zero of the energy at the level of the plateau of Fig. 9.44(a) as shown in Fig. 9.44(b), and drop the $U = 0$ level that was shown in Fig. 9.43. Energy band diagrams in most electronics and electrical engineering textbooks are drawn as shown in Fig. 9.44(b). The energies are shown in electron volts (eV) and we will now discuss the different energy levels shown.

Fermi Energy in Metals

If relatively few electrons reside in an otherwise empty conduction band, ample unoccupied states are available into which the electrons can move. To know about the amazingly large number of available states in an energy band spanning 1 eV, refer to "Density of States" in Appendix B. The Fermi energy is the term used to describe the top of the collection of electron energy levels at absolute zero temperature. So, at absolute zero, the free electrons pack into the lowest available energy states and build up a "Fermi sea" of electron energy states.

At 0 K, the top of the free electron energy distribution of a metal is defined as the Fermi energy, E_{Fm} measured from the bottom of the sea of free electrons (not "inner" electrons) and is shown in Fig. 9.44(b). At 0 K, it is not possible for an electron to escape from the metal unless it is given an amount of energy equal to E_B shown in Fig. 9.44(b) and the maximum energy possessed by an electron is only E_{Fm}. It is necessary to supply an additional amount of energy equal to the difference between E_B and E_{Fm} in order to make this escape possible.

For electrons to experience acceleration in an applied electric field, they must be able to move into new energy states. There must be empty states (allowed energy states which are not already occupied by electrons) available to the electrons. In metals, the valence and the conduction bands either are only partially filled as shown in Fig. 9.45(a) or overlap as shown in Fig. 9.46.

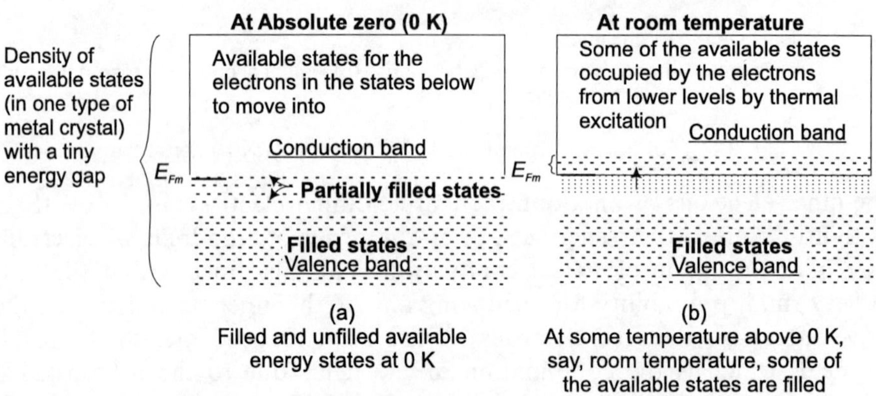

Fig. 9.45 Metal with partially filled valence and conduction bands (a) at absolute zero (0 K) (b) at room temperature

Note that all potential energy levels of the electrons are negative (Section 9.7), but in the case of metals and semiconductors, for discussions of the energy levels of the electrons in the valence and conduction bands and not the "inner" electrons (Appendix B), the zero energy level is conveniently taken to be the bottom of the sea of electrons.

The Fermi energy E_{Fm}, for a metal, is the surface of the electron sea as shown in Fig. 9.45(a) where, at absolute zero, no electrons will have enough energy to rise above the surface. One of the remarkable things about the Fermi energy (we remind ourselves that the Fermi energy is defined only at 0 K) is how large it is compared to the energies which electrons could gain by ordinary physical interactions with their environment. The amount of energy available as a result of the temperature of the material is on the order of the average thermal energy, for which kT= 0.026 eV at 300 K (say, room temperature) is a representative number. This is very small compared to the Fermi energy E_{Fm} of 7 eV for copper.

Note: In the model used to calculate the Fermi energy, it is assumed that there is no interaction among electrons so the potential energy of each of the electrons is zero and that all their energy is kinetic. Therefore, the Fermi energy is positive. See http://web.archive.org/web/20150421013249/http://people.duke.edu/~ad159/files/p112/27.pdf andhttp://web.archive.org/web/20040824021212/http://www.ph.utexas.edu/~phy317n/PHY317L/Lectures/Lecture20/L20examples.pdf.

This tells us that thermal energy can interact with *only a tiny fraction* of the electrons (roughly .026/7 or about 0.4% of the energy range), since the overwhelming majority of the electrons are separated from the top of the Fermi sea by much more than thermal energy.

Visit http://web.eng.fiu.edu/npala/Applets/education/semicon/fermi/functionAndStates/functionAndStates.html or http://jas.eng.buffalo.edu/education/semicon/fermi/functionAndStates/functionAndStates.html and observe this by varying only the temperature beginning 0 K to around 300 K of the material. Observe how the Fermi function changes shape in the left figure. Also observe the electrons (.) occupying or vacating quantum states (__).

Note that while the Fermi level (Fermi Energy E_{Fm} for a metal) remains at the level at 0 K, there are more electrons occupying higher energy states supplied from the partially filled conduction band shown by the arrow in Fig. 9.45(b). The Fermi level (Fermi Energy E_{Fm} for a metal) varies from substance to substance but is constant with temperature for a given substance. In the applet above, one can shift the Fermi level (Fermi Energy E_{Fm} for a metal) independently to that corresponding to different substances.

It should be noted that the Fermi energy E_{Fm} is not an "average" of the distribution but is governed by the Fermi-Dirac statistics that takes into account the complex bottom heavy nature of the electron energy state occupation distribution. See more in *Introduction to Solid State Physics* by Charles Kittel, 7th edition, Wiley India Pvt Ltd.

Metals with Partially Filled States have High Conductivity

Metals in which the energy states are only partially filled have a high electrical conductivity, since the empty levels (though only a tiny fraction of the Fermi energy itself) provide *head room* for electrons to gain energy and move about freely and return to the lower level on losing energy after a collision. This is an echo of the simple start-stop model of electron conduction in a metal.[31] Visit http://hyperphysics.phy-astr.gsu.edu/hbase/solids/fermi2.html for a discussion of Fermi energies for metals and see Appendix B "Density of States".

In a metal with overlapping valence and conduction bands as shown in Fig.9.46, above the Fermi level, energy levels are empty (at absolute zero), and can accept excited electrons elevated in energy by an applied electric field. Such metals possess, not surprisingly, a high mobility 'μ'.

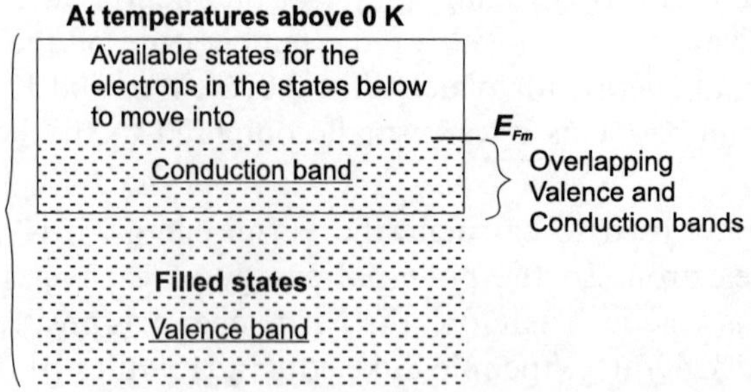

Fig. 9.46 A metal with overlapping valence and conduction bands

Certain semiconductor materials, such as Gallium Arsenide (GaAs) have satellite valleys in the conduction band that are above the central valley (in the conduction band), and the two are separated by about 0.3 eV. When a powerful electric field is applied to a sample of the GaAs material, conduction electrons are transferred into the satellite valley from the central valley, and since the satellite valley has a much larger (available) density of states than the central valley, the electrons scatter to occupy these states.[30] When the applied field is increased, the electrons scatter less into the lower central valley and remain in the satellite valley because of the high density of available states there. The effect is called the *transferred-electron* mechanism or the Gunn effect. In the satellite valley, their mobility is much lower and their drift velocity does not increase with an increase in applied field, the electrons seemingly preferring to scatter between the closely spaced available states (large density of states) and is a cause for a negative resistance region with applied field above a critical field.

This peculiar behavior of the GaAs semiconductor is utilized in making oscillators in the microwave frequency range. A full treatment of this effect is available in Ref. [30].

Substances with Completely Filled Bands make Poor Conductors

The diamond structure is such that the valence band is completely filled with electrons at 0 K and the conduction band is empty as shown in Fig. 9.47. The band gap energy is typically 5 eV or more. It is not possible to move an electron to a higher energy state without supplying a very large amount of energy. At ordinary temperatures, such a material is a very good insulator and a very poor conductor of electric current. There can be no charge transport within the valence band, since no empty states are available into which electrons can move.[31]

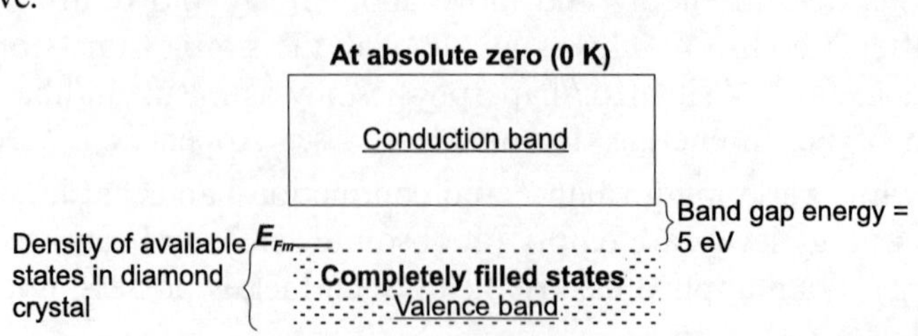

Fig. 9.47 Energy band structure of a diamond crystal (valence and conduction bands)

In a manner of speaking they are boxed in so within the valence band that there is no *head room* for movement to a higher level, being blocked by the band gap (where no electron can occupy any energy level) to move into the conduction band.

Fermi Level in Semiconductors

Since the Fermi level is a representative energy mathematically arrived at, of a distribution of energies in a population of electrons of atomic nuclei, it may not necessarily be an energy in which one should always expect an electron to be present. For example, the Fermi level of a semiconductor made from pure silicon (intrinsic) will be in the middle of the band gap. This is like finding sometimes that the average of marks awarded in a class test, is a fraction even when the possible total of marks a student can be awarded, is a whole number.

Figure 9.45(a) is the energy level diagram for metals or conductors to be more precise. The energy gap is seen to be tiny and if this width of the energy gap (also called the *forbidden energy region* because no mobile carriers can be found in this energy gap) between the valence and conduction band edges is of the order of 1 electronvolt (eV), it is called a semiconductor. Germanium, has an energy gap of 0.72 eV at 27°C and Silicon has an energy gap of 1.1 eV at 27°C.

For semiconductors, the conduction band edge is the reference for electrons. All electrons in the conduction band edge have the potential energy of the conduction band edge. An electron at the conduction band edge has a kinetic energy of zero. The higher it is in the conduction band edge, the larger is its kinetic energy, the density of states being so large that it hardly matters for the potential energy. For holes, the same reasoning holds. Holes at the valence band edge have a kinetic energy of zero. Those higher in the valence band just gain kinetic energy.

Fermi Energy of Metals and Fermi Level of Semiconductors

As was shown earlier in the Fermi (energy) level for metals, at zero Kelvin (0 K), there is no thermal energy and electrons fill the band from the bottom till the Fermi *energy* (7 eV for Copper, a *metal* which is not a semiconductor) which defines a sort of cutoff point beyond which electrons do not occupy any energy level.

The Fermi energy is typically taken to mean the *bottom* of the conduction band while, the Fermi level which is also an energy level, is more a measure of the concentration and not strictly the energy of the electrons. The Fermi *level* shifts with doping types and concentrations and also with temperature.

The reader is cautioned to be careful with the usage of the terms Fermi level and Fermi energy since some authors use these interchangeably leading to confusion in their interpretation.

The Work Function

Having learned in some detail how conduction band electrons gain energies from an applied field in both metals and semiconductors, and given the large number of available metals, making the right choice to prepare a metal to semiconductor contact ohmic requires some knowledge of the work function of the metal, a property of the metal which we will now describe.

The work function represents the minimum amount of energy that must be given to the fastest-moving electron at the absolute zero of temperature in order for this electron to be able to escape from the metal, because there is no probability of finding an occupied quantum state of energy greater than the Fermi energy level at the absolute zero of temperature.

The work function is denoted $q\phi_m$ for a metal where q is the magnitude of the electronic charge and ϕ_m the potential difference in volts between the Fermi Energy for metals (and the Fermi level in the case of semiconductors) and the vacuum level.

The work function $q\phi_m$ is shown in Fig. 9.44(b) and is given by $q\phi_m = E_B - E_{Fm}$.

The region outside the metal is now at a potential equal to E_B, the height of the potential-energy barrier in electron volts. As the temperature varies, the probability of occupation of energy states by the electrons also varies, but given a metal and its composition the Fermi Energy does not vary, even at high temperatures when only a few electrons will have large values of energy.

In the case of semiconductors, device designers prefer to work with the electron affinity '$q\chi$' (Fig. 9.19) rather than the work function, because the work function varies based on doping concentration while the bottom of the conduction band just inside the semiconductor E_C from which the electron affinity $q\chi$ is measured, does not.

Ohmic Contacts

The metal to *n*-type semiconductor contact

Recollect from Section 9.3 that when a *p-n* junction is formed, the Fermi levels must equilibriate and this was achieved by charges diffusing across the junction. Similarly, when a metal and a semiconductor are brought in contact, the Fermi level (or strictly speaking, electrochemical potential) of the two must be equal in thermal equilibrium, as shown in Fig. 9.48 where a metal is shown before and after a contact is made with an *n-type* semiconductor.

The metal and the semiconductor can have different work functions, denoted $q\phi_m$ and $q\phi_s$ respectively as shown in Fig. 9.48(a). The figures also show the electron affinity χ which is measured from the vacuum level to the bottom of the conduction band just *inside the semiconductor*.

After contact as shown in Fig. 9.48(b), electrons will flow from the material with the smaller work function (the metal) to the material with the larger work function and electron affinity (*n-type* semiconductor) until the Fermi levels equilibrate. As a result, the metal takes on a slight positive charge and the semiconductor will become slightly negative. The top semiconductor energy levels are occupied (the electrostatic potential is lowered, i.e., more negative because of the stronger negative quality of the atoms in the semiconductor as a result of the electron transfer, and the potential energy is raised, i.e., the potential energy is *less* negative). As long as the selected metal (carefully chosen) has a smaller work function than that of the *n*-type semiconductor, there will result an ohmic contact and a tiny contact potential.[30]

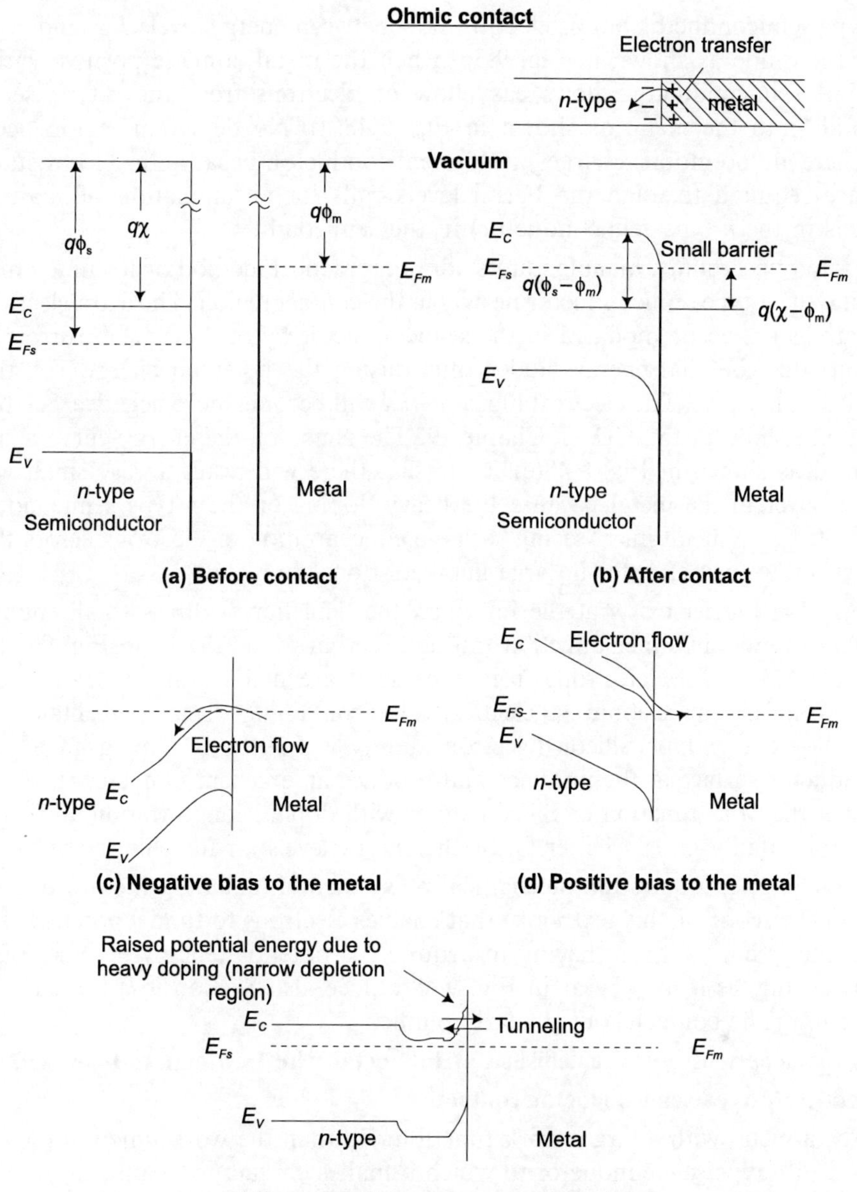

Fig. 9.48 Energy levels of a metal to *n*-type semiconductor contact

In this case, the barrier to electron flow between the metal and the semiconductor is small and easily overcome by a small forward bias to the diode as shown in Fig. 9.13 when the *n*-type semiconductor is positive with respect to the metal contact. As shown in Fig. 9.48(c) with the metal side made negative with respect to the *n*-type semiconductor the electrons from the metal flow easily into the n-type semiconductor. (**Note:** The reader is cautioned not to view the figures as if the electrons move from outside the metal into the *n*-type semiconductor but as electron flow between energy levels E_{Fm} and E_C). A reverse bias to the diode as shown in Fig. 9.36, when the metal is made positive with respect to the *n*-type semiconductor allows easy flow of electrons from the *n*-type semiconductor "downhill" into the metal as shown in Fig. 9.48(d). No depletion region occurs (region where there are no mobile charges) in the semiconductor, because the electrostatic potential difference required to align the Fermi levels calls for accumulation of majority carriers (electrons in the *n*-type semiconductor) in the semiconductor.

To avoid uncertainty, manufacturers adopt a practical method of forming ohmic contacts by doping the *n*-type semiconductor heavily in the contact region. The extra electrons from the donor atoms are accommodated in the semiconductor by pushing them further up (mutual repulsion) the potential energy ladder thus raising the potential energy, i.e., the potential energy is less negative. The electrostatic *potential* will become more negative. See the animated power point slides in the CD of Chapter 9. The shape of the energy curve at the junction is changed as shown in Fig. 9.48(e). Note that there will result a very small width of the depletion layer in the metal because the heavy doping of the *n*-type semiconductor would have "fooled" the metal into "seeing" a large concentration of electrons across the interface, thus preventing their transfer in large numbers from the metal.

Thus, if a barrier exists at the interface, the depletion width is small enough to allow electrons to tunnel in either direction through the barrier as shown in Fig. 9.48(e). See Fig. 9.45(b) also as to how electrons from the metal are made available. For example, Gold (whose Work function is about 5.2 electron volts) containing a small percentage of Antimony can be alloyed to *n*-type Silicon (electron affinity is 1.39 eV), forming an n+ layer at the semiconductor surface at the interface and produce an excellent ohmic contact. It should be noted that the work function of Silicon varies with doping concentration and type, therefore the electron affinity gives a better fix of the energy levels for matching with the metal.

"Tunneling" is a quantum mechanical phenomenon (detailing the process of tunneling is beyond the scope of this textbook) that enables electrons to tunnel potential barriers that are extremely thin[30] without having to acquire the potential energy necessary to overcome the barrier, tiny as it may be as in Fig. 9.48(c) (see also Section 9.7) for a forward biased *p*-*n* junction. The contact is very nearly ohmic.

How is the supply of holes replenished at the metal wire to *p* material contact?

The metal to *p*-type semiconductor contact

Similarly, a metal with a larger work function $q\phi_m$ than the work function $q\phi_s$ and electron affinity of a *p*-type semiconductor to which it makes contact will cause easy hole flow. This is seen in Fig. 9.49(a), which should be viewed keeping in mind that a hole flow corresponds to the opposite flow of electrons in the valence band (Section 9.7 "How holes surmount the potential barrier").

As long as the metal has a larger work function than the work function $q\phi_s$ and electron affinity of the *p*-type semiconductor, there will result an ohmic contact and a tiny contact potential.[30]

After contact as shown in Fig. 9.49(b), electrons will flow at the valence energy level from the material with the smaller work function (the *p*-type semiconductor) to the material with the larger work function (metal) until the Fermi levels equilibrate. As a result, the *p*-type semiconductor takes on a slight positive charge and the metal will become slightly negative, the excess electrons residing as a sheet at the interface in the metal. The potential energies in the *p*-type semiconductor are lowered to more negative and the electrostatic potential is raised, i.e., less negative because of the stronger positive quality of the atoms in the semiconductor as a result of the electron transfer to the metal. Because the electron transfer at the junction, is at the valence energy level, it may be seen as a transfer of holes from a metal such as, Aluminium to the *p*-type semiconductor.[30]

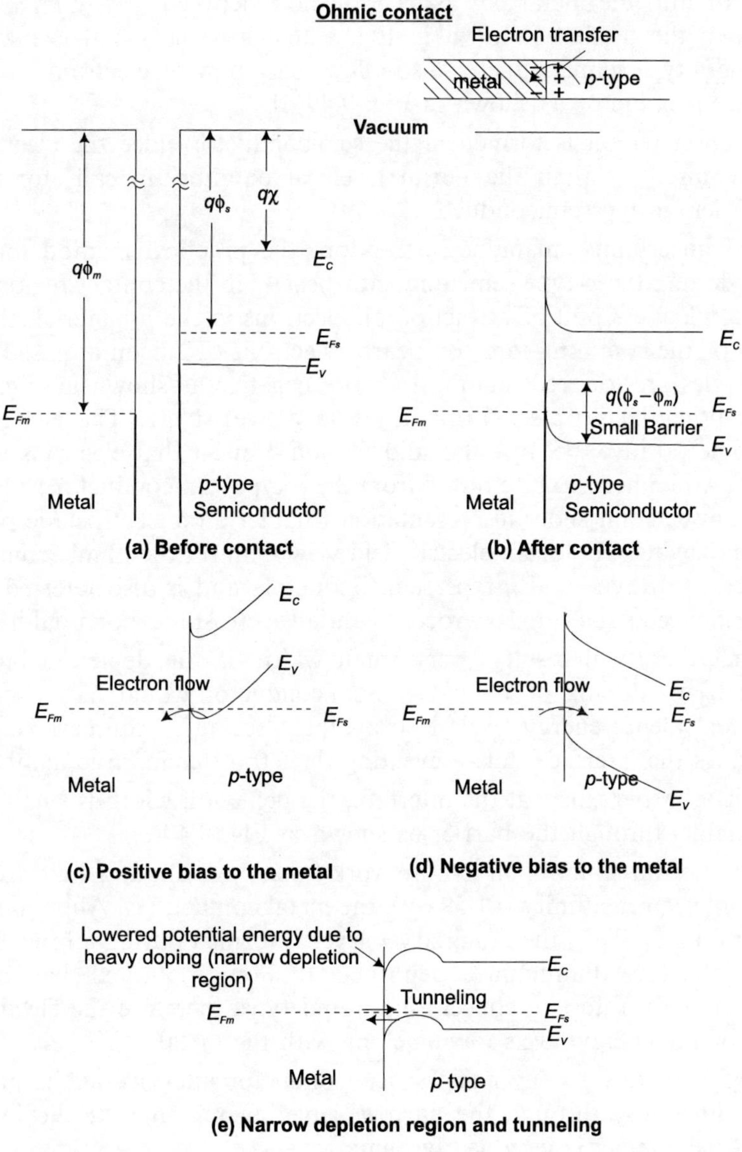

Fig. 9.49 Energy levels of a metal to *p*-type semiconductor contact

In this case, the barrier to electron flow between the metal and the semiconductor is small and easily overcome by a small forward bias to the diode as shown in Fig. 9.13 when the metal side of the contact is made positive with respect to the *p*-type semiconductor. As shown in Fig. 9.49(c) with the metal made positive with respect to the *p*-type semiconductor, electrons from the *p*-type semiconductor (valence energy levels) can flow rather easily "downhill" into the metal overcoming the potential barrier tiny as it may be. (Note: The reader is cautioned not to view the figures as if the electrons move from *outside* the *p*-type semiconductor into the metal but as electron flow between *energy levels* E_V and E_{Fm}). A reverse bias to the diode as shown in Fig. 9.36 when the metal is made negative with respect to the *p*-type semiconductor also allows easy flow of electrons from the metal into the *p*-type semiconductor as shown in Fig. 9.49(d).

No depletion region is formed in the semiconductor since the electrostatic potential difference required to align the Fermi levels at equilibrium calls for accumulation of majority carriers in the semiconductor.

To avoid uncertainty, manufacturers adopt the practical method for forming ohmic contacts by doping the *p*-type semiconductor heavily in the contact region. The additional holes exert a relentless pull of attraction on electrons in the *p*-material, thereby increasing the strength of the confining force on nearby electrons of silicon atoms. The p-side system electron energies are lowered tending to more negative as shown in Fig. 9.49(e) and the electrostatic potential is raised[30] (made less negative). **Note:** The electrostatic potential cannot become positive, because if it did, it would mean that electrons will no longer be confined and would need to be emitted from the p-type semiconductor ! The reader is urged to view the Power Point slides in presentation Chapter 9 on electrostatic potential included in the accompanying CD for an electric field viewpoint on the ohmic contact development for both metal to *p*-type and *n*-type semiconductors and is also referred to the article in http://academic.reed.edu/chemistry/roco/Potential/electrostatic_potential.html.

Note that there will result a very small width of the depletion layer in the metal because the heavy doping of the *p*-type semiconductor would have "fooled" the *p*-type semiconductor valence energy level electrons into "seeing" a comparable concentration of electrons across the interface thus preventing their transfer in large numbers to the metal.

Hence, if a barrier exists at the interface, the depletion width is small enough to allow carriers to tunnel through the barrier as shown in Fig. 9.49(e).

In the case of an Aluminium (whose work function is about 4.1 electron volts) wire on *p*-type silicon (electron affinity is 1.39 eV), the metal contact, i.e., Aluminium also provides the acceptor dopant. Thus the required *p*+ layer is formed during a brief heat treatment of the contact after the Aluminium is deposited. It should be noted that the work function of Silicon varies with doping concentration and type, therefore the electron affinity gives a better fix of the energy levels for matching with the metal.

Thus, by forming a *p*+ layer at the semiconductor interface at the junction, electrons can tunnel either way through the narrow depletion region near the junction (See Fig. 9.49(e)) and the contact is very nearly ohmic.

Measuring the Potential Barrier V_o

It is not possible to measure the potential barrier V_o with a voltmeter, and to understand why this is so, we need consider the effect of the contacts to the semiconductor material as shown in Fig. 9.50.

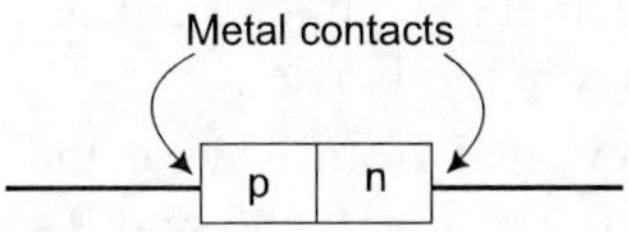

Fig. 9.50 A *p-n* junction diode showing the metal contacts

Having understood how a contact potential comes into existence at a metal to semiconductor contact, we are now in a position to understand why it is not possible to measure V_o using a voltmeter.

If an applied voltage V to a *p-n* junction were set to zero as shown in Fig. 9.51, the *p-n* junction would be short-circuited.

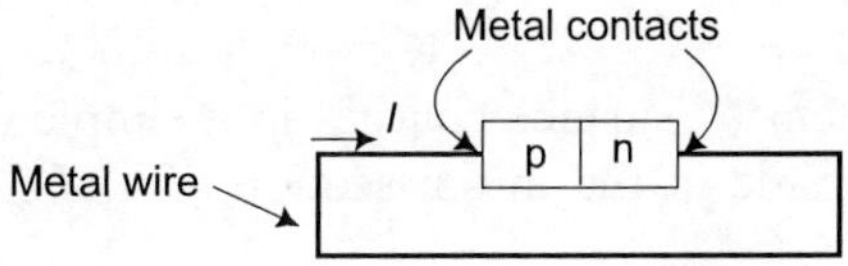

Fig. 9.51 A *p-n* junction diode short-circuited

No current can flow, because if there was a current, the metal wire would become heated and since there is no external source of energy available, the energy required to heat the metal wire and contacts would have to be supplied by the *p-n* bar. The semiconductor bar therefore, would have to cool off. Clearly, under thermal equilibrium (there can be no source of heat other than that of the surroundings with which thermal equilibrium of the diode is established), the simultaneous heating and cooling of the bar is impossible, and we conclude that the current $I = 0$.

Since under short-circuit conditions, the sum of the voltages around the closed loop must be zero, the junction potential V_o must be *exactly compensated* by the metal-to-semiconductor contact potentials at the metal to semiconductor ohmic contacts.

If in an attempt to measure V_o we connect a voltmeter (say, a moving-coil type) across the *p-n* junction, the voltmeter would register '0' volts as shown in Fig. 9.52.

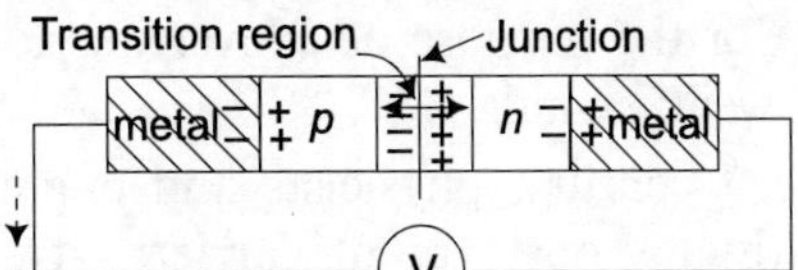

Fig. 9.52 Voltmeter connected across a diode to measure V_o will read '0'.

The circuit process is as follows: The voltmeter will register a non-zero voltage if a current I_{vm} flows in it in the direction indicated, or electron current I_{electron} in the opposite direction. At the metal to *p*-type semiconductor contact, the electrons are prevented by the small barrier (See Figs. 9.49b and 9.49e if the *p*-type is heavily doped) set up by the accumulated electrons on the metal contact surface. The voltmeter does not supply a forward biasing voltage and lower the barrier.

Also, the electrons are prevented from leaving the positive terminal of the voltmeter toward the metal by the barrier (See Fig. 9.48b and 9.48e if the *n*-type is heavily doped) at the metal to *n*-type semiconductor contact. Therefore, no current can flow in the voltmeter. Since the current is zero, the voltmeter will register '0' volts. Reversing the voltmeter connections will still result in zero current and the voltmeter will register '0' volts again. The wire can be cut without changing the situation, and the voltage drop across the cut must remain zero, due to junction potential compensation. In other words, it is not possible to measure the contact difference of potential of a *p-n* junction directly with a voltmeter.

SUMMARY

A review of the development of surface charges in a simple circuit consisting of a battery, wires and a resistor was made in the initial section of the chapter, and then with a diode replacing the battery.

A semiconductor is the basic building material of solid-state electronic devices, characterized by a gap in the energy band between the valence and conduction bands. The conduction band is unfilled if the temperature of the semiconductor is 0 Kelvin. When its temperature is raised, the bonds of the semiconductor atoms break, and hole-electron pairs are formed with electrons occupying the conduction band, and holes occupying the valence band. We learned that the hole-electron pairs simultaneously begin to recombine as well, yet, there remains a net concentration of hole-electron pairs at any instant of time.

Physicists and engineers recognized that the electrons could conduct electricity and so could the holes.

We learned that in copper, steel and other metallic wires, conduction is due to electrons and current flows in either direction. They were thus unsuitable to control current direction. Vacuum diodes afford a means of controlling current flow in one direction by a negative plate (cathode) which emits electrons that are collected by a positive plate (anode). Current flow in the reverse direction is prevented because the anode does not emit electrons.

Electronic circuits built for the purpose of converting ac to dc relied on vacuum diodes and triodes for amplifying weak signals.

In the middle of the last century, physicists and engineers developed semiconductor materials with predominantly any one type of carrier; *p*-type (hole) and *n*-type (electron). The dominant or majority current in the *n*-type is due to electrons while, in the *p*-type, the majority carrier is the hole.

When the two types are joined forming a metallurgical junction, a *p-n* junction is formed. We learned how holes and electrons on either side of the junction diffuse (a process it may be noted is due to a gradient or variation in concentration levels of a particular type of carrier in a semiconductor material) and recombine, at and around the junction, leaving a region around the junction which is devoid of mobile electrons and holes. This region is called the transition or depletion region, whose width is about 10^{-4} cm which corresponds to a few atomic layers.

The formation of the depletion region is related to the *pn* junction seeking an equilibrium energy level; a level that is uniform everywhere in the diode, such that there is no further migration of charge. This we understood with the example of how when two steel tumblers of water at different temperatures are brought in contact, the temperatures will settle to a uniform value in both the tumblers, when further exchange of energy stops.

In the case of the *p-n* junction however, the Fermi level which represents the characteristic energy of the system (noting that the electron energies are not uniform but spread in bands) will equilibrate and as a consequence, an electric field pointing from the *n-* side to the *p*-side across the junction develops, preventing further charge migration. Associated with the field is a potential barrier V_o.

When a diode is forward biased, surface charges develop and the electric field associated with V_o is reduced in magnitude, and the mobile charges in the semiconductor regions gain energy by the applied electric field. The process of how the electrons and holes gain both potential and kinetic energy was studied in some detail.

The carriers from the bulk regions cross the junction, and then recombine with charges of opposite polarity. Since the movement of holes constitutes a current in the *p*-side up to the point when it recombines with an electron, likewise, the movement of electrons constitutes a current in the *n*-side up to the point when it encounters a hole to recombine. We learned that there is a current everywhere within a forward biased diode either due to holes or due to electrons.

The hole current, which is essentially a movement of the electron at the valence level and not the conduction band level, of course, exists only in the *p*-material and not in the terminals or the *n*-side of the diode.

We digressed briefly to study the potential and kinetic energies of electrons in the conduction and valence band using potential energy vs. distance (radial distance from the nucleus) diagrams and noted that the spread of energies is over a few tens of electron volts with the energy band-gap between the conduction and valence bands of about an electron volt. Very importantly, we learned that the potential energy vs. distance (radial distance from the nucleus) diagram could be drawn because there is attraction between the positive nucleus and electrons occupying orbital spaces around them and because of the electric fields surrounding the electrons and protons. We also learned that the inner electrons (which are electrons other than those occupying the conduction and valence bands) are not shown in the graphs because they possess large energies (of the order of kilo electron volts for the innermost electrons), making them strongly attached to the nucleus. This makes it difficult to supply energy using moderate voltages applied in common electric and electronic circuits to make the inner electrons participate in conduction.

Then, we learned how electrons surmount the potential energy barrier noting carefully, the differences in the way holes gain energy (with detailed illustrations) from the manner in which electrons do, with detailed illustrations of the process of establishing a current when a diode is forward biased.

We then reached a very crucial aspect concerning diode operation; what happens when the forward bias voltage exceeds V_o?

We learned that the potential barrier V_o can never be made zero. The forward bias merely lowers the barrier potential which is typically around 0.7 volts for a silicon diode (it can vary and its value depends on the semiconductor material and the doping concentrations). With the help of a numerical as illustration, we learned that as the applied voltage is increased and approaches V_o, the diode current begins to rise very rapidly after a value called the cut-in voltage, is crossed. As the applied voltage is further increased, the current begins to rise rapidly. The junction resistance, flooded with mobile charges all around, drops. This happens while there is little or an insignificant drop in the bulk regions of the diode. However, when the applied voltage is quite close to the value of V_o, with small increments of it, the current grows extremely rapidly. At this point, the voltage drop across the bulk neutral regions begins to grow and across the junction, the drop never quite reaches the barrier voltage V_o.

We gained an intuitive insight of the process with the analogy of a water check valve used to control the flow of water from a pump to an overhead tank. The valve assembly consists of a cone with a hole at the vertex fitted snugly into a pipe. A ball is held above the hole (orifice) and exerts pressure by a spring attached (the other end of the spring is anchored on a perforated plate fixed a little distance above the cone) against the walls of the cone and the orifice so that the entry of water is blocked. If water flows, driven by a pump, through the orifice, the flow exerts a pressure on the ball which rises against the spring tension, thus opening out the blocked cone and water flows between the surface of the ball and along the walls of the cone, through the perforated plate and fills the pipe. The compression of the spring by the water pushing against the ball is like the lowered barrier potential. More the flow rate more is the compression of the spring. In a similar manner, more the applied voltage, the smaller the barrier. The spring does not give way and recovers when the water flow is stopped when the pump stops. Likewise, the potential barrier does not become zero with applied voltage and recovers when the applied voltage is removed.

One of the other important observations which we made concerning the controlling portion of a diode is that it is almost entirely controlled by the processes at the junction viz. the barrier potential (lowered or raised), the injection amounts and the recombination process. And, all these processes occur within the tiny width of the junction which is a mere 10^{-4} cm, where the overall diode length is of the order 1.3 to 1.8 cm.

In a check valve we observed that where the length of the control segment in the valve maybe a couple of centimeters, the length of the piping within which it controls the flow of water, could exceed a few tens of meters.

After studying how a diode conducts in the forward direction, we then focused on learning what happens to the current, when a diode is reverse biased. During a reverse bias, we learned, electrons migrate away from the junction and recombine with positive charges on the positive terminal of the supply, and holes migrate away from the junction to recombine with electrons on the negative terminal of the supply. The process is extremely rapid and exposes more immobile ions around the junction, and more importantly, we observed that the reverse bias voltage drops across the junction. Another important aspect we also learned was that there are still plenty of mobile charges available in the bulk neutral *p*- and *n*- side regions. The few additional immobile ions exposed around the junction are sufficient enough to drop the negative bias voltage; the electric field in the transition region is now stronger which corresponds to the large potential drop. There is hardly any current possible through the diode, because there is a slightly wider yet thin depletion region, a little more than 10^{-4} cm now with the reverse bias, containing no mobile charges. Yet, since hole-electron pairs are continuously being generated in the bulk neutral regions, a few wander near the edges of the junction and are swept by the electric field in the depletion region. The direction of conventional current is from the *n*-type to the *p*-type as we established of this tiny current which is called the reverse saturation current, I_o. It is tiny and therefore, we can say that in a forward bias, a diode conducts and in a reverse bias, it blocks current.

This property, we learned, enabled the use of diode as a rectifier which can convert ac voltages to constant or steady dc voltages. When a diode is used in a rectifying circuit, care should be taken to ensure that the applied reverse voltage does not exceed its Peak Inverse Voltage rating. The rating tells us that with this voltage applied, the junction breaks down due to the powerful field and the diode will short circuit and could lead to eventual burn out.

One other application of a diode is to limit the high voltage which is generated when the current of an inductor coil is abruptly terminated. This was illustrated with the example of a relay coil driven by a transistor. Without the free-wheeling diode connected across the inductor coil of the relay, the high voltage generated when the transistor current is switched off, could damage the transistor, by breaking down its collector base junction. The diode conducts from the instant of transistor switch OFF and dissipates the stored magnetic energy of the coil by effectively shorting the coil through its low forward voltage.

We had learned that a depletion region could be formed between a *p*-type and *n*-type semiconductor. A depletion region therefore, can form between the metallic leads attached to the diode's *p*-type and *n*-type materials. There is an exchange of energy when contact is made, and a depletion layer is formed and is associated with a barrier potential that causes rectification at the diode leads. In low power circuits this is quite troublesome and we learned that manufacturers prepare ohmic contacts or contacts that do not rectify ac voltages or signals.

To study the fabrication and operation of ohmic contacts required the study of metal to *p*-type and metal to *n*-type junctions.

Metals, we learned are giant structures of atoms held in place by the delocalized electrons in what is known as non-directional metallic bonds unlike the covalent bonds that bind semiconductor atoms.

To enable making the right choice of metals given a material and type of semiconductor requires that the energy levels of both are matched. This required knowledge of the work function of metals which we proceeded to examine in some detail beginning with a description of the potential energy graphs of nuclei in a metallic crystal.

From the graph we learned that there is a potential energy barrier near the surface just outside the last nucleus. An inner electron, because of its strong attraction to the nuclei lacks the energy necessary to move into the valence band and remains therefore, bound. On the other hand, an electron in the conduction band being in a region of lower potential roams around freely. Yet, it lacks the energy required to overcome the potential barrier and break out of the surface and is one of the reasons, it may be said, why you cannot pour electrons out by shaking a piece of metal!

Metals, are characterized by the Fermi Energy E_{Fm} which lies at the top of the electron energy distribution, when the metal is held at 0 K. We then learned that this energy level itself is quite large, and is around 7 eV measured with the zero energy reference at the bottom of the conduction band. Thermal energy interacts with only a tiny fraction of the electrons (roughly .026/7 or about 0.4% of the energy range).

Metals in which the energy states are only partially filled can make good conductors, because these states provide head room for the electrons to gain energy and move about freely and return to the lower level on losing energy after a collision. A study of energy levels in diamond, an insulator was made in which all the electrons remain in the valence band and there is a band gap between the valence and conduction bands of the order of a few electron volts. The band gap is too large for electrons to cross into from the valence band.

A brief study of the Fermi level of semiconductors was provided from which we learned that the Fermi level lies roughly in the middle of the band gap of Silicon and Germanium which are two well-known semiconductor materials, the band gap itself being of the order of an electron volt.

The work function is denoted $q\varphi_m$ for a metal where q is the magnitude of the electronic charge and φ_m the potential difference in volts between the Fermi level and the vacuum level.

The electron affinity $q\chi$, we learned, is the preferred parameter to consider when selecting suitable semiconductors to make an ohmic contact because it does not change with doping concentrations.

The detailed discussion on the Fermi Energy of metals and the Fermi level of semiconductors was made to enable us understand the fabrication and operation of ohmic contacts.

We then studied, in some detail, the processes when a metal is joined to an *n*-type semiconductor whose electron affinity is greater than the work function $q\varphi_m$ of the metal. After contact is made, electrons flow from the metal to the *n*-type semiconductor and no barrier is formed since electrons are the majority carriers in the semiconductor; the Fermi levels equilibrate to form a nearly ohmic or nonrectifying contact. We then learned that manufacturers dope the *n*-type material in the contact region heavily to prevent any depletion region formation in the metal. And, the contact is nearly ohmic and even if a barrier exists at the interface, it would be so thin that electrons can tunnel through to the other side.

We then studied the metal to *p*-type semiconductor contact and learned that as long as the metal has a larger work function than the work function $q\varphi_s$ and electron affinity of the *p*-type semiconductor, there will result an ohmic contact and a tiny contact potential. The processes of electron transfer however, unlike in the case of contact with *n*-type semiconductor, are from the *p*-type to the metal at the valence energy level. Once again, manufacturers dope the *p*-type material in the contact region heavily to prevent any depletion region formation in the metal. And, the contact is nearly ohmic and even if a barrier exists at the interface, it would be so thin that electrons can tunnel through to the other side.

The potential barrier V_o of a diode cannot be used to provide power nor can it be measured by voltmeters and we learned that the reason this is so is due to metal to semiconductor contact potentials.

Questions

1. What happens to a silicon bar when its temperature is raised from 0° K?
2. How do the energies of the electrons distribute when a *p-n* junction is formed and explain how the equilibrium Fermi level is made constant throughout.
3. Describe the nature and strength of the electric field in the transition region and the electrostatic potential barrier which is formed when there is dynamic equilibrium.
4. Describe in detail the processes when a diode is forward biased with special emphasis on how the potential barrier is lowered and the process of recombination to enable conduction.
5. What is meant by binding energies of "inner electrons"? Why are the inner electron energies not shown in energy band diagrams?
6. With reference to the potential and kinetic energies of electrons during forward bias explain in detail how the holes and electrons surmount the potential barrier.
7. Is the potential barrier reversed when a diode is forward biased? With the help of a suitable analogy explain how the potential barrier can never be made 0 volts.
8. Name and describe the 3 processes when a *p-n* junction is formed. Explain these processes briefly.
9. Describe the process of the influence of built-in electric field on the energies of electrons with the help of energy band diagrams and derive the Einstein relation from it.
10. What are the possible origins of *built-in* electric fields in semiconductors and in *p-n* junctions in particular?
11. Beginning from an expression of the total current density of a *p-n* junction, derive the expression $p_{po} = p_{no}\, e^{\frac{qV_O}{kT}} = p_{no}\, e^{\frac{V_O}{V_T}}$ where p_{po} and p_{no} are the hole thermal equilibrium concentrations in the *p* and *n* neutral regions respectively. Explain briefly why this expression is referred to by some authors as the Boltzmann relationship of kinetic gas theory.

12. Planck's law appears in the expression in the Density of States and the solutions to Schrodinger's Wave Equation. Would you call this particular application of the law i) Classical Electrodynamics ii) Quantum Electrodynamics iii) Quantum Mechanics or iv) Quantum Field Theory? Explain. Hint: See Section 6.6.
13. What is *low-level injection*? If this condition is not met which expressions of a junction diode will not become applicable?
14. Is the diffusion current component of holes uniform through a *p-n* diode? Is the diffusion current component of electrons uniform through a *p-n* diode? How is the conservation of current principle upheld when there is current in a diode?
15. Beginning with the expression $p_{po} = p_{no}\, e^{\frac{qV_O}{kT}} = p_{no}\, e^{\frac{V_O}{V_T}}$ and with a bias voltage V applied derive the diode current equation $I = I_O\,(e^{\frac{V}{V_T}} - 1)$.
16. What is meant by cutin voltage of a diode? Can it be derived? Explain the difference between cutin voltage and the potential barrier V_o.
17. Name 3 major processes that occur during and after the formation of a *p-n* junction. Explain them briefly.
18. Is the potential barrier reversed when a diode is forward biased? With the help of the diode equation $I = I_O\,(e^{V/V_T} - 1)$ explain how the potential barrier can never be made 0 volts.
19. Explain where the voltage drops when a diode is sufficiently forward-biased into conduction and thereon for large forward bias voltages exceeding V_o.
20. Explain the processes that occur when a diode is reverse biased.
21. Describe the operation of a half-wave rectifier and how it can be used to convert a sinusoidal voltage to a steady dc voltage.
22. Explain how the current and voltage ratings, in particular, the peak inverse voltage ratings are important in selecting diodes for rectifier operation.
23. Physicists normally refer to the concentration of electrons by the term Fermi level. What is Fermi energy? Define Fermi Energy of a metal.
24. With the help of energy level diagrams explain why metals in which the energy states are only partially filled have high conductivity.
25. Explain why diamond is an insulator with the help of energy level diagrams.
26. Draw the potential energy graph of a few nuclei at the surface of a metal. Mark clearly in the graph the position of an "inner" electron and the location of an electron which participates in the electric conduction process. In which portion of the metal is there a sharp variation of potential and which portion can we call "equipotential"?
27. Can electrons exist at the same energy level? How do the millions and millions of electrons occupy energy levels. Indicate approximately how many energy levels are present say in a small energy interval. Are these levels available typically in the thousands or hundreds of thousands or millions? What are these finely split energy levels usually called?

28. In a sketch of Fig. 9.45 mark the '0' reference of the Fermi Energy E_{Fm} for metals.
29. Draw the "plateau" in the energy band diagram and describe the potential barrier at the surface of a metal. What energy level does the plateau signify? Explain its importance.
30. What are ohmic contacts? Why are they necessary in rectifier diodes? Explain.
31. When a metal and *p*-type semiconductor are joined which energy levels equilibrate? How is this achieved? Describe with the help of energy level diagrams. What should be the work function of the *p*-type semiconductor to form a nearly ohmic contact with the metal?
32. When a metal and *n*-type semiconductor is joined which energy levels equilibrate? How is this achieved? Describe with the help of energy level diagrams. What should be the work function of the *n*-type semiconductor to form a nearly ohmic contact with the metal?
33. Manufacturers usually dope the semiconductor heavily at the semiconductor to metal contact to achieve a reliable ohmic contact. How does this help? Explain with energy level diagrams.
34. Describe the change in energy levels at the metal to semiconductor contact that was made ohmic (both for *n*- and *p*-type) during a forward bias and explain how the nature of the ohmic contact enables a free movement of electrons and holes in both directions.
35. Explain why it is not possible to measure the potential barrier V_o using voltmeters.

Additional practice questions are available in the CD with this book.

10

Wave Propagation, Antennas and Waveguides

This chapter is about electromagnetic waves, their propagation, production and interaction with matter. Electromagnetic waves (em waves) are time-varying fields and can exist in a range of frequencies as in radio and television and as in light. Electromagnetic waves are not some strange mixture of the two fields, electric and magnetic, but propagate distinctly separate and in a coordinated manner. We have not used the word "move" to describe their travel through space because the fields themselves do not move. It is the *pulsations* of the electric and magnetic fields that travel through space and other medium.

The chapter begins by tracing the history of the discovery of electromagnetic waves and one section is devoted to the mechanism of their propagation through space and the properties em waves should have, to propagate. The next few sections deal with the mechanism of generation of electromagnetic waves by the motion of charges.

In the sequence of courses in physics of optics, the modern trend is to study electromagnetic waves first, followed by a reading of optics since light is an electromagnetic wave. Therefore, the chapter deals with a few fundamental processes involving light (as a form of em wave), and the phenomena of refraction and reflection. An important application of the phenomenon of internal reflection is the use of "fiber optics" for long-distance high-speed Internet connections. These phenomena are also found in radio frequency waves (as radiated by antennas and propagated by ground waves and sky waves and their interactions with atmospheric layers above the earth), though a treatment is not included being beyond the scope of the textbook. It is however recommended that these phenomena be understood in this chapter, since they find direct application in topics related to propagation of electromagnetic waves through the atmospheric layers and there are very few textbooks on antennas and wave propagation that deal with these processes.

Then, the effect of em waves in the form of light on matter and how light refracts, reflects and scatters by reradiation is discussed. The manner of generation of light by incandescent bulbs, is slightly different from the manner of production of radio frequency em waves by antennas, and this is discussed in one section. Following this, the emission of photons by energy level translations of atoms, and the discovery made by Planck (which heralded the birth of Quantum Physics) that the energy emitted by blackbodies is not continuous but transferred in multiples of a small amount, is described. In the same section, the complex process of the production of sparks in ionized air by the recombination of electrons with ions, is described.

Later sections deal with the aspect of acceleration of charges and how this produces electromagnetic radio frequency waves, and the manner of their production by antennas.

One section is devoted to a description of the principle of operation of Waveguides which are used to transmit electromagnetic waves in the radio frequency range. This section is placed near the end of the chapter, so that the reader having read the phenomenon of reradiation and reflection in earlier sections, will find it easy to understand waveguide operation.

At the end of the chapter, links to authoritative lectures on Relativity and to videos of the electric fields of moving and interacting particles by Hermann Härtel are provided.

Before reading this chapter, it is recommended that the topics "superposition of fields" and "Faraday's law" alongwith other relevant sections in Chapter 1 and Chapter 5 that treat the nature, properties and effects on matter of electric and magnetic fields, be reviewed.

10.1 THE DISCOVERY OF ELECTROMAGNETIC WAVES

While all discoveries in science were made and the laws governing them were formulated by experiments and observations, two stand out among the few for having been predicted mathematically first. Electromagnetic waves is one among the two. The presence of one of the planets was first predicted from mathematical considerations and *later* verified by observations. Unexpected changes in the orbit of Uranus led Alexis Bouvard to deduce that its orbit was subject to gravitational perturbation by an unknown planet. The unknown planet was subsequently observed with a telescope by Johann Galle within a degree of the position predicted by Urbain Le Verrier and was named Neptune.

Seeking an explanation to the piling up of charge on the plates of capacitors, Maxwell noticed that Ampere's law (that predicted a magnetic field around a current) fails and needed to add a term to Ampere's mathematical expression to set things right. That single term led to the prediction of electromagnetic waves and that *a changing electric field induces a magnetic field*, just as from Faraday's law a non-coulomb electric field is associated with a changing magnetic field (Faraday's law, Chapter 5).

Now, the two fields could be seen to be able to propagate as electromagnetic waves through space without the presence of sources or external charges.

Radio waves are waves of electromagnetic energy and were first produced deliberately by Heinrich Hertz in 1887, using electrical circuits which produced oscillations at a much lower frequency than that of visible light. Hertz also developed ways to detect these waves, and produced and characterized what were later termed radio waves and microwaves. Light is also a form of electromagnetic wave.

With the discovery of radio and other high frequency waves, mankind had a means to communicate with each other across vast distances without the need to travel.

10.2 WAVES OF ELECTROMAGNETIC FIELDS TRAVELING THROUGH SPACE

In Chapter 5 we showed how a time-varying magnetic field is *associated* with an electric field. James Clerk Maxwell, while studying currents and fields of capacitor charging circuits, predicted that a time-varying electric field is associated with a magnetic field and thence fixed Ampere's circuital law by adding a term [22]. This explained the continuity of current through a capacitor as the field between its plates varies while charging. It is difficult to verify Maxwell's ingenious "guess" because the unavoidable capacitor charging current becomes the source of the observed magnetic field (remembering that the field penetrates through matter) and the gap between the plates of a capacitor is very tiny. Therefore, the real currents may mask the effects of the time-varying electric field.

The confirmation of Maxwell's guess is indirect in the explanation he was able to provide of the existence of light and radio as electromagnetic waves. It can be shown by applying Maxwell's equations that time-varying electric and magnetic fields can exist in empty space, without any charges or currents nearby, and that their speed of propagation is given by $v = \dfrac{1}{\sqrt{\mu_o \varepsilon_o}}$, where μ_o, ε_o are the permeability and permittivity of empty space. It turns out that $v = 3 \times 10^8$ meters/second and is the measured value of the speed of light 'c'! Thus an electromagnetic wave, which comprises configurations of time-varying electric and magnetic fields of any frequency, propagate through empty (free) space at the speed of light, denoted by the symbol 'c'.

The speed is applicable to all frequencies of electromagnetic waves traveling in free space; very low frequency to radio to Microwave to light to X-rays and beyond, to Gamma rays.

Propagation

(Chabay and Sherwood 2011, used by permission)

The propagation of electromagnetic wave means the spreading of its constituent electric and magnetic fields into empty space or into new regions of space. Recollect from Chapters 1 and 2 where we stated that electric and magnetic fields don't move.

Then how do fields propagate? Why do we not directly connect the generation of electromagnetic waves with their propagation?

The fact is that electromagnetic waves propagate in space *independent* of the sources of the fields. One method of solving for differential equations like Maxwell's equations, which describe the relationships between electric and magnetic fields, both stationary as well as time-varying, is by taking derivatives of the guessed function and plugging them into the differential equation. This is one of the most common schemes for solving a differential equation, and it can be used to solve integral equations as well. We could consider various configurations of time-varying electric and magnetic fields, in the absence of any charges, and we would find that most of them do not satisfy Maxwell's equations and hence are not possible. However, if we add the assumption that the region of nonzero field is "traveling" through space at a velocity $\vec{v}$, we can find a solution that works. This idea which we had introduced in Section 5.22 about how a Transverse Electromagnetic (TEM) wave is initiated in a transmission line can be applied to the case of waves in space. You may recognize this phenomenon – a disturbance moving through space – as a wave. A similar idea that the region where the fields are nonzero must move through space – is the key to finding time-varying electric and magnetic fields that satisfy all of Maxwell's equations. In practical applications, the wave is sometimes launched from a transmission line to an opened out transmission line, which is an antenna. We will learn more about antennas in later sections.

To simplify the analysis, we will first consider the simplest possible wave: a single propagating pulse. To understand this, consider a slab (or pulse) of electric field propagating to the right, as shown in Fig. 10.1(a).

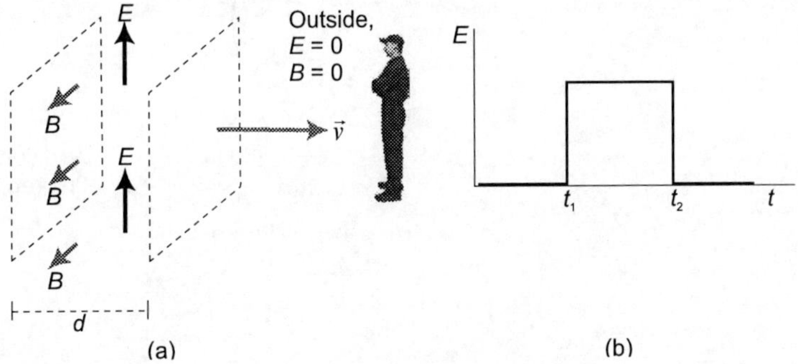

Fig. 10.1 (a) A slab of space contains electric and magnetic fields, with no fields outside the slab. The slab moves to the right with speed *v*. (b) Magnitude of electric field observed at a particular location in space, as a function of time, as the slab passes. Note that at the times t_1 and t_2, the electric field at this location is changing ($dE/dt \neq 0$)

Consider a rectangular "slab" of space, of thickness d and very large height and depth, filled with a uniform electric field $\vec{E}$ pointing up and a uniform magnetic field $\vec{B}$ pointing out of the page (Fig. 10.1(a)). The electric and magnetic fields are zero outside the slab. We imagine the field-filled slab moving to the right with an unknown speed v, by which we mean that at a later time the slab-shaped region of space that contains fields is somewhere to the right of the present position (and at that time the fields are zero in the original position).

Time-varying Fields

If you stand at a location to the right of this region, as shown in Fig. 10.1(a), for a while you don't experience any electric or magnetic fields. Then, as shown in Fig. 10.1(b), the fields at your location change. For a short time you can detect an upward-pointing electric field, and a magnetic field coming out of the page. After the slab passes by, you again experience no electric or magnetic fields. You might describe your experience by saying that a short "pulse" of fields passed by. As the edges of the pulse pass a location, there are time-varying fields at that location. (In a real pulse, the magnitude of the field would not change instantaneously, so dE/dt and dB/dt at the edges of the pulse would not be infinite).

We say that in our example the fields "propagate" to the right, or that the "direction of propagation" is to the right.

Strictly speaking, the fields don't move. Rather it's just that at one instant of time there are fields in a particular slab-shaped region of space, and at a later time there are fields in a different slab-shaped region of space, to the right of the original region (Fig. 10.2(a)).

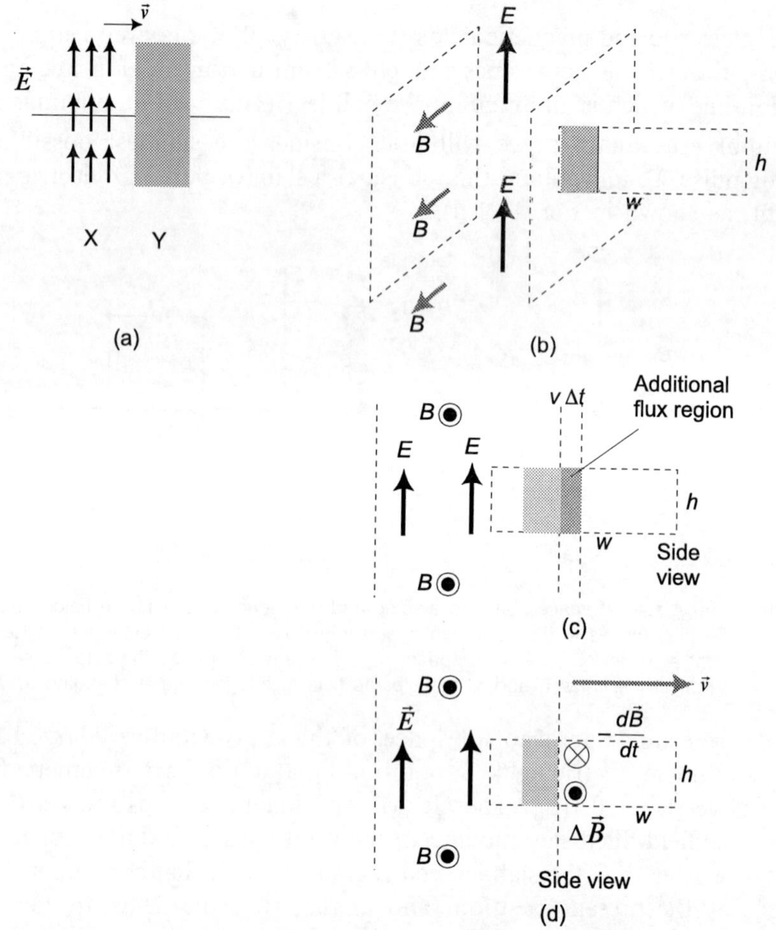

Fig. 10.2 (a) At a given instant the electric field is nonzero in the region X. In the next instant, the region of nonzero field will propagate to the shaded area. For clarity magnetic field is not shown here (b) Applying Faraday's law to a rectangular loop partly inside the slab. The lower diagram (c) shows a side view. (d) The direction of electric field is consistent with Faraday's law.

At a given instant the field exists in the slab of region X and not in the empty region marked Y.

In the next instant, region Y of zero field is filled with a non-zero electric field as predicted by Maxwell's equations. For clarity only the electric field is shown and not the magnetic field. The electric field in the region X is "curly" in the sense that it has a non-zero curl even though the electric field lines shown in region X do not close all the way around.

We will now show that this configuration of electric and magnetic fields is consistent with two of Maxwell's equations viz. Faraday's law and the Ampere-Maxwell law. The proof that the field configuration is consistent with Gauss's law and Ampere's law is left as an exercise for the reader who should remember that there are neither static charges nor currents in the region of space.

Faraday's law Applied to a Moving Slab of Time-varying Electric and Magnetic Fields

(Chabay and Sherwood 2011, used by permission)

Consider a rectangular path of height h and width w fixed in space partly inside the slab and partly outside as shown in Fig. 10.2 (b).

We begin by applying Faraday's law which relates a changing magnetic flux to a non-Coulomb electric field:

$$\oint \vec{E} \cdot d\vec{l} = -\frac{d\phi_B}{dt} \quad \text{(Section 5.12)} \tag{1}$$

$$\oint \vec{E} \cdot d\vec{l} = -\frac{d}{dt}\int \vec{B} \cdot \hat{n}\, dA \tag{2}$$

The slab is moving to the right at speed v, and the rectangular path is fixed in space. In a time Δt, the area of the magnetic field B is uniform within the slab, but increases by an amount $(v\Delta t)h$ in the rectangular path.

Since the magnetic field B is uniform within the slab, the magnetic flux increases by an amount $B(v\Delta t)h\cos 0° = Bvh\Delta t$ as shown in Fig. 10.2(c). We take the cosine of the angle the flux component makes with the right side of the box and this is seen to be 0°.

Therefore, the rate of change of magnetic flux inside the chosen path is

$$\frac{d\phi_{magnetic}}{dt} = Bvh \tag{3}$$

By Faraday's law, the rate of change of magnetic flux is equal to emf around the path, which is a path integral of the electric field (Section 5.12).

Keeping in mind that part of the fixed path is outside the moving slab, and since there is a parallel component of nonzero electric field only along the piece of length 'h', we have

$$|\text{emf}| = \left|\oint \vec{E} \cdot d\vec{l}\right| = Eh\cos 0° \tag{4}$$

Equating the expressions (3) and (4) we get

$$Eh = Bvh, \quad \text{or}$$

$$E = vB \tag{5}$$

as the relationship between E and B in the slab.

In the region ahead of the slab, the magnetic field changes from being zero to pointing out of the page, so $\Delta\vec{B}$ points out of the page and $-\frac{d\vec{B}}{dt}$ points into the page (Fig.10.2(d)).

By applying Lenz's law (Section 5.13) with the right thumb pointing in the direction of $-\frac{d\vec{B}}{dt}$, the fingers of the right hand curl clockwise, which is the sense of the electric field around the path.

Thus, our chosen configuration of moving fields is consistent with Faraday's law if $E = vB$.

Ampere-Maxwell law Applied to a Moving Slab of Time-varying Electric and Magnetic Fields

The Ampere-Maxwell law relates a changing electric flux to a magnetic field, or

$$\oint \vec{B} \cdot d\vec{l} = \mu_o \left[\Sigma I_{inside\ path} + \varepsilon_O \frac{d\phi_{electric}}{dt} \right] \tag{6}$$

In this situation there are no currents. Because the field containing slab is "moving", the electric field at a location in space changes with time, and by the Ampere-Maxwell law (Section 5.22) this changing electric field must be associated with a curled magnetic field.

To consider the Ampere-Maxwell law, we draw a rectangular path fixed in space, that is horizontal (so this rectangle is perpendicular to the electric field) as shown in Fig. 10.3(a).

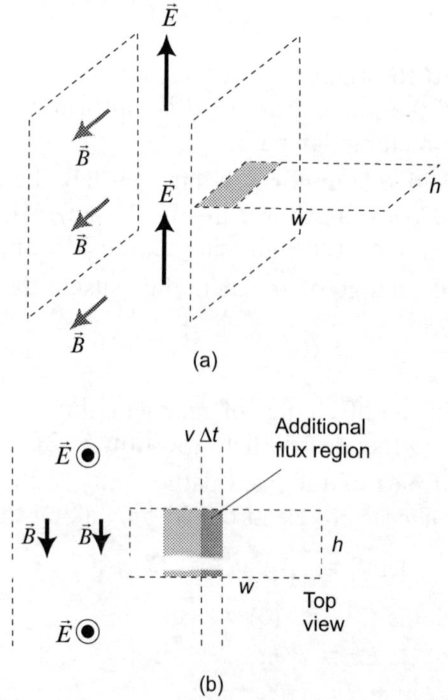

Fig. 10.3 (a) Applying the Ampere-Maxwell law to the slab. The lower diagram (b) shows a view from above.

Figure 10.3(b) includes a view from above, looking down on the upward-pointing electric field.

In a time Δt the area of electric field inside our chosen path increases by an amount $(v\Delta t)h$, and since the electric field E is uniform within the slab, the electric flux increases by an amount $E(v\Delta t)h\cos 0° = Evh\Delta t$. Therefore, the rate of change of electric flux inside our chosen path is this:

$$\frac{d\phi_{electric}}{dt} = Evh \tag{7}$$

The only portion of the path along which there is a parallel component of nonzero magnetic field is along the piece of length h, so we have this:

$$\oint \vec{B} \cdot d\vec{l} = Bh\cos 0° = Bh \tag{8}$$

Application of the Ampere-Maxwell law will establish another algebraic relation between E and B. There is no current I, so the path integral of the magnetic field from equation (6) is equal to $\mu_o \varepsilon_o$ times the rate of change of electric flux, $Bh = \mu_o \varepsilon_o Evh$, so we have the following relationship between the B and E in the slab:

$$B = \mu_o \varepsilon_o vE \tag{9}$$

Since we have shown that our chosen configuration of fields is possible if the region of nonzero field is traveling with speed v, it is interesting to find what v must be.

Solving the algebraic equations (5) and (9) we find

$$v^2 = \frac{1}{\mu_o \varepsilon_o} \tag{10}$$

Evaluating this numerically we obtain this:

$$v = \sqrt{\frac{1}{\mu_o \varepsilon_o}} = \sqrt{\left(\frac{1}{4\pi\varepsilon_o}\right)\left(\frac{4\pi}{\mu_o}\right)}$$

$$v = \sqrt{(9\times10^9\,\mathrm{N\cdot m^2/C^2})(1\times10^7\,\mathrm{A/T\cdot m})}, \text{ or}$$

$$v = 3 \times 10^8 \text{ m/s}$$

This is a spectacular result: the pulse is predicted to advance at a speed that is the same as the known speed of light!

By the application of Faraday's law and the Ampere-Maxwell law to the moving slab we have shown that a pulse of radiation must contain, both time-varying electric and magnetic fields, so that the pulse can move forever through otherwise empty space. The transverse electric field is accompanied by a transverse magnetic field perpendicular to the electric field, with the cross-product $E \times B$ in the direction of propagation.

If the source that produced the electromagnetic waves is sinusoidal, the field strength variations in both time and space will be sinusoidal too.

Consider one end of a rope fastened to a wall and its free end is moved up and down. If a snapshot of the rope is taken, it will be seen to have a wavelike shape. One useful way to understand the propagation of the fields in space is to imagine replacing the varying displacement of the rope from the datum line with the varying *strength* of the fields. There are no sinewaves in space of electric and magnetic fields. One representation of the variation of field strength in time *and* distance of a propagating wave in free space could be a sinewave. We shall see below how a sinusoidal wave can be produced by a charge oscillating continuously sinusoidally.

A detailed treatment with proof using Gauss's law of how an electromagnetic wave propagates through space is available in Ref. [2].

When in 1862 Maxwell first arrived at the result that the field pattern of an electromagnetic wave travels with speed c, the constant c in his equations expressed only a relation among electrical quantities as determined by experiments with capacitors, coils and resistors. To be sure, the dimensions of this constant $\frac{1}{\sqrt{\varepsilon_o \mu_o}}$ were those of velocity, but its connection with the actual speed of light had not yet been recognized.

Maxwell wrote, "The speed of light had most recently been measured by Fizeau in 1857. The velocity of transverse undulations in our hypothetical medium, calculated from the electromagnetic experiments of MM Kohlrausch and Weber, agrees so exactly with the velocity of light calculated from the optical experiments of M. Fizeau, that we can scarcely avoid the inference that *light consists in the transverse undulations of the same medium which is the cause of electric and magnetic phenomena*". The italics are Maxwell's [22]. The medium was called "aether" whose existence was disproved by the Michelson-Morley experiment (1887 and later by other scientists in the 1920's) [Refer Section F.1, Appendix F]. This led to the proposal of several theories to explain charged particle interaction and wave propagation; Weber's Electrodynamics (1875, without the field concept), the "field" theory of classical electrodynamics due to Maxwell-Lorentz (around 1905, described in this text), Page's Emission theory (1924) of massless moving elements and the more recent Quantum Electrodynamics leading to Quantum Field theories. See Sections F.1 and F.5, Appendix F for a discussion of these theories.

The Units of the constant $\frac{1}{\sqrt{\varepsilon_o \mu_o}}$

The constant $\frac{1}{\sqrt{\varepsilon_o \mu_o}}$ is known from Maxwell's equations to represent the speed of light '*c*'.

The units of ε_o in the SI system equals $8.85 \times 10^{-12}\ \frac{\text{Farad}}{\text{m}} = \frac{\text{Coulombs}^2}{\text{Newtons} \cdot \text{m}^2}$.

The units of μ_o in the SI system equals $4\pi \times 10^{-7}\ \frac{\text{Tesla} \cdot \text{m}^2}{\text{Coulomb} \cdot \text{m / sec}}$, where the units of Tesla is $\frac{\text{Newton}}{(\text{Coulomb / sec}) \cdot \text{m}}$. Therefore, the units of μ_o maybe expressed as

$$\frac{\text{Newton} \cdot \text{m}^2}{(\text{Coulomb / sec}) \cdot \text{m} \cdot \text{Coulomb} \cdot \text{m / sec}} = \frac{\text{Newton} \cdot \text{sec}^2}{\text{Coulomb}^2}.$$

The product $e_o \mu_o$ therefore has units $\frac{\text{Coulombs}^2}{\text{Newtons} \cdot \text{m}^2} \cdot \frac{\text{Newton} \cdot \text{sec}^2}{\text{Coulomb}^2} = \frac{\text{sec}^2}{\text{m}^2}$.

The constant $\frac{1}{\sqrt{\varepsilon_o \mu_o}}$ therefore has units $\frac{1}{\sqrt{\frac{\text{sec}^2}{\text{m}^2}}} = \sqrt{\frac{\text{m}^2}{\text{sec}^2}} = \frac{\text{m}}{\text{sec}}$ which is a unit of velocity.

The constant $\frac{1}{\sqrt{\varepsilon_o \mu_o}}$ therefore is $\frac{1}{\sqrt{8.85 \times 10^{-12} \times 4\pi \times 10^{-7}}}$ and this works out to 299863380 m/s or 3×10^8 m/second.

Note: It is a little remarkable that while discoveries related to electricity and magnetism were being made, the atom and its structure still remained to be discovered. Remember that JJ Thomson discovered the electron and its properties only in 1897, the atomic nucleus was discovered in 1909 and the Bohr model was proposed in 1913.

Properties of Electromagnetic Waves[22]

An electromagnetic wave has the following properties

i. The field pattern travels with speed c.

ii. At every point in the wave at any instant of time, the electric and magnetic field strengths are equal. In CGS units, **B** is expressed in gauss and **E** in statvolts/cm; but **B** and **E** have the same dimensions (1 statvolt = 1 erg/esu where esu is electrostatic unit; 1 gauss = 1 dyne/esu). If the electric field strength is 10 statvolts/cm, the associated magnetic field strength is 10 gauss. Note: 1 esu = 1 electrostatic unit.

iii. The electric field and the magnetic field are perpendicular to one another and to the direction of travel, or propagation.

Other Waveforms and Superposition of Waves

The waviness of the sinusoidal function has nothing to do with the essential property of wave motion which is the propagation unchanged of a form or pattern – *any* pattern.[22] The radio frequency waves that are broadcast by your local TV or radio station are not sinewaves and yet they have traveled from the station to your receiver. This is possible because the electric field **E** and magnetic field **B** are linear and the superposition of two solutions is also a solution. Moreover, the signals of different channels travel through space without affecting each other.

It is remarkable that an electromagnetic wave signal travels through space with independent existence without charges (as we shall see in Section 10.10), and deposits the energy it carries into the antenna in your TV or radio receiver!

10.3 PRODUCING RADIATION BY ACCELERATED CHARGES

(Chabay and Sherwood 2011, used by permission)

Having seen how electromagnetic waves can propagate through space we now show how such a wave can be produced. Electromagnetic waves cannot be produced by charges at rest or when they are moving with uniform velocity. Electromagnetic radiation is produced by accelerating and by decelerating charges. For the moment we shall see how an accelerated charge can produce a propagating electromagnetic wave.

A positive charge q is shown sitting at rest (Fig. 10.4).

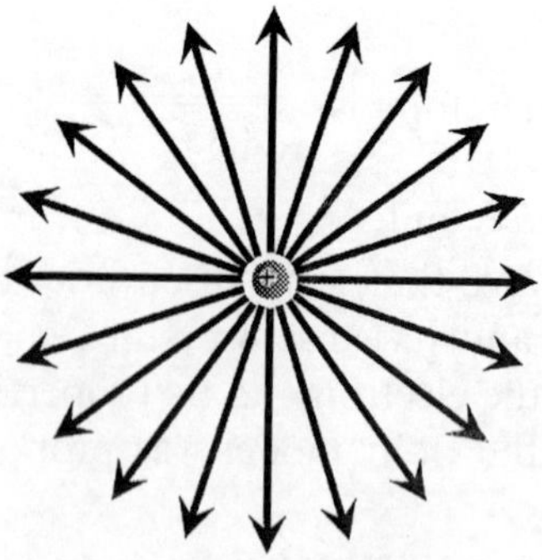

Fig. 10.4 The directions of electric field surrounding a stationary positive charge

The electric field of this charge points radially outward in all directions indicated by the electric field lines.

If the charge is kicked downward, it picks up a little bit of speed and then coasts downward at that slow speed since there is no friction. It had momentarily accelerated from rest and attained a slow velocity.

If we are able to take a snapshot of the direction of the electric field throughout this we will see that the snapshot has the peculiar appearance shown in Fig. 10.5.

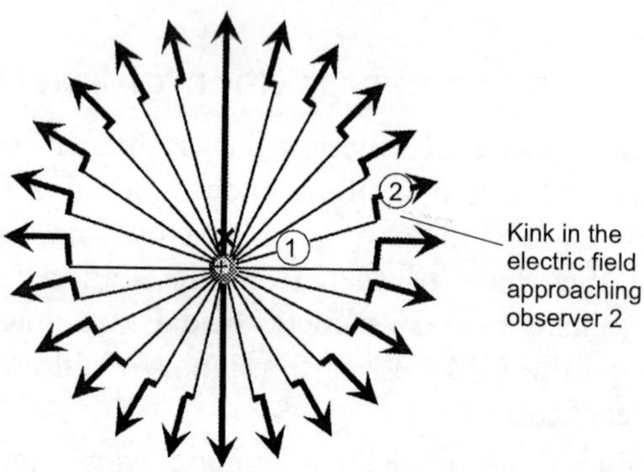

Fig. 10.5 Directions of electric field shortly after the charge was kicked

An observer stationed at location (1) sees an electric field that points away from the present location of the slowly moving charge.

Since information about the change in the electric field has not yet reached observer 2, because the change in the electric field propagates at the speed of light, it takes a while for the new field to appear at observer 2's position.

Between observer 1 and observer 2 there is a kink (a sharp twist or bend) in the electric field direction. The kink which is a sideways or transverse electric field will be detected by observer 2 only after a time $t = r/c$ where r is the distance of observer 2 from the charge.

At some later time 't', this kink will be at location (2) as shown in Fig. 10.6. The original location of the charge is marked '×'.

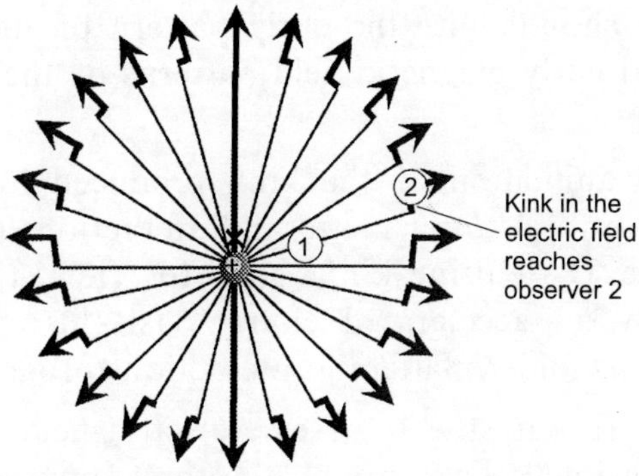

Fig. 10.6 Information of the changed charge position has reached observer 2

The sphere of radiation is expanding and outside this sphere there is no magnetic field because the charge was initially at rest. The expanding sphere carries energy in the radiating electromagnetic fields. Because the energy spreads out over the sphere, the magnitudes of the energy bearing electric and magnetic fields fall off as $1/r$ where r is the distance of a far away point.

The transverse electric field (transverse because it is perpendicular to the direction of the field of the charge at rest) is accompanied by a transverse magnetic field perpendicular to the electric field.

Figure 10.6 is a two-dimensional slice through the three-dimensional phenomenon that is shown in Fig. 10.7, with the pulse headed out of the page and into the page as well as moving in the plane of the diagram.

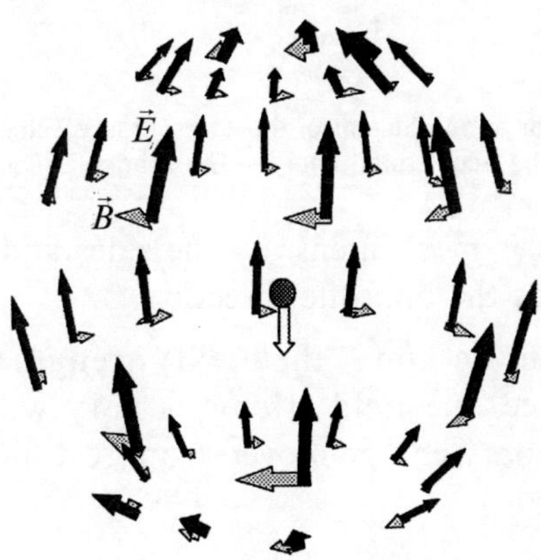

Fig. 10.7 Radiation expands in a spherical pattern from a charge (center) that has been briefly accelerated downward

Applying the curly product rule E×B in the direction of propagation, a radiative magnetic field must point out of the page in Figs. 10.6 and 10.7 and into the page on the left of these figures. This magnetic field is shown by the lightly shaded block arrows. The reader should note the curly pattern of the magnetic field and discern four such individual curly magnetic field patterns of their associated electric fields in Fig. 10.7.

To view a nice animation of the kink produced by an accelerated charge visit http://www.glowscript.org/#/user/matterandinteractions/folder/matterandinteractions/program/23-radiate2D-fieldline-kink and to view the radiation (electric and magnetic fields) by an accelerated charge visit http://www.glowscript.org/#/user/matterandinteractions/folder/matterandinteractions/program/23-radiation-3D.

Figure 10.8 shows only the transverse electric field sometime after the kick and is only a two-dimensional slice through the three-dimensional phenomenon shown in Fig. 10.7.

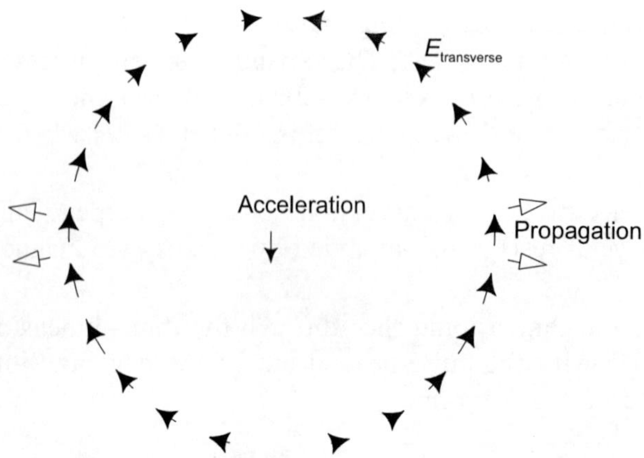

Fig. 10.8 A snapshot at one instant of the transverse electric field, which propagates away from the accelerated charge. The magnetic field is not shown.

Note how the radiation is most intense to the sides and there is zero radiation in the direction of acceleration and the opposite direction.

At a particular instant in time, the field strength varies with location. At a particular location in space, the field strengths vary with time. Such variations are produced by either a stationary charge moved from rest or by a continuously oscillating charge.

Sinusoidally Varying Electromagnetic Wave

If a ball carrying a positive charge is hanging from a spring and was to move sinusoidally, it will emit continuous radiation and the patterns of electric and magnetic field will be as shown in Fig. 10.9 after the spring begins to oscillate.

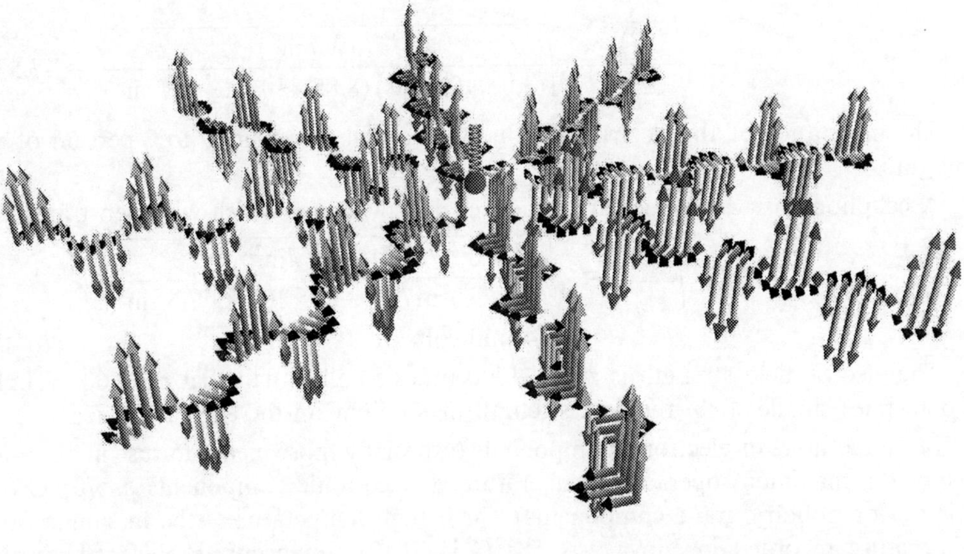

Fig. 10.9 Electromagnetic radiation from a charged ball moving sinusoidally in the $\pm y$ direction

A charge moving sinusoidally in the $\pm y$ direction emits sinusoidal electromagnetic radiation, which propagates outward from the source. For clarity, this radiation is shown only in the xz plane, and the $1/r$ fall off in magnitude is not shown. The electric fields are represented by shaded block arrows and the magnetic fields by shaded block arrows with black arrowheads. Also see plot in Q.10.2 in Additional Practice Questions for all chapters in the CD.

To get a "feel" of the strength of the electric and magnetic fields in the sun's rays, we compute their strengths given the time-averaged Poynting vector also called Irradiation or Insolation, at the upper surface of the earth's atmosphere. The vector is 1.5×10^3 W/m^2. In some literature, this is mentioned as 1.367×10^3 W/m^2.

The time-averaged Poynting vector is related to the amplitude of the electric field by

$$\langle S \rangle = \frac{c}{2}\varepsilon_o E_o^2 \text{ where, } c \text{ is the speed of light}$$

The amplitude of the electric field E_o is

$$E_o = \sqrt{\frac{2\langle S \rangle}{c\varepsilon_o}} = \sqrt{\frac{2\,(1.35\times 10^3 \text{ W}/\text{m}^2)}{(3.0\times 10^8 \text{m}/\text{s})\,(8.85\times 10^{-12} C^2/\text{N.m}^2)}}$$

$$= 1.01 \times 10^3 \text{ V/m}$$

Note that the field in the radiation is about 1 kV/m.

The average Poynting vector on unit area of the Earth facing the Sun, is calculated and found to be roughly 684 W/m^2 for which $E_o = 0.72 \times 10^3$ V/m.

If a 60 W bulb illuminates a room of area 5 m $\times$ 4 m = 20 m^2 area the power density will be approximately $\dfrac{60\,\text{W}}{20\,\text{m}^2}$ or 3 W/m^2. Substituting this value in the formula for the amplitude of the electric field, we get

$$E_o = \sqrt{\frac{2\langle S\rangle}{c\varepsilon_o}} = \sqrt{\frac{2(3\text{W}/\text{m}^2)}{(3.0\times10^8\text{m/s})(8.85\times10^{-12}C^2/\text{N.m}^2)}} = 47.53\text{V/m}$$

The magnitude of the electric field due to the bulb is a mere 5 to 7 percent of that of sunlight.

A cellphone has an incident radiation of 10^{-6} watts/m^2 which will then translate to

$$E_o \sqrt{\frac{2\langle S\rangle}{c\varepsilon_o}} = \sqrt{\frac{2(10^{-6}\text{W}/\text{m}^2)}{(3.0\times10^8\text{m/s})(8.85\times10^{-12}\text{C}^2/\text{N.m}^2)}}$$

$$= 27.4 \text{ millivolts/m}$$

The electric field strength is about 1700 times smaller than that of a 60 W bulb and 26,666 times smaller than the field strength of sunlight on the earth's surface.

Manufacturers of electronic components (especially those meant for use in automobiles) specify the maximum operating temperature of electronic components as 70° C or even 125° C (for military grade components). Such high temperatures (the maximum ambient temperature recorded in Libya was 58° C in 2012) can occur in enclosed interiors of automobile engine compartments parked in the sun and furnace rooms. It will be instructive to measure the temperature of a thin sheet of metal say, a couple of square feet in area, exposed to direct sunlight outdoors (with the wind blocked). Also, observe the effect on the temperature of the plate when a gentle wind is allowed to blow over it.

Rays

(Chabay and Sherwood 2011, used by permission)

The radiation traveling in a particular direction of propagation is often described as a ray and represented by a single line in the direction of propagation, perpendicular to wavefronts (Fig. 10.10(a).

Fig. 10.10 (a) Planar segments of wavefronts far from the source (b) A ray describes a section of a wavefront propagating in a given direction

Far from the source, segments of wavefronts appear nearly planar, and parallel to each other. A line called a ray is often used to describe a section of a wavefront propagating in a given direction.

There are not really discrete rays- the radiation emitted by an oscillating charge is spatially continuous- imagining particular rays can be useful in understanding the geometric behavior of light as it travels through objects such as lenses and prisms.

10.4 RE-RADIATION OR "SCATTERING"

(Chabay and Sherwood 2011, used by permission)

Reradiation or Scattering is a very important phenomenon to know since it occurs almost always with all radiation perhaps barring the em waves of space communication signals above the earth's atmosphere where, the waves travel almost freely unhindered from and towards satellites and space vehicles.

Consider once again the charged ball hanging from a spring, exposed to electromagnetic radiation and therefore oscillating at the radiation frequency (Fig. 10.11).

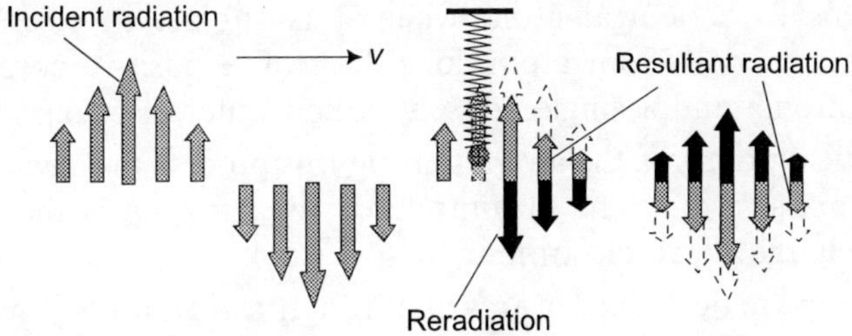

Fig. 10.11 Reradiation by an accelerated charged ball

Since a charge moves by a force exerted by an electric field, the ball that carries a positive charge will oscillate sinusoidally and will sinusoidally accelerate noting that the tangent representing velocity to a sinusoidal shape of the displacement vs. time graph has a slope that varies with time at every instant of time. Won't the ball radiate? since it is accelerating? Yes.

The ball will reradiate since the oscillating ball is a sinusoidally moving (and sinusoidally accelerating) charged object. The reradiation by the charged ball redistributes the energy in new directions. The reradiation is often called "scattering" because part of the original energy moving to the right has been scattered into other directions.

The reradiation by the accelerated charge provides a mechanism for decreasing the energy of the original radiation. To the right of the charged ball, the outgoing electromagnetic radiation is the superposition of the original fields plus the fields radiated by the accelerated ball. If an incoming radiation points upward, it accelerates a positively charged ball upward, and to the right of the ball the reradiated electric field points downward.

The incident radiation has lost momentum having imparted some of it to the ball. Thus, we see that after an interaction with matter the electric and magnetic fields are smaller in magnitude. As a result, the net outgoing electric field to the right of the ball in Fig. 10.11 is smaller than the original incoming electric field and we say, the outgoing field is "attenuated". Also refer "Attenuation of a wave" in Section 5.22.

The energy in the electromagnetic radiation, which is proportional to E^2, is smaller after giving kinetic energy to the ball.

10.5 LIGHT PROPAGATION THROUGH A MEDIUM

(Chabay and Sherwood 2011, used by permission)

The most important aspect of the interaction between electromagnetic radiation and matter is the effect the electric field has on the charged particles inside atoms and molecules. Of course, the magnetic field must be present in electromagnetic radiation, but usually only the electric field need be considered when we try to understand the main effects of electromagnetic radiation on matter. It will be useful to recollect that electric and magnetic fields are "alterations" of space, which penetrate through matter that includes the protons, electrons and neutrons of atoms and that the principle of superposition is applicable to the fields which we learned in Chapters 1, 2 and 5.

What happens to a neutral atom when it is subject to electromagnetic radiation? Consider a simple model of an atom consisting of a positive core (nucleus plus tightly bound inner electrons) and a single, loosely bound outer electron.

The electric field polarizes the atom. During this process the electron moves farther than the much more massive ion. The magnitude of the force eE is the same on both objects where 'e' is the charge of an electron.

When electromagnetic radiation travels through a region filled with a medium such as glass, air, water, metal, plastic and for that matter anything other than vacuum, it may interact with charged particles in the medium. The electric field in the incoming radiation accelerates charged particles in the medium, causing these accelerated charges to emit electromagnetic radiation as well. The magnitude of the electric force on a proton is the same as that on an electron, but the acceleration of the low-mass electron is much larger, so it is the electrons that contribute almost all of the reradiation.

Therefore, at any location inside the medium, the net radiative electric and magnetic fields are the *superposition* of the incident (initial) sinusoidally oscillating fields *and* the reradiated oscillating fields. As indicated schematically in Fig. 10.12, this sum can be quite complex, especially in dense medium such as a solid material, which is composed of many charged particles.

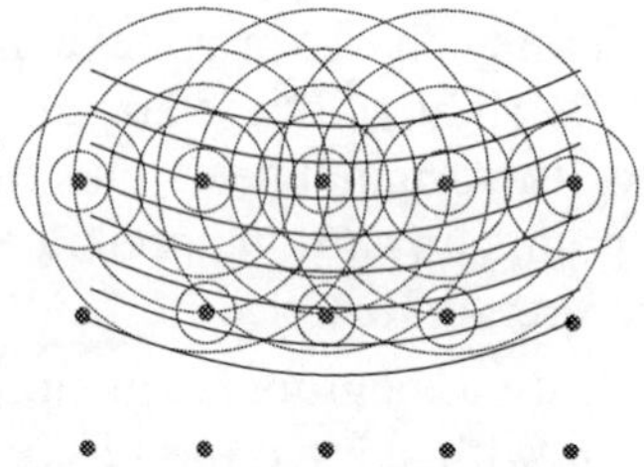

Fig. 10.12 Gray dots represent atoms in a solid medium. The solid curves represent wavefronts in the incident radiation. The dashed curves represent wavefronts in the reradiation from charges in the atoms that were accelerated by the incident radiation. The net field at any location is the sum of all fields at that location.

The behavior of light in a material or other electromagnetic radiation such as radio frequency waves depends very much on the structure of the material at the atomic level.

Opaque Materials

One possible result of the superposition is that the incident wave and re-radiated waves may superimpose in just such a way that troughs in one wave occur at the same locations as crests in the other wave. In this case, the net amplitude is zero, and no electromagnetic radiation emerges from the other side of the medium. This is called destructive interference. In this case the material is opaque. This is how we were able to explain why the cardboard blocked light in Section 4.7. In the experiment in Section 4.7 with a light bulb and cardboard, destructive interference caused the *net fields* of the constituent electric and magnetic fields of light wave to be reduced so drastically in strength or *attenuated* that we say, the light is blocked by the cardboard. Remembering that the fields had an independent existence when they were incident on the cardboard, the cardboard molecules continue to accelerate by the electric field of the light. The interfering reradiation exists as long as the light exists. There is no means by which the net fields can regenerate as long as the cardboard blocks the light because on the other side of the cardboard the independently existing fields are too weak to produce and propagate more powerful electric and magnetic fields anymore. To get back the light on the other side of the cardboard, you have to remove the source of the interfering (reradiated) electromagnetic radiation, viz. the cardboard.

Opacity depends on the wavelength of the light and the nature of the material.

Transparent Materials

In a material that is transparent to electromagnetic radiation of a particular wavelength, incident radiation and reradiation add constructively, so the amplitude of the waves inside the material is not zero. However, this addition process can result in changes in the wavelength of the net waves and in the speed at which a particular wave crest travels through the medium.

Index of Refraction

(Chabay and Sherwood 2011, used by permission)

Most commonly, in ordinary solid materials, the incident wave and the reradiated waves combine to create a resultant net wave whose crests travel through the medium at a speed much slower than c. Because the frequency f of the reradiation is the same as that of the incident radiation, the wavelength inside the medium is less than the wavelength in vacuum: $v = \lambda/T = \lambda f$, so $\lambda = v/f$, and smaller v means shorter λ because f is unchanged. The frequency is unchanged because the "beat" or time period of the pulsations of the em fields in the medium are determined by the incoming radiation.

The index of refraction of a medium is defined as the ratio of c, the speed of a wavefront in a vacuum to v, the speed of a wavefront in the medium, both measured by the time for one crest to travel a given distance:

$$n = \frac{c}{v}$$

The index of refraction for a 589 nanometer wavelength (yellow sodium light) of vacuum is 1, air is 1.00029, water is 1.33 and that of plexiglass is 1.51.

The fact that the speed of a wavefront is different in different materials leads to bending of a ray when a wave moves from one medium into another. To understand how light slows down in a medium, see Q.10.4 in the file "Additional Practice Questions for all chapters" in the CD.

The index of refraction of most materials is not exactly the same for light of different wavelengths. This variation is due to differences in how light of different wavelengths interacts with the charged particles that compose the material. As a result of the wavelength dependence of index of refraction, a glass prism can be used to show that white light in fact contains a mixture of all colors of light.

Since the index of refraction depends on the wavelength, light of different wavelengths is bent by different amounts as it passes from air into glass, and again when it passes from glass back into air. The speed of the light color waves on emerging from the glass is '*c*'.

A Marching Band Changes Direction

There is a useful analogy that helps explain what happens to wavefronts and rays of light at the surface between the two materials. Think of a marching band that is marching quickly across grass to the beats of a drummer. When it comes to a muddy part of the field it has to slow down. Their feet continue to stamp the muddy part to the beat of the drummer (the frequency is unchanged) who doesn't change his beat.

In Fig. 10.13 all members of each marching rank reach the mud at the same time, and the pileup on reaching the mud makes the ranks ("wavefronts") closer together in the mud than on the grass.

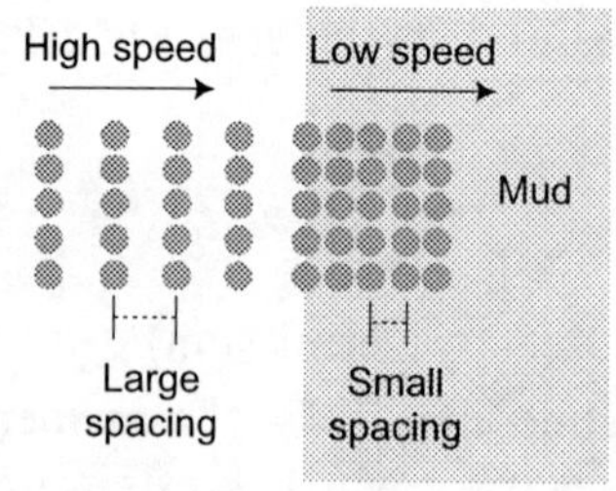

Fig. 10.13 When members of a marching band encounter mud and slow down, the ranks get closer together.

The marching rhythm of ranks on the grass and in the mud is the same. This is analogous to the decreased wavelength and unchanged frequency when light passes from air into glass.

When the band approaches the mud at an angle, the members slow down to allow the other members to come up to the muddy field. This is depicted in Fig. 10.14.

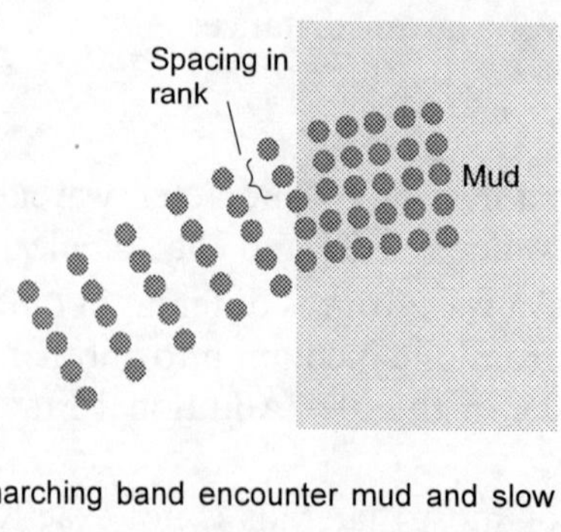

Fig. 10.14 When members of a marching band encounter mud and slow down, the ranks get closer together.

When the first band members in a rank encounter the mud and slow down, there are other marchers in their rank who are still on the grass and still moving quickly. Again the marchers don't miss a step to the beat of their drummer. If the marchers are required to maintain the same spacing within their rank as before, the ranks ("wavefronts") in the mud end up being at a different angle than the ranks still on grass.

The direction of march changes; the outgoing "ray" is in a different direction than the incoming "ray".

We now examine the phenomenon in the case of light.

Bending of Light

(Chabay and Sherwood 2011, used by permission)

In Fig. 10.15 light spreads out in air from a sinusoidally oscillating point source and encounters a slab of glass.

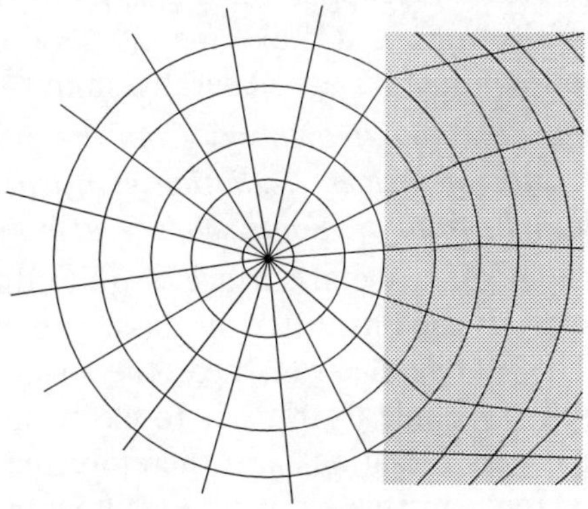

Fig. 10.15 Light in air encounters a slab of glass.

The rays indicate the direction of propagation, and the curving arcs are the wavefronts, perpendicular to the rays. The wavefronts, indicated by the curving arcs, get closer together in the glass, as a result of the superposition of the fields of the point source and the additional fields reradiated by accelerated charges in the glass. The wavelength λ of the radiation is shorter inside the glass than it was in the air. Because the frequency and period of the oscillation don't change, the speed of a wavefront in the glass is slower than it was in air (speed = λ/T). The rays as seen in Fig. 10.15 bend inward at the surface between the air and glass. The bending of rays at an interface between two materials is called "refraction". There is a nice article on the process of light slowing down when it travels through matter in the link http://www.madsci.org/posts/archives/may98/893732585.Ph.r.html.

Reflection

(Chabay and Sherwood 2011, used by permission)

In addition to the refraction (bending of light), there is also some reflection of the light in situations like that shown in Fig. 10.15. The accelerated electrons in the glass radiate in all directions, and in addition to making the wave-length shorter in the glass, they produce light in backward directions. It looks as though this "reflected" light bounces off the surface, but in fact this new light is emitted by all of the accelerated charges throughout the glass. The reflected light may be quite weak; you have probably noticed that reflections off a sheet of glass can be quite dim.

When an electromagnetic wave (say, a radio frequency wave) encounters an electrical conductor, the electric field causes currents to flow.[22] This generally results in energy being dissipated within the conductor at the expense of the energy in the wave. If the resistivity of the reflector is near zero, the incident wave will be totally reflected.

If the resistivity of the reflector is not zero, the amplitude of the reflected wave will be less than that of the incident wave. Aluminium, for example, reflects visible light, at normal incidence, with about 92 percent efficiency.[22]

The reflectivity of most metals for visible light (as in mirrors made by applying a reflective surface of tin, silver and copper on a glass substrate) is essentially due to the highly mobile conduction electrons that make metals good conductors of steady current. It is no accident that good conductors are generally shiny.[22]

What can be said about the reradiated energy from the inner electrons of the metal since the electric and magnetic fields of incident radiation can penetrate through matter? Why do we see reflection of visible light from a smooth surface at only one angle?[2]

Reflection of visible light off a metal surface is quite different from the reflection off transparent insulating materials such as water or glass. The mobile electrons in a metal are very easy to accelerate. The mobile electrons near the surface gain a lot of energy, so we conclude that little energy is available to transfer to electrons far from the surface. It must be the case that the intensity of radiation (and therefore the intensity of the electric field) inside the metal is small. How can this happen? Presumably there is so much re-radiation by accelerated electrons near the surface of the metal that there is very little net field in the interior of the metal. There are no significant interference effects between layers at different depths from the surface. Of course there are the usual interference effects involving electrons in different locations on the surface, so that there is strong intensity only at the "reflection" angle, and metals make good mirrors.

Copper and bronze alloy mirrors have been produced in China and India from around 2000 BC. Mirrors with glass as their component are said to have been invented in Sidon (modern-day Lebanon) in the first century AD. The Romans invented a method for creating crude mirrors by coating blown glass with molten lead. The silvered-glass mirrors found throughout the world today first got their start in Germany almost 200 years ago.

In 1835, German chemist Justus von Liebig developed a process for applying a thin layer of metallic silver to one side of a pane of clear glass. This technique was soon adapted and improved upon, allowing for the mass production of mirrors.

Energy can also be absorbed when an electromagnetic wave meets nonconducting matter (insulators or dielectrics). Little of the light that strikes a black rubber tyre is reflected, although the rubber is an excellent insulator for low-frequency electric fields. Here the dissipation of the electromagnetic energy involves the action of the high-frequency electric field on the electrons of the molecules of the material.[22]

A few insulators transmit electromagnetic waves with very little absorption. The transparency of glass for visible light, with which we are so familiar, is really a remarkable property. In the purest of glass fibers used for optical transmission of audio and video signals, a wave travels as much as a kilometer, or more than 10^9 wavelengths, before most of the energy is lost. However transparent a material medium may be, the propagation of an electromagnetic wave within the medium differs in essential ways from propagation through the vacuum.[22]

The glass fibers used in fiber optic communications are designed in such a way that the light rays traveling through the glass undergo total internal reflection at the surface of the fiber rather than escaping. This is achieved by using glass with an appropriate index of refraction.[2]

10.6 THE FIELD DUE TO A MOVING CHARGE

The previous sections in this chapter discussed electromagnetic waves, the principle of their generation and propagation emphasizing the aspects related to the effect on them by matter as when light is incident on various substances.

In this section we study the fields due to moving charges. The fields associated with a charge that is accelerated and the fields due to several charges (initially accelerated) in uniform motion provide the key to understanding the generation of electromagnetic waves by antennas and light sources like incandescent lamps. A single charge moving at a uniform speed is shown in Fig. 10.40. The aspect of the field of a particle which has been moving at constant speed forever and whose past history determines the field observed would especially be useful. An aid to visualizing such a field is provided in "Field of a moving single charge" in Appendix F. The present discussion concerns a single charge that is accelerated. Consider an electron which starts from rest, and suddenly acquires the speed v at $t = 0$. The field of the electron will look something like the one shown in Fig. 10.16.[22]

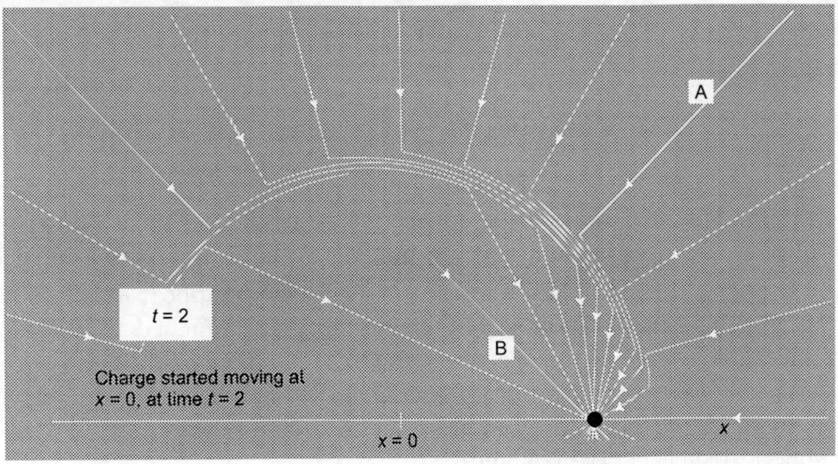

Fig. 10.16 An electron initially at rest in a laboratory frame is suddenly accelerated at $t = 0$ and moves with constant velocity thereafter. This is how the electric field looks at the instant $t = 2$ all over the laboratory frame.

In Fig. 10.16, at A, the field resembles the far field of a charge stationary at the origin. At B, the field resembles that of a charge moving with uniform velocity described in Section 10.12. Note the thin spherical shell between A and B. This is what was shown as a kink in the field in Figs. 10.5 and 10.6. The shell simply expands with speed c, its center remaining at $x = 0$. The arrowheads on the field lines indicate the direction of the field when the source is a negative charge, as we have been assuming. The field lines seem to remind us of ribbon streamers on a stick waved by a gymnast; but of course, the fields themselves do not travel but do seem to appear like the ribbon streamers. There is no difference between the transverse field described in Section 10.3 and in this section. A detailed proof of the shape and presence of the shell-shaped brush of field lines can be shown to be consistent with Gauss's law[2,22].

What will the field of a uniformly moving electron look like if it were stopped?

Figure 10.17 shows the field of an electron which had been moving with uniform velocity until $t = 0$, at which time it reached $x = 0$ where it was abruptly stopped.

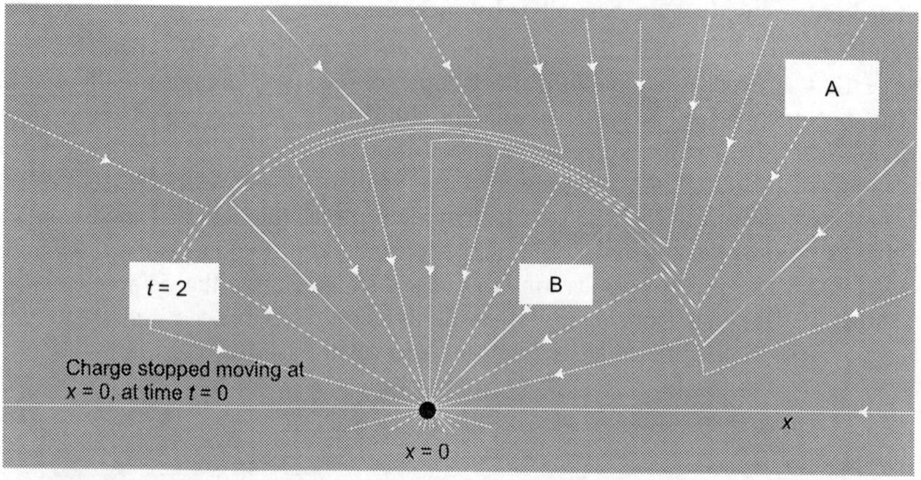

Fig. 10.17 A charge that was moving with uniform velocity is stopped.

Now the news that it was stopped cannot reach, by time t, any point farther than ct from the origin. The field outside the sphere of radius $R = ct$ must be that which would have prevailed if the electron had kept on moving at its original speed as in Fig. 10.40.[22]

Again, the fields themselves do not travel but do seem to appear like the shape the wave-like ribbon streamers would assume if the stick they are attached to is abruptly stopped but the remarkable aspect in the snapshot view of Fig. 10.17 are the throw of the lines at A, where the far field resembles that of a charge moving with uniform velocity described in Section 10.12. At B the field resembles that of a charge stationary at the origin. Note the thin spherical shell between A and B.

The shape can be explained by invoking Gauss's law[22]. The circular shape of the electric field that expands as shown in Figs. 10.16 and 10.17 is described by Härtel as a "circular wave". Circular waves, produced by massless moving elements along lines of electric force and magnetic force (Ref. [39]) are used as an aid to visualization of the process of self-induction and transmission line wave propagation by Härtel in his article "A Qualitative Approach to Electricity" available in the Teacher's Guide in the CD with this book.

10.7 RADIATION PRODUCED BY ACCELERATING AND DECELERATING CHARGES

For the generation of a *radiating* electromagnetic wave it was seen in Secs. 10.3 and 10.6 that charges must accelerate (or decelerate). The oscillating and therefore accelerating single charge radiating electromagnetic waves shown in Figs. 10.5 to 10.7 would be so weak that no practical use can be made from it. Figure 10.18(a) depicts the electric field lines produced by an oscillating charge at some instant.

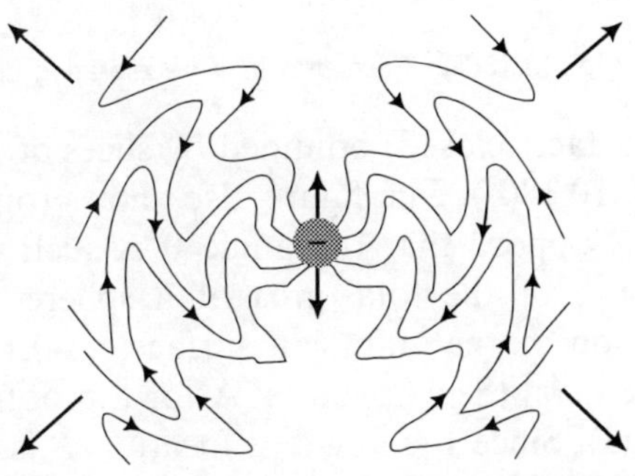

Fig. 10.18(a) Electric field lines of an oscillating point charge

Current in a wire as was seen in Section 1.18 is a start-stop motion of millions and millions of electrons everywhere within the wire with a drift superimposed, the drift being uniform. The interior of the wire is neutral leaving little scope for the generation of curly electric fields. If however, a large number of electrons can be made to accelerate (or decelerate) by an oscillatory motion, the generation of a curly electric field is inevitable. To produce radiation that can propagate over large distances, we need to be able to accelerate and decelerate a whole lot of charges. What better than to consider moving millions and millions of free (conduction band) electrons in small conducting rods, the electrons being mobile.

We will now show how the transverse electric field is proportional to the accelerating charge. The proof is based on the derivation in *Electricity and Magnetism*, Ref. [22] and fine details in *Electric and Magnetic Interactions*, [2010 edition].

Consider a negative point charge q initially moving at speed $v_o << c$ which then stops, decelerating uniformly for duration t_o. At a time $T >> t_o$ after this happens, the pulse of radiation (the "kink" in the electric field) has reached a radius of $R = cT$ as shown in Fig. 10.18(b).

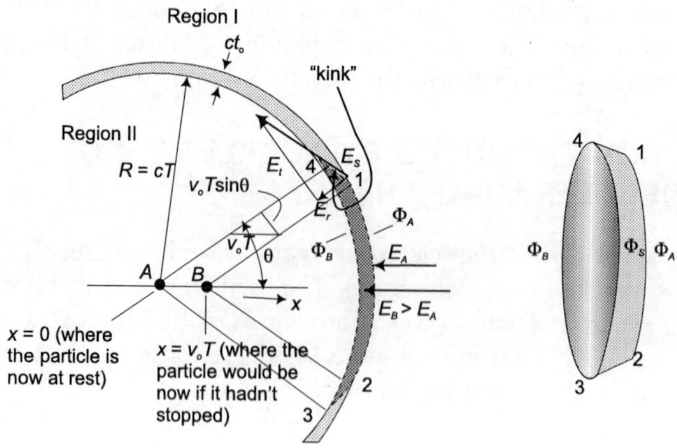

Fig. 10.18(b) Field due to a decelerating charge

Choose a Gaussian surface (closed) bounded by slices of two concentric spheres shown with a dotted line in Fig. 10.18(b). The figure also shows (on the right) a 3-D view of the closed Gaussian surface. Suppose that the angle θ is such that the right portion of the surface represents one-tenth of the total area of a sphere around location *A*. Then the flux Φ_A on the right portion of the Gaussian surface is $-0.1q/\varepsilon_o$, since the total flux on a complete sphere would be $-q/\varepsilon_o$ (See Appendix A for a proof of how the flux on a surface equals the charge enclosed). Since the angle at location *B* is also θ so the flux on the left side of the Gaussian surface is $+q/\varepsilon_o$ (because the unit vector is in the same direction as the electric field lines), and $|\Phi_A|=|\Phi_B|$. Because *B* is closer than *A*, the electric field component $E_B > E_A$, but the right side of the Gaussian surface has less area than the left side. These two effects compensate each other to make the magnitude of the flux on the left side equal to the magnitude of the flux on the right side.

Since $\Phi_A + \Phi_B = 0$, (because the flux "in" should equal the flux "out" from Gauss's law) it must be that the flux Φ_s on the other, ring-shaped portion of the Gaussian surface is also zero, and given the symmetry of the situation it must be that the electric field $\vec{E}_s$ on that surface is everywhere parallel to the surface.

From the Fig. 10.18(b), note that $\vec{E}_s$ has a tangential component $\vec{E}_t$ that is parallel to a perpendicular direction of the (static) electric field of the charge $-q$, which is what is expected for a radiative field and this perpendicular component is also called the *transverse* component.

The triangle in Fig. 10.18(b) for which the upper line is the hypotenuse is shown enlarged in Fig. 10.18(c), and this triangle is similar to the triangle formed by $\vec{E}_s$ and its components.

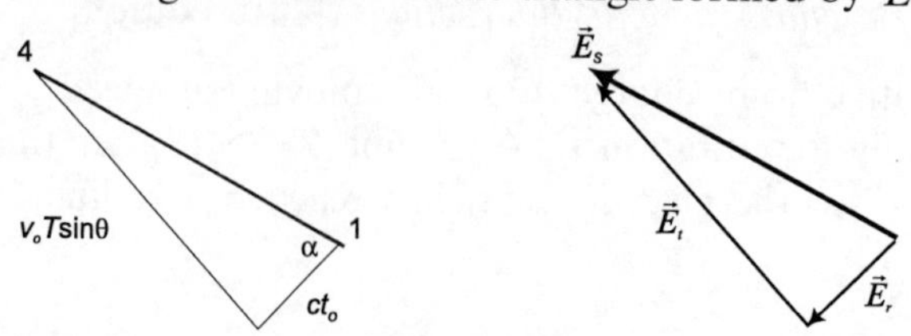

Fig. 10.18(c) Two similar triangles. The triangle on the left is enlarged from the slanted surface in Fig. 10.18(b).

From tan α in Fig. 10.18(c) we obtain a relationship between the tangential and radial components of $\vec{E}_s = \vec{E}_t + \vec{E}_r$ that we will need in a moment:

$$\tan\alpha = \frac{\vec{E}_t}{\vec{E}_r} = \frac{v_o T \sin\theta}{ct_o} \tag{11}$$

E_r must have the same value within the shell thickness that it does in the region II near point 4 which can be shown by drawing a thin Gaussian surface that is just inside and just outside the inner surface of the shell, with very small areas and therefore negligible flux on the top and bottom in the diagram of Fig. 10.18(d).

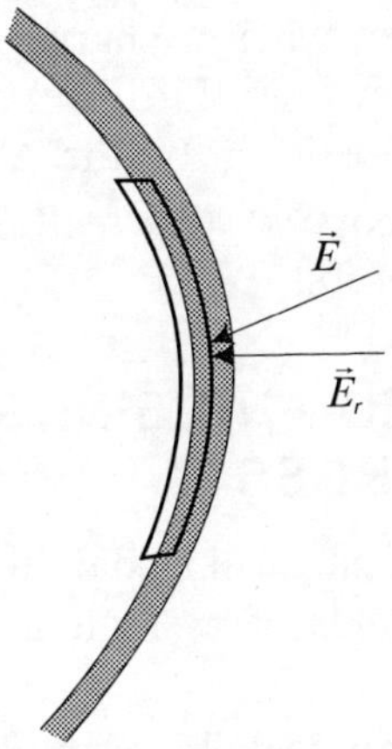

Fig. 10.18(d) A close-up look at a small section of the expanding spherical shell.

Because this Gaussian surface includes no charge, the magnitude of the flux $E_r\Delta A$ on the left and right sides of this Gaussian surface must be equal, which proves that the radial component of the electric field doesn't change as we cross from just inside the inner surface of the shell to just outside the inner surface of the shell. In particular, the radial component of $\vec{E}_s$ must be the same as the electric field of a charge at location B (and we use the approximation that since $v_o T << cT$, the distance from B to the spherical shell is approximately R):

$$E_r = \frac{1}{4\pi\varepsilon_o}\frac{-q}{R^2} \tag{12}$$

Substituting this result into our earlier equation (11) and simplifying, noting that $R = cT$ (so $T = R/c$) and $\frac{v_o}{t_o}$ is the acceleration a:

$$E_t = \frac{1}{4\pi\varepsilon_o}\frac{-q}{R^2}\frac{v_o T\sin\theta}{ct_o}$$

$$= \frac{1}{4\pi\varepsilon_o}\frac{-qa\sin\theta}{c^2 R} \tag{13}$$

i.e., the tangential field component strength is proportional to the acceleration (or deceleration) '*a*' of the charge.

Note how the transverse "kink" field (E_t) falls off with distance as $1/R$ and not as $1/R^2$.

The Kinetic Energy of Charges When they Radiate

It would seem that the energy of charges as they accelerate and produce radiation should in some way reduce. And indeed it is true. We have seen how an accelerating charge radiates and this radiation carries off energy, which must come at the expense of the particle's kinetic energy. A bullet that leaves the barrel of a gun produces recoil and this is predicted by Newton's third law. But the mechanism of reduction in energy of the accelerated charges as they radiate is more than that due to a reaction force.

A stationary charge cannot exert a force on itself (See Appendix A "The electric field due to a point charge"). However, when a charge is accelerated and radiates energy, the rule breaks down (and so does Newton's third law) and the charge does exert a force upon itself. An expression for the force is given by the Abraham-Lorentz formula.

The details are beyond the scope of this textbook and a derivation with explanation is due to David J. Griffiths in Ref. [8].

10.8 UNIFORMLY MOVING AND SINUSOIDALLY OSCILLATING CHARGES

In Section 10.3 it was seen how a single sinusoidally oscillating charge produces radiation. In Section 10.6 was shown the fields of a single accelerated charge and of a decelerated charge.

In a later Section 10.10, we will describe a simplified dipole antenna construction, and the zones of fields surrounding it. The dipole is constructed using conducting rods in which plenty of mobile (or free) electrons are available for the production of electromagnetic radiation.

To produce powerful electromagnetic waves which are of practical importance and for applications in TV and radio broadcasts, telecommunications and in radar, antennas are designed using conducting rods.

Before studying fields due to antennas, it is instructive to study in some detail the fields due to currents of sheets of charges. In Sections 1.17, 1.18 it was shown how conduction band electrons move *randomly* within the lattice of a metal and that they make nearly 10^{14} collisions per second with the lattice ions in conductors. While they do accelerate to speeds of about 1/200th the speed of light the randomness in their motion leads them nowhere. The radiative losses due to collisions of the electrons with the ions in ordinary wires are negligible[8, Chapter 11, Problem in "Point Charges"]. When a field is applied to a conductor, they drift slowly in a preferred direction and this drift is superimposed on their start and stop motion.

When dealing with currents in antennas it is the *drift* of the millions and millions of electrons that is most important and when we speak of *accelerating* charges to produce powerful radiations of electromagnetic waves that possess the properties listed in Section 10.2, it is the *drift* of the millions and millions of electrons that needs to increase (or decrease) with time to produce *time-varying* currents.

Accelerating millions of electrons that creates a time-varying current can produce radiating electromagnetic waves. Readers familiar with the fields surrounding configurations of charges especially the field due to an infinite sheet of charge, complex notation and complex integrals will find it easy to follow the proofs and derivation of the field patterns in this section and therefore a review of these topics is recommended before reading on.

We attempt to create a "kink" in an electric field line by moving charges from rest with uniform velocity. Since changes in the electric field travel at the finite speed of light, it should be a possible radiation mechanism by which some changes can be produced by charges moving with uniform velocity *from rest*.

Suppose we have an infinite sheet of charge located in the *yz*-plane, initially at rest, with surface charge density σ coulombs/meter^2 as shown in Fig. 10.19. The sheet is *not* neutral but charged.

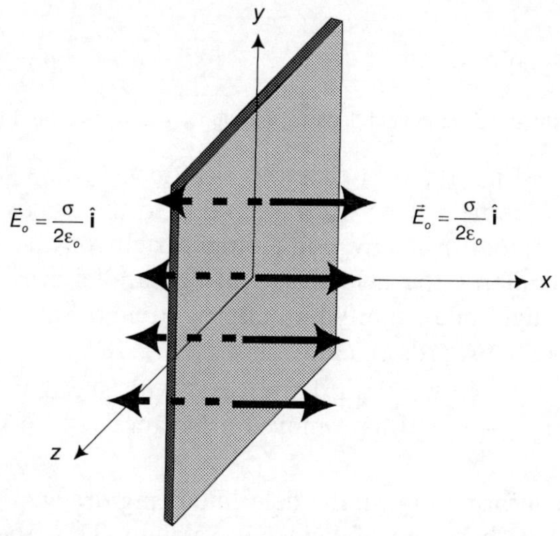

Fig. 10.19 Electric field due to an infinite sheet of charge density σ coulombs/meter^2

The charge *density* σ of the sheet is a surface charge density, which is expressed in coulombs/meter^2 and not in coulombs/meter^3.

The surface charge when the sheet is stationary will give rise to a remarkably uniform static electric field $\vec{E}_o$ that extends to infinity to the right and to the left. It is difficult to create such a uniform field !.[32] To view a simulation of such a field visit http://www.falstad.com/vector3de/ and in the "Field Selection" box choose "infinite plane" and the display to Field vectors.

By the application of Gauss's law [32] or Coulomb's law it can be shown that the field is

$$\vec{E}_o = \begin{cases} +\left(\dfrac{\sigma}{2\varepsilon_o}\right)\hat{\mathbf{i}}, & x > 0 \\ -\left(\dfrac{\sigma}{2\varepsilon_o}\right)\hat{\mathbf{i}} & x < 0 \end{cases} \tag{14}$$

The field $\vec{E}_o$ (Fig. 10.19) does not contain a distance variable and $\hat{\mathbf{i}}$ is the unit vector in the *x*-direction.

At $t = 0$, the sheet of charge which was at rest since $t < 0$ is pulled downward with constant velocity $\hat{\mathbf{v}}_o = -v\hat{\mathbf{j}}$ where $\hat{\mathbf{j}}$ is a unit vector in the y-direction. This is shown in Fig. 10.20(a).

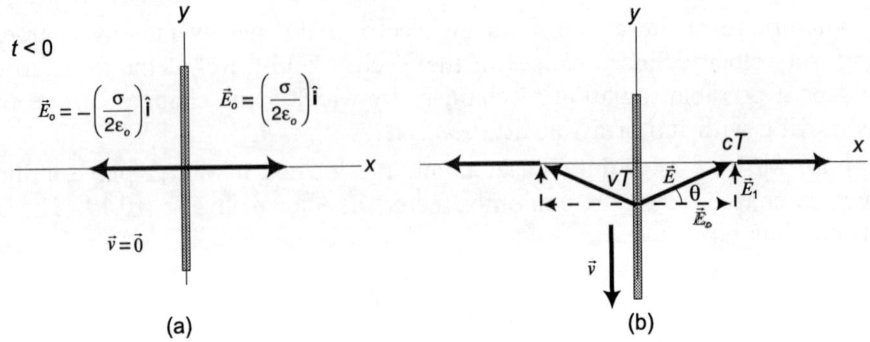

Fig. 10.20 Electric field lines (a) through $y = 0$ at $t < 0$, and (b) at $t = T$.

It can be seen in Fig. 10.20(a) that the electric field line originates from charge at the foot of the line. The field lines will be parallel to the x-axis and of equal magnitude (signifying uniform strength) at any point on the right of the sheet and all lines point towards the right. Similarly, the field lines will be parallel everywhere on the left of the sheet and of equal magnitude (signifying uniform strength) at any point in the left of the sheet and all lines point towards the left.

Since the sheet is pulled downward from rest with speed v, the foot of the field line which was at $y = 0$ at $t = 0$ will have moved a distance $y = -vT$ down the y-axis at time $t = \mathrm{T}$.

Assume that the information of the field line being dragged downward will propagate outward from $x = 0$ at the speed of light c in vacuum. Thus the portion of the field line (remembering that it is the *change* in the electric field that propagates at the speed of light 'c') located at distances beyond cT, i.e., $x > c\mathrm{T}$ along the x-axis doesn't know the charges are moving, and thus has not yet begun to move downward.

The field line therefore appears at time $t = T$ as shown in Fig. 10.20(b), straight out beyond $x = c\mathrm{T}$. Note that the foot of the field line is a distance $y = -vT$ down the y-axis and we guess the orientation of the field line for $x < c\mathrm{T}$ by drawing a straight line that connects the foot of the moving charge to the position on the field line that we know about at time T; $|x| = cT$, the modulus taking care of both field lines, one on the left of the sheet and the other on the right of the sheet.

One may wonder why when there are so many field lines rooted to charges on the plate, then this one field line that we have picked, is inclined. The visualization of field lines like ribbon streamers on a stick that is waved (Section 10.6) would be useful. Application of Gauss's law to the field line can confirm this.[8,22] This is not different from the field line that we drew of a single charge that was accelerated or stopped in Section 10.6, and as a matter of fact, all the field lines of all charges on the sheet will be inclined.

By pulling down on the charged sheet, we have created a perturbation in the electric field shown by $\vec{E}_1$ in addition to the static field $\vec{E}_o$. Thus, the total field for $0<|x|<cT$ is

$$\vec{E} = \vec{E}_o + \vec{E}_1 \tag{15}$$

As shown in Fig. 10.20(b), the field (total) vector also must be parallel to the field line connecting the foot of the field line and the position of the field line at $|x| = cT$. Hence

$$\tan\theta = \frac{E_1}{E_o} = \frac{vT}{cT} = \frac{v}{c} \tag{16}$$

where $E_1 = |\vec{E}_1|$ and $E_o = |\vec{E}_o|$ are the magnitudes of the fields, and θ is the angle with the x-axis. Using Eq. (16), the perturbation field is

$$\vec{E}_1 = \left(\frac{v}{c}E_o\right)\hat{j} = \left(\frac{v\sigma}{2\varepsilon_o c}\right)\hat{j} \tag{17}$$

where we have used $E_o = \sigma/2\varepsilon_o$. Note how an electric field perturbation has been generated and the expression tells us how large the perturbation field $\vec{E}_1$ is for a given speed of the sheet of charge, v.

The direction of $\vec{E}_1$ is such that the forces it exerts on the charges in the sheet resist the motion of the sheet. That is, there is an upward electric force on the sheet when we try to move it downward.

Once the charged sheet is in motion, a sheet of surface current density (current per unit length)

$\hat{\mathbf{K}} = -\sigma v\hat{j}$ is present. From Ampere's law therefore, a magnetic field appears in addition to $\vec{E}_1$.

$\hat{\mathbf{K}}$ is the conventional symbol used for sheet currents and is not to be confused with the unit vector $\hat{k}$.

The strength of this magnetic field is given by

$$\vec{B}_1 = \begin{cases} +(\mu_o\sigma v/2)\,\hat{k}, & x>0 \\ -(\mu_o\sigma v/2)\,\hat{k}, & x<0 \end{cases} \tag{18}$$

where, $\hat{k}$ is a unit vector in the z-direction.

As one moves from negative to positive values of x, (across the current sheet) the magnetic field changes direction. The configuration of the field due to the downward current is shown in Fig. 10.21 for $|x| < cT$.

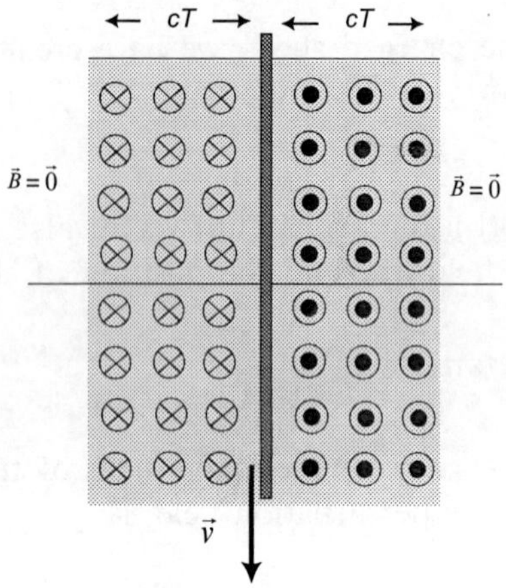

Fig. 10.21 Magnetic field at $t = T$

Note from the figure that the information that the charged sheet has started moving, producing a *current* sheet and associated magnetic field, can only propagate outward from $x = 0$ at the speed of light c. Therefore, the magnetic field is still zero, $\vec{B} = \vec{0}$ for $|x| > cT$.

It is seen that the magnetic field $\vec{B}$ generated by the current sheet is perpendicular to E_1 with a magnitude $B_1 = E_1/c$, as expected and constitutes a transverse *electro-magnetic* wave.

The perturbation electric and magnetic fields carry energy and the fields carry exactly the same amount of energy flux to the left as the energy flux they carry to the right since the magnetic field $\vec{B}_1$ changes direction across the plane $x = 0$.

The energy flowing to the right is given by

$$\vec{\mathbf{S}} = \frac{1}{\mu_o}(\vec{\mathbf{E}}_1 \times \vec{\mathbf{B}}_1) = \left(\frac{1}{\mu_o}\right)\left(\frac{v\sigma}{2\varepsilon_o c}\hat{\mathbf{j}}\right) \times \left(\frac{\mu_o \sigma v}{2}\hat{\mathbf{k}}\right) = \left(\frac{v^2\sigma^2}{4\varepsilon_o c}\right)\hat{\mathbf{i}} \tag{19}$$

This is half the total energy and the other half is the energy flowing to the left. So the total energy flux carried off by the perturbation electric and magnetic fields we have generated is exactly equal to the rate of work per unit area to pull the charged sheet down against the tension in the electric field. Thus we have generated perturbation electromagnetic fields that carry off energy at exactly the rate that it takes to create them.

Where does the energy carried off by the electromagnetic wave come from?

This energy is provided by the external agent who originally "shook" the charge (on the infinite sheet) and had to do work to get it moving against the drag of $\vec{E}_1$ that resists the motion of the sheet.

An exactly analogous situation exists when one asks where the energy carried by a wave on a string comes from. The agent who originally shook the string to produce the wave had to do work to shake it against the restoring tension in the string, and that agent is the ultimate source of energy carried by a wave on a string.

Fields of Oscillating Charges on an Infinite Plane

An infinite plane full of source charges, can be made to oscillate so all charges or electrons oscillate together, all having the same amplitude and phase. We will in our discussion ignore the presence of the static field due to the infinite charged sheet (or plane) since due to the principle of superposition, it will carry in all our expressions for the total field and can be dropped to reveal the changing field components.

Shown in Fig. 10.22 is a sheet of oscillating charges in the *x*-*y* plane.

To compute the field at a point P far away on the *z*-axis, assume there are σ charges per unit area of the plane, and that each one of them has a charge q.

All of the charges move with simple harmonic motion, with the same direction, amplitude, and phase. Now, let the motion of each charge, *with respect to its own average position*, be $x_o \cos \omega t$. Using complex notation, the motion (its displacement) can be described by the real part of $x_o e^{i\omega t}$.

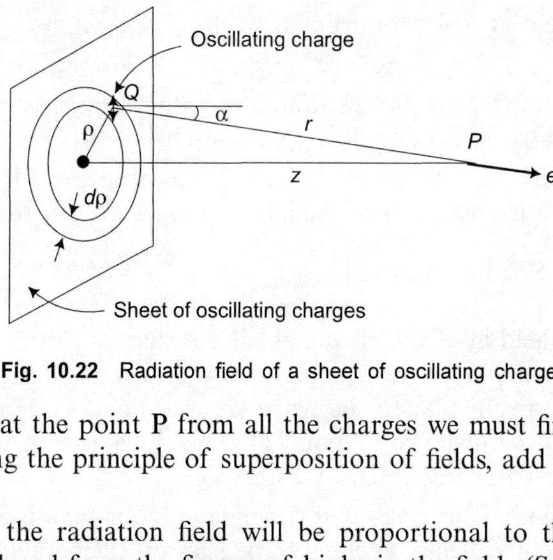

Fig. 10.22 Radiation field of a sheet of oscillating charges.

To find the field at the point P from all the charges we must find the field there from *each q*, and then using the principle of superposition of fields, add the contributions from all the charges.

The strength of the radiation field will be proportional to the acceleration of the charge, as can be deduced from the figures of kinks in the fields (Section 10.6, Fig. 10.17) of charges that accelerate from rest and stopped when moving with uniform velocity.

The radiative electric field of an oscillating point charge at a point far away is given by

$$(\text{Electric field at P from charge at Q}) \approx \frac{-q}{4\pi\varepsilon_o c^2} \frac{d^2 e_{r'}}{dt^2} \tag{20}$$

where $e_{r'}$ is the unit vector in the direction from the point P where the **E** is being measured.

This is the same formula for the transverse field that was derived with reference to the Fig. 10.18(b) with the difference that the acceleration '*a*' of the particle in Eq. (13) is replaced by $\frac{d^2 e_{r'}}{dt^2}$, the acceleration of the unit vector.

If the charge is moving in a very small motion and it is laterally displaced $x(t)$, then the angle that the unit vector $\boldsymbol{e}_{r'}$ is displaced is x/r. Since r is practically constant, the x component of $\frac{d^2 e_{r'}}{dt^2}$ is the acceleration of x itself at an earlier time. Therefore, the electric field in the x-direction at P is

$$E_x(t) = \frac{-q}{4\pi\varepsilon_o c^2} a_x\,(t - r/c) \tag{21}$$

where, a_x is the acceleration. The acceleration of the charges executing simple harmonic motion is obtained by differentiating the displacement $x_o e^{i\omega t}$ twice and is $-\omega^2 x_o e^{i\omega t}$ (and is the same for every charge).

Taking into account the finite time of propagation of changes in the field, the electric field at the point P due to a charge at the point Q is proportional to the acceleration of the charge q, at the earlier time $t' = t - r/c$, where r/c is the time it takes the waves to travel the distance r from Q to P.

Therefore, the field at P is proportional to the acceleration a_x and is

$$-\omega^2 x_o e^{i\omega\,(t-r/c)}. \tag{22}$$

The electric field at large distances from a radiating charge is inversely proportional to distance r as was shown in Section 10.7 equation (13).

Using the value for the acceleration a_x in (22) as seen from P in formula (21) for the electric field at large distances from a radiating charge, we get, therefore

$$(\text{Electric field at P from charge at Q}) \approx \frac{q}{4\pi\varepsilon_o c^2}\,\frac{\omega^2 x_o e^{i\omega(t-r/c)}}{r} \tag{23}$$

To get the total field at P, the effects of all the charges in the plane should be added. The acceleration a_x should be really $a_x \cos\alpha$ where α is the angle the unit vector $e_{r'}$ subtends with the perpendicular to the plane sheet at point Q. However, point P is so far away compared with the distance of point Q from the axis that the cosine factor can be left out which would be nearly equal to 1 anyway.

To our approximation the field at P depends only on the distance r, so all charges at the same r produce equal fields. Therefore, the sum of all the fields of those charges in a ring of width $d\rho$ and radius ρ is first computed.

Then, following a procedure quite similar to the one followed in Appendix A, "Field of Uniformly Charged Disks" replacing the summation by an integral taken over all ρ, the total field can be obtained due to all the charges on the disk of infinite radius, the sheet being of infinite extent can be considered to be circular with no loss of generality.

The number of charges in the ring is the product of the surface area of the ring, $2\pi\rho d\rho$, and σ, the number of charges per unit area. Thus,

$$\text{Total field at P} = \int \frac{q}{4\pi\varepsilon_o c^2}\,\frac{\omega^2 x_o e^{i\omega(t-r/c)}}{r}\cdot \sigma 2\pi\rho d\rho \tag{24}$$

This integral should be evaluated from $\rho = 0$ to $\rho = \infty$ to cover an infinite circular plane sheet carrying charge of density σ (charges per unit *area* of the plane). The variable

t is to be held fixed while evaluating the integral, so the only varying quantities are ρ and r.

Leaving out all the constant factors, *including the factor* $e^{i\omega t}$, for the moment, the integral of (24) is

$$\int_{\rho=\infty}^{\rho=0} \frac{e^{-i\omega(r/c)}}{r}\rho d\rho \tag{25}$$

To do this integral we need to use the relation between r and ρ:

$$r^2 = \rho^2 + z^2. \tag{26}$$

We take the differential of this equation, knowing that z is independent of ρ, and we get

$$2rdr = 2\rho d\rho$$

Replacing $\rho d\rho$ by rdr in the integral (25) and simplifying, we get

$$\int_{r=z}^{r=\infty} e^{-i\omega(r/c)}\, dr \tag{27}$$

The limits of r are not the same as the limits of ρ. When $\rho = 0$, $r = z$, so the limits of r are z to infinity. Therefore, the integral in (27) evaluates to

$$-\frac{c}{i\omega}\left[e^{-i\infty} - e^{-(i\omega/c)z}\right] \tag{28}$$

where ∞ is used in place of $(\omega/c)\infty$, since they both just mean a very large number.

Now $e^{-i\infty}$ can just be taken to be zero. To see why this is so the reader is referred to details in Chapter 30, Section 7, of Feynman, Leighton, and Sands, The Feynman Lectures on Physics, Vol 1, Addison-Wesley, 1963. The integral of (27) therefore evaluates through (28) to

$$\frac{c}{i\omega}\left[e^{-(i\omega/c)z}\right] \tag{29}$$

Equation (29) is the evaluation of integral of equation (25). Now, combining this with the factors we had left out of equation (24), we get

$$\text{Total field at P} = -\frac{\sigma q}{2\varepsilon_o c} i\omega x_o e^{i\omega(t-z/c)} \tag{30}$$

because, $1/i = -i$, and where, σ is the number of charges per unit area of the plane.

It is interesting to note that $(i\omega x_o e^{i\omega t})$ is just equal to the velocity of the charges (let us remember that we had begun with the field due to the acceleration of a single charge $-\omega^2 x_o e^{i\omega t}$, and it turns out that the resultant field due to *all* the charges in the sheet, is proportional to the *velocity* of the charges), so that we can also write the equation for the field as

$$\text{Total field at P} = -\frac{\sigma q}{2\varepsilon_o c} \times [\text{velocity of charges}]_{\text{at } t-z/c}, \tag{31}$$

Although formula (31) is valid only for distances far from the plane of oscillatory charges, it turns out that the formula (30) or (31) is correct at any distance z, even for $z < \lambda$.

The electric field generated by the oscillation of the current sheet is shown in Fig. 10.23 for the instant when the sheet is moving down and the perturbation electric field is up.

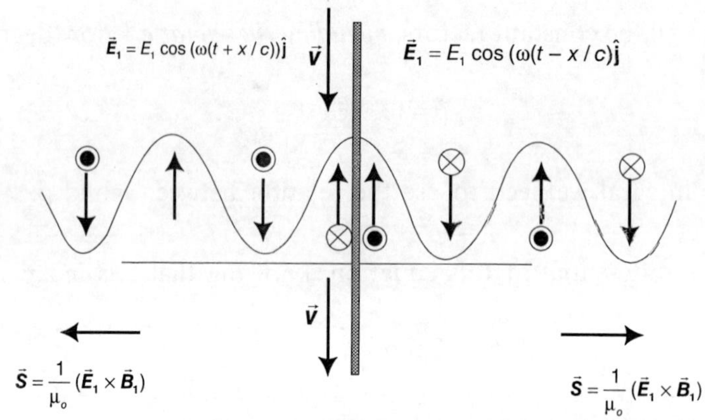

Fig. 10.23 Electric field generated by the oscillation of a current sheet.

The magnetic fields, which point into or out of the page, are also shown. The construction shown assumes that the "feet" of the electric field lines move with the charges, and that information propagates at c shows that electromagnetic waves can be generated by shaking a plane of charge sinusoidally. The wave generated has electric and magnetic fields perpendicular to one another, and transverse to the direction of propagation, with the ratio of the electric field magnitude to the magnetic field magnitude being equal, one of the properties required by an electromagnetic wave to exist and propagate through free space.

The energy flux $\vec{\mathbf{S}} = \frac{1}{\mu_o}(\vec{\mathbf{E}}_1 \times \vec{\mathbf{B}}_1)$ carried off by the wave comes from the agent who shakes the charges, and thereby generates the electromagnetic wave. For more complicated geometries, these statements become much more complicated in detail, but the overall picture remains as presented here.

To view an interactive applet that simulates the electromagnetic radiation generated by an oscillating sheet of charge visit http://ocw.mit.edu/ans7870/8/8.02T/f04/visualizations/light/09-planewaveapp/09-planewaveapp320.html.

The reader should be careful in visualizing electromagnetic wave propagation as *pulsations* of the electric and magnetic fields that travel through space and note that the fields do not move from point to point in space. See "Propagation" Section 10.2.

Fabricating a set up with a charged infinite sheet and moving such a sheet to produce radiation is a difficult task.

In Section 10.7, it was seen that an accelerated charge has the necessary "push" (and pull) to generate electromagnetic waves. Powerful forces if applied over short intervals of time to accelerate these can convert energy very rapidly to produce powerful electromagnetic waves comprising strong electric and magnetic fields and therefore suitable for use in wireless broadcasts and communications.

We will learn soon how to produce such waves especially for propagation to regions where there was previously no existing field which was not the case with the infinite sheet of charge, which had a static electric field set up extending all the way to infinity!

Assume an infinite plane sheet of charge was moving with uniform velocity since the beginning of time before we marked $t = 0$, then at $t = 0$ and even after $t = 0$. Will the charge produce a kink in the electric field as was shown in Fig. 10.20(b)?

To see what happens read the note in Section 10.12.

10.9 RADIATION BY LIGHT BULBS, PHOTON EMISSION AND SPARKS IN AIR

A light bulb is made by positioning carefully a helically wound tungsten wire in a sealed glass bulb which either is evacuated or filled with an inert gas (Helium or Argon) that won't react with the filament. Such a bulb is also called an incandescent lamp. Incandescence is the emission of electromagnetic radiation (including visible light) from a hot body as a result of its temperature.

If an open tungsten wire were to be heated by passing a current through it, it will oxidize and melt or burn away. Filling the glass envelope of a light bulb with an inert gas helps prevent oxidization and extends the normal life of the filament.

When an incandescent light bulb is energized a current in the wire filament is established and the temperature of the filament rises in the manner that was described in Section 1.18, Fig. 1.26. The conduction band electrons gain kinetic energy and pass this to the vibrating lattice ions (atomic cores) causing an increase in the intensity of vibrations of the cores. The atomic cores of the tungsten filament vibrate very intensely and are the cause of the heat produced.

When the temperature to which the filament is heated is around 2,500°C the vibrating atomic cores which are really oscillating charges (Section 10.3) emit visible electromagnetic waves.

The voltage source could be either dc or ac, but the mechanism of production of electromagnetic radiation remains essentially the same.

When an incandescent light bulb is switched ON and kept ON for a long time, won't the filament melt and burn off? (Reproduced by permission, Chabay and Sherwood, 2010 edition).

A quantitative analysis by applying the energy conservation principle to the energy conversion mechanisms in the wire will help us understand why the filament does not melt. For each electron that moves through a tungsten filament wire of length L (say about 150 ohms resistance which is roughly the hot resistance of a 100 Watt/120 Volt light bulb), due to an electric field E, the electric potential energy of the system (electron plus wire) decreases by an amount $e\Delta V = eEL$ where ΔV is the potential difference across the wire. This loss of potential energy is equal to the gain in kinetic energy of the electron, which is dissipated in collisions with the atoms in the wire. As the wire's temperature increases, the rate of energy transfer to the surroundings increases in the form of emission of light. The temperature of the wire stops rising once the temperature is high enough that the rate of emitted energy (as light) has become equal to the rate of energy dissipation in the wire.

Photon Emissions

(Chabay and Sherwood 2011, used by permission)

A photon is a particle that has only kinetic energy – its rest energy is zero. A photon can never be at rest; it may be thought of as a packet of energy traveling at the speed of light. A microscopic object such as an atom can gain energy by absorbing a photon; if this happens, all of the photon's energy is transferred to the object. Conversely, an object can lose energy by emitting a photon with a particular amount of energy.

The general term applied to photons is "electromagnetic radiation", or simply light, and their production can be understood by considering the electronic energy levels for a hydrogen atom shown in Fig. 10.24.

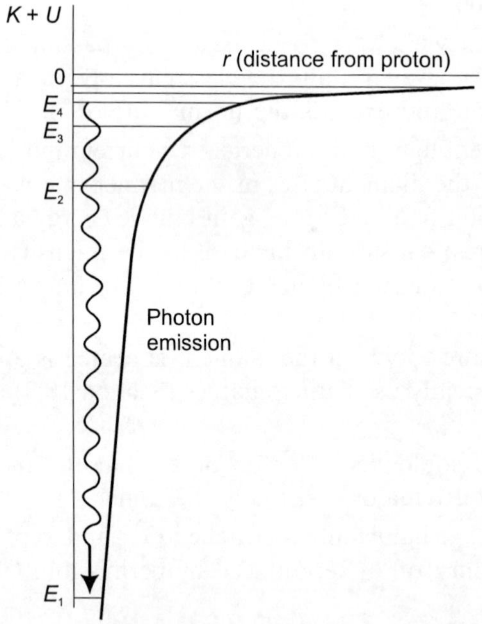

Fig. 10.24 Electronic energy levels for a hydrogen atom. An atom can drop to a lower energy level by emitting a photon whose energy corresponds to the difference between the energies of the levels. By convention a squiggly line with arrowhead indicates a photon.

The potential energy curve describes the potential energy of the electron-proton system as a function of separation. These energy levels denote different bound states of the electron + proton system. We had described for such a system in Section 9.7 (refer Fig. 9.20), the variation in potential energy levels with distance of an electron from the nucleus or proton in the case of the hydrogen atom.

In Fig. 10.24 the horizontal lines represent the possible bound energy states for the hydrogen atom; bound states with other values of $(K + U)$ where K = kinetic energy and U = potential energy are not observed. Unbound states $(K + U \geq 0)$, however are not quantized. (See "What is temperature?" in Appendix B on quantization levels of atomic energies).

When a hydrogen atom drops from a higher energy level E_4 to a lower energy level E_1 it emits a photon with the associated energy $E_4 - E_1$ (Fig. 10.24). A collection of hydrogen atoms can emit photons whose energies correspond to the differences in energy between any two electronic energy levels of the hydrogen atom. The light emitted by excited atomic hydrogen is called the "hydrogen emission spectrum", and the bright lines of light of particular energies in the spectrum are called "spectral lines".

This is the process of emission of photons that makes light emitting diodes (LEDs) emit colored light (see Section 9.9, *recombination* in forward-biased diodes).

The Birth of Quantum Physics

Physicists made several attempts to model atomic vibrations to explain the shape of the radiation vs. wavelength curves of blackbody radiation. The *irregular* motion of electrons, atoms, and molecules as the temperature is raised, is called "heat" and is more, the more rapid the motion (see "Chemical bonds can be modeled as springs" in Appendix B). Irregular thermal motion produces irregular oscillatory electron motion (the electron being lighter than the nucleus), representing a continuous spectrum of frequencies. Each oscillation at a particular frequency emits and also receives electromagnetic radiation. See also "How is heat produced in conductors and resistors?" in Appendix B.

Blackbody radiation refers to the spectrum of light emitted by any heated object; common examples include the heating element of a toaster and the filament of a light bulb. The spectral composition of a heated body depends on the materials of which the body consists, and is not the case for an ideal radiator or absorber which absorbs and thus emits radiation of *all frequencies equally and fully*. A radiator/absorber of this kind is called a blackbody. A good approximation of a blackbody is a piece of coal or, better yet, a cavity in a piece of coal that is visible through a small opening. As the piece of coal is heated to higher and higher temperatures, one first observes a dull red glow, followed by a change in colour to bright red; as the temperature is increased further, the colour changes to yellow and finally to white.

The thin curved lines in the figure are a plot of measurements of the spectral emission of blackbodies (blackbody radiation intensity vs. wavelength) at three different temperatures.

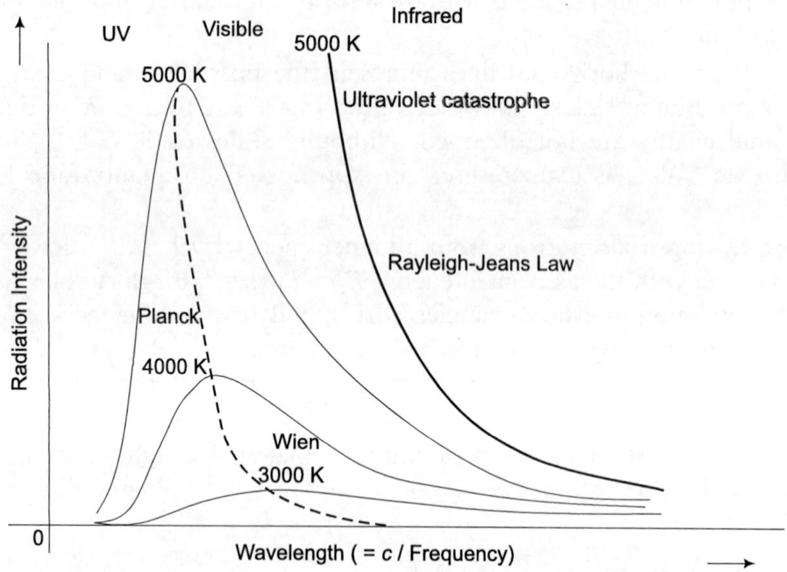

Black body radiation intensity vs. wavelength.

The Rayleigh–Jeans Law predicted an energy output that diverges towards infinity as wavelength approaches zero (as frequency tends to infinity) but the measured emission agreed with the Rayleigh–Jeans law at low frequencies (thick curved line in the figure).

When the temperature of a blackbody radiator increases, the overall radiated energy increases and the peak of the radiation curve moves to shorter wavelengths. The product of the peak wavelength and the temperature is a constant and is called "Wien's Displacement Law" (dashed curved line).

The inconsistency between observations and the predictions of the Rayleigh-Jeans Law, a law based on classical physics, is commonly known as the ultraviolet catastrophe.

Upto the end of the 19th century, none of the theories could explain the peaks in the radiation vs. wavelength curves nor why the *colour* of the light given off by a hot object changes with temperature (dashed curved line).

The resolution of the ultraviolet catastrophe was a foundational aspect of the development of quantum mechanics in the early 20th century.

Max Planck hypothesized that the equations of motion for light describe a set of harmonic oscillators, one for each possible frequency from which he derived an approximate mathematical function for the blackbody radiation spectrum.

As an aid to visualization of the model of a blackbody as a cavity filled with photons bouncing back and forth, see the Fig. 2.1 in the interesting and informative article *Mathematical Physics of Blackbody Radiation* by Claes Johnson, a link to which is available in the Links_to_animations_articles file in the CD.

Max Planck's (1858-1947) first formula for the emitted energies was based on ideas he borrowed from Boltzmann and Maxwell's statistical approach to thermodynamics (see "Quantum Mechanics, Probabilities And Orbitals of Electrons" and "Density of States" in Appendix B about statistical approaches to analyse the Energies and their distribution of large numbers of particles). Also, see Note in "Computing Pn(0)" in Appendix E for the Maxwell-Boltzmann distribution of energies of gas particles. But, since the curves generated by Planck's first formula still did not fit the observed radiation vs. wavelength data, Planck put forth the additional condition that the vibrational atomic energies are not a continuous infinitely divisible quantity, but a discrete quantity composed of an integral number of finite equal parts.

According to Jan Faye, *Copenhagen Interpretation of Quantum Mechanics*, "In 1900 Max Planck found that he could correctly predict the properties of blackbody radiation if he assumed that transfers of energy in the form of electromagnetic radiation were quantized – that is, an energy transfer could take place only in fixed amounts, rather than having a continuous range of possible values. He discovered that the radiation energies are separated by the value $h\nu$, where ν is the frequency and h is a new constant, the so-called Planck constant" (See Chapter "Energy Quantization" in Ref. [41]). "According to classical physics (Rayleigh-Jeans Law) the intensity of this continuous radiation would grow unlimitedly with growing frequencies, resulting in what was called the ultraviolet catastrophe. But Planck's suggestion was that if blackbodies only exchange energy with the radiation field in a proportion equal to $h\nu$ that problem would disappear". The fact that the absorption and the emission of energy is discontinuous is in conflict with the principles of classical physics. Jan Faye's article is published in *The Stanford Encyclopedia of Philosophy* (Fall 2014 Edition), Edward N. Zalta (ed.). See Links_to_animations_articles for a link to the article.

The principle of Planck's law, formally proposed in 1901, is that electromagnetic energy could be emitted only in quantized form, in other words, the energy could only be a multiple of an elementary unit: $E = h\nu$ where h is Planck's constant and ν is the frequency of the radiation. This relation is used in Planck's energy distribution function:

The power emitted per unit area of the blackbody, per unit solid angle that the radiation is measured over, per unit frequency is

$$\frac{2h\nu^3}{c^2} \cdot \frac{1}{e^{\frac{ch}{k\lambda T}} - 1},$$

where $\dfrac{1}{e^{\frac{ch}{k\lambda T}} - 1}$ is the Planck distribution function.

Also, see Chapter Phonons II in "Introduction to Solid State Physics" by Charles Kittel, Wiley, 7th edition and articles by Boya and Gearhart on how Max Planck derived the distribution function, links to which are available in the Links_to_animations_articles file in the CD.

The blackbody radiation values calculated using Planck's distribution function accurately matched experimental data (thin curved lines) in the radiation vs. wavelength plot at three different temperatures.

A few years later Albert Einstein used this discovery in his explanation of the photoelectric effect. He suggested that light waves were quantized, and that the amount of energy which each quantum of light could deliver to the electrons of the cathode, was exactly $h\nu$."

Also, according to Bohr, quantum mechanics, as formulated by Heisenberg, was a *rational generalization* of classical mechanics (physics), when the quantum of action (Planck's constant) and the spin property were taken into account.

In one sense, Planck's law can be considered as a unification of the wave and particle model of light in which, the wave property explains interference patterns and the particle model explains how a fixed amount of energy can be delivered to a system by a photon. For a nice explanation of this behavior of light, see the topic "Unifying the two models" in the Section "The Wave Model vs. The Particle Model of Light" in the Chapter "Waves and Particles" in the textbook *Matter and Interactions II, Electric and Magnetic Interactions* by Chabay and Sherwood, Wiley, USA, 2011.

The discreteness in electromagnetic radiation, which may not be physically observable because of the extreme tininess of the Planck constant embodied in Planck's law, heralded the birth of Quantum Physics. The reader should be careful in noting that the energy levels of the electron orbitals can only take on discrete values (Fig. 10.24) and this should not be confused with Planck's law which states that the *emitted* radiation energy is quantized. The discretization of *energy levels* of electrons is described by Rydberg's formula (single hydrogen atom) and solutions to Schrodinger's wave equation, both of which have included in them Planck's constant Ref. [17]. This discretization of the energy levels is what Bohr had referred as a "rational *generalization* of classical mechanics when the *quantum* of action (Planck's constant) and the spin property were taken into account".

I urge the reader to view a lecture by Prof. Leonard Susskind, Stanford University, USA on aspects of the probability approach to "Classical Randomness" and "Quantum Randomness" and its implication or effects on the conservation of energy with illustrative examples, a link to which is provided in the Links_to_animations_articles file in the CD.

For comprehensive descriptions of the ideas exchanged across the disciplines of mechanics and electricity and magnetism by renowned scientists Kirchhoff, Boltzmann, Planck, Rayleigh-Jeans see *The Theory of Electrons* by H. A. Lorentz, pages 63 to 97. Lorentz's lectures in the book trace the evolution beginning from the classical electrodynamics then developing along the lines of classical thermodynamics to the statistical interpretation of thermodynamics (second law) leading to Planck's law. A link to this textbook is given in the Links_to_animations_articles in the CD.

The terminology of Electrodynamics and the origin of Quantum Electrodynamics (QED) and Quantum Field Theory (QFT) that was introduced to explain certain specific aspects of the interaction between an electron and radiation, is described in Section 6.6. More details about QED and QFT are provided in Section F.5, Appendix F.

Note The word "photon" was adopted by Arthur H. Compton around 1925 from suggestions made by other scientists. In the period the photon was coined, light was described by a semiclassical model and not QED. That is, light was described by Maxwell's equations but it's emission and absorption by material objects (e.g. black body radiation) was in quantized amounts. However, it is the theory of Quantum Electrodynamics (QED) which has explored the nature and properties of the photon.

Readers should note that the description of circuit processes in this textbook treats the electromagnetic field *classically*, which is considered a limit of large quantum numbers as we will learn in Section F.5, Appendix F.

The Frequency and Wavelengths of Visible Light

The speed of light in free space is a physical constant defined as 299,792,458 meters per second. It is often denoted by the symbol *c* - originating from the Latin word *celeritās* (speed).

A typical human eye will respond to wavelengths from about 390 to 700 nm (1 nanometer (nm) = 10^{-9} meters). In terms of frequency, this corresponds to a band in the vicinity of 430–790 THz (1 TeraHertz (THz) = 10^{12} Hz).

It should be noted that such high frequencies of radiation are produced by the high frequency vibrations of the lattice ions in the tungsten filament of a light bulb when energized.

Sparks in Air

(Chabay and Sherwood 2011, used by permission)

You may occasionally see a spark when you flip a switch, unplug a power cord, or when you separate clothing that has stuck together in the dryer or pull a blanket away from a sheet in the winter. During a spark, charge is transferred from one object to another. How can such sparks occur in air?

We will describe briefly one of the models that explains the origin of sparks in air though a full description of the phenomenon and the models physicists use to explain it are beyond the scope of this textbook.

Supposing that we have two small metal spheres, one negatively charged and one positively charged as shown in Fig. 10.25(a). There is an electric field *E* in the space between the two spheres due to the charge on them.

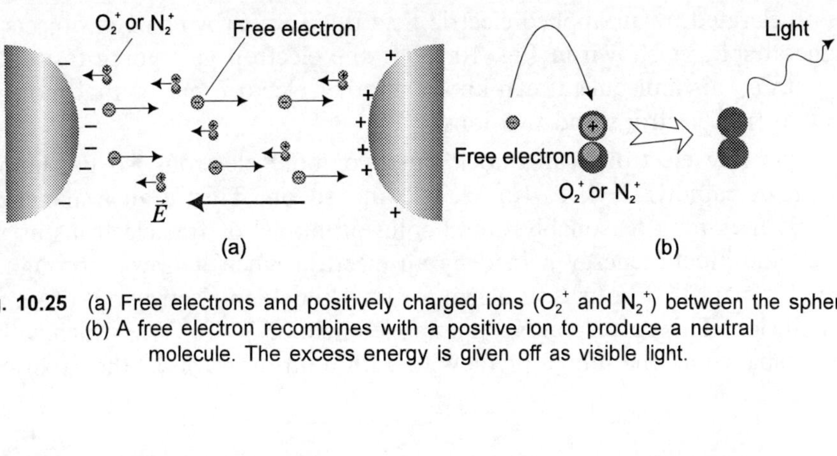

Fig. 10.25 (a) Free electrons and positively charged ions (O_2^+ and N_2^+) between the spheres. (b) A free electron recombines with a positive ion to produce a neutral molecule. The excess energy is given off as visible light.

When the two spheres are brought close together (but not touching), there is a spark between them. How does this happen?

If many of the oxygen and nitrogen molecules in air became ionized as shown in Fig. 10.25(a), we would have a gas composed of charged particles, all of which are free to move.

Assuming that air can somehow become ionized, with plenty of mobile electrons and ions, electrons drift toward the positive ion. The positive ions in air drift (much more slowly) toward the negative electron. Occasionally a free electron comes near enough to a positive ion (not necessarily the original ion) to be attracted and recombine to form a neutral molecule. When the free electron and the ion recombine, there is a transition from the high-energy unbound state (free electron and ion) to a low-energy bound state, with the emission of a photon as shown in Fig. 25(b).

It is important to understand that the emission of light goes on simultaneously with charge conduction through the ionized air. As long as the applied field is big enough, neutral molecules are continually being ionized, at the same time that some free electrons and ions are recombining to make neutral molecules (with the emission of light).

How does the air become ionized?

It is observed experimentally that an electric field intensity of about 3×10^6 V/m is sufficient to ionize air and make it a conductor. However, the electric field strength required to pull the outer electron out of an atom is of the order of 1.4×10^{11} V/m. The enormous difference between the experimentally observed field of 3×10^6 V/m and the value necessary to ionize an atom means that there is no possibility of direct atomic ionization by the applied electric field.

However, dramatic events such as "muons" produced by cosmic rays, and fast moving charged particles such as electrons, positrons, or alpha particles (helium nuclei) emitted by radioactive isotopes present in trace quantities in materials in and around you can knock out electrons from atoms. There are fast-moving charged particles passing through your body at this very moment, ionizing some atoms and molecules in your body! The reader can find detailed descriptions of the process in Ref. [2]. The ionization due to muons and energetic particles emitted by trace radioactive isotopes is not extensive, because if it were, air would be a good conductor at all times.

If there happens to be a single free electron in the air (due to muons or radioactivity), it can be accelerated by an applied electric field due to nearby charged objects such as the charged metal spheres shown in Fig. 10.25. If the electron gets going fast enough before colliding with an air molecule, it can knock another electron out of that molecule, so now there are two free electrons and two ions.

These two free electrons can knock out two more electrons, so the number of free electrons grows rapidly: 2, 4, 8, 16, 32, 64, and so on. This is an avalanche, or "chain reaction". By making a reasonably sound collision model of free electrons in air, it can be shown[2] that the kinetic energy gained by an electron when it travels through a potential difference of a mere 14.4 Volts is sufficient for it to knock an electron out of an air molecule during a collision. This corresponds to an applied field of 3×10^7 V/m which will be required to initiate a spark and the model predicts to within an order of 10 the experimental value of 3×10^6 V/m for the applied field to cause ionization of air. The most probable process is that the first ionization event is by one of the dramatic events of a muon or particles from radioactive decay and the chain reaction follows.

The emission of light is a sort of side effect of the ionization of the air when the neutral molecules are continuously being ionized, at the same time that some free electrons and ions are recombining emitting light in the form of sparks and making neutral molecules.

10.10 ELECTROMAGNETIC RADIATION BY ANTENNAS

Antennas are structures with wires and rods assembled in specified shapes which when excited by time-varying voltages are capable of transmitting (or producing) electromagnetic radiation essentially in the radio frequency bands spanning 3 KHz to 300 GHz. (1 Giga Hertz = 10^9 Hz.)

Antennas are usually made with very low resistance wires and the mechanism of radiations they produce are vastly different from the mechanism of light (electromagnetic waves) generation (Section 10.9).

The radio frequency electromagnetic (em) waves produced by antennas are based on the process of accelerating charges in large numbers, electrons to be more precise. The mechanism is different from the mechanism of generation of light by light bulbs which use high resistance tungsten wire and at high temperatures the atomic core vibrations emit em waves. The antenna is capable of emitting em waves of the frequency of the carrier frequency voltage unlike the high frequencies emitted by the tungsten wire of the light bulb where light radiations of frequencies ranging in the TeraHertz are hard to control precisely.

The Hertzian (or Ideal) Dipole

In Section 10.6 it was shown how a kink in an electric field that propagates outward is produced when a single charge is accelerated from rest and when a charge moving with uniform velocity is stopped. Before we begin the discussion on the mechanism of radiation by an antenna, it is useful to study the radiation produced by an ideal or Hertzian dipole.

Hertz's first experimental apparatus to create electromagnetic waves consisted of two brass knobs separated by a small distance. Both knobs were connected to an induction coil driven by an oscillator. When the coil current was interrupted sparks were produced across the knobs. This served as a transmitter, and for the receiver, a loop of wire also connected to two brass knobs was kept a few yards from the transmitter. When the transmitter was switched ON and a series of sparks were produced between the transmitter knobs, then electromagnetic waves spread out from the transmitting knobs and currents were induced in the receiver loop of wire and sparks were produced between the receiver knobs.

An oscillating charge sheet produces electromagnetic waves as was shown in Section 10.7. Time-varying currents will be shown to hold the key to producing electromagnetic waves from stationary conducting rods because such currents are made by charges that are accelerated (or decelerated).

A tiny length of a very thin wire carrying a time-varying current is a current element and if the ends of the wire are terminated on two tiny disks or spheres, the charges of the current element will accumulate and disperse from them. If the current of the element is time-varying then the current maybe written $\frac{dI_z}{dt} = q_l \frac{dv_z}{dt} = q_l\, a_z$ where, q_l is the charge (in coulombs) per unit length, and v_z is the velocity of the charges (meters/sec) and $dv_z/dt = a_z$ is the acceleration of the charge in (meters/sec^2). If the current is sinusoidal, then the rate of *change* of current $\frac{dI_z}{dt}$ is given by a tangent drawn at '*t*'. Such slope lines are drawn for the input and output voltage of a capacitor subject to a sinusoidal voltage in the sketch below Fig. 2.38, and similar slope lines can be drawn on the current plot. During the first quarter of the positive half-cycle of the current, slope lines pointing upward would represent *acceleration* of charges while during the second quarter of the positive half-cycle, slope lines pointing downward will indicate *deceleration*.

If the distributed capacitance of the wire is negligible (between wire and ground "earth", see Section 5.22) compared to the capacitance between the spheres, then the current in the wire can be assumed to be uniform (See Section 2.9, Fig. 2.31 and the resulting steady current when a "ramp" voltage is applied to a capacitor, and conversely, a steady current to a capacitor produces a ramp voltage across the capacitor). That is, the charge accumulation at the ends can be given by $\frac{dq}{dt} = I \cos \omega t$ where, I represents the steady current amplitude in the wire along its length, however tiny, and cos ωt is the variation imposed on I with respect to time and represents the accelerating (or decelerating) charges.

We had studied in Sections 2.10 and 2.11 that the reactance of a capacitor varies in inverse proportion to the frequency of the applied voltage and to the capacitance. Therefore, more powerful oscillations and radiations due to strong currents are possible if the frequency is increased.

The thin wire with a time-varying current terminated on tiny spheres is practically impossible to realize but it formed the bare essential oscillatory filamentary current (I) element (dl) for Hertz's mathematical approach towards computing the electric and magnetic fields of radiation by a dipole. In his honor, it was named the Hertzian Dipole and is shown in Fig. 10.26.

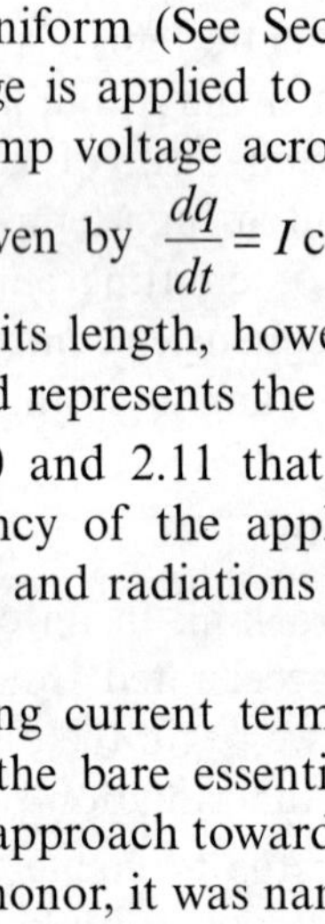

Fig. 10.26 The Hertzian Dipole.

If the current element forms part of a complete circuit, charges will not accumulate at its ends if the current is uniform throughout the circuit, for the current from one element flows into the next. If the situation is depicted using Hertzian dipoles as in Fig. 10.27(a), then the positive charge at the end of one dipole is just canceled by an equal amount of charge at the opposite end of the adjacent dipole.

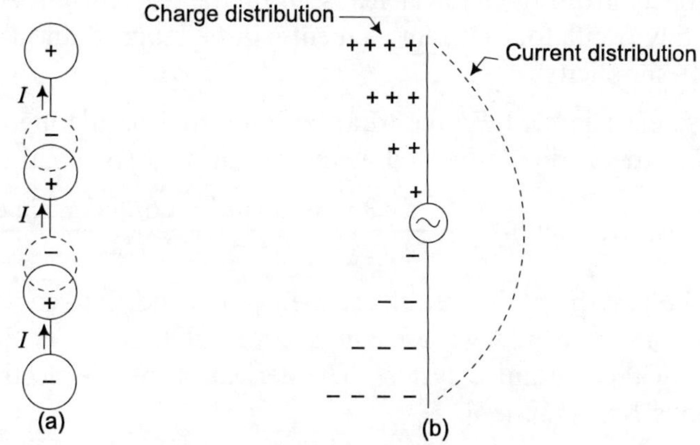

Fig. 10.27 (a) Chain of Hertzian dipoles equivalent of an antenna (b) Current and charge distribution on a linear antenna.

In this case, the electrostatic field that would vary as $1/r^3$ where r is the distance from the dipole will vanish, leaving only induction and radiation fields which we shall soon see in Eq. 32.

However, if the current along the length of the circuit or antenna is not uniform and assumes the distribution shown in Fig. 10.27(b), this could be represented as a chain of current elements, or Hertzian dipoles, having slightly different amplitudes. In this case the adjacent charges do not completely cancel, and there is an accumulation of charge on the surface of the wire, as indicated in Fig. 10.27. These surface charges are responsible for a relatively strong component of electric field strength normal to the surface of the wire. A description of practical dipole antennas is provided later in this section and we will see that the current distribution shown in Fig. 10.27(b) is that of a half-wave dipole antenna.

The astute reader may have noted something peculiar about the distribution. It is drawn along the *length* of the dipole and there does not seem to be anything suggestive of a *time variation* of current from the plot. As a matter of fact, what the distribution represents is a standing wave of current and the plot masks the dynamics of the charge movement. Tangent lines to the current distribution curve, therefore, do not represent the acceleration (deceleration) of charge necessary to produce radiation. This will become clear soon, when we plot the current in a real dipole antenna vs time (Fig. 10.33) and the process which leads to the standing wave of current is described.

The electric and magnetic fields due to the Hertzian (oscillatory) dipole can be computed assuming either an infinitesimal electric dipole or an infinitesimal current element (the Hertzian dipole).

If an oscillating electric dipole, the derivation can be made beginning $\vec{p}(t) = \vec{p}_o \cos \omega t$ where, $\vec{p}(t)$ is a time-varying electric dipole moment qdl of charge q separated by a distance dl.

Such a dipole can be created by making a point charge undergo simple harmonic motion.[8] Alternatively, from the equivalence seen of the tiny current element that feeds the electric dipole, a tiny oscillatory current element can be realized and is preferred by several authors due to its simplicity.

The transverse electric field E_θ due to an infinitesimal oscillating current element such as the Hertzian (or Ideal) dipole located at the origin of a spherical co-ordinate system is

$$E_\theta \; (\textit{transverse electric field}) = \frac{Idl \sin\theta}{4\pi\varepsilon}\left(\frac{-\omega \sin \omega t'}{rv^2} + \frac{\cos \omega t'}{r^2 v} + \frac{\sin \omega t'}{\omega r^3}\right) \tag{32}$$

where, dl is the length of the current element, ω is the frequency of oscillations, t' is the retarded time or $t - r/c$ that we used in Section 10.8, and r is the distance from the origin of the spherical co-ordinate system. The derivation of the electric field can be found in several textbooks Ref. [16], [35].

The $\mathbf{1/r^3}$ term is the *quasi steady-state electrostatic field* of the classical electrostatic dipole formed by the surface charges shown in Fig. 10.27(b). See the derivation in Appendix B "Electric Field due to a dipole" of the field surrounding a dipole and its variation as $1/r^3$ with distance.

The $\mathbf{1/r^2}$ term is the induction field which is due to the magnetic field strength (its origin is the quasi steady-state magnetostatic field) varying with distance as $1/r^2$ obtained from the Biot-Savart law (See Section 5.2) and keeping in mind that the changing magnetic field is associated with a non-coulomb electric field that in turn would vary as $1/r^2$. Since the area of a sphere increases at the rate of r^2, then the quasi steady-state field electrostatic (dipole) and the induction field (magnetostatic origin) both become insignificant a small distance away measured radially around the dipole from the origin, since they vary at the rates of $1/r^3$ and $1/r^2$.

It is the $\mathbf{1/r}$ term which remains significant over large distances and is the term which represents the radiative field. The radiative field diminishes in strength or attenuates proportional to the $1/r$ term and this is because the expanding pulsations of the electric field energy spread over a large spherical surface whose area enlarges as the distance r increases from the radiating source. Also see "Attenuation of a wave" in Section 5.22.

Hertz was careful enough to place the receiving wire loop with the brass knobs far enough to shut out the induction field and the quasi steady-state electrostatic field from the transmitting antenna and observed the receiver's response to the radiation field.

Beginning from the microscopically tiny elementary oscillatory current element and by applying Maxwell's equations including the equations wherein Maxwell's added term for the time varying electric field that is associated with a magnetic field and the equation for a time-varying magnetic field that associates with an electric field (Faraday's law), Heinrich Hertz was able to predict and confirm by measurements the presence of the three component fields of the oscillatory current element. He also thus showed by the Hertzian dipole construct that the oscillating current element and an oscillatory electric dipole are equivalent[16] in terms of the fields they produce which is evident from the elegant mathematical expression for the field in Eq. (29).

Hertz's analysis is the backbone on which rests the analysis and design of all antennas that were constructed since his time and in modern antenna designs.

The Hertzian dipole is the fundamental construct used in realizing practical antennas which connect chains of them capable of producing strong radiation for application in radio broadcasts, radio frequency communication links, satellite communication and radar.

The Hertzian dipole radiation mechanism and derivations of the expression of the electric field can be found in several textbooks and affords a clearer picture of the fields that comprise practical antenna radiation processes which will be discussed next.

Radiation from a Dipole Antenna

A common form of the antenna is a dipole shown in Fig. 10.28 made from two short rods (conducting) that are separated by a small distance and connected to a sinusoidal signal source.

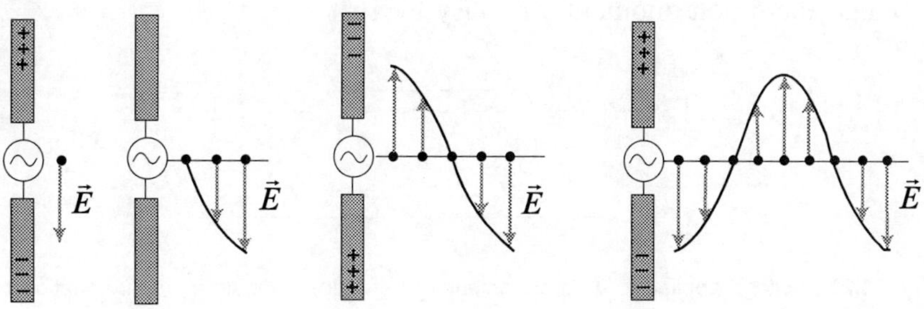

Fig. 10.28 Electric fields produced by an electric dipole antenna.

At time $t = 0$, the ends of the rods are charged so that the upper rod has a maximum positive charge and the lower rod has an equal negative charge. The electric field near the antenna points downward. Treating the antenna like a capacitor, the current will be as shown in Fig. 2.32. The charges begin to decrease and vanish momentarily at $t = T/4$. Subsequently, the polarities of the rods are reversed with negative charges continuing to accumulate on the upper rod and positive charges on the lower until $t = T/2$, when maximum charge accumulation is attained. At this instant, the electric field near the rod points upward and as the charges continue to oscillate between the rods, electric fields are produced and changes (waves) in these fields (and *not* the fields themselves) propagate at the speed of light. There is a current in the dipoles (that exhibits a capacitor like current behavior) and this current sets up a magnetic field encircling the rods.

The behavior of the fields near the antenna is expected to be very different from that far away from the antenna. Shown in Fig. 10.28 is the far field electric field pattern and is representative of the field patterns picked up by radios and mobile phones situated far away from the broadcasting and transmitting antennas.

It is difficult to pick the process of production of electromagnetic radiation due to the accelerating and decelerating electrons (*time-varying* drift current) in the rods from among the considerable jumble of fields, (i) the fields surrounding the charges, and (ii) the dipole (resultant) field (Section 1.14, "*Electric Field of a dipole*"). The radiating field should be invested with the properties which are listed in Section 10.2.

Consider an electric dipole whose dipole moment varies in time according to

$$\vec{\mathbf{p}}(t) = p_o\left[1+\frac{1}{10}\cos\left(\frac{2\pi t}{T}\right)\right]\vec{\mathbf{k}} \tag{33}$$

Remember an electric dipole moment $\vec{\mathbf{p}}$ of a dipole is given by $\vec{\mathbf{p}} = q\vec{\mathbf{d}}$, where $\vec{\mathbf{d}}$ is the displacement vector pointing from the negative charge to the positive charge and $\vec{\mathbf{k}}$ is a unit vector in the direction of propagation. What equation (23) tells us is that the strength of the charge on the dipole varies (ignoring for the moment the mechanism by which such a variation is produced) if the distance between the opposite charges is a constant. We assume that the charges accumulate on the tips of the rods and that this distance between the tips is held constant, the rods being held stationary.

If a sinusoidal voltage of strength and amplitude shown in Fig. 10.29 is applied to a dipole antenna, the dipole moment will vary by 10%.

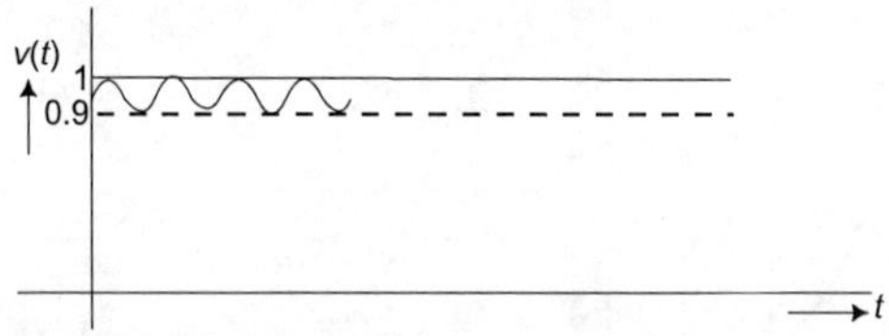

Fig. 10.29 Voltage applied to a dipole antenna to produce a dipole moment variation of 10%.

Figure 10.30 shows the fields at one instant of time. Close to the dipole, the field line direction and thus the Poynting vector is first outward and then inward, corresponding to

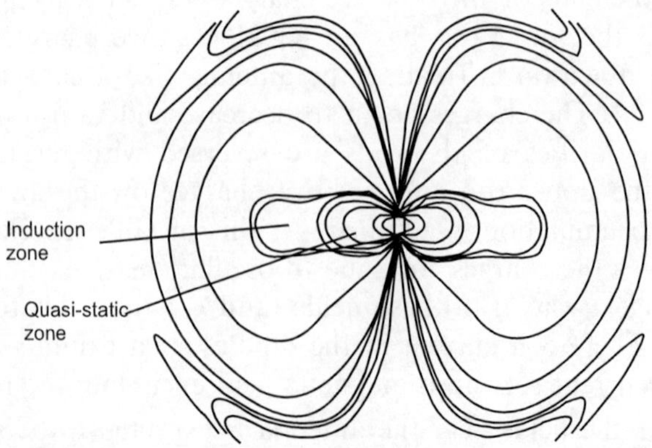

Fig. 10.30 Radiation from an electric dipole in one plane whose dipole moment varies by 10% by the applied voltage of Fig. 10.29.

energy flow outward as the quasi-static dipolar electric field energy is being built up, and energy flow inward as the quasi-static dipole electric field energy collapses. This can be seen in Fig. 10.31 which shows a simplified view of the electric and magnetic fields of an electric-dipole (half-wavelength) antenna and the Poynting vector.

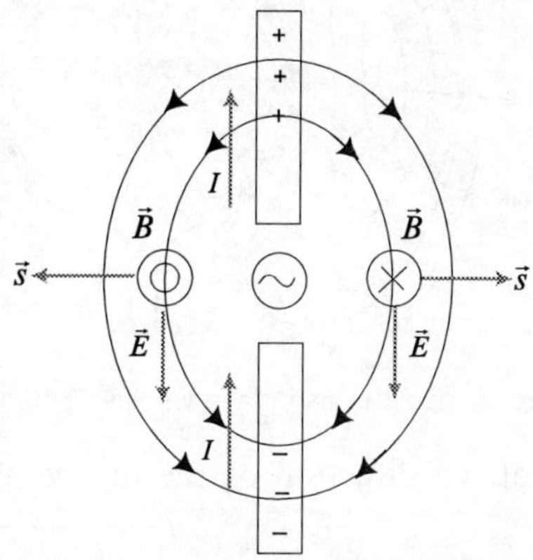

Fig. 10.31 Electric and magnetic field lines produced by an electric dipole antenna.

The dipole *does not reverse* but the time-varying nature of the fields due to variation (by 10%) of the strength of the charges will produce radiation of some sort because a time-varying electric field (not necessarily reversing) is associated with a magnetic field and a time-varying magnetic field (not necessarily reversing) is associated with an electric field.

Even though the energy flow direction changes sign in these regions, there is still a small time-averaged energy flow outward. This small energy flow outward represents the small amount of energy radiated away to infinity. Outside of the point at which the outer field lines seemingly detach from the dipole and move off to infinity, the direction of the pulsating field lines (due to Faraday's law and the Ampere-Maxwell term) and thus the direction of the electromagnetic energy flow, is always outward. This is the region dominated by radiation fields, which carry energy outward to infinity. The "loops" of em wave radiation field lines in the radiation zone are not depicted in Fig. 10.30 but may be seen in the visualization in the MIT website below.

For visualizations (animated) of the fields in the quasi-static zone, the induction field zone and the radiation zone visit http://ocw.mit.edu/ans7870/8/8.02T/f04/visualizations/light/01-DipoleRadiation/01-Dipole_320.html and view the Dipole radiation, fields near the dipole (quasi-static zone) and a 3D field line representation of the entire phenomenon.

Figure 10.32 shows the electric and magnetic fields of an electric dipole characterized by

$$\vec{\mathbf{P}}(t) = p_o \cos\left(\frac{2\pi t}{T}\right)\vec{\mathbf{k}} \tag{34}$$

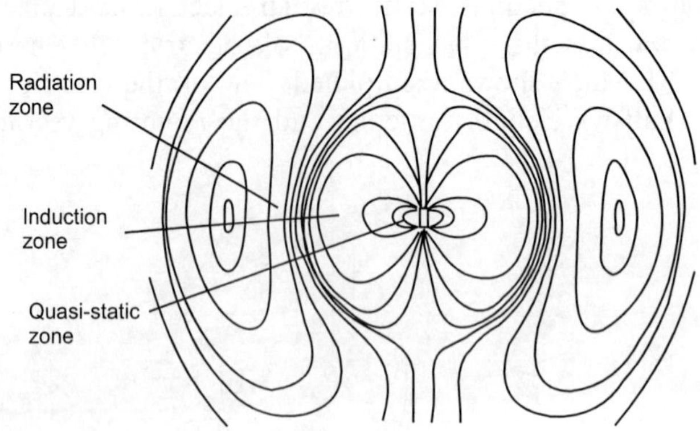

Fig. 10.32 Radiation from an electric dipole in one plane whose dipole moment completely reverses with time.

Equation (34) shows that the direction of the dipole moment varies between $+\vec{\mathbf{k}}$ and $-\vec{\mathbf{k}}$ or the dipole reverses.

It seems easier to nudge the radiative fields from the dipole when the dipole moment is reversed.

In Fig. 10.32 the radiation field lines are shown seemingly breaking and propagating away from the near field (quasi-static and induction) zones of the dipole when the field is reversed. The "loopy" pulsating electric field lines are self-propagating and possess the properties of propagating waves (Sec 10.2).

Note Some authors state explicitly that radiation field lines "detach" and race away from the antenna as the wave travels outward and one lecture note (MIT, USA) states that "most antennas reversibly link radiation fields to currents flowing in wires".

In Chapter 3, Ref. [36] Assis has pointed out contradictions in the interpretation of field ideas by several scientists especially in the context of whether the fields can independently carry energy and momentum and conform to Newton's Third Law which requires that the fields absorb the reaction *independent* of their source, say, when they interact with a charged particle. Read the topic "Fields and Momentum Conservation" in Appendix B and the topics "Coulomb interaction" and "Magnetic Interaction" in Appendix F for more details on this aspect of fields.

In this textbook, I have suggested that the radiating field lines "seemingly" detach and propagate away, without further qualification.

Recollect from Section 10.7 and the few paragraphs at the beginning of Section 10.8 that strong radio frequency electromagnetic waves can be produced by accelerating large numbers of electrons. Antennas do just that.

Unlike the charged infinite plane sheet that was made to oscillate sinusoidally to generate electromagnetic waves, the antenna is a stationary structure and the mechanism of accelerating a large number of electrons is by the application of a sinusoidally oscillating voltage to the conductive antenna rods. The voltage produces a very vigorous electron motion back and forth which is necessary to launch loopy electric and magnetic fields that can propagate through space.

A simplified antenna structure of a quarter-wave "dipole" wire transmitting antenna, is shown in Fig. 10.33, which transmits a signal applied to its two ends.

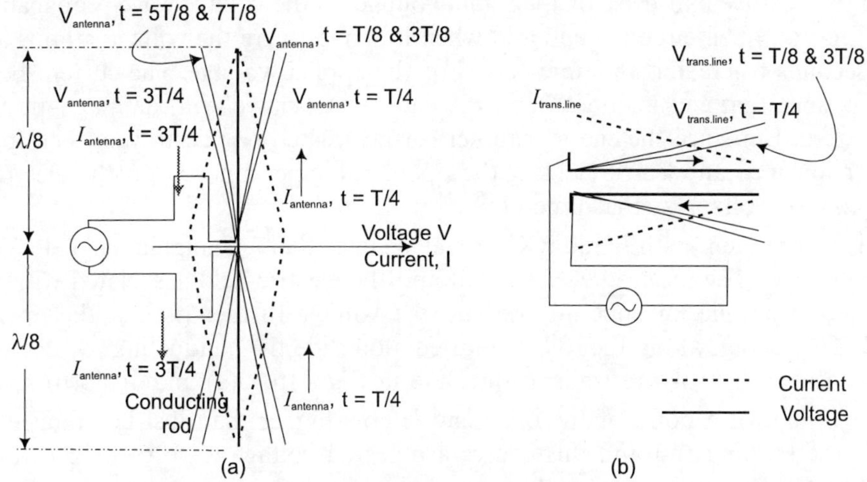

Fig. 10.33(a) A dipole wire antenna and its voltage and current distribution during the positive and negative half-cycles of a sinusoidal voltage (b) A two-wire transmission line depiction of the antenna voltage and current distribution.

The conducting rods of the antenna assumed to be of negligible resistance are arranged as shown in Fig.10.33(a). The rods are excited by a sinusoidally oscillating voltage in the radio frequency range by a source connected to the antenna rod terminals in the middle of the structure, one terminal of the source to one rod and the other terminal to the other rod.

The current distribution along the length of the antenna will be as shown in Fig. 10.33 and the current is not uniform along its length. In order to understand the shape of the current distribution curve shown, it is useful to review Section 2.21 on how a single wire connected to one terminal of a battery attains the potential of the terminal and Section 5.22 on how the currents and voltages along a long transmission line may behave peculiarly. As a matter of fact, the antenna may be viewed as a two-wire transmission line (Fig. 10.33(b)) that has been unfolded to form an antenna. If we had a transmission line with very little spacing in between the lines, the currents being oppositely directed there will result no radiation from the line and this will be satisfactory for transmission line designs. When shown unfolded (Fig. 10.33(a)) as if for an antenna, the effect of the current in the rods reinforces the fields which an antenna design requires.

When a sinusoidal signal voltage is applied to such a transmitting antenna (or the transmission line) then currents flow in its wires and the voltage of the line assumes that of the voltage source. And since the source voltage is varying sinusoidally, a quasi-steady state can be assumed for the voltages and currents at each and every instant and along each and every point along the line (or each rod of the antenna) where the electrons in motion cause some of them to "condense" and "disperse" on the surface as shown in Fig. 10.27(b) and are responsible for the non-uniform current and for a voltage wave to propagate down the line as shown in Fig. 10.33(a).

Positive Half-cycle of Signal Voltage

During the positive half-cycle of the signal voltage, as the voltage wave propagates down the line, it finds an open-circuited end and when it arrives there the voltage attains a maximum at T/8 seconds where T is the time period of the applied voltage. The current becomes zero however, since that end is "open" and the energy associated with its magnetic field should become zero. But since the energy can neither be created nor destroyed, then it must effect a conversion into an electric field at the open end superimposed on the electric field due to the applied voltage at that time (T/8).

This is consistent with Faraday's law which states that a changing magnetic field induces an electric field. The induced electric field and the electric field associated with the voltage of the wave combine or superimpose and the voltage further rises at the open end until $t = T/4$. The voltage along the rods is plotted alongside the conducting rods of the antenna and alongside the two-wire transmission line in Figs. 10.33(a) and 10.33(b) respectively.

The voltage at a point of the open end is now higher than just before a little distance towards the sending end and this causes a reflected voltage wave to be generated. This is not unlike the "fringe" electric field generated after charges accumulate on the plates of a capacitor and the fringe field takes control and drags charges away from the plate as we learned in Section 2.9 and depicted in Fig. 2.29 when a sinusoidal voltage is applied to a capacitor. The forward voltage (due to the signal source which does not stop generating the sinusoidal voltage!) and the return voltage combine to produce a maximum at the open end.

The reflection of the voltage from the open end of the line in turn causes a current to return up the line towards the signal source. The voltage is nearly zero at the signal source end, its peak at $t = T/4$ having been nullified by the returning voltage wave.

The result of the forward and reflected wave combining during the first positive (half) cycle of the applied sinusoidal voltage will result in the antenna voltage taking a shape as depicted in the snapshot views at $t = T/8$, $T/4$ and $3T/8$ in Fig. 10.33(a). The voltage curve of the transmission line is shown in Fig. 10.33(b). The voltage of the antenna forms a standing wave.

The current is a maximum at the source end at $t = T/4$ of the antenna during the positive cycle and zero at the open end because of the open circuit. The current in the *other* rod is similarly maximum at the source end and zero at the open end simultaneously.

There is a continuity of current through the source the maximum occurring at the source end of both the rods when $t = T/4$. Note the direction of the current I_{antenna} in both the rods! It was seen in Section 5.22 that the current in a transmission line is a differential current. Since the antenna maybe viewed as an opened out transmission line, the current is a differential current, i.e. the current in one rod if moving to the open end then the current in the other returns from the open end to the source. This is evident from the transmission line equivalent of Fig. 10.33(b). This should put to rest fears if viewing the current in one rod that the current should result in a large voltage change of the rod since it appears as if the current is continuously pumping charge through the rod.

Negative Half-cycle of Signal Voltage

The sinusoidal signal source voltage reverses when $t > T/2$ and this voltage wave has a maximum magnitude at the open end of the antenna when $t = 3T/4$.

Full Cycle of Signal Voltage

Clearly from the Fig. 10.33 it is seen that during a full cycle of the signal sinusoidal voltage of period T, the voltage swings alternately every half-cycle (for a period of $T/2$) between being a positive maximum to a negative maximum at the end of the antenna rods.

The current in the antenna is what we will be focusing on to understand how the antenna radiates.

Shown in Figs. 10.33(a) and (b) are the currents in the transmission line and the antenna during the first positive cycle of the sinusoidal signal voltage, on the right of the antenna and above the transmission line. The current at the ends of the transmission line is zero at all times since the current cannot go anywhere from an open-ended line (or antenna rod).

Near the signal source the current swings alternately every half-cycle (for a period of $T/2$) between being a positive maximum to a negative maximum. Like the voltage, the current in the antenna also forms a standing wave.

See the power point presentation "Chapter_10" with animations showing the standing waves of voltage and current of a Half-Wave Dipole antenna in the accompanying CD. The same can also be viewed at https://en.wikipedia.org/wiki/Dipole_antenna.

The radiation will be strong at the rods' ends near the voltage source because the current which consists of a large number of electrons will be accelerated very strongly when they undergo a direction reversal. The radiation from a $\lambda/4$ or quarter wave dipole antenna oriented along the z-axis (into the page) is depicted in Fig.10.34 to which is applied a sinusoidally oscillating voltage (not shown).

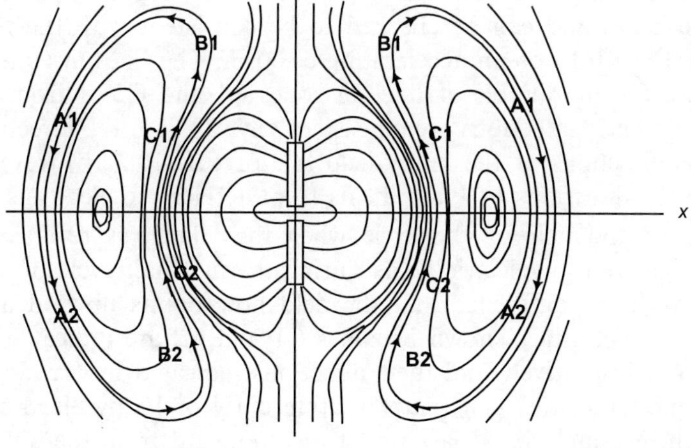

Fig. 10.34 Radiation from a quarter-wave antenna in one plane over the full period.

The slope of the current curve plotted against distance as depicted in Fig. 10.33(a) suggests that there are points where radiation does occur other than at the ends. There is a *velocity* change along the major portion of the antenna rods since there is a reduction in the magnitude of the current gradually moving from the source towards the open end, but note that the current is a standing wave which means that at every instant during the cycle the current undergoes an overall reduction in magnitude since it has to undergo a phase reversal at the end of one half of a cycle.

During the first half of the sinusoidal voltage cycle, the mechanism of production of the electric field is as described in Section 10.8 by the initially pushed and moving charged infinite plane sheet with the difference that the rods of the antenna are stationary and it is the applied voltage and accompanying fields that cause the charges in the antenna rods to move and produce currents. The source end of the rod and the long section of the antenna produce electric field lines pointing downward as shown in the zone A1 in Fig. 10.34 in the right hand portion of the loopy field lines. The field line direction is consistent with the direction of current and considering that an electron in the upper rod will move downward opposite to conventional current I_{antenna} at $t = T/4$ that is shown in Fig. 10.33. From Fig. 10.16 of an accelerated electron it may be seen that the direction of the field in the kink points in the same direction as the electron acceleration.

At the open end of the rods there is a sharp cut-off of current signifying a large velocity change and a reversal but there is little radiation given the small magnitude of the current at the ends.

While the source end (of one rod) and a small portion in addition from this end generates a significant portion of radiation (zone marked A1), the radiation at the open end contributes in providing a small "lick" as shown in the zone marked B1 in Fig. 10.34 of the left and right segments of the upper half of a loopy electric field.

The function of the other rod is to contribute likewise the lower half of a loopy electric field to join to the loopy pattern made by the upper rod. The corresponding zones are marked A2 and B2 in Fig. 10.34. The direction of the electric field lines is indicated by arrows in these zones and can be checked to be consistent with the field line directions in the kinks of Fig. 10.16 of an accelerating electron. The lines in zones A1 and A2 may have originated from the jumble of lines in zones C1 and C2 or from somewhere nearer the antenna rods. Their directions are consistent anyway with the directions we will obtain by applying the directions of the kinky field line of Fig. 10.20(b) (reversed for electrons) or by applying the directions of kinky fields from accelerated electrons of Fig. 10.16.

During the second half of the cycle when the voltage is reversed, the descriptions given of the production of electric fields during the first half-cycle of the applied voltage are applicable with the difference that the field now points upward as depicted in Fig. 10.34. The electric field lines shown in zones C1 and C2 are indicated for the upper rod and the lower rod respectively and their directions indicated by arrows. The directions of the field lines produced are such that pulsating curly or loopy electric field patterns are generated which seemingly break away and race outward from the antenna into space at the speed of light c. And, since the loops in turn spread around a sphere their strength diminishes proportional to 1/(distance from the antenna). The field pattern shown in

Fig. 10.34 is a two-dimensional slice of a 3-dimensional field pattern of a real world antenna. The full field pattern in 3D can be visualized as a figure of revolution around the *y*-axis.

That is how the process of radiation is enabled by an antenna and that is how from among the maze of fields in the quasi-static zone and the induction zone, the pulsating radiating fields apparently break away and radiate into space.

As was mentioned in Section 5.22 special antennas make use of the properties of quarter-wave transmission lines that are shorted at one end to maximize the impedance to differential-mode currents and operate with common-mode currents (Fig. 5.41). These antennas are called folded dipoles and were designed by mixing transmission-line theory and antenna theory elegantly. Details of folded dipoles are in Refs. [16] [35], and in MIT lecture links in the CD with this book.

Propagation of Radio Frequency Electromagnetic Waves in Space

It would be useful to review the propagation of electromagnetic waves through space in "Propagation" in Section 10.2, because that is how the electromagnetic waves produced by antennas in the radio frequency range propagate; by a series of pulsations of the electric and magnetic fields orthogonal to one another. They propagate at the speed of light '*c*'.

The photo shows a half-wave dipole antenna with a usable frequency range of 1 to 4 GHz (1 GHz = 10^3 MHz).

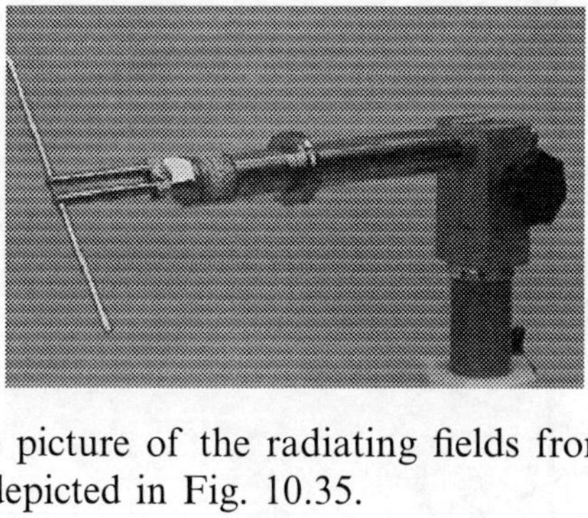

Image courtesy:
"Half Wave Dipole" by
Schwarzbeck Mess-Elektronik

A complete picture of the radiating fields from a dipole antenna ($l \leq \lambda/2$) in the far-field region is depicted in Fig. 10.35.

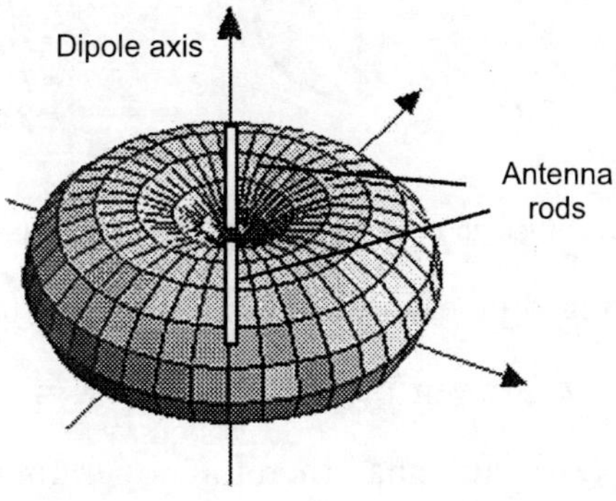

Fig. 10.35 The radiation pattern (electric field in the far-field region) of a dipole antenna. The antenna is assumed positioned vertically along the dipole axis.

Note that the magnitude of the antenna current is finite. If we consider a negligibly small antenna resistance energised by a sinusoidally oscillating voltage there should have been a large current, at least momentarily. Recollect that there is a reflected voltage from the open end. Now, the applied (field) voltage and the reflected (field) voltage combine to produce a resultant (field) voltage that limits the current not quite different from the case of the capacitor charge and discharge currents which we learned in Section 2.9.

As a matter of fact the reflected field follows the applied voltage so closely it leaves little room for the current to increase indefinitely.

We have not considered so far the length of the antenna rod and its effect on the overall current profile. As depicted in Fig. 10.33 the length of the rod used is $\lambda/8$. What this means is that the voltage maximum at the open end (when the maximum amount of charge accumulates) occurs when the voltage from the initial time considered to be the start of the sinusoidal oscillation at t = 0 attains a value corresponding to $T/4$ since the source voltage continues to grow between t = $T/8$ to t = $T/4$ where T is the time period of the oscillation.

If we had a rod that was $\lambda/4$ long then the voltage maximum at the open end would occur when the voltage of the source attains a value corresponding to $T/4$.

This can be seen in Fig.10.36 which shows the current profiles for rods of various lengths in terms of the wavelength of the applied sinusoidal voltage.

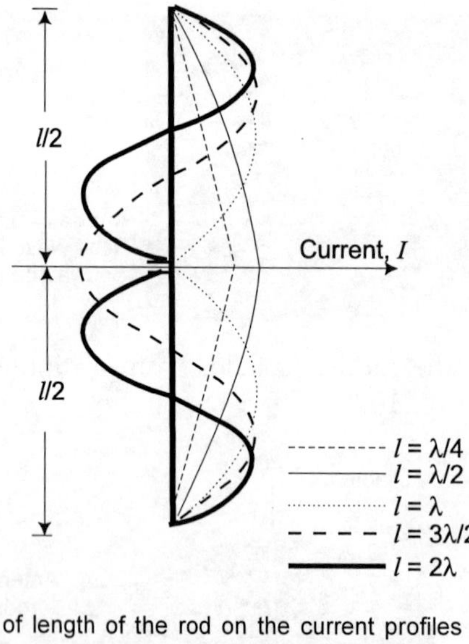

Fig. 10.36 Effect of length of the rod on the current profiles of a dipole antenna.

Consider a transmitting dipole antenna operating at 100 Megahertz or 100×10^6 Hz.

The free space wavelength λ is given by $c/f = \dfrac{3\times10^8}{100\times10^6} = 3$ meter.

For a quarter wave ($\lambda/4$) antenna, therefore operating with a signal frequency of 100 Megahertz the length of each rod is, $\lambda/8$ = 3/8 meter = 0.375 meter = 37.5 cm. The magnitude of the current will be a maximum when the time of the wave is T/8 sec = $1/8f$, because

$T = 1/f$, and for the 100 Megahertz signal is $1/8 \times 100 \times 10^6$ Hz, i.e., 1.25 nanoseconds from the start of the excitation voltage *plus* the time it takes to travel to the end of the rod which is 0.375 m/3×10^8 m/second = 1.25 nanoseconds. In other words, it would take 2.5 nanoseconds for the voltage maximum to travel to the open end of the rod assuming the speed of the wave is *c*, the free space speed of the wave, which is not an unreasonable assumption. More reasonable values of wave velocity '*v*' for transmission lines is 0.6*c*.

If the length of each antenna rod were to be increased to $\lambda/4$ to make a half-wave antenna, the voltage maximum will occur at $T/4$ at the end of the rod, i.e after 1.25 nanoseconds +2.5 nanoseconds = 3.75 nanoseconds. The maximum current near the source end expectedly will be larger than the maximum current of a $\lambda/8$ antenna rod and is depicted in Fig.10.34.

If the length of each antenna rod were to be increased to half of a wavelength (i.e 1-1/2 meter for a 100 MHz signal) to make a *full wavelength* antenna, there will be two current maxima along the length of the antenna, one in each rod around its middle, as shown in Fig. 10.36.

If the length of the antenna were to be made 3 $\lambda/2$ and more, then there will be a negative maximum of current alongwith a positive maximum in each rod. The effect is a cancellation of fields and a weakening due to superposition of oppositely directed fields at points in space around the antenna. It is therefore preferred to use antenna rods of lengths of $\lambda/2$ or less to maximize the strength of the radiating fields.

Radiation Resistance

In Section 10.7, it was seen that the accelerated charge has a self-force which reduces its kinetic energy, the energy that is carried away by the radiation. The action it seems is similar in nature to the collisions conduction band electrons undergo in a conductor when subject to an electric field, and they slow down when exchanging kinetic energy with the lattice ions. Thus, we can associate a resistance called the 'Radiation Resistance' with the radiation by a large number of electrons in antennas in which the electrons exchange kinetic energy with the radiated fields and slow down. While the resistance of the wires and rods that are used to make dipole antennas maybe of the order of a few milliohms to a fraction of milliohms, the radiation resistance associated with the loss kinetic energy when the electrons radiate fields is of the order of a few ohms. See links to MIT lectures in the accompanying CD.

The propagation description of em waves is essentially a classical view (since the time of Hertz's and Michelson-Morley's experiments in 1888) which is demonstrably correct. However, with the advancement of Quantum theories, this viewpoint has changed considerably. Electromagnetic waves, described by Maxwell's equations, are photons, which are the force-particles of Quantum Electrodynamics (Sections 4.2, 6.6 and 10.10).

In the book, Classical Electrodynamics, John David Jackson mentions that the classical notion of the electric and magnetic fields (as lines of force or as vectors in space) is a limit of large quantum numbers of a quantum-mechanical description in terms of real or virtual photons, and that in the "domain of macroscopic phenomena and even some atomic phenomena, the discrete photon aspect can usually be ignored or at least glossed over". See Section F.5 "Theories of interactions between charge carriers", Appendix F for a calculation of the number of photons in a propagating electromagnetic wave of a Frequency Modulation signal in free space.

While sound or waves in water are fluctuations (or differences) in the densities of the medium (air, solid material, water), the photons are particles which are excitations of a quantum field. So the "medium" where photons propagate is just space-time which is still there, even in most abandoned places in the universe and in absolute vacuum.

The reader may refer to "Introduction to Elementary Particles" by David J. Griffiths for a comprehensive treatment. In Leigh Page's Emission theory, charges when accelerated produce em waves by the moving elements of Ref. [39], by producing an intrinsic magnetic field in addition to the changing electric field. The intrinsic magnetic field is distinguished from an apparent magnetic field due to steady currents. While the apparent field vanishes with a moving observer, the intrinsic field will not. The field theories described above are one among several interpretations. For more on these theories see the end of Section 10.12 and "Fields and Momentum Conservation" in Appendix B.

Receiving Antenna

Receiving antennas are constructed in a manner similar to transmitting antennas, but of course, without the signal source since the receiving antenna is designed to receive electromagnetic waves from a transmitting antenna and convert them into emfs or signals. A demonstration of the operation of a receiving antenna is shown in Fig. 10.37.

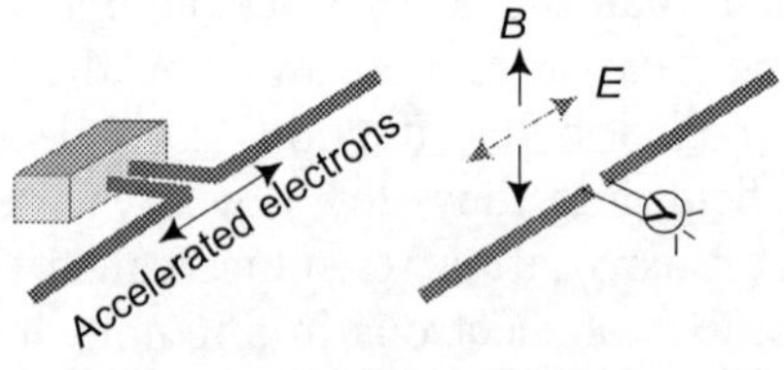

Fig. 10.37 Radio transmitter and receiver: The radiative electric field drives current through the bulb.

A transmitting dipole (like the one in Fig. 10.35) is connected to a high-frequency ac voltage supply at 300 megahertz (300 × 10^6 Hz). The radio receiver or receiving antenna is made of two metal rods connected by a flashlight bulb.[2]

When positioned as shown in the figure, the bulb lights. For the radiation produced by accelerated electrons in the transmitter that propagates through the right, the oscillating electric field is horizontal, and it accelerates the mobile electrons in the receiving antenna, driving them back and forth through the flashlight bulb, and lighting the bulb.

Also see the power point presentation "Chapter_10" showing the operation of a Receiving Dipole antenna with animations in the accompanying CD.

To view a nice demonstration of an electromagnetic wave from an antenna lighting an electric bulb, visit https://www.youtube.com/watch?v=4xF1Fq2wB1I.

Attenuation of Propagating Electromagnetic Waves

When electromagnetic radio waves from a radiating source such as an antenna propagate through space, the strength of the electric field (and consequently the associated magnetic field strengths) will diminish. This is attributable to a few phenomena (See "Attenuation of a wave", Section 5.22) and among these the dominant ones are due to the wave energy being spread over an expanding spherical surface at large distances from the antenna as was described in "The Hertzian (or Ideal) dipole. This attenuation is because of the $1/r$ proportionality of the radiative field strength. See notes below Eq. 32. The other dominating factor is the presence of objects (walls, trees and also the earth) that absorb wave energy which by the phenomenon of scattering cause the field strength to diminish and this is called attenuation.

Hence, the field components of an electromagnetic wave received by an antenna will surely be smaller in strength than near the radiating source. For a detailed description of the phenomenon of reradiation or scattering see Section 10.4 and also see "Attenuation of a wave" Section 5.22.

The receiving antennas which capture audio and video signal modulated high-frequency em waves in radios, television sets, cellphones and satellite communication systems operate by the same principle as was demonstrated in Fig. 10.37 with signal amplifiers replacing the bulb. The amplifier converts the weak signals from far away broadcasting stations into stronger signals (after demodulation) to operate the voice coils of headphones and loudspeakers and video picture rendering devices.

The transmitting and receiving antenna energy exchange processes may be analysed using the Poynting Theorem (Section 4.2) and the energy carried by the electromagnetic wave can be computed using the Poynting Vector (Section 4.2).

10.11 ELECTROMAGNETIC WAVE TRANSMISSION BY WAVEGUIDES

While an antenna radiates waves in all directions in space, some antennas are designed to radiate in specific directions by arranging several of them in arrays and the superposition of fields concentrates the fields in those directions. However, the power in the fields at receiving antennas may still not be sufficient even if one were to ignore the losses due to reradiation and scattering in the space between the transmitter and the receiver. Besides, one has to think of how to *get* the large amounts of high frequency signal power generated by sources to the antennas (especially high power antennas) and other electronic equipment.

We had studied in Section 5.22 how guided waves between a pair of perfectly conducting parallel planes convey energy from a voltage source say, a DC supply or a sinusoidal generator from the source to the load. Examples of lines which can guide waves are the twin-lead cable and the coaxial cable.

While two-wire lines such as the twin-lead cable and the coaxial cable can support any of the three modes (TE, TM and TEM) for propagating em waves (Section 5.22), the preferred mode of propagation in these cables is the Transverse Electromagnetic Mode (TEM) mode at frequencies below 300 MHz.

Engineers frequently use devices which can guide electromagnetic waves in the radio frequency range to couple high frequency signal sources to the cooking chamber in microwave ovens, radar, antennas and in printed circuit boards. Some of the common structures are coaxial cables, two-wire and microstrip transmission lines and hollow conducting (single-conductor) waveguides.

Waveguides cannot support the TEM mode of propagation (See Ref. [16]). The modes of wave propagation supported by waveguides are the Transverse Electric (TE) and the Transverse Magnetic (TM). Notwithstanding the fact that waveguides have a lower cutoff frequency, they find application in transmitting high frequency radiowaves and microwaves. Visit http://www.ece.rutgers.edu/~orfanidi/ewa/ch09.pdf for the theory, transmitted powers and typical attenuation factors of waveguides. The frequency range of operation of waveguides is between 100 megahertz (100×10^6 Hz) upto 40 gigahertz (40×10^9 Hz). Manufacturers do offer waveguides which operate in higher frequencies above 40 GHz for special applications.

As the name suggests, waveguides *guide* electromagnetic waves and these are confined inside the waveguide due to their reflection from the waveguide wall. They are *not* used to transmit light which is also an electromagnetic wave in the infrared to visible range roughly 1 Terahertz (10^{12}) to 790 Terahertz (790×10^{12} Hz). For the transmission of light, optical fibers are used. Optical fibers rely on the phenomenon of bending of light and total internal reflection for confining waves within the fiber. Modern optical waveguides use a dielectric slab of three dielectric layers with different refractive indices.

Most of the phenomena of electromagnetic wave interactions with matter viz. superposition of waves, radiation by accelerated charges, reradiation and reflection described in the Sections 10.1 to 10.5 are observable as the electromagnetic waves propagate in a rectangular waveguide with metallic walls described next.

The wave is usually launched by a small coaxial cable feeder (used like an antenna) through the wall and protruding a small distance into the center of the waveguide, the outer conductor of the coax being terminated on the wall itself as shown in the transverse electric (TE) wave in Fig. 10.38.

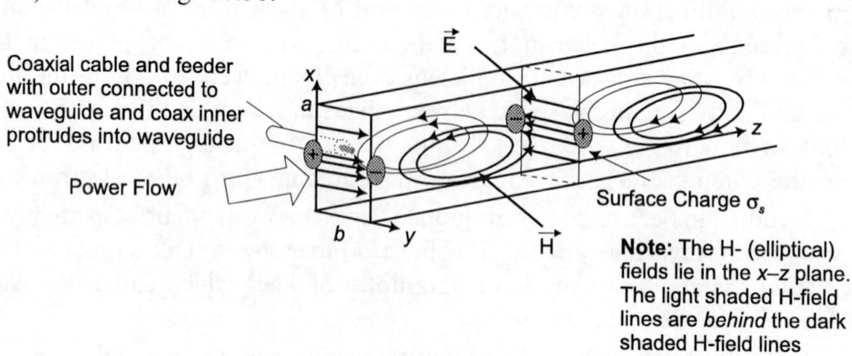

Fig. 10.38 A rectangular waveguide showing surface charges on the walls and the electric and magnetic fields.

Refer http://www.radio-electronics.com/info/antennas/waveguide/waveguide-launchers-couplers-transitions.php.

The solution of the time-varying form of Maxwell's equations gives the electric and magnetic field components of waves propagating in free-space, the theory of which is described in several textbooks.[8,16,33]

In waveguides, the solutions to the time-varying form of Maxwell's equations gives the component electric and magnetic fields with appropriate boundary conditions at the conducting walls of the waveguides. For example, an electric field along any of the perfectly conducting walls of the waveguide should be zero is one such boundary condition. This would mean that there can be no component electric and magnetic fields outside the waveguides with a large percentage of the input electromagnetic wave energy being either reflected back and forth between the walls and a small percentage dissipated by induced currents in its walls. Refer to Section 10.4 "Reradiation" on how free charges accelerated on the wall dissipate energy while simultaneously attenuating the waves outside the waveguide. Also, refer to Section 10.5 and how incident and reradiated waves superimpose to produce net fields inside the waveguide material. Superposition is described in Sections 1.6 and 2.2.

The usual convention for the field orientation is to align the edges of rectangular waveguide walls along the x and y-axes with the direction of propagation in the direction of the z-axis as shown in Fig. 10.38. The solutions to the Transverse Electric (TE) and Transverse Magnetic ("transverse" means the field direction is perpendicular to the direction of propagation or power flow; Figs. 10.38 and Fig. 10.39 show the Transverse Electric mode) modes of propagation which occurs when the frequency of the signal is above a certain *cutoff* frequency that is determined by the dimensions a and b of a rectangular waveguide (Fig. 10.38).

One set of solutions does not give a component of $\vec{H}$ along the direction (z–) of wave propagation but does give a component of $\vec{E}$ in the direction of propagation ($\vec{E}_z$) and therefore the net electric field $\vec{E}$ is no longer transverse to the direction of propagation. Thus, such a wave is said to be Transverse Magnetic (TM). It is the $\vec{E}_z$ component of the TM wave that enables it to travel in the direction (z–) of propagation unlike the case of a Transverse Electromagnetic (TEM) wave where there is no component of both the electric and magnetic field along the direction of propagation and yet the wave travels due to its initial "push".

Similarly, in a Transverse Electric (TE) wave, it is the $\vec{H}_z$ component of the magnetic field that enables the TE wave to travel in the direction (z–) of propagation.

The solutions are developed assuming perfectly conducting walls. This makes the derivation of the electric and magnetic field components in the three directions viz., x, y and z relatively simple by using the appropriate boundary conditions. This is justified by the fact that the metal (of the wall) losses ($I^2_{\text{surface}}\, R_{\text{surface}}$) will be small assuming that the currents flowing as shown in Fig. 10.39 in conductors of finite conductivity are essentially the same as those for the perfect-conductor case with near zero electric field components near the top and bottom surfaces (y-z planes) of the walls, or that the losses are small and the fields and currents are affected only slightly thereby.

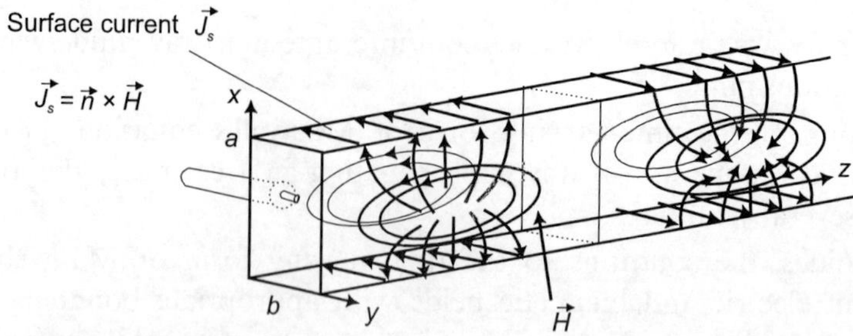

Fig. 10.39 A rectangular waveguide showing free surface currents $\vec{J}_S = \vec{n} \times \vec{H}$.

The electric field component "*terminates*" seemingly on electrons on the walls (surface charges shown in Fig. 10.38) and the magnetic field component *induce* currents in the walls of the waveguide. The surface current J_s (current per unit width in the z-direction) is a current which adjusts itself to a finite value and equal to a finite magnetic field just *outside* the surface. The surface current is produced as a result of the prediction made in the Ampere-Maxwell relation $\oint \vec{H} \cdot d\vec{s} = \int \left(\dot{\vec{D}} + \vec{J}\right) \cdot d\vec{a}$ where $d\vec{s}$ is an elemental length of the closed path is equal to the conduction current plus the time derivative of the electric displacement through any surface $d\vec{a}$ bounded by the path (Section 5.22 and Section 8.28). If however, the walls have perfect conductivity, there can be no electric field, the displacement $\dot{\vec{D}} = \varepsilon\dot{\vec{E}}$ term will be zero, and as we learned in Section 5.22 with reference to Fig. 5.36(a), the magnetic field and surface current J_s will be parallel to the surface, but perpendicular to each other and given by $\vec{J}_s = \vec{n} \times \vec{H}$.

The reader is cautioned once again not to visualize the fields *traveling* through the waveguide. Fields never *travel*; their pulsations travel and these being wavelike in nature, it is said that the wave propagates. See Section 10.2 "Propagation".

The electric field component of the wave still retains its essential *non-conservative* nature (see Sections 5.12 and 5.13) despite it being terminated on charges on the walls, the field component having been launched by the coaxial cable from a time-varying high frequency signal. When there is no phase difference spatially along the z-direction of the electric and magnetic wave components, the waves pick that solution which makes the incident waves to reflect and bounce back and forth between the walls of the guide. The waves dissipate their energy in the guide walls and a sharp decrease in the strength of the electric and magnetic fields occurs due to the effects of reradiation and scattering. This happens to signals below a frequency called the cutoff frequency and the wave does not pass through the waveguide.

Above the *cutoff* frequency, waveguides pass high frequency signals. The waves continue to reflect back and forth on the walls of the waveguide but now a (spatial) phase difference appears and the waves seemingly glide along the z-direction (Figs. 10.38 and 10.39), the direction in which the wave propagates.[16,33]

For nice animations of wave propagation in waveguides visit http://mason.gmu.edu/~jdilles/classes/ece305/ehwave.html and http://www.met.reading.ac.uk/clouds/maxwell/waveguide.html.

The total power carried by the fields in the z-direction is obtained by integrating the z-component of the Poynting vector (Section 4.4) over the cross-sectional area of the guide which may be a dielectric like air.

Above the cutoff frequency, the energy of the wave is dissipated by the surface currents of the conducting walls of real waveguides which have a finite though large conductivity. The power loss which occurs in the conducting walls will be small compared with the power transmitted in the dielectric which could be air in ordinary waveguides between the walls. There are several loss mechanisms in a waveguide and amongst them the significant ones are the *Ohmic* losses in the dielectric and the other the surface currents due to the presence of magnetic fields at the surface. The surface currents are calculated by

$$\vec{J}_s = \vec{n} \times \vec{H}$$

where $\vec{n}$ is the normal to the conductor as shown in Fig. 5.36 (Section 5.22) of a perfectly conducting parallel plane transmission line. The ohmic losses in the conducting walls may be calculated by multiplying the square of J_s with the surface resistance R_s of the walls and integrating over all the walls. See www.ece.rutgers.edu/~orfanidi/ewa/ch09.pdf.

The signals above the cutoff frequency lose energy and the fields their strength by the loss mechanisms above but, the attenuation is small compared to that when the signal frequency is below the cutoff frequency.

See also "Attenuation of a wave" in Section 5.22 on how the electromagnetic wave is attenuated above the cutoff frequency.

Below the cutoff frequency, the energy of the wave dissipates heavily by penetrating currents in the conducting walls of real waveguides which have a finite though large conductivity (See "Electric field and current flow within the conductor" in Chapter "Guided Waves" Ref. [16]). The nulls of the electric field component of the wave no longer occur near the top and bottom surface of the walls and there is thus a strong current produced which is dissipated resulting in the wave being attenuated rapidly.

See https://www.youtube.com/watch?v=FN2Ih6ntQFo of a wave below cutoff frequency in a waveguide. See also "Attenuation of a wave" in Section 5.22 on how the electromagnetic wave is attenuated below the cutoff frequency.

A full treatment of these and other related phenomena is beyond the scope of this textbook and there are several textbooks that treat these in detail.[16,33]

10.12 THE FIELD OF A SINGLE MOVING CHARGE

What is the field of a uniformly moving charge? Figure 10.40(a) shows the field **E** of a charge Q at rest in an inertial frame F (See Appendix F). Figure 10.40(b) shows the field as observed in an inertial frame F' moving with uniform speed v. The field component parallel to the direction of motion is observed as $E'_{\parallel}=E_{\parallel}$ i.e it remains the same, while the component perpendicular is observed to be $E'_{\perp} = \gamma E_{\perp}$ where γ, the Lorentz factor is >1 and $\theta' > \theta$ (see Eq. 15, Appendix F). Figure 10.40(c) corresponds to the view from frame F', and shows how the field concentrates in a pancake perpendicular to the direction of motion (speed of 0.866c).

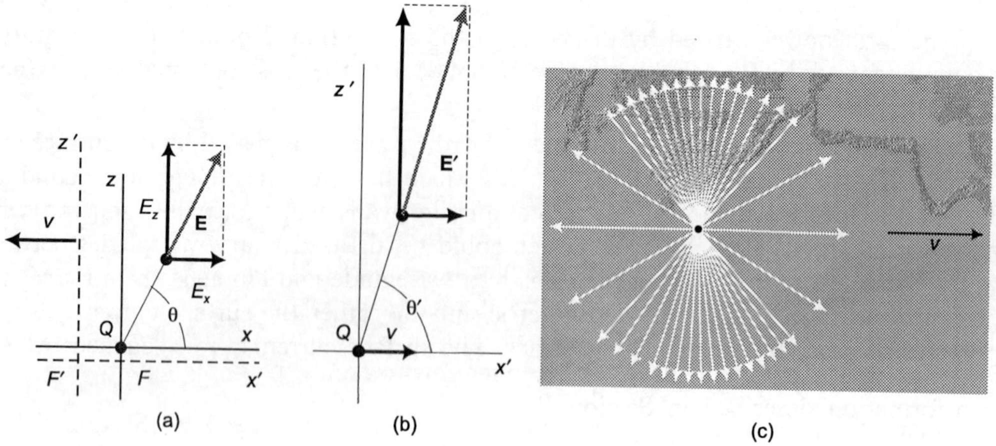

Fig. 10.40 (a) Electric field of a point charge (a) in a frame in which the charge is at rest and (b) in a frame in which the charge moves with constant velocity (c) Another representation of the field of a uniformly moving charge.

A derivation of expressions for the magnitude of the electric field is provided in Refs. [8] and [22]. While Griffiths' derivation is based on retardation considerations, Purcell's derivation makes use of the Lorentz transformation set of equations.

Note that we have not included the magnetic field of the moving charge in Fig. 10.40. For a mapping of the pattern of electric field and the magnetic field of a charge in uniform motion at relativistic velocities see the topic "How Fields Transform" in the chapter Magnetic Field in Ref. [22].

Two nice videos that depict the field pattern around a charge moving with constant velocity (assuming it was moving since the beginning of time) at two different velocities is available in the links here http://web.mit.edu/8.02t/www/802TEAL3D/visualizations/electrostatics/MovingChargePosElec/movingChargePosElec.htm and http://web.mit.edu/8.02t/www/802TEAL3D/visualizations/electrostatics/MovingChargeNegElec/movingChargeNegElec.htm.

Beginning with the description for a charge with uniform velocity of the enhancement of the field perpendicular to the velocity and a reduction in the forward and backward direction (Fig. 10.40), Härtel has explained the repulsion between two steady-current carrying wires. It will be instructive to read this description which is available in Sections 3.2.3 - 3.4.3 in the article "Prof Härtel A qualitative approach to electricity A Guide to visualization of electrodynamics" in the folder "Teacher's Guide" in the CD with this book (In Härtel's article, the interaction between the current-carrying conductors Fig. 3.5 in Section 3.2.3 appears before the detailed descriptions of the interaction). Prof. Härtel's description is a simpler version of that made by Purcell in Ref. [22] for the force of repulsion between two wires carrying steady-currents.

Härtel has made videos based on Leigh Page/s Emission Theory of Electrodynamics of a charge moving with uniform velocity (after being accelerated) and of circular waves (essentially the patterns depicted in Figs. 10.16 and 10.17) which may be seen here http://

portia.astrophysik.uni-kiel.de/~hhaertel/Videos/videos-uk-fl.htm. The videos are also available in the folder "Videos by Prof Härtel" in the CD alongwith this book. It would be highly instructive to view these videos. Note the difference in the field pattern of a charge moving since the beginning of time with uniform velocity and that of a particle which was first accelerated before settling down to move with uniform velocity.

It would be useful to become familiar with the Lorentz transformation and the visualization in "Field of a single moving charge" in Appendix F. In Section 6.6 we learned that if you were to run past a stationary charge you would detect a magnetic field along with an electric field, and if you stood still, you would detect the electric field but not a magnetic field. These phenomenon may be explained by the Lorentz transformation described in Sections F.2 to F.3 in Appendix F and "Magnetic Forces" Ref. [22]. A set of lectures by Prof. Ramamurti Shankar, Yale University introducing Special Relativity and the Lorentz transformation are here https://www.youtube.com/watch?v=pHfFSQ6pLGU&index=12&list=PLFE3074A4CB751B2B. There are more links in the Links to animations articles in the CD.

Also, L. Page's theory of Emission by moving elements (see Section 4.2 and Ref. [39]) describes quite precisely the production of em waves. The circular shape of the electric field that expands as shown in Figs. 10.16 and 10.17 is described by Härtel as a "circular wave". Circular waves are produced by massless moving elements emitted along lines of electric force and magnetic force (Ref. [39]) by charges when accelerated. It would be instructive to review Section 3.5.2 and then to read Sections 3.8 on Transmission Line wave propagation and Section 3.9 on Electric Circuits and Transmission Line in the article by Härtel in the "Teacher's Guide" in the CD.

Note: We had posed a question in Section 10.8 whether a kinky electric field line will be produced if an infinite sheet of charge moves with uniform velocity since the beginning of time and at $t = 0$ and forever thereafter. The answer is 'no'. The plane sheet of charge (Section 10.8) was *pulled* from *rest* and then began moving with uniform velocity and so the kink in the electric field line was produced (Fig. 10.20(b)).

SUMMARY

In the beginning of the chapter, we learned that only two major discoveries of science are attributed to mathematics; i) the detection of the planet Neptune, and ii) the existence of a time-varying magnetic field associated with a changing electric field.

We then learned that electromagnetic waves are a combination of pulsating electric and magnetic fields, and that they propagate through space as what we call radio waves and light. The permittivity ε_o and the permeability μ_o of free space were constants that were determined by making very accurate measurements more than a century ago and while these measurements were made on static fields, Maxwell proved that the speed of light (consisting of time-varying electric and magnetic fields) in vacuum is a constant that is given by these same fundamental constants $\frac{1}{\sqrt{\varepsilon_o \mu_o}}$ which is 3×10^8 meters/second.

The speed is applicable to all frequencies of electromagnetic waves; audio to radio to Microwave to light through X-rays and beyond to Gamma rays, and is the calculated speed anywhere in the Universe even today!

We then studied in detail the manner of propagation of electromagnetic waves, by assuming a slab of electric field travelling through space. Of course, the field does not travel, we learned, but a new field free region is filled by it at a later instant. The slab also consists of a magnetic field which points in a direction perpendicular to the electric field.

We saw how in the slab, a time-varying magnetic field which produces (or associates with) an electric field, and the time-varying electric field which produces (or associates with) a magnetic field, the electric and magnetic fields as a wave, move into unfilled regions. This manner of propagation it was shown, is consistent with the two main laws of wave propagation, Faraday's law and the Ampere-Maxwell law.

The electromagnetic (em, for short) wave has three properties (i) it travels with speed c, (ii) the strengths of **E** and **B** are equal (expressed in statvolts and Gauss respectively), and (iii) the fields are perpendicular to one another and to the direction of wave travel.

We also learned that the fields obey the principle of superposition of waves while noting that the radio frequency waves received by television sets are not sinewaves and yet, they have traveled from the transmitter to the receiver.

The em waves can travel without any charges around through free space; they can exist independent of charges. Radiation or the generation of electromagnetic waves is performed by accelerated charges. This important concept we learned by giving a stationary positive charge, a kick in one direction. Since the changes in the field configuration take a finite time to propagate because the speed of light is finite, the field develops a kink in the field lines in the region that demarcates the field when the charge was stationary, with the field corresponding to its new position after being kicked.

The kink in the field lines or the transverse field expands outward. We then viewed a pattern of the field vectors as they were radiated by the accelerated positive charge in three dimensions.

A sinusoidally accelerating charge emits sinusoidal electromagnetic radiation, which propagates outward from the source, and this was important to understand, because most sources of radiation oscillate for long periods of time either sinusoidally or as additions of various frequencies of sinusoidal waves with varying phases.

It was observed after making simple calculations that while a cellphone has an electric field about 27.4 millivolts/meter strong incident on it, the strength of the field due to the sun incident on the earth's surface is about 720 Volts/meter.

This makes the solar radiation field strength nearly 26,666 times more powerful than the field strength of the em waves received in a cellphone from a tower.

We observed that the wavefronts of radiation far away from sinusoidal sources appear planar or parallel to each other, being part of ever expanding large spherical wavefronts.

All radiation, perhaps barring the em waves of space communication signals above the earth's atmosphere, we learned, undergo the phenomenon of reradiation or scattering and is a very important phenomenon to know.

We observed that a charged ball hanging from a spring, oscillates at the frequency of an incident wave and therefore emits radiation redistributing the incident energy in all directions. The original incident radiation superimposes on the emitted radiation of the ball and is a mechanism for decreasing the energy of the original radiation, the incident radiation losing some of its momentum having imparted it to the ball.

The example also helped us make a proper interpretation of the term "attenuation" of a wave as a decrease of the amplitude or magnitude of coherent or incoherent electromagnetic waves which we had studied in Section 5.22.

Next, we studied the propagation of light through opaque and transparent media, in which only the electric field in em waves was considered in the study of their interaction with matter even though the magnetic field also is present. The electric field which penetrates through matter affects the charged particles of atoms and molecules that make up matter.

The electric field polarizes an atom, as we had seen in Chapter 1 and when em radiation travels through glass, air, water, metal it interacts with their charged particles causing them to accelerate and emit radiation, or reradiate. The electron being a very light (least heavy) constituent of an atom, it contributes almost all of the reradiation. Since all matter is made up of millions and millions of atoms, their millions and millions of loosely bound electrons reradiate and the superposition of their reradiated fields makes the resultant fields very complex. And, what is true of the effect of light on matter, similar considerations apply to em waves in the radio frequency range also on matter.

We then studied the effect of light on opaque materials and learned that they block light because of destructive interference, i.e., the incident and reradiated wave superimpose in a manner such that the magnitude of the emerging reradiated wave is reduced. And, there is no means to raise the magnitude of the transmitted light other than by removing the obstacle such as the piece of cardboard that was blocking the light.

In transparent materials, the incident and reradiated waves add constructively, so inside the material, the waves are not zero. However, the process can result in their slowing down.

A simple analogy of a marching band on grass that comes to a muddy part of the field was useful to understand the complex process of the slowing down of light (or for that matter any em wave of any frequency) in a dense medium. And, since the frequency of the light wave is the same as the original wave, the wavelength reduces.

That the speed of a wavefront is different in different materials leads to a bending of the rays and this is known as refraction. In the marching band analogy, when the band approaches the muddy portion at an angle, the ranks in the mud slow down to allow the soldiers in the grass to catch up; similarly, the wavefronts slow down in the medium so that portions of the wavefront outside the medium can catch up with them. And, thus the rays bend in the medium and the bending angle is dependent on the wavelength.

A glass prism can be used to show that white light in fact contains a mixture of all colors of light.

We also learned that the speed of the light color waves on emerging from the glass is '*c*'.

Next we learned that if an em wave encounters a good conductor say, a metal, the electric fields causes currents to flow and a portion of the energy of the em wave is dissipated as heat when the electrons collide with the lattice ions. If the resistivity of the reflector is near zero, the incident wave will be totally reflected and that is why we observe that good conductors are generally shiny.

Mirrors, we noted, are made by applying a thin layer of metallic silver to one side of a pane of clear glass.

It was then pointed out that optical fibers used in high-speed communication links, are exceptionally transparent and are designed in such a way that the light rays through the glass undergo total internal reflection.

We then studied in some more detail with diagrams, the fields made by moving charges particularly when an electron is suddenly accelerated and when a charge moving with uniform velocity is suddenly stopped.

Next, an expression for the strength of the transverse electric field due to a single charge was obtained by a detailed analysis using Gauss's law. We found that the transverse field is proportional to the acceleration of the charge as expected, but inversely proportional to the distance from the charge; the field thins as it propagates outwards.

Preparatory to understanding how an antenna which is used to produce em waves with useful power, we studied the production of radiation by moving from rest an infinite sheet of charge.

The field of a stationary infinite sheet of uniform charge is static, uniform, points away from the sheet perpendicular to the sheet on both the sides, and also extends up to infinity.

A moving sheet of charge constitutes a current sheet, and from Ampere's Law, a magnetic field appears in addition to the electric field. After deriving an expression for the strength of the electric field in a kink produced as the sheet moved, we learned that it is a fraction v/c times the static field where, v is the velocity of the sheet. It was observed that the total field at a point P due to *all* the accelerating charges in the sheet, is proportional to the *velocity* of the charges, a rather interesting consequence of the principle of superposition of the fields.

A light bulb radiates light because it's filament, when heated up by the current to around 2500°C, makes the vibrating atomic cores (oscillating charges) emit electromagnetic waves in the visible frequency range.

We then learned an important aspect of the energy conservation of the system of the light bulb; when the heat dissipated to the surroundings of the filament equals the loss of potential energy of an electron (eEL) where E is the electric field in the filament and L the filament length, then a dynamic equilibrium is established and the filament temperature stops rising.

Photons are packets of electromagnetic radiation. The electron energy diagram of hydrogen was used to illustrate the production of a photon. It was shown that the electron energies are quantized and that whenever a hydrogen atom drops from a higher energy level to a lower energy level, it emits a photon. The light in the form of photons emitted by the excited atomic hydrogen is the hydrogen emission spectrum and we learned that it is this process of emission of photons which makes light emitting diodes emit colored light. We then learned about an important discovery made by Planck that blackbody radiation involves the transfer of energy in multiples of fixed amounts. His important discovery of the formula $E = h\nu$ and prediction of the distribution of radiation intensity with respect to wavelength of radiation was used in several important theories, Schrodinger Wave Equation whose solutions describe atomic orbitals and their quantization into energy levels.

The processes which produce sparks we learned, are fairly involved. It is observed that when the strength of the electric field in air exceeds 3×10^6 V/m, air is ionized and mobile charges are produced. However, the electric field strength required to pull an electron out of the atom is 1.4×10^{11} V/m, which suggests that dramatic events such as muons and fast moving charged particles such as electrons, positrons or alpha particles present in trace quantities around us, initiate the generation of a few ionized particles. The ionized particles are accelerated by powerful electric fields near charged objects around us (though not having enough energy to be fatal to us), and can collide with an air molecule nearby to knock another electron out. The two fast moving electrons can now knock out more electrons from more molecules when they collide with them and there is a cascading effect, producing more numbers of fast moving electrons.

The production of the spark occurs when the free electrons and ions recombine.

The process of Electromagnetic Radiation by Antennas in the radio frequency bands spanning 3 kHz to 300 GHz was discussed, which we learned, is based on radiation by accelerating large number of electrons.

We then learned that the Hertzian dipole is a basic oscillatory filamentary current element and forms the basis for analysis of currents and radiation patterns of almost all antennas. A linear antenna may be viewed as a chain of Hertzian dipoles with tiny disks at its' ends.

The current distribution of a linear antenna is normally plotted as a sinusoid. This current is not static, but varies with time, the positive maximum at the center swinging through zero to a negative maximum, and then back from the negative maximum through zero to a positive maximum and the cycle repeats.

Such a time-varying current represents an acceleration of the charges and a strong em wave is produced around the linear antenna.

The expression for the electric field around such a dipole (a linear antenna) which is excited by a sinusoidal signal contains three terms, each being proportional to i) $1/r^3$, ii) $1/r^2$ representing the induction field, and iii) $1/r$ which is the radiation field.

We learned that as the currents oscillate in the rods of a half-wave dipole antenna, waves of the electromagnetic fields race away from the antenna. The field patterns break away from the antenna and have an independent existence as they propagate farther and farther away.

Next, we studied the process of radiation by a quarter-wave dipole antenna constructed by two rods separated by a small distance. The antenna, we learned can be treated as an opened out transmission line. If the rods were oriented like a transmission line however, there can be no radiation, the radiation of each half of the line will cancel out the radiation from the other half, because of the currents in the two rods directed in opposite directions, and the rods being near each other. So, the purpose of studying the antenna as a transmission line was meant to illustrate the currents in them.

A detailed study of the generation of the fields of a quarter-wave Transmitting antenna during one full cycle revealed how the fields lift off from the antenna rods and then loop around due to the formation of small licks at the open ends of the antenna rods during a full cycle. This forms loopy field lines which then race away into the space around the antenna.

We then learned that in an antenna there is a loss of energy to the radiated fields and the effect it produces is similar to that of a real resistance, where electrons lose kinetic energy when they collide with lattice ions, setting the ions into motion. This resistance effect in antennas when they radiate energy is termed Radiation Resistance.

Receiving antennas are similar in construction to Transmitting Antennas with the difference that an amplifier or a detector is used in place of the signal generator in the latter.

We learned how the electric field component of an electromagnetic wave produces a current in the rods of the antenna when the field is polarized in the direction of the antenna rods. It was shown how a low watt bulb can be lit by the current when the antenna is positioned and properly oriented near a transmitting antenna.

We next studied how electromagnetic waves are attenuated or lose their strength as they propagate through space. The em wave strength drops in proportion to 1/distance from the antenna since the wave expands into wider and wider regions. The other reason, we learned is due to the presence of objects such as walls, trees and also the earth that absorb wave energy (dissipated in their resistance when the wave induces currents in them) and some of it is scattered in new directions.

Waveguides are single conductor hollow rectangular or circular tubular structures. While two-conductor lines are normally arranged to transmit energy by the TEM mode, waveguides it was seen, do not support the TEM mode. They are dimensionally configured to support TE or TM modes of propagation in the radio frequency range. They cannot transmit light, which is usually done by optical fibers.

The em wave in the TE mode is usually launched in a waveguide by means of a coaxial cable at the closed end of the guide with the center conductor protruding into the chamber and the shield terminated on the outer wall. The coax wave launcher behaves like an antenna.

The component fields are mostly confined within the conducting walls of the waveguide and the solutions for them are made assuming perfectly conducting walls. The surface currents, we were reminded, are the currents associated with the magnetic fields (assuming perfectly conducting walls).

We learned that the electric fields in the waveguides are essentially non-conservative but they seemingly terminate on surface charges on the walls. The waves essentially reflect from the walls and in the TE mode, a magnetic field component in the direction of wave travel along the guide is observed.

Next, we studied in some detail the processes which cause the wave in the TE mode to be attenuated sharply below the cutoff frequency. Above the cutoff frequency, the waves carry energy inside the guide as the waves glide along the inner walls. There is some attenuation above the cutoff frequency and this is due to the surface currents which in slightly resistive walls, cause losses of the transmitted energy.

The losses below the cutoff frequency are due to heavy penetration of the powerful electric fields whose nulls no longer occur near the walls in this frequency region and the currents penetrate the walls more resulting in heavy dissipation.

The final section of the chapter discussed the electric field of a single charge moving with uniform velocity. The charge as it moves at relativistic speeds will make the electric field lines that were uniformly divergent in all directions when the charge was static, to concentrate in a direction perpendicular to the direction of motion.

At the end of the chapter, we learned that the electric field and the magnetic field are very closely related, and the magnetic field is a relativistic effect of the electric field. The Lorentz transformation provides one explanation of how when one runs past a stationary charge, one observes a magnetic field in addition to the electric field. The Special Theory of Relativity draws upon the Lorentz transformation between inertial frames of reference to explain the occurrence of magnetic forces as electric phenomenon. We were then introduced The Emission theory of Electromagnetism by Leigh Page. This theory, uses the concept of massless moving elements issued by accelerated charges, whose loci are Faraday's lines of electric and magnetic lines of force, and are continuously emitted as electromagnetic circular waves composed of expanding circular electric and magnetic field lines.

Questions

1. Write expressions of Faraday's law and Ampere-Maxwell law.
2. Electromagnetic fields don't move. Then, how do electromagnetic waves propagate through space? Describe the process of propagation.
3. Visualize a pulse of electric and magnetic fields propagating through space. Using Faraday's law and the Ampere-Maxwell law derive expressions relating the electric field and magnetic fields.
4. What are the properties of propagating electromagnetic waves through free space?
5. Two phenomena enable the propagation of electromagnetic waves through free space. What are these two phenomena (which are included in Maxwell's equations)? Describe the phenomena and name the physicists who discovered them.
6. Beginning with the Lorentz transformation equations, explain Lorentz contraction. State how this is different from the common observation of lengths of moving trains from a railway platform.
7. A charge moves with uniform velocity v in the negative x direction. Sketch the electric field E in an inertial frame in which the charge is at rest and in an inertial frame which moves with the charge. Describe one useful visualization of the electric field in the moving frame. (Hint: See Section F.3 in Appendix F).

8. Using the example of a test charge moving with uniform velocity parallel to a current carrying wire, show that magnetism is relativistic.
9. Is the electric field real? Do the concepts of E and B fields precisely explain the operation of circuits and circuit components? Explain.
10. Describe with sketches the production of radiation by an accelerated charge.
11. A charge is made to oscillate continuously. Sketch the electric and magnetic fields that are radiated.
12. When an electromagnetic wave interacts with matter, the charges in matter reradiate. Describe this process which leads to "attenuation" of the incoming wave.
13. When light is incident on glass, describe the processes which leads to its slowing down in glass.

    What processes cause light to be blocked by opaque materials?
14. Explain how smooth and shiny surfaces reflect more light than dull surfaces.
15. Dipole antennas are stationary structures that can radiate electromagnetic waves in the radio frequency range. Describe the process of radiation by antennas.
16. With the help of sketches describe how a perturbance in the electric field of an infinite charged sheet is produced.
17. The electrons in an infinite charged sheet are oscillated. Derive the expression for the radiative field.
18. Describe the production of light by electric filament bulbs and how this is different from its production by photon emission in light emitting diodes (LEDs).
19. Do photons (Classical electrodynamics) have a rest energy? Explain. Hint: See "Photon Emission" Section 10.9.
20. Give examples of phenomena which are analysed by statistical methods. Give examples of phenomena that can be described by equations which predict the future unambiguously or deterministically. Hint: See "Quantum Mechanics, Probabilities And Orbitals of Electrons" in Appendix B.
21. Is quantization an integral part of the Boltzmann distribution expression for the energies of particles? Explain.
22. Planck borrowed the concept from the Maxwell-Boltzmann distribution of energies of gas particles to explain which physically observable phenomenon? How did Planck apply his law which fit observed data? Explain and write the expression that Planck finally derived.
23. Briefly describe the phenomenon of blackbody radiation and how its intensity varies with wavelength. Describe the theories that were made to explain the shape of the radiation intensity vs. wavelength curve and why they failed to explain it's shape. Then, describe how Max Planck formulated a theory that correctly predicted the shape of the curve.

    Name two expressions in atomic physics to which Planck's law was applied.
24. Briefly trace the development of Quantum Electrodynamics from Classical Electrodynamics.

25. What mechanism could possibly trigger the production of sparks in air? Describe it.
26. The Hertzian dipole though practically impossible to realize, is a useful construct to understand the principle of radiation of electromagnetic waves. Describe the dipole and how it can explain the equivalence of radiation by an oscillatory electric dipole and an oscillatory current element.
27. Sketch the radiation pattern produced by a quarter-wave antenna over one full period and explain how the electromagnetic wave breaks away to propagate freely through space.
28. Is a standing wave current pattern of an antenna stationary? Explain the processes which lead to the shapes of the standing wave patterns of voltage and current waveforms of these antennas.
    Sketch the current patterns of dipole antennas of lengths $\lambda/4$, $\lambda/2$ and λ.
29. What is the order of the resistance of dipole antennas made of wires and rods? State the order of radiation resistance of a dipole antenna and explain the difference between antenna resistance and radiation resistance.
30. What are the force carriers (particles) in Quantum Electrodynamics? Describe briefly their action to produce static and dynamic electric fields.
31. What loci do the moving elements in L. Page's Emission Theory trace? Describe briefly the types and configuration of fields they produce when moving with uniform velocity and when accelerated.
32. Explain why John David Jackson considered the discrete photon aspect of electric and magnetic fields **E** and **B** "can be ignored or at least glossed over".
    Calculate the number of photons per second per meter^2 in the radiation from a cellphone antenna operating at 1 GHz with a radiating power of 300 mW at a distance 1 kilometer from the phone. Does the discrete nature of the photons become evident from their number? Hint: See Section 2.1.3, *Applied Quantum Mechanics,* A. F. J. Levi, Cambridge University Press, 2003.
33. Describe the mechanism of electromagnetic wave propagation in waveguides. Which are the two laws which predict electromagnetic wave propagation in the space between its walls. Are these processes different from the mechanism of propagation in free space?
34. Waveguides cannot support the TEM mode. Explain why this is so with supporting arguments referring to the magnetic field lines and associated current types. Hint: See "Impossibility of TEM wave in Wave Guides" in the Chapter Wave Guides in Ref. [16].
35. What phase condition makes the wave propagate along the direction of the waveguide. Is this a spatial or time dependent condition?
36. Describe the processes which make the waves dissipate their energy in the walls of the waveguide.
37. Show how accelerating charges produce *circular* waves and sketch the start of wave propagation on a transmission line. Hint: See Hartel's article in "Teacher's Guide" in the CD.

Additional practice questions are available in the CD with this book.

APPENDICES

Appendix A

From "Electric and Magnetic Interactions" by Chabay and Sherwood, 1995, used by permission.

Electric Field

The electric field at a location is defined by the following relation

$$\vec{F}_{\text{electric on } q} = q\,\vec{E}$$

where $\vec{F}$ is the force acting on charge q and $\vec{E}$ is the value of the electric field at the location of charge q as shown in Fig. A.1. The electric field therefore has units of newtons per coulomb or N/C.

$$\vec{E} = \frac{\vec{F}}{q} \tag{1}$$

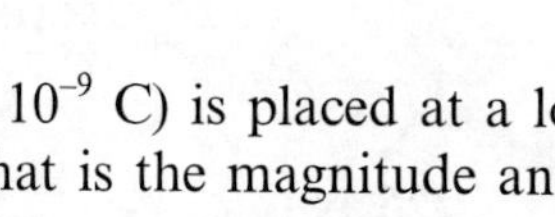

Fig. A.1 Force $\vec{F}$ acting on a charge q due to the electric field $\vec{E}$ at the location of q.

Example

If a +2 nC (nanocoulomb = 1×10^{-9} C) is placed at a location and it experiences a force 3×10^{-7} Newtons to the right, what is the magnitude and direction of the electric field at that point?

$$\vec{E} = \frac{\vec{F}}{q} = \frac{3\times10^{-7}\text{ N}}{2\times10^{-9}\text{ C}} = 150 \text{ N/C}$$

The electric field points to the right because the charge is positive.

The Electric Field due to a Point Charge (See "Point Charges" in Appendix B)

If there is a charge q_1 at some location in space, we know from Coulomb's law that if we place another charge q_2 at a location a distance 'd' from q_1, the other charge q_2 will experience a force whose magnitude is as shown in Fig. A.2.

$$\vec{F}_{\text{on } q2} = \frac{1}{4\pi\varepsilon_o}\frac{q_2 q_1}{d^2}$$

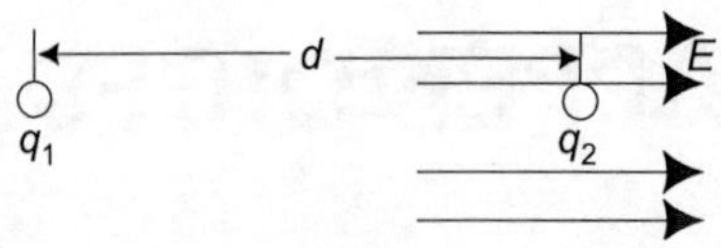

Fig. A.2 Electric field at location q_2 due to a charge q_1.

This can be rewritten as

$$\vec{F}_{\text{on } q2} = q_2 \left[\frac{1}{4\pi\varepsilon_o} \frac{q_1}{d^2} \right] = q_2 \vec{E}$$

Note: In Appendix B in the topic "The Idea of Momentum of a field" and in Section F.1, Appendix F in the topic "Coulomb Interaction", we will learn that the Coulomb forces between two charged particles do not act at the same point in space and that each carrier experiences a force due to its interaction with the field, but locally, there is no reaction back onto the field. This reaction force is found when the force on the opposite charge carrier is taken into account.

Note: The electric field of a stationary charge cannot exert a force on itself. It is the electric field due to q1 that exerts a force on q2. IT IS AN EXTERNAL FIELD THAT EXERTS THE FORCE ON q2.

So the electric field of point charge q_1 has magnitude

$$\vec{E} = \left[\frac{1}{4\pi\varepsilon_o} \frac{q_1}{d^2} \right] \tag{2}$$

What is the electric field $\vec{E}$, 20 cm away from a positive point charge of one nano-coulomb ($+1 \times 10^{-9}$ Coulomb).

You may use $\quad \dfrac{1}{4\pi\varepsilon_o} = 9 \times 10^9$ N-m²/C².

Magnitude of $\quad \vec{E} = \dfrac{1}{4\pi\varepsilon_o} \dfrac{q_1}{d^2} = \left[\dfrac{9 \times 10^9 \text{ Nm}^2}{\text{C}^2} \right] \left[\dfrac{10^{-9} \text{ C}}{(0.2 \text{ m})^2} \right]$

$$= 225 \text{ N/C}$$

The direction of $\vec{E}$ is radially outward away from the positive charge as shown in Fig. A.3.

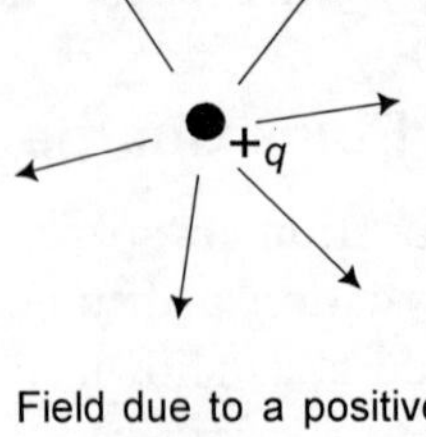

Fig. A.3 Field due to a positive charge.

The Electric Field of a Dipole

Fig. A.4 An electric dipole.

A dipole is composed of equal and unlike charges separated by a distance.

What is a dipole?

Matter contains both positive and negative charges (protons and electrons). Dipoles occur frequently in nature. Certain molecules occur in nature as dipoles with inherent polarization and are called permanent dipoles, for example, water molecules and HCl molecules. For example, a single HCl molecule is an electric dipole, the H end is somewhat positive and the Cl end is somewhat negative. A molecular dipole could be either an *induced* dipole created by external charges or a permanent dipole such as the water molecule.

Is an atom a dipole? See Appendix B for an answer to this.

The analysis of a single dipole provides a basis for analyzing more complicated forms of matter.

Superposition of Electric Fields

A very important property of electric fields is a consequence of the superposition principle which states: "The net electric force on a charged object is the vector sum of the individual electric forces on the object due to all other charged objects. Each individual interaction is *unaffected* by the presence of other charges." What this means is that the force between one particle and another is *unaffected* by the presence of a third particle. As a consequence of this principle, the net electric field due to two or more charges is *unaffected* by the presence of other charges. Superposition as a vector sum holds true for forces in general, so it also holds true for fields, because field is a force per unit charge.

Force on a Point Charge by a Dipole (Chabay & Sherwood, 1995, used by permission)[2]

The *net charge* of a dipole is zero, so we might expect that the force exerted by a dipole on a charged object having charge Q would be zero (see Fig. A.5a), using Coulomb's law:

Fig. A.5(a) Illustrating the force exerted by a dipole on a charge Q.

$$F_{\text{on } Q \text{ by dipole}} = \frac{1}{4\pi\varepsilon_o}\frac{(dipole\ charge)Q}{d^2}$$

$$= \frac{1}{4\pi\varepsilon_o}\frac{(0)(Q)}{d^2} = 0\ !$$

However, if we draw a diagram of the situation, we realize that this may not be the case, since one of the ends of the dipole is *closer* to Q than the other.

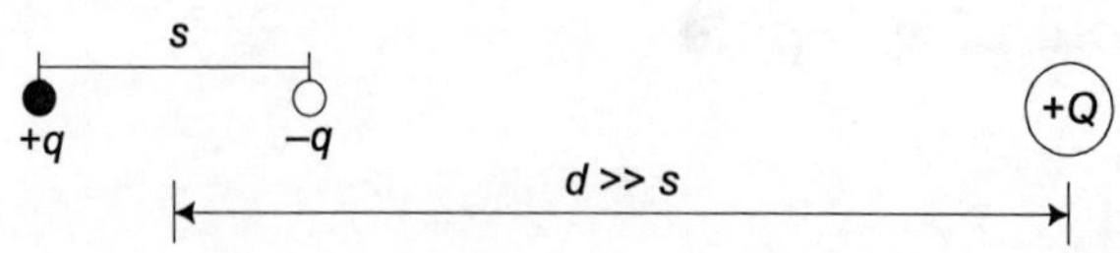

If the distance 'd' is much larger than the dipole separation 's', perhaps this difference won't matter, and the net force will be zero, then. Just to be sure, let us first draw two vectors showing the force on Q by the negative end of the dipole, and the force on Q by the positive end of the dipole as shown in Fig. A.5(b).

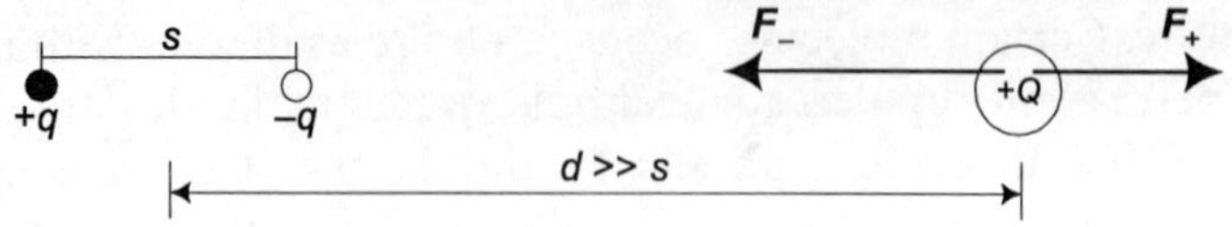

Fig. A.5(b) The forces F_- and F_+ due to the two ends of the dipole on Q.

Since the two force vectors point in opposite directions, the magnitude of the net force is the difference of the magnitude of the two individual forces:

$$\mathrm{F}_{\text{on } Q \text{ by dipole}} = F_{\text{on } Q \text{ by –ve end}} - \mathrm{F}_{\text{on } Q \text{ by +ve end}}$$

Do the two forces have exactly the same magnitudes?

No ! The magnitude of F_+ is less than the magnitude of F_- because the positive end is farther away from charge Q by the separation 's'. The net force on Q due to the dipole is small, and points to the left.

Therefore, the *net force will be one of attraction* as shown in Fig. A.6.

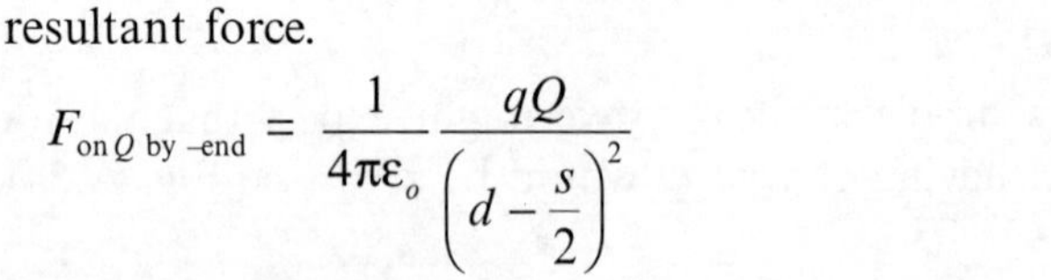

Fig. A.6 Showing the resultant force on Q due to the dipole.

Let us calculate the resultant force.

$$F_{\text{on } Q \text{ by –end}} = \frac{1}{4\pi\varepsilon_o}\frac{qQ}{\left(d - \frac{s}{2}\right)^2}$$

$$F_{\text{on } Q \text{ by +end}} = \frac{1}{4\pi\varepsilon_o}\frac{qQ}{\left(d + \frac{s}{2}\right)^2}$$

$$F_{\text{net on } Q} = \frac{1}{4\pi\varepsilon_o}\frac{qQ}{\left(d - \frac{s}{2}\right)^2} - \frac{1}{4\pi\varepsilon_o}\frac{qQ}{\left(d + \frac{s}{2}\right)^2}$$ ...from the principle of superposition

$$= \frac{1}{4\pi\varepsilon_o} qQ \left[\frac{1}{\left(d - \frac{s}{2}\right)^2} - \frac{1}{\left(d + \frac{s}{2}\right)^2} \right] \quad \text{...common factor}$$

$$= \frac{1}{4\pi\varepsilon_o} qQ \left[\frac{\left(d + \frac{s}{2}\right)^2 - \left(d - \frac{s}{2}\right)^2}{\left(d - \frac{s}{2}\right)^2 \left(d + \frac{s}{2}\right)^2} \right] \quad \text{...common denominator}$$

$$= \frac{1}{4\pi\varepsilon_o} qQ \left[\frac{\left(d^2 + 2d\frac{s}{2} + \left(\frac{s}{2}\right)^2\right) - \left(d^2 - 2d\frac{s}{2} + \left(\frac{s}{2}\right)^2\right)}{\left(d - \frac{s}{2}\right)^2 \left(d + \frac{s}{2}\right)^2} \right] \quad \text{...expand numerator}$$

$$= \frac{1}{4\pi\varepsilon_o} qQ \left[\frac{2ds}{\left(d - \frac{s}{2}\right)^2 \left(d + \frac{s}{2}\right)^2} \right] \quad \text{...simplify numerator}$$

Upto now we have done straightforward algebra, but in order to gain some physical insight into the nature of this force, we will make an important approximation.

We have already stated that $d >> s$, so

$\left(d - \frac{s}{2}\right)^2 \approx d^2$ and $\left(d + \frac{s}{2}\right)^2 \approx d^2$, and we can replace the denominator by d^4.

$$F_{\text{net on } Q} = \frac{1}{4\pi\varepsilon_o} qQ \left[\frac{2ds}{\left(d - \frac{s}{2}\right)^2 \left(d + \frac{s}{2}\right)^2} \right] = \frac{1}{4\pi\varepsilon_o} qQ \left[\frac{2ds}{d^4} \right].$$

How good is the approximation?

Let us take an example.

Let $d = 1000$ and $s = 1$. Calculate the value of $\left(d - \frac{s}{2}\right)^2 \left(d + \frac{s}{2}\right)^2$ exactly using your calculator.

$(1000 + 0.5)^2 (1000 - 0.5)^2 = [9.999995 \times 10^{11}]$. (Some calculators can't handle enough significant figures to do this calculation accurately.)

What value did you get using d^4?

$(1000)^4 = 1.0 \times 10^{12}$.

What is the percent error? $\left[\frac{1.0 \times 10^{12} - 9.999995 \times 10^{11}}{9.999995 \times 10^{11}} \right] = 0.0000005 = 0.00005\%$.

Note that even if d is only 100 instead of 1000 the error is only 0.005%. So if $d >> s$, we can be confident that $\left(d+\frac{s}{2}\right)^2 \approx d^2$ is a very good approximation.

Why cannot we use $s = 0$ in the numerator? How much is *ds*? You cannot put zero here! Coulomb's law is applicable to point charges only, not dipoles!

So, using the approximation that $\left(d-\frac{s}{2}\right)^2 = \left(d+\frac{s}{2}\right)^2 = d^2$ when $d >> s$, and canceling d in the numerator and denominator, we get:

$$F_{\text{on } Q \text{ by dipole}} = \frac{1}{4\pi\varepsilon_o} qQ \frac{2s}{d^3}$$

Electric Field due to a Dipole

(Chabay & Sherwood, 1995, used by permission)[2]

In the previous section, we calculated the force exerted by a permanent dipole on a charge placed along the x-axis. We will now calculate the electric field E_1, due to the dipole at a distance x from the centre of the dipole.

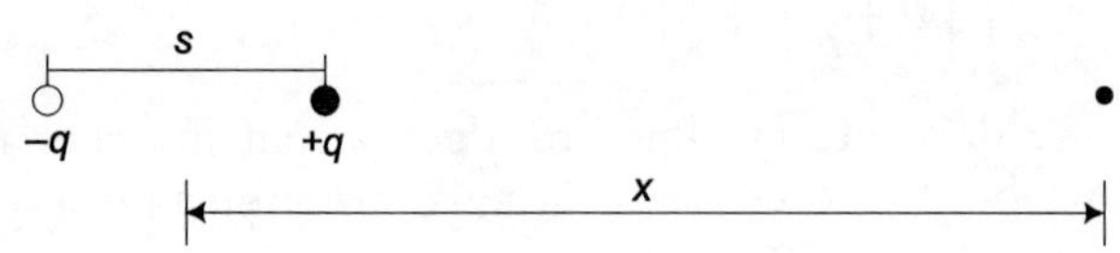

Fig. A.7 Electric field due to a dipole at a distance x from the dipole.

The net field E_1 is horizontal to the right, because at this location, the field E_+ of the nearer $+q$ charge is larger than the field E_- of the more distant $-q$ charge. The magnitude is $E_1 = E_+ - E_-$.

Let us find a simplified expression for E_1.

Using the formula for $\vec{E} = \left[\frac{1}{4\pi\varepsilon_o}\frac{q_1}{d^2}\right]$ due to a point charge q_1 at an observation location distance 'd' away from the point charge applied to our above system of a dipole:

$$E_1 = E_+ - E_- = \frac{1}{4\pi\varepsilon_0}\frac{q}{\left(x-\frac{s}{2}\right)^2} - \frac{1}{4\pi\varepsilon_0}\frac{q}{\left(x+\frac{s}{2}\right)^2} \qquad \text{...superposition}$$

$$= \frac{q}{4\pi\varepsilon_o}\left[\frac{1}{\left(x-\frac{s}{2}\right)^2} - \frac{1}{\left(x+\frac{s}{2}\right)^2}\right] \qquad \text{...common factor}$$

$$= \frac{q}{4\pi\varepsilon_o}\left[\frac{\left(x+\frac{s}{2}\right)^2-\left(x-\frac{s}{2}\right)^2}{\left(x-\frac{s}{2}\right)^2\left(x+\frac{s}{2}\right)^2}\right] \quad \text{...common denominator}$$

$$= \frac{q}{4\pi\varepsilon_o}\left[\frac{\left(x^2+sx+\frac{s^2}{4}\right)-\left(x^2-sx+\frac{s^2}{4}\right)}{\left(x-\frac{s}{2}\right)^2\left(x+\frac{s}{2}\right)^2}\right] \quad \text{...expand numerator}$$

$$= \frac{1}{4\pi\varepsilon_o}\left[\frac{2qsx}{\left(x-\frac{s}{2}\right)^2\left(x+\frac{s}{2}\right)^2}\right] \quad \text{...simplify numerator (3)}$$

If we are far away from the dipole, as is normally the case when we interact with molecular dipoles, the magnitude of the field can be further simplified.

This situation is depicted as $x >> s$, then

with $$E_1 = \frac{q}{4\pi\varepsilon_o}\left[\frac{2sx}{\left(x-\frac{s}{2}\right)^2\left(x+\frac{s}{2}\right)^2}\right]$$

from Eq. (3) above

$x - \frac{s}{2} \approx x$ and $x + \frac{s}{2} \approx x$ as well, so

$$E_1 = \frac{q}{4\pi\varepsilon_o}\left[\frac{2sx}{(x)^2\,(x)^2}\right] = \frac{q}{4\pi\varepsilon_o}\frac{2sx}{x^4}$$

$$= \frac{1}{4\pi\varepsilon_o}\frac{2qs}{x^3} \qquad (4)$$

Notice how the electric field drops away rapidly, proportional to $1/x^3$ at distances along the x-axis that are far away from the dipole.

Next, consider the electric field $\vec{E}_2$ at a location y along the y-axis as shown in Fig. A.8.

Note that the distance from either charge to the location y is, by Pythagorean theorem,

$$\left[y^2+\left(\frac{s}{2}\right)^2\right]^{1/2}$$

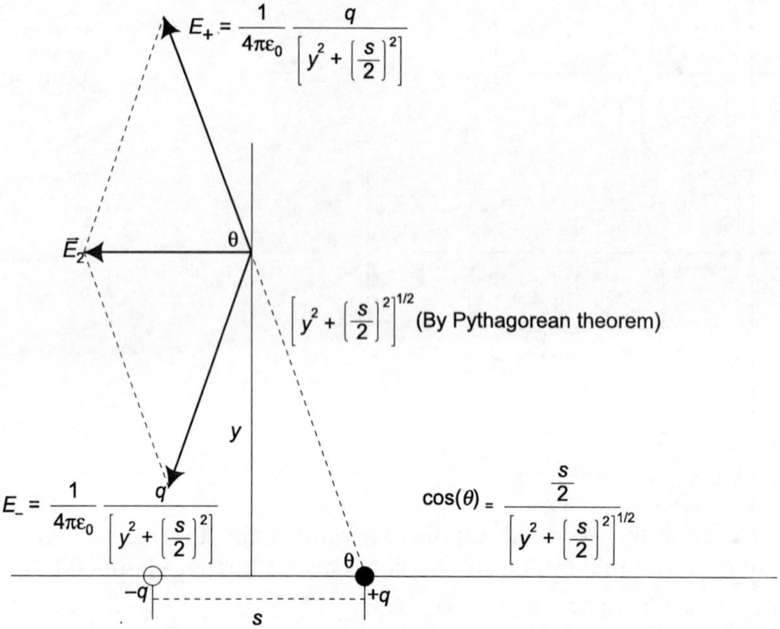

Fig. A.8 Vector diagram for calculating the field at a location *y* along the *y*-axis.

The net field $\vec{E}_2$ is the vector sum of $\vec{E}_+$ and $\vec{E}_-$. The vertical components of the two fields $\vec{E}_+$ and $\vec{E}_-$ cancel, and the net field $\vec{E}_2$ is therefore horizontal to the left.

Note that both $\vec{E}_+$ and $\vec{E}_-$ have an *x*-component that involves $\cos(\theta) = \dfrac{\frac{s}{2}}{\left[y^2+\left(\frac{s}{2}\right)^2\right]^{1/2}}$.

The magnitude is $E_2 = E_+ \cos\theta + E_- \cos\theta = 2E_+\cos\theta$.
From Fig. A.8 above we compute E_2 as:

$$E_2 = 2\frac{1}{4\pi\varepsilon_o}\frac{q}{\left[y^2+\left(\frac{s}{2}\right)^2\right]}\left[\frac{\left(\frac{s}{2}\right)}{\left[y^2+\left(\frac{s}{2}\right)^2\right]^{1/2}}\right]$$

$$= \frac{1}{4\pi\varepsilon_o}\frac{qs}{\left[y^2+\left(\frac{s}{2}\right)^2\right]^{3/2}}$$

If we consider points far away along the *y*-axis where $y >> s$, then

$$E_2 = \frac{1}{4\pi\varepsilon_o}\frac{qs}{\left[\left(y+\frac{s}{2}\right)\right]^{3/2}} \approx \frac{1}{4\pi\varepsilon_o}\frac{qs}{\left[y^2\right]^{3/2}} \text{ (since } y >> s)$$

$$= \frac{1}{4\pi\varepsilon_o}\frac{qs}{y^3}$$

Since $\vec{E}_2$ points to the left (in a direction that is opposite to the normal positive x-direction), the x component of the vector E_2 is negative. See Fig. A.9.

The electric field far away from the dipole along the y-axis drops off rapidly proportional to $1/y^3$.

The reason we consider the cases when $y >> s$ is because while the stretch of the atomic dipoles are tiny in orders of magnitude of 10^{-12} m in most circuit applications, the points where the field is of significance in conduction processes may be at distances several molecular distances away say, at 10^{-9} m or more.

Note that 10^{-9}m >> 10^{-12}m.

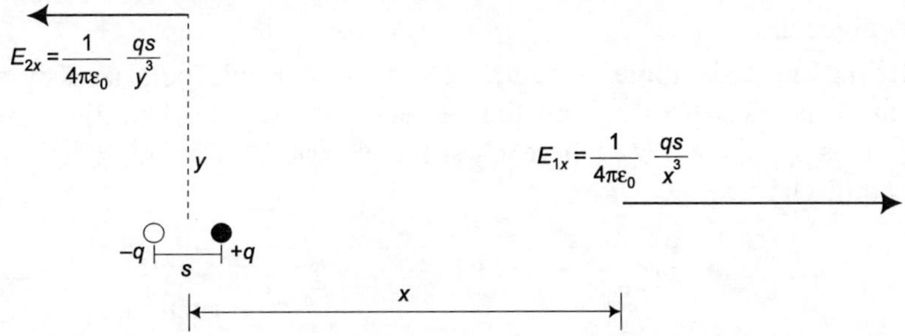

Fig. A.9 The field strengths along the x-axis and y-axis of a dipole.

The electric field at both locations is proportional to the product qs, called the "dipole moment" and denoted by $p = qs$. Far away from a dipole, the same electric field could be due to a small q and a large s, or a large q and a small s. The only thing that matters is the "dipole moment" qs.

Moreover, this is the quantity that is measurable for molecules such as HCl and H_2O that are permanent dipoles, not the individual values of q (the amount of charge on an end) and s (the stretch of the molecule).

Force on a Point Charge due to a Permanent Dipole

(Chabay & Sherwood, 1995, used by permission)[2]

We can easily calculate from the electric field calculations above the force acting on a point charge at distances away from a dipole of separation s as depicted in Fig. A.10.

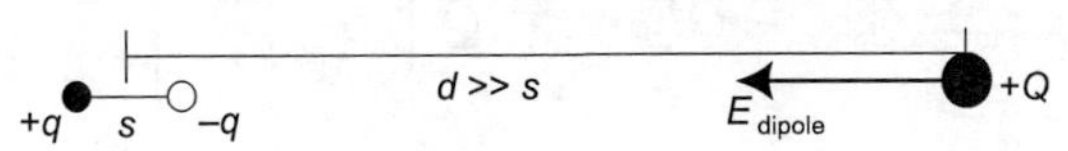

Fig. A.10 Force acting on a point charge due to a dipole.

Assume that there is a point charge $+Q$ located a distance $d >> s$, then
Direction of $F_{\text{on } +Q \text{ by dipole}}$ is to the left;
From Eqs. (1) and (4)

$$F_{\text{on } +q \text{ by dipole}} = QE_{\text{dipole}} = Q\frac{1}{4\pi\varepsilon_o}\frac{2qs}{d^3}$$

Force on a Point Charge due to an Induced Dipole

(Chabay & Sherwood, 1995, used by permission)[2]

When a small electric field E is applied to an atom or molecule that is not a permanent dipole, it is found experimentally and theoretically that the charge separation s (and the induced dipole moment $p = qs$) is proportional to E. Double the electric field E, double the stretch s. This proportionality does not hold for very large electric fields such as the electric fields in the electromagnetic radiation in a tightly focused beam of a powerful laser. But under most conditions it is an excellent approximation to say that the dipole moment $p = qs = \alpha E$, where α is a constant that expresses the polarizability of the particular kind of atom or molecule.

Consider an induced dipole, a neutral atom or molecule polarized by a point charge $+Q$ that is at a distance $d >> s$ from the center of the dipole. The dipole moment of the induced dipole is $p = qs = \alpha E_Q$, where E_Q is the electric field at the location of the dipole, due to the point charge $+Q$.

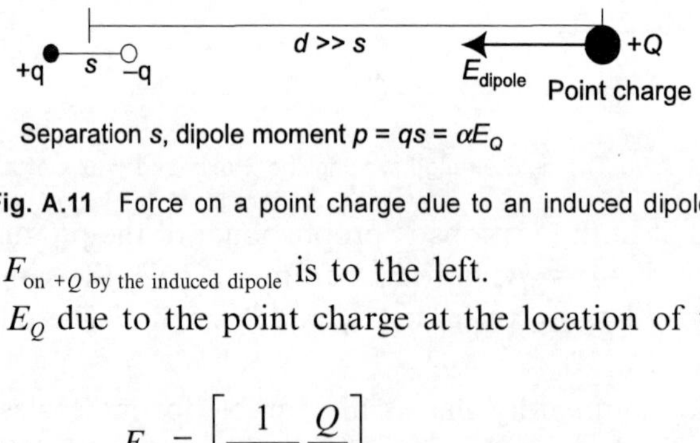

Fig. A.11 Force on a point charge due to an induced dipole.

The direction of $F_{\text{on } +Q \text{ by the induced dipole}}$ is to the left.

The electric field E_Q due to the point charge at the location of the dipole is given by Eq. (2)

$$E_Q = \left[\frac{1}{4\pi\varepsilon_o}\frac{Q}{d^2}\right] \tag{5}$$

Therefore,

$$F_{\text{on } +Q \text{ by dipole}} = QE_{\text{dipole}} = Q\left[\frac{1}{4\pi\varepsilon_o}\frac{2qs}{d^3}\right] \text{ using expression (4)}$$

Since $\quad p = qs = \alpha E_Q;$

Therefore, $F_{\text{on } +Q \text{ by induced dipole}} = QE_{\text{dipole}} = Q\left[\frac{1}{4\pi\varepsilon_o}\frac{2qs}{d^3}\right]$

$$= Q\frac{1}{4\pi\varepsilon_o}\frac{2}{d^3}\left[\alpha\frac{1}{4\pi\varepsilon_o}\frac{Q}{d^2}\right]$$

$$= \left[\frac{1}{4\pi\varepsilon_o}\right]^2 \frac{2\alpha Q^2}{d^5}$$

Notice that the force F on $+Q$ by an induced dipole falls away at the rate of $1/d^5$.

By Newton's third law, *the magnitude of the force exerted by Q on the dipole is the same*.

Amount of the Electron Cloud Shift in the Carbon Atom of Small Bits of Paper that is Attracted to a Charged Plastic Pen

The electric field of the charged pen shifts the electron cloud of the carbon atoms in the small bits of paper a small amount, creating an induced dipole (ignoring the effects that the other induced dipoles in the paper have on this carbon atom).

When the pen is a height h above the paper, the carbon atom's outer shell of four electrons (charge –4e) shifts an unknown distance s relative to the nucleus. The inner shell of 2 electrons shifts a negligible amount relative to the nucleus, because the inner electrons on the average are much closer to the nucleus and much more strongly attracted to the nucleus.

Assume that the pen carries a charge of about $-Q = -10^{-7}$ coulomb and height h = 1 cm, we will calculate how far the carbon atom's outer shell of four electrons must be shifted, s, in order that the upward electric force of attraction between the pen and the carbon atom equals the downward gravitational force on the carbon atom.

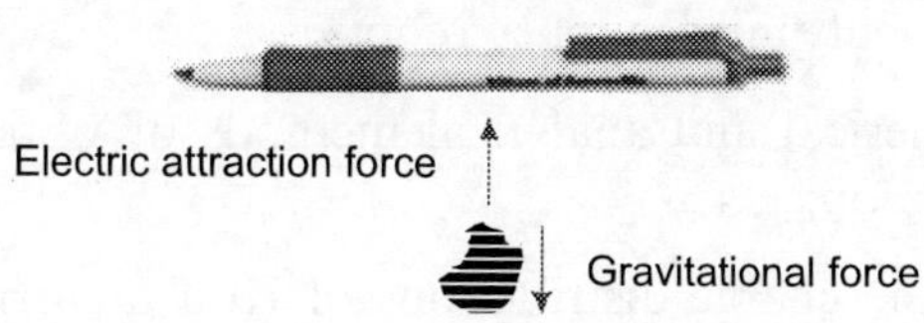

The mass of the carbon atom is (12×10^{-3} kg/mole)/(6×10^{23} atoms/mole).

Force on the carbon atom by the pen will be given by $F_{\text{on Q by dipole}} = \frac{Q}{4\pi\varepsilon_o}\frac{2qs}{h^3}$ where Q is on the plastic pen and the dipole is in the bit of paper, using expression (4) for the electric field at the pen due to the dipole charge q.

The downward force due to gravity on the carbon atom will be

$$mg = \frac{12\times10^{-3}}{6\times10^{23}} \times 9.81 \text{ newtons.}$$

When the pen is h = 1 cm away, the force of attraction of the pen should be slightly greater than the downward force of gravity in order that the pen is able to lift the bit of paper. Therefore,

$$\frac{12\times10^{-3}}{6\times10^{23}} \times 9.81 = \frac{Q}{4\pi\varepsilon_o}\frac{2qs}{h^3} = \frac{10^{-7}}{4\pi\varepsilon_o} \times \frac{2\times4\times1.6\times10^{-19}\times s}{(0.01)^3}$$

[**Note:** $\frac{1}{4\pi\varepsilon_o} = 9 \times 10^9$ Nm²/C²]

Therefore,

$$s = \frac{12\times10^{-3}\times9.81}{6\times10^{23}}\times\frac{1}{9\times10^{9}}\times\frac{(0.01)^3}{10^{-7}\times2\times4\times1.6\times10^{-19}}$$

$$= \frac{19.62}{72}\times 10^{-15}\text{ meter}$$

$$= 0.72\times 10^{-15}\text{ meter.}$$

The distance the electron cloud shifts turns out less than 10^{-15} meters.

This is roughly the radius of a single proton. The radius of the electron cloud of the carbon atom is about 10^{-10} meters, or about 100,000 times larger than a proton.

The shift is remarkably *tiny* to have such a marked effect. This is another example of the fact that electric interactions are much stronger than gravitational interactions.

Electric Field due to Distributed Charges

(Chabay and Sherwood, 2010, used by permission)[2]

Several textbooks describe the procedure to obtain the electric field of large numbers of point charges distributed over large areas, but few textbooks discuss mathematical techniques for adding up the contributions to the electric field by dividing the charge distribution into a large but finite number of pieces.

We will discuss numerical and analytical methods of obtaining the electric field at a few locations around the objects.

Here, we will treat the charge distributions of (i) a uniformly charged thin rod, (ii) a uniformly charged thin circular ring, (iii) a uniformly charged disk, and (iv) two uniformly and oppositely charged disks spaced a small distance apart (called a "capacitor").

The electric field due to two individual charged particles can be calculated as the superposition of the fields due to each particle. In a similar way, we could calculate the net electric field due to three or four charged particles.

But how can we calculate the electric field due to the superposition of the fields of millions of individual charged particles?

For this, we apply the superposition principle, and add the contributions of all the point charges by following the general steps in the calculation of the field due to objects carrying distributed charges.

1. Divide the charged object into small pieces. Make a diagram and draw the electric field $\Delta\vec{E}$ contributed by one of the pieces.

2. Choose an origin and axes. Write an algebraic expression for the electric field $\Delta\vec{E}$ due to one piece.
3. Add up the contributions of all pieces, either numerically or symbolically.
4. Check that the result is physically correct.

Electric Field of a Uniformly Charged Thin Rod

(Chabay and Sherwood, 2010, used by permission)

As an example of how to calculate the electric field due to large numbers of charges, let us consider the electric field of a uniformly charged thin rod.

The rod might, for example, be a glass rod that was rubbed all over with silk, giving it a nearly uniform distribution of positive charge on its surface. We will consider a thin rod of length L and total positive charge Q.

The field of a point charge was spherically symmetric, since a point charge itself has spherical symmetry. Since a rod is shaped like a cylinder, we should expect cylindrical symmetry in the electric field of a uniformly charged rod as shown in the figure.

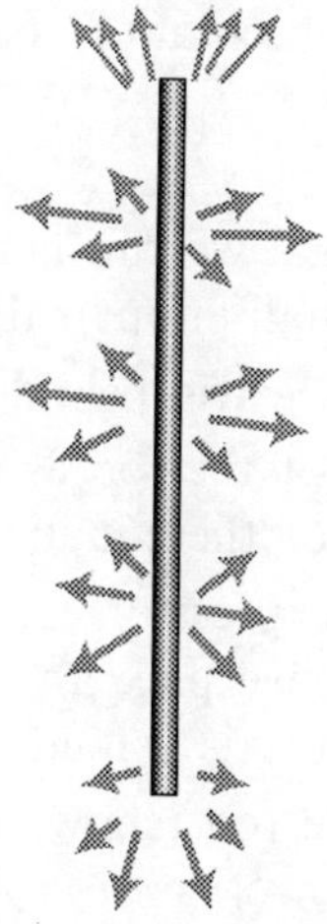

The electric field at locations near a charged thin rod. Near the middle of the rod the electric field vectors lie in a plane perpendicular to the rod.

The process of finding the electric field due to charge distributed over a macroscopic object has four steps.

Step 1: Divide the distribution into pieces; draw $\Delta\vec{E}$

To apply the superposition principle, we imagine cutting up the thin rod into very short sections each with positive charge ΔQ, as shown in the figure.

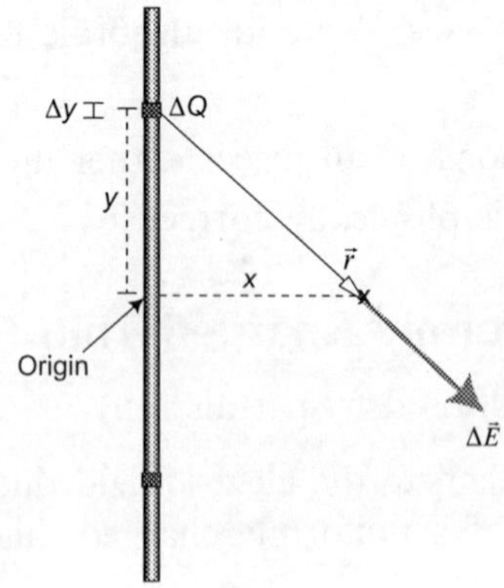

A uniformly charged thin rod.

Here ΔQ is a small portion of the total charge Q of the rod, which contributes $\Delta\vec{E}$ to the net field at the observation location. We can treat the piece ΔQ as though it were a point charge, which should be a fairly good approximation as long as its size is small compared to the distance to the observation location.

In the figure we have picked a representative piece of the rod that is not at a "special" location (in this case, neither at an end nor at the middle), and drawn its contribution $\Delta\vec{E}$ to the net field at our chosen observation location.

Assumptions

We assume that the rod is so thin that we can ignore the thickness of the rod. We choose the piece of the charge ΔQ to be small enough that it can be modeled as a point charge.

Step 2: Write an expression for the electric field due to one piece

We will approximate each piece of the rod as a point charge. Therefore, the expression for $\Delta\vec{E}$ will have the familiar form of the equation for the electric field of a point charge.

In order to write an algebraic expression for the contribution to the electric field of one representative piece of the rod, we pick an origin at the center of the rod, a choice that makes the algebra easier. The x-axis extends to the right, and the y axis extends up.

The location of one piece of the rod (shown in the figure) depends on y.

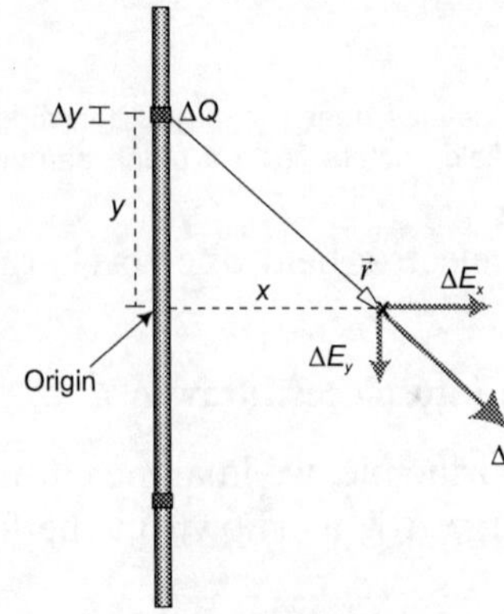

A uniformly charged thin rod showing choice of origin and axes.

The coordinates of the observation location should remain, but we will sum (integrate) over all pieces of the rod, so the coordinates of the piece of the rod should not remain. However, in the expression we want to integrate it is okay to have variables representing the location of the rod segment; they are called "integration variables" and will disappear after we do the sum.

A piece of the rod is a distance y from the origin. The length of the piece is Δy, a small increment in the integration variable y, shown in the figure. Our invented variable will not appear in the final result.

The vector $\vec{r}$ points from the source (the representative little piece of charge ΔQ) to the observation location. We can find the components of $\vec{r}$ from the diagram in the figure.

$$\vec{r} = \langle obs.loc. \rangle - \langle source \rangle = \langle x, 0, 0 \rangle - \langle 0, y, 0 \rangle$$
$$= \langle x, -y, 0 \rangle$$

Note that –y, the y component of $\vec{r}$, is negative (as it should be).

$$|\vec{r}| = [x^2 + (-y)^2]^{1/2}$$

Unit vector $$\hat{r} = \frac{\vec{r}}{|\vec{r}|} = \frac{\langle x, -y, 0 \rangle}{[x^2 + (-y)^2]^{1/2}}$$

The scalar part of the field, due to this piece, is:

$$\frac{1}{4\pi\varepsilon_o}\frac{\Delta Q}{r^2} = \frac{1}{4\pi\varepsilon_o}\frac{\Delta Q}{[x^2 + (-y)^2]}$$

We can now write the vector $\Delta\vec{E}$:

$$\Delta\vec{E} = \frac{1}{4\pi\varepsilon_o}\frac{\Delta Q}{[x^2 + (-y)^2]}\frac{\langle x, -y, 0 \rangle}{[x^2 + (-y)^2]^{1/2}}$$

$$\Delta\vec{E} = \frac{1}{4\pi\varepsilon_o}\frac{\Delta Q}{(x^2 + y^2)^{3/2}}\langle x, -y, 0 \rangle$$

The components of $\Delta\vec{E}$ are then:

$$\Delta\vec{E}_x = \frac{1}{4\pi\varepsilon_o}\frac{x\Delta Q}{[x^2 + (-y)^{3/2}]}$$

$$\Delta\vec{E}_y = \frac{1}{4\pi\varepsilon_o}\frac{-y\Delta Q}{[x^2 + (-y)^{3/2}]}$$

$$\Delta\vec{E}_z = 0$$

ΔQ and the Integration Variable

Which things are constants and which things are variables in the expressions for ΔE_x and ΔE_y?

You should have identified y as a variable, since it differs for each small piece. We need to rewrite the expression in a form in which it is easy to add up all the ΔE_x's and ΔE_y's due to all the pieces of the rod. This means expressing everything in terms of one integration variable, related to the coordinates of the piece; in this case y is the integration variable.

In particular, we need to express the charge ΔQ in terms of the integration variable y. The rod is uniformly charged with a total charge Q (positive or negative), so the amount of charge on a section of length Δy is equal to

$$\Delta Q = \left(\frac{\Delta y}{L}\right) Q$$

since $\Delta y/L$ is the fraction of the whole rod represented by Δy. Alternatively, there is a linear charge density of Q/L in coulombs/meter, so on a length of Δy meters there is an amount of charge $\Delta Q = (Q/L)\Delta y$. The important thing is to realize that ΔQ is a small fraction of the total charge Q, and that you must express ΔQ in terms of Δy.

Algebraic Expression for $\Delta \vec{E}$

Putting it all together, we have an expression for the x component:

$$\Delta \vec{E}_x = \frac{1}{4\pi\varepsilon_o} \frac{Q}{L} \frac{x}{(x^2+y^2)^{3/2}} \Delta y$$

A similar approach yields an expression for the y component:

$$\Delta \vec{E}_x = \frac{1}{4\pi\varepsilon_o} \frac{Q}{L} \frac{-y}{(x^2+y^2)^{3/2}} \Delta y$$

A note on the symbol Δ

We have used the Greek letter Δ (capital delta) in two ways in setting up the problem. ΔE or ΔQ refers to a small contribution to a total quantity. Δy refers to a change in the integration variable y, which determines the location of the piece of the rod currently under consideration. You will need to use the symbol Δ in both these ways when setting up problems like this one.

Step 3: Add Up the contributions of all the pieces

Before performing the actual summation, we should think about what we expect the answer to be. Since we have chosen an observation location on the midline of the rod, we should expect the y component of the final result be zero, as indicated in the figure.

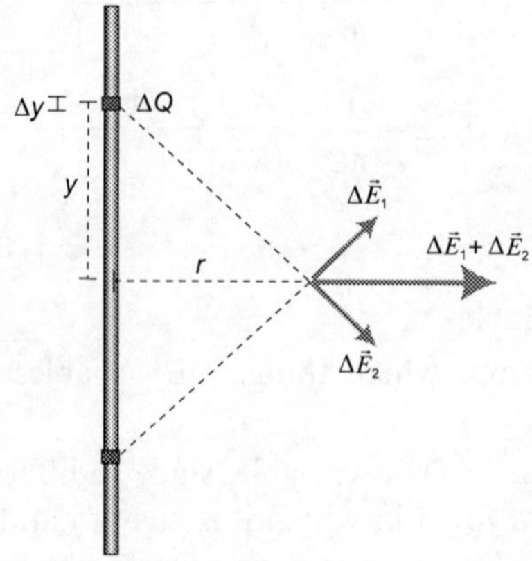

$\Delta \vec{E}$ is the contribution to the total field at location $(x, 0, 0)$ made by a small piece of the rod of length Δy.

x Components

Each ΔQ contributes ΔE_x to the net field. If we number each piece 1, 2, 3, and so on, we have

$$\Delta E_x = \Delta E_{x1} + \Delta E_{x2} + \Delta E_{x3} + \cdots$$

It is standard practice to write such sums in a more compact form using a Σ (the Greek capital sigma stands for "Summation"):

$$E_x = \Sigma \Delta E_x = \sum \frac{1}{4\pi\varepsilon_o} \frac{Q}{L} \frac{x}{(x^2 + y^2)^{3/2}} \Delta y$$

y Components

$$E_y = \Sigma \Delta E_y = \sum \frac{1}{4\pi\varepsilon_o} \frac{Q}{L} \frac{-y}{(x^2 + y^2)^{3/2}} \Delta y$$

At this point we have to decide how to add up these terms. Each term in this summation is different. Although we can choose to make all the pieces the same length Δy and take Δy out of the sum, y itself is different for every piece along the rod. How can we add up all the contributions of all the pieces?

Numerical Summation

One way to add up all the different contributions would be to divide the rod into 10 slices as shown in the figure, calculate the contribution of each of those 10 slices (using our formula with $\Delta y = L/10$), and add up these 10 numbers. Of course that would be only an appropriate result, because each slice isn't really a point particle, but it might be good enough for many purposes. We will do this calculation for the case of $L = 1$ meter, $\Delta y = 0.1$ meter, $Q = 1$ nanocoulomb (1 nano unit $= 10^{-9}$ unit), and $x = 0.05$ meter. For each slice we'll take y to be at the center of the slice, and we'll use a calculator to evaluate each term in the summation. The figure shows a diagram of the 1 meter long rod divided into 10 slices.

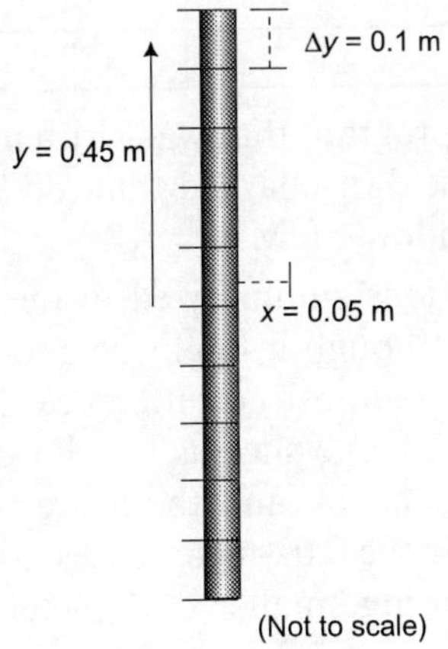

Division of the rod into 10 equal slices.

The results are shown in the Table 1 including the sum for all pieces. As expected, we see the sum of the y components is zero.

Contributions to $\vec{E}$

Table 1

y	ΔE_x	ΔE_y
+0.45	0.484	−4.363
+0.35	1.018	−7.217
+0.25	2.715	−13.577
+0.15	11.384	−34.153
+0.05	127.279	−127.279
−0.05	127.279	127.279
−0.15	11.384	34.153
−0.25	2.715	13.577
−0.35	1.018	7.127
−0.45	0.484	4.363

It would be possible to get a more accurate value by cutting the rod into more slices, but it would be tedious to do the calculations by hand. A computer program was used to do these calculations for various numbers of slices, and the results of these calculations are shown in Table 2.

Effect of number of slices

Table 2

# slices	E_x
10	285.764
20	353.764
50	358.214
100	358.214

A listing of the MATLAB program that was written to do these calculations is available in the CD ROM file "Electric field of charged thin rod". The program can be directly copy pasted into the Command Window of MATLAB.

Judging from the computations summarized in the Table 2, our calculations with just 10 slices was not very accurate (though it might be good enough for some purposes), while taking more than 50 slices makes almost no difference. The accuracy depends on how good an approximation it is to consider one slice as a point charge. With 50 slices, each slice Δy is 1/50th of a meter long (0.02 meters), and the nearest slice is 0.05 meters away from the observation location. Apparently the 0.02-meter slices are adequately approximated by point charges at their centers, since using smaller 0.01 meter slices (100 slices) gives practically the same result.

How do we know that our computer summations are correct? For 10 slices they agree with the calculator results. For larger and larger numbers of slices the computer results approach a constant value, which is expected. Later we will discuss additional ways to check such work.

Summation as an Integral

A major disadvantage of adding up the contributions numerically is that we don't get an analytical (algebraic) form for the electric field. This means that we can't easily answer such questions as, "How does the electric field vary with r near a rod?" For a point charge we know that the field goes like $1/r^2$, and far from a permanent dipole it goes like $1/r^3$. Is there a way to do the summation to get an algebraic rather than a numerical answer for the rod? That is what integral calculus was invented to do.

The key idea of integral calculus applied to problems like ours is to imagine taking not 50 or 100 slices but an infinite number of infinitesimal slices. We let the number of slices $N = L/(\Delta y)$ increase without bound, and the corresponding slice length $\Delta y = L/N$ decreases without bound. We take the limit as Δy gets arbitrarily small:

$$E_x = \lim_{\Delta y \to 0} \frac{1}{4\pi\varepsilon_o}\frac{Q}{L} x \sum \frac{1}{(x^2+y^2)^{3/2}} \Delta y$$

This limit is called a "definite integral" and is written like this:

$$E_x = \frac{1}{4\pi\varepsilon_o}\frac{Q}{L} x \int_{-L/2}^{+L/2} \frac{1}{(x^2+y^2)^{3/2}} dy$$

The integral sign ∫ is a distorted S standing for Summation, just as the Greek sigma Σ stands for summation. It is important that in this context you think of an integral as a sum of many contributions.

The integration variable y ranges from $y = -L/2$ (the bottom of the rod) to $y = +L/2$ (the top of the rod), so these are limits on the definite integral. If you have drawn an appropriately labeled diagram, you should be able to read the correct limits of the integration variable off the diagram. This is another benefit of drawing a well-labeled physics diagram.

Our small length Δy has now changed into dy, an "infinitesimal" increment in y that must appear in the integrand. This is why Δy must appear in the algebraic expression for the contribution of one piece of charge distribution.

Evaluating the Integral

Most of the physics in this problem went into setting up the integral. Evaluating the integral is simply mathematics. In some cases it is easy to evaluate the integral. In this particular case, the integral is not a very simple one, but if you would like to exercise your skill at integration, give it a try! Otherwise, it can be found in tables of integrals in mathematical handbooks and in some calculus textbooks. Some calculators can do such integrals, or one can use a tool such as Maple or Mathematica. Looking up the result in a table of integrals, we get the following:

$$E_x = \frac{1}{4\pi\varepsilon_o}\left(\frac{Q}{L}x\right)\left[\frac{y}{x^2\sqrt{x^2+y^2}}\right]_{-L/2}^{+L/2}$$

$$E_x = \frac{1}{4\pi\varepsilon_o}\left[\frac{Q}{x\sqrt{x^2+(L/2)^2}}\right]$$

Note that as we expected this result does not contain the integration variable y, which was simply a variable referring to the coordinates of one piece of the rod and was necessary in setting up the summation.

The y component of the electric field can also be found by integration. In this situation the result comes out to zero, as expected.

$$E_y = \frac{1}{4\pi\varepsilon_o}\frac{Q}{L}\int_{-\frac{L}{2}}^{+\frac{L}{2}}\frac{-y}{(x^2+y^2)^{\frac{3}{2}}}dy$$

$$E_y = 0$$

Replace *x* with *r*

Because the rod, and its associated electric field, are cylindrically symmetric, the axis we called the x axis could have been rotated around the rod by any angle, and we would have obtained the same answer. To indicate this, we replace x with r in our result (you may recognize this as converting to cylindrical coordinates):

Electric Field of a Uniformly Charged Thin Rod ....contd

$$E_y = \frac{1}{4\pi\varepsilon_o}\left[\frac{Q}{r\sqrt{r^2+(L/2)^2}}\right]$$

at a location a distance r from the midpoint of the rod along a line perpendicular to the rod. Q is the total charge on the rod, and L is the length of the rod. The direction is either radially away from or toward the rod (depending on the sign of Q). See the figure.

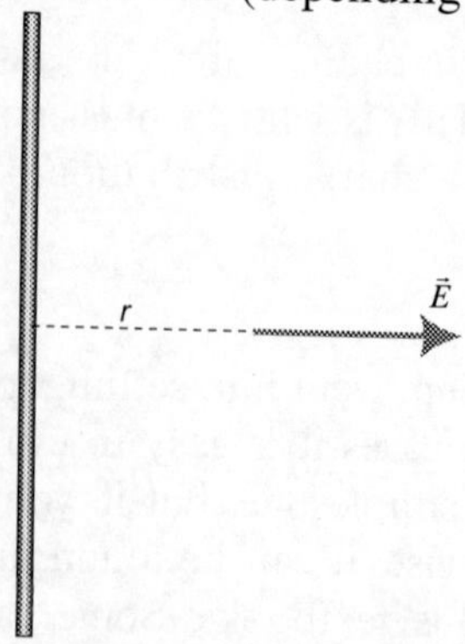

Electric field of a uniformly charged thin rod, near the midpoint of the rod.

Step 4: Check the result

Because there are many opportunities to make mistakes in this procedure, it is extremely important to check the result in as many ways as you can. Different checks provide information about different kinds of possible errors.

Direction: First, is the direction qualitatively correct? We have the electric field pointing straight away from the midpoint of the rod, which is correct, given the symmetry of the situation. The vertical component of the electric field should indeed be zero.

Units: Second, do we have the right units? The units should be the same as the units of the expression for the electric field for a single point particle:

$$\frac{1}{4\pi\varepsilon_o}\frac{Q}{r^2}$$

We easily verify our answer does have the right units, since

$$\frac{1}{r\sqrt{r^2+(L/2)^2}} \text{ has the same units as } \frac{1}{r^2}$$

Special case: $r >> L$ (very short rod, or very far from rod)

If r is very much larger than L, the distant rod looks almost like a point charge, so the net field ought to look like the field of a point charge. Does it?

$$\text{If : } r >> L\text{, then } \frac{1}{r\sqrt{r^2+(L/2)^2}} \approx \frac{1}{r\sqrt{r^2+0}} = \frac{1}{r^2}.$$

Special case: $L << r$

For another special case, suppose that r is fairly near the rod, but the rod is so short as to look almost like a point charge. In that case $L << r$, and we again get $E \rightarrow \frac{1}{4\pi\varepsilon_o}\frac{Q}{r^2}$, as we should.

Compare to numerical calculation: Earlier we carried out a numerical calculation for the case of $Q = 1 \times 10^{-9}$ C, $L = 1$ meter, and $r = 0.05$ meter, and we found that if we used 50 or more slices the result was $E_x = 358.214$ N/C.

The analytical solution gives

$$E = \frac{1}{4\pi\varepsilon_o}\left[\frac{Q}{r\sqrt{r^2+(L/2)^2}}\right]$$

$$= \frac{1}{4\pi\varepsilon_o}\left[\frac{1\times10^{-9}}{0.05\sqrt{0.05^2+0.5^2}}\right]$$

$$= 358.213 \text{ N/C}$$

which agrees very well with the numerically calculated value.

Checking a Numerical Solution

Similar techniques can be used to check a numerical integration done on a computer. For example, set the length of the rod to be very short or the distance to be very large, and the numerical integration should give a result equal to what you calculate by hand for a point charge.

Infinitesimals and Integrals in Science

In pure mathematics, an infinitesimal quantity is conceived to be arbitrarily close to zero. In science, however, this may not make sense. For example, in the case of the uniformly charged thin rod it makes no physical sense to choose dz to be smaller than an atomic diameter. It would enormously complicate the summation if we had to take into account the detailed distribution of charged particles inside each atom. All that is required is that dz be small compared to the significant dimensions of the situation (r and L). We want each slice dz to be an adequate approximation of a point charge, and dz = 0.02 m may be good enough ! We can consider the mathematical infinitesimal dz to be an idealization of what a scientist means by an infinitesimal quantity, which is a quantity that is "small enough" within the desired and possible precision of measurements and analysis.

Oddly, the result of doing an integral is approximate, not exact, because at the microscopic level the actual charge distribution is not in fact uniformly distributed. In Section 1.14, we found that on a charged piece of tape the individual electrons or ions were far from each other on an atomic scale. However, if you are many atomic diameters away from a "uniformly" charged rod, the mathematical integral gives an excellent approximation to the actual electric field.

There is another important sense in which our analysis is approximate. If a rod is an insulator, and has been charged by rubbing it with a cloth, the charge distribution will be only approximately uniform, even on a macroscopic scale, since charge transfer cannot be accurately controlled. If the rod is a metal, charge will tend to pile up on the ends rather than being uniformly distributed. Nevertheless, modelling a charged rod as a *uniformly* charged rod in practice can be a useful model. We just need to remember that our analytical solution is approximate, not exact.

Special case: A very long rod

A very important special case, which we will refer to often, is the case of a rod that is very long (or alternatively a rod very close to the observation location).
In either case $L >> r$.

Before reading further, try to show that if $L >> r$, the electric field is approximately:

$$\frac{1}{4\pi\varepsilon_o}\frac{2(Q/L)}{r}$$

$$E = \frac{1}{4\pi\varepsilon_o}\left[\frac{Q}{r\sqrt{r^2+(L/2)^2}}\right]$$

If $L >> r$, then $\dfrac{1}{r\sqrt{r^2+(L/2)^2}} \approx \dfrac{1}{r\sqrt{(L/2)^2}} = \dfrac{1}{r(L/2)}$, so

$$E \approx \frac{1}{4\pi\varepsilon_o}\left[\frac{2(Q/L)}{r}\right]$$

The distance dependence is illustrated in the figure. This also holds for a short rod if we are very close, so that $L >> r$.

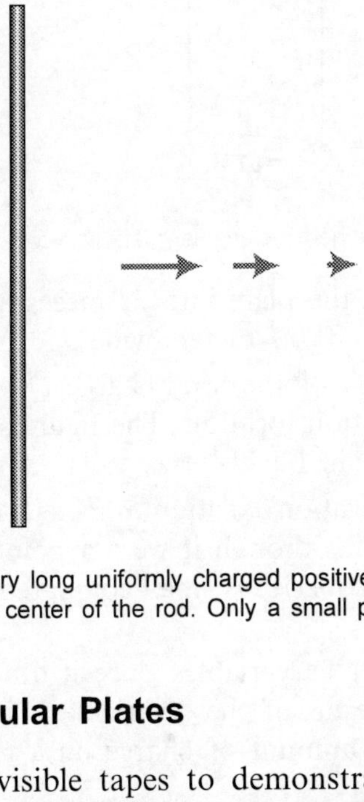

The electric field of a very long uniformly charged positive rod, shown at locations along a line very near the center of the rod. Only a small portion of the rod is visible.

Electric Field of Rectangular Plates

In Chapter 1 we had used invisible tapes to demonstrate the attractive force of two oppositely charged strips of tape brought near one another.

The electric field at locations on a line from the midpoint of charged tapes can be expressed by forming an expression for the electric field contributions of small pieces of a charged rectangular plate.

The figure shows a model of a rectangular plate L meters long and W meters wide (area $A = L \times W$ meter^2) with a total charge Q in which the plate is divided into a number of pieces Δz wide and Δy high, each piece carrying a charge ΔQ.

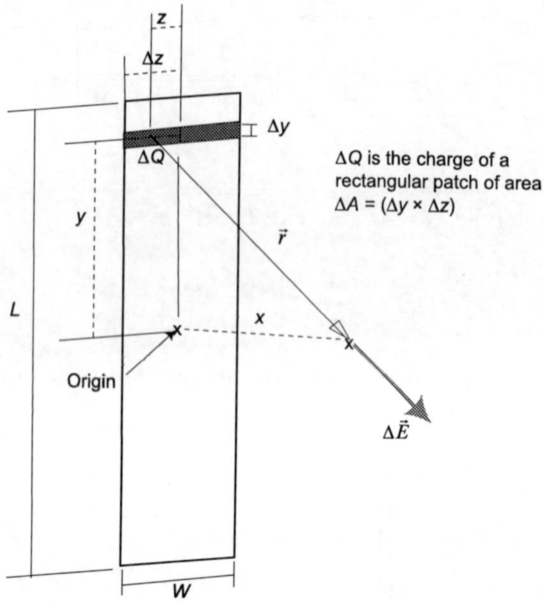

A uniformly charged rectangular plate with two rectangular patches.

If we divide the width of the plate into M pieces and the length of the plate into N pieces, then each piece is $\Delta z = W/M$ meters wide and $\Delta y = L/N$ meters long.

Here ΔQ is a small portion of the total charge Q of the plate which contributes $\Delta \vec{E}$ to the net field at the observation location. The figure shows two pieces of the plate $W/2$ meters wide and Δy meters high. Each piece c[illegible] small amount of charge ΔQ.

We have ch[illegible] observation location x meters from the center of the plate. The piece of charge ΔQ is treated as though it were a point charge, which should be a fairly good approximation as long as its size is small compared to the distance to the observation location.

We have identified z and y as variables since it differs for each small piece. We need to express the charge ΔQ in terms of these variables. Since the plate is uniformly charged with a total charge Q, so the amount of charge on a rectangular patch of width Δz and height Δy is equal to

$$\Delta Q = \left(\frac{\Delta z \times \Delta y}{W \times L}\right) Q$$

since $\dfrac{\Delta z \times \Delta y}{W \times L} = \dfrac{\Delta A}{A}$ is a fraction of the whole area of the plate. Alternatively, there is a surface charge density of $Q/(W \times L)$ in coulombs per meter^2, so on a patch of area ΔA meter^2, the amount of charge $\Delta Q = \left(\dfrac{Q}{W \times L}\right) \Delta z \Delta y$.

We will briefly outline the procedure to be followed to obtain expressions for the electric field components and leave the details for the reader.

To obtain the net electric field $\Delta\vec{E}$ we follow the procedure we adopted to write expressions for the electric field components of a uniformly charged thin rod while assuming that the source point is located at the center of each rectangular patch carrying the charge ΔQ. The vector $\vec{r}$ and the unit vector $\hat{r}$ are written using the Pythagorean theorem in three dimensions $d^2 = x^2 + y^2 + z^2$. We can then write expressions for the electric field components: $\Delta\vec{E}_x$, $\Delta\vec{E}_y$ and $\Delta\vec{E}_z$ of $\Delta\vec{E}$ due to each piece ΔQ,

Using the expressions, a computer can be programmed to obtain the electric field at the observation location by numerically adding the contributions made by each piece to the individual components of the electric field.

A listing of MATLAB programs for computing the efield of thin rods and invisible tape "MATLAB_programs_efield_thin_rod_and_invisible_tape" is provided in the CD accompanying this book.

Uniformly Charged Ring (Chabay & Sherwood, 1995, used by permission)[2]

Field of a single ring at a point on the axis

Step 1: Cut up the charge distribution into pieces (Δl) and draw ΔE.

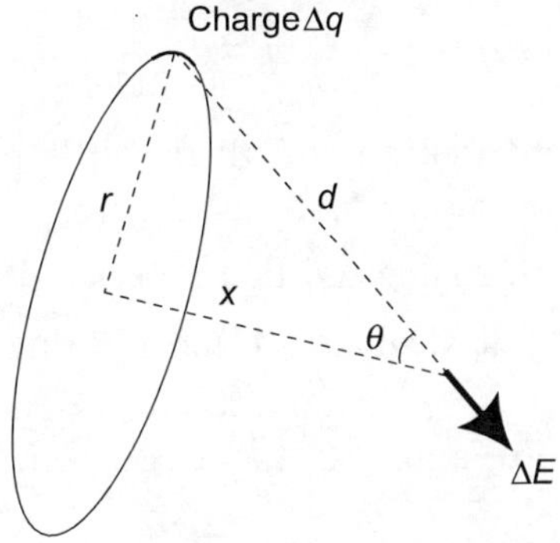

Fig. A.12 Electric field along the axis of a charged ring.

We cut up the thin ring into very short sections each of length Δl with positive charge Δq. Here Δl is a small portion of the total circumference $2\pi r$, and Δq is a small portion of the total charge q of the ring.

Step 2: We invent an integration variable to refer to each piece. We use the variable "l" measured circularly around the ring with Δl a small increment of this integration variable. There is a small amount of charge Δq on a piece of length Δl. We show the components of ΔE parallel to the axis (ΔE_x) and perpendicular to the axis ($\Delta E_\perp$):

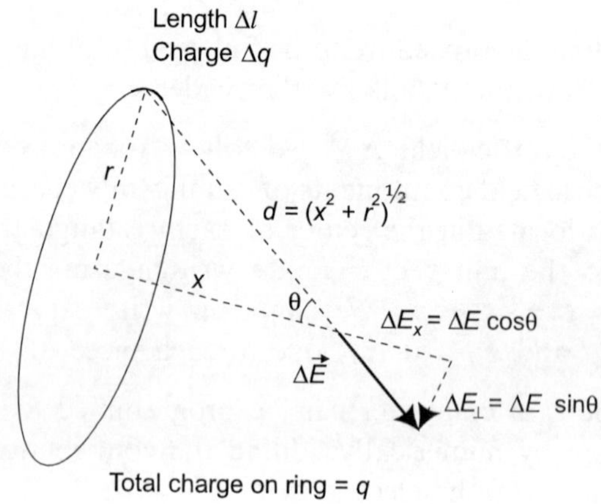

At a distance along the axis from the center of the ring, the little piece of charge Δq is a distance $d = (x^2 + r^2)^{1/2}$ away, and it makes a field $\Delta \vec{E}$ of magnitude

$$\Delta E = \frac{1}{4\pi\varepsilon_o}\frac{\Delta q}{d^2} = \frac{1}{4\pi\varepsilon_o}\frac{\Delta q}{(x^2+r^2)}$$

Since the ring is uniformly charged with a total charge q, and $\dfrac{\Delta l}{2\pi r}$ is the fraction of the whole ring represented by Δl, so the amount of charge on a section of length Δl is

$$\Delta q = \left(\frac{\Delta l}{2\pi r}\right) q = \left(\frac{q}{2\pi r}\right)\Delta l$$

When we add up the contributions of all the little pieces that make up the ring, the components perpendicular to the axis add up to zero.

For each ring segment with charge Δq_1 that makes a field $\vec{E}_1$ whose y-component is $\vec{E}_{1y}$, there is a piece Δq_2 (located 180 degrees around the ring from Δq_1) that makes a field $\vec{E}_2$ whose y-component $\vec{E}_{2y}$ is equal in magnitude and opposite in direction to $\vec{E}_{1y}$. So the components all add up to zero, and only the x-component is non-zero.

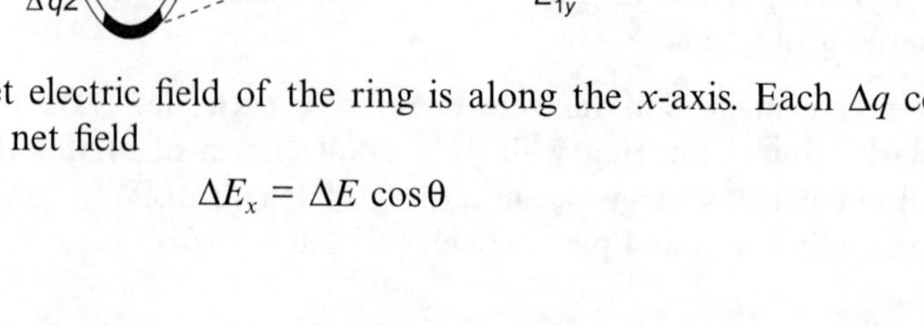

And, the net electric field of the ring is along the x-axis. Each Δq contributes $\Delta E_x = \Delta E\cos(\theta)$ to the net field

$$\Delta E_x = \Delta E \cos\theta$$

Writing $\cos\theta$ in terms of x and r

$$\cos\theta = \frac{x}{d} = \frac{x}{\left(x^2 + r^2\right)^{1/2}}$$

Now put it all together.

$$\Delta E_x = \frac{1}{4\pi\varepsilon_o} \frac{\Delta q}{\left(x^2 + r^2\right)^{1/2}} \cos\theta$$

$$\cos\theta = \frac{x}{d} = \frac{x}{\left(x^2 + r^2\right)^{1/2}}$$

and $$\Delta q = \left(\frac{q}{2\pi r}\right)\Delta l$$

So, $$\Delta E_x = \left[\frac{1}{4\pi\varepsilon_o} \frac{q}{2\pi r} \frac{\Delta l}{(x^2 + r^2)} \frac{x}{(x^2 + r^2)^{1/2}}\right]$$

$$= \frac{1}{4\pi\varepsilon_o} \frac{q}{2\pi r} \frac{x}{(x^2 + r^2)^{3/2}} \Delta l$$

Step 3: Add up all the contributions of all the pieces. The net field lies along the axis and is the summation of ΔE_x contributed by each little piece of the ring:

$$E_x = \frac{1}{4\pi\varepsilon_o} \frac{q}{2\pi r} \frac{x}{(x^2 + r^2)^{3/2}} \sum \Delta l$$

The integration variable "*l*" runs around the circumference of the ring, and Δl is an increment of the integration variable, corresponding to a short step along the circumference of the ring. If we go all the way around the ring, we will have accounted for every small piece of the ring, and adding up all the steps, we get the circumference of the ring:

$$\sum \Delta l = 2\pi r.$$

More formally, if we let $\Delta l \to dl$ (an infinitesimally small increment with no bounds), we can integrate

$$\int dl = 2\pi r.$$

Adding up all the *dl*'s gives us the total length (the circumference). So, the result which does not contain the integration variable *l*, is:

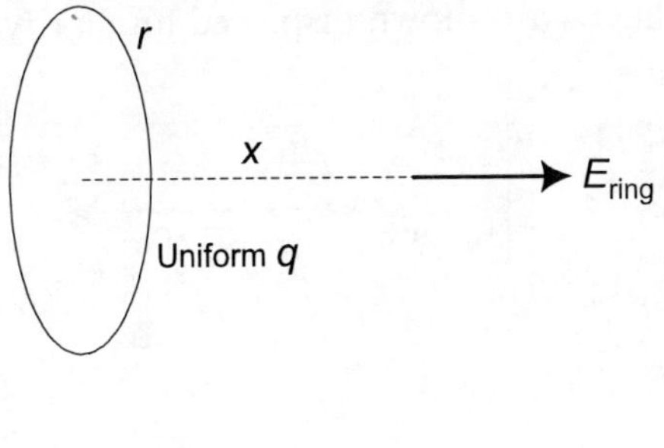

$$E_{\text{ring}} = \frac{1}{4\pi\varepsilon_o} \frac{qx}{(x^2 + r^2)^{3/2}} \tag{6}$$

for a ring of radius r and charge q, a distance x from the center, along the axis.

Special case

If $x = 0$, were at the center of the ring, the electric field should be zero, since then the contributions of all portions of the ring point inward and cancel each other. Setting $x = 0$ in the expression given by Eq. (6), we indeed find $E_{\text{ring}} = 0$.

If x is very much larger than r, the distant ring looks almost like a point charge, so the net field ought to look like the field of a point charge.

If $x >> r$, then $x^2 >> r^2$

$$\frac{x}{(x^2 + r^2)^{3/2}} = \frac{x}{(x^2)^{3/2}} = \frac{x}{x^3} = \frac{1}{x^2}, \quad \text{so}$$

$$E_{\text{ring}} = \frac{1}{4\pi\varepsilon_o} \frac{q}{x^2}.$$

The above expression does indeed look like the electric field of a point charge when we compare this with expression (2).

Field of Two Rings (Chabay & Sherwood, 1995, used by permission)[2]

Two rings of radius of 5 cm are 20 cm apart and concentric with a common axis. The ring on the left carries a uniformly distributed charge of +35 nC and the ring on the right carries a uniformly distributed charge of –35 nC.

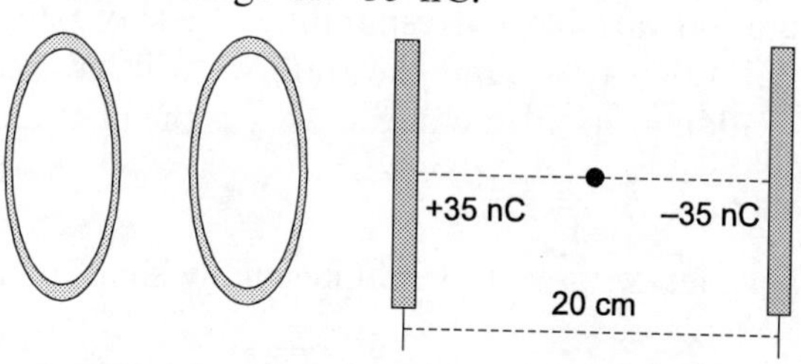

What is the magnitude and direction of the electric field on the axis, halfway between the two rings?

The net field is the vector sum of the field due to ring 1 and the field due to ring 2.

The vectors $\vec{E}_1$ and the $\vec{E}_2$ are shown displaced for clarity, they are equal and *collinear*.

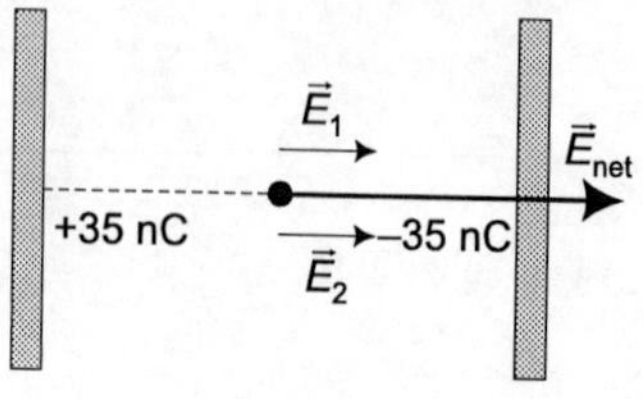

The magnitude of the net field E_{net} is $E_1 + E_2 = 2E$.
Using the expression

$$E_{ring} = \frac{1}{4\pi\varepsilon_o}\frac{qx}{(x^2+r^2)^{3/2}}$$

We can compute

$$E_{net} = 2\cdot\frac{1}{4\pi\varepsilon_o}\frac{(35\times10^{-9}\,\text{C})(0.1\,\text{m})}{\left((0.1\,\text{m})^2+(0.05\,\text{m})^2\right)^{3/2}} = 4.5\times10^4\ \text{N/C}$$

If the rings both carry a charge of +35 nC, i.e., like charges are present on both the rings, then the magnitude and direction of the electric field midway between the rings will be zero.

Field of Uniformly Charged Disks

(Chabay & Sherwood, 1995, used by permission)[2]

We now consider the important case of a uniformly charged disk made from insulating material say, glass that has been rubbed with silk. We will see in the next section that two charged metal disks can form a "capacitor", a device that is important in electric circuits.

Field of a Uniformly Charged Disk, at a Point on the Axis

Step 1: Cut up the charge distribution into pieces and draw ΔE. We will calculate the electric field of a uniformly charged disk of radius R.

We cut up the uniform disk into rings, each contributing some known electric field ΔE.

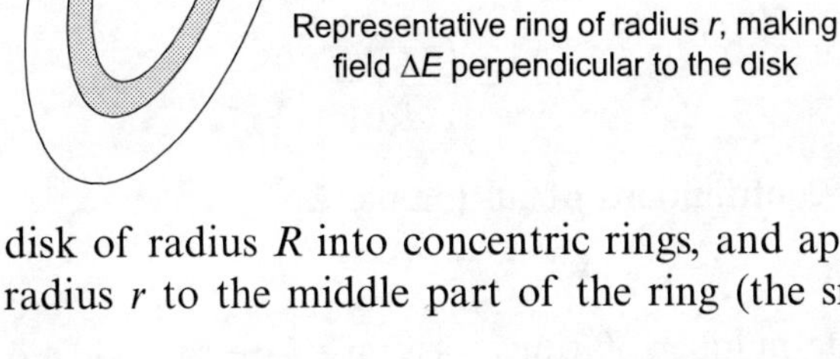

Let us divide the disk of radius R into concentric rings, and approximate each ring as having some average radius r to the middle part of the ring (the smallest ring is actually a small disk).

Each ring has some amount of positive charge ΔQ (not necessarily the same on each ring). Each ring contributes an amount ΔE to the electric field perpendicular to the disk.

Step 2: Write an expression for the electric field due to one piece.

We have already invented an integration variable: the radius r of a ring. (The distance x from the center of the disk to the observation point is a given constant, but r is a variable that we invented to refer to pieces into which we cut the disk). Using our previous result for the electric field of a single ring of charge, at a location on the axis, we can write: (See expression (6))

$$\Delta E = \frac{1}{4\pi\varepsilon_o} \frac{\Delta q x}{(x^2 + r^2)^{3/2}}$$

where Δq is the amount of charge on the ring, and r is the radius of the ring. Both Δq and r will be different for different rings.

What is Δq for a particular ring in terms of the integration variable r?

Imagine dividing the disk into a very large number of rings, and consider one ring with radius r and very narrow width Δr.

If you cut such a ring and roll it straight, you get an approximately rectangular strip of length of $2\pi r$ (the circumference of a circle) and width Δr, so this area is

$$\Delta A = 2\pi r \Delta r$$

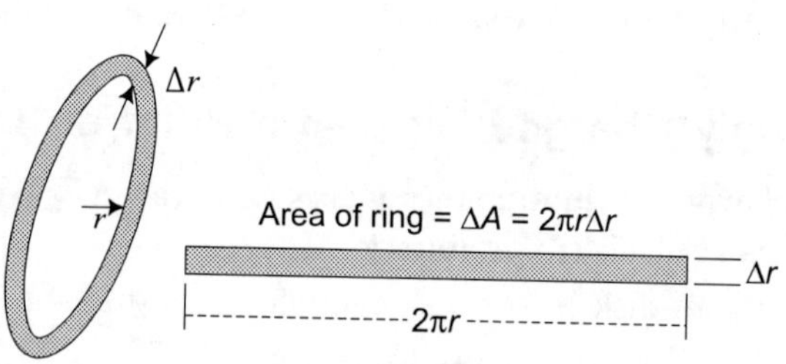

The total charge Q on the disk is spread over a total area A ($= \pi R^2$), so the uniform charge density is Q/A (coulombs per square meter). Therefore, $\Delta q = (Q/A)(2\pi r\Delta r)$, since the ring area is $2\pi r\Delta r$.

Now that we know the amount of charge on a ring, we can write an algebraic expression for the contribution of one ring, in terms of integration variable r:

$$\Delta E = \frac{1}{4\pi\varepsilon_o}\left(\frac{Q}{A}\right)\frac{x}{(x^2 + r^2)^{3/2}}(2\pi r\Delta r)$$

Step 3: Add up the contributions of all the pieces

Numerical summation

If we consider a finite number of rings—say five—we can add up the electric fields of all five rings to get the approximate field of the disk.

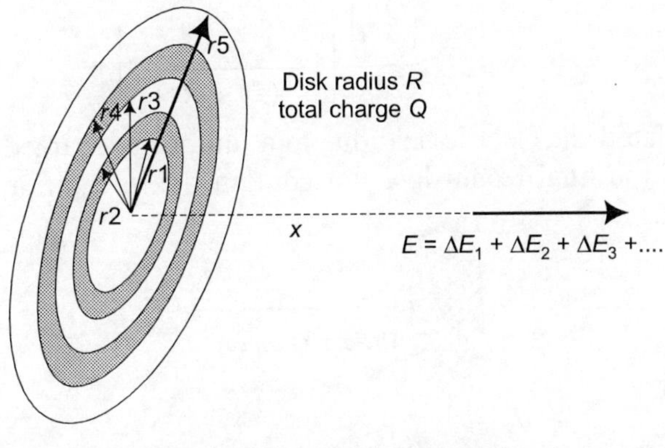

The total field E is the summation of the field contributions of the 5 rings:

$$E = \sum_{n=1}^{n=5} \frac{1}{4\pi\varepsilon_o}\left(Q/A\right)\frac{x}{(x^2 + r_n^2)^{3/2}}(2\pi r_n \Delta r)$$

This result will not be exact, because we need an average radius r_n for the nth ring. We could evaluate the summation numerically to get an approximate result for a particular disk. We would get a more accurate result if we divided into ten rings, or a hundred, or a thousand rings. If we divide the disk into an infinite number of infinitesimally narrow rings, we get an exact result, and we consider that procedure next.

Integration

Let the number of rings become very large, so Δr becomes dr, an infinitesimally small thickness. Remember, that the integral sign $\int$ is just a distorted S standing for "Summation".

Since $$dA = 2\pi r dr,$$

$$E = \int \frac{1}{4\pi\varepsilon_o}\frac{x}{(x^2 + r_n^2)^{3/2}}\left(Q/A\right)(2\pi dr)$$

Since r ranges from 0 to R,

$$E = \int_0^R \frac{\left(Q/A\right)}{2\varepsilon_o}\frac{x}{(x^2 + r^2)^{3/2}} r dr = \frac{\left(Q/A\right)}{2\varepsilon_o} x \int_0^R \frac{r}{(x^2 + r^2)^{3/2}} dr$$

Let $u = x^2 + r^2$, so that $du = 2rdr$ and $dr = du/2r$ and rewrite this integral

$$E = \frac{\left(Q/A\right)}{2\varepsilon_o} x \int_{r=0}^{r=R} \frac{1}{2}\frac{1}{u^{3/2}} du = \frac{\left(Q/A\right)}{2\varepsilon_o} x \left[-\frac{1}{u^{1/2}}\right]_{r=0}^{r=R}$$

$$E = \frac{\left(Q/A\right)}{2\varepsilon_o} x \left[-\frac{1}{(x^2 + R^2)^{1/2}} - \frac{-1}{(x^2)^{1/2}}\right]$$

$$= \frac{\left(Q/A\right)}{2\varepsilon_o}\left[1 - \frac{x}{(x^2 + R^2)^{1/2}}\right] \tag{7}$$

We have calculated the electric field due to a uniformly charged disk of radius R, along its axis. Note that the final result does not contain the integration variable r:

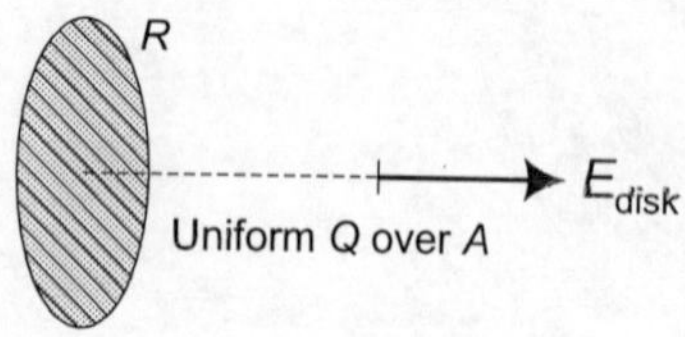

$$E_{\text{disk}} = \frac{\left(Q/A\right)}{2\varepsilon_o}\left[1 - \frac{x}{(x^2 + R^2)^{1/2}}\right]$$ perpendicular to the center of the disk ($A = \pi R^2$).

Electric field near the center of a uniformly charged disk

A special case turns out to be of considerable interest. Let us see what the electric field is like *near* the center of a uniformly charged disk (x small, but not zero). Suppose that we are so close to the disk that x is very small ($x << R$), or equivalently that we are some distance away but the disk is very large ($R >> x$).

Since, $E_{\text{disk}} = \frac{\left(Q/A\right)}{2\varepsilon_o}\left[1 - \frac{x}{(x^2 + R^2)^{1/2}}\right]$, then with $R >> x$,

$(x^2 + R^2)^{1/2} \approx R$ so

$$E_{\text{disk}} = \frac{\left(Q/A\right)}{2\varepsilon_o}\left[1 - \frac{x}{R}\right] \tag{8}$$

Further, if x/R is very very small, $\left[1 - \frac{x}{R}\right]$ reduces to 1, so we have approximately

$$E_{\text{disk}} = \frac{\left(Q/A\right)}{2\varepsilon_o}\left[1 - \frac{x}{R}\right] = E_{\text{disk}} = \frac{\left(Q/A\right)}{2\varepsilon_o}$$ near the center of a disk.

The result is approximately nearly independent of the distance x and is true near any large uniformly charged plate, not just a circular one.

What does this result mean?

Assuming a disk of radius R = 20 cm and the total charge is distributed uniformly all over the disk is $Q = 6 \times 10^{-6}$ C and that $\varepsilon_o = 9 \times 10^{-12} \text{C}^2/\text{N–m}^2$, we find the electric field 1 mm from the center of the disk using expression (7) to be 2.639×10^6 N/C and that 3 mm from the center of the disk it is only very slightly smaller: 2.613×10^6 N/C.

This electric field is nearly strong enough to cause breakdown in air – the breakdown field strength is roughly 3×10^6 N/C. The process of ionization of air at these field strengths is described in some detail in Chapter 10.

You find that the field at a distance of 1 mm from the center of the disk and the field at a distance of 3 mm are nearly the same.

Here is a way to make a "nearly uniform field" in a small region; an electric field that doesn't vary much in direction or magnitude in that region.

Does this contradict Coulomb's law that the electric field varies as $1/d^2$?

Coulomb's law describes the electric field of a point charge. A uniformly charged disk is not a point charge, so its field does not necessarily look like $1/d^2$.

The field of the disk is calculated by integrating the contributions of point-like pieces of the disk over the whole surface of the disk. The contribution of each point-like piece does follow Coulomb's law, *but the sum does not.*

Electric Field Strength near a Charged Invisible Tape

In the Section 1.14, we demonstrated that invisible tapes normally take on an excess charge of about 10 nanocoulombs. We can use the approximate expression for the field near the center of a charged disk to compute the electric field strength near the midpoint of a charged tape. Plugging in the values of Q = 10 nC, L = 0.2 meter and Width = 0.01 meter, we obtain:

$$E_{tape} = \frac{\left(Q/A\right)}{2\varepsilon_o} = \frac{10 \times 10^{-9} \times 9 \times 10^{9} \times 4 \times \pi}{2 \times 0.2 \times 0.01} = \frac{90 \times 4 \times \pi}{0.004}$$

$$= \frac{90 \times \pi}{0.001} = \frac{282.7433}{0.001} = 282743.3 = 2.827 \times 10^5 \text{ N/C.}$$

This is a very rough approximation. See also "Electric Field of Rectangular Plates" for one model of a charged invisible tape to obtain expressions for it's electric field components.

Computer Programs and Graphical Rendition of Electric Field of Distributed Charges

A listing of a program that can be run (click "Run this program" to run it) that graphs the electric field due to a charged disk is available in this link http://www.glowscript.org/#/user/matterandinteractions/folder/matterandinteractions/program/15-Edisk.

This uses the browser (GlowScript) version of VPython. More demo programs are available at matterandinteractions.org/student.

At vpython.org, in the documentation section, is full documentation including a tutorial. Or go to glowscript.org and click "VPython Help" for the similar documentation on the browser version of VPython, where you will also find a link to extensive tutorials on using GlowScript VPython in mechanics, by Rhett Allain.

Two Uniformly Charged Disks: A Capacitor (Chabay & Sherwood, 1995, used by permission)[2]

Consider two uniformly (oppositely) charged metal disks placed very near each other (separation or gap distance s), carrying charges of $-Q$ and $+Q$.

This arrangement is called a "capacitor" and such devices are used in electric and electronic circuits.

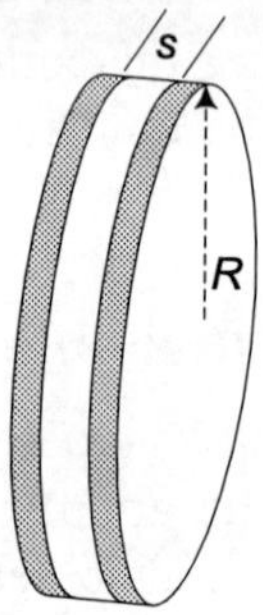

Two oppositely charged metal disks very near other

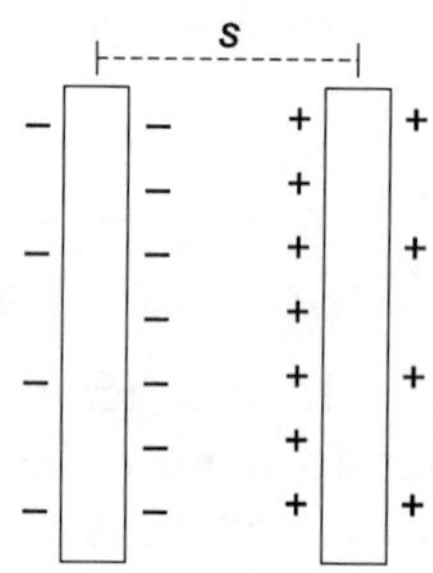

Side view of the 2 disks

A single charged metal disk cannot have a truly uniform charge density, because the mobile charges tend to push each other to the edge of the disk. Nevertheless, our results for disks made of insulating material are approximately correct for metal disks, especially in this two disk configuration if the disks are very close together ($s << R$). Due to the attraction by the neighboring disk, almost all of the charge is uniformly distributed on the inner surfaces of the disks, with very little charge on the outer surfaces of the disks.

We'll calculate the strength of the electric field at locations near the center of the disks, both inside and outside the capacitor.

Step 1: Cut up the charge distribution into pieces and draw $\Delta\vec{E}$.

We cut up the distribution into just two pieces because we know the electric field made by a single uniformly charged disk.

The diagram below shows a side view of the region near the center of the disks, blown up. The disks are very close together ($s << R$), and extend up and down beyond the boundaries of the diagram. Drawn in the diagram at each of the locations labeled 1, 2 and 3 are arrows labeled E_+ and E_- showing the contributions of each disk to the net electric field at that location.

Note: The tails of the arrows are shown at the location points' different vertical levels only for the sake of clarity.

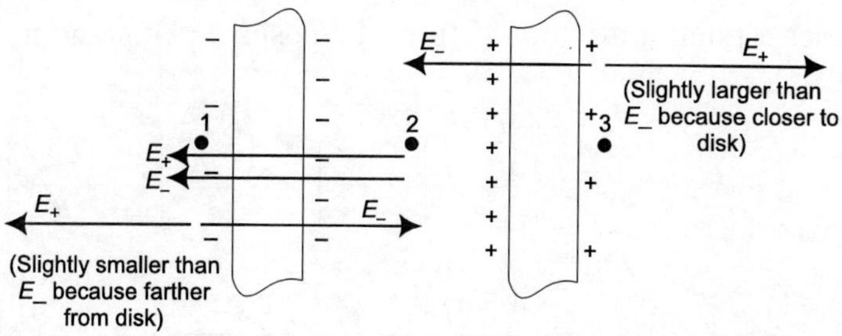

At locations 1 and 3, $E_+ \neq E_-$ because the distances to the two disks are slightly different. On the diagram below are drawn the "*net*" electric field at points 1, 2 and 3. The vector lengths reflect the relative magnitudes of the electric fields.

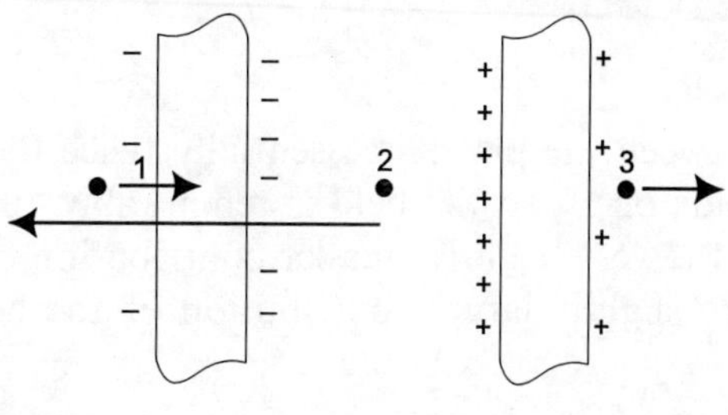

The above figure is important to understand the behavior of capacitors.

Step 2: Write an expression for the electric field due to one piece.

Assuming a uniform charge density on each disk (and ignoring the small amount of charge on the outer surfaces of the disks) we know that the field E_{disk} near the center of a single disk is approximately using Eq. (8)

$$E_{\text{disk}} = \frac{\left(Q/A\right)}{2\varepsilon_o}\left[1-\frac{x}{R}\right]$$

We're assuming that the plate separation s is very small ($s << R$), and that locations 1 and 3 are very near the disks ($x << R$).

Step 3: Add up the contributions of all the pieces

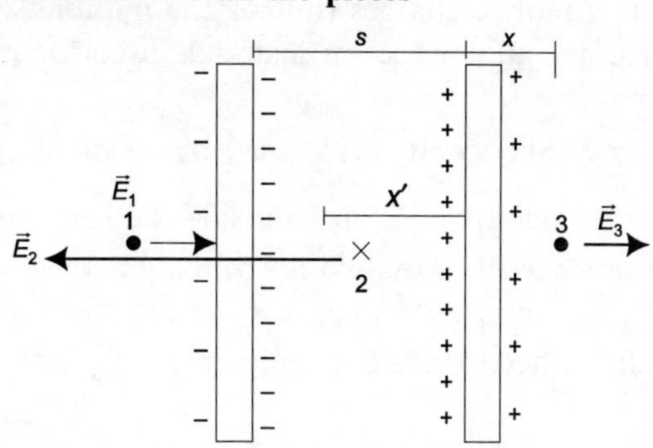

We calculate the magnitude of $\vec{E}_2$ inside the gap as the superposition of the field of the negative disk and the field of the positive disk, both of which point to the left inside the gap. Consider a point a distance x' from the positive plate (and a distance $s - x'$ from the negative plate).

$$E_2 = \frac{\left(Q/A\right)}{2\varepsilon_o}\left[1 - \frac{x'}{R}\right] + \frac{\left(Q/A\right)}{2\varepsilon_o}\left[1 - \frac{s - x'}{R}\right]$$

$$= \frac{\left(Q/A\right)}{2\varepsilon_o}\left[1 - \frac{x'}{R} + 1 - \frac{s - x'}{R}\right] = \frac{\left(Q/A\right)}{2\varepsilon_o}\left[2 - \frac{s}{R}\right]$$

$$= \frac{(Q/A)}{\varepsilon_o}\left[1 - \frac{s/2}{R}\right] \approx \frac{(Q/A)}{\varepsilon_o}$$

to the left, because s<<R.

The electric field between the plates is essentially twice the field due to one plate. The magnitude hardly depends on x', so the field is remarkably "uniform". As you move away from the positive plate, the very slightly smaller contribution of the positive plate is nearly compensated by the very slightly larger contribution of the negative plate.

Capacitance

The potential difference between the plates will be $V = Es = \frac{Qs}{\varepsilon_o A}$. The straight multiplication *Es* was possible because the field *E* between the plates is *uniform* from one plate to the other. Therefore, the capacitance which is the amount of charge held in the plates divided by the voltage $\left(\frac{Q}{V}\right)$ is given by $\frac{\varepsilon_o A}{s}$. The reader may appreciate that capacitance depends entirely on the *geometry* of the distribution of charges on the conductors. Since all conductors have an *enormous* store of mobile charges, their *number and distribution* on capacitor plates is what determines its capacitance and not its resistivity. It is to be regarded that the conductive material of the plates enabled the capacitor to *guide and hold* the distribution (tight packing) of the few mobile charges (out of the millions available in the conductor) on its plates. This is true for all conductive materials used for making capacitors.

Fringe Field (Chabay & Sherwood, 1995, used by permission)[2]

Next, let us calculate the magnitude $\vec{E}_3$ just outside the capacitor.

Location 3 is a distance x from the main positive charges on the inner surface which makes a field to the "right", and a distance $x + s$ from the main negative charges, which make a field to the "left". Therefore, the magnitude of $\vec{E}_3$ is the "difference" of the fields of the two disks:

$$E_3 = \frac{\left(Q/A\right)}{2\varepsilon_o}\left[1 - \frac{x}{R}\right] - \frac{\left(Q/A\right)}{2\varepsilon_o}\left[1 - \frac{x + s}{R}\right]$$

$$= \frac{\left(Q/A\right)}{2\varepsilon_o}\left(\frac{s}{R}\right) \text{ to the right}$$

Similarly, we find in the same way at location 1 just to the left of the capacitor, we have

$$E_1 = \frac{\left(Q/A\right)}{2\varepsilon_o}\left[1-\frac{x}{R}\right] - \frac{\left(Q/A\right)}{2\varepsilon_o}\left[1-\frac{x+s}{R}\right]$$

$$= \frac{\left(Q/A\right)}{2\varepsilon_o}\left(\frac{s}{R}\right) \text{ to the right}$$

The most important thing to note is that the field outside the capacitor is very small compared to the field inside the capacitor. Inside the gap, the fields of the two disks *add*, but outside the gap the fields of the two disks, subtract.

By calculating the ratio of the outside field to the inside field, we can show that the ratio is very small if $s << R$.

$$\frac{E_{outside}}{E_{inside}} = \frac{E_3}{E_2} = \frac{\dfrac{Q/A}{2\varepsilon_o}\left(\dfrac{s}{R}\right)}{\dfrac{Q/A}{\varepsilon_o}} = \frac{s}{2R} \rightarrow 0 \text{ if } s << R. \tag{9}$$

The small field outside the capacitor is often called the "fringe field". We have calculated the **fringe field** at a location outside the capacitor but very close to it, where the fringe field hardly changes with distance x. (If however, you are very far away, so $d >> R$, the capacitor looks like an electric dipole, and the fringe field falls off like $1/d^3$.)

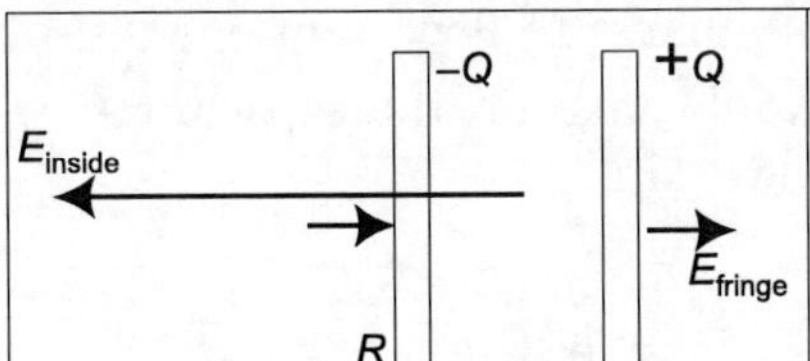

Two-disk capacitor: plate area $A = \pi R^2$, near center

$$E_{inside} = \frac{Q/A}{\varepsilon_o}$$

$$E_{fringe} = \frac{Q/A}{\varepsilon_o}\left(\frac{s}{2R}\right)$$

The results we obtained apply not only to circular capacitors but also to capacitors with rectangular or other shapes or plates, where A is the area of one of the plates. The same restriction applies as with circular capacitors; the plate separation s must be much smaller than the width or height of a plate. As long as the gap is very small, E_{inside} is nearly equal to $\frac{(Q/A)}{\varepsilon_o}$ near the center of the plates. $E_{outside}$ near the center of the plates depends on the shape of the plates, but is always much smaller than E_{inside} by a factor of about (s/L) where L is the width or height of a plate (or diameter $2R$ for circular plates).

We saw that the electric field inside the capacitor near the center of the plates is remarkably uniform across the gap. It is also true that the magnitude and direction of the electric field are practically the same everywhere in the gap; not just near the center of the plates. Only if you get very near the outer edges of the plates does the electric field vary much from its center value. Here is a close-up view of the gap in a parallel plate capacitor, not necessarily circular, far from the outer edges of the plates (the small fringe field is not shown).

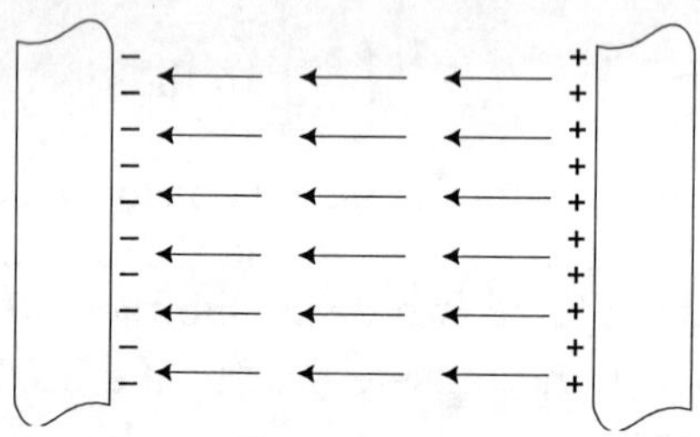

Note that despite the $1/d^2$ character of electric interactions among point particles, the electric field due to a distribution of a large number of charges can have a very different distant dependence. Here the electric field is uniform in magnitude and direction throughout a region.

The small *fringe field* plays an important role in the discharge of capacitors and in the current pattern that results whenever a capacitor is charged when a voltage is applied to it.

Example: If the electric field in air exceeds roughly 3×10^6 newtons/coulomb, the air breaks down and a spark forms. For a two-disk capacitor of radius 50 cm with a gap of 1 mm, what is the maximum charge (positive and negative) that can be placed on the disks without a spark forming (which would permit charge to flow from one disk to the other?)

We can calculate how much charge it would take on the capacitor ($+Q$ and $-Q$) to produce a field sufficiently strong to breakdown air:

$$E = \frac{\left(Q/A\right)}{\varepsilon_o} = \frac{Q}{\pi R^2 \varepsilon_o} \quad \text{because } A = \pi R^2$$

So,

$$Q = \pi\varepsilon_o R^2 E$$

$$= 9 \times 10^{-12}\ \frac{C^2}{Nm^2}\ \pi(0.5m)^2 \left[3\times10^6\ \frac{N}{C}\right]$$

$$= 2.08 \times 10^{-5} C$$

Under these conditions, what is the strength of the fringe field just outside the center of the capacitor?

The fringe field just outside the capacitor is approximately (using expression (9) with E_{fringe} being the field $E_{outside}$:

$$E_{\text{fringe}} = \frac{s}{2R}\,E_{\text{inside}}$$

$$= 0.5\left[\frac{0.001\text{m}}{0.5\text{m}}\right] 3\times10^{6}\ \text{N/C} = 3\times10^{3}\ \text{N/C}$$

We also see that E_{fringe} is a fraction ($s/2R$) times the electric field inside E_{inside} and in this example is about 1/*thousand times* E_{inside}.

It is interesting to note that if the radii of the disks are infinitely large, then the fringe field using the *ideal approximation* (see Section 2.11) is zero.

A Spherical Shell of Charge

(Chabay & Sherwood, 2011, used by permission)

Spherical shells of charge produce patterns of electric field that are special and distinct from those produced by rings and plates of charge we saw earlier. We will learn how this is due to the unique spatial distribution of charge having *spherical* symmetry.

A sphere with charge spread uniformly over its surface produces a surprisingly simple pattern of electric field. For brevity, we will use the phrase "a uniform spherical shell" to refer to "a sphere with charge spread uniformly over its surface", since the charge itself forms a thin, shell-like spherical layer.

We will qualitatively graph the electric field pattern of the uniform spherical shell, by dividing the shell into six regions, as shown in Fig. A.13.

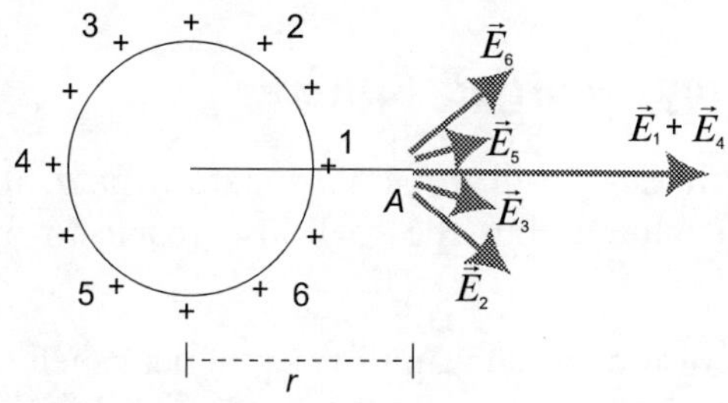

Fig. A.13 Contributions to the net electric field outside a charged sphere.

$\vec{E}_2$ means "the electric field due to the charges in the region 2 of the spherical shell". As long as we are rather far from a region of distributed charge we can approximate the electric field of that region as being due to a point particle with the total charge of the region.

Direction: Consider the field at location A outside the shell, a distance r from the center of the sphere. By symmetry, the net electric field is horizontal and to the right, because the vertical component of the net field cancels for pairs of fields such as $\vec{E}_2$ and $\vec{E}_6$, or $\vec{E}_3$ and $\vec{E}_5$. It is clear that the net field (the superposition of the contributions of all the source charges) is radially outward from the center of the sphere.

Magnitude: It is somewhat surprising that the magnitude of the net field turns out to be

$$E = \frac{1}{4\pi\varepsilon_o}\frac{Q}{r^2}$$

where Q is the total charge on the sphere and r is the distance to location A from the center of the sphere (not from the surface of the sphere!). In Fig. A.14 note that the distance from location A to some regions is less than r, while the distance from location A to other regions is greater than r.

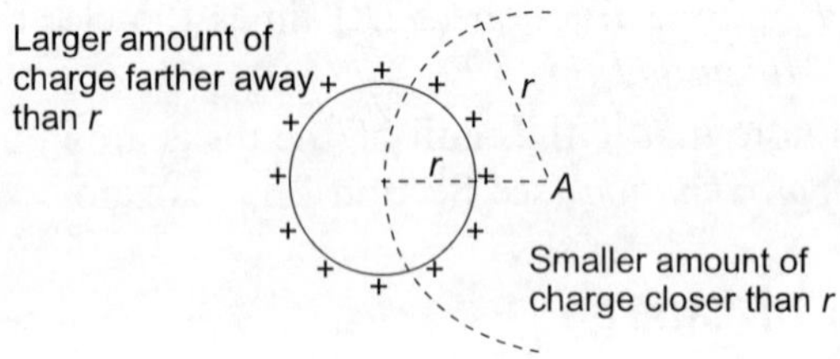

Fig. A.14 Some charge is closer than r to point A, but more of the charge is farther away.

Because of the $1/r^2$ distance dependence for point charges, a smaller amount of charge closer than r makes as big a contribution to the net field as the larger amount of charge that is farther away than r.

Equivalence to a Point Charge at Center of Sphere

It is an extraordinary aspect of the $1/r^2$ behavior of Coulomb's law that the net effect of all these charges comes to be simply equivalent to placing one point charge Q at the center of the sphere. This is not an obvious result!

Force on a Uniformly Charged Sphere

A uniformly charged spherical shell not only makes a field outside that looks as though it were made by a point charge, but the shell also reacts to outside charges as though it were a point charge.

Consider the interaction between a uniform spherical shell and an outside point charge. The shell exerts a force on the point charge as though both were point charges. By the reciprocity principle for electric forces (Newton's third law), the real point charge must exert an equal and opposite force on the shell, and so it exerts a force on the shell as though the shell were a point charge. Hence from the outside a uniform spherical shell "looks" just like a point charge, both as a source of electric field and when it reacts to external fields.

See also "The Electric Field due to a Point Charge" at the beginning of this appendix.

Inside a Uniform Spherical Shell of Charge

(Chabay & Sherwood, 2011, used by permission)

Now consider a location B inside the shell, as indicated in Fig. A.15.

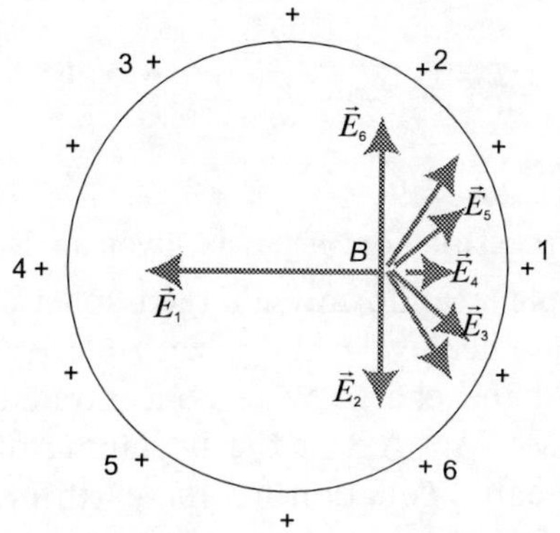

Fig. A.15 Contributions to the net electric field at a location inside a uniformly charged spherical shell add up to zero (the spherical shell of Fig. A.13 is magnified for clarity).

A small number of charges near location B (region 1) make a large contribution to the left, while a large number of charges far away from location B (regions 3, 4, and 5) each make small contributions to the right. Because the surface area of a portion of the sphere is proportional to r^2 while the electric field contributed by this region is proportional to $1/r^2$, these contributions exactly cancel each other. The electric field at location B due to charges on the surface of the sphere turns out to be exactly zero. Although intuitively plausible, this is not an obvious result.

The resultant field of a uniform spherical shell of charge is shown in Fig. A.16.

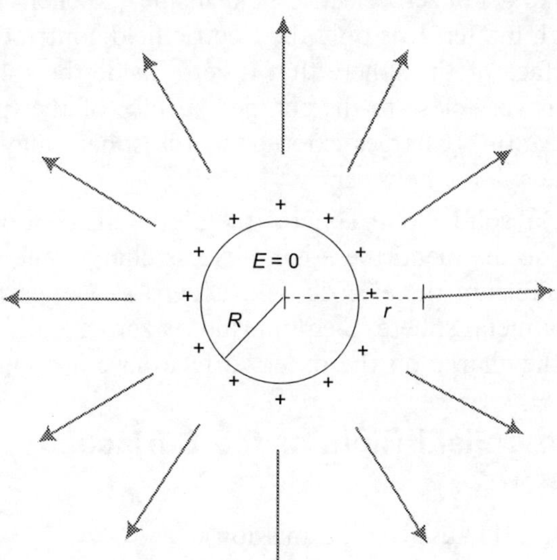

Fig. A.16 The field of a uniform spherical shell of charge.

The electric field at a distance r from the center of the uniformly charged spherical shell of radius R (Fig. A.16) carrying a charge Q is given by

$$\vec{E}_{sphere} = \frac{1}{4\pi\varepsilon_o}\frac{Q}{r^2} \qquad \text{for } r>R \text{ (outside the sphere)} \qquad (10)$$

$$\vec{E}_{sphere} = 0 \qquad \text{for } r<R \text{ (inside the sphere)} \qquad (11)$$

A derivation of the expression for $\vec{E}_{sphere}$ is given in Refs. [2] and [8].

Equation (10) shows that a uniformly charged spherical shell looks from the outside like a point charge but on the inside has a zero electric field. The uniformly charged shell appears as though all the charge were concentrated into a point at the center of the spherical shell ! Also, see Fig. A.3 at the beginning of this Appendix. Equation (10) enabled us to write the potential equation for the earth in Section 2.21 and use it to find the capacitance of the earth.

An alternative proof that the electric field inside a uniformly charged sphere is zero can also be made using Gauss's law which is given later in this Appendix.

The resultant zero electric field inside uniformly charged spheres has interesting implications. For example, if the uniformly charged shell is filled with plastic, the charges on the surface do not polarize the molecules of the plastic, because the surface charges contribute zero field.

Warning: *E* Is Not Always Zero Inside A Charged Sphere

Do not overgeneralize this [illegible] of uniformly charged spherical shells. Other charges in the Universe may make a nonzero electric field inside the shell. Recollect that an electric field penetrates through matter. It is only the electric field contributed by charges uniformly distributed on the surface of the sphere that is zero inside the sphere. Even if the location of a plastic molecule is very close to the charged surface of the sphere, E inside the sphere is still zero at that location. (Charges external to the sphere may contribute to a nonzero electric field inside the sphere, however.)

Excess charge on a solid metal sphere (metals being conductors) arranges itself on the outer surface. Inside the metal there is no excess charge, and there is no field, because inside a metal in equilibrium the electric field is zero as we learned in Sections 1.13 and 1.14. The field inside a metal sphere in equilibrium is zero even if external charges polarize the metal and make the charge on the outer surface have a nonuniform distribution.

What is the Electric Field Right at the Surface of a Charged Sphere?

(Chabay & Sherwood, 2011, used by permission)

The electric field is zero just inside the surface of a uniformly charged spherical shell, but the field is nonzero just outside the surface. You might wonder what the field is right at the surface. The physical reality is that the electric field is highly variable in direction and magnitude right at the surface, because the individual excess point charges (electrons and lattice ions) on the surface are many atomic diameters away from each other. Typically about only one in a million atoms on a surface has an extra charge as we saw in "Fraction of Surface Atoms with Excess Charge" in Section 1.14. That density of surface charge is enough to produce powerful potentials on charged spheres! It is only when you get some thousands of atomic diameters away from the surface, the field outside the surface looks like the field of a point charge at the center of the sphere, and the field inside the surface is zero.

A Note on the Units of the Electric Field

The electric field may also be expressed in volts/meter. If we write the units of volts in joules (J) per coulomb (C), then since joules (J) are expressed in Newton (N)-meter (m), this becomes evident:

$$\frac{V}{m} = \frac{J}{C-m} = \frac{N-m}{C-m} = \frac{N}{C}$$

Energy Stored in Stretched Springs and Unlike Charges that are Separated

In case of a real spring anchored at one end to a wall with a mass attached to the other end, you have to apply a force that is *non-constant* to slowly pull the mass to a distance x_2 away from x_1 where x_1 is the initial rest distance of the mass from the wall. This is because the spring exerts a force that is k times x (where k is the spring stiffness) that is opposite to the force you exert to pull the mass. The work you did will be stored in the spring as "elastic potential energy" and will be $\int_{x_1}^{x_2} kxdx = \frac{1}{2}kx_2^2 - \frac{1}{2}kx_1^2$. Similarly, if you pull a positive charge q_2 that is a distance r_1 away from a (fixed) negative charge $-q_1$ you have to apply a non-constant force of magnitude $\frac{1}{4\pi\varepsilon_o}\frac{q_1q_2}{r^2}$ that is opposite to the force exerted by the charge $-q_1$. If you move it to a new distance that is r_2 away, the work done will be stored as electric potential energy and will be $\int_{r_1}^{r_2} \frac{1}{4\pi\varepsilon_o}\frac{q_1q_2}{r^2}dr = \frac{q_1q_2}{4\pi\varepsilon_o}\left[\frac{1}{r_1} - \frac{1}{r_2}\right]$.

The observant reader would have noted that the force required to cause separation of an object of mass m from the earth (to raise it) to a few metres above the earth will be *constant* (mg) and the force required to pull the mass attached to the spring *increases* with distance (kx) of separation from the wall (within reasonable distance limits before the spring snaps !). The force required to pull the positive charge away from the negative charge *decreases* (proportional to $1/r^2$) with increasing separation distance r.

Energy Stored in Capacitors

Suppose that at some point in the process of charging a capacitor, the charge on the positive plate is q, so that the potential difference is q/C. Then, the work you must do to transport the next piece of charge, dq, is $dW = \left(\frac{q}{C}\right)dq$. The total work necessary, then to go from $q = 0$ to $q = Q$, is $W = \int_0^Q \left(\frac{q}{C}\right) dq = \frac{1}{2}\frac{Q^2}{C}$, or since $Q = CV$, $W = \frac{1}{2}CV^2$, where V is the final potential of the capacitor.

Gauss's Law and its Application

(Chabay and Sherwood, 2011, used by permission)

In Section 1.10 it was seen that electric field lines drawn by physicists and engineers around a charge configuration indicate patterns of the electric field over large expanses of space. The field *lines* form a continuum of field *vectors* which are used to depict the electric field strength at different points in space.

While Coulomb's law tells us how to derive the electric field if the charges are given, with Gauss's law we can determine how much charge is in any region if the field is known. Gauss's law was formulated by Carl Gauss in 1835 and is applicable to magnetic and gravitational fields as well. The law is a powerful analytical tool that can solve complicated problems involving charge configurations that possess cylindrical, planar or spherical symmetry.

Electric flux

Electric flux is a quantitative measure of the amount and direction of electric field over an entire surface. Like the electric field the electric flux doesn't travel; it simply exists (See Section 2.3, "Fields do not travel"). The quantitative definition of electric flux has the following properties: (i) The sign of the electric flux is related to the angle the electric field makes with the surface. The angle factor is parameterized as the cosine of an angle between the angle and the outward-going normal, a unit vector $\hat{n}$ perpendicular to the surface pointing outward from the closed surface. See the figure. (ii) If the field strength is doubled by doubling the amount of electric charge enclosed by the surface, the flux is doubled.

It is reasonable to assume that if a small closed surface encloses an amount of charge Q, the field strength on its' surface will be stronger than on a large closed surface that encloses the same amount of charge Q. Therefore, the area A of a surface is included into the definition. Thus, electric flux is defined as $\vec{E} \cdot \hat{n}\, A$. The magnitude and direction of $\vec{E}$ may vary from location to location on a large surface, leading to the more appropriate definition of electric flux over a surface as $\vec{E} \cdot \hat{n}\, \Delta A$ where $\Delta A = \Delta x \Delta y$.

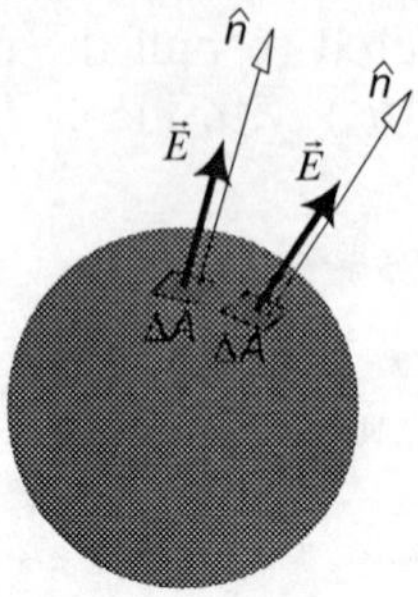

Gauss's law states that the total flux of any closed surface surrounding a charge is equal to the charge enclosed and is stated as

$$\sum_{\text{Closed surface}} \vec{E} \cdot \hat{n}\Delta A = \frac{\sum q_{inside}}{\varepsilon_o}$$

where, $\hat{n}$ is a unit vector normal to the surface. (**Note:** The unit vector $\hat{n}$ of the tiny patch ΔA is *defined* perpendicular to the patch). What this means is that if we consider a closed spherical surface surrounding a positive point charge $+Q$ as shown in the figure, then the total flux on the surface is equal to the charge enclosed.

It is very important to understand that the spherical surface is a purely mathematical construction and cannot be made out of any material.

We divide the spherical surface into small pieces ΔA and evaluate $\vec{E} \cdot \hat{n}\, \Delta A$ for each piece as shown in the figure. Because the closed spherical surface is centered on the point charge, the angle θ between $\vec{E}$ and $\hat{n}$ is zero at all locations on the surface. Therefore $\vec{E} \cdot \hat{n} = E \cos\theta = E$ since $\cos\theta = +1$.

The magnitude of the electric field at all locations on the sphere is simply

$$E = \frac{1}{4\pi\varepsilon_o}\frac{Q}{r^2}$$

Hence, $E \cos\theta\, \Delta A$ is simply equal to $E\Delta A$, since $(\cos\theta = +1)$, and E is constant, so the net flux over the entire surface is given by

$$\sum_{\text{Closed surface}} \vec{E} \cdot \hat{n}\Delta A = \frac{1}{4\pi\varepsilon_o}\frac{Q}{r^2}(+1)\,(4\pi r^2) = \frac{Q}{\varepsilon_o}$$

Note: In Section 2.27, we learned that the flux density $D = \dfrac{Q}{4\pi r^2}$, where we used r in place of the distance d. An alternate statement of Gauss's law is then

$$\psi = \Sigma_{closed\ surface}\ \varepsilon_o \vec{E} \cdot \hat{n}\, \Delta A = \frac{1}{4\pi}\frac{Q}{r^2}\ (+1)(4\pi r^2) = Q$$

or $\psi = Q$, where we have used the relation $\vec{D} = \varepsilon_o \vec{E}$.

The expression we arrived at is based on geometry and is therefore a geometry-based proof rather than an algebra-based proof.

The size of the surface doesn't matter nor does its shape. This is the speciality of Gauss's law. It tells us that the electric flux due to a charge is divergent (note the *spread* of the flux that we took into account when we cut up the surface into small patches of ΔA and that the electric field has a direction normal to the surface at every point on it), and simply exists everywhere. Remember that the electric field penetrates matter but that need not bother us also since the surface is a mathematical construct and on any surface the distribution simply spreads evenly and smoothly without throwing any surprises in a manner such that the total on any closed surface equals the charge inside the surface.

We must choose a surface that encloses all the charges when using it to evaluate the flux of charges that may not be distributed symmetrically.

The fact that the shape of the surface does not matter can be easily understood by assuming different surfaces (cubes, cones, tetrahedrons and others) around the charge and visualizing the flux present on them.

Application of Gauss's Law in Circuits

(Chabay and Sherwood, 2011, used by permission)
We had assumed that the charge enclosed by the spherical surface was static and a static equilibrium condition exists. However, Gauss's law expresses a completely general relationship between charges and electric field, so it applies to electric circuits, too.

Proof that the Interior of a Current Carrying Wire is Neutral

We will now demonstrate its application to confirm that the interior of a wire is neutral. The figure shows a portion of a current-carrying wire of constant cross-section *A* and uniform composition.

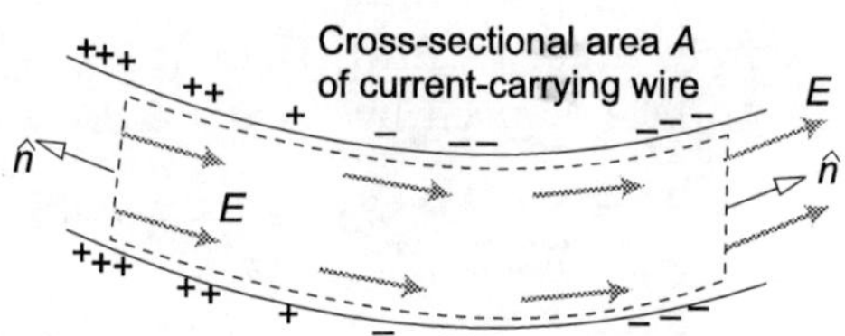

In the steady-state, excess charges arrange themselves on the surface of such a wire so that the electric field *E* has the same magnitude at different locations along the wire, and is everywhere parallel to the wire. The dashed tube-shaped Gaussian surface nearly fills the wire.

We will now use Gauss's law to prove that there is no excess charge in the interior of the wire.

Along the sides of the tube the electric field is parallel to the surface and contributes no flux. Assume that the area of cross-section of the wire is *A*. The right end of the tube contributes an amount of flux *EA*, but the left end contributes –*EA*, so the total electric flux on the Gaussian surface is zero, which means that there must not be any net charge inside the Gaussian surface; all excess charge is on the surface of the wire.

That's it and that completes the proof. There are no special mathematical expressions and manipulations required apart from our assumption that the current in the wire is due to the electric field that acts on all the mobile electrons in the wire and that this electric field owes its existence to the "excess" charges on the surface !

Proof that Field and Current Inside the Wire are Uniform

Not only must the electric field follow the wire, but it must also be uniform across a cross-section of the wire. Suppose for a moment that this isn't true – that instead at different locations in the cross-section of a straight wire the drift speed and therefore the electric field are different as in the figure.

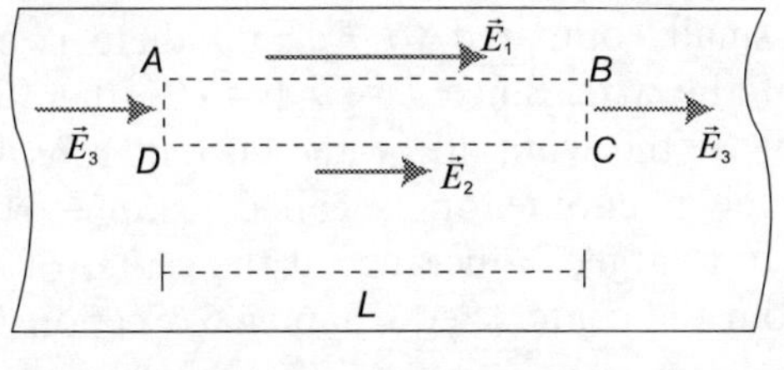

Could E vary at different locations in a uniform wire?

Consider the potential difference

$$\Delta V_{ABCDA} = -\int_A^A \vec{E} \cdot d\vec{l}$$

along the round-trip path $ABCDA$ in the figure (from A to B to C to D and back to A).

Since the fields are set up by point charges on the surface, the requirement that the round-trip potential difference (See Sections 2.19 and 5.11) be zero means that $\vec{E}_1$ and $\vec{E}_2$ have to be equal. Therefore the electric field must be uniform both along the length of the wire and also across the cross-sectional area of the wire. Since the drift speed is proportional to E, we find that the current is indeed uniformly distributed across the cross-section. This result is true only for uniform cross-section and uniform material, in the steady state.

Skin Effect (Chabay and Sherwood, 2011, used by permission)

Note: It is recommended that this topic be read after reading Chapter 5 of this book since knowledge of non-Coulomb forces is a pre-requisite to understand the Skin effect.

The current is not uniformly distributed across the cross-section in the case of high-frequency (non-steady-state) alternating currents, because time-varying currents can create non-Coulomb forces. It is not true that the field should be uniform in an AC circuit, because the magnetic field made by the AC current varies sinusoidally with time. The magnetic field inside the loop $ABCDA$ is out of the page as seen in the figure.

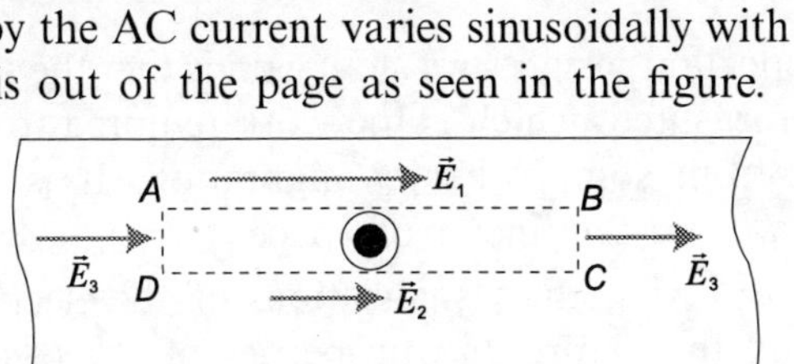

An AC current produces a time-varying magnetic field in the wire, and $\vec{E}_1 > \vec{E}_2$.

When the current is increasing, the magnetic field is increasing out of the page, and there will be a clockwise curly non-Coulomb electric field (Section 5.13). Going around the path *ABCDA*, the path integral $\int_A^A \vec{E}\cdot d\vec{l} = E_1L - E_2L = d\phi/dt$, which is positive. Therefore $E_1 > E_2$. The bigger $d\phi/dt$, the larger the difference between E_1 and E_2.

The higher the AC frequency of the current, the bigger is $d\phi/dt$, so that at very high frequencies E_2 is very small compared to E_1, and there is significant electric field only very near the surface of the wire. Since $J = I/A = \sigma E$, there is almost no current density anywhere in the interior of the wire: all of the current runs very near the surface of the wire. This high-frequency phenomenon is called "skin effect". The effect can also be explained as being due to a larger inductance at the center of the wire in comparison with the inductance away from the center. See Chapter 5, Section 5.22 "Skin effect".

Several authors[32] refer to Gauss's law in the following form

$\Psi = Q_{enc}$ where, Ψ is the total electric flux through any closed surface (i.e. closed surface) '*S*' and is given by $\Psi = \oint d\psi = \oint_s \vec{D}\,.\,d\vec{s}$ where *ds* is the variable of integration over the domain '*S*'. Since, from Section 2.27, $\vec{D} = \varepsilon_o \vec{E}$, we can write Gauss's law as $\oint_S \frac{\vec{D}}{\varepsilon_o}\cdot d\vec{s} = \frac{1}{\varepsilon_o}\oint \vec{D}\cdot d\vec{s}$, or alternatively, we may write,

$$\frac{1}{\varepsilon_o}\Sigma_{Closed\ surface}\ \vec{D}\cdot\hat{n}\,\Delta A = \frac{1}{\varepsilon_o}\Sigma_{Closed\ surface}\ \varepsilon_o\,\vec{E}\cdot\hat{n}\Delta A = \frac{1}{4\pi\varepsilon_o}\frac{Q}{r^2}\,(+1)\,(4\pi r^2) = \frac{Q}{\varepsilon_o}.$$

Application of Gauss's law to Charge Distributions and Energy Stored in Fields

We will apply Gauss's law to find the electric field of different charge distributions in space. We should not concern ourselves unduly about where and *how* such charge distributions occur but that if they did, then what would their fields look like. The study will provide valuable insights into what the spread of the fields are like and more importantly from the unified viewpoint of electrostatics and circuits, shed light on the behavior of circuits in special situations.

One of the most important insights we shall gain is that the *total* energy of a distribution only can be physically measured, which is the work required to bring the charge into some configuration, starting from some other configuration. It is impossible to measure the electrostatic energy in a cubic centimeter of space.

It is recommended to review the visualizations of the electric field in Sections 1.3 and 1.6 and to keep in mind the nature and properties of the electric field, especially (i) the electric field as an *alteration* of space, (ii) the field of a charge *fills* the entire universe, (iii) it penetrates through matter, (iv) the field obeys the principle of superposition which enables us to obtain the *resultant* field given a charge configuration at various points in space, and (v) the field surrounding an individual charge in a configuration always exists at every point in space even if it is observed that the resultant of the charge configuration is different from that of that individual charge.

We will begin by determining the electric field due to an infinite sheet of charge which carries a constant surface charge density σ coulombs/cm^2 as in Fig. A.17.

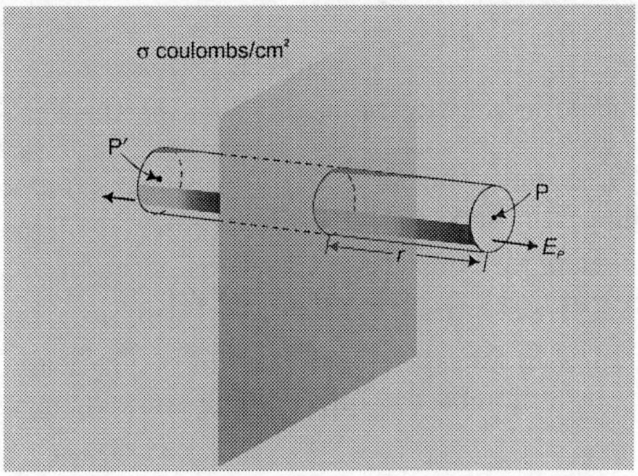

Fig. A.17 Gauss's law applied to find the field due to an infinite flat sheet of charge.

We had seen earlier in the appendix that the electric field of a very large uniformly charged disk does not vary much in direction or magnitude and is given by $\frac{(Q/A)}{2\varepsilon_o}$, where replacing Q/A by σ, we get, $\frac{\sigma}{2\varepsilon_o}$. We can obtain the same result by applying Gauss' law. Two important facts should be recognized before arriving at an expression for the intensity of the field on either side of the infinite sheet carrying charge of density σ coulombs/cm^2 (Fig. A.17);

i) the electric field on either side of the sheet must surely point perpendicular to the plane of the sheet, there being no other unique direction.

ii) due to symmetry, the field must have the same magnitude and the opposite direction at two points P and P′ equidistant from the sheet on opposite sides.

The electric field is the resultant of all the charges in the sheet and the resultant and magnitude is determined by applying the principle of superposition of fields.

These can be inferred from the expressions of the field of uniformly charged disks derived earlier in this appendix which we repeat;

i) the field of the very large uniformly charged disk is along the axis and does not vary much in direction or magnitude, and

ii) If the disk were to become infinitely large then we would arrive at much the same field in regions close and far away from the disk, an independence that is also borne by the fact that the field given by $\frac{\sigma}{2\varepsilon_o}$ does not contain a distance r term.

Draw a cylinder as in Fig. A.17 with P on one side and P′ on the other cross-section area A. Since the flux is outwards at both ends and found only at the ends, then if E_P

denotes the magnitude of the field at P, and $E_{P'}$ the magnitude at P′ the outward flux is $(AE_P + AE_{P'}) = 2AE_P$. The charge enclosed is σA and hence from Gauss's law,

$$2AE_P = \frac{1}{\varepsilon_o}\sigma A \tag{12}$$

or

$$E_P = \frac{\sigma}{2\varepsilon_o} \tag{13}$$

Note that the change in the electric field across the sheet is $\frac{\sigma}{\varepsilon_o}$.

The field strength is independent of r, the distance from the sheet and extends to infinity on either side of the sheet !

Spherically Symmetrical Distribution of Charge

We can apply Gauss's law to find the electric field of a spherically symmetrical distribution of charge of density ρ coulombs per unit volume somewhere out in space as shown in Fig. A.18.

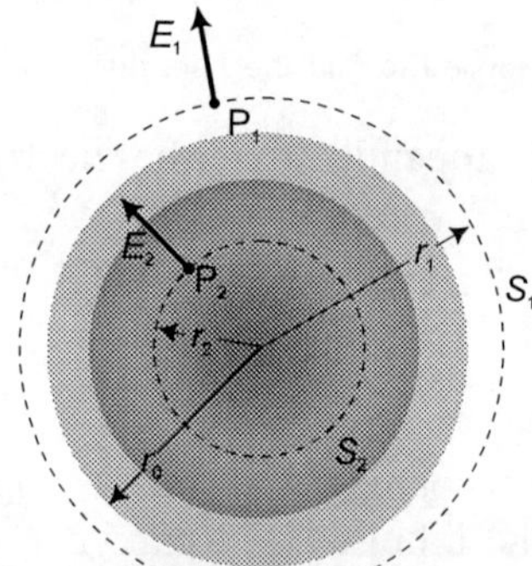

Fig. A.18 A spherically symmetrical distribution of charge.

We are not concerned on how such a distribution was formed nor held in place but that the charge is held in place somehow. The charge density ρ depends only on the radius from the center and is high at the center and zero beyond r_o.

Because of the spherical symmetry, the electric field at any point must be radially directed because no other direction is unique. The field magnitude E_1 must be the same at all points on a spherical surface S_1 of radius r_1, for all such points are equivalent. The flux through this surface S_1 is therefore simply $4\pi r_1^2 E_1$, and by Gauss's law this must be equal to $\frac{1}{\varepsilon_o}$ × (the charge enclosed by the surface)

Therefore,

$$E_1 = \frac{\textit{Charge inside } S_1}{4\pi\varepsilon_o r_1^2} \tag{14}$$

Comparing this with the field of a point charge, we see that the field at all points on S_1 is the same as if all the charge within S_1 were concentrated at the center.

The electric field is the resultant of all the charges in the distribution and the shape of the field lines is determined by applying the principle of superposition of fields.

The same comment applies to a sphere drawn inside the charge distribution. The field at any point on S_2 is the same as if all charge within S_2 were at the center, and all charge outside S_2 absent.

From the above observations we can infer that the field inside a "hollow" spherical charge distribution of Fig. A.19 with a surface charge density σ coulombs/cm^2, is zero.

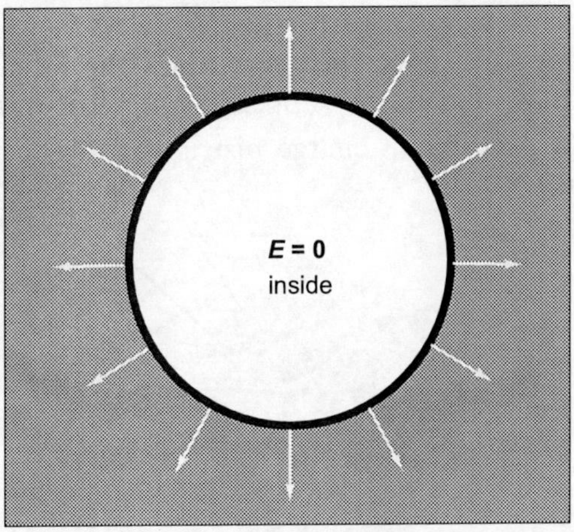

Fig. A.19 A "hollow" spherical distribution of charge.

The fact that $E = 0$ can be readily established by drawing a spherical surface on the inner surface of the sphere and noting that since the charge enclosed is *zero*, therefore there cannot be any electric field inside the shell.

We had shown that $E = 0$ inside a uniform spherical shell in the topic "Inside a Uniform Spherical Shell of Charge" earlier in this appendix.

We can now estimate the field *just* outside a hollow spherical surface.

A Hollow Spherical Distribution of Charge

Consider a spherical surface with uniform charge density σ coulombs/cm^2 as shown in Fig. A.19. Outside the sphere, the field is $\frac{Q}{4\pi\varepsilon_o r^2}$, where Q is the total charge $4\pi r_o^2\sigma$ on the sphere.

The electric field *just* outside the sphere can be found by considering that E_P for an infinite flat sheet with constant surface charge density σ is given by $E_P = \frac{\sigma}{2\varepsilon_o}$ (Eq. 13) and that the change in the electric field is

$$2 \times E_P = 2 \times \frac{\sigma}{2\varepsilon_o} = \frac{\sigma}{\varepsilon_o} \tag{15}$$

Therefore, the change in E, from one side of the layer of the hollow spherical surface of Fig. A.19 to the other, is equal to $\frac{\sigma}{\varepsilon_o}$, or the field just outside the spherical surface is

$$E = \frac{\sigma}{\varepsilon_o}. \tag{16}$$

Force on a Layer of Charge

With charges distributed in spherical shells, we have seen that the electric field inside the sphere is zero. It is instructive to learn what the electrical force is that make up such distributions. We are not really concerned about how these charges were distributed in the first place and for all that we care they appear as if freeze-dried into position !

The sphere of Fig. A.20 has a charge distributed over its surface with the uniform density σ coulombs/cm^2.

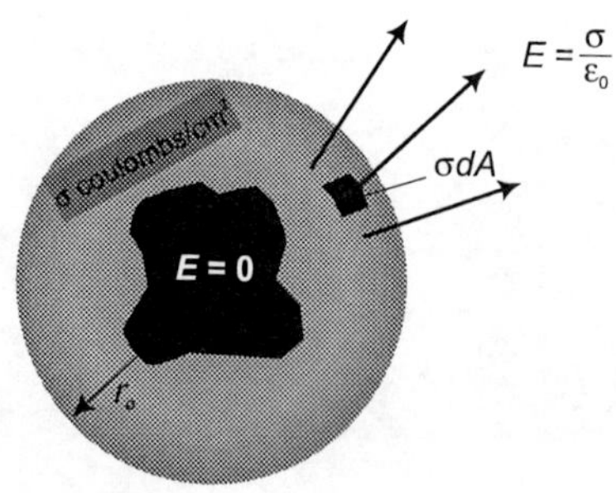

Fig. A.20 A spherical surface with uniform charge density σ.

What is the electrical force experienced by the charges that make up this distribution?

Consider the force on a small element of charge dq, such as a small patch of area dA with charge $dq = \sigma dA$. Consider, separately, the force on dq due to all the other charges in the distribution, and the force on the patch due to the charges within the patch itself.

$$\mathbf{E} = \mathbf{E}_{\text{patch}} + \mathbf{E}_{\text{other}} \tag{17}$$

Since Coulomb repulsion between the charges within the patch is just another example of Newton's third law, the patch as a whole cannot push on itself and this force $\mathbf{F}_{\text{patch}}$ is surely zero. The charge cannot exert a force on itself as mentioned in Appendix A "The electric field due to a point charge". This is similar to not being able to lift yourself by standing in a basket and pulling up on the handles.

This simplifies the problem for it allows us to use the entire electric field **E**, *including* the field due to all charges in the patch, in calculating the force **dF** on the patch of charge dq:

$$d\mathbf{F} = \mathbf{E} \cdot dq = \mathbf{E}\sigma dA$$

But which **E** shall we use, the field $E = \frac{\sigma}{\varepsilon_o}$ outside the sphere (see "A hollow spherical distribution of charge" above this topic), or the field $E = 0$ inside?

In the magnified view of the layer of surface charge, since real charge layers do not have zero thickness, we can consider a distribution say, as in Fig. A.21.

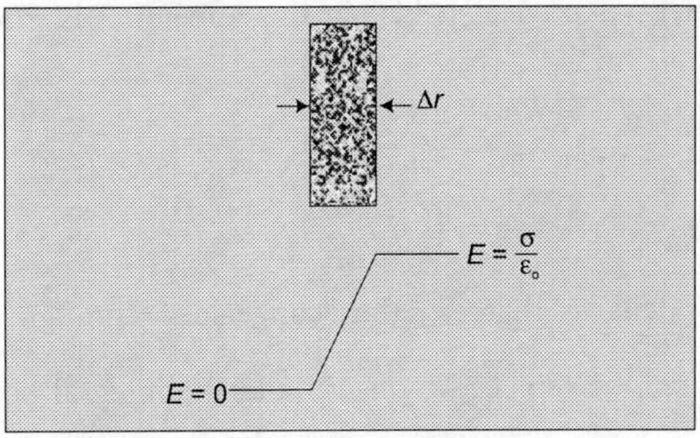

Fig. A.21 The net charge in field at a charge layer depends only on the total charge per unit area.

We have considered only one charge distribution through the thickness of the layer among several possible[22] and the analysis that follows is applicable to any distribution. We will now find the other field $\mathbf{E}_{\text{other}}$ of the layer.

To make it more general, we can let the field on the left be E_1 (rather than zero, as it is inside the sphere), with E_2 the field strength on the right. The condition imposed by Gauss's law for given σ, is in each case

$$E_2 - E_1 = \frac{\sigma}{\varepsilon_o} \tag{18}$$

Let us look carefully within the layer where the field is changing continuously from E_1 to E_2 and there is a volume charge $\rho(x)$ extending from $x = 0$ to $x = x_o$, the thickness of the layer. Consider a much thinner slab of thickness $dx << x_o$, which contains per unit area an amount of charge $\rho\, dx$ as shown in Fig. A.22.

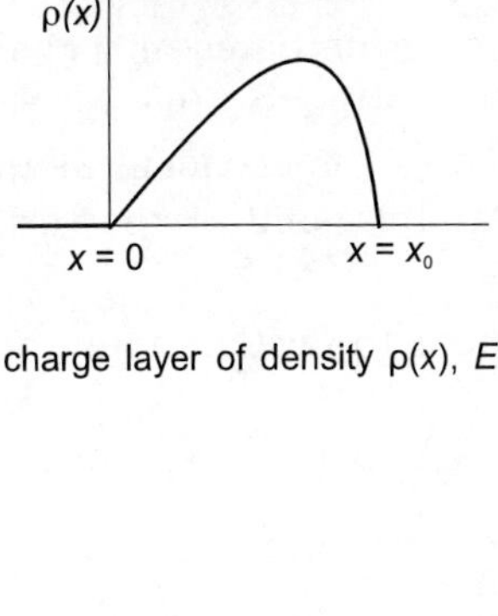

Fig. A.22 Within the charge layer of density $\rho(x)$, $E(x + dx) - E(x) = \dfrac{\sigma}{\varepsilon_o}$.

The change in E through the layer is plotted in Fig. A.22 and the change in electric field intensity (Refer change in field across a sheet "Application of Gauss's law to charge distributions and energy stored in fields" in this appendix) is $E(x + dx) - E(x) = \dfrac{\sigma}{\varepsilon_o}$.

The force on it is

$$dF = E\sigma \tag{19}$$

and thus, the total force per unit area of the charge is

$$F = \int_0^{x_o} E\,\sigma \tag{20}$$

But Gauss's law tells us that dE, the change in E through the thin slab is just $\dfrac{\sigma}{\varepsilon_o}$.

Hence, σ in Eq. 20 can be replaced by $\varepsilon_o\, dE$ and the integral becomes

$$F = \varepsilon_o \int_{E_1}^{E_2} E\, dE$$

$$= \frac{\varepsilon_o}{2}(E_2^2 - E_1^2) \tag{21}$$

The result in Eq. 21 after being factored, can be expressed as

$$F = \frac{\varepsilon_o}{2}(E_2 - E_1)(E_2 + E_1)$$

$$= \frac{\varepsilon_o}{2}\frac{\sigma}{\varepsilon_o}\ (E_2 + E_1)$$

where, we have used $\quad E_2 - E_1 = \dfrac{\sigma}{\varepsilon_o} \quad$ (Eq. 18).

Hence, the total force per unit area of the charge is given by,

$$F = \frac{\sigma}{2}\ (E_1 + E_2) \tag{22}$$

where, we have changed the order of the electric field for clarity.

Since, for a sphere with uniform charge density σ, $E_1 = 0$ (inside) and $E_2 = \dfrac{\sigma}{\varepsilon_o}$ (outside), we find that the force exerted by a patch on its surface is

$F = \dfrac{\sigma}{2}\ (E_1 + E_2) = \dfrac{\sigma}{2}\ (0 + \dfrac{\sigma}{\varepsilon_o}) = \dfrac{\sigma^2}{2\varepsilon_o}$. The direction of the force will be along a normal outward to the patch.

Thus, for a given σ, the force per unit area on a charge layer is determined by the mean of the external field on one side and that on the other which also answers our question as to which E we must consider. This is independent of the thickness of the layer, as long as it is small compared to the total area, and of the variation $\rho(x)$ in charge density within the layer.

Gauss's law and its applications to several more static charge distributions are available in Ref. [2].

Energy Stored in the Fields

There is no way of identifying the energy stored in a particular cubic centimeter of space. Only the total energy is physically measurable, that is, the work required to bring the charge into some configuration. We had calculated the energy stored in a capacitor by computing the change in potential energy. In the example that follows, we will compute the work done to compress a charged spherical shell to calculate the energy stored in the field.

Suppose a spherical shell of charge say, a charged rubber balloon, of surface density σ coulombs/cm^2 is compressed slightly from an initial radius of r_o to a smaller radius $r_o - dr$ as in Fig. A.23.

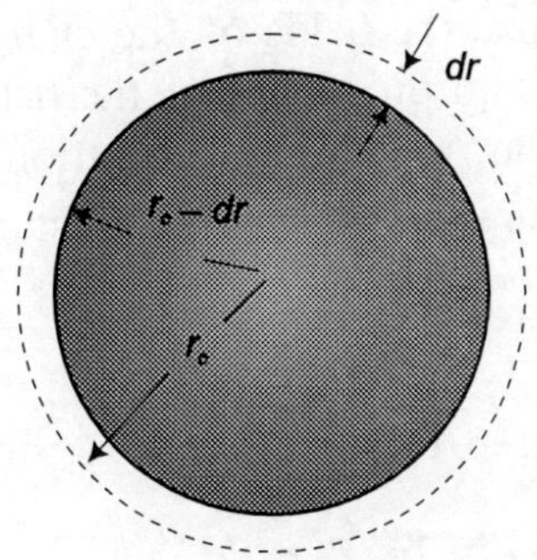

Fig. A.23 Shrinking a spherical shell or a charged balloon.

Outside the sphere, the electric field is $\frac{Q}{r^2}$, where Q is the total charge on the sphere equal to $4\pi r_o^2\sigma$. From Gauss's law we found that *just* outside the surface of the sphere, the field is $\frac{\sigma}{\varepsilon_o}$ (Eq. 16).

The direction of the electrical force on an element of charge on the sphere is outward whether the surface charge is positive or negative as was demonstrated with reference to Figs. A.20 to A.22. Since the charges do not fly off the sphere, that outward force must be balanced by some inward force which holds the charge carriers in place.

If the charged rubber balloon is of 10 cm radius with 6.66×10^{-9} coulombs of negative charge with the charge spread as uniformly as possible on its surface, it forms a surface charge of density $\frac{6.7 \times 10^{-9}}{4 \times \pi \times 10 \times 10} = 5.3 \times 10^{-12}$ coulombs/cm^2.

From Eq. 22, the repulsive force of the charge is $\frac{\sigma}{2}(E_1 + E_2)$ and with $E_1 = 0$ for a spherical shell of charge (Figs. A.3 or A.4) and because E_2 is $\frac{\sigma}{\varepsilon_o}$, this force is $\frac{1}{2\varepsilon_o}\sigma^2$ and it acts normal to the surface.

A closer look at the distribution of charge on the 10 cm radius balloon reveals the "graininess" of their distribution. There are $\frac{6.7\times10^{-9}}{1.6\times10^{-19}}$, or roughly 4×10^{10} electrons attached to the rubber film of the balloon which would mean that there are about 30 million extra electrons per sq. cm.

However, if we could look at one of these extra electrons, we would find roughly 10^{-4} cm – an *enormous* distance on an atomic scale – from its nearest neighbor. The electron would be stuck, electrically stuck, to this molecule of rubber. The rubber molecule would be attached to adjacent rubber molecules, and so on. If we could pull on one of the electrons, the force will be transmitted to the whole piece of rubber.

When attempting to visualize the fields of the charged rubber balloon, the reader will do well to recollect the nature and properties of the field of a charge that was enumerated under "Application of Gauss's law to charge distributions and energy stored in fields" above.

Now, when we compress the shell slightly by a distance *dr*, the total work done is given by surface area of sphere × force per unit area × $dr = 4\pi r_o^2 \left(\frac{1}{2\varepsilon_o}\sigma^2\right) dr$. This represents an increase in energy *dU* required to assemble the system of charges and is,

$$dU = 2\pi \frac{r_o^2}{\varepsilon_o}\sigma^2 dr \tag{23}$$

Notice how the electric field *E* has been changed. Within the shell of thickness *dr*, the resultant field before compressing the shell was zero and is now, $\frac{\sigma}{\varepsilon_o}$. In effect, we have *created* a field of strength $E = \frac{\sigma}{\varepsilon_o}$ filling a region of volume $4\pi r_o^2 dr$. Of course, an electric field can never be created. It is the resultant of all the charges that were displaced from their old positions.

Beyond r_o, the field is unchanged and drops at the rate of $1/r^2$. (Remember that the electric field of a spherical shell is equivalent to that of placing all of its' surface charge at the center of the spherical shell).

If we substitute $\sigma = \varepsilon_o E$ from Eq. 16 in Eq. 23, we get

$$dU = 2\pi \frac{r_o^2}{\varepsilon_o}\varepsilon_o^2 E^2 dr$$

We arrange the terms slightly to gain insight into the contribution the field makes in the calculation of the total energy;

$$dU = \varepsilon_o E^2\, 2\pi r_o^2 dr \tag{24}$$

Multiplying and dividing the right hand side by 2, Eq. 24 may be written thus

$$dU = \frac{\varepsilon_o}{2} E^2\, 4\pi r_o^2 dr \tag{25}$$

The potential energy *U* of a system of charges, which is the total work required to assemble the system of charges, can be calculated from the electric field itself simply by assigning an amount of energy ($\frac{\varepsilon_o}{2}$ E^2) dv to every volume element dv (i.e $4\pi r_o^2 dr$) and integrating over all space where there is electric field.

$$U = \frac{\varepsilon_o}{2} \int_{\substack{Entire\\Field}} E^2\, dv \tag{26}$$

E^2 is a scalar quantity: $E^2 \equiv \mathbf{E}\cdot\mathbf{E}$.

We do know that there is a field far away outside the compressed balloon also and energy is contained in the field there too. This goes to show that while we do have a formula (Eq. 26) for the *total* energy of the configuration and that it is stored *everywhere* outside the sphere (we used the field *just* outside the sphere in our formula), we cannot conceivably obtain the energy contained in one cubic centimeter of space *anywhere* outside the sphere.

When two charged particles interact, then the total energy calculated using Eq. 26 brings into highlight the energy of individual particles. We had seen an instance of this in Section 1.7 when two point charges interacted. We will explore this using one more example of the interaction between charged particles.

Consider two charged particles, a proton and a negative pion, for instance. Let $\mathbf{E}_{pr}$ be the electric field of the proton, $\mathbf{E}_{pi}$ that of the pion. The total field is $\mathbf{E} = \mathbf{E}_{pr} + \mathbf{E}_{pi}$, and $\mathbf{E}\cdot\mathbf{E}$ is $E_{pr}^{\ 2} + E_{pi}^{\ 2} + 2\mathbf{E}_{pr} \cdot \mathbf{E}_{pi}$. According to Eq. 26 the total energy in the electric field of this two-particle system is

$$\begin{aligned} U &= \frac{\varepsilon_o}{2} \int E^2\, dv \\ &= \frac{\varepsilon_o}{2} \int E_{pr}^2\, dv + \frac{\varepsilon_o}{2} \int E_{pi}^2\, dv + \varepsilon_o \int E_{pr} \cdot E_{pi}\, dv \end{aligned} \tag{27}$$

The value of the first integral is the self-energy of the proton and is a constant of nature and is not changed by moving the proton around. The same goes for the second integral, involving the (negative) pion's electric field alone. It is the third integral with the cross-term that directly concerns us, for it expresses the energy required to assemble the system given a proton and a pion as constituents.

We will no longer be able to ignore the first integral if the two particles were to interact strongly that the electrical structure of one is distorted by the presence of the other. That may happen if the particles were a distance 10^{-13}cm apart. At shorter distances, for strongly interacting particles like the proton and the pion, nonelectrical forces dominate the scene anyway.

But, since we assume our particles to belong to a system of discrete elementary charged particles interacting at distances far greater than 10^{-13}cm, therefore, we can omit the self-energy terms like the first two integrals of Eq. 27.

We did just that, in effect, when we replaced the actual distribution of discrete elementary charges (the electrons on the rubber balloon) by a perfectly continuous charge distribution, which we called 'σ'.

We now have an expression for the total energy of the proton and pion over *all space* and not the energy of one cubic centimeter somewhere.

To illustrate the use of Eq. 27 and its versatility we will consider one final case of an assembly made up of two individual charge configurations (sub-assemblies made of not individual charge particles but charge configurations).

Consider an assembly of two individual sub-assemblies of concentric spherical shells (maybe one rubber balloon inside another), of radii a and b, b being the radius of the outer spherical shell.

Suppose that the inner shell carries a charge $+q$ and the outer one a charge $-q$ (both of them uniformly distributed over the surface).

We can proceed to obtain the total energy in two ways; (i) from the known field of the total assembly, and (ii) from the individual fields of each sub-assembly (each spherical shell of charge).

Method I: from the field of the total assembly

Using Gauss's law, we learn that the electric field inside the shell (Fig. A.24) whose radius is a is zero. We also learn from Gauss's law that there cannot be a field outside the shell whose radius is b because if we draw a spherical Gaussian surface S outside this shell the total charge enclosed is $+q$ and $-q$ which sums to '0' *assuming* a *perfectly* continuous charge distribution. See also "Point Charges" in Appendix B.

We had learned that the field outside (called the *fringe* field) two disks separated by a distance s, a capacitor, is given by $\frac{(Q/A)}{2\varepsilon_o}\left(\frac{s}{R}\right)$. The field outside the disks has a dependence on the size of the disks (radius R). While noting that the fields of charge configurations of objects is dependent on their geometry, the field outside the disks can become near zero if the radius of the disks is large, extremely large but, the field outside the uniform equal and oppositely charged concentric spherical shells of finite radii is zero. This is due to the unique geometrical shape of spheres and the $1/r^2$ proportionality of the field of a uniformly charged spherical shell.

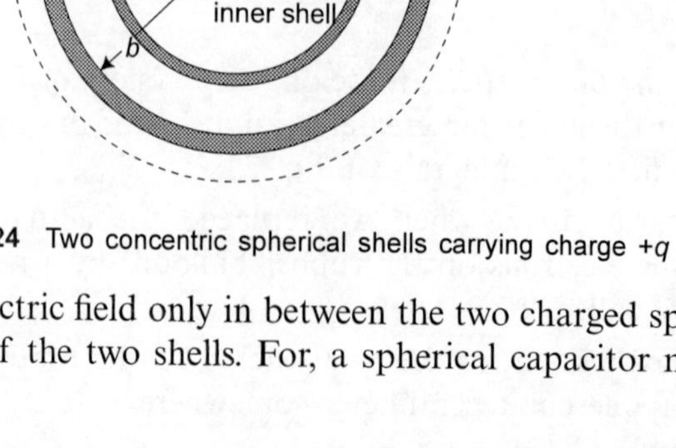

Fig. A.24 Two concentric spherical shells carrying charge $+q$ and $-q$.

There can be an electric field only in between the two charged spherical shells and is the field of the assembly of the two shells. For, a spherical capacitor made of two concentric hollow spherical conductors a fringe field will appear in the wires connected to the shells (a hole is necessary in the outer spherical conductor for threading the wire to the inner conductor) which break the spherical symmetry of the assembly of the two conductors.

Using Eq. 16, we can write for the total energy U of the system of two shells

$$\mathrm{U} = \frac{\varepsilon_o}{2} \int \vec{E}^2 \, dv \tag{28}$$

The electric field $\vec{E} = \frac{1}{4\pi\varepsilon_o} \frac{q}{r^2}$ for $a<r<b$, zero elsewhere.

The volume dv is surface area of the sphere × incremental change in radius

$$= 4\pi r^2 dr$$

Substituting in Eq. 28, we get

$$U = \frac{\varepsilon_o}{2} \left(\frac{q}{4\pi\varepsilon_o} \right)^2 \int_a^b \left(\frac{1}{r^2} \right)^2 4\pi r^2 dr$$

$$= \frac{q^2}{8\pi\varepsilon_o} \int_a^b \frac{1}{r^2} dr$$

$$= \left[\frac{q^2}{8\pi\varepsilon_o} \left(\frac{1}{a} - \frac{1}{b} \right) \right] \tag{29}$$

Method II: from the individual fields of each sub-assembly

To use the method we need the energy of a charged spherical shell of total charge q and radius R.

Inside the sphere $E = 0$; outside,

$$\mathbf{E} = \frac{1}{4\pi\varepsilon_o} \frac{q}{r^2} \hat{r} \quad \text{so} \quad E^2 = \frac{q^2}{(4\pi\varepsilon_o)^2 r^4}$$

To use Eq. 26 and find the total energy, we find an expression of the volume of an infinitesimal volume element dv. See Ref. [8] for an expression in spherical coordinates: $dv = r^2 \sin\theta \, dr d\theta d\varphi$.

The energy stored is equal to the work done to assemble the charges on the spherical shell given by Eq. 26, and is

$$W_{total} = \frac{\varepsilon_o}{2(4\pi\varepsilon_o)^2} \int_{outside} \left(\frac{q^2}{r^4} \right) (r^2 \sin\theta \, dr d\theta d\varphi)$$

$$= \frac{1}{32\pi^2\varepsilon_o} q^2 4\pi \int_R^\infty \frac{1}{r^2} dr = \frac{1}{8\pi\varepsilon_o} \frac{q^2}{R}.$$

We have two charge spherical shell sub-assemblies and obtain

Work done to assemble shell carrying charge $+q$: $W_1 = \frac{1}{8\pi\varepsilon_o} \frac{q^2}{a}$

Work done to assemble shell carrying charge $-q$: $W_2 = \frac{1}{8\pi\varepsilon_o}\frac{q^2}{b}$

Electric field due to shell carrying charge $+q$: $\vec{E}_1 = \frac{1}{4\pi\varepsilon_o}\frac{q}{r^2}\hat{r}$ $(r>a)$,

Electric field due to shell carrying charge $-q$: $\vec{E}_2 = \frac{1}{4\pi\varepsilon_o}\frac{-q}{r^2}\hat{r}$ $(r>b)$.

So, using Eq. 27

$$\varepsilon_o \int \vec{E}_1 \cdot \vec{E}_2 \, dv = -\varepsilon_o \left(\frac{1}{4\pi\varepsilon_o}\right)^2 q^2 \int_b^\infty \frac{1}{r^4} 4\pi r^2 \, dr$$

$$= \frac{-q^2}{4\pi\varepsilon_o b}$$

Therefore, the total work done (W_{total}) to assemble the charge configuration comprising two spherical shells carrying charges $+q$ and $-q$ will be

$$W_{total} = W_1 + W_2 + \int \vec{E}_1 \cdot \vec{E}_2 \, dv$$

$$= \frac{1}{8\pi\varepsilon_o} q^2 \left(\frac{1}{a} + \frac{1}{b} - \frac{2}{b}\right)$$

$$= \frac{q^2}{8\pi\varepsilon_o}\left(\frac{1}{a} - \frac{1}{b}\right) \tag{30}$$

Eqs. 29 and 30 tell us that either way the total energy calculated remains the same.

In Section 2.7, we had seen how the electric field where there was none, grew in strength between the plates of a capacitor as it was charged by a battery. A chemical energy conversion in the battery caused charges to be separated and these charges produced the field strength to overcome the restraining influence of the growing fringe field. This caused the displacement of charges on the capacitor plates and resulted in the storage of energy in the electric field between the plates of the capacitor.

When the charged balloon or charged spherical shell was compressed, the compressing agent provided the energy that overcame the restraining influence of the forces on the layers of charge to convert energy and increase the electric field in the region dr in Eq. 23.

Magnetic Field

Careful experimentation shows that a moving point charge not only makes an electric field but also makes a magnetic field (Section 2.27). We had shown (Section 5.2) that the Biot-Savart law for a single moving source charge gives the magnetic field B at the observation location by

$$\Delta\vec{B} = \frac{\mu_o}{4\pi}\frac{q\vec{v}\times\hat{r}}{r^2}.$$

The Biot-Savart Law for a Short Thin Length of Wire

(Chabay and Sherwood, 2011, used by permission)

We don't often observe the magnetic field of a single moving charge. It would be useful to remember that the magnetic field produced by a moving charge is a *different* alteration of space from that which surrounds a stationary charge which is called the electric field. The magnetic field (Section 5.1) obeys the principle of superposition and magnetic fields penetrate through matter.

Usually we are interested in the magnetic field produced by a large number of charges moving through a wire in a circuit. The superposition principle is valid for magnetic fields, so we need to add up the contributions of the individual charges in the magnetic field.

Let's calculate the magnetic field due to a bunch of moving positive charges contained in a small thin wire of length Δl and cross-sectional area A.

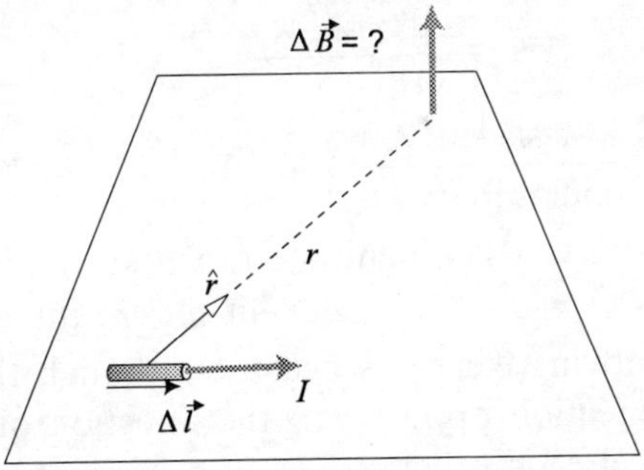

Magnetic field contributed by a short thin length Δl of current-carrying wire.

If there are n moving charges per unit volume, in this small volume $A\Delta l$ there are $nA\Delta l$ moving charges. We will measure the magnetic field at a location far enough away from this small volume that each moving charge produces approximately the same magnetic field at that location.

The net magnetic field of all the moving charges in this small volume, far from the volume, is

$$\Delta \vec{B} = \frac{\mu_o}{4\pi} \frac{I\Delta\vec{l} \times \hat{r}}{r^2}$$

where $\Delta\vec{l}$ is a vector with magnitude Δl (the length of the segment of wire) pointing in the direction of conventional current I, r is the distance of the observation point from the thin length of wire, and $\hat{r}$ is a unit vector in the direction from source to the observation point.

This results from the fact that the volume $A\Delta l$ contains $n(A\Delta l)$ moving charges, so the magnetic field is $n(A\Delta l)$ times as large as the magnetic field of one of the charges. Each moving charge has a charge q, and the conventional current is $I = |q|\ (nA\bar{v})$ where, $\bar{v}$ is the average drift speed. Collecting terms, we find the result given above.

We have an alternative form of the Biot-Savart law for the magnetic field contributed by a short thin length of current-carrying wire (see figure below):

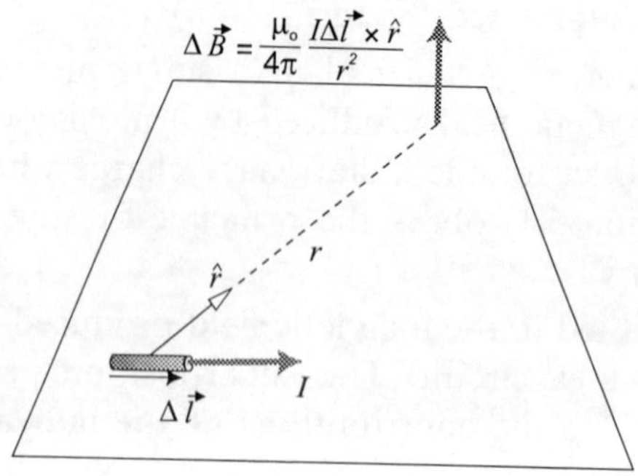

Magnetic field contributed by a short thin length Δl of current-carrying wire.

THE BIOT-SAVART LAW FOR A SHORT THIN LENGTH OF WIRE

$$\Delta\vec{B} = \frac{\mu_o}{4\pi}\frac{I\Delta\vec{l}\times\hat{r}}{r^2}$$

where $\frac{\mu_o}{4\pi} = 1\times10^{-7}\ \frac{\text{tesla}\cdot\text{m}^2}{\text{coulomb}\cdot\text{m/s}}$ exactly. $\Delta\vec{l}$ is a vector in the direction of the conventional current I, and whose magnitude Δl is the length of the segment of wire. $\hat{r}$ is a unit vector that points from the wire segment *toward* the observation location.

The law works for moving electrons because although they have the opposite charge, they also move in the direction opposite to the direction of conventional current. These two changes in sign cancel each other.

It is important to keep in mind that there is really only one Biot-Savart law, not two. Always remember that

$$\Delta\vec{B} = \frac{\mu_o}{4\pi}\frac{I\Delta\vec{l}\times\hat{r}}{r^2} \tag{31}$$

is simply the result of adding up the effects of many moving charges in a short thin length of wire, each of which contributes a magnetic field

$$\Delta\vec{B} = \frac{\mu_o}{4\pi}\frac{q\vec{v}\times\hat{r}}{r^2} \tag{32}$$

The key point is that there are $nA\Delta l$ electrons in a short length of wire, each moving with average speed $\bar{v}$, so that the sum of all the $q\bar{v}$ contributions is $|q|(nA\bar{v})\Delta l = I\Delta l$.

The Magnetic Field of Current Distributions

(Chabay and Sherwood, 2011, used by permission)

Using the same four-step approach that we used to find the electric field of distributed charges (Appendix A, "Electric field of a uniformly charged thin rod"), we will find the magnetic field of distributed charges.

1. Cut up the distribution into pieces; draw $\Delta\vec{B}$ for a representative piece.
2. Write an expression for the magnetic field due to one piece.
3. Add up the contributions of all the pieces.
4. Check the result.

Applying the Biot-Savart Law: A Long Straight Wire

Using the procedure outlined above, we can calculate the magnetic field near a long straight wire. This will make it possible to predict the compass needle deflection that you observe when you bring a wire near your compass, and you can compare your prediction with your experimental observation.

The long straight wire is one of the few cases that we can calculate completely by hand. Except for a few special cases, calculating the magnetic field due to distributed currents is often best done by computer. The basic concepts you would use in a computer program are the same as we will use here, but the computer would carry out the tedious arithmetic involved in the summation step of the procedure.

Step 1: Cut up the distribution into pieces and draw $\Delta\vec{B}$

We will consider a wire of length L as shown in the figure.

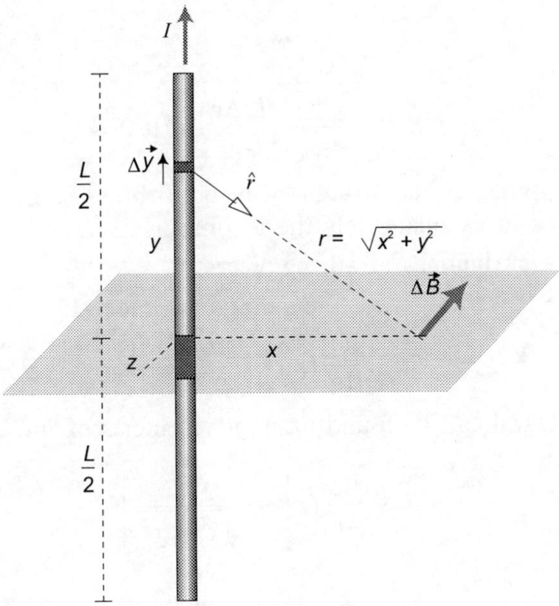

Cut the wire into short pieces and draw the magnetic field contributed by one of the pieces.

We cut up the wire into very short sections each of length Δy, where Δy is a small portion of the total length. We will calculate the magnetic field a perpendicular distance x from the center of the wire. Using the right-hand rule, we can draw the direction of $\Delta\vec{B}$ for this "piece" of current.

Step 2: Write an expression for the magnetic field due to one piece

Origin: Center of wire

Vector $\vec{r}$: $r = \langle x, 0, 0\rangle - \langle 0, y, 0\rangle = \langle x, -y, 0\rangle$

Magnitude of $\vec{r}$: $r = [x^2 + (-y)^2]^{1/2} = [x^2 + y^2]^{1/2}$

$$\textit{Unit vector } \hat{r} = \frac{\vec{r}}{|\vec{r}|} = \frac{\langle x, -y, 0\rangle}{\left[x^2 + y^2\right]^{\frac{1}{2}}}$$

Location of piece: Depends on y (so y will be the integration variable)

$\Delta\vec{l}$ in terms of y : $\Delta\vec{l} = \Delta y\ \langle 0, 1, 0\rangle$

Magnetic field due to one piece: (using Eq. 31)

$$\Delta\vec{B} = \frac{\mu_o}{4\pi}\frac{I\Delta\vec{l}\times\hat{r}}{(x^2+y^2)} = \frac{\mu_o}{4\pi}\frac{I\Delta y\langle 0,1,0\rangle}{(x^2+y^2)}\times\frac{\langle x,-y,\ 0\rangle}{[x^2+y^2]^{1/2}}$$

Evaluate cross product:

$$\langle 0, 1, 0\rangle \times \langle x, -y, 0\rangle = \langle 0, 0, -x\rangle$$

Note: The cross product of two vectors is

$\vec{A} \times \vec{B} = (A_yB_z - A_zB_y,\ A_zB_x - A_xB_z,\ A_xB_y - A_yB_x)$

For example $\langle 2, 3, 4\rangle \times \langle 5, 6, 7\rangle = \langle 3\cdot 7 - 4\cdot 6,\ 4\cdot 5 - 2\cdot 7,\ 2\cdot 6 - 3\cdot 5\rangle$

$= \langle -3, 6, -3\rangle$

Expression for $\Delta\vec{B}$

$$\Delta\vec{B} = -\frac{\mu_o}{4\pi}\frac{Ix\Delta y}{[x^2+y^2]^{3/2}}\langle 0,0,1\rangle$$

Components: Every piece of the straight wire contributes some magnetic field in the $-z$ direction, so we need to calculate only the z component.

Step 3: Add up the contributions of all the pieces

By letting $\Delta y \to 0$ we write the sum as an integral:

$$\vec{B} = \frac{\mu_o}{4\pi}Ix\int_{y=-L/2}^{y=+L/2}\frac{dy}{(x^2+y^2)^{3/2}}$$

Fortunately, this integral can be found in standard tables of integrals:

$$B = \frac{\mu_o}{4\pi}Ix\left[\frac{y}{x^2\sqrt{x^2+y^2}}\right]_{y=-L/2}^{y=+L/2}$$

$$B = \frac{\mu_o}{4\pi}Ix\left[\frac{L/2}{x^2\sqrt{x^2+(L/2)^2}} - \frac{-L/2}{x^2\sqrt{x^2+(L/2)^2}}\right]$$

$$B = \frac{\mu_o}{4\pi}\frac{LI}{x\sqrt{x^2+(L/2)^2}} \tag{33}$$

An extremely important case is that of a very long wire ($L >> x$), or equivalently, the magnetic field very near a short wire ($x << L$).

If $L >> x$, $\sqrt{x^2 + (L/2)^2} \approx L/2$, which leads to $B \approx \dfrac{\mu_o}{4\pi}\dfrac{2I}{x}$.

In summary, the figure shows the pattern of magnetic field a distance r from the center of a straight wire of length L.

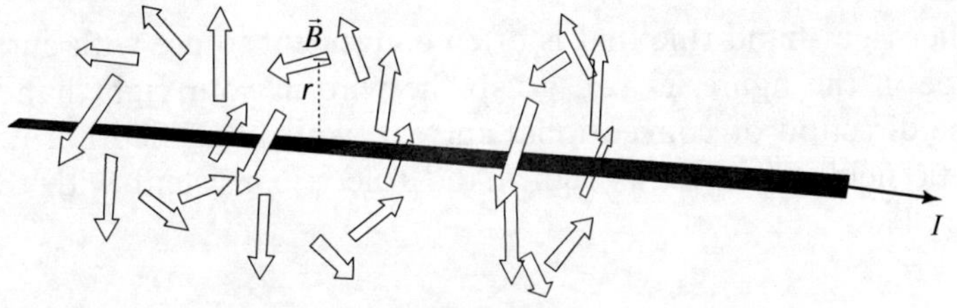

Magnetic field of a straight current-carrying wire, at selected locations a distance r radially outward from the wire.

Because of the symmetry of this field, it does not matter where we draw our x axis– the field will be the same all around the rod.

To indicate this, we replace x in our formula with r, the radial distance from the rod. Although the magnitude of the field is constant at constant r, the direction of the field is different at each angle, since the field curls around the wire. To express this as a vector equation we would have to include a cross product in the expression: it's simpler to get the direction using the right-hand rule.

MAGNETIC FIELD OF A STRAIGHT WIRE

$$B_{wire} = \frac{\mu_o}{4\pi}\frac{LI}{r\sqrt{r^2 + (L/2)^2}}$$

for length L, conventional current I, a perpendicular distance r from the center of the wire.

$$B_{wire} \approx \frac{\mu_o}{4\pi}\frac{2I}{r} \qquad \text{if } L >> r$$

Historically, the result for a very long straight wire was first obtained by the French physicists Biot and Savart. Their names have come to be associated with the more fundamental principle (Biot-Savart law) that leads to this result.

Step 4: Check the result

Direction: In the figure the right-hand rule is consistent with the diagram.

Units:

$$\left(\frac{\text{T}\cdot\text{m}}{\text{A}}\right)\frac{(\text{m}\cdot\text{A})}{\text{m}\cdot\text{m}} = \text{T}$$

Far away ($r >> L$):

$$B_{wire} \approx \frac{\mu_o}{4\pi}\frac{LI}{r^2} = \frac{\mu_o}{4\pi}\frac{\left|I\Delta\vec{l} \times \hat{r}\right|}{r^2}$$

which is indeed the magnetic field contributed by a short wire of length Δl.

Another Right-Hand Rule

There is another right-hand rule that is often convenient to use with current-carrying wires. As you can see in the figure, if you grasp the wire in your right hand with your thumb pointing in the direction of conventional current, your fingers curl around in the direction of the magnetic field. Clearly this right-hand rule is consistent with the one we used for the Biot-Savart law.

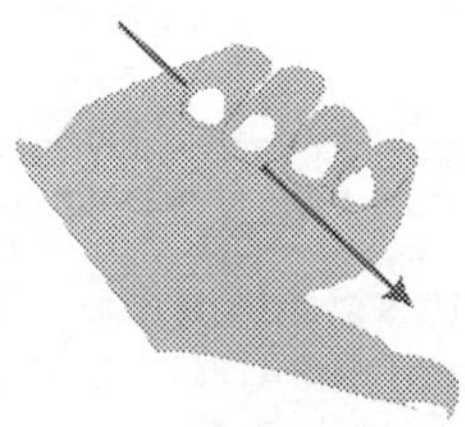

The thumb of the right hand points in the direction of conventional current flow *I*, and the fingers curl around in the direction of the magnetic field.

A Circular Loop of Wire

(Chabay and Sherwood, 2011, used by permission)

We will calculate the magnetic field of a circular loop of wire that carries a conventional current *I*. We take the easiest case – magnetic field at any location along the axis of the loop, which is a line going through the center and perpendicular to the loop.

This calculation is important for two reasons. First, many scientific and technological applications of magnetism involve circular loops of current-carrying wire. Second, the calculation will also lead into analysis of atomic current loops in bar magnets.

Step 1: Cut up the distributions into pieces and draw $\Delta\vec{B}$

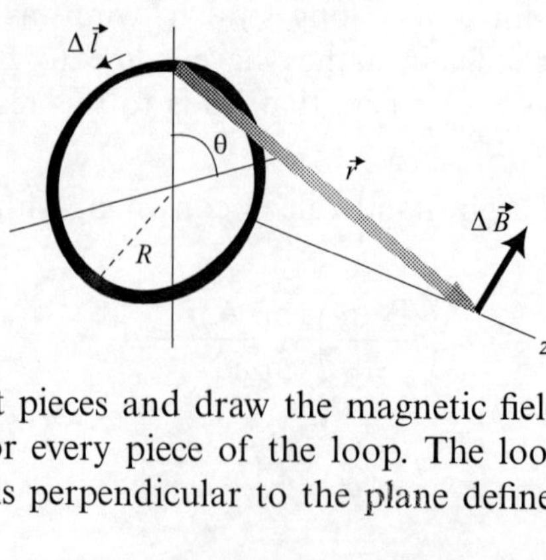

Cut the loop into short pieces and draw the magnetic field contributed by one of the pieces. Note that $d\vec{l} \perp \hat{r}$ for every piece of the loop. The loop is in the *xy* plane, with *x* to the right and *y* up. $\Delta\vec{B}$ is perpendicular to the plane defined by $\Delta\vec{l}$ and $\vec{r}$.

We cut up the loop into very short sections each of length Δl (a small portion of the total circumference $2\pi R$); a side view of the loop is shown in the figure.

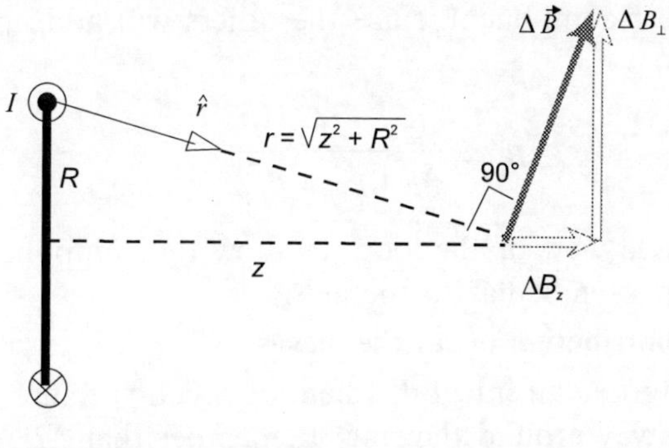

Side view of the current-carrying loop. Again note that $d\vec{l} \perp \hat{r}$ for every piece of the loop.

Determine the direction of $\Delta\vec{B}$ with the right-hand rule. Note that the angle between $\Delta\vec{l}$ and $\hat{r}$ is 90° for every location on the ring.

Step 2: Write an expression for the magnetic field due to one piece

To work out $\Delta\vec{B}$ for an arbitrary $\Delta\vec{l}$ will be algebraically messy, since

$$\Delta\vec{l} = \langle R\cos(\theta + d\theta), R\sin(\theta + d\theta), 0\rangle - \langle R\cos\theta, R\sin\theta, 0\rangle$$

and

$$\vec{r} = \langle 0, 0, z\rangle - \langle R\cos\theta, R\sin\theta, 0\rangle$$

However, we can simplify the problem considerably by noticing the symmetry of the situation. Because of the circular symmetry of the ring, the ΔB_x and ΔB_y contributed by one piece will be canceled by the contributions of a piece located on the other side of the loop.

Furthermore, the ΔB_z contributed by each piece of the ring will be exactly the same. This allows us to select one piece for which ΔB_z is easy to calculate, and use this value for every piece in the sum. We will select the piece shown in the diagram, located on the y axis.

Origin: Center of loop

Vector $\vec{r}$: $\vec{r} = \langle\text{obs.loc.}\rangle - \langle\text{source}\rangle = \langle 0, 0, z\rangle - \langle 0, -R, z\rangle$

Magnitude of $\vec{r}$: $r = [R^2 + z^2]^{1/2}$

Unit vector $\hat{r} = \dfrac{\vec{r}}{r} = \dfrac{\langle 0, -R, z\rangle}{\left[R^2 + z^2\right]^{1/2}}$

Location of piece: Depends on θ (so θ will be the integration variable)

$\Delta\vec{l}$: $|\Delta\vec{l}\,| = \langle -R\Delta\theta, 0, 0\rangle$ the magnitude of $\Delta\vec{l}$ is $R\Delta\theta$.

Magnetic field due to one piece: (using Eq. 31)

$$\Delta\vec{B} = \frac{\mu_o}{4\pi}\frac{I\Delta\vec{l}\times\hat{r}}{r^2} = \frac{\mu_o}{4\pi} I \frac{\langle -R\Delta\theta, 0, 0\rangle \times \langle 0, -R, z\rangle}{\left[R^2 + z^2\right]^{3/2}} \tag{34}$$

Evaluate the cross product:

$$\langle -R\Delta\theta,\ 0,\ 0\rangle \times \langle 0,\ -R,\ z\rangle = \langle 0,\ zR\Delta\theta,\ R^2\Delta\theta\rangle$$

We need only the z-component, since the others will add up to zero:

Expression for ΔB_z:

$$\Delta B_z = \frac{\mu_o}{4\pi}\frac{IR^2\Delta\theta}{\left[R^2+z^2\right]^{3/2}}$$

In the figure of the side view of the loop we show the component ΔB_z along the axis and the component $\Delta B_\perp$ perpendicular to the axis.

Step 3: Add up the contribution of all the pieces

We can express the sum as an integral, where θ, which specifies the location of a piece of the ring, runs all the way around the ring. Remember that ΔB_z contributed by each piece is the same.

$$B_z = \int_0^{2\pi}\frac{\mu_o}{4\pi}\frac{IR^2 d\theta}{(z^2+R^2)^{3/2}} = \frac{\mu_o}{4\pi}\frac{IR^2}{(z^2+R^2)^{3/2}}\int_0^{2\pi} d\theta$$

The integral $\int_0^{2\pi} d\theta$ is just 2π. Here is the result (see figure)

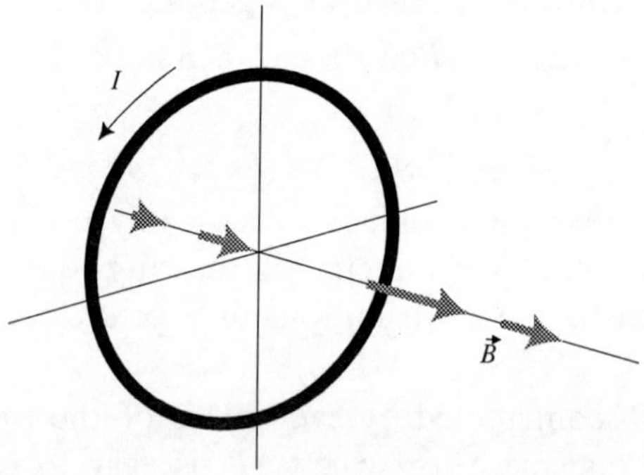

Magnetic field of a circular loop of current-carrying wire.

MAGNETIC FIELD OF A LOOP

$$B_{loop} = \frac{\mu_o}{4\pi}\frac{2\pi R^2 I}{(z^2+R^2)^{3/2}} \tag{35}$$

for a circular loop of radius R and conventional current I at a distance z from the center, along the axis.

Step 4: Check the result

Units:

$$\left(\frac{\text{T}\cdot\text{m}}{\text{A}}\right)\frac{(\text{m}^2\cdot\text{A})}{(\text{m}^2)^{3/2}} = \text{T}$$

Direction: Check several pieces with the right hand rule.

Special case: Center of the loop, where the magnetic field is especially easy to calculate from scratch.

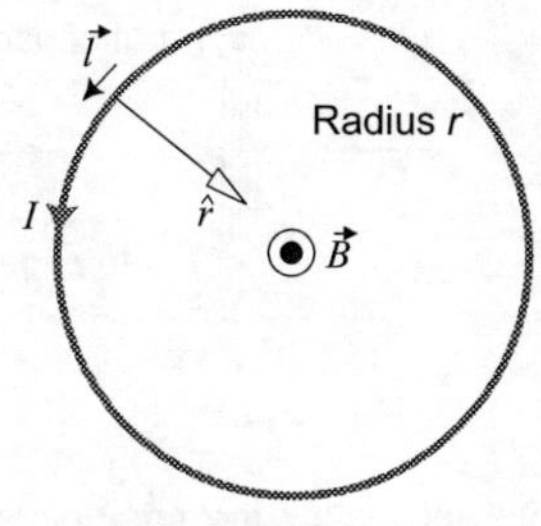

Magnetic field at the center of a loop of current-carrying wire.

By the right-hand rule, $\Delta\vec{l} \times \hat{r}$ points out of the page for every piece of loop, and its magnitude is simply Δl, since $\Delta\vec{l} \perp \hat{r}$. The magnitude of $\hat{r}$ is 1, and sin (90°) = 1.

We have this:

$$B = \sum \frac{\mu_o}{4\pi}\frac{I\Delta l}{R^2} = \frac{\mu_o}{4\pi}\frac{I(2\pi R)}{R^2} = \frac{\mu_o}{4\pi}\frac{2\pi I}{R}$$

because the sum of all the Δl's is just $2\pi R$, the circumference of the ring.

Qualitative Features of the Magnetic Field of a Loop of Wire

If z is very much larger than the radius R, the magnetic field is approximately equal to

$$\frac{\mu_o}{4\pi}\frac{2\pi R^2 I}{z^3}$$

The key to this result is that if $z >> R$, $(z^2 + R^2)^{3/2} \approx (z^2)^{3/2} = z^3$. We see that the magnetic field of a circular loop falls off like $1/z^3$.

A Coil of Many Loops

The figure shows a photo of a coil made of many loops.

A coil made of many loops.

The figure below shows the end view of a coil containing many loops and the pattern of magnetic field along the axis and at points above and below the coil. The size and direction of the arrows represent the strength and direction respectively of the magnetic field.

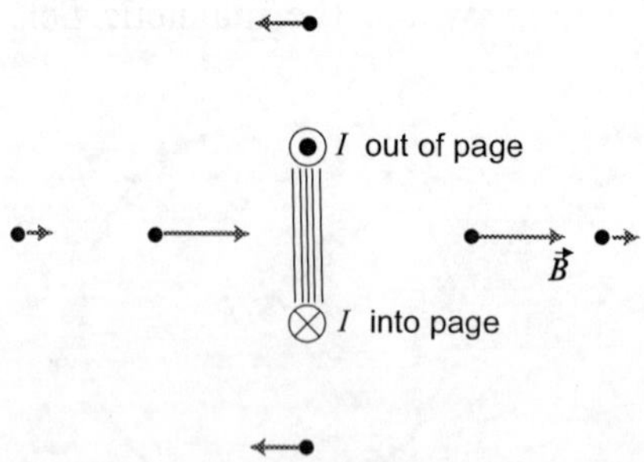

The magnetic field above and below the coil points in the opposite direction to the field along the axis (shows end view of a coil containing many loops).

Far above the coil, the closer upper part of the coil contributes a larger magnetic field to the left than does the slightly farther away lower part of the coil to the right. A detailed calculation shows that this is also true a short distance above the coil. Compare the pattern of magnetic field with the pattern of *electric* field around an electric dipole (figure in Section 1.14).

Magnetic Field at other Locations outside the Loop

The magnetic field at other locations outside the loop is more difficult to calculate analytically, but the magnetic field has a characteristic dipole pattern as shown in the figure, which is the result of a computer calculation that added up all the contributions of many short sections of the loop.

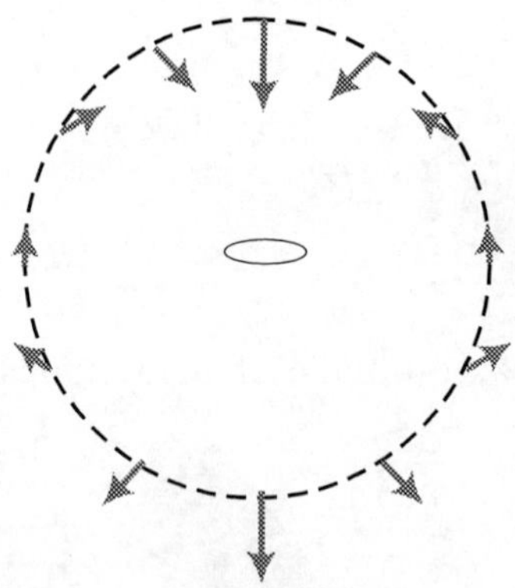

The magnetic field of a current loop (which lies in the *xz* plane, viewed edge-on), at locations outside the loop, in a plane containing the axis of the loop.

Magnetic Field of A Solenoid

(Chabay and Sherwood, 2011, used by permission)

The analysis in this section offers one more example of how to apply the Biot-Savart law. It shows that the magnetic field is nearly uniform along the axis of a solenoid, far from the ends. A solenoid is a coil that is much longer than its radius.

Step 1: Cut up the distribution into pieces and draw $\Delta\vec{B}$

We consider a solenoid of length L that is made up of N circular loops wound tightly right next to each other, each of radius R as shown in the figure.

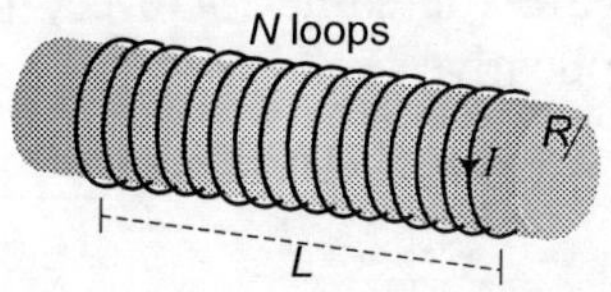

A solenoid of length L with N loops of radius R carrying current I.

We consider each loop as one piece. The conventional current in the loops is I. What will be the direction of the magnetic field $\Delta\vec{B}$ contributed by each of the loops at any location along the axis of the solenoid in the figure above?

All of the contributions inside the solenoid will point to the right in the diagram.

Step 2: Write an expression for the magnetic field due to one piece

Origin: Center of solenoid

Location of one piece: Given by z, so the integration variable is z; the distance from the loop to the observation location A is $d - z$ (see figure).

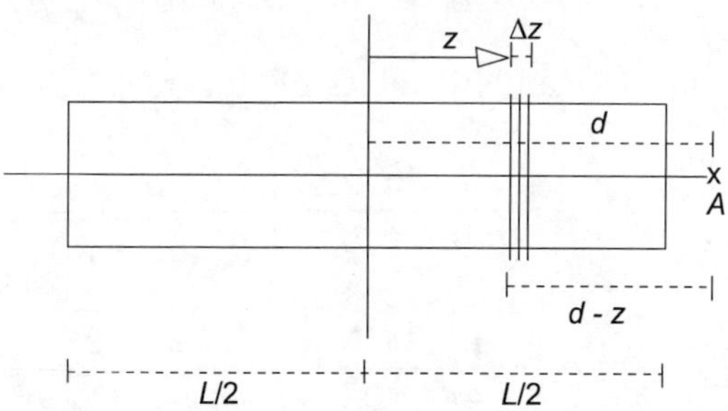

Defining an integration variable z for summing the contributions of many loops.

Our result will be valid for any location along the axis, inside or outside the solenoid.

How many closely packed loops are contained in a short length Δz of the solenoid?

There are N/L loops per meter, so the number of loops in a length Δz is $(N/L)\,\Delta z$. We know the magnetic field made by each loop along its axis.

The loops contained in the section Δz of the solenoid in the figure above contribute a magnetic field at the observation location in the z direction of this amount: (Using Eq. 31 and see Eq. 34)

$$\Delta B_z = \frac{\mu_o}{4\pi}\frac{2\pi R^2 I}{[(d-z)^2 + R^2]^{3/2}}\frac{N}{L}\Delta z$$

Step 3: Add up the contributions of all the pieces

The net magnetic field lies along the axis and is the summation of all the ΔB_z contributed by all the loops:

$$B_z = \sum \Delta B_z = \sum \frac{\mu_o}{4\pi} \frac{2\pi R^2 IN}{L} \frac{\Delta z}{[(d-z)^2 + R^2]^{3/2}}$$

Many of these quantities are the same for every piece and can be taken outside the summation as common multiplicative factors:

$$B_z = \frac{\mu_o}{4\pi} \frac{2\pi R^2 IN}{L} \sum \frac{\Delta z}{[(d-z)^2 + R^2]^{3/2}}$$

This can be turned into an integral, with z ranging from $-L/2$ to $+L/2$:

$$B_z = \frac{\mu_o}{4\pi} \frac{2\pi R^2 IN}{L} \int_{z=-L/2}^{z=+L/2} \frac{dz}{[(d-z)^2 + R^2]^{3/2}}$$

Let $u = d - z$, in which case $dz = -du$, and therefore the limits on the integration run from $u = d - (-L/2)$ to $u = d - (+L/2)$:

$$B_z = \frac{\mu_o}{4\pi} \frac{2\pi R^2 IN}{L} \int_{u=d+L/2}^{u=d-L/2} \frac{-du}{[u^2 + R^2]^{3/2}}$$

$$B_z = \frac{\mu_o}{4\pi} \frac{2\pi R^2 IN}{L} \int_{u=d-L/2}^{u=d+L/2} \frac{du}{[u^2 + R^2]^{3/2}}$$

The integral can be found in standard tables of integrals:

$$B_z = \frac{\mu_o}{4\pi} \frac{2\pi R^2 IN}{L} \left[\frac{u}{R^2 \sqrt{u^2 + R^2}} \right]_{u=d-L/2}^{u=d+L/2}$$

$$B_z = \frac{\mu_o}{4\pi} \frac{2\pi IN}{L} \left[\frac{d + L/2}{\sqrt{(d+L/2)^2 + R^2}} - \frac{d - L/2}{\sqrt{(d-L/2)^2 + R^2}} \right] \quad (36)$$

In this expression, d is the distance from the center of the solenoid.

The figure below shows a plot of this expression for B_z for a particular solenoid, as a function of the distance from the center of the solenoid.

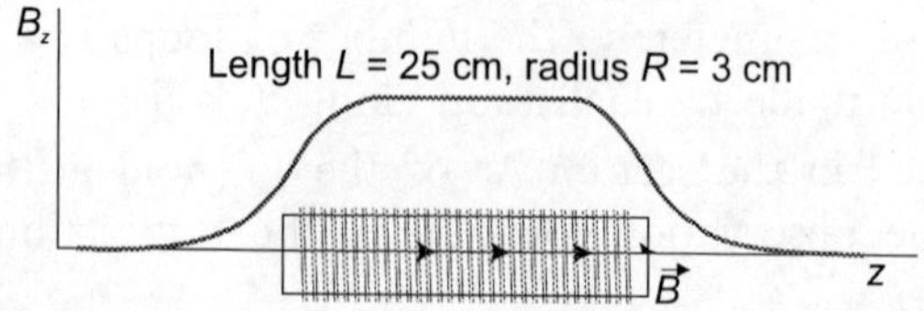

Notice that the magnetic field is nearly uniform inside the solenoid, as long as you are not too near the ends. The field is even more uniform for solenoids that are longer and/or thinner than this one.

If the radius R is small compared with the length $L (R << L)$, show that at the center of the solenoid the magnetic field has a remarkably simple value:

$$B_z \approx \frac{\mu_o NI}{L}$$

With $d = 0$ and $R << L$, the quantity in square brackets above reduces to

$$\frac{L/2}{\sqrt{(L/2)^2}} - \frac{-L/2}{\sqrt{(-L/2)^2}}$$

which is equal to 2, from which follows the simple form $B_z \approx \mu_o NI/L$.

The function plotted above shows that this is also the approximate magnetic field inside the solenoid along much of the axis, not too close to the ends. Outside the solenoid, the magnetic field falls off rapidly. At a large distance the field falls off like $1/r^3$, since the solenoid then looks like a simple collection of current loops.

MAGNETIC FIELD INSIDE A LONG SOLENOID

$$B_z \approx \frac{\mu_o NI}{L}$$

inside a long solenoid, with the radius of the solenoid << L.

Step 4: Check the result

Does our final result make sense? In particular, do we have the right units? Comparing with the magnetic field for a single current element,

$$\frac{\mu_o}{4\pi}\frac{I\Delta\vec{l} \times \hat{r}}{r^2}$$

we easily verify that our answer does have the right units, since

$$\frac{NI}{L} \text{ has the units as } \frac{I\Delta l}{r^2}$$

(Remember that $\hat{r}$ is dimensionless).

Faraday's Law and Moving Coils or Magnets

We had seen in Section 5.16 that a time-varying current in a coil induces an emf in the same coil. You will also find an emf if you move a bar magnet toward or away from a coil, since this creates a time-varying magnetic field (and magnetic flux) inside the coil.

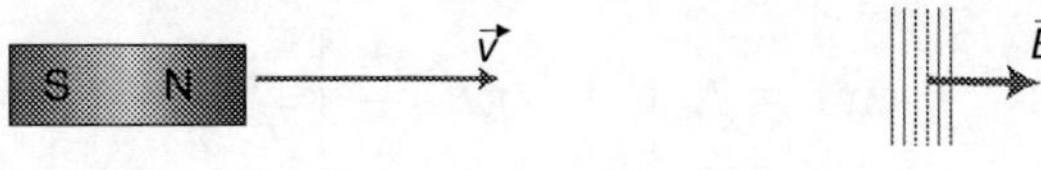

Moving a magnet toward a coil creates a time-varying magnetic field inside the coil.

You can also get an emf by rotating the bar magnet since this changes the flux through the coil as long as you are rotating and therefore changing the magnetic field (and the magnetic flux) in the region of space surrounding the second coil.

Rotating a bar magnet produces a time-varying magnetic field inside the coil.

These various experiments give additional confirmation that the emf in one or more loops of wire is associated with a time-varying magnetic field, and numerically equal to the rate of change of the magnetic flux enclosed by the loops:

$$\text{emf} = -N\frac{d\phi_{mag}}{dt} = -N\frac{d}{dt}\left(\sum \vec{B}_1 \cdot \hat{n}\,\Delta A\right)$$

where N is the number of turns in the coil, ΔA is the area of a small patch of the coil and $\hat{n}$ is a unit vector in a direction perpendicular to the face of the coil. See Sections 7.4 to 7.6.

INDUCTANCE OF A SOLENOID

(Chabay and Sherwood, 2011, used by permission)

It will be useful to read Faraday's law (Sections 7.4 to 7.6) before reading this section. We found that the magnetic field inside a solenoid of length d (we will represent the length of the solenoid by d instead of L since we will need L to represent a constant that is widely used in circuit analysis that include inductors) is $B = \mu_o NI/d$. Using this expression we can find out how much emf is generated in one loop of the coil if the current changes at a known rate dI/dt.

We have this

$$\text{emf} = \left|\frac{d\phi_{mag}}{dt}\right| = \frac{d}{dt}\left[\frac{\mu_o NI}{d}\pi R^2\right] = \frac{\mu_o N}{d}\pi R^2\frac{dI}{dt} \qquad \text{(one loop)}$$

where πR^2 is the area A of the face of one loop of the solenoid (See Section 7.6), assuming each loop has the same face area. See Sections 7.4 to 7.6 also on Faraday's law.

There are N loops, and each loop encloses approximately the same amount of time-varying flux. Therefore, what is the net emf along the full length of the solenoid?

Evidently, we add up all the individual one-loop emfs in series to obtain the emf from one end of the coil to the other:

$$\text{emf} = N\left[\frac{\mu_o N}{d}\pi R^2\frac{dI}{dt}\right] = \frac{\mu_o N^2}{d}\pi R^2\frac{dI}{dt} \qquad \text{(entire solenoid) (37)}$$

It is standard practice to lump the many constants together and write that the self-induced emf is proportional to the rate of change of the current, with proportionality constant L:

$$|\text{emf}_{\text{ind}}| = L\left|\frac{dI}{dt}\right| \tag{38}$$

The proportionality constant L is called the "inductance" or "self-inductance" of the device, which is called an "inductor".

Evidently the (self-) inductance of a solenoid is this:

$$L = \frac{\mu_o N^2}{d}\pi R^2 \tag{39}$$

The unit of inductance (volts-seconds/ampere) is called the "henry" in honor of the 19th-century American physicist Joseph Henry, who simultaneously with Michael Faraday in 1831 discovered the effects of time-varying magnetic fields.

Electric Potential, Potential Difference and Potential Energy

If we know the value of the potential at location A in a system of charged particles, we can use this to find what the **potential energy of the system** would be if we placed a charged object at that location. Do remember that there is a complex pattern of electric field in the system.

$$U_A = qV_A$$

This leads us to write an expression for **electric potential** defined as potential energy per unit charge:

$$V_A = U_A/q$$

Using the expression for V from Section 2.21 we get the familiar expression for the electric potential energy U of two particles (noting that potential energy is only defined for a multiparticle system and is zero for a single particle) with charges q_1 and q_2 separated by a distance r:

$$U = q_1\left(\frac{1}{4\pi\varepsilon_0}\frac{q_2}{r}\right) = \frac{1}{4\pi\varepsilon_0}\frac{q_1 q_2}{r}$$

The potential difference ΔV can be positive or negative, and the sign determines whether a particular charged particle will gain or lose energy in moving from one place to another in a system. An increase in potential energy $q\,\Delta V$ is associated with a decrease in kinetic energy, whereas a decrease in potential energy $q\,\Delta V$ is associated with an increase in kinetic energy.

Direction of Path Same as Direction of Electric Field: ΔV is Negative

If the direction of the path from the initial location A to final location B is the same as the direction of the electric field, the potential difference is negative (Section 2.21).

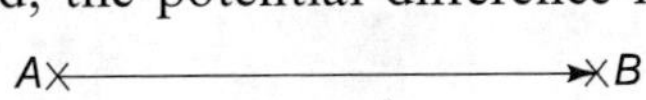

Direction of path same as direction of electric field: Potential Difference is Negative.

Direction of Path Opposite to Direction of Electric Field: ΔV is Positive

On the other hand, the path from the initial location B to final location A is opposite to the direction of the electric field, and so, the potential difference is positive (Section 2.21).

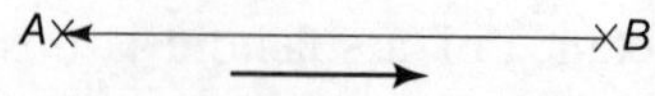

Direction of path opposite to direction of electric field: Potential Difference is positive.

Appendix B

Comparison of Electric and Gravitational Forces (Chabay & Sherwood, 1995, used by permission)[2]

To get a feel for the strength of electric interactions, let us compute the magnitude of the electric force between two particles, and compare it to the magnitude of gravitational forces.

The radius of an atom is of the order of 10^{-10} meters. The electric force of attraction between a proton and an electron that are this far apart (as they might be in a hydrogen atom consisting of 1 proton and 1 electron) is

$$F = \frac{1}{4\pi\varepsilon_o}\frac{e^2}{d^2} = \left(9\times10^9\,\frac{\text{Nm}^2}{\text{C}^2}\right)\frac{(1.6\times10^{-19}\,\text{C})\,(1.6\times10^{-19}\,\text{C})}{(10^{-10}\,\text{m})^2} = 2.3\times10^{-8}\ \text{N}$$

The force does not seem to make much sense unless a meaningful comparison is made with some other force such as a gravitational force.

We can calculate the gravitational force on a hydrogen atom due to the earth:

$$F_{\text{grav}} = mg = (1.7\times10^{-27}\ \text{kg})(9.8\ \text{N/kg}) = 1.7\times10^{-26}\text{N}$$

How does the magnitude of the electric force binding the hydrogen atom together compare to the gravitational force exerted by the entire earth on the hydrogen atom?

$$\frac{F_{\textit{electric on electron by proton}}}{F_{\textit{gravitational on atom by earth}}} = \frac{2.3\times10^{-8}\ \text{N}}{1.7\times10^{-26}\ \text{N}} = 1.3\times10^{18}$$

So, the *electric* force exerted by the proton on the electron is about 10^{18} times larger than the *gravitational* force exerted by the earth on the hydrogen atom.

The electric force of the outermost electron on the proton of an atom in a copper bar will be small, say, 2.3×10^{-8} Newtons by a similar calculation as above, and a hammer and sharp chisel that exerts a force of 100 Newtons *could* prise electrons from the bar. It should be noted that the force actually transmitted to individual atoms of the bar would be quite small, however sharp the chisel point is. The chisel point would be several atomic diameters thick resulting in a stress or force per unit area transmitted to be of the order of 10^{-8} N.

Observing static electric phenomenon of this kind between metals require careful and controlled experimentation and have not been done, but it has a fair chance to be proven given the fact that if two optically flat metallic pieces were brought in contact, it would be diffcult to prise them apart on account of electron transfer.[24]

The force that holds *inner* electrons to the nucleus is far greater than the force of 2.3×10^{-8} Newtons and these cannot be affected much by the forces exerted by the hammer and chisel.

The force exerted by a voltmeter probe will still be tinier and will not get "inside" a surface charge on a conducting wire !

See Appendix D for a very simple description with images of the forces at work at the atomic level when an object is touched.

The gravitational attraction *between the proton and electron is* (Chabay & Sherwood, 1995, used by permission)[2]

$$F = G\frac{m_e\, m_p}{d^2} = \left(6.7\times10^{-11}\,\frac{\text{Nm}^2}{\text{kg}^2}\right)\frac{(9\times10^{-31}\text{ kg})\,(1.7\times10^{-27}\text{ kg})}{(10^{-10}\text{ m})^2} = 1.0 \times 10^{-47}\text{N}$$

How does the magnitude of the electric attraction between the proton and the electron compare with the magnitude of the gravitational attraction between the proton and the electron?

$$\frac{F_{\textit{electric on electron by proton}}}{F_{\textit{gravitational on electron by proton}}} = \frac{2.3\times10^{-8}\text{ N}}{1.0\times10^{-47}\text{ N}} = 2.3 \times 10^{39}$$

So, the *electric* force exerted by the proton on the electron is about 10^{39} times larger than the *gravitational* force exerted by the proton on the electron.

Stable Atoms (Made to Nature's Specifications)

Why does an electron not fall into the nucleus?

If one takes a positive nucleus and a stationary negative electron, the nucleus will attract the electron, and the electron will start moving towards the nucleus until the electron falls into the nucleus. This cannot be a model for an atom. The electrons cannot be *hanging stationary* above the nucleus. They must be moving. We can think of circular motion, just as the planets move around the sun. The planets (uncharged objects) are attracted towards the center of the sun, by gravity, in a manner which constantly *changes their direction,* but *not magnitude of tangential velocity* given to them since the time that they were created. If halted, the planets would fall in towards the sun. As a matter of fact, they are going too fast to actually do this. This circular motion protects the planets against falling into the Sun. And, also note there is no friction where the planets orbit the sun, because there is no atmosphere, and there is no friction where the electron orbits the nucleus as well ! Therefore, there is no loss of energy of the frictional type. So one might think, this would work for electrons as well. But……

The Model Works for Planets but does not work for electrons! Why?

Classical mechanics teaches us that if a *charged* object (such as the electron) moves with an acceleration (in other words, the magnitude or the direction of its velocity is changing in time), this object RADIATES ELECTROMAGNETIC WAVES![25]

If an electron (which is a charged object) would be doing a circular motion around a nucleus, the direction of its velocity would be changing, so it would radiate out energy; therefore it would lose energy, which means it would spiral down to the nucleus.

Classical mechanics **does not** give a satisfactory answer, as to why atoms are stable.

The answer is found in Quantum Mechanics (QM).

1. If an electron is in the electric field of a nucleus, the electron can occupy only certain energy levels. When it is sitting on one of these energy levels, it does not radiate, it does not lose energy. QM shows the way how to calculate these energy levels. Electrons are allowed to move around the nucleus ONLY IN CERTAIN SPECIFIC orbits, called Sommerfeld orbits and the enhancements of Schrödinger and Pauli. In these orbits, the electrons DO NOT RADIATE!!!! So, they are stable. And again, the orbits are not definite.
2. It is possible for electrons to change energy levels, but they have to either absorb, or emit a quantum of energy. This energy is an INTEGER number multiplied by a Planck's constant.
3. It is *now* known that the electrons are not localized in certain, well defined and precise positions or orbits around the nucleus, but occupy *orbital spaces* as seen in the images below of simulations on a computer of these: the *s*-orbital (left image) and *p*-orbital (right image).

In fact, there is a finite "nonzero" probability of finding an electron anywhere in our universe.

Quantum Mechanics, Probabilities and Orbitals of Electrons

Quantum mechanics gives a prediction and shows how to calculate the probability of finding an electron at a given space point of our universe. Universe? Yes, solutions to the electron wave equation are available at several points upto the ends of the universe! And probability?

We have learned that the electron is very close to the nucleus (say, within a 10^{-10} meter radius of the nucleus), and this is so because an electron prefers to be so near the nucleus where it will be in stable equilibrium. There is a *high probability* of it being near the nucleus than elsewhere in the universe, just as at the end of the day after your energies are spent in college or in a playground or a movie theatre or in a shopping mall, there is a *high probability* of finding you in the comfort of your room! High probability? Not sure? Yes, it is entirely probable that you may go away to a relative's house and spend the night!

In the interpretation of large numbers of events or data gathered about physical phenomena, as your movements in the course of a day, mathematicians use statistical methods. These phenomenon cannot be described by equations that predict the future unambiguously and *deterministically*, as you would with equations of motion, that you use to *determine* the position and velocity of an object after a certain time dropped from a height with an initial velocity. See also Summary of Section 2.4 of Chapter 2 in Ref. [30].

When studying physical phenomenon that involve a large number of parameters, the more the number of observations of events or experimental inferences, better are the statistical models and their predictions. The results of experiments on systems comprising large numbers of events or particles will then be probabilistic with expectation values.

With the development of the atomic structure of matter and the study of electrical phenomenon, physicists found it necessary to formulate statistical laws to describe the energies of electrons.

Since it was not possible to experimentally *determine* the movement of electrons, physicists used a process of elimination to predict the likelihood of electron positions around the nucleus and in the entire universe; for example, for an object lying on a table it is not entirely possible for one of the electrons of its atoms to be in the corner of the room, so a very tiny probability is assigned to that position, and one metre away the probability is a tiny fraction more, and so on.

That is why the orbitals of electrons are expressed using probabilities. The solutions to Schrodinger's Wave Equations give the wavefunctions or orbitals which with the application of boundary conditions allows *eigenfunctions* as valid solutions, with corresponding *eigenenergies*, determined by allowed *quantum* numbers. Results of physical measurements are no longer *deterministic* (as in classical mechanics), but *probabilistic*, with an expectation value (Ref. [17]).

The electron cloud model represents a sort of history of where the electron has probably been and where it is likely to be going. Imagine, as the electron moves it leaves a trace of where it was. This collection of traces quickly begins to resemble a cloud.

For an animation of the quantum mechanical model of the hydrogen atom visit http://dwb4.unl.edu/chemAnime/BOHRQD/BOHRQD.html and click on "Quantum Mechanical Model". Visit http://www.av8n.com/physics/orbitals.htm#htoc6 for an animation of electron position from positional probability data extracted from orbitals.

In semiconductor physics, how is it possible to obtain an expression for the number of electrons and holes in an intrinsic semiconductor using such probabilistic models?

In semiconductor physics, the origin of the Fermi-Dirac probability function can be traced to probabilistic considerations. We first associate a density of possible states or energy levels with the probabilistic occupancies of these states by electrons and holes to determine the number of holes in the valence band and the number of electrons in the conduction band, at a given thermal equilibrium temperature.

An integration is performed over a band of energies to "capture" those electrons and holes that have *probable* occupancies into the *verifiable* number of conduction electrons and valence holes that fill these energy bands.

This is similar to the summation of the number of students who have obtained equal grades over the entire range of possible grades to obtain the total number of students who appeared in an examination.

In semiconductors, the numbers of electrons in the conduction band and the number of holes in the valence band are essential to determine the conductivity of the material which is used in the derivation of the volt-ampere relationship and the current components of *p-n* junctions.

Energy Levels of Electrons

To a good approximation, we can identify the total energy of an electron with the size of its orbit. That is, we can think of each radius on the left of the figure below as equivalent to an energy level of the figure on the right.

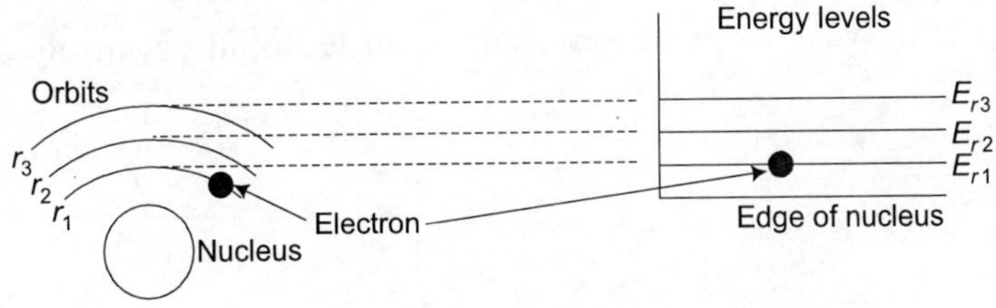

Energy level is proportional to orbit size (radius)

Electrons in the smaller orbit r_1 are on the first energy level E_{r1}; electrons in the second orbit r_2 are on the second energy level E_{r2}, and so on.

Inner Electrons and the Binding Energies of Electrons

The amount of energy that is required to be given to the electron to pull it away from the attractive (Coulombic) force of attraction to the nucleus is called the binding energy. The calculation of the binding energy can be made exactly both by the Schrödinger equation and the relativistic level by the Dirac equation. However, when more than one electron is present in the orbital space around a nucleus one must further consider the electrostatic repulsion which arises between the electrons.

The binding values for the innermost electron in Copper (29 electrons, 29 protons and 35 neutrons) is 8.983 keV and that of the innermost electron in Silicon (14 electrons, 14 protons and 14 neutrons) is 1.844 keV. The outermost electron of Copper has a binding energy of 7.7 eV and the outermost electron of Silicon has a binding energy of 8.15 eV. Note the huge difference in energy levels between the innermost and the outermost electrons for both Copper and Silicon. It is as if the "springs" holding the outermost electrons and the nucleus are thin and soft, which is to be expected given the larger distance from the nucleus these occupy than the inner electrons and the comparatively weak electric binding forces of the outer electrons. The conduction band levels available for the conduction band electrons to move freely given by the density of states are immense and these are separated by fractions of electron volts. Descriptions of the binding energy of electrons is given here http://www.chembio.uoguelph.ca/educmat/atomdata/bindener/elecbind.htm.

Should the binding energy be positive or negative?

There is some confusion in the way physicists refer to the binding energy. Some prefer to show it as positive while others seem to think it should be negative. I recommend that if the electron binding energy is referred as an energy level, then in keeping with the convention of Section 9.7, it should be shown negative. On the other hand, if the binding energy is being referred to as the energy that should be supplied to the electron with the purpose of removing it completely from the atom, it should be shown positive.

Is an atom a dipole?

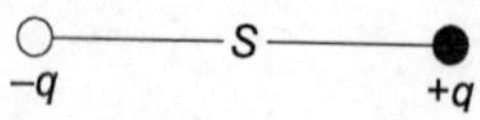

The electric dipole moment (product of the charge in coulombs '*q*' and the separation distance '*s*' = *qs*) is caused from an electric dipole (a separation between positive and negative charge).

Picture an atom with its positively charged nucleus and negatively charged electrons.

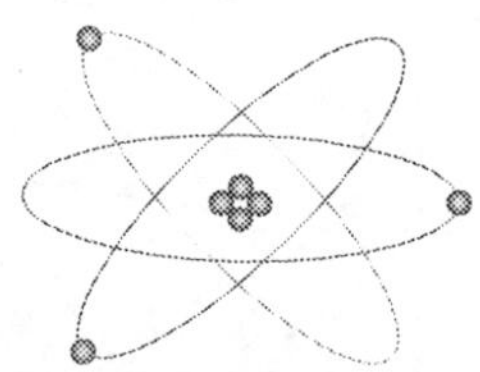

The separation between the positive and negative charge is constantly changing direction (i.e., electric dipole is a vector), and the average (if you wish) is just zero.

A second way is to imagine the electron as a "cloud" that surrounds the atom, then it *appears* that the dipole moment is zero. This is essentially a classical model.

With an asymmetric distribution of charge, one can *expect* a dipole to appear. But even with a perfectly symmetric charge distribution (assuming point particle and a stable atom in the ground state) there can result an electric dipole moment (*qs*), quantum mechanically.

However, the electric dipole moment (*qs*) of an atom is yet vanishingly small due to symmetries. This is the cause for near perfect field cancellation around the atom and the resultant negligibly tiny electric field outside that it qualifies to treat the atom as being "neutral" and not as a dipole.

In metals, where millions of atoms are packed closely together, there is a resultant electric field around the atoms that is not zero, yet vanishingly small....so negligibly small, that it qualifies to treat the interior of an isolated metal as being *neutral*, not forming significant dipoles.

These tiny insignificant dipoles, however, produce "noise" emfs that superimpose on small signals transmitted in sensitive electronic circuits !

How is a molecule a dipole?

Consider the water molecule. A single H_2O molecule is an electric dipole, the H end is somewhat positive and the O end is somewhat negative.

A molecular dipole could be either an induced dipole created by external charges *or* a *permanent* dipole such as the water molecule.

In a molecule there is an electric field *around the component atoms* in so far as treating the atoms independently.

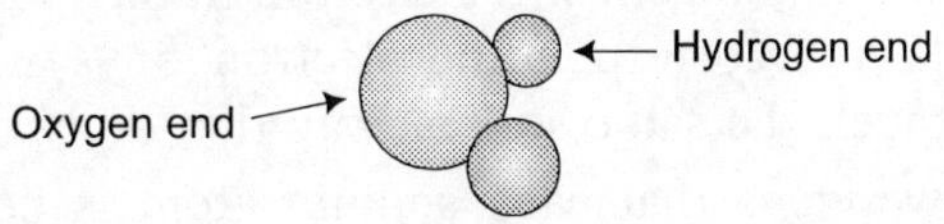

Molecules bond when their electron "clouds" overlap substantially enough that there are in fact electrical fields within each atom induced by the radial fields between the electron "cloud" and nucleus of the other.

To view the electric field (resultant) surrounding a pair of unlike charges…that *grows* (uncanceled fields of each charge) around the pair when pulled apart to form a dipole and *collapses* when brought near each other, visit http://ocw.mit.edu/ans7870/18/18.013a/textbook/HTML/chapter28/section04.html. On the interactive exercise in the webpage there, make all charges = 0, except q_1 and q_2, to view the field due to these two charges.

You can even drag to join them together…and the field collapses totally around the pair…..however, note that this is not possible to do *practically* with a real nucleus and its valence electron (a pair of unlike charges)….nature prevents you from doing this ! There will always be a tiny…ever so tiny resultant electric field around the atom!

Will it so happen that if the atoms were in a place where the temperature is extremely cold (say, in the depths of outer space) the electrons will move so slowly and will "fall" into the nucleus?

Quantum mechanics states that among all the possible energy levels an electron can sit in the presence of a nucleus, there is one, which has THE MINIMAL energy. This energy level is called the "*ground state*". So, even if atoms are in a *very very cold* environment, QM prohibits electrons from falling to the nucleus. The worst that can happen is that they will sit in their *ground states*.

Metallic bonding and Metallic structures (a special kind of great importance to all engineers and scientists)

Unlike the ionic compounds (sodium chloride) and atoms joined together by the classic covalent bond (diamond and graphite) that gives rise to their structures, the manner in which the atoms are arranged in a lattice of the crystal, for example, in a grain of salt and in a diamond in your ring, a bond *may not always determine* a structure. We consider a metallic structure, which is of special importance to mechanical and electrical and electronic engineers, where the bond does not determine its structure.

This is due to the manner in which the *atoms were packed* when the metal was created, whether occurring naturally or man-made as in a furnace.

The structure of the metal is determined by the fact that each atom *tries to be as close to as many other atoms as possible*. Metals are well known to be solids (except for Mercury!). The bonds between atoms in metals can *loosely be described as covalent bonds* (due to sharing electrons), except that the metal atoms do not just share electrons with 1, 2, 3 or 4 neighbors, as in covalent bonding, but *with many atoms.*

Take the metal Sodium that has the electronic structure $1s^2 2s^2 2p^6 3s^1$. Sodium has an electrical conductivity that is about a quarter of that of aluminium.

When sodium atoms come together, the electron in the 3*s* atomic orbital of one sodium atom shares space with the corresponding electron on a neighboring atom to form a molecular orbital—in much the same sort of way that a covalent bond is formed.

The difference, however, is that each sodium atom is being touched by *eight* other sodium atoms—and the sharing occurs between the central atom and the 3*s* orbitals on all of the eight other atoms. And each of these eight is in turn being touched by eight sodium atoms, which in turn are touched by eight atoms—and so on and on, until you have taken in all the atoms in that lump of sodium.

Therefore, this kind of bonding is given a special name "Metallic Bonding".

All of the 3*s* orbitals on all of the atoms overlap to give a vast number of molecular orbitals which extend over the whole piece of metal. There have to be huge numbers of molecular orbitals, of course, because any orbital can only hold two electrons.

The electrons can move freely within these molecular orbitals, and so each electron becomes detached from its parent atom.

The electrons are said to be delocalized. The metal is held together by the strong forces of attraction between the positive nuclei and the delocalized electrons. They must be holding the metal together....else, there would not have been pieces of metal...there would be only metal ions in the universe! No metal pieces would have been possible! It could be that nature decided in some inexplicable way not to make an ideal (noble gas style) bonding in favor of a structure, that we could hammer, bend and shape to any object we desire!

Metals tend to have high melting points and boiling points suggesting strong bonds between the atoms.

Metals are therefore, giant structures of atoms held together by metallic bonds; "Giant" means that large numbers of atoms are involved—depending on the size of the bit of metal.

Density of States

To get a feel for the number of tiny miniscule orbital spaces and the number of electrons nature accommodates in them, we calculate the number of states that are *allowed* (*allowed* quantum states based on the solutions to Schrödinger's Wave equation and Planck's constant, see Appendix IV in Ref. [30]) between two closely spaced energy levels, radially separated in a metal crystal lattice by a few Ångströms (1 Ångström = 10^{-7}mm). The metal crystal lattice comprises millions and millions of atoms packed together that causes their electron orbitals to overlap, delocalizing them.

These are the states that the electrons (conduction band) *can* occupy.

Note that if these states were not *allowed* states, there would be a chaotic distribution of electrons in atoms and molecules with the possibility of shocking charges around us with unimaginable catastrophic consequences to the order of the material universe!

An estimate of the allowable *density of states*, number of states per electron volt per cubic meter for electrons in tungsten lying between 8.5 and 8.6 eV can be made. See Ref. [17], Chapter 3, Problem 3.9. After calculations are made, it is found that there are $6.82 \times 10^{27} \times (8.55)^{1/2} = 1.99 \times 10^{28}$ states per electron volt per cubic meter.

Notice how in the tungsten lattice there are 1.99×10^{28} states (orbital spaces) available inside a few Ångströms (1 Ångström = 10^{-7}mm) of radial separation distance, finer than the finest of rulings of a diffraction grating used in optics experiments!

Each individual state (orbital space) within the 8.6−8.5 = 0.1 eV band of energies is distinctly defined by demarcated solutions to the wave equation. Calculation shows that this band accommodates as many as 2×10^{27} electrons in one meter cube of tungsten. Note that a mere $\dfrac{2\times10^{27}}{1.99\times10^{28}}$ or *one-tenth* of the spaces is occupied, leaving 9 empty states (orbital spaces) available for each electron to move into should it gain or lose energy.

This is what makes electrical conduction possible in the metal.

The delocalized electrons are the conduction band electrons or the outermost electrons that gained just enough energy to become free and roam the lattice freely.

Chemical Bonds can be Modeled as Springs

(Chabay and Sherwood, 2010, used by permission)

Blocks of metal say, copper, aluminium or silicon, do not spontaneously fall apart or evaporate, so the atoms of the block must be attracting each other. The atoms also resist attempts to compress the block. Evidently, the atoms in a block of metal are normally at just the right distance from each other, neither too close not too far away. Two atoms linked by a chemical bond behave in a manner very similar to two macroscopic balls attached to the ends of a very low mass spring as shown in the figure, of a ball-spring model of two linked atoms.

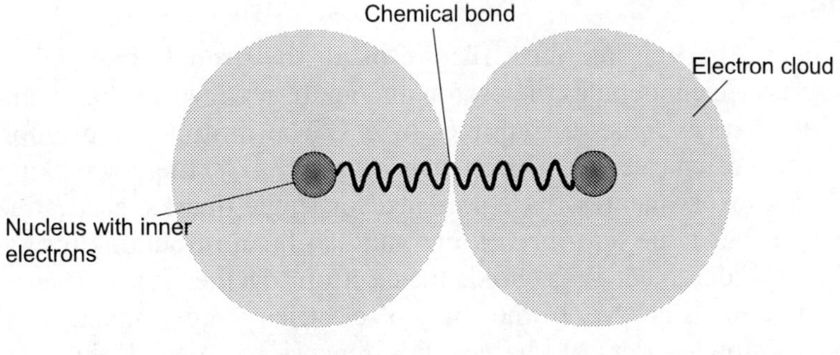

The ball-spring model for two atoms connected by a chemical bond represented by a spring

Many solid objects of metals are crystals of regular arrays of atoms and these may be represented by tiny balls (representing the nuclei with the inner atoms) connected to their neighboring atoms in a lattice with tiny springs. For a description of the ball-spring model of solid objects, the reader may refer "Modern Mechanics" by Ruth W. Chabay and Bruce A. Sherwood Ref[41].

Energy Stored in a Two Electron Configuration

The rest energy of the electron 0.511 Mega electronvolts or 0.511×10^6 eV was calculated using Einstein's famous mass-energy relation: $E = mc^2$ where, m = rest mass of the electron $= 9.109 \times 10^{-31}$ kilogram and c = the speed of light $= 3 \times 10^8$ m/sec. We assume that the electron is at rest for the calculation, an impossible achievement! We have idealized the electron as a point charge. See "Point Charges" below.

Using $E = mc^2$, we get, $E = 9.109 \times 10^{-31} \times (3 \times 10^8)^2 = 82 \times 10^{-15}$ Joules. Since 1 Joule $= 6.2 \times 10^{18}$ electronvolts, therefore, 82×10^{-15} Joules $= 82 \times 10^{-15} \times 6.2 \times 10^{18}$ electronvolts $= 0.508 \times 10^6$ electronvolts or approximately, 0.51 Mega electronvolts(eV).

If we could *compress* two electrons together to a distance of say, 10^{-10} meters, then the potential energy (*PE*) stored in this configuration is given by $PE = kqQ/r$

where, $q = Q = 1.6 \times 10^{-19}$ Coulombs, $r = 10^{-10}$ meters, and $k = \dfrac{1}{4\pi\varepsilon_o} = \left(9 \times 10^9 \dfrac{\text{Nm}^2}{\text{C}^2}\right)$.

Substituting these values in the relation for the *PE*, we obtain. $PE = 23 \times 10^{-19}$ [illegible]

Using, 1 Joule $= 6.2 \times 10^{18}$ electronvolts, we get the $PE = 14.2$ electronvolts(eV).

This is the interaction energy of the two electrons separated by a distance of 10^{-10} meters, or in other words, the energy required to assemble this configuration.

While it does not make sense to compare the energy due to the field interaction of the two electrons, or pontential energy, with their rest energy, the comparison puts into perspective the tininess of the amount of the potential energy; the two electrons have a total *rest* energy of approximately 10^6 electronvolts while the energy of their *interaction* a distance 10^{-10} meters apart, is a *mere* 14.2 electronvolts.

Point Charges

There is evidence that the internal structure of all the strongly interacting particles called hadrons—a class which includes the proton and neutron—involves basic units called quarks, whose electric charges come in multiples of e/3. The proton, for example, is made with three quarks, two of charge 2/3e and one with charge −1/3e, thereby making it positive with charge e. Several experimenters have searched for single quarks, either free or attached to ordinary matter, but none have been found, and so far no fractionally charged matter has been conclusively identified. What holds the electron together is as mysterious as what fixes the precise value of its charge. In the study of electricity and magnetism, charged particles are treated as simply carriers of charge, with dimensions so small that their extension and structure is for most purposes quite insignificant. Even heavy nuclei have their electric charge distributed over a region smaller than 10^{-11} cm. For early physicists, a "point charge" (a mathematical point having no dimensions with a charge) remained an abstract notion.

Today we are on familiar terms with atomic particles. The graininess of electricity is so evident in modern descriptions of nature that it is found that a point charge is less of an artificial idealization and is more a smoothly varying *distribution* of charge density. See "Animation-An electron as a point charge" in Appendix B - Links to animation pdf in CD. When such smooth charge distributions are postulated, these may be thought of as averages over very large numbers of elementary charges, in the same way that a macroscopic density of a liquid is defined, without regard to its *lumpiness* on a molecular scale.[22]

Resistivity of Metals

If the metal crystal lattice were to be perfect, geometrically perfect and flawless (free from lattice imperfections) an electron will move freely with a mean time between collisions of *infinity*.

But it has been observed experimentally that the mean free time in a piece of sodium metal between collisions is 3×10^{-14} sec and the metal begins to exhibit resistance to the flow of electrons. This finite mean time is the reason for the 10^{14} collisions per second an electron experiences as was described in Section 1.18 "Neutrality in current carrying conductors".

The progress of the electron-wave like movement (of conduction band electrons) that interacts with large regions in the crystal is *interrupted* by irregularities in the array or so-called lattice imperfections and that is one of the reasons for the observed mean time and the resistivity of metals.[22] While atomic arrangements are characterized by translational periodicity in three dimensions in ideal crystals, real crystalline materials are rarely perfect.

Several types of lattice imperfections are present in metallic crystals. Lattice imperfections are classified into: (i) point defects (ii) line defects (iii) planar defects and (iv) bulk or volume defects. Among these the most significant lattice imperfections that contribute to the resistivity of metals are the point defects.

Atoms in a material are arranged in a periodic manner. A pure copper crystal structure is face-centred cubic (FCC) and copper atoms occupy all the sites of the FCC lattice. Occasionally, a few lattice positions remain vacant. A missing atom from a normal lattice site in a crystal is called a "vacancy" as shown in the figure.

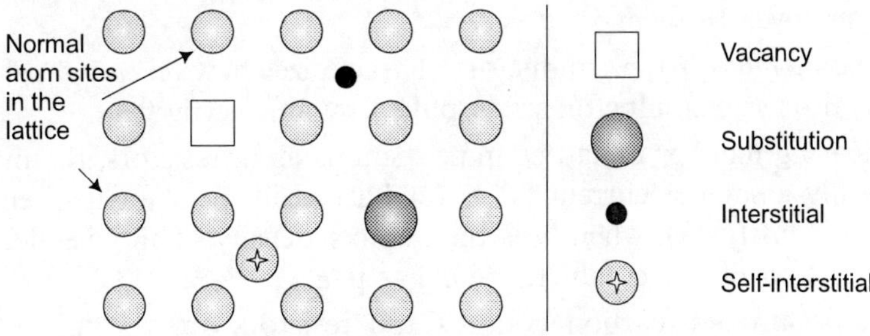

Vacancies influence the property of the conductivity of materials. Sometimes, instead of a vacancy an impurity atom may occupy a regular lattice site. The impurity atoms can be accommodated either at normal lattice sites or in the interstices between the matrix (lattice site) atoms.

If the impurity atom of nearly the same size as the lattice's original matrix atom occupies a site as in the figure above, it is called "substitution". Sometimes if the impurity atoms are very tiny they may occupy interstitial sites between the atoms. In steel (iron-carbon alloys), for example, carbon atoms occupy interstitial positions as shown in the figure above. Another defect occurs when the lattice's original atom occupies an interstitial site. Such a defect is called self-interstitial defect as seen in the figure above.

Point defects such as vacancies and interstitials influence the electrical conductivities of materials. The defects of one of the above types or several in combination may be present in materials.

The other significant contributor to resistivity is the vibrations of the lattice ions at elevated temperatures. The vibrations are random in nature which makes the lattice at any moment slightly irregular geometrically, and the more so, the higher the temperature.

It is this effect which further impedes the progress of the electrons inside the lattice structure and makes the resistivity of a pure metal to increase as the temperature is raised.

If a real piece of copper is cooled below about 20 kelvin, we may expect to attain very low resistivity. But this does not happen; as a matter of fact, there is some residual resistivity due to the lattice defects.

Electron Mobility

For conductors, $v(drift) = \mu E$ where μ is the electron mobility and E the applied electric field strength. While the words "electron mobility" (the dictionary meaning of mobility is "the ability to move physically" and a usage example is "a knee operation has restricted his mobility) may seem to suggest that the electron can at different times be more mobile than at other times, the mobility is *not* a characteristic of an electron; rather, it is the environment it is *in*, which makes it's movement "tight" or "free", i.e., with less or more mobility respectively.

An electron can travel very fast and a good example is the hydrogen atom which is in all our water. A calculation shows that the electron is traveling at about 2,200 kilometers per second. That's less than 1% of the speed of light, but it's fast enough to get it around the Earth in just over 18 seconds.

So, electron *mobility* (μ) in conductors characterizes how *quickly* an electron can move through a metal or semiconductor, when pulled by an electric field.

With increasing number of lattice imperfections as in resistors, the mobility decreases and consequently a smaller current "i" is established in them for a given applied electric field, because $i = nAv(drift)$, where n is the number density of mobile electrons and A the area of cross-section of the conductor and $v = \mu\ E_{net}$.

In electronic circuits, carbon composition resistors are the most commonly used components. The carbon atom has six electrons, 4 of the electrons are in its valence shell (outershell). There are three types of carbon, fullerenes, diamonds, and graphite. The difference between fullerenes, diamonds, and graphite is in their molecular structure.

Graphite has a layered molecular structure and there are very weak bonds between the layers, which gives graphite its slippery feel as opposed to diamonds which have rigid bonds between their atoms which allows very little or no movement.

Graphite has many forms and uses in our everyday life. Natural graphite can be found in China, Mexico, Canada, Ukraine, and Brazil, and it is used as a lubricant, filler for flexible graphite, artistry tools, KEVLAR, and of course, pencils. It has a very low resistivity ranging from 9 to 40 μΩm.

The different forms of graphite accounts for the range in resistivities. Some graphite is manufactured while some is mined from the above locations. Because of its layered molecular structure, graphite has parallel and perpendicular resistivities, depending on how the graphite is oriented in a circuit.

Graphite is an interesting material since it displays properties of *both* metals, and nonmetals. However, like a metal, graphite is a very good conductor of electricity due to the mobility of the electrons in its outer valence shells.

The resistive element consists of finely ground carbon (graphite) particles with refractory filling. These are bonded together using a synthetic resin binder and the mixture is then either compressed into shape and fired in a kiln or coated onto a substrate such as glass and then fired. The composition of the mixture determines the resistance value.

Carbon Film Resistors have many of the same characteristics as carbon (graphite) composition resistors and are made by coating ceramic rods with a mixture of carbon materials. This material is applied to these rods in a variety of means, the ones most familiar are dipping, rolling, printing, or spraying the rods in the appropriate solution.

These rods are cut to the length required for a specific size of resistor. These rods can then be spiral cut to a wide range of values. The original method of spiraling these was done with grinding wheels on a machine similar to a lathe. Modern processes use lasers that are programmed to cut to specific values.

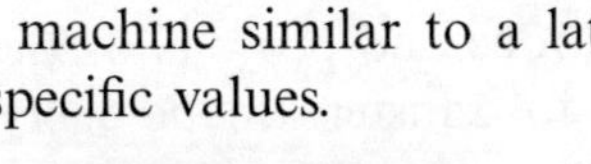

How is Heat Produced in Conductors and Resistors?

Heat is a form of energy and an *increase* in the *random* kinetic or vibrational energy of lattice ions manifests itself as heat and a rise in the temperature of the conductor. The heat energy of a substance is determined by how active its atoms and molecules are. A hot object is one whose atoms and molecules are excited and show rapid movement. A cooler object's molecules and atoms will be less excited and show less movement. Also refer Section 2.23 "The process of capacitors sharing charges" for a description of heat, its production and dissipation as e-m waves in conductors (wires).

For a nice animation of the vibration of atoms with temperature visit http://group.chem.iastate.edu/Greenbowe/sections/projectfolder/flashfiles/thermochem/eqilibrium-v1.html.

Now that we have understood heat, let us understand what types of collisions occur in nature. A collision can be inelastic which is what most collisions in nature are and fortunately (else, there would not be the electric light bulb!) for us, occurs when an electron collides with a lattice ion.

So what is different between an inelastic collision and an elastic collision? An inelastic collision can result in the production of heat while an elastic collision (if done to perfection!) cannot.

The nearest thing to an *elastic* collision we can produce is the collision of two billiard balls. The smoothness of their surfaces, the smoothness of the billiard table surface (very little friction) and the point-like area of contact when they collide ensures that kinetic energy is conserved (almost) along with momentum. Therefore, after the balls collide their kinetic energies macroscopically change but are *conserved* and most importantly they move at the same temperature they were before collision.

Another example is the gravity assisted slingshot that spacecraft use to alter the spacecraft's path and movement using their relative movement and the gravity of a planet. The linear momentum gained by the spaceship is equal in magnitude to that lost by the planet, though the planet's enormous mass compared to the spacecraft makes the resulting change in its speed negligibly small. Typically gravity assist is used to save propellant, time, and expense. Note that there is no physical contact between the spacecraft and the planet!

With an electron and lattice ion collision, which is inelastic, momentum will be conserved but the kinetic energies will not. So, where is the missing energy?

The electron in an inelastic collision with a lattice ion acquires some extra kinetic energy, as well as momentum, between collisions. A collision, or at most a few collisions, redirects its momentum at random (conserving it) but does not necessarily restore the kinetic energy to normal. Some of the missing energy is radiated away due to acceleration and lost movement of the neutrons and protons of the lattice ions. For kinetic energy to be conserved, the electron has to transfer kinetic energy to the lattice ion that deflects it. Again, note that the collision between the electron and the lattice ion is not a physical contact due to strong electron repulsions! And since it has a smaller mass than the ion it collides with, like a tiny marble colliding say, with a cricket ball, the electron will continue to accumulate extra energy until its average kinetic energy is so high that its average loss of energy equals the amount gained between collisions. In this way, by first "heating itself up", the work done by the electrical force driving the electrons is eventually passed on to the rest of the medium (to the lattice ions of the conductors and then to the air molecules with which the conductors are in contact) as random kinetic energy (vibrations), or heat.[22]

What is temperature?

Most of us would have done several calculations in secondary school and pre-University Physics level involving temperature and heat. Seldom do we pause to understand what temperature is ….apart from some vague notion that temperature is somehow related to energy!

In this appendix we had seen that the chemical bonds between atoms can be modeled as springs. Max Planck, a German physicist, found that he could explain the spectrum of light emitted through a small hole in a hot furnace by assuming that the transfers of energy in the form of electromagnetic radiation were quantized-that is, an energy transfer could take place only in fixed amounts.

By solving the quantum mechanical Schrödinger equation for a system involving an atom and a chemical bond, it can be shown that the spacing ΔE between the allowed, quantized energies has the following value $\Delta E = \frac{h}{2\pi}\omega_o$, where ω_o is the angular frequency of the oscillator, and h is Planck's constant. From the classical ball-spring oscillator it is known that $\omega_o = \sqrt{\frac{k_s}{m}}$ where k_s is the spring constant and m is the mass of the ball attached to the spring.

Therefore, classically the energy E can have values as

$$E = E_o,\ 1h\omega_o + E_o,\ 2h\omega_o + E_o,\ 3h\omega_o + E_o \quad \ldots\text{and so on.}$$

In both classical and the quantum picture, the angular frequency $\omega_o = \sqrt{\frac{k_s}{m}}$ of a particular oscillator is fixed and is the same whether the energy of the oscillator is large or small. It can be said that the frequency is determined by the parameters k_s and m, while the energy is determined by the initial conditions or rather to the amplitude of the oscillations ($E = \frac{1}{2}k_sA^2$) where, A is the amplitude of oscillation.

This is fine as far as quantization is concerned into energy levels E_o, $1h\omega_o + E_o$, $2h\omega_o + E_o$, $3h\omega_o$ ....of equal spacing.

This is quite different from the uneven spacing of energy levels found in the hydrogen atom, and implies that in the quantum world, the amplitude can have only certain *specific* values. Other suggested models are "square well" that predict uneven spacing between energy levels that come nearer to the levels of real atoms.

The above description of the energy levels of the atom (nucleus and the electrons) and their quantization was meant to give the reader an intuitive understanding of the topic in quantum theory. See also "The Birth of Quantum Physics" in Section 10.10. For a more comprehensive treatment of the quantum model, the reader is referred to "Modern Mechanics", Ruth W. Chabay and Bruce A. Sherwood, Wiley, 2011 Ref. [41].

When studying temperature, one has to consider an object comprising several atoms and maybe millions of them and their chemical bonds.

So what really is temperature? Temperature *partitions* energy. If you put the magnitude of energy of a single atom on a scale, then temperature can be used to "scribe" the energy (kinetic, the incessant movement of atoms and molecules) scale into partitions of constant energy!

The **Boltzmann constant** (k or k_B) is the physical constant relating temperature to energy.

It is named after the Austrian physicist Ludwig Boltzmann, who made important contributions to the theory of *statistical* mechanics (because millions of atoms and molecules are involved), in which this constant plays a crucial role. Boltzmann's constant k is a bridge between macroscopic and microscopic physics; a proportionality factor that scales empirical temperature to the microscopic mean kinetic energy.

The microscopic properties of molecules introduces Boltzmann's constant

$$pV = NkT$$

where, p = pressure, V = Volume, N is the number of molecules of gas, and k is Boltzmann's constant 1.380×10^{-23} J/K, T = absolute temperature in K.

Thermionic Emission

Thermionic emission is the heat-induced flow of charge carriers from a surface or over a potential-energy barrier. This occurs because the thermal energy given to the carrier overcomes the forces restraining it. The charge carriers may be electrons or ions, and in older literature are sometimes referred to as "thermions".

In vacuum tubes, a **hot cathode** is a cathode electrode which emits electrons due to thermionic emission. Hot cathodes are the main source of electrons in electron guns in cathode ray tubes, electron microscopes, vacuum tubes, and fluorescent lamps.

Heating the metal vibrates the lattice structure vigorously increasing its kinetic energy which is passed on to the electrons by collisions which causes a copious emission of electrons. In the analogy of the cricket ball (lattice ion) and marble (electron) collision (see Section1.18), heating a metal causes the heavier cricket balls to transfer their energy to the lighter electrons. (Applying an electric field to cause emission of electrons is very difficult because of the tremendous field strength and uniformity in strength with distances that will be required to cause the electrons to be ripped off).

After emission, a charge will initially be *left behind in the emitting region that is equal in magnitude and opposite in sign* to the total charge emitted. But if the emitter is connected to a battery, then this charge left behind will be neutralized by charge supplied by the battery, as the emitted charge carriers move away from the emitter, and finally the emitter will be in the same state as it was before emission.

The thermionic emission of electrons is also known as thermal electron emission. Hot cathodes may be either **directly heated**, where the filament itself is the source of electrons, or **indirectly heated**, where the filament is electrically insulated from the cathode; this configuration minimizes the introduction of hum when the filament is energized with alternating current. The filament is most often made of tungsten.

With indirectly heated cathodes, the filament is usually called the *heater* instead. The cathode for indirectly heating is usually realized as a nickel tube which surrounds the heater.

Einstein Relation

To gain an intuitive insight into the Einstein relation let us perform a thought experiment. Supposing we remove a bunch of electrons from a bunch of positive charges in a neutral body B_1 a small distance x, such that the potential energy of the system is a value P. Mark the energy P and the voltage V of the separated charges on a scale.

Change the temperature of a body B_2 to a temperature T from 0 K and mark the energy of the body kT on a scale, where k is Boltzmann's constant.

Fix the temperature T of the body B_2 and separate the charges of the body B_1 to a distance until their potential energy $P = kT$ the energy of the body B_2.

Proceed to do this for several values of the temperature T of the body B_2, plotting T and kT against corresponding P and voltage V of the separated charge centers always maintaining $P = kT$ for the two bodies B_1 and B_2.

We will observe that the relationship between the voltage of the charge pairs in body B_1 and the temperature of the body B_2 is linear, and when the temperature of the body is 25°C then the corresponding voltage (V_T) of the separated charge centers is 25.7 millivolts. The relationship between the voltage and the energy will be governed by $V_T = kT/e$ and V_T is called the *volt-equivalent of temperature*.

In the above processes of charge separation and the increase of the kinetic energy of the particles, the values we would have recorded are essentially *statistical* averages because of the large number (millions and millions!) of particles involved in the above thought experiment.

For example, consider a long semiconductor bar that is at equilibrium in which no net current flows. The role of thermal energy is to knock charges around (with an average energy kT, but having a typical Gaussian-like distribution). Some with higher energy (the mobile charges have a mobility μ) can fly out farther against the electric field created from separated charge centres.

Thus, there is a tug-of-war between the thermally knocked charge diffusing out, with drift induced by the electric field that results from the charge centers separating from each other.

Since, both diffusion D and mobility μ are *statistical* thermodynamic phenomenon, D and μ are not independent. The relation between them is given by the Einstein relation:

$$D_p/\mu_p = V_T \text{ (for holes)}$$

and

$$D_n/\mu_n = V_T \text{ (for electrons)}$$

that relates the diffusion constant D and mobility μ to the *volt-equivalent of temperature*, V_T.

A formal derivation of the relation is provided in Appendix E in "The third process of the influence of the electric field on the energies of electrons" (p. 731).

Fields and Momentum Conservation

Are fields necessarily the carriers of energy and momentum?

Beginning in the mind of Faraday as a region of space which is influenced by a charge, the "field" was defined around charges with the charges as it's source, using various constructs by several scientists. Faraday originally depicted the electric and magnetic fields as electrostatic and magnetic lines of force and Maxwell changed Faraday's lines of force to tubes of force.

Aether, on the other hand (not to be confused with Ether, which is an organic chemical compound), was modeled to be an omnipresent, perfectly *elastic* medium having zero density and was considered a medium for light propagation or light wave disturbances, longitudinally (not transversely). The principle of propagation was similar to the idea of propagation of sound or pressure wave fluctuations in a material medium like air. Aether, is a constituent of space itself, pervading it, and not something belonging to a charge.

Sir Isaac Newton (1642 – 1727) in the Principia makes a mention of Aether when describing the force of attraction between bodies "from the action of the bodies themselves, as tending mutually to or agitating each other by spirits emitted; or whether it arises from the action of the aether or of the air, or of any medium whatsoever". Newton mentions the aether in several places and includes it in addition to air, water and mercury as media.

Sir Joseph Larmor regarded "electricity" (or charges) as consisting of so-called isolated point singularities in the aether, that is, points at which E and H become infinite in a certain manner. See H. Bateman, *The structure of the Aether, Am.* Math. Society, 1915. Soc. 21 (1915) 299-309.

Maxwell, Faraday and Lorentz believed in the existence of two types of aether; a luminiferous one to carry light waves and a non-luminiferous one to carry electromagnetic radio waves. Maxwell believed in any case that the electromagnetic undulations were carried by the medium (or aether). See the last few slides in the power point presentation of Chapter 6 in the CD about the aether theory. When it was shown that the velocity of light c is the same as electromagnetic waves, Maxwell discarded the luminiferous aether and retained the nonluminiferous aether. It would seem that these scientists did not think it right to make massless and invisible lines of force to carry the electric and magnetic fields' undulations, and so introduced the idea of an elastic medium aether to do that. There is a difference in the two ideas, namely Aether, and a field. Aether, was considered to be present all the time, as a constituent of space, pervading it and not belonging to a charge, while a field surrounds the charge. It is important to note the difference in the idea, to understand what Assis's Ref. [36] objections are, to several ideas concerning the field theory.

After the Michelson-Morley experiment disproved the existence of aether, it was replaced by the "fields" themselves as carriers of light and em waves. See the end of Section F.1 in Appendix F about the Michelson-Morley experiment and for links to webpages on animations of the experiment see the Links_to_animations_articles file (Appendix F) in the CD.

Amongst the accepted constructs for depicting electric and magnetic fields are: as *lines of force* surrounding a charge, or as *field lines* (useful for visualizing a *field of vectors*), or as a *field of vectors* at each and every point in space surrounding a charge.

A simple thought experiment will illustrate why the idea of a field came into prominence.

Consider the interaction between two charges (or masses). If the first charge or mass is moved, then the force acting on the second charge or mass immediately (or rather instantly), responds and this is what Coulomb's law and Newton's Law states. However, according to the theory of Special Relativity, the maximum speed with which information can propagate through the Universe is the speed of light in vacuum. So, if the first charge or mass is moved then there must always be some time needed for a light signal to propagate between the two charges or masses, before the second charge or mass responds. Consider what this means if one of the charges (or masses) was suddenly annihilated. The second charge or mass only finds out about this some time later. *During* this time interval, the second charge or mass experiences an electrical or gravitational force which is as if the first charge or mass were still there. So, during this period, there is an action but no reaction, which violates Newton's third law of motion. Therefore, action at a distance is not compatible with relativity, and, consequently, Newton's third law of motion is not strictly true.

In Section 4.2, we stated "We can also compute a momentum called the *electromagnetic momentum* that the fields themselves carry". In Section 8.2 "Momentum" of Griffiths's "Introduction to Electrodynamics" (Ref. [8]), an experiment of the interaction between two charges moving perpendicular to each other is described. It presented a difficulty in explaining the conformance to Newton's Third law when their forces were found not to be equal and opposite.

Most scientists would then say, "Never mind, because conservation of momentum holds, anyway. And, if there is some difficulty there, it is because it is a *hidden* momentum in the fields". See Section 12.2.4 "Relativistic Dynamics", Ref. [8].

Then what had happened to Newton's third law? According to Griffiths, "the proof of conservation of momentum rests on the cancellation of internal forces" (the forces that need to be accounted in the third law). Then, this leaves the system with the momentum of particle and field, reminiscent of the cancellation of the force of action by a reaction, which we saw in "Nature of a field", Section 1.10, when after an object is hit once, it settled down to move with uniform velocity, or when billiard balls after a collision move with uniform velocity, the forces of impulse and collision, having diminished afterwards.

Regarding the location of the so-called internal forces during a charged particle interaction with a field, Härtel (2017) comments "who knows where the origin of the force is located".

In the Electrodynamic case, "only when the field momentum is added to the mechanical momentum of the charges is momentum conservation restored"[8]. This is a crucial aspect of the application of conservation principle in Classical electrodynamics.

Going further, the fields provide a convenient means to explain electromagnetic propagation embodied in Maxwell's equations. Also read the note in Section 10.10 on the radiation from an electric dipole (below Fig. 10.32) about how the radiating field lines "seemingly" break away carrying energy and momentum with them, from an antenna.

The Idea of Momentum of a Field

The set of electrodynamic equations formulated by Maxwell, in one sense marked the beginning of defining the electric field as a "real" physical entity, though Maxwell himself did not think so, since he thought some medium say, aether was mediating the interaction between charged particles, magnets and so forth. After the Michelson-Morley experiment disproved the existence of an aether, the field itself began to be defined as an object like a particle with properties of acquiring energy and momentum. EM fields can transmit the influence of one charge to the other, mediating the interaction (Ref. [8]). Field theories were postulated by theoretical physicists as answers to questions which are summarized in the figure.

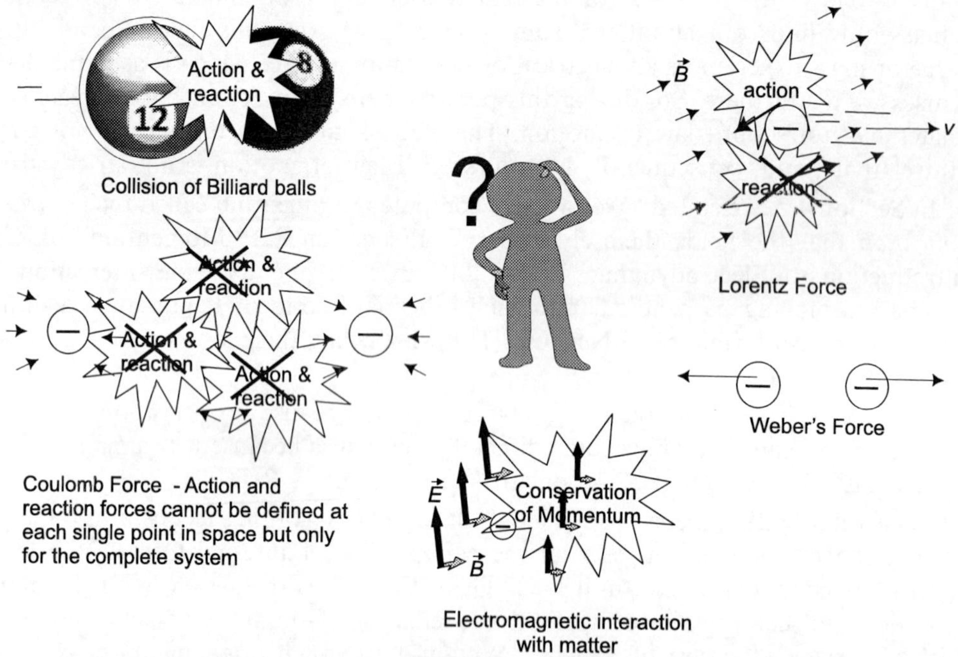

The law of conservation of momentum precedes the law of action and reaction.

These are fundamental aspects of Physics and concern the conservation of energy and momentum. The conservation of momentum is usually explained with the example of colliding billiard balls. There is an action force and a corresponding reaction force at the point of contact. As may be seen in Section F.1, Appendix F on the Coulomb and Magnetic interactions, according to Härtel "The Coulomb forces between two charged carriers are equal and opposite, but they do not act at the same point in space. Each charge carrier experiences a single force due to its interaction with the field; but locally, there is *no reaction* back onto the field", and for the magnetic interaction, "a single force acts on a moving charge carrier within a magnetic field which results in an acceleration of the charge carrier perpendicular to its velocity. There is no local reaction back onto the magnetic field". This may be expected since the fields are abstractions with zero mass. Similar difficulties arise when explaining the loss of momentum due to reradiation when an electromagnetic wave interacts with a charged particle in an object.

For a presentation of the idea of momentum of a field, the author recommends viewing four introductory video lectures by Professor Leonard Susskind, Stanford University on Classical Mechanics. Beginning with a discussion of the Lagrangian function, the Action principle leading to Newton's formula, $F = ma$, the lectures progress with numerous examples by way of illustration, to explain the laws of physics of equilibrium (an alternate formulation of Newton's equations), relations between symmetry principles and conservation of energy, leading to the ideas of energy conservation and deriving the equations of motion from the Lagrangian. Following lectures introduce the Hamiltonian function, and how it is different from the Lagrangian. In the next two lectures, Prof. Susskind discusses the motion of particles in a magnetic field and the vector potential leading to the Lorentz force law. In the last part of the lecture, Prof. Susskind uses the Hamiltonian to derive the conservation of Energy principle of particles moving in a magnetic field. Links to these video lectures are available in the Links_to_animations_articles file in the CD accompanying the book.

For a mathematical derivation of the Lorentz force from the Lagrangian equations of motion, the Hamiltonian and the Field Momentum, see Appendix G, "Vector Potential, Field Momentum, and Gauge Transformation" in *Introduction to Solid-State Physics*, Charles Kittel, Wiley-India, 7th ed.

In their article, The Coulomb force and the Ampere force, Parry Moon, MIT, USA and Domina Eberle Spencer, University of Connecticut wrote in 1954: "In most engineering applications, Maxwell's equations are perfectly satisfactory; but there are occasional ambiguities that have never been resolved. Thus there is a growing feeling that electromagnetic theory should be re-examined from the beginning. The classical experiments of Coulomb on the force between stationary charges were reported in 1785 and led to the well-known Coulomb equation. Gauss (1835) extended the equation to include the effect of motion of the charges. Grassmann (1845), Weber (1846), Helmholtz (1873), Riemann (1875), and others made contributions in the same spirit. Ampere's equation (1823) for the force between current elements [See Appendix F], is likewise free of the field concept. All this work was founded directly on experimental data, without the introduction of a magnetic field".

Proofs of energy and momentum conservation have been put forward based on Weber's force laws applied to electrodynamics and the interaction of charged particles *without* the ideas of fields. These are mentioned by Maxwell himself in the Article 856 of his Treatise: "Weber proceeded to apply his formula to the explanation of the production of electric currents by electro-magnetic induction". Further, "Weber's law is also consistent with the principle of the conservation of energy, in so far that a potential exists, and this is all that is required for the application of the principle by Helmholtz and Thomson".

Dispelling Misconceptions about Action-At-A-Distance Theories

The primary objection to action-at-a-distance theories is its *seemingly* built-in instantaneous action property, which does not conform to the maximum speed limit of motion of any object or the maximum speed limit of propagation of information in the Universe, which according to the postulate of the Theory of Special Relativity, is the speed of light, c.

Kirchhoff and Weber worked independently of one another and predicted the existence of periodic modes of oscillation of the electric current propagating at light velocity in a conducting circuit of negligible resistance (Ref. [37]).

In the article, *On the First Electromagnetic Measurement of the Velocity of Light by Wilhelm Weber and Rudolf Kohlrausch* (2003), Assis states "Although in this derivation the interaction between any two charges is given by Weber's action at a distance law, the collective behavior of the disturbance propagates at light velocity [not instantly] along the wire. This is somewhat similar to the propagation of sound waves derived by Newton or the propagation of signals along a stretched string obtained by d'Alembert. In all these cases classical Newtonian mechanics was employed, without time retardation, without displacement current and without any field propagating at a finite speed. Although the interaction of any two particles in all these cases was of the type of action at a distance, the *collective* behavior of the signal or disturbance did travel at a finite speed". See also "Speed of the Voltage Wave" in Section 5.22, where it was shown how the speed of propagation is increased when the internal inductance of the wires is reduced by the skin effect. Of course, it can never under any circumstance, exceed the finite ultimate speed of light, *c*.

According to Härtel, "At the end of discussions of action-at-a-distance theories in present-day textbooks, authors summarize that they are false and should be ignored. This is wrong, because it is not that the theories are false, but that the *interpretation* of such theories is false. The authors of these theories did not claim that changes occur immediately with *no time lag*. These theories leave the question about transition processes open."

The first to derive the telegraphy equation (sometimes referred as "telegraph" equation) was Weber and Kirchhoff in 1857 based on Weber's electrodynamics, which is the action-at-a-distance theory of Weber.

Assis, in *Circuit theory in Weber electrodynamics* (1997) had presented a derivation of the equation describing the current flow in a circuit with self-inductance based on Newton's second law plus the Weber force. In Weber's approach, "the self-inductance can be treated as a measure of the *effective* average inertial mass of the conduction electrons", according to Assis. A description of the derivation is made by Assis (2005) and a link to the article is available in the Links_to_animations_articles file in the CD.

Weber's Fundamental force law (Ref. [36,38]) for charged particles, is an action-at-a-distance theory, without the idea of a field. According to it, the interacting forces and energies depend only on the relative radial distance, velocity and acceleration between them. Weber's force on the line connecting the two interacting charges, obeys Newton's action and reaction law. It complies with the principles of the conservation of linear momentum, angular momentum and energy.

A detailed proof of this important aspect with a nice description is available in Chapter 3, "Weber's Electrodynamics" by A. K. T. Assis (Ref. [42]). Also, in Chapter 4 of "Weber's Electrodynamics", Assis has dismissed the contention of several scientists that since, Weber's force was developed from Ampere's force based on Fechner's hypothesis which states "the positive and negative charges in metallic wires move in opposite directions with equal velocities", therefore, Weber's Electrodynamics is wrong.

Assis shows "if we assume only Weber's force and the neutrality of the current elements we can still derive Ampere's force even when Fechner's hypothesis is wrong, as is the case in the usual metallic conductors [in which the mobile charges are only the electrons]. The proof of this Section overcomes this limitation pointed out against Weber's electrodynamics as it is based on general assumptions more general than the particular case specified by Fechner's hypothesis." Assis proceeds to give the proof in Chapter 4 of "Weber's Electrodynamics" (Ref. [42]).

From Weber's force, the Coulombic force, Gauss's Law and Ampere's force laws are readily derived. See Section F.5, Appendix F for a description of these and other theories of particle interactions.

The Idea of Relativity

Isaac Newton (1642 – 1726) imagined the velocity of a body in relation to absolute space and measured by absolute time. To know more about his ideas of absolute space and absolute time see Section 1.2, "Laws of Motion" in Ref. [36].

In contrast to this idea, it would be useful to get an idea about relativity in the physical sense. Imagine a Universe or empty space with no objects but one. Would it be possible to say that the object is i) at rest, ii) in translational motion, or iii) spinning around some axis? It would not be possible to state exactly since there is no reference point for observation (even you cannot be present as an observer !). For example, in i) and ii), since there is no second object no displacement between the object and the absent object is possible. In the case of iii), since the space around is empty and all space around is the same, it would be difficult to say that the object is spinning around some axis. Therefore, when describing motion or its absence, defining a frame of reference, which could be the earth, the galaxies, or maybe a laboratory bench and so on, becomes necessary.

The Maxwell-Lorentz's force $\vec{F} = q\vec{E}(\vec{r}) + q\vec{v} \times \vec{B}(\vec{r})$ (Eq. 11 in Appendix F), was developed by Maxwell between 1861 and 1873, being also presented by Lorentz in 1895. It describes the force on a test charge due to electromagnetic fields generated by other source charges. It could be written in the form of a force of interaction between point charges due to the works of Lienard, Wiechert, Schwarzschild and Darwin. When this is done there also appear velocities and accelerations between the charges and the Aether (as thought by Lorentz) or between the charges and inertial frames of reference or observers (interpretation introduced by Einstein). Once more it is not the velocity and acceleration *between* point charges which are relevant, but their motion relative to something *external* to them. This led to multiple definitions of the reference for the velocity in Lorentz's force law. Examples are i) Relative to the average velocity of the microscopic charges (normally electrons) which generate the magnetic field, ii) Relative to the magnetic field itself, iii) Relative to the detector of the magnetic field, iv) Relative to any inertial frame of reference, v) Relative to an arbitrary frame of reference, which does not need to be inertial, and vi) Relative to the aether.

According to Assis "The introduction of physical forces which depend on the state of motion of the observer has created many problems and paradoxes for the explanation of several simple phenomena of nature. Unfortunately it has been part of theoretical physics ever since that time. No experiment has suggested or forced this new interpretation. This whole interpretation arose from Einstein's mind. The usual expression for the magnetic force might have been maintained, interpreting $\vec{v}$ as the velocity of the test charge relative to the magnet or relative to the current carrying wire (See Fig. 3.3 in Härtel's "A qualitative approach to electricity" in the Teacher's Guide in the CD), without any contradictions with experimental data. Instead of adopting this reasonable point of view, Einstein decided to change the interpretation of this velocity. He created an enormous confusion with this new interpretation [by bringing in inertial frames and observers] which has plagued theoretical physics ever since 1905".

Emission Theory, Quantum Electrodynamics and Classical Electrodynamics

Leigh Page's Emission Theory (Ref. [39]), assumes the continuous emission by charges, of massless moving elements, which *do not* carry energy, with a velocity *c*. Their loci map the lines of force. Energy and momentum, according to the theory, are conserved in the classical sense. For visualizations of the interaction of accelerating charge carriers, Härtel adopted Page's theory (see Teacher's Guide in CD).

Quantum Electrodynamics subscribes to a field theory in which the field is quantized and photons are carriers of energy and momentum. Quantum Electrodynamics or Electromagnetism (as some authors would call it) is electromagnetism on the microscopic or atomic scale. See Sections 4.2, 6.6 and 10.10 and "Introduction to Elementary particles" by David J. Griffiths.

A nice lecture (Modern Physics; Quantum Mechanics) that introduces the Quantum character of Physics was delivered by Prof. Leonard Susskind, Stanford University, USA, a link to which is provided in the Links_to_animations_articles file in the CD.

In contrast, it may be said that Classical Electrodynamics (or Classical Electromagnetism as some authors would call it) is electromagnetism on length-scales much larger than the atomic scale.

Note that among these theories, in this textbook, Classical Electrodynamics (or classical field theories) is used to explain the operation of circuits and Antennas and the process of Wave Propagation (including the visualizations in the article by Härtel in the Teacher's Guide in the CD). According to these theories, energy and momentum are conserved in the classical sense that is, in the fields.

According to Purcell (Ref. [22]), two facts make Classical Electromagnetism important in modern physics; i) Special Relativity required no revision of classical electromagnetism, and ii) quantum modifications of the electromagnetic forces have turned out to be unimportant down to distances less than 10^{-10} centimeters, 100 times smaller than the atom.

Appendix C

SURFACE VOLTAGE AND FIELD STRENGTH OF INSULATORS

By definition, insulators do not have a voltage. Before we understand how the charge distribution inside an insulator prevents us from analyzing an insulator's voltage profile, it is instructive to see what really makes a conductor (made of copper, aluminium or steel or any metal) an equipotential object. The description following is reprinted with permission from *Compliance Engineering*, "Surface Voltage and Field Strength: Part I, Insulators", September/October 2001. Copyright © 2001 Canon Communications LLC. The original article is by Niels Jonassen, DSc of the Technical University of Denmark and is available for viewing at http://incompliancemag.com/article/surface-voltage-and-field-strength-part-1-insulators/

Figure C.1 shows an insulated conductor A, which is a conductor surrounded by insulation to prevent current leakage or short circuits, with a charge q.

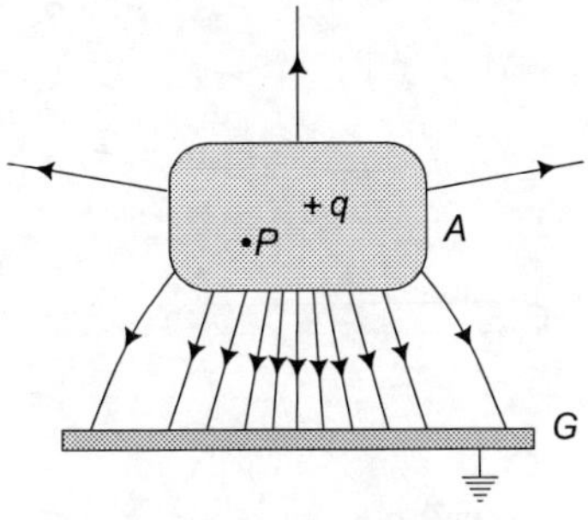

Fig. C.1 Charged conductor.

The charge will automatically distribute itself on the surface of the conductor in such a way that the field in the interior of the conductor will be zero, the field will be perpendicular to the surface, and the integral of the field strength E from any point P *in or on the conductor* to a ground point G is constant and given by

$$V = \int_{P}^{G} E \cdot da \tag{1}$$

where V is the voltage or potential of the conductor.

The potential will progressively decrease at points farther and farther away in space from the surface of the conductor.

Interpret the equation 1 like this: it would take more work to push a test unit positive charge a distance Δa from some point in space towards the conductor....against the repulsive force of the positive charge (encapsulated in the field E....also recollect from Chapter 1 the visualization using *springs* that push like charges apart) on the conductor.... as the test unit charge gets nearer ...you would need to do more work to push a distance Δa towards the conductor against the repulsive force .....and nearer.....more work.....until you reach the conductor.

V will remain the same whichever way you approach because the field strength around the conductor has been adjusted by a trick of nature....thin from far away, like infinity, and thick from the grounded surface which is nearer!

The voltage V and the charge q are proportional, and q is

$$q = C.V \tag{2}$$

where C is the capacitance of the insulated conductor and is determined by the conductor's size and shape, and its placement relative to other conductors and ground.

The charged system stores an electrostatic energy of

$$W = \frac{1}{2}\ C.V^2 \tag{3}$$

which can be dissipated in a single discharge or current pulse. Figure C.2 shows a charged insulator B.

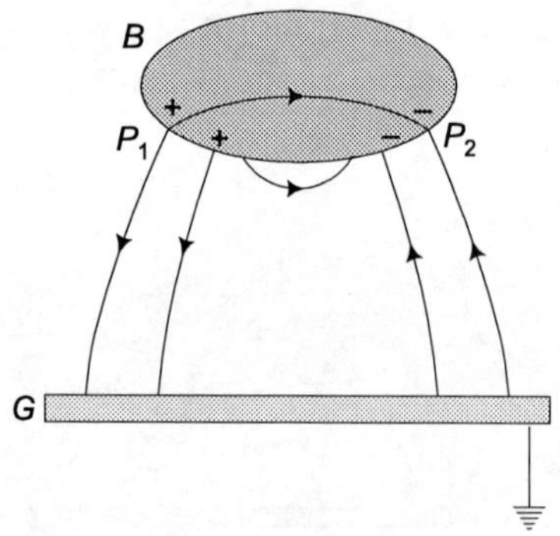

Fig. C.2 Charged insulator.

The field conditions here are very different from those at a charged conductor.

The polarity of the charge may be different from point to point, the field in the interior may be different from zero, the field is not necessarily perpendicular to the surface, and the integral of the field strength from a point on or in the insulator to ground is usually different from point to point.

In Fig. C.2, the integrals of the field strength for P_1 and P_2 are

$V_{P1} = \int_{P1}^{G} E \cdot da > 0$ and $V_{P2} = \int_{P2}^{G} E \cdot da < 0$ respectively which can be vastly different. V_{P1} and V_{P2} are the surface voltages (or surface potentials) of the two points.

In general, the surface voltage of an insulator will vary from point to point, as will the potential of any point in the interior.

It is therefore not possible to characterize a charged insulator with a potential of single value.

In other words, an insulator *does not have a single value of potential throughout its volume and surface.*

There are cases in which the surface of an insulator has a constant surface voltage.

But apart from such instances, rare as they are, there is only one situation in which all points in and on an insulator can be ascribed a well-defined (but unmeasurable) voltage. This is shown in Fig. C.3.

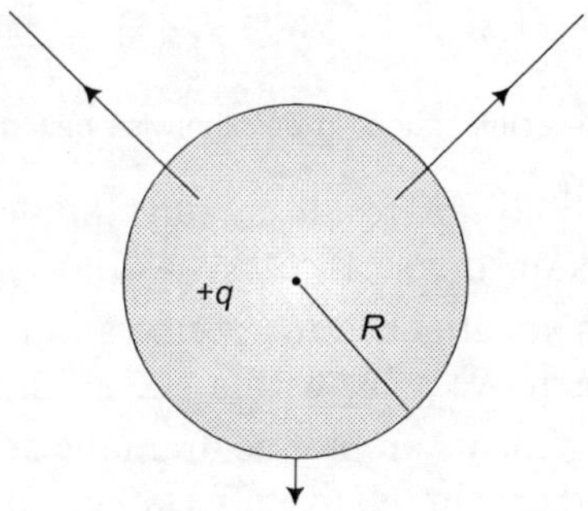

Fig. C.3 Uniformly charged, spherical insulator.

If a spherical insulator with radius R and uniform charge q as in Fig. C.3, is placed infinitely far (a distance much greater than R) from any conductors, the sphere would have a voltage of

$$V_{\text{sphere}} = \frac{q}{4\pi\varepsilon_o R} \tag{4}$$

However, this very theoretical situation is the only case in which it makes sense to talk about the voltage of an insulator. Similarly, the concept of an insulator's capacitance is meaningless.

Although it is possible to get a discharge from a charged insulator, the discharge will always be partial, and the energy dissipated can neither be related to the total charge nor be related to any kind of voltage.

In other words, voltage and capacitance are quantities of a conductor, not an insulator.

Dielectric in a Uniform Electric Field

When a sphere of a homogeneous linear dielectric material is placed in an otherwise uniform field E_o, the electric field flux lines will be as shown in Fig. C.4.

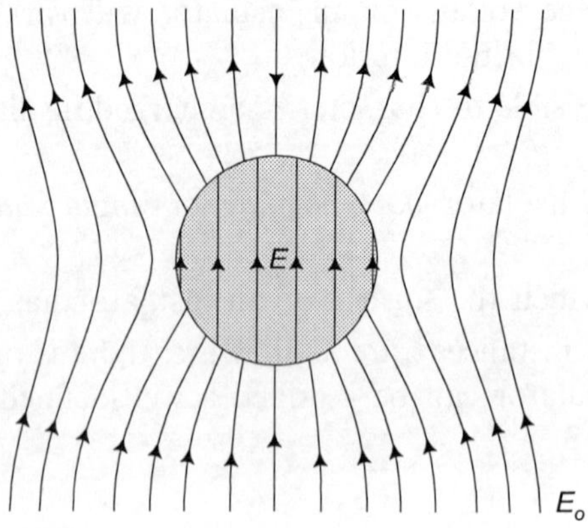

Fig. C.4 Dielectric sphere in an otherwise uniform electric field E_o.

Recollect how polarization can result in tiny dipoles (atomic or molecular) in the dielectric material as well as on the surface as was described in Section 1.14.

In dielectrics, all charges are attached to specific atoms or molecules—they're on a tight leash, and all they can do is move a bit within the atom or molecule.

Such microscopic displacements are not as dramatic as the large scale rearrangement of charge in a conductor, but their cumulative effects account for the characteristic behavior of dielectric materials.

The reason there is a field inside the sphere is due to the fact that the cancellation of the external field E is only partial.

Physicists study polarized material that is, an object containing a lot of microscopic diploes lined up using the concept of *bound* volume charges and surface charges.

The field of a polarized object is identical to the field that would be produced by a certain distribution of *bound* charges and this equivalence makes it easier to compute the fields both inside and outside objects that are regular and symmetrical in shape such as rectangular boxes and spheres. Also, refer Section 2.27.

You may have noticed that the field in Fig. C.4 is *uniform* in the interior of the sphere. The reader is referred to "Introduction to Electrodynamics" by David J. Griffiths for a calculation of the internal field

$$E = \frac{3}{\varepsilon_r + 2} E_o$$ where ε_r is the relative permittivity of the dielectric.

The effect of "cancellation" of the electric field within the dielectric is what makes manufacturers build capacitors of small dimensions with a dielectric between its plates that can store larger amounts of energy.

When the space between the plates of a capacitor is filled with a dielectric, the dielectric will reduce E and hence also the potential difference V.

Accordingly, the capacitance $C = Q/V$ is increased by a factor of the dielectric constant $C = \in_r C_{\text{vacuum}}$ where $\in_r$ is the relative permittivity of the dielectric and C_{vacuum} is the capacitance of the capacitor without the dielectric. Also refer Sections 1.14 and 2.26.

The formation of the dipoles being not so well-defined in the bulk in dielectrics, is one among the many reasons why it is difficult to characterize the no-load output voltage (without a filter capacitor) of a half-wave rectifier circuit made using a diode when subjected to an input sinusoidal voltage.

There is a transition region in the diode that is devoid of mobile charges, which when treated as a dielectric that is already polarized, (that caused the barrier potential as described in Section 9.4) makes the computation of the electric field and the potential quite difficult.

Conductor in a Uniform Electric Field

In sharp contrast to the effect of a dielectric placed in a uniform electric field, when an *uncharged* conducting sphere is placed in a uniform electric field, the field due to the induced charge (only surface, no volume charge) completely cancels E_o within the sphere.

This effect is a result of the movement of mobile charges in a conductor. This is shown in Fig. C.5. Also refer Sections 1.13 and 1.14.

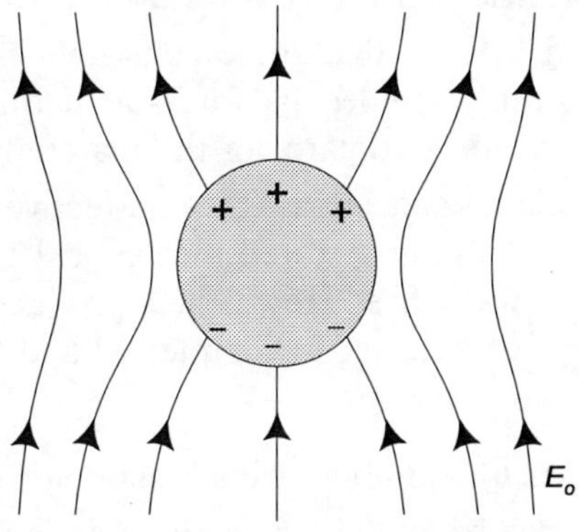

Fig. C.5 Uncharged conducting sphere is placed in an otherwise uniform field E_o.

What are insulators and how do they prevent electric shocks?

An **insulator**, also called a *dielectric*, is a material that resists the flow of electric current. An insulating material has atoms with *tightly* bonded valence electrons. These materials are used in parts of electrical equipment, also called *insulators* or *insulation*, intended to support or separate electrical conductors without passing current through themselves. The term insulator is also used more specifically to refer to insulating supports that attach electric power transmission wires to utility poles or pylons.

Some materials such as glass, paper or Teflon are very good electrical insulators. A much larger class of materials, for example rubber-like polymers and most plastics are still "good enough" to insulate electrical wiring and cables even though they may have lower bulk resistivity. These materials can serve as practical and safe insulators for low to moderate voltages (hundreds, or even thousands, of volts).

It is *not true* that an electrical insulator such as the one that is used on an electrician's tester, "blocks" the voltage from being felt by its user; in reality, the voltage on the insulated handle is difficult to calculate and measure and is rarely blocked entirely. It is an insulator's high resistivity to *current* (to less than a few thousandth of a milliamp), that enables it to prevent electric shocks to personnel, because it is the currents within the range of a fraction of a milliamp to a few milliamps that is the cause of electric shocks leading to fibrillation and fatalities.

A piece of wood, which is an insulator, in contact with a live wire, will have an internal and surface electric field profile (and voltages at various points in its interior and on its surface) that is *same*, whether *dry or wet*. A person who touches the surface of the wood will not receive a shock when the wood is dry, but may feel a shock when the wood is wet because of the availability of mobile charge carriers that cause currents in the wood and through his body.

Lossy Dielectrics

A lossy dielectric is a medium in which an electromagnetic wave loses power as it propagates due to poor conduction (see Ref. [32]). In other words, a lossy dielectric is a partially conducting medium (imperfect dielectric or imperfect conductor) with σ (conductivity) $\neq 0$, as distinct from a lossless dielectric (perfect or good dielectric) in which σ (conductivity) $= 0$.

The reader is cautioned to note that some texts define lossy dielectrics differently: A lossy dielectric exhibits loss due to molecular forces as if this is a spring with stiffness that the electric field has to overcome in polarizing the material.

Both the effects produce a loss of energy (the conductive loss being dissipative "kinetic" whereas the molecular spring loss is a stored mechanical type) and a consequent loss in power of the electromagnetic wave. See "How is heat produced in conductors and resistors" Appendix B, "Reradiation or Scattering" in Chapter 10 and "Energy in Dielectric Systems" below.

Why does a balloon charged by rubbing in one's hair not give a shock?

The voltage of the balloon may be quite large, perhaps several thousands of volts and would be several thousand joules of energy if the charge were 1 coulomb. But, the amount of charge that is transferred on to the balloon is typically less than a millionth of a coulomb. A high voltage means a lot of energy only if a lot of charge were involved. We also see from this an important difference between electric potential energy and electric potential.

Energy in Dielectric Systems

In Section 2.26, we had seen how an air-filled capacitor can store more energy if a dielectric is inserted between the plates. The electric field polarizes the dielectric, and the voltage source will need to pump up more charge with the dielectric in place, to attain the same voltage. But, note that the expression (26) we derived in Appendix A "Energy Stored in the Fields" $U = \frac{\varepsilon_o}{2}\int_{\substack{Entire\\Field}} E^2\, dv$ will not include the work involved in stretching and twisting the dielectric molecules. If we picture the positive and negative charges as held together by tiny springs (recollect the spring compress-release and stretch-release exercises that we did in Section 1.4 !), the "spring" energy or energy of the separated charges is given by $1/2kx^2$ (See "Chemical bonds can be modeled as springs", Appendix B and "Energy stored in stretched springs and unlike charges that are separated" in Appendix A).

In dielectrics, the energy loss due to the bound charge (see Section 2.27) as the dipoles twist and stretch, maybe indistinguishable from the loss due to free charge conduction that is quantified by σ (See "Lossy Dielectrics" above).

According to Griffiths Ref. [8], the energy in a dielectric given by $U = \frac{1}{2}\int_{\substack{Entire\\Field}} \vec{D}\cdot\vec{E}\,dv$ where, dv is an infinitesimally small volume in space around the charge configuration, will give us the total energy $W_{tot} = W_{free} + W_{bound} + W_{spring}$, where W_{free} is the electrostatic energy of the free charge, W_{bound} is the electrostatic energy of the bound charge and W_{spring} is the molecular "spring" energy. On the other hand, the expression $U = \frac{\varepsilon_o}{2}\int_{\substack{Entire\\Field}} E^2 dv$ would have given us the total energy stored in the fields as if the charge system was assembled together including the polarization of the molecules (to produce bound charge), leaving out the "spring" energy, the energy expended to overcome spring stiffness before the molecules reach the final equilibrium *polarized* positions. We did get an indication of several of these energy exchange processes in Section 4.2 equations (1) and (5).

For a more detailed treatment see in Ref. [8] Chapter 4, "Linear Dielectrics" and in Ref. [32], Chapter "Wave Propagation in Lossy Dielectrics".

Induced Charge Configurations

We had seen several examples of induced charges in Chapters 1 (Fig. 1.14), 2 (Fig. 2.9 and Fig. 2.63) and 3 (Fig. 3.40). The charge configurations were shown rather arbitrarily with no great degree of accuracy. Induced charge configurations vary with the geometry of conductor shapes, and are quite complex.

See Fig. C.1 in this Appendix. There are a few excess negative charge on the upper surface where the field lines terminate (the negative charges are assumed and not shown on the ground plane). If the ground plane were to be replaced by an uncharged (ungrounded) conducting disc of finite dimensions, then there will surely be positive charge on the lower side of the disk in addition to the negative charge on the upper surface, produced by charge separation.

To know the distribution of the positive charge, the reader is referred to the Fig. 3.10 in the topic "Some simple systems of conductors" in Chapter 3, Electric Field around Conductors (Ref. [22]).

Appendix D

DO WE REALLY TOUCH AN OBJECT

Before we discuss what really happens when we touch an object, let us understand the four fundamental forces in nature.

There are four known fundamental forces and these are listed below *in the order in which these occur*: beginning from the innermost regions of an atom (inside the nucleus) and moving outwards to encompass the entire universe.

1. **Weak Force:** This causes Beta decay (the conversion of a neutron to a proton, an electron and an antineutrino) and various particles (the "strange" ones) are formed by strong interactions but decay via weak interactions. The weak force **is a short range force** whose range is of the order of 10^{-16} centimeters. ***Not of much concern to electrical and electronics engineers**.
2. **The Strong Force:** This force binds neutrons and protons together in the cores of atoms. The *strong interaction* is very strong, and like the weak force, is **short-ranged**. It acts only over ranges of order 10^{-13} centimeters and is *responsible for holding the nuclei of atoms together*. It is basically attractive, but can be effectively repulsive in some circumstances. ***Not of much concern to electrical and electronic engineers**.
3. **Electromagnetic:** This acts between electrically charged particles or pieces of matter carrying an electrical charge. Electricity, magnetism, and light are all produced by this force and it has **infinite range**. **It can be attractive or repulsive,** and acts only between pieces of matter carrying electrical charge. ***of concern to scientists, mechanical, civil, chemical, metallurgical, electrical and electronic engineers**.

   This force can be separately studied in electrostatics and magnetostatics, but is always treated together as a *fundamental force*.

   Taken singly, the fields store energy: electric energy between the plates of a charged capacitor and magnetic energy in the field of a bar magnet or magnetic energy in the field surrounding a steady current carrying shorted *pure* inductor.

   Together, they transport energy in circuits and in radiation. For other theories on charged particle interactions without field concepts, see Ref. [36].
4. **Gravity:** This force acts between all mass in the universe and like the electromagnetic force it also has **infinite range**. ***of concern to all engineers and scientists**. The *gravitational force* is weak, but very long ranged. Furthermore, it is always attractive.

Fundamental forces are also known as fundamental *interactions*, and are the interactions in physical systems that don't appear to be reducible to more basic interactions. The fundamental interactions gravitational, electromagnetic, strong nuclear, and weak nuclear are characterized on the basis of the following four criteria: the types of particles that experience the force, the relative strength of the force, the range over which the force is effective, and the nature of the particles that mediate the force.

The weak and electromagnetic interactions have been unified under electroweak theory (Glashow, Weinberg, and Salaam were awarded the Nobel Prize for this in 1979).

Grand unification theories attempt to treat both strong and electroweak interactions under the same mathematical structure; attempts to include gravitation in this picture have not yet been successful.

Important Note: All the other forces are derived from these four fundamental forces; for example, frictional forces and reaction forces that prevent you from falling from your chair when you sit to read these notes are electromagnetic in origin!

You may use one type of force to overcome another! **Example:** An electromagnet uses electromagnetic forces to lift a car which is resting on the ground due to the force of gravity. Thank nature for permitting us to do that!

For a gentle introduction to new theories in the advanced topics of Quantum Electrodynamics and Quantum Field on the nature of these forces see video lectures here https://www.youtube.com/watch?v=YHTt6LGW9s0 (in three parts) and here https://www.youtube.com/watch?v=mBFDp-ppG6I.

[illegible] an object is touched? At the atomic level? What's in a Touch?

When two physical surfaces make contact, we say they are 'touching' each other. This contact may make it appear that two different surfaces *are* actually touching, but in fact, *no atoms of any one object can ever touch the atoms of another*![23, 24]

The man on ***the left*** is leaning on a wall, which his jacket is 'touching'. At least, it seems that way. But when you look very closely at the two surfaces 'in contact', and understand a little about what atoms are like, you discover that *no jacket atoms are touching any wall atoms at all!*

Here's a simplified explanation of why this is so:

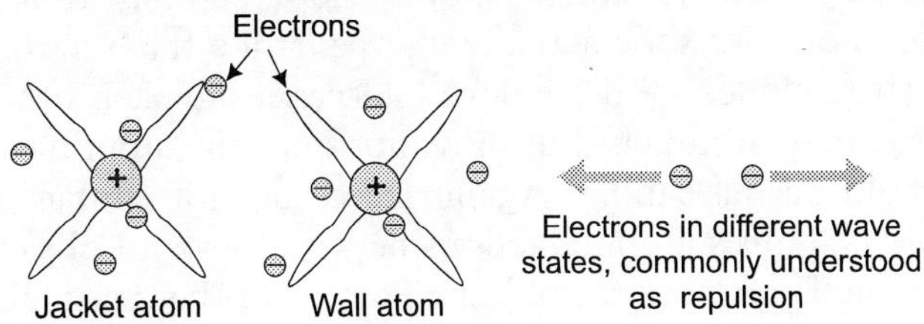

An atom could be described as a positively charged nucleus, surrounded by a cloud of negatively charged electrons. The *tangential velocity* keeps the negatively charged electrons in stable orbits, or actually orbitals, around the positively charged nucleus and prevents them from falling into the nucleus.

When two atoms are forced into 'contact' or interact closely (here, the jacket touching the wall), a force due to *electromagnetism*, keeps negatively charged electrons from actually making contact with each other. The only interaction is electromagnetic and weak force. There is NO repulsive force keeping them apart. It is the fact that the electron must be in a particular wave state that keeps them apart. The separation is a quantum mechanics effect. Electrons are point particles and cannot touch because they do not have surfaces. An infinitely small object has a surface area = 0. In reality the electron clouds penetrate each other and overlap, so in a sense electrons occupy the same space, but that is not the same thing as touching.

Push as hard as you want; you cannot make two electrons come into contact. The force you apply is about 20 newtons resulting in a stress (force/the area of contact) not even close to 2.3×10^{-8} newtons that is the force of attraction between the electron and the proton as we saw in Appendix B. The electron bound to one atom in "contact" with another atom has a negligibly small probability of being found at the location of the nucleus of the near atom. So the only thing actually 'in contact' is *the electromagnetic repulsive force between the electrons*, and there is always space between them! When the letter blocks are placed one above the other, electrons in the contact surface atoms of the upper and lower blocks overlap such that the 'normal' force of the lower block counters (by repulsive interactions of overlapping electrons) the force (say, gravitational or hand pressure on the upper block) exerted by the upper block.

It's as if you are pushing with similar poles of two very strong bar magnets; you can push one around with the other, but can never force them together.

Even standing on the floor, you are really supported by a very thin layer of electromagnetic force; no atoms of your foot actually touch atoms of the floor! The electron clouds of your shoe actually interpenetrate the electron clouds of the floor atoms, and the electron clouds of the blocks interpenetrate each other (overlapping orbitals), so one might say they do touch. But in reality since the atom has no definite surface the idea of touch could be misleading. Again, forces do not contact each other! Forces are not physical objects. Forces are interactions between physical objects and so can be considered as pushes or pulls on physical objects. Forces can be considered to add together and are independent of each other.

But wait a minute ... what about when you cut something!?!!

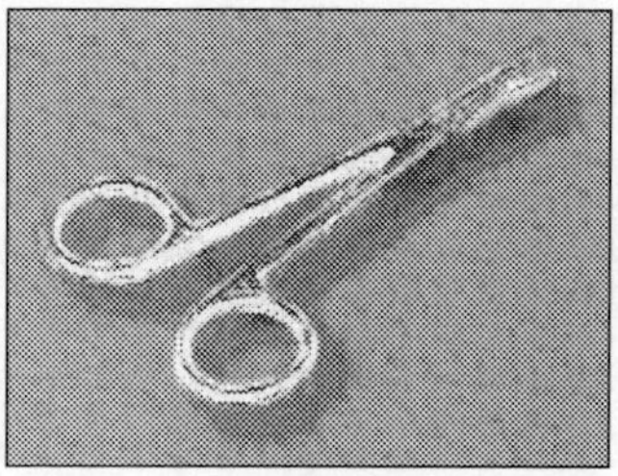

Don't the scissor blades have to touch what they're cutting, in order to cut?

No, they don't. Atoms can't touch, remember? They just push against each other with an electromagnetic force. That's what happens with the scissors:

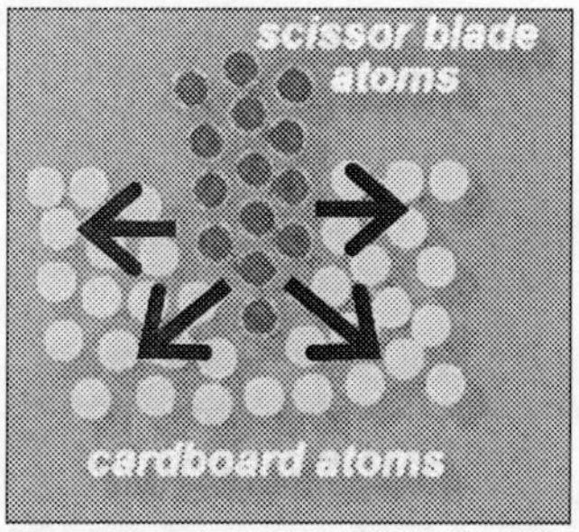

Even the sharpest knife doesn't 'cut' ... it pushes apart the atoms of the material it's being used on. The atoms of the knife or scissors blade push electromagnetically against the atoms of cardboard (see the figure above), spreading them apart. Even when the blade is withdrawn, and the material rejoins (as with flesh cut by a scalpel), the sealing is because of the positive and negative attraction between atoms and molecules with net charges (chemistry!) ... but this is not strong enough to force electrons together, just as it wasn't in the original uncut material.

How do we feel things if we are not touching them at the atomic level?

You may be wondering how you can *feel* things, if you don't actually make contact with anything. For example, when you pick something up in your fingers, you can certainly 'feel' the contact!

What's happening? The caveman certainly felt the club used by the cavewoman when it made contact with his head! Let's examine what actually took place when that happened.

Atoms of the club pushed against atoms of his head, and probably moved them around a lot (electromagnetically)! The nerve endings sense the compression of interatomic bonds in the skin on his head when the club hit it Ref. [40].

Buried in the skin are nerve cells or (sensory) neurons that gather and transmit electrochemical signals. Neurons have axons which are long, thin, cable-like projections of the cell that carry electrochemical messages (nerve impulses or action potentials of the order of a few tens of millivolts) along the length of the cell. The nerve endings are dendrites which by a method of changing action potentials carry messages between the billions of (receptor) neuron dendrites in our bodies, eventually reaching the brain, where other cells interpret this electrical signal as 'pain'. At no time did any atoms or electrons actually *touch* each other; the only 'contact' was the electromagnetic force (positive and negative) between particles!

Friction is the effect of atoms pushing against each other. On a smooth surface, with few bumps, there is little friction. Most of the atoms can slide right by each other without actually pushing *against* each other. But, no surface is totally free of friction. Watch the man skiing on snow.

Man skiing on snow

If you examine any surface under a powerful enough microscope, no matter how smooth it appears to the naked eye, there will be bumps. This will cause atoms to come face-to-face.

If the forces between one set of atoms (snow, in this case), are weak, those atoms get pushed out of the way (by the ski atoms, in this case). But this uses up energy. It's called 'friction'.

If the atoms are joined by strong forces, as in rock, then they *can't* easily be pushed away. The forces between these atoms and your ski atoms, supposing that you were skiing on rock, might be enough to bring you to a stop!

In either case, it's atoms pushing against each other, at a distance, (electromagnetically), that causes friction. Generally, the energy "dissipated" by friction remains as molecular motion or what is called heat.

When we hit an object, the physical force felt would be the normal force, this force is an electromagnetic force, the electrons in the object repelling the electrons in the hand.

When a batsman hits a cricket ball, the force involved is electromagnetic; the electrons in the bat repelling the electrons in the ball.

The idea of touch appears to be so simple using classical mechanics concepts but is the quantum mechanical concept consistent with the classical descriptions?

While it is true that theoretically electrons can not touch because of the repulsion, and of course the nuclei do not touch each other, the surfaces of the atoms (defined in some mathematical fashion) can touch and actually penetrate each other. Forces are interactions between two objects.

All atoms are not solid and have no definite surfaces. One can make a mathematical definition of where the surface is located, but that is subject to interpretation. Actually when you touch something, bonds form between the two surfaces so in a sense sub-microscopically the two surfaces penetrate each other slightly. So sub-microscopically there is no gap.[24]

Remember that outside an atom there is no *perceptible* static electrical field because the forces due to the protons and electrons are balanced on any nearby object. But if you think about it classically there is only a repulsive force if the "edges" of the atoms overlap each other.

The model that you use for physics depends on the scale; at a macroscopic level when objects touch, use classical mechanics but sub-microscopically use quantum mechanics and here classical concepts such as touch can no longer be applied.

Note

The electrons in an atom have a probability distribution which is non-zero at the nucleus, that is, there is a nonzero (though small) probability of finding an atomic electron inside the nucleus of the atom.

It is possible to make electrons interact with the protons in a nucleus. This is accomplished artificially by making them interact due to their high momentum and energy at relativistic speeds in a cyclotron.

One of the things that can happen when the electron hits a proton is that they can combine to create a neutron. This process also occurs in a neutron star, where the intense gravitational pressure collapses all the atoms of the star, so that all the electrons combine with all the nuclear protons, forming a solid soup of neutrons.

Appendix E

THE P-N JUNCTION DIODE CURRENT

The expression for *p-n* junction diode current is

$$I = I_o \left(e^{\frac{V}{V_T}} - 1 \right) \quad \text{where,}$$

$$I_o \equiv \frac{AqD_p p_{no}}{L_p} + \frac{AqD_n n_{po}}{L_n} \text{ for an applied voltage V.}$$

In the expression for I_o, '*A*' is the area of cross-section of the junction, D_p (D_n) is the diffusion constant for holes (electrons), L_p (L_n) is the diffusion length for holes (electrons) and p_{no} (n_{po}) is the thermal equilibrium concentration of holes (electrons) in the *n*-type (*p*-type) material.

The above expression for current *I*, being different from the Ohm's law expression $J = \sigma E$ and $V = IR$, which expresses Ohm's law in terms of the voltage and current in a conductor of resistance *R* with a voltage across it, is a source of confusion to several students.

It would be useful to remember that in a simple circuit comprising a battery and a series of say, resistors, it is the total resistance offered that determines the current, neglecting drops in connecting wires whose resistance is assumed to be very small. In an unbiased *p-n* junction diode, there is a thin depletion or space-charge region with few or no *mobile* charges causing the diode end-to-end resistance to be large. To get the diode to conduct by applying a forward bias with a battery and connecting wires, mobile carriers from the neutral bulk regions should somehow *bridge* this thin depletion region and it should be appreciated that the processes which will be described here for this region, will be the significant factor that will decide the overall forward current. Among the assumptions to derive the current, we neglect the resistance of the bulk neutral *p*- and *n*-regions considered to be very small and also voltage drops in connecting wires.

Further, the assumptions made in the derivation are (i) the depletion-layer thickness is negligible (ii) there is no carrier generation and recombination in the depletion region.[17]

To gain a proper perspective of the *p-n* junction diode current-voltage relation, it is useful to view the 3 different processes viz. diffusion, recombination, and the influence of an electric field on the energies of electrons. Figure E.1 shows the concentration levels of holes at a distance x from the junction into the n-type material of a forward-biased p-n junction.[17]

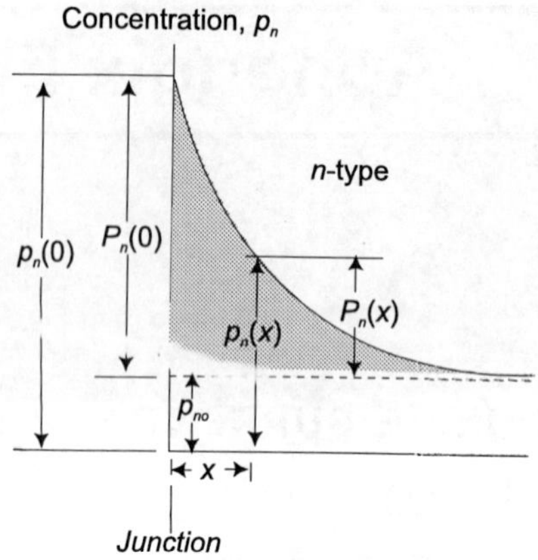

Fig. E.1 Defining the several components of hole concentration in the **n** side of a forward-biased p-n ju[illegible] [illegible]gram is not to scale since $p_n(0) >> p_{no}$.

[illegible] Diffusion Process

The diffusion hole current is derived considering the *first process of diffusion,* when an arbitrary bunch of injected holes $P_n(0)$ at $x = 0$ moves into regions where there are fewer holes (in the n-type material) as a result of a *thermodynamic statistical* phenomenon. When a closed perfume bottle is opened in one corner of a closed room, the scent is [illegible] throughout the room. If there is no convection or other net motion of the air, the scent spreads, by diffusion. Similarly, carriers in a semiconductor diffuse in a carrier gradient by random thermal motion and scattering from the lattice and impurities.

The hole current is *not* given by $I = qnAv$ where 'n' is the number of holes, 'A' the area of cross-section and 'v' is the drift velocity. Instead it is given by

$$J = -qD_p \frac{dp(x)}{dx} \tag{1}$$

where, $\frac{dp}{dx}$ is the concentration gradient along 'x' meters, p is the concentration of holes per m^3, D_p is the diffusion constant in m^2/second and q is the electronic charge. Thus, J is in amperes per m^2.

The fewer the holes to the right of the junction, when injection is initiated whatever the means of injection, more is the depth of the penetration by the holes into the n-type region.

Since, D_p in (1) is expressed in square meters per second, a time rate of change gives the expression for 'J' in (1) the units of current density.

2. The Recombination Process

From the continuity equation, the concentration gradient $\frac{dp}{dx}$ is evaluated, in which a factor $e^{-\frac{x}{L_p}}$, determined by the *second process of recombination* of holes with the electrons, appears. This accounts for the variation in concentration of holes along the x-axis in the n-material leading to an expression

$$P_n(x) = P_n(0)\, e^{-\frac{x}{L_p}} \tag{2}$$

Note how $P_n(x)$ has an exponential (decreasing) of the fraction of distance (from $x = 0$) to the diffusion length L_p which is the distance into the n-type semiconductor at which the injected concentration falls to $1/e$ of its value at $x = 0$.

This accounts for the exponentially decreasing hole concentration curve in Fig. E.1 showing the various hole concentrations.

Taking the derivative of the hole concentration $P_n(x)$ expression (2) and substituting in (1), we obtain the expression for the hole current (multiplying by 'A' the area of cross-section of the semiconductor) as

$$I_{pn} = \frac{AqD_pP_n(0)}{L_p}\, e^{-\frac{x}{L_p}} \tag{3}$$

Note from Eq. (3) that the current component is not constant with distance, as in a uniform conductor, but varies with distance.

Also, the expression does not include the electric field or a drift velocity of holes but includes the *hole concentration* $P_n(0)$, *the Diffusion constant* D_p in square meters per second and *a divisor* L_p *in meters.*

3. The Third Process of the Influence of the Electric Field on the Energies of Electrons

The key difference between a metal and a semiconductor lies in n, the number of available charge carriers (say, conduction band electrons) per unit volume, and the slightest potential difference produces some current. In a metal there are always lots of mobile charges, because the energy band is only partially filled (See Appendix B "Density of States") and changing the temperature hardly changes the number density of mobile charges (conduction band electrons). In a partially filled band, we can add a small amount of energy to electrons near the top of the occupied levels, moving them to nearby unoccupied levels[31] (also see Section 9.15 "Fermi level in metals). No more than two electrons can occupy one electronic state (the two electrons must differ in their spin directions in order to obey the Pauli exclusion principle).

Such behavior means that at least some of the electrons in a metal are truly free, not bound to the atoms. If they were bound we would have to apply some field to free some up, and then more field to increase the current. This explains the large conductivities of metals. See the table below of the resistance of a copper conductor of length 0.5 cm and area of cross-section 10^{-4} cm^2. In a metal, n is a fixed number (one mobile electron or hole per lattice ion or atomic core).

In a semiconductor, at very low temperatures there are no mobile electrons or holes and the material acts nearly like an insulator for small applied fields. However, the electrons are not tightly bound to the atomic cores and it is possible to free up some electrons by applying a large enough electric field, so n is *variable*, and increases with increasing potential difference. In the table below is shown the resistivity and resistance of silicon and of n-type silicon for samples of length 0.5 cm and area 10^{-4} cm^2. Among Copper, Silicon and n-type Silicon, Copper has the highest conductivity (lowest resistivity) and n-type Silicon has medium resistivity and yet its conductivity is still quite large. Silicon has a high resistivity and a lower conductance compared to n-type Silicon, as can be expected.

Table of Resistivity and Resistance of Copper, Silicon and n-type Silicon samples of equal dimensions l = 0.5 cm and area $10^{-4}cm^2$

Material	Resistivity 'ρ'	Resistance = $\rho \dfrac{l}{A}$ ohms	Remarks
Copper	10^{-6} Ω-cm	0.005 ohms	Low resistance; High conductance
n-type Silicon	0.1 Ω-cm	500 ohms	Medium resistance; high conductance
Silicon	2.5×10^5 Ω-cm	1250 x 10^6 ohms	High resistance; low conductance

Raising the temperature can also free up some electrons (or holes), and the effect of increasing n (number of available charge carriers) more than compensates for the higher rate of collisions with the atomic cores, so that for a semiconductor, the resistance decreases with increasing temperature.[31]

If an electric field was to be applied to either of the three samples (isolated), whether copper, silicon or n-type silicon, internal currents will cause carriers to move and the net field inside will be made 'zero' nearly instantly. See Section 1.13 "The Electric Field inside a metal" and Section 1.14 "Electric Fields in matter". The fact that the resistivity is vastly different for the three materials does not restrict their ability to be polarized in the presence of the external field and to make the net electric field inside the samples to be zero.

As a matter of fact, it is this process which keeps the electric field zero in the bulk regions of an open-circuited p-n junction (Section 9.5) and indicated in Fig. 9.12 by the tiny amount of excess charges on the surfaces at the edges of the p-n junction diode.

The injected hole concentration $P_n(0)$ at the n-side which was taken arbitrarily in the beginning should now be replaced by its actual value that will take into account the donor concentration in the n-side.

To determine the value $P_n(0)$ we need to first establish the Einstein relation for V_T (the thermal voltage), which is present in the p-n junction diode current expression.

When discussing the motion of carriers of a semiconductor (electrons in the conduction band and holes in the valence band) in an electric field, the *third process of its influence on the energies of electrons* in the band diagrams should be considered. Figure E.2(a) shows the energy band structure of an intrinsic piece of semiconductor where E_c is the lowest energy level in the conduction band, E_v is the uppermost energy level in the valence band and E_i is the intrinsic energy level.

The change in potential energy of electrons in an electric field is shown in Fig. E.2(b) and since electrons drift in a direction opposite to the field, we expect the potential energy for electrons to increase in the direction of the field, as we have depicted in Fig. E.2(b).[30]

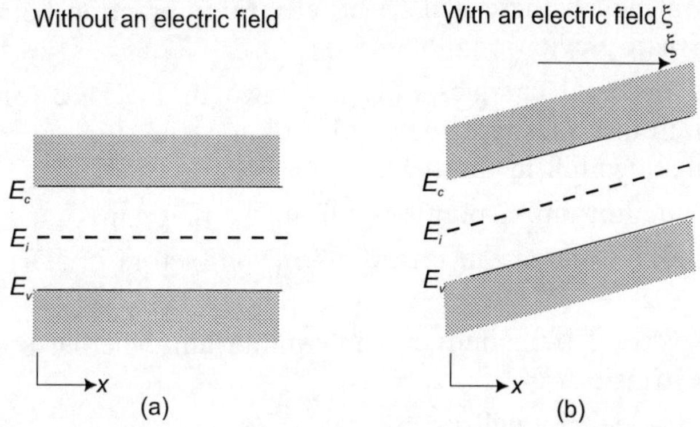

Fig. E.2 Energy band diagram of a semiconductor in an electric field.

In an intrinsic semiconductor (*no* trivalent or pentavalent impurities added), the E_F level lies near the middle of the band gap and this level is designated E_i.

We have not commented on the source of the electric field, as yet.

Assuming an electric field $\xi(x)$ in the x-direction, the energy bands are drawn to show the change in potential energy for electrons in the direction of the field. The electrostatic potential $V(x)$ varies in the opposite direction and is related to the electron potential energy $E(x)$ ($\xi(x)$ is the electric field) depicted in the figure, by $V(x) = \dfrac{E(x)}{-q}$.

From the definition of electric field,

$$\xi(x) = -\frac{dV(x)}{dx} \tag{4}$$

The electric field $\xi(x)$ can be related to the electron potential energy in the band diagram by choosing some reference in the energy band for the electrostatic potential. The spatial variation in $V(x)$ for Eq. (4) is of interest[30] indicative of the electron energy variations caused by separation of electrons from their lattice ions, and the effect of the electric field is most on the valence and conduction band electrons.

Choosing E_i as a convenient reference, the electric field can be related to this reference by

$$\xi(x) = -\frac{dV(x)}{dx} = -\frac{d}{dx}\left[\frac{E_i}{-q}\right] = \frac{1}{q}\frac{dE_i}{dx} \tag{5}$$

There is a nice description with animation of electric potential energy and electric potential in the webpage https://kaiserscience.wordpress.com/physics/electromagnetism/

electric-fields-and-potential/. The reader should remember that the energies discussed here are in a crystal with its millions and millions of lattice ions and millions and millions of valence electrons and millions and millions of conduction band electrons that occupy allowable energy levels. Also, electrons in the real world do not *stick* to protons as shown in the animation in the article which is written with a view to explain the concepts of potential energy and potential. And, the reader should also visualize the electric field surrounding the charges without which there will be no electric interaction between the electrons and the proton other than gravitational!

The variation of band energies $E(x)$ as drawn in Fig. E.2 (on the right) is correct, because the diagram indicates electrons and we know the slope in the bands must be such that electrons drift downhill in the field.

Therefore, at equilibrium, ξ points uphill in the diagram.[30]

We will see how E_f can *vary* in intrinsic semiconductors in "Origins of built-in electric field" later.

Consider the piece of semiconductor in equilibrium, whether pure (intrinsic) or doped with impurities (extrinsic).

The total current density will be

$$J_p(x) = \underbrace{q\mu_p\, p(x)\,\xi(x)}_{\text{Drift}} - \underbrace{qD_p \frac{dp(x)}{dx}}_{\text{Diffusion}} \tag{6}$$

where, μ_p is the mobility of holes. The negative sign for diffusion is because the net motion of holes due to diffusion is in the direction of decreasing hole concentration.

An **important** result of expression (6) is that minority carriers (holes in a *n*-type semiconductor or electrons in a *p*-type semiconductor) can contribute significantly to the current through *diffusion*. Since the drift term in (6) is proportional to carrier concentration, minority carriers seldom provide much drift current. On the other hand, diffusion current is proportional to the gradient of concentration.[30] This is a point to be noted because as we shall see later it is the sharp *gradient* in concentration across a *p*-*n* junction with its large ξ that will decide the overall quantum of current dose in a forward-biased diode.

At equilibrium, no net current flows in a semiconductor. Diffusion, is a statistical thermodynamic process, which triggers a charge separation exposing immobile charges, and drift is due to the presence of the built-in electric field due to the charge separation, which exerts a restraining tug on the diffusing charges. The reader is also advised to review the notes on "Separation of charges" at the end of Section 1.10 and visualize the growth of the field, when unlike charges are separated. The net current being zero, we can set expression (6) for J_p equal to zero for an open-circuited semiconductor in equilibrium, to get

$$\xi(x) = \frac{D_p}{\mu_p}\frac{1}{p(x)}\frac{dp(x)}{dx} \tag{7}$$

The concentration of holes is obtained by multiplying the effective density of states by the probability of *vacancy* state for holes.[30]

The intrinsic hole concentration p_i may be written

$$p_i = N_v e^{-\frac{(E_i - E_V)}{kT}} \tag{8}$$

where, N_v is the density of available number of states per electron volt per cubic meter, *not* acceptor concentration (See Appendix B "Density of States"), and $e^{-\frac{(E_i - E_V)}{kT}}$ is the probability of vacancy of a state for holes between the intrinsic energy level and the edge of the valence band.[30]

The thermal equilibrium concentration, p_o is given by

$$p_o = N_v e^{-\frac{(E_F - E_V)}{kT}} \tag{9}$$

where E_F is the Fermi energy level. Eq. (9) is valid whether the material is intrinsic or doped *near the middle of the band gap.*

Therefore, for a material with impurities the thermal equilibrium concentration of holes may be written

$$\begin{aligned} p_o &= p_i e^{\frac{(E_i - E_F)}{kT}} \\ &= n_i e^{\frac{(E_i - E_F)}{kT}} \end{aligned} \tag{10}$$

because the intrinsic hole and electron concentrations are equal.

Substituting $p(x)$ in (7) with p_o, we obtain

$$\xi(x) = \frac{D_p}{\mu_p} \frac{1}{n_i e^{\frac{(E_i - E_F)}{kT}}} \cdot n_i e^{\frac{(E_i - E_F)}{kT}} \cdot \frac{1}{kT}\left[\frac{dE_i}{dx} - \frac{dE_F}{dx}\right]$$

The equilibrium Fermi level does not vary with 'x' so $\dfrac{dE_F}{dx} = 0.$

$$\therefore \qquad \xi(x) = \frac{D_p}{\mu_p} \cdot \frac{1}{kT}\left[\frac{dE_i}{dx}\right] \tag{11}$$

Setting this equal to $\xi(x)$ from (5), we obtain

$$\frac{D_p}{\mu_p} = \frac{kT}{q} = V_T \tag{12}$$

which is Einstein's relation for the thermal voltage V_T.[30] V_T is a constant and is precisely 25.9 mV at 300 K room temperature. See the end of Appendix B on the Einstein Relation.

The key processes when a semiconductor, is in equilibrium with no applied voltage are

(i) when thermal fluctuations begin a diffusion current, an electric field is also set up being the result of separation of electrons from the lattice ions and their potential energy varies, giving atoms a more positive potential quality (less negative) with the electrons moving "downhill" in Fig. E.2(b), and

(ii) this electric field in turn, redistributes the carriers (electrons) by drift.

Origins of Built-in Electric Field

An **important** result of the *balance* of drift and diffusion at equilibrium (E_F constant) in semiconductors is that built-in fields *accompany* gradients in E_i (see Eq. (5)).[30] Such gradients in the bands can arise when the band gap varies due to changes in alloy composition. An example of an alloy used in semiconductor device fabrication is AlGaAs (Aluminium Gallium Arsenic).

More commonly, built-in fields result from doping gradients (built-in fields due to this factor are more common due to doping inaccuracies during manufacture). For example, a donor distribution variation with distance causes a gradient in n_o, the thermal equilibrium concentration of electrons in *n*-type material, which must be balanced by a built-in electric field $\xi(x)$.

See Ref. [30] for the field strength ξ calculation of an intrinsic silicon sample with an exponential impurity distribution at equilibrium, in which ξ is found to be 259V/cm. Ref [18] shows a calculation of ξ for [illegible] abrupt *p-n* junction [illegible] (no applied voltage) to be 40 kV/cm. While such large field strengths look alarming, when traversed over minute distances ([illegible]icrometers) across a *p-n* junction, the voltage (potential barrier) will be a mere fraction of a [illegible] volt.

In metals such as copper and silver, the existence of an electron sea where delocalized electrons roam freely in the crystal lattice precludes the existence of concentration gradients.

In *p-n* junctions, where hole (electron) concentrations in the *p*(*n*)-type region are greater than that of holes(electrons) in the *n*(*p*)-type region, diffusion causes carrier migration to produce *built-in* electric fields, giving rise to the built-in barrier potential V

Computing $P_n(0)$

In the next step towards obtaining an expression for the excess injected holes $P_n(0)$ [illegible] forward-biased *p-n* junction diode, we combine Eqs. (5), (7) and (12) to [illegible]

$$-\frac{q}{kT}\frac{dV(x)}{dx} = \frac{1}{p(x)}\frac{dp(x)}{dx} \tag{13}$$

Equation (13) can be solved by integration over the potential on either side of the junction V_p and V_n and the hole concentration at the edges of the transition region, p_p and p_n. Integration gives

$$\frac{-q}{kT}\int_{V_p}^{V_n} dV = \int_{p_p}^{p_n}\frac{1}{p}\,dp$$

$$-\frac{q}{kT}(V_n - V_p) = lnp_n - lnp_p = ln\frac{p_n}{p_p} \tag{14}$$

The potential $V_n - V_p$ is the contact potential (barrier potential) V_o and is simply

$$V_o = \frac{kT}{q}\,ln\,\frac{p_p}{p_n} \tag{15}$$

Written in the exponent form

$$p_p = p_n e^{\frac{qV_o}{kT}} \tag{16}$$

and for an open-circuited junction at thermal equilibrium,

$$p_{po} = p_{no} e^{\frac{qV_o}{kT}} = p_{no} e^{\frac{V_o}{V_T}} \tag{17}$$

where p_{po} and p_{no} are the hole thermal equilibrium concentrations in the p and n neutral regions respectively.

Note: Some authors refer to expression (16) as the Boltzmann relationship of kinetic gas theory.[17] There are several types of distributions and the commonly used ones are normal distributions (Gaussian or bell shaped curve) applied for example to the grades scored by students in a test, Maxwell-Boltzmann distribution for gas particles, and Bose-Einstein for photons. In the topic "Quantum Mechanics, Probabilities And Orbitals of Electrons" in Appendix B, a qualitative description of the need for statistical approaches to analyze the Energies and their distribution of large numbers of particles, is provided. In a Boltzmann distribution of gas particles (neutral), the number n_2 of gaseous atoms in state E_2 is related to the number n_1 in E_1 (with $E_2 > E_1$) at thermal equilibrium by $\frac{n_2}{n_1} = e^{-(E_2 - E_1)/kT}$ assuming the two levels have an equal number of states. The exponential term $e^{-(E_2 - E_1)/kT}$ is commonly called the Boltzmann factor, in which k is Boltzmann's constant. Electrons in solids obey Fermi-Dirac statistics developed in late 1920's (Maxwell-Boltzmann statistics applicable in the classical regime cannot be applied to electrons because of the limit imposed by the Heisenberg uncertainty principle for a particle's position and momentum), and this exponential appears in the denominator of the Fermi-Dirac distribution function $f(E) = \frac{1}{1 + e^{(E - E_F)/kT}}$. The expressions (8), (9), (10) and (16) of hole concentrations at energy levels E_i, E_V and E_F of an open circuited junction at thermal equilibrium bear a similarity to the Boltzmann relation of gaseous atoms in states E_2 and E_1 indicated above, because $(E - E_F)$ is most of the time vastly greater than kT and so, the Fermi-Dirac distribution function $f(E)$ can be approximated by $f(E) \approx e^{-(E - E_F)/kT}$. Authors therefore, sometimes refer to the expression (16) as the Boltzmann relation of kinetic gas theory.[30]

Forward Biasing the *p-n* Junction

If the junction is forward biased by an applied voltage V, surface charges develop around the circuit and the dynamic equilibrium that was established without an applied voltage (open-circuit) is disturbed by the forward bias voltage. The barrier potential is reduced like the compressed spring in the check valve analogy in Chapter 9, and the electric field strength in the junction is reduced (it's action explained by the principle of superposition of potential, Section 3.1 "Superposition of Potential"), the transition region width decreases

(See Section 9.6), and a new equilibrium needs to be established in the steady-state to favor a large diffusion component of current. The drift component is I_o and seems very tiny and stays that way.

With the forward bias, holes in the *p*-side and electrons in the *n*-side gain sufficient *potential* energy *and kinetic* energy from the electric field (now, a resultant of the applied and the built-in) and fill available states to cross over the lowered barrier. See Figs. 9.23 to 9.28.

The equilibrium ratio of hole concentration on each side of the junction given by expression (the Boltzmann relation, Eq. 16) becomes, with a bias voltage V, when (16) is *still* applicable for *low-level injection* (to be explained shortly)

$$p_{po} = p_n(0)\, e^{\frac{(V_o - V)}{V_T}} \tag{18}$$

where, p_{po} is the thermal equilibrium hole concentration throughout the *p*-region, and $p_n(0)$ is the hole concentration at the right-hand edge of the depletion region as shown in Fig. E.1.

Combining this with (17), we get

$$p_n(0) = p_{no}\, e^{\frac{V}{V_T}} \tag{19}$$

In some textbooks[17], this is called the law of the junction.

Since, $P_n(0) = p_n(0) - p_{no}$, Eq. (19) can be written

$$P_n(0) = p_{no}\left[e^{\frac{V}{V_T}} - 1\right] \tag{20}$$

The injected hole concentration $P_n(0)$ into the *n* side at the junction is the factor $e^{\frac{V}{V_T}}$ minus 1 times the equilibrium concentration of holes on the *n*-side. It shows how the concentration $P_n(0)$ injected *adjusts* itself to suit the available concentration of holes in the *n*-type region and the level of forward bias voltage V.

It is instructive to calculate the injected hole concentrations for a *p-n* junction diode with different applied forward voltages. With the impurity concentrations $N_A = 2.5 \times 10^{15}$ cm^{-3} and $N_D = 5 \times 10^{16}$ cm^{-3} at 300°K, the intrinsic concentration $n_i = 1.5 \times 10^{10}$ (see the numerical in Section 9.8), we note that

$$p_{no} = \frac{n_i^2}{N_D} = \mathbf{4500}\ \text{cm}^{-3}, \text{ a very small concentration.}$$

For a forward bias voltage of **0.3 Volts**, the injected hole (excess) concentration $P_n(0)$ using Eq. (20) is $\mathbf{4827 \times 10^4}$ cm^{-3}.

For a forward bias voltage of **0.6 Volts**, the injected hole (excess) concentration $P_n(0)$ using Eq. (20) is $\mathbf{5.177 \times 10^{13}}$ cm^{-3}.

Note the large increase in excess hole concentration when the bias voltage was increased by a mere 0.3 volts to 0.6 volts.

And thus, the depletion region is "bridged" by the injected holes (electrons) from the *p*(*n*) sides by the forward bias, its resistance is lowered, and diode conduction is established.

The injection is termed *low-level injection,* because the injected hole concentrations ranging 4827×10^4 cm^{-3} – 5.177×10^{13} cm^{-3} are very small in comparison with the donor concentration in the *n*-side (N_D) 5×10^{16} cm^{-3}.

Substituting P_n (0) from (20) in (3), the hole current I_{pn} (0) is

$$I_{pn}(0) = \frac{AqD_p p_{no}}{L_p}\left(e^{\frac{V}{V_T}} - 1\right) \tag{21}$$

with x set equal to zero. The conservation of current principle enabled us to make this substitution and drop the $e^{-\frac{x}{L_p}}$ term since it does not matter where, along x, we compute the current.

But, the diffusion current *does* vary with distance due to the process of recombination as is evident from the hole concentration graph Fig. E.1 and Eq. (3), and other components of current (which we will not deal with here and are available in several textbooks) in the junction, combine to make the current uniform in the semiconductor in conformance with the law of current conservation. Thus, we could drop the x from the expression of the injected (excess) hole concentration.

The electron current I_{np} (0) crossing the junction into the *p*-side is obtained from Eq. (21) by interchanging n or p, or

$$I_{np}(0) = \frac{AqD_n n_{po}}{L_n}\left(e^{\frac{V}{V_T}} - 1\right) \tag{22}$$

where, L_n is the diffusion length for electrons.

The total diode current I is the sum of I_{pn} (0) and I_{np} (0), or

$$I = I_o\left(e^{\frac{V}{V_T}} - 1\right) \tag{23}$$

where
$$I_o \equiv \frac{AqD_p p_{no}}{L_p} + \frac{AqD_n n_{po}}{L_n} \tag{24}$$

Summarizing, some of the key processes that are involved in *p-n* junction currents are the process of diffusion, the process of recombination of electrons and holes that diffused across the junction, and the effects of the electric fields built-in due to gradients in energy bands on electron potential energies, alongwith the diffusion of carriers as the *p-n* junction is formed and also when the junction is forward biased. All these processes harmoniously combine (adjusting themselves all the time) to ensure that the steady-state conservation of current principle is upheld.

In a forward bias, holes are injected into the *n*-side bridging the depletion region, electrons are injected into the *p*-side bridging the depletion region. Under constant thermal equilibrium conditions, the holes and electrons move on and recombine in, near and around the depletion region. The stream of holes in the *p*-region and stream of electrons everywhere in the crystal and in the connecting wires move on and on, recombining around the junction, and the processes continue for the duration of the forward bias. This is the forward current of the diode.

The expression for the *p-n* junction diode current

$$I = I_o \left(e^{\frac{V}{V_T}} - 1\right) \tag{25}$$

indicates that the *p-n* junction simply absorbs an applied forward voltage to lower the potential barrier and the current is decided mostly by the diffusion component of current that crosses a junction subtracting out the tiny ever present (because of p_{po} and p_{no} in (24)) "drift" reverse saturation current for carriers swept by the electric field in the junction as can be seen in the $-I_o$ term when the right-hand side of expression (25) is multiplied out.

The current expression includes in I_o, the constant D_p, which is expressed in the units square meters per second and a divisor L_p in meters giving a coulombs per second "flow"; there is no mobility 'μ' appearing explicitly in the expression but is implicit in V_T which is due to the built-in electric field. And this field, exerts a "pull" on the separated charges of diffusion.

Again, since the *resistance* of the depletion region is changed by the injected dose of holes and electrons (see Eq. (20)), we cannot expect the Ohm's law relations $J = \sigma E$ and $V = IR$ to apply for varying bias voltages, since the Ohm's law expressions assume a constant R and σ for the device, such as a resistor.

During a forward bias, the forward current is entirely *drift*, elsewhere in the diode.

The applied voltage seemingly regulates the "height" of the potential barrier establishing a new equilibrium condition for every value of the applied voltage whether forward or reverse.

The expression gives the total current through the diode for either a forward or a reverse bias that is fairly accurate for currents for forward biasing voltages upto the barrier potential V_o. For the current with biasing voltages beyond V_o, refer Sections 9.8 and 9.9

For a large reverse bias, the only current is [illegible] reverse saturation current I_o that drifts with the field in the junc[illegible]. And if you examine expression (24) minutely, note how I_o the drift[illegible] reverse saturation current depends on the *concentration* of minority holes (electrons) in the *n*(*p*) side. This should explain the tininess of its magnitude!

The expression (25) encapsulates the behavior of a *p-n* junction device that allows current to flow in the forward direction and blocks current in the reverse direction barring the tiny reverse leakage current I_o.

Appendix F

LORENTZ FORCE LAW AND RELATIVISTIC NATURE OF MAGNETISM

F.1 THE AETHER VS. FIELDS VS. ACTION-AT-A-DISTANCE THEORIES

The descriptions of the electric and magnetic fields in this textbook are based purely on interpretations by most authors of texts on Classical Electricity and Magnetism. When an uncharged comb is brought near tiny bits of paper, there is definitely no attraction between the two. However, when the comb is charged, the bits of paper are attracted. The experiment indicates that an "action at a distance" event occurred. The charged comb surely exerts a "force" that caused the tiny bits of paper to move towards it. I then mentioned that there was "something" or "something happened" between the comb and the bits of paper, which was called "the electric field".

Michael Faraday (1791 – 1867) utilized the word "field" for the first time on November 7, 1845, in his Diary. However, long before this time he already utilized analogous expressions such as "magnetic curves," "lines of magnetic force," or "lines of force"[38]. In a paper published in 1851, he defined the field as a region of space around the bodies he was investigating.

Definitions of the field were made by several scientists, J. C. Maxwell, for example, since Faraday's first use of the term to explain that "something" that we indicated in the first paragraph of this Appendix and in Chapter 1.

My intention was neither to ascribe any *material* aspect (electric field) to the space between the comb and the paper nor to ascribe a *property* which I described as an "alteration of space" to space itself. It could as well have been described as "an action at a distance".

We are used to observing actions in our daily lives such as contact forces, when objects collide with other objects or apply a force to them by contacting them. The term "action at a distance" is applicable in the context of theories of *gravity and electromagnetism* which describe how an object responds to the influence of distant (non-contacting) objects. For example, an object falling to the ground causes the earth to move (imperceptibly) to meet the object, according to the law of Universal gravitation. A charged particle attracts (or repels) another charged particle, depending upon their signs, from a distance, according to Coulomb's law.

In Relational Mechanics (Ref. [36]), **A. K. T. Assis** lists some of the definitions, meanings and properties associated with the field concept. While several interpretations of charged particle interactions are propounded by physicists, I must emphasize that the field concept with its use as an aid to visualization of circuit processes described in the chapters of this textbook are sufficiently accurate and precise for understanding and analyzing circuit operations.

Why does the force of magnetic origin warrant so much attention in the study of Electricity? **Charles-Augustin de Coulomb** (1736 – 1806), famous for the Coulomb's law, was the first physicist who, in the year 1785, explained the laws of attraction and repulsion between electric charges and magnetic poles, although he did not find any relationship between the two phenomena. The forces between electric charges and magnetic poles *appeared* straightforward to scientists at the time as was described in the "Electric field due to a point charge" and "Force on a point charge by a dipole" in Appendix A.

Coulomb Interaction

Yet, according to Härtel, "In the case of the Coulomb interaction between two particles, Newton's Third Principle cannot be applied in a completely analogous way to the mechanical case of colliding objects. The Coulomb forces between two charged carriers are equal and opposite, but they do not act at the same point in space. Each charge carrier experiences a single force due to its interaction with the field; but locally, there is *no reaction* back onto the field. This reaction force is found when the force on the opposite charge carrier is taken into account. This reaction force, again, is the result of an interaction of field and charge with no reaction back onto the field." See Section 1.4.1 of article "*A Qualitative Approach to Electricity*", in the folder "Teacher's Guide" in the CD with this book.

I have noted several students take the "action-at-a-distance" aspect of a coulombic force for granted without knowledge of the underlying implications.

According to Härtel in a communication (March 2017) to the author, "A *local* scale would be to look only on a field-vector and a point charge at the same location, from where you derive: $\vec{F} = q\vec{E}$. The *complete system* includes both charges, the charge as origin of the field and the test charge to measure the force, and possibly the space in between. Who knows where the origin of the force has to be located."

In Sections 2.3 and 2.4, "A Qualitative Approach to Electricity: A Guide to visualizing Electrodynamics using Symbols and Images" (available in the CD), Härtel discusses ways to represent the Coulomb interaction between two particles. As an aid to proper visualization of the Coulomb interaction, I urge the reader to sketch the interaction depicted using ellipses and vectors in Figs. 2.4, 2.5, 2.6 and 2.8 in the Guide.

It was **Hans Christian Oersted** (1777 – 1851) who observed in 1820, the deflection of a compass due to the presence in its neighborhood of a long straight wire carrying a constant current during a classroom demonstration. He wrote a short work of 4 pages describing his observations, in Latin, which he sent to several scientists.

Call the magnetic force "rotatory" or "orthogonal" as many of the scientists did, **Oersted's** discovery gave "birth" to the science of electromagnetism, that is, the systematic study of the relation between electric and magnetic phenomena which was made by **André-Marie Ampère** (1775–1836), **Michael Faraday** (1791 – 1867), **Jean-Baptiste Biot** (1774 – 1862), **Félix Savart** (1791 – 1841), **James Clerk Maxwell** (1831 – 1879) and **Wilhelm Eduard Weber** (1804 – 1891) for the most part of the 19th century, spanning nearly 80 years.

Magnetic Interaction

Again, Härtel remarks: "In the case of the magnetic interaction, a single force acts on a moving charge carrier within a magnetic field which results in an acceleration of the charge carrier perpendicular to its velocity. There is no local reaction back onto the magnetic field. Such a single force or a single acceleration is a contradiction to Newton's Third Principle. For the magnetic interaction, this principle is only fulfilled when the reaction of the moving particle with the *complete* system is taken into account. The principle is, however, not valid on a local scale."

Härtel in a communication (March 2017) explains the magnetic interaction: "In the *local* perspective, you have an interaction between a line of magnetic force and a moving point charge. But when looking at the *complete system*, you see the interaction between the moving charges as origin of the magnetic field and the single moving charge".

In Sections 3.3 and 3.4, "A Qualitative Approach to Electricity: A Guide to visualizing Electrodynamics using Symbols and Images" (available in the CD), Härtel discusses ways and means to represent the fields of charges at rest, in motion and the interactions between two particles moving at right angles and the interaction between a current and an electron. As an aid to proper visualization of the interaction of charges at rest and in motion, I urge the reader to sketch the fields and the interactions depicted using dashed circles, ellipses and vectors in Figs. 3.6, 3.7, 3.8, 3.9 and 3.10 in the Guide.

Recollect from Chapters 5, 6, 7 and 8, in which the magnetic force $q\vec{v}\times\vec{B}$, and Faraday's law $\oint E_{NC}\cdot d\vec{l} = -\frac{d}{dt}\int \vec{B}\cdot\hat{n}dA$, which associated an electric field with a changing magnetic field, explained the effects we observed. It is when Newton's third law and the laws of conservation of energy and momentum are applied that the real meaning behind the search for intrinsic phenomenon to corroborate them by scientists, begins to make sense.

During all those 80 odd years of the 19th century, none of the scientists were aware of the structure of the atom. They used several terms to describe the current, for example, the "flow" of electrical "fluids". And yet, Ampere used compass needles and a chemical battery in a large wooden box to demonstrate (Ref. [38]) that there was a continuity of current *inside* the battery !

The first phase of **André-Marie Ampère's** (1775–1836) researches began in September 1820 and continued to January 1821.

According to Ampere, the following definitions should assist in distinguishing the usage of terms

- Electrostatic phenomena: Forces and torques between electrified bodies which are at rest relative to one another.

- Magnetic phenomena: Forces and torques between magnets, together with the torques exerted by the earth on magnets (orientation of compass and dip needles).
- Electromagnetic phenomena: Forces and torques between a current-carrying conductor and a magnet, together with the forces and torques exerted by the earth on current-carrying conductors
- Electrodynamic phenomena: Forces and torques between current-carrying conductors.

It is appropriate for reasons which we shall see a little later in this appendix, to introduce the work of **Wilhelm Eduard Weber** (1804 – 1891). **Weber's** electrodynamics, was the main electromagnetic theory in Germany during the third quarter of the 19th century (after around 1875), which does *not utilize* the concept of lines of force, lines of magnetic induction, nor lines of magnetic field $\vec{B}$. Weber's electrodynamics deals only with the direct interaction between charges and Weber's force depends only on the distances, relative radial velocities and relative radial accelerations of the interacting charges (Ref. [36]). The concepts of electric and magnetic fields do not need to be introduced in Weber's electrodynamics, which was also discussed at length in the last chapter of **Maxwell's** Treatise.

A study of the research work of all the above-mentioned scientists reveals nothing that suggests that the magnetic or the electric field was a "discovery" or that the magnetic field "produced" the force during the interaction between current elements. There is nothing sequential either in the formulation of a "field" theory and the Lorentz force law. It was Ampere who initially considered *molecular* currents as being located in planes orthogonal to the straight line connecting the magnetic poles of a magnet. Ampere never worked with the concept of a "magnetic field".

Ampere's main experimental and theoretical works on electrodynamics, the second phase of his research, was from 1821 to 1827, during the period in which he built several apparatus and performed experiments with them systematically to arrive at the final form of the force between current elements *later* to be embodied in the famous Lorentz force law. He utilized straight wires, wires wound in helices, suspended wires, astatic coils (which Ampere developed to eliminate the magnetic action of the earth in delicate experiments performed for studying the force between current elements), levers, pivots and numerous devices in his apparatus and performed numerous experiments, checking and double-checking results even as he prepared notes and wrote equations with sketches and presented them in the Academy of Sciences of Paris over this period (Ref. [38]).

1. FORCE BETWEEN TWO CURRENT ELEMENTS

Ampere's force d^2f between two current elements ids and $i'ds'$ separated by a distance r is

$$d^2 f = \frac{ii'dsds'}{r^n}(\sin\alpha\sin\beta\cos\gamma + k\cos\alpha\cos\beta) \tag{1}$$

with $n = 2$ and $k = -\frac{1}{2}$, Eq. 2.1 in "Ampere's Electrodynamics"(Ref. [38]). The constant $k \neq 0$, represents the existence of a net force between two collinear and parallel current elements and played a crucial role in Ampere's deduction of his force law which conformed to Newton's Third Law in all situations.

The currents i and i' are in units introduced by Ampere (called nowadays the electrodynamic system of units). They are related to the International System of Units by $ii' \Leftrightarrow \frac{\mu_o}{2\pi} II'$. The angles α, β and γ are defined as depicted in Fig. 2.15 in Ref. [38].

Ampere's force $d^2F_{ji}^A$ exerted by a current element $I_j\vec{r}_j$ localized at $\vec{r}_j$ on another current element $I_i\vec{r}_i$ localized at $\vec{r}_i$ can be expressed in the International System of Units utilizing modern vector notation and is given by

$$d^2\vec{F}_{ji}^A = \frac{\mu_o I_i I_j}{4\pi}\frac{\hat{r}_{ij}}{r_{ij}^2}\left[3\left(d\vec{r}_i \cdot \hat{r}_{ij}\right)\left(d\vec{r}_j \cdot \hat{r}_{ij}\right) - 2\left(d\vec{r}_i \cdot d\vec{r}_j\right)\right] = -d^2\vec{F}_{ij}^A \quad (2)$$

according to Eq. 5.1 in "Inductance and Force Calculations in Electrical Circuits" by M.d.A. Bueno and A.K.T. Assis (Ref. [40]) and it is observed that this expression always satisfies the principle of action and reaction, i.e., $d^2\vec{F}_{ji}^A = -d^2\vec{F}_{ij}^A$. The superscript '$A$' of the force indicates Ampere's expression.

The force expression Eq. 5.1 in Ref. [40], is equivalent to that given by Eq. 2.16 in "Ampere's Electrodynamics" (Ref. [38]), which is

$$d^2\vec{F}_{I'ds' \text{ on } Ids}^A = -\frac{\mu_o II'}{4\pi}\frac{\hat{r}}{r^2}\left[2\left(d\vec{s} \cdot d\vec{s}'\right) - 3\left(\hat{r} \cdot d\vec{s}\right)\left(\hat{r} \cdot d\vec{s}'\right)\right] = -d^2\vec{F}_{Ids \text{ on } I'ds'}^A. \quad (3)$$

The equivalence is evident after properly redefining the currents, current elements, displacement vectors and multiplying the expression within square brackets by the –1 outside the brackets.

The derivation of the expression (Ref. [38]) in the International System of Units is made from Ampere's original expression, which in the electrodynamic system of units is $\frac{ii'\,dsds'}{r^n}(\cos\varepsilon + h\cos\theta\cos\theta')$ Eqs. 2.3 and Eq.7.1 in Ref. [38].

In Ref. [38], the equivalence of the expression for force (Eq. 2.1), $\frac{ii'dsds'}{r^n}(\sin\alpha\sin\beta\cos\gamma + k\cos\alpha\cos\beta)$ and the expression (Eq. 2.3), $\frac{ii'dsds'}{r^n}(\cos\varepsilon + h\cos\theta\cos\theta')$ is first established by noting that α = θ and β = θ′ from Fig. 2.15 and $h = k - 1$ given in Eq. 2.6.

Then, after making the substitution $\cos\varepsilon = \sin\alpha\sin\beta\cos\gamma + \cos\alpha\cos\beta$ (Eq. 7.2), in Eq. 2.3, Eq. 2.1 is obtained. This proves the equivalence of Eqs. 2.1 and 2.3 in Ref. [38].

The derivation of the force expression in the International System of Units (Eq. 2.16) above follows from Eq. 2.15 (which is the same as Eq. 2.3) to 2.21 in Ref. [38] making the substitution $n = 2$.

Note: To interpret the angle ε, refer to Fig. 7.18 (Ref. [38]). When interpreting Fig. 7.18 (Spherical triangle) and Eqs. 7.2 to 7.7, especially with reference to the line OA, visualize another line *parallel* to OA with its element *ds*, somewhere ahead along OC, was slid back to the center 'O' of the sphere. I urge the reader to obtain the equation in the International System of Units following the steps as described above.

The Biot-Savart formula (though they never used the idea of a magnetic field) for the magnetic field produced by an element of a current *i.ds* at a point distant *r* from it in a direction making an angle θ with the element, is $\frac{Ki.ds\sin\theta}{r^2}$ in the electrodynamic system of units and the field lies in the direction perpendicular to both *ds* and *r*. The factor *K* is introduced to enable make any subsequent adjustment of units, according to R. A. R. Tricker. See Chapter III Section E "The Design of Ampere's Basic Experiments", in *Early Electrodynamics - The First Law of Circulation*, R. A. R. Tricker, a link to which is available in the Links_to_animations_articles file in the CD.

Grassmann's Force Expression

The Biot-Savart formula for the magnetic field at $\vec{r}_i$ due to a current element $I_j\vec{r}_j$ localized at $\vec{r}_j$ can be expressed in the International System of Units as $\frac{\mu_o}{4\pi}\frac{I_j\,d\vec{r}_j\times\hat{r}_{ij}}{r_{ij}^2}$. The force on a second element $I_i\vec{r}_i$, according to **Hermann Günther Grassmann** (1809-1877) (Eq. 5.10 in "Inductance and Force Calculations in Electrical Circuits" by M.d.A. Bueno and A.K.T. Assis (Ref. [40])), is given by

$$d^2\vec{F}_{ji}^G = I_i d\vec{r}_i\times\frac{\mu_o}{4\pi}\frac{I_j d\vec{r}_j\times\hat{r}_{ij}}{r_{ij}^2} = \frac{\mu_o I_i I_j}{4\pi}\frac{1}{r_{ij}^2}\left[\left(d\vec{r}_i\cdot\hat{r}_{ij}\right)d\vec{r}_j - \left(d\vec{r}_i\cdot d\vec{r}_j\right)\hat{r}_{ij}\right] \tag{4}$$

where the triple vector product identity **a**× (**b**×**c**) = **b**(**a**·**c**) – **c**(**a**·**b**) is used to expand the second expression.

It can be observed that Grassmann's force does not satisfy the principle of action and reaction (third law of Newton) in general (see details in Chapter 5, Section 5.2 "Grassmann's Force" in "Inductance and Force Calculations in Electrical Circuits" by M.d.A. Bueno and A.K.T. Assis Ref. [40])), i.e., $d^2\vec{F}_{ji}^G \neq -d^2\vec{F}_{ij}^G$ where the superscript '*G*' of the force indicates Grassman's expression.

Over several years, scientists Faraday and Biot were in constant communication with Ampere and were exchanging notes and analyzing results especially on the action of forces between current carrying wires and magnets. In numerous instances, Ampere did not accept the hypotheses of Biot and Faraday on the origins of the forces, which seemed to him contrary to Newton's Third Law.

According to Tricker, "In applying the Biot-Savart formula to parts of circuits, however, we have to remember that Newton's third law does not apply" and "On the other hand, the use of Ampere's formula obviates this difficulty, since the mechanics he assumes agrees with that of Newton on this point, and makes the line joining the elements the line of action of the forces." But, "When integrated *round the complete circuit* generating the magnetic field, however, the formulae of both Ampère and Biot and Savart [utilized in Grassman's expression] give identical results", as we will see below.

2. FORCE BETWEEN A CURRENT-CARRYING CLOSED CIRCUIT *S* ON A CURRENT ELEMENT

Ampere's net force $d\vec{F}$ exerted by a closed circuit *S* carrying a current *i* on a current element $i'd\vec{s}'$ in electrodynamic units is given by

$$d\vec{F} = i'd\vec{s}' \times \left(\frac{i}{2} \oint_S \frac{d\vec{s} \times \vec{r}}{r^3} \right) \tag{5}$$

according to Eq. 10.24 in Ref. [38]. This equation is shown in Ref. [38] to have been obtained by integrating Eq. 7.16 (which is obtained by way of interpreting the elements of Eq. 2.1 from Fig. 7.18) in steps illustrated in Section 10.1 in Ref. [38].

Ampere found that the net force was always *orthogonal* to the direction of the vector of length $d\vec{s}$ as seen from the formula for $d\vec{F}$ above. He concluded that this net force belonged to a certain plane which he called the directing plane. The straight line orthogonal to this directing plane and passing through the midpoint of the vector of length $d\vec{s}'$ was initially called normal to the directing plane. In the Th´eorie Amp`ere, he called it "directrix" (Ref. [38]).

For the sake of illustrating the appearance of a vector, later to be called the magnetic field, we adopt one of the original forms of Ampere's force law[38].

Ampere's Force $d\vec{F}$, exerted by a closed circuit *S* carrying a current *I* and acting on a current element $i'd\vec{s}'$, which does not belong to this closed circuit, can be written in the International System of Units as

$$d\vec{F}_{d\vec{s}'} = \frac{\mu_o}{2\pi}\frac{II'}{2} d\vec{s}' \times \vec{D} = I'd\vec{s}' \times \left(\frac{\mu_o}{4\pi} \oint_S I \frac{d\vec{s}' \times \vec{r}}{r^3} \right) \tag{6}$$

where, *r* is the distance between the *midpoints* of the elements of lengths $d\vec{s}$ and $d\vec{s}'$ carrying currents *I* and *I'* respectively. The directrix $\vec{D}$ is $\left(\oint_S \frac{d\vec{s}' \times \vec{r}}{r^3} \right)$ according to Eq. 10.23 in Ref. [38].

The magnetic field $d\vec{B}$ at a point r_1 due to a current element $Id\vec{s}$ at point r_2 is defined by the following in the International System of Units

$$d\vec{B} \equiv \frac{\mu_o}{4\pi} \frac{Id\vec{s} \times \vec{r}}{r^3} \tag{7}$$

in which the vector $\vec{r}$ points from $\vec{r}_1$ to $\vec{r}_2$, that is, from the midpoint of element 1 to the midpoint of element 2, and *r* represents the distance between these two points. Ampere never worked with the concept of a magnetic field. However, it is possible to express the integrated results of Ampere in terms of the magnetic field[38]. The magnetic field $\vec{B}$ due to a closed circuit *S* of arbitrary form by integrating Eq. (7) along *S*, is

$$\vec{B} = \frac{\mu_o I}{4\pi} \vec{D} = \frac{\mu_o I}{4\pi} \oint_S \frac{d\vec{s}' \times \vec{r}}{r^3} \tag{8}$$

Note how the magnetic field strength drops away as $1/(\text{distance})^3$ as we found for the magnetic field of a circular loop at a point located far away from the loop in Appendix A.

Thus, the force exerted, determined by Ampere's Eq. (6) by the closed circuit *S* acting on the current element $I'd\vec{s}'$ could then be written by combining with Eq. (8) as

$$d\vec{F} = I' d\vec{s}' \times \vec{B} \tag{9}$$

Equations (7) and (8) are due to Maxwell's definition of the magnetic field. Maxwell finally replaced this current element $I'd\vec{s}'$ by $q\vec{v}$, where q is the charge of the electrical body and $\vec{v}$ it's velocity.

The magnetic force $\vec{F}_m$ acting on this charged body moving in a magnetic field would then be written as

$$\vec{F}_m = q\vec{v} \times \vec{B} \tag{10}$$

Hendrik Antoon Lorentz (1853 – 1928) also obtained this relation in 1895 and in the textbooks of classical electromagnetism, the electromagnetic force (called the Lorentz force law) at a position vector $\vec{r}$ on a point particle moving with velocity $\vec{v}$ is

$$\vec{F} = q\vec{E}(\vec{r}) + q\vec{v} \times \vec{B}(\vec{r}) \tag{11}$$

where, $\vec{E}(\vec{r})$ and $\vec{B}(\vec{r})$ represent the electric and magnetic fields respectively at the location $\vec{r}$ from the point charge q.

Grassmann's Force Expression

In the International System of Units, Ampere's force $d\vec{F}_{1j}^{A}$ exerted by a closed circuit Γ_1 carrying a current I_1 on a current element $I_2 d\vec{r}_j$ of the circuit Γ_2 is given by

$$d\vec{F}_{1j}^{A} = -\frac{\mu_o I_1 I_2}{4\pi}\left(3\oint_{\Gamma_1}\frac{\hat{r}_{ij}}{r_{ij}^2}\left(d\vec{r}_i\cdot\hat{r}_{ij}\right)\left(d\vec{r}_j\cdot\hat{r}_{ij}\right) - 2\oint_{\Gamma_1}\frac{\hat{r}_{ij}}{r_{ij}^2}\left(d\vec{r}_i\cdot d\vec{r}_j\right)\right) \tag{12}$$

according to Eq. 5.23 in Ref. [40], and Grassman's force by

$$d\vec{F}_{1j}^{G} = \frac{\mu_o I_1 I_2}{4\pi}\left(\oint_{\Gamma_1}\frac{\hat{r}_{ij}}{r_{ij}^2}\left(d\vec{r}_i\cdot d\vec{r}_j\right) - \oint_{\Gamma_1}\frac{1}{r_{ij}^2}\left(d\vec{r}_j\cdot\hat{r}_{ij}\right)d\vec{r}_i\right) \tag{13}$$

according to Eq. 5.24 in Ref. [40].

Proof of the equivalence of Ampere's and Grassmann's force expressions for the force exerted by a closed circuit on a current element external to it is given in detail in Ref. [40]. It would be instructive for students to read and understand the proof which requires the application of conditions of exact differentials. The reader is referred to theorems for exact differentials and path independence of integrals in the Links_to_animations_articles file in the CD. Such a reading will enable to clear confusion regarding the accuracy and use of Grassmann's force which is based on the Biot-Savart magnetic field formula and Ampere's force in different situations.

Force and Magnetic Field in Textbooks

According to classical electromagnetic theory, a current element generates a magnetic field which can act on another current element. The expression satisfied by this magnetic field due to a current element is usually called Biot-Savart's law, although Biot and Savart did not work with the field concept (Ref. [38]).

Based on the interpretation of a Bridge experiment (Sections 8.2 and 21.2 in Ref. [38]) Ampere would certainly be against three specific consequences of this *magnetic interaction*:

(a) The absence of a net force between two *collinear and parallel* current elements, but which is present according to Ampere's force law, hinted by the presence of the product of cosines term with $k \neq 0$ in Eq. (1); (b) the primitive couple due to this magnetic field (Biot-Savart law) generated by a current element, as discussed in Section 20.2, Ref. [38]; and (c) the inequality between action and reaction in some specific situations, as discussed in Section 20.3, Ref. [38].

Although, Ampere's force expression Eqs. (1 & 2) is accurate universally in all situations from the standpoint of Newton's Third Law, Grassmann's force expression Eq. (4) (though it violates Newton's third law in certain situations) is considered sufficiently accurate for the force between current elements, and is adopted in most textbooks.

In modern textbooks, the magnetic field is given by

$$d\vec{B}(\vec{r}) = \frac{\mu_o}{4\pi} \frac{I' d\vec{s}' \times \hat{r}}{r^2} \tag{14}$$

The equivalence of Eq. (14) to Eq. (7) is seen by using the unit vector definition, i.e., $\hat{r} \equiv \vec{r} / r$. According to Tricker, "Ampere's directrix possesses the same direction as what is now called the magnetic induction."

Therefore, it may be said that the magnetic field appearing in the Lorentz force law (Eq. 10) is based on the interpretation of Ampere's term for the directrix (Eqs. 6 and 7) expressed in the International System of Units.

Sometimes textbooks present the formula expressing the magnetic field $\vec{B}(\vec{r})$ at a point $\vec{r}$ of an inertial coordinate system, with the field generated by a point charge q' which is passing through the origin while moving with velocity $\vec{v}'$. This field is given by

$$\vec{B}(\vec{r}) = \frac{\mu_o}{4\pi} \frac{q' \vec{v}' \times \hat{r}}{r^2} \tag{15}$$

Specifically, Ampere's and Grassmann's expressions of the force between a current carrying circuit S and a current element are shown to be equivalent as discussed previously here in Ref. [40].

It should be emphasized that according to Ampere's original conceptions, this force resulted from a direct action between the closed conductor S carrying a current I and the current element $I' d\vec{s}'$. That is, according to Ampere this direct net force was not mediated through a "magnetic field". According to Ampere, the net force was the result of a direct *action at a distance* between current elements (Ref. [38]). The net force being *orthogonal* as seen from Eq. (6) captured the imagination of many scientists and it is this phenomenon that led to several interpretations of the interaction of charged particles especially particles in motion.

Ampere did not think of the magnetic field; Faraday mentioned it, Maxwell formalized it, and Einstein added a new dimension to it.

Maxwell believed in a material medium and wrote "The undulatory theory of light also assumes the existence of a medium. We have now to shew that the properties of the electromagnetic medium are identical with those of the luminiferous medium".[36]

Einstein dismissed the existence of a "luminiferous aether" assumed till 1905, as being the "medium" in which light traveled as a wave (see "Fields and Momentum Conservation" in Appendix B), until the Michelson-Morley experiment designed to detect it, discounted its existence. As we will soon see, this led to the formal establishment of the "field" interpretation of fundamental charged particle interactions[36]. For an animation and intuitive understanding of the assumptions made for the Michelson-Morley experimental setup, see the "Links to animations_articles" in the CD.

It is indeed the painstaking research undertaken by **Ampere** to realize an expression for the force between the straight line joining the midpoints of two arbitrarily spatially oriented current elements, and that this obeyed Newton's third law, viz., the principle of action and reaction, which eventually led to the formal expression of the Lorentz force. His steadfastness in the belief that the force must obey Newton's third law enabled him to point out several anomalies in the theories propounded by renowned scientists Faraday, Biot and Savart in the theory of electrodynamics and directed his researches towards the expression for this force.

All the scientists including Faraday, Maxwell, Biot and Savart later expressed their admiration for the accuracy in Ampere's work in treatises and papers.

The theory of atomic structure matured in 1913 with Niels Bohr's publication, nearly 77 years after Ampere died. Nowadays, students who learn topics in Electricity and Magnetism would be reading and visualizing circuit processes with a good exposure to the atomic structure with electrons occupying orbital spaces around the stationary lattice ion and the aggregation of atoms as molecules in conductors and insulators, far different from Ampere's world of Electricity and Magnetism and yet, his Force equation has withstood the test of time !

Weber's Force Expression

Weber in 1846 gave an expression for the force, exerted by an electric charge q_j on another charge q_i separated by a distance r_{ij}, which is given using vectorial notation in the International System of Units by

$$\vec{F}_{ji}^{W} = \frac{q_i\, q_j}{4\pi\varepsilon_0} \frac{\hat{r}_{ij}}{r_{ij}^2}\left(1 - \frac{\dot{r}_{ij}^2}{2c^2} + \frac{r_{ij}\, \ddot{r}_{ij}}{c^2}\right).$$

See Section 3.2, Chapter 3, Weber's Electrodynamics by A. K. T. Assis (Ref. [42]).

Note from the expression that Coulomb's force is a particular case of Weber's Force, obtained when the charges are at rest, relative to one another. The velocity and acceleration which appear are only the radial velocities and accelerations between the two charges.

Ampere's Force between current elements can be easily derived from Weber's Force expression between charges (See Section 4.2, Chapter 4, Ref. [42]).

Maxwell's equations can be derived from Weber's force: Gauss's Law, Lorentz's force and Faraday's Law as shown in Ref. [42].

Weber's Force expression, it is seen, represents a *unification of electrostatics and electrodynamics*.

F.2 THE FIELDS OF MOVING CHARGES ACCORDING TO THE THEORY OF SPECIAL RELATIVITY

The Lorentz transformation is the starting point for *understanding* the Lorentz force law (Eq. 11) and leading to a description of the relativistic nature of Magnetism.

Lorentz first postulated coordinate transformation of events in inertial frames in the period 1892-1904 assuming the luminiferous aether for light propagation. Einstein in 1905, derived them in the Theory of Special Relativity with the aether made superfluous by the Michelson-Morley experiment. Also see the note in Q.10.2 in the Additional Practice Questions for all chapters, Einstein's paper "On the Electrodynamics of Moving Bodies" and links to video lectures in the CD.

Most essential is the concept of an inertial frame of reference for space time events and the transformation of the coordinates of an event from one inertial frame to another. A frame of reference is a coordinate system laid out with measuring rods and provided with clocks. It is used to describe the position, and also the velocity and acceleration, of an object.

An *inertial* frame of reference has a uniform velocity or a coordinate system which is not accelerating. That is, it is moving at a uniform speed in a straight line, or it is standing still. Understand that when something is standing still, it has a velocity of zero. In an inertial frame of reference the laws of physics of inertia (tendency to resist change in velocity) and Newton's laws of motion hold with the absence of fictitious forces.

A *non-inertial* frame of reference does not have a uniform velocity and is accelerating. The law of inertia does not hold in it. When in a moving automobile, if the brakes are suddenly applied, you will be thrown forward and there is no real force (a force whose origin is unknown sometimes termed "fictitious" in older textbooks) that pushed you. The car is a non-inertial frame of reference since it was slowing down or decelerating due to the braking action, and the law of inertia no longer holds if such a non-inertial frame is used to judge your motion.

When reading topics related to special relativity and frames of reference (inertial and non-inertial), it is important to make note of the "location" of the observer. For instance, when discussing an object dropped inside a moving train using special relativity, the observer may be i) on the ground (outside the train), ii) the object, or iii) the moving train. Each observer makes different observations. See a nice video on inertial vs. non-inertial frames of reference and a 1960 educational film in the links provided in the "Links to animations_articles" in the CD.

For an accurate description of inertial frames, read "The Principle of Relativity", Chapter 1 followed by "Inertial Frames", Chapter 2 in "Modern Mechanics" (Ref. [41]).

In the following discussion of the Lorentz force law, we use the inertial frame of reference.

Time is measured by a *local* clock that is *stationary* in the frame. The clocks belonging to the frame are all *synchronized* (see Ref. [22] for how this is done). The frame is one in which an isolated body, free from external influences moves such that the history of the particle's motion is a sequence of *events*. An event is located in space and time by its coordinates x, y, z, t in some chosen reference frame. If, for example, a sequence of events has the property that $x = v_x t$, $y = v_y t$, $z = v_z t$, at every time t, with v_x, v_y and v_z constant, then it describes motion in a straight line at constant speed with respect to this frame.

Two frames F and F' can differ in several ways. One frame can be simply displaced with respect to the other, the origin of coordinates in F' being fixed at a point F which is not at the F coordinate origin. As for the timing of the events, if F and F' are not moving with respect to one another, a clock stationary in F is stationary also in F'. All the clocks in F can be set to agree with all the clocks in F'. Such frames do not evoke special interest. Suppose now that the origin of frame F' is *moving* relative to the origin of frame F. Assume two observers are stationed in the two frames F and F' each equipped with clocks that have been synchronized and measuring rods to measure displacements. Their description of a sequence of events will differ.

We will consider the case in which F is an inertial frame and F' is a frame which is moving relative to F at constant velocity in a straight line. In that case F' is also an inertial frame, as deduced from Newton's first law also sometimes called the Law of Inertia. The frames should not be *accelerating* !

Lorentz Transformation

Consider two events, A and B, observed in an inertial frame F. The displacement of one event from the other is given by the four numbers

$$x_B - x_A, \quad y_B - y_A, \quad z - z_A, \quad t_B - t_A \tag{16}$$

The same two events could have been located by giving their coordinates in some other frame F' which is moving with respect to F. Suppose F' is moving with respect to F in the manner indicated in Fig. F.1.

The spatial axes of F' remain parallel to those in F, while, as seen in F, the frame F' moves with speed v in the positive x direction.

Event A, as observed in F', occurred at x'_A, y'_A, z'_A, t'_A the last of these numbers being the reading of a clock belonging to (that is, *stationary in*) F'. The space-time displacement or interval between events A and B in F' is not same as in F. Its components are related to those in F by the Lorentz transformation (Ref. [22]):

$$\begin{aligned} x'_B - x'_A &= \gamma(x_B - x_A) - \beta\gamma c(t_B - t_A) \\ y'_B - y'_A &= y_B - y_A \\ z'_B - z'_A &= z_B - z_A \\ t'_B - t'_A &= \gamma(t_B - t_A) - \beta\gamma(x_B - x_A)/c \end{aligned} \tag{17}$$

The space (displacement) on the LHS is measured in frame F' and the space (displacement) and time on the RHS of this equation is measured in the frame F. In these equations, c is the speed of light, $\beta = v/c$, and *Lorentz factor* $\gamma = 1/\sqrt{1-\beta^2}$. The inverse transformation has a similar appearance — as it should if no frame is unique. It can be obtained from Eq. (17) simply by exchanging the primed and unprimed symbols and reversing the sign of β. This will be shown later in the "Important note on Inverse Lorentz Transformation". The universal constant c appears in these formulas as a limiting velocity, approached by an energetic particle but never exceeded.

A physical interpretation of the Eq. (17) can be made by observing two events A and B that are *simultaneous* in F. It is observed that $t_B - t_A = 0$ will make the events simultaneous. But that does not make the events simultaneous in F', i.e., $t'_B - t'_A = 0$ unless $x_B = x_A$. This fundamental "relativity of simultaneity" should not be confused with the obvious fact that an observer not equally distant from two simultaneous explosions will receive light flashes from them at different times. The times t'_A and t'_B are recorded by *local* clocks at each event, clocks stationary in the moving frame F'.

Lorentz contraction

Consider a rod stationary in F', which is parallel to the x' axis and extends from x'_A to x'_B. Its length in F' is just $x'_B - x'_A$. The rod's length as measured in frame F is the distance $x_B - x_A$ between two points in the frame F that its ends pass *simultaneously* according to clocks in F. For these two events, then, $t_B - t_A = 0$. With this consideration the first of the Lorentz transformation equations above gives us (Ref. [22])

$$x_B - x_A = (x'_B - x'_A)/\gamma \tag{18}$$

where, $\beta = v/c$, and the Lorentz factor, $\gamma = 1/\sqrt{1-\beta^2}$.

This is the famous *Lorentz contraction* which makes the lengths between fixed points in F', if parallel to the relative velocity of the frames, as judged by observers in F, to be shorter by the factor $1/\gamma$. The statement is true if F' and F are interchanged as we will see in the "Important note on Inverse Lorentz Transformation".

Lengths *perpendicular* to the relative velocity measure the *same* in the two frames. There are no common examples for observing this phenomenon as we will learn by putting in a few numbers.

Consider two inertial frames one stationary and the other moving as shown in Fig. F.1(a).

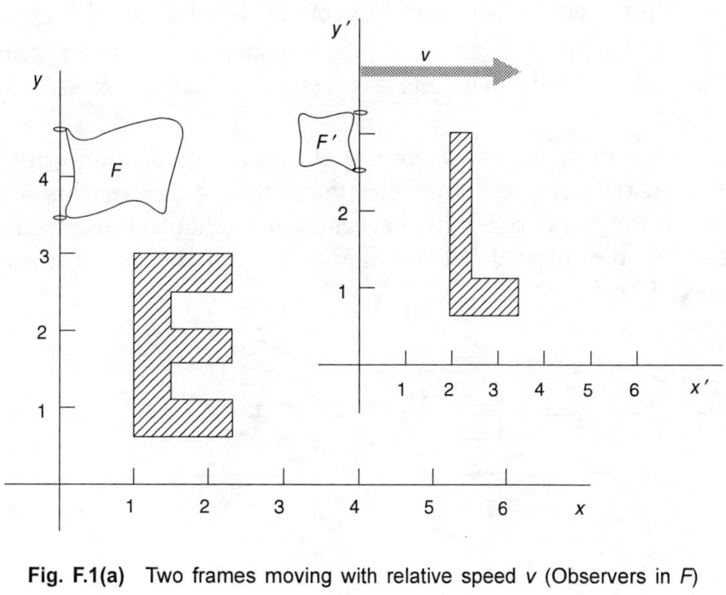

Fig. F.1(a) Two frames moving with relative speed v (Observers in F)

The "*E*" is stationary in frame *F*. The "*L*" is stationary in frame F'. In this example, $\beta = v/c = 0.866$ (a relatively *very high* speed) and γ therefore, equals 2. Figure F.1(a) shows where everything was, as determined by observers in *F* at a particular instant of time *t* according to clocks in *F*. We can only say, as stationary observers in *F* that the "*L*" in F' does look strangely shortened along the base and the tall segment of "*L*" appears thinner $\left(\frac{1}{\gamma} = \frac{1}{2}\right)$.

Now, let us switch to frame F' and note what the observers here will record. What the observers in F' at a particular instant of time t' (according to clocks in F') will observe is depicted in Fig. F.1(b).

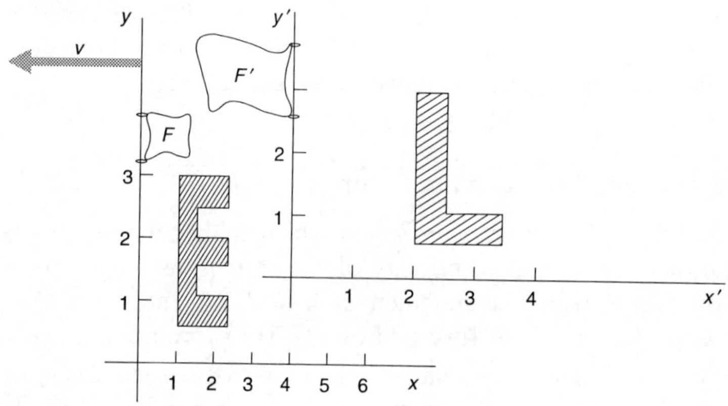

Fig. F.1(b) Two frames moving with relative speed *v* (Observers in F')

The letter "*L*" appears to be normal to the observers in F' but, the letter "*E*" in frame *F* appears shortened to them and the vertical portion of "*E*" appears thinner.

The phenomenon causes pronounced effects when the relative speeds are *relativistic*, i.e., close to the speed of light. The effect is less pronounced at speeds much slower than the speed of light.

The reader is cautioned not to confuse Lorentz contraction with observations of moving trains from railway platforms. An approaching train appears longer because the light received from the last carriage left earlier than the light received simultaneously from the front engine. In the interval it takes the light from the last carriage, the train itself moves a distance $L' - L$:

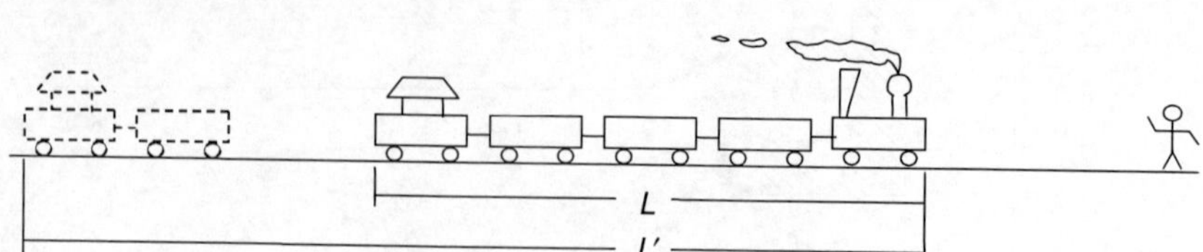

So approaching trains appear *longer*, by a factor (1 – v/c^{-1}), i.e., $L' = \dfrac{L}{1 - v/c}$. By contrast, a train going away from you looks *shorter* by a factor (1 + v/c^{-1}). This is reminiscent of the Doppler Effect[8]. See the Links_to_animations_articles.pdf file in Appendix F for links to animation on the Doppler Shift and also to articles on the Lorentz contraction.

The contraction predicted by the Lorentz transformation is different as can be seen from the factor in our observation of lengths of trains, namely v/c. The Lorentz transformation, which predicts the Lorentz contraction of rods in a moving frame when observed from a stationary frame, is a more *fundamental* connection between space and time coordinates of an event as measured by an observer in each frame.

Important Note on Inverse Lorentz Transformation

According to the principle of relativity, there is no privileged (or unique) frame of reference, so the transformations from F' to F must take exactly the same form as the transformations from F to F'. The only *difference* is F' moves with velocity $-v$ relative to F (i.e., the relative velocity has the same magnitude but is oppositely directed).

Consider the two frames F and F' in Fig. F.1(a), in which F' is moving at velocity v in the *positive* x direction of F. Given a space time event that occurs at (ct, x, y, z) in frame F, the Lorentz transformation Eq.(17) helps us to find the space-time coordinates (ct', x', y', z') of that event in frame F'. If, however, you know the event occurs at (ct', x', y', z'), the inverse Lorentz transformation helps us find the space-time coordinates (ct, x, y, z) of that event in frame F.

The Lorentz transformation for the x coordinates (Fig. F.1(a)) is given by the first equation of Eq.(17):

$$x'_B - x'_A = \gamma(x_B - x_A) - \beta\gamma c(t_B - t_A)$$

Recollect from the interpretation of Eq. (17) that the space (displacement) on the LHS is measured in frame F' and the space (displacement) and time on the RHS of this equation is measured in the frame F. From the first postulate of special relativity the laws of physics in frame F' must be the same as those in frame F so to find $x_B - x_A$ we can use:

$$x_B - x_A = \gamma(x'_B - x'_A) - \beta'\gamma c(t'_B - t'_A)$$

where $\beta' = \dfrac{v'}{c}$ in which v' is the velocity of frame F in the frame F'. Because of the assumed directions in Fig. F.1(a), v' is $-$ v, thus

$$\beta' = \frac{v'}{c} = \frac{-v}{c} = -\beta .$$

Therefore,
$$x_B - x_A = \gamma(x'_B - x'_A) + \beta\gamma c(t'_B - t'_A) \tag{19}$$

This is the inverse Lorentz transformation for the x-coordinate, and everything on the LHS is measured in frame F and on the RHS in frame F'.

The Lorentz transformation Eq. (17) is used when going from frame F to F' (Fig. F.1(a)) and the inverse transformation is used when going from frame F' to frame F (Fig. F.1(b)).

Time dilation

Consider one of the clocks in F'. It is moving with speed v through the frame F. If t'_A is its reading as it passes one of the local clocks in F, the local clock reads at that moment t_A. Later the moving clock passes another F clock. At that event the local clock F reads t_B, and the reading of the moving clock is recorded as t'_B. The two events are separated in the F frame by a distance $x_B - x_A = v\,(t_B - t_A)$. Substituting this in the fourth equation of the Lorentz transformation, Eq. (17), we obtain at once

$$t'_B - t'_A = \gamma(t_B - t_A)(1-\beta^2) = (t_B - t_A)/\gamma \tag{20}$$

According to the moving clock, less time has elapsed between the two events than is indicated by the stationary clocks in F. This is called *time dilation* which figures in the "twin paradox". It has been verified in many experiments, including one in which an atomic clock was flown around the world (Ref. [22]).

While a skeptical reader may dismiss the Lorentz contraction and time dilation as mere conjecture requiring elaborate experiments to corroborate, physicists and electrical engineers will not easily dismiss them. Let us see how they view this. Before we see the phenomenon applied to electricity and magnetism we will need to establish a formula for the addition of velocities and how they transform from one inertial reference frame to another.

Formula for the addition of velocities

Suppose an object is moving in the positive x direction in frame F with velocity u_x. What is its velocity in the frame F'? To simplify matters let the moving object pass the origin at $t = 0$. Then its position in F at any time is simply $x = u_x t$. To simplify further, let the space time origins of F and F' coincide. The first and the last of the Lorentz transformation equations (Eq. 17) are

$$x' = \gamma x - \beta\gamma ct \qquad \text{and} \qquad t' = \gamma t - \beta\gamma x/c$$

By substituting $u_x t$ for x on the right side of each and dividing the first by the second (Ref. [22]), we get

$$\frac{x'}{t'} = \frac{u_x - \beta c}{1 - \beta u_x / c} \tag{21}$$

On the left we have the velocity of the object in the F' frame, u'_x. The formula is usually written with v instead of βc.

$$u'_x = \frac{u_x - v}{1 - u_x v / c^2} \tag{22}$$

By solving Eq. 22 for u'_x it can be verified that the inverse is

$$u_x = \frac{u'_x + v}{1 + u'_x v / c^2} \tag{23}$$

and that in no case will these relations lead to a velocity, either u_x or u'_x, larger than c.

Having established a few basic relations using the Lorentz transformations of events A and B in two inertial frames, we proceed to investigate the electric field measured in different frames of reference. If charge is to be invariant under a Lorentz transformation, the electric field has to transform in a particular way. If an observer in a certain inertial frame F measures an electric field **E** as so many statvolts/cm, at a given point in space and time, what field will be measured at the same space-time point by an observer in a different inertial frame F'?

In the frame F in (Fig. F.2(a)) there are two stationary sheets of charge of uniform density σ and $-\sigma$ esu/cm^2 (1 esu = 3.335×10^{-10} Coulombs), respectively.

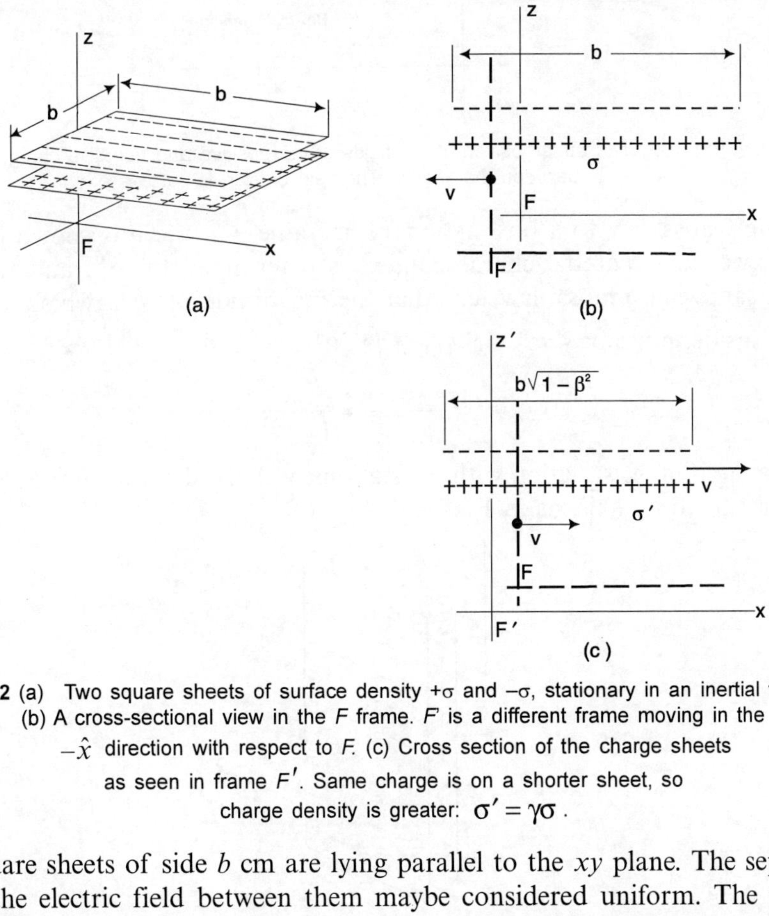

Fig. F.2 (a) Two square sheets of surface density $+\sigma$ and $-\sigma$, stationary in an inertial frame F. (b) A cross-sectional view in the F frame. F' is a different frame moving in the $-\hat{x}$ direction with respect to F. (c) Cross section of the charge sheets as seen in frame F'. Same charge is on a shorter sheet, so charge density is greater: $\sigma' = \gamma\sigma$.

The square sheets of side b cm are lying parallel to the xy plane. The separation is so small that the electric field between them maybe considered uniform. The magnitude of the field measured by an observer in the F frame is $4\pi\sigma$ (in CGS units, not SI units that we used to derive for the field between the plates of a capacitor in Appendix A). Now consider an inertial frame F', which moves toward the left, with respect to F, with velocity v. To an observer in F', the charged "squares" are no longer square. Their x' dimension is contracted from b to $b\sqrt{1-\beta^2}$, where β stands for v/c. But total charge is invariant, that is, independent of reference frame, which we shall not prove, so the charge density measured in F' must be *greater* than σ in the ratio γ, that is $1/\sqrt{1-\beta^2}$.

Figure F.2 (b) shows the system in cross section as seen in F and Fig. F.2(c) as seen in F'.

The superposition of the fields of the positive and negative charges in the F' frame is shown in Fig. F.3.

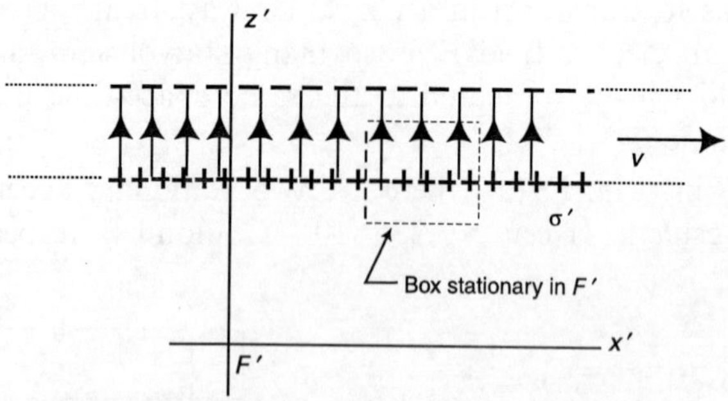

Fig. F.3 The superposition of the fields of the positive and negative charge between the sheets. The field outside is zero.

Applying Gauss' law to a box stationary in frame F', the box shown in cross section in Fig. F.3, we note that the charge content is determined by σ', and the field is zero outside the sandwich. Gauss' law tells that the magnitude of E_z', which is the only field component inside, must be $4\pi\sigma'$, or $4\pi\sigma/\sqrt{1-\beta^2}$

$$E_z' = \frac{E_z}{\sqrt{1-\beta^2}} = \gamma E_z \tag{24}$$

Consider a different situation with the stationary charged sheets in the frame F oriented perpendicular to the x axis, as in Fig. F. 4.

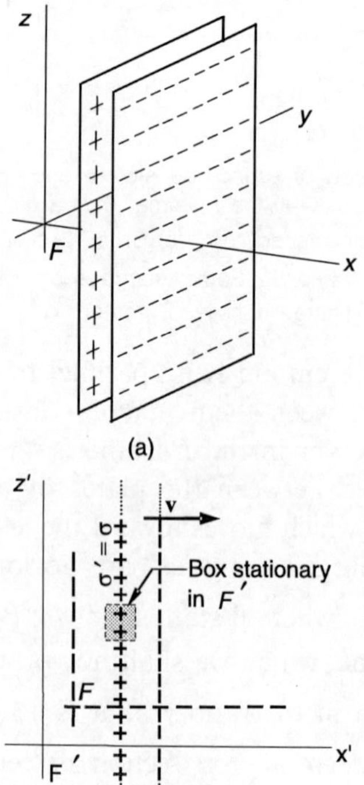

Fig. F.4 The electric field in another frame of reference (relative velocity parallel to field direction). (a) in reference frame F. (b) Cross-sectional view in reference frame F'.

The observer in F now reports a field in the x direction of magnitude $E_x = 4\pi\sigma$. In this case, the surface charge density observed in the frame F' is the same as that observed in F. The sheets are not contracted; only the distance between them is contracted (Fig. F.4 (b)), but that doesn't enter into the determination of the field. This time we find by applying Gauss' law to the box stationary in F':

$$E'_x = 4\pi\sigma' = 4\pi\sigma = E_x \tag{25}$$

Summarizing, charges at rest in frame F are the source of a field E. Let frame F' move with speed v relative to F. At any point F, resolve E into a longitudinal component $E_{\parallel}$ parallel to v and a transverse component $E_{\perp}$ perpendicular to the direction of v. At the same space-time point in F', the field E' is to be resolved into $E'_{\parallel}$ and $E'_{\perp}$, $E'_{\parallel}$ being parallel to v and $E'_{\perp}$, perpendicular thereto. Hence,

$$\begin{aligned} E'_{\parallel} &= E_{\parallel} \\ E'_{\perp} &= \gamma E_{\perp} \end{aligned} \tag{26}$$

Note the increased strength of the perpendicular component of the electric field observed in F'.

This is applicable to charges stationary in F.

F.3 FIELD OF A MOVING SINGLE CHARGE AND THE RELATIVISTIC NATURE OF MAGNETISM

In Chapter 10, Section 10.12 was shown the field of a *single charge* moving with uniform velocity (Fig. 10.40). By considering a system of a charge Q stationary in F and an observer in a frame moving in F', it is shown (Ref. [22]) that the electric field is relativistically compressed into a thin pancake (more powerful in strength) in a direction perpendicular to the direction of motion, which becomes thinner with increased strength, as the velocity increases (See Eqs. 24, 25 and 26). An observer in F' will report that "the electric field in his vicinity was pointing, exactly radially from the origin" (see proof in Ref. [22]). "How can an observer a mile away know where the particle is at the same instant?". Prof. Purcell answers that the information that the particle has been moving at constant speed forever, ...has been available for a long time. "It is the *past history* of the particle that determined the field observed".

One useful visualization of this statement is to imagine tiny vectors (representing the electric field in the manner in which the fields are depicted as waves in the animation here http://electronicsgurukulam.blogspot.in/search/label/Transmission%20Lines%20and%20 Waveguides) surrounding the charge and extending to infinity, which continuously keep changing their direction (and strength) *always pointing towards the charge* as it moves. Visualize the information of field vector strength and direction changes (not the vectors themselves because the field can never travel!), traveling at the speed of light, *rippling* through the densely populated field vectors, continuously in the space surrounding the moving single charge.

In Section 3.3 of the Guide to visualization of electrodynamics (available in the Teacher's Guide in the CD), Härtel has used this result to demonstrate the interaction of two electrons moving at right angles and between a single charge carrier and a current.

Several moving charges

If charges (*several* single charges) in F are moving, the prediction of the electric field in F' involves knowledge of *two* fields in F, the electric and magnetic.

If the postulates of relativity are valid, if electric charge is invariant, and if Coulomb's law holds, then, the effects we commonly call "magnetic" are bound to occur. This can be explained by a simple system (Ref. [22]).

In the lab frame shown in Fig. F.5(a), with spatial coordinates x, y, z, there is a line of positive charges (or ions), at rest and extending to infinity in both directions.

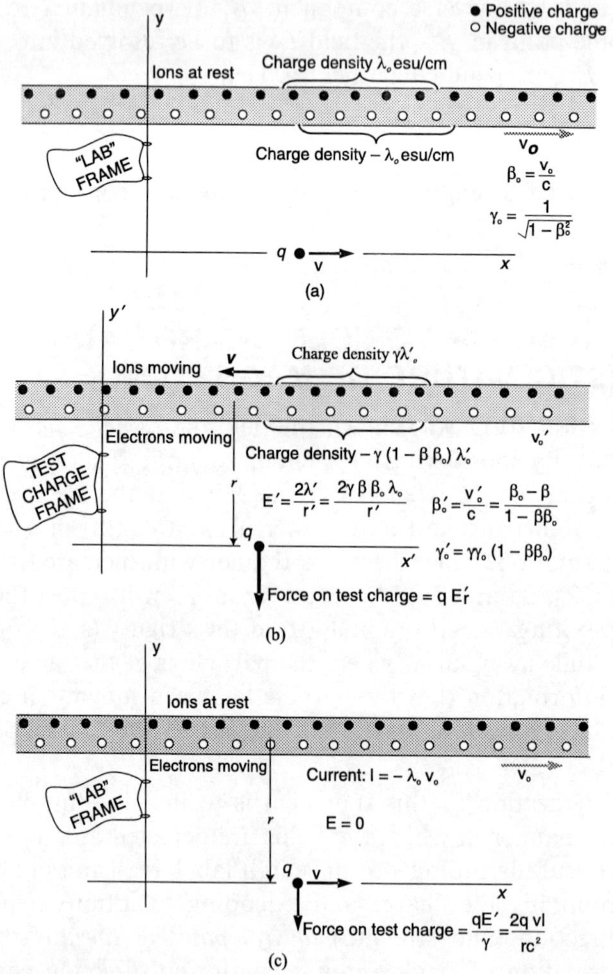

Fig. F.5 A test charge q moving parallel to a current in a wire. (a) The current consists of electrons moving to the right with speed v_o. The net charge on the wire is zero and there is no electric field outside the wire. (b) In the frame in which the test charge is at rest, the positive ions are moving to the left with speed v and the electrons are moving to the right with speed v'_o. The linear density of positive charge is greater than the linear density of negative charge. The wire appears positively charged, with an external field E'_r which causes a force qE'_r/γ on the stationary test charge q. (c) That force transformed back to the lab frame has magnitude qE'_r/γ

There is also a line of negative charges that are electrons moving to the right with a speed v_o. In a real wire the electrons would be intermingled with the ions but are shown separated for clarity. The linear density of positive charge is λ_o in esu/cm which is equal in magnitude to the linear density of negative charge. The net charge on the wire is zero and from Gauss' law there can be no flux from a cylinder that contains no charge, so the electric field must be zero everywhere outside the wire and so the test charge q at rest near this wire experiences no force whatsoever.

Suppose the test charge is not at rest in the lab frame but is moving with speed v in the x direction. Transform to a frame moving with the test charge, the x', y' frame in Fig. F.5(b). The test charge q is here at rest, but something else has changed. The wire appears to be charged ! Observe carefully in Fig. F.5(b) that the positive ions are closer together, and the electrons farther apart. Because the lab frame in which the positive ions are at rest is moving with speed v, the distance between positive ions as seen in the test charge frame is contracted by $\sqrt{1-\beta^2}$, or $\sqrt{1-v^2/c^2}$, or $1/\gamma$ (see Fig. F.1(a) and Eq. 18). The linear density of positive charge in this frame is correspondingly greater and will be $\gamma\lambda_o$.

The density of negative charge (electrons) takes a little longer to calculate, for the electrons were already moving with speed v_o in the lab frame in which their density was $-\lambda_o$. Their linear density in the lab frame, which was $-\lambda_o$, had already been increased by a Lorentz contraction of factor γ_o. So, in the electrons' own rest frame, the negative charge density must have been $-\lambda_o/\gamma_o$.

We need the speed of the electrons in the test charge frame in order to calculate their density there. To find that velocity v_o', in Fig. F.5(b), we must add the velocity $-v$ to the velocity remembering to use the formula for addition of velocities from Eq. 22. Let $\beta_o' = v_o'/c, \beta_o = v_o/c$, and β = v/c. Then

$$\beta_o' = \frac{\beta_o - \beta}{1 - \beta\beta_o} \tag{27}$$

The corresponding Lorentz factor γ_o', obtained from Eq. 27 with a little algebra, is

$$\gamma_o' = (1 - \beta_o'^2)^{-1/2} = \gamma\gamma_o(1 - \beta\beta_o) \tag{28}$$

The total linear density of charge in the wire in the test charge frame λ', can now be calculated (Ref. [22]):

$$\lambda' = \gamma\lambda_o - \frac{\lambda_o}{\gamma_o}\underbrace{\gamma\gamma_o(1 - \beta\beta_o)} = \gamma\beta\beta_o\lambda_o \tag{29}$$

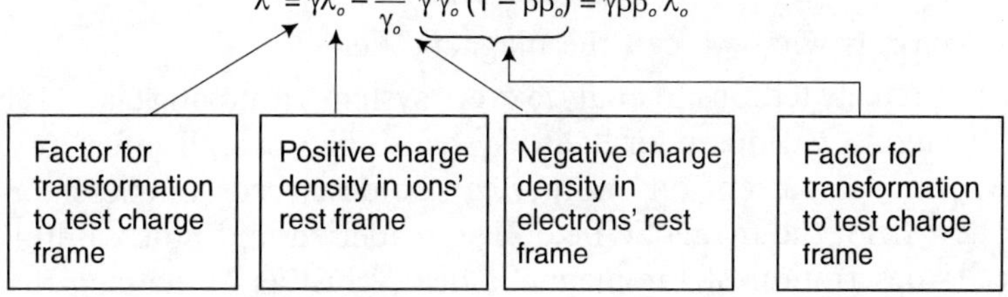

The wire is positively charged. Gauss' law guarantees the existence of a radial electric field E_r' given by the formula for the field of any infinite line charge (See Appendix A "Special case: A very long rod" in "Electric field of a uniformly charged thin rod"):

$$E_r' = \frac{2\lambda'}{r'} = \frac{2\gamma\beta\beta_o\lambda_o}{r'} \tag{30}$$

At the location of the test charge q, this field is in the $-y'$ direction. The test charge will experience a force

$$F_y' = qE_y' = -\frac{2q\gamma\beta\beta_o\lambda_o}{r'} \tag{31}$$

Since we wish to know the force experienced in the lab frame and not the test charge frame, then let us check this from Fig. F.5(c). If its value is qE_y' in the rest frame of the test charge, observers in the lab frame will report a force smaller by the factor $1/\gamma$. Since $r = r'$, the force on our moving test charge, measured in the lab frame is

$$F_y = \frac{F_y'}{\gamma} = -\frac{2q\beta\beta_o\lambda_o}{r} \tag{32}$$

Now $-\lambda_o v_o$ or $-\lambda_o\beta_o c$ is just the total current I in the wire, in the lab frame, for it is the amount of charge flowing past a given point per second. We'll call the current positive if it is equivalent to positive charge flowing in the positive x direction. Therefore, the current in this example is negative, and we can therefore write:

$$F_y = \frac{2I}{rc^2} q v_x \tag{33}$$

Note that the force on the moving test charge does not depend separately on the velocity or density of the charge carriers (in the wire) but only on the product, $\lambda_o\beta_o$ in our example, that determines the charge transport. In other words, we can have a certain current say, as small as 3.3 milliamps and it does not matter whether this current is composed of high-energy electrons moving with 99 percent of the speed of light, or of electrons in a metal executing nearly random thermal motions with a slight drift in one direction. The force exists when there is a current. Furthermore, the force on the test charge is strictly proportional to the velocity of the test charge v.

Tremendous is the effect of electric interaction, indeed; a miniscule *apparent* increase in charge density given tiny drift speeds in wires, is sufficient to produce significant electric effects!

The force on the test charge F_y is a force proportional to v_x (Eq. 33) and can be written $\boldsymbol{v} \times \mathbf{B}$, if $\mathbf{B}$ is a vector in the $\hat{z}$ direction whose strength is $\frac{2I}{rc^2}$,

$$F_y = q\,(v \times \mathbf{B}) \tag{34}$$

That $\mathbf{B}$, a vector, is what we call the magnetic field.

It would be extremely tedious to analyze every system of moving charges by transforming back and forth among various coordinate systems. The overall effect of a current on a moving charge, and one current on another can be described completely and concisely by introducing a new field, the magnetic field. The reader should note carefully the choice of words in Purcell's description of the magnetic field, "Just as we defined the electric field **E** as the vector force on unit charge at rest, so we can *define another field* **B** by the *velocity-dependent* part of the force that acts on a charge in motion". Prof. Purcell does not make an explicit reference to a physical entity **B**.

From the Eqs. 33 and 34, it is evident that a stationary charge will not experience a force near the current-carrying wire, but a moving charge will. Therefore, we say magnetism is relativistic.

F.4 THE LORENTZ FORCE LAW

At some instant t a particle of charge q passes a point (x, y, z) moving with velocity $\boldsymbol{v}$. At that moment the force on the particle (its rate of change of momentum) is **F**. The electric field at that time and place is known to be **E**. Then the magnetic field at that time and place is defined as the vector **B** which satisfies the vector equation

$$\mathbf{F} = q\mathbf{E} + \frac{q}{c} v \times \mathbf{B} \tag{35}$$

This is called the Lorentz force law in the CGS system, and is

$$\mathbf{F} = q\mathbf{E} + qv \times \mathbf{B} \tag{36}$$

in the SI system.

Repulsion of the wires in Chapter 6 (Fig. 6.1)

In Chapter 6 (Fig. 6), we had demonstrated the application of the Lorentz force law to explain the mutual repulsion of conductors. We will now show how Eq. 33 can be applied to explain the same effect.

The Fig. F.6 shows two wires carrying current in opposite directions in the lab frame in Fig. F.6(a). Assume the wires are uncharged in the lab frame. Then there is no electrical force from the opposite wire on the positive ions which are stationary in the lab frame.

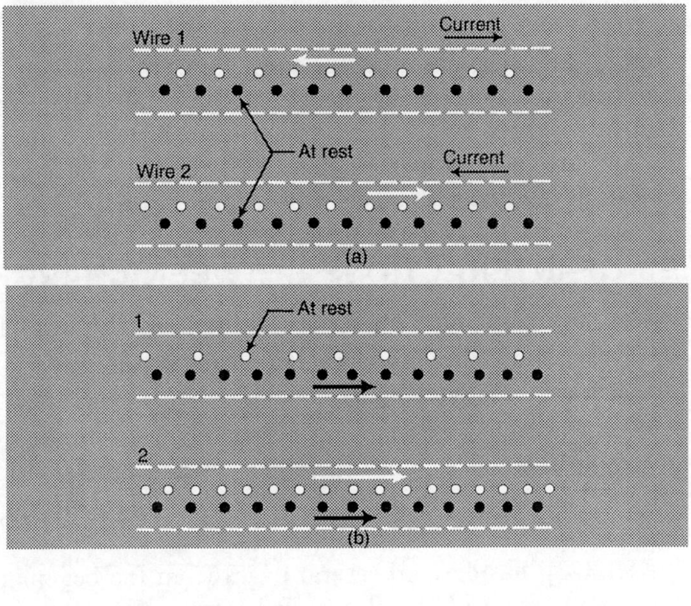

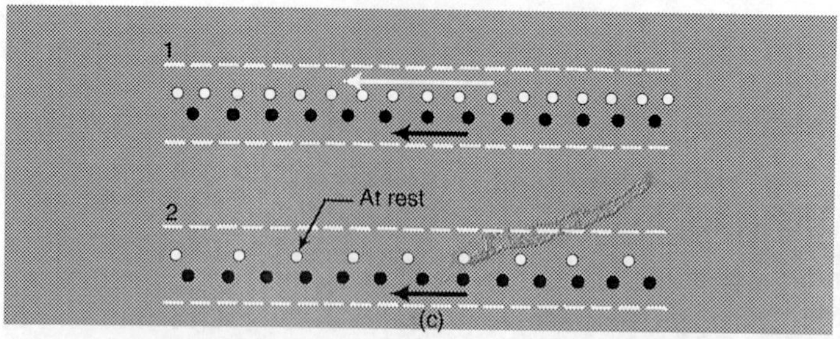

Fig. F.6 (a) Lab frame with two wires carrying current in opposite directions. As in metal wire, current is due to motion of negative ions (electrons) only. (b) Rest frame of electrons in wire 1. Note that in wire 2 positive ions are compressed, but electron distribution is contracted even more.(c) Rest frame for electrons in wire 2. Just as in (b) the other wire appears to these electrons at rest to be negatively charged.

Transferring to a frame in which one set of electrons is at rest (Fig. F.6(b)), we find that in the other wire the electron distribution is Lorentz-contracted more than the positive ion distribution. Because of that the electrons at rest in this frame will be repelled by the other wire. But when we transfer to the frame in which those other electrons are at rest (Fig. F.6(c)), we find the same situation. They too will be repelled. These repulsive forces will be observed in the lab frame as well, modified only by the factor γ.

We conclude that the two streams of electrons will repel one another in the lab frame. The stationary positive ions, although they feel no direct electrical force from the other wire, will be the indirect bearers of this repulsive force if the electrons remain confined within the wire. So the wires will be pushed apart producing tension in the wires, as in Fig. 6.1, until some external force such as those in the supports at the ends of the wires, balances this tension.

Purcell (Ref. [22] Chapter 5, *The Fields of Moving Charges*) has described the experiment of Fig. F.6 and mentions that there is no effect on the force of repulsion if we put a sheet of metal between the two wires (no shielding effect is produced). It will be instructive to reason why this is so. **Hints:** the property of an electric field to penetrate matter, transformation properties of the field are local properties Ref. [22], Chapter 6, Section How The Fields Transform, and location of observers when transforming fields.

F.5 THEORIES OF INTERACTIONS BETWEEN CHARGE CARRIERS

I refer the reader to the topic "Some properties of the magnetic field" in Ref. [22].

In view of the fact that a stationary charge does not experience a force near a current carrying wire, but a moving charge will, we conclude that magnetism is relativistic.

In the case when a test charge is moving perpendicularly towards or away from the current carrying wire, it is possible to use the effect we had described in Section 10.12 to study the interaction. The details are available in "The fields of moving charges" in Ref. [22].

Prof. Purcell (Ref. [22]) has devoted several pages from the beginning of his textbook before he takes on a discussion of the electric field itself. He answers the question as to whether the electric field is something real or whether it is merely a numerical value assigned to a force we measure. His first observation, and I quote, "First, since it (concept of a field) works, it doesn't make any difference." And later, "To visualize an electric field, you need to associate a vector, that is, a magnitude and direction, with *every* point in space".

This should in some way comfort the reader in knowing that one of the several interpretations several physicists make, that it is an "alteration of space" and adopted in this text in several chapters is sufficient and precise for the study of circuits and visualization of circuit processes.

The concepts of the fields described in Classical Electromagnetism are useful in the analysis of circuit processes and the reader should not dismiss these outright. Computations using field quantities are sufficiently accurate and precise in most applications of circuit theory though the idea as to how the fields themselves carry momentum is still the subject of study by physicists.

For example, Leigh Page of Yale University (Ref. [39]), propounded the "Emission Theory of Electromagnetism" in which the loci of massless moving elements issued out by charges continuously trace the path of the lines of force of Faraday. Page's Emission theory uses the two main postulates of Special Relativity (i. laws of physics are same in all inertial frames, and ii. the speed of light in free space has the same value c in all inertial frames of reference) to prove all of Maxwell's equations, the Lorentz force law and electromagnetic waves.

Härtel has used Page's circular waves produced by accelerating charges to help students visualize interactions intuitively due to changing currents (self- and mutual-induction). The Guide (reproduced by permission, Hermann Härtel) is available in "Teacher's Guide" in the CD with this book. Also available in the Guide are Härtel's visualizations of electrodynamics of charges moving with constant velocity and with acceleration in Section 3.4 (near current carrying wires), Section 3.5 of induction, Sections 3.6 and 3.7 of wave propagation, Section 3.8 of circuits and Section 3.9 of transmission lines. The visualizations in the Guide are sufficiently accurate and precise for an undergraduate course that covers topics on circuits and transmission line theory.

Note: Leigh Page's Emission theory should not be confused with other emission theories. Some Emission theories obey the first postulate of special relativity by having no preferred frame for light transmission. And, instead of applying the second postulate of the Theory of Special Relativity which states that the speed of light in empty space with a definite velocity c that is independent of the state of motion of the emitting body, say that light is emitted at a nominal speed of c with respect to, or relative to the emitting object and obeying the usual laws of Newtonian mechanics. This means that we can expect light to be moving towards us with a speed that is offset by the speed v of the distant emitter ($c \pm v$). On the other hand, Leigh Page's emission theory is in conformance with both the postulates of the Theory of Special Relativity, because Page's theory was formulated beginning with the assumption of the postulates of the Theory of Special Relativity.

Edward Purcell (Chapter 5 Ref. [22], *From Oersted to Einstein*), quotes Leigh Page "Viewed from another standpoint, the fact that we have been able, by means of the principle of relativity, to deduce the fundamental relations of electrodynamics from those of electrostatics, may be considered as some confirmation of the principle of relativity".

In Section 10.9, we learned that when Planck quantized the energy emitted by black bodies, the idea for which can be traced to a statistical interpretation of the second law of thermodynamics by Maxwell and Boltzmann, Quantum Physics began to influence scientists to seek solutions to describe atomic orbitals (See the topic "Quantum Mechanics, Probabilities and Orbitals of Electrons" in Appendix B). The influence can be very clearly seen in the Schrödinger Wave Equation and also in the expression of Density of States of conduction band electrons in conductors, where Planck's constant *h* makes it's appearance denoting the quantization of momentum and energy levels (See "Introduction to Solid State Physics" by Charles Kittel, Wiley, 7th Ed and Ref. [17]). Planck's law helped to provide an explanation of black-body radiation, of photoluminescence, of the photoelectric effect, and of the ionization of gases by ultraviolet light. This was the period when the semiclassical model of electrodynamics became popular and the word "photon" was increasingly used to replace the words "light quantum" which Einstein used to describe the quantization of light waves (See Note in Section 10.9, Chapter 10).

With these beginnings and seeking an explanation to experimental observations of radiative transitions of electrons, Paul Adrien Maurice Dirac (1902 – 1984), in the late 1920s found it "necessary not only to quantize material particles, but also the *electromagnetic field*", thus commencing research into the area now called Quantum Electrodynamics and Quantum Field Theory. Details are in the article "Recent Developments in the theory of the electron" by Victor F. Weisskopf of MIT, Cambridge, Reviews of Modern Physics, Volume 21, Number 2, April 1949, a link to which is provided in the Links to [illegible]_articles in the CD accompanying this book. The refinement of Quantum Electrodynamics was made later in 1948 by Tomonaga, Schwinger, and Feynman.

In contrast to Page's theory using massless moving elements to describe the fields, Quantum Electrodynamics (Sections 4.2, 6.6, 10.9 and 10.10) uses the concept of virtual and real photons that mediate the interactions between charges and according to this theory the *field* itself is quantized (See "Introduction to Elementary particles" by David J. Griffiths).

According to John David Jackson [Section I.1, Maxwell's Equations in Vacuum, Fields and Sources, Introduction, *Classical Electrodynamics*, Wiley, 3rd edition, Chapter 1], "the concept of **E** and **B** is a classical notion", and "can be thought of as the classical limit (limit of large quantum numbers) of a quantum mechanical description in terms of real or virtual photons". Jackson states "In the domain of macroscopic phenomena and even some atomic phenomena, the discrete photon aspect can usually be ignored or at least glossed over". For example, "An isotropic FM antenna with a power of 100 watts at 10^8 Hz (100 MHz) produces an rms electric field of only 0.5 mV/m at a distance of 100 kilometers, but this still corresponds to a flux of 10^{12} photons/cm^2.s." For computing the number of photons/area-second see Section 2.1.3 in "Applied Quantum Mechanics" by A. F. J. Levi, Cambridge University Press, 2003.

The article "Michelson-Morley experiment: length contraction?" in physicspages traces how the failed experiment led Lorentz and Fitzgerald to invoke ideas of length contraction and time dilation from Special Relativity to obtain the null result of the experiment and to keep the aether model intact though without it's drag, leading later to the introduction of "fields" in electromagnetism. Also see Section 15.6 of Relational Mechanics (Ref. [36]) on the Michelson-Morley Experiment. The article "Riddle of Inertia" by Peter Graneau, in Wireless World is important to understand the concept of inertial forces and a reading will prove instructive.

Links to the physicspages article and the article by Graneau are available in the Links_to_animations_articles in the CD.

In the beginning of this Appendix, we mentioned one other interpretation of interaction among charged particles made using Relational Mechanics (Ref. [36]). Borrowing concepts from Weber's electrodynamics for analyzing interactions between charged bodies, several physicists have been contributing to this important subject. Relational mechanics utilizes the concept of a force as an *interaction between material bodies*. In relational mechanics a force does not describe, for instance, an interaction of a body with "space," the interaction of a body with the observer, of a body with the frame of reference, nor the interaction of a body with a "field"[36].

Further, according to the theory, 'The sum of all forces of *any* nature (gravitational, electric, magnetic, chemical, etc.) acting on any body is always zero in all frames of reference.' This postulate of relational mechanics does not necessitate the introduction of the concepts of gravitational/electric/magnetic fields, nor inertia, nor the concept of inertial mass nor the concept of an inertial frame of reference, and this is it's main advantage compared to classical mechanics. Assis strongly argues against the rationale of Lorentz and Fitzgerald in introducing the ideas of fields and compounding a simple problem. In Relational Mechanics (Ref. [36]), see also Section 16.6 "General Comments", Section 17.4 "Properties of Weber's Potential Energy and Force as Applied to Electromagnetism and Gravitation", Section 25.2 "Electromagnetism", Chapter 26 "Conclusion", and Appendix A. In these topics, it is shown that the postulates of relational mechanics are valid in *all* frames of reference both, inertial and non-inertial.

The student, who has read through the chapters of this textbook which utilizes the field concept in all the chapters, may find the discussions in Relational Mechanics (Ref. [36]) a bit unsettling, since it challenges some of the concepts assimilated, concerning fields especially. In the chapter "The Electric Field", Griffiths [Ref. 8] states, "It is even possible, though cumbersome, to formulate classical electrodynamics as an "action-at-a-distance" theory, and dispense with the field concept altogether". Such is the nature of the action at a distance theories and physical observations that I urge the student, in the spirit of inquiry, to read Relational Mechanics without prejudice, since it provides an enrichment of scientific knowledge. According to Assis[37], "But the primary reality for us is the force or interaction between material bodies (generating their relative accelerations relative to inertial frames), and not the abstract field concepts existing in space independent of the presence of a charged test particle which can detect the existence of these fields."

Books and articles by Andre K. T. Assis are available in Assis' homepage http://www.ifi.unicamp.br/~assis.

References

1. Ruth W. Chabay and Bruce A. Sherwood, *"Unified treatment of electrostatics and circuits"*, Center for Innovation in Learning and Dept. of Physics, Carnegie Mellon University.
2. Ruth W. Chabay and Bruce A. Sherwood, *"Electric and Magnetic Interactions"*, John Wiley, USA, 1995.
3. Sherwood Parker, "*Electrostatics and Current flow*", American Journal of Physics, Vol. 38, June 1970.
4. Paul G. Hewitt, "*Conceptual Physics*", Little, Brown and Company, Boston, USA.
5. Paul Falstad, "*Electrostatics Applets*", http://www.falstad.com/vector2de/
6. Hermann Häertel, *"Constant electric current and the distribution of surface charges - 1"*, Institute for Science Education, University Kiel.
7. MIT Open Course Ware, Massachusetts Institute of Technology, Massachusetts, USA.
8. David J. Griffiths, "*Introduction to Electrodynamics*", Prentice-Hall, India, 1999.
9. Norris W. Preyer, *"Surface charges and fields of simple circuits"*, American Journal of Physics, Volume 68, No. 11, November 2000.
10. J.D. Jackson, *''Surface charges on circuit wires and resistors play three roles"*, American. Journal of Physics, Volume 64, pp. 855-870, 1996.
11. William J. Beaty, http://amasci.com/miscon/eleca.html.
12. Lecture presentations from the Massachusetts Institute of Technology, Boston, USA
13. David Haliday, Robert Resnick, Jearl Walker, *"Physics*", Wiley India, 2001.
14. The Physics Classroom Tutorial – Glenbrook South Physics (http://www.glenbrook.k12.il.us/gbssci/phys/phys.html)
15. Edward Hughes, "*Electrical Technology*", Longman, 7th edition, 1995.
16. Edward Jordan and Keith Balmain, "*Electromagnetic Waves and Radiating Systems*", Prentice-Hall of India, 2nd edition, 2003.
17. Jacob Milman, Christos Halkias, Satyabrata Jit, "*Electronic Devices and Circuits*", Tata McGraw-Hill, India.
18. Bart Van Zeghbroeck, http://ecee.colorado.edu/~bart/book, "*Principles of Semiconductor Devices*", 2007.
19. Richard F.W. Bader, "*An introduction to the electronic structure of atoms and molecules*", http://www.chemistry.mcmaster.ca/esam/Chapter_1/intro.html, Department of Chemistry, McMaster University, Hamilton, Ontario, Canada.
20. Paul Malvino, "*Electronic Principles*", Tata McGraw-Hill, India.
21. George Kennedy and Bernard Davis, "*Electronic Communication Systems*", Tata McGraw-Hill, India.
22. Edward Purcell, "*Electricity and Magnetism*", Tata McGraw-Hill, India.
23. Worsley School Online, Alberta, Canada, "*What's in a touch*". https://www.scribd.com/document/24059107/What-is-Touch.

24. John M. Clement and John Denker, "[Phys-l] *Fields and touch*", Archives of forum replies https://web.archive.org/web/20100808134424/https://carnot.physics.buffalo.edu/archives/2010/2_2010/msg00132.html Also see …../msg00129.html and …../msg00132.html.
25. Fermilab, U.S. Dept. of Energy, "*Inquiring minds*" http://www.fnal.gov/pub/inquiring/questions/bob.html.
26. Elementary Physics II, Boston University, "*Electric energy and potential*" http://physics.bu.edu/~duffy/PY106/Potential.html.
27. Physics and Astronomy, University of St. Andrews, "*Energy and moving electrons*" http://www.st-andrews.ac.uk/~www_pa/Scots_Guide/audio/part6/page2.html.
28. Steve Wirt, "*Introduction to Magnetism*" http://www.regentsprep.org/Regents/physics/phys03/cintromag/
29. Lee, Y.C. and Mei, W.N. (1996), *Self-inductance of a long wire: infinite multiple loop approach*, International Journal of Mathematical Education in Science and Technology, 27:1,79-86.
30. Ben G. Streetman and Sanjay Banerjee, "*Solid State Electronic Devices*", Pearson Prentice Hall, 2006.
31. Ruth W. Chabay and Bruce A. Sherwood, "*Electric and Magnetic Interactions*", 2011, John Wiley, 3rd Edition, USA.
32. Mathew N.O. Sadiku, "*Elements of Electromagnetics*", Oxford University Press, 3rd Edition, 2006.
33. John D. Ryder, "*Networks, Lines and Fields*", Prentice-Hall of India, 2nd Edition, 2003.
34. Millman and Taub, "*Pulse, Digital and Switching Waveforms*", Tata McGraw-Hill, India, 2000.
35. John D. Krauss, Ronald J. Marhefka, Ahmad S Khan, "*Antennas and Wave Propagation*", 2012, Tata McGraw Hill, India (4th edition).
36. Andre Koch Torres Assis, "*Relational Mechanics and Implementation of Mach's Principle with Weber's Gravitational Force*", 2014, C. Roy Keys Inc., Quebec, Canada
37. Andre Koch Torres Assis and Julio Akashi Hernandes, "*The Electric Force of a Current - Weber and the surface charges of resistive conductors carrying steady currents*", 2007, C. Roy Keys Inc., Quebec, Canada.
38. Andre Koch Torres Assis and J. P. M. C. Chaib, "*Ampère's Electrodynamics-Analysis of the Meaning and Evolution of Ampère's Force between Current Elements, together with a Complete Translation of His Masterpiece: Theory of Electrodynamic Phenomena, Uniquely Deduced from Experience*", 2015, C. Roy Keys Inc., Quebec, Canada.
39. Page, L.,"*The Emission Theory of Electromagnetism*", Connecticut Academy of Arts and Science, Vol. 26, pp. 213-243.
40. M. d. A. Bueno and A. K. T. Assis, *Inductance and Force Calculations in Electrical Circuits*, Nova Science Publishers, Hauppauge, NY, USA.
41. Ruth W. Chabay and Bruce A. Sherwood, "*Modern Mechanics*", John Wiley, USA, 2011.
42. A. K. T. Assis, *Weber's Electrodynamics*, 1994, Springer Science+Business Media Dordrecht.

Images

1. **Chapter 1**
   Structure of metals and conductors
   mysite.verizon.net/kdrews47/solids/metal.gif
   Electric field of a point charge
   http://www.falstad.com/vector3de/ (reproduced with permission)
2. **Chapter 2**
   Potential and equipotential lines of a dipole
   http://www.falstad.com/vector3de/ (reproduced with permission)
   Electric field lines of a capacitor
   http://www.falstad.com/emstatic/ (reproduced with permission)

Index

Page numbers in bold font type point to formal description of the topic

D